WEBSTER'S NEW WORLD™

Telecom Dictionary

WEBSTER'S NEW WORLD™

Telecom Dictionary

Ray Horak

Wiley Publishing, Inc.

Webster's New World Telecom Dictionary

Published by
Wiley Publishing, Inc.
10475 Crosspoint Boulevard
Indianapolis, IN 46256
www.wiley.com

Published simultaneously in Canada

ISBN: 978-0-471-77457-0

Manufactured in the United States of America

10 9 8 7 6 5 4 3 2 1

Library of Congress Cataloging-in-Publication Data

Horak, Ray.
Webster's New World telecom dictionary / Ray Horak.
p. cm.
ISBN 978-0-471-77457-0 (pbk.)
1. Telecommunication—Dictionaries. I. Title.
TK5102.H65 2007
621.38203—dc22

2007024232

For general information on our other products and services please contact our Customer Care Department within the United States at (800) 762-2974, outside the United States at (317) 572-3993 or fax (317) 572-4002.

To Margaret

for whom my love is infinite and eternal

ABOUT THE AUTHOR

Ray Horak is president of The Context Corporation, an independent consultancy specializing in telecommunications and data communications systems and networks. Well-known for translating highly technical subject matter into plain-English, commonsense terms, he has written hundreds of articles and columns on various aspects of telecommunications. This is his fifth book for Wiley. He teaches public and private seminars on various telecommunications subjects and is a popular speaker at industry events. He also serves as an expert consultant and witness in litigation involving intellectual property matters such as patent infringement and trademark/service mark infringement as well as product/service misrepresentation.

Technical Editor

William A. Flanagan is President of Flanagan Consulting, an independent consultancy specializing in converged networking for voice, data, and video. There he analyzes enterprise WANs and carrier networks, defining technical architectures and writing procurement-related documentation. He also assists equipment vendors to define product feature sets and market positioning. Bill is a physicist by training, a frequent speaker at conferences and seminars, and the author of dozens of articles and eleven books on networking.

Advisory Board

Rick Luhmann has over 20 years of experience in media and publishing, with a focus on call centers and telecom. He was editor in chief (1987) of *Teleconnect Magazine* and founding editor in chief (1993) of *Computer Telephony Magazine*. In the late 1990s he moved into online publishing, as co-founder of the telecom and networking portal CommWeb for CMP Media. He currently is Director, Online Business Development, ICMI Group, CMP Technology.

Brett Parker has a background of more than 20 years as an independent consultant with a concentration in the design, securing, and administration of complex networks and internetworks for business and government. He currently works as a professional IT administrator for a local government agency in northwest Washington.

Gene Retske is a respected telecommunications authority, with 30 years of industry experience and over 12 years as a professional author and journalist. He is the author of two highly regarded books on telecommunications and over 800 articles that have appeared in publications around the world, including *Newsweek*. Retske is currently editor in chief of *The Prepaid Press*, the leading trade publication for the prepaid services industry.

Mark Simon is a senior technology executive with 20 years of successful experience in information technology and business leadership. Recognized in the areas of aligning business and information technology, technology strategies, security strategies, communicating business needs into technology requirements and building organizations with a commitment to customer service and a can-do culture. Mark is an active board and advisory board member, senior strategy and operations consultant to numerous early/mid stage companies.

Dave Thomas is a partner at Hogan & Hartson, LLP, an international law firm based in Washington, D.C. During his 17 years of practice as counsel to the nation's top cable companies, cable and telecommunications trade associations, and other competitive communications organizations, Dave has become a nationally recognized authority in pole attachments and communications plant deployment issues and has handled a wide range of regulatory, litigation, and transactional matters involving competitive wireless and wireline telecommunications providers, broadcasters, and satellite communications companies.

CREDITS

Acquisitions Editor
Jenny Watson

Development Editor
Ami Frank Sullivan

Special Editorial Help
Rebekah Gholdson

Technical Editor
Bill Flanagan

Editorial Manager
Mary Beth Wakefield

Production Manager
Tim Tate

Vice President and Executive Group Publisher
Richard Swadley

Vice President and Executive Publisher
Joseph B. Wikert

Project Coordinator, Cover
Lynsey Osborn

Proofreader
Sossity Smith

Compositor
Kate Kaminski, Happenstance Type-O-Rama

Anniversary Logo Design
Richard Pacifico

CONTENTS

ACKNOWLEDGMENTS

A work of this magnitude is not a simple matter. It is the culmination of more than 35 years' combined experience in telecommunications and incorporates knowledge drawn from thousands of articles and Web sites and hundreds of books. Most of those sources are very recent, but some of the books are more than 100 years old and still relevant for context, if not for their descriptions of contemporary cutting-edge technologies. My thanks to all those authors who went before me and variously put ink to paper or fingers to a keyboard.

Neither is a work of this magnitude an individual effort. I owe a great deal to Bill Flanagan, who served as Technical Editor, providing a great deal of guidance across a wide range of technologies and applications. Bill is perhaps as knowledgeable as anyone across the full range of subject matter covered in this book and has written eleven excellent books, from which I drew extensively. He went through every word of every draft of this work as if his life depended on making me get every single Hz, bit, and byte correct. Bill is a patient and skilled collaborator with a great sense of balance and a wonderful sense of humor.

My thanks to the members of the advisory board, each of whom provided valuable guidance in his area of expertise. I am fortunate indeed to have the guidance and support of Rick Luhmann, Brett Parker, Gene Retske, Mark Simon, and Dave Thomas.

My thanks also to the team at Wiley. This book would not exist, or at least I would not have written it, had Vice President Joe Wikert not gotten involved and done the right thing. Ami Frank Sullivan, Senior Development Editor, patiently and competently shepherded me through the development process. She and Rebekah Gholdson also read every word of the manuscript and did the early copy editing, which must qualify them for some sort of special award in editorial circles. Patience and positive attitude seem to be common virtues at Wiley.

We all worked together to make this book as good as it is, and we think it is very good. Ultimately, of course, the responsibility for any errors and omissions is mine, and mine alone. If you have any suggestions for improving it, please address them to me (ray@contextcorporation.com).

INTRODUCTION

We live in a complex world that increasingly is defined by information technology, by which I do not mean classical data processing but rather the creation, storage, and distribution of all forms of information, including audio, text, image, and video. In combination, such information becomes multimedia in nature. Computers of one sort or another are used to capture and create much of that information, which is remarkable in itself, but the value of the data is largely dependent on our ability to share it with others. Robert Metcalfe, who invented the Ethernet local area network in 1973, clearly recognized the value of networking resources. Metcalfe's Law states that the value of a telecommunications network is proportional to the square of the number of users (n^2) of the system. (I think Alexander Graham Bell clearly recognized the same relationship roughly 100 years earlier, but he failed to state it in the form of a law, or even a theory.) Information, of course, is our most valuable resource and telecommunications systems and networks allow us to share it with others.

In the 97 years between Bell's invention of the telephone and Metcalfe's invention of the LAN, there was a lot of technological progress, and the pace has increased markedly since. Copper wires have given way to glass fibers in the WAN backbone, and fiber optics is now making its way into the local loop. Coaxial cable has been obsoleted in the LAN by twisted pair, glass and plastic optical fiber, and now RF technologies. Wireline technologies have yielded to or are supplemented by wireless in many applications, not only in the LAN, but also in the MAN and WAN. Some estimates now place the number of cellular telephones worldwide at over two billion, which means that there are more cellular telephones than landlines. Since its invention in 1877 and for well over 100 years, circuit switching was the sole method by which telephone calls were connected, but is now rapidly being replaced by packet switching. There seems to be no question that Voice over Internet Protocol (VoIP) technologies will obsolete the traditional circuit-switched PSTN. Since its origins in the late 1960s as a closed network for academics and intellectuals working on projects for the U.S. military, the Internet has been commercialized and made available to the general public. The Internet now comprises over 60,000 networks connecting nearly 395 million host computers in more than 150 countries. On a daily basis, the Internet handles more than 84 billion e-mail messages and total traffic of approximately 5,175 petabits. It is in large part due to the popularity of the Internet and the World Wide Web that so many millions of miles of optical fiber have been deployed and that the available bandwidth has reached such incredible and even indescribable proportions.

All of these and many other relevant technologies build in some way on those that came before and each adds to the vocabulary of telecommunications a set of terminology, along with the seemingly requisite abbreviations, acronyms, contractions, initialisms, and portmanteaux. Many of these have multiple definitions, sensitive to historical or technological context. *Broadband*, for example, has one set of definitions in a WAN context, but quite another in the LAN domain. *Carrier* also has several definitions, as do *buffer* and *ATM*, and there are at least four kinds of *cells*. Some acronyms really aren't acronyms at all — ISO comes to mind. The origin of some terms is fascinating, with *bug* and *ping* being good examples. Some definitions in this book are very short, such as *plug*, whereas some are more like mini-tutorials, such as *SONET* or *frame relay*. Some definitions are highly relevant, whereas others are only marginally so, and still others have no relevance to telecom whatsoever, but I find them interesting. *Rules of engagement*, for example, is a term I find fascinating and I think you will, as well. Also fascinating are the cross-references from *rules of engagement* to *Geneva Convention* and *warrior's code*. Most terms are cross-referenced, which you will find to be of great value. Some terms are just for fun, so I hope you have a sense of humor. Check out *euphemism* and *OCD* as examples. If you don't find them funny, just skip over them and forgive me, but please don't get the impression that I take this book lightly. I take my job very seriously. I also enjoy what I do, and I try to have a little fun with it now and then.

I hope that you enjoy the book and find it valuable. This is not an open-source dictionary. I wrote every word of it and I am solely responsible for its content, which is how I know that it is correct and objective. However, please feel free to contact me if you have a suggestion for a correction or perhaps an additional term. I plan many more editions of this book and intend for each to be bigger and better.

Ray Horak
1500A East College Way, PMB 443
Mount Vernon, WA 98273
United States of America
ray@contextcorporation.com
Tel: 360.428.5747
Fax: 360.416.3378

WEBSTER'S NEW WORLD™

Telecom Dictionary

Symbols

Octothorp, octothorpe, pound sign, hash mark, or square. There is disagreement about the origin of the symbol, although many agree that it was invented by scientists at Bell Telephone Laboratories (Bell Labs) in the early 1970s to designate one of the special function keys on a touch-tone telephone keypad. (The other is the *, or asterisk.) The # is found in the lower right-hand corner of the standard dual tone multifrequency (DTMF) keypad grid. See also *DTMF*.

* Asterisk. The symbol used to designate one of the special function keys on a touch-tone telephone keypad. (The other is the #, or octothorpe.) The * is found in the lower left-hand corner of the standard dual tone multifrequency (DTMF) keypad grid. See also *DTMF*.

. **1.** Pronounced *dot*. In an e-mail address, the symbol used to separate the organization name from the domain. E-mail addresses follow the convention *user@organization.domain*. My e-mail address, for example, is *ray@contextcorporation.com* (pronounced *ray at context corporation dot com*), which is the e-mail address for Ray Horak (user) at The Context Corporation (organization providing the connection), a *com*mercial enterprise. See also *domain*. **2.** Period. A punctuation mark used to terminate sentences or sentence fragments that are neither interrogatory (?) nor exclamatory (!).

/ **1.** Forward slash. A symbol used to separate the parts of a directory path in UNIX and Linux. Internet addresses use the forward slash convention for describing directory and path names. **2.** Virgule. A punctuation mark that separates alternatives, with SONET/SDH and hybrid fiber/coax as examples.

// **1.** Double forward slash or double slash. In a Web address, the notation used in conjunction with a colon (:) to separate the Internet protocol (e.g., Hypertext Transfer Protocol, or HTTP) from the uniform resource locator (URL). My company address, for example, is *http://www.contextcorporation.com*. See also *protocol*, *URL*, and *WWW*.

@ Pronounced *at*. In an e-mail address, the symbol used to separate the user name from the organization name. E-mail addresses follow the convention *user@organization.domain*. My address, for example, is *ray@contextcorporation.com* (pronounced *ray at context corporation dot com*), which is the e-mail address for Ray Horak (user) at The Context Corporation (organization providing the connection), an enterprise in the *com*mercial domain. The @ was selected in 1971 by Raymond Tomlinson, who is widely credited for inventing a method of transmitting e-mail between computers, and was used to separate the user name from the machine name. At the time, Tomlinson was working as an engineer for Bolt Beranek and Newman (BBN) and was engaged in the development of the Advanced Research Projects Agency Network (ARPANET), the predecessor to the Internet.

™ TradeMark. Intellectual property comprising a word, phrase, logo, or other graphic symbol, sounds, or colors used by a manufacturer or seller to distinguish its products. See also *trademark*.

© The symbol for copyright, referring to the exclusive legal right of an author or publisher to publication, production, or sale of the rights to an original literary, dramatic, musical, or artistic work. I, Ray Horak, am the author of this original work, to which Wiley owns the copyright. Please don't even think about violating their copyright. Thank you. Have a nice day.

Δ The Greek letter *delta*, Δ, is the symbol for change or difference, as in delta modulation.

λ The Greek letter *lambda*, λ, is used to denote wavelength. See *wavelength.*

μ The Greek letter *mu*, μ, is used to denote one-millionth, as in μm, a micrometer, or micron. See *micron.*

μ-law (mu-law) A voice companding technique specified in the ITU-T G.711 Recommendation for pulse code modulation (PCM). This technique is used in North America and areas under North American influence. See *mu-law.*

π The Greek letter *pi*, the rough equivalent of 3.14. See *pi.*

Ω The Greek letter *omega*, **Ω**, is used as the symbol for ohm. See *ohm* and *Ohm's Law.*

Numbers

0B+D Referring to a variation of the ITU-T specification for ISDN basic rate interface (BRI), also known as basic rate access (BRA). BRI supports two bearer (B) channels and one data (D) channel. Intended for applications that do not require an information-bearing channel for voice or data transmission, 0B+D supports zero bearer (B) channels and one data (D) channel. The D channel is designated for signalling and control purposes, but also can be used for low-speed packet data applications such as transaction processing, credit card verification, and telemetry. See also *1B+D*, *BRI*, and *ISDN.*

0080 A dial prefix for toll free services in some countries. See also *800 Service*, *area code*, and *toll free service.*

020 A dial prefix for toll free services in some countries. See also *800 Service*, *area code*, and *toll free service.*

0500 A dial prefix for toll free services in some countries. See also *800 Service*, *area code*, and *toll free service.*

0800 A dial prefix for toll free services in some countries. See also *800 Service*, *area code*, and *toll free service.*

1A The first key telephone system (KTS), the 1A was a hardwired system comprising components physically wired together by hand. Developed by the Bell System and first marketed in 1938, the 1A was an electromechanical system. See also *1A1*, *1A2*, and *KTS.*

1A1 The A hardwired electromechanical key telephone system (KTS) comprising components physically wired together by hand. Developed by the Bell System and first marketed in 1953, 1A1 systems superseded the 1A systems, adding a few features, including line status lamps that lit steadily to indicate a line in use and flashed to indicate a line on hold. 1A1 systems were superseded, in turn, by 1A2 systems. See also *1A*, *1A2*, and *KTS.*

1A2 A modular electromechanical key telephone system (KTS) comprising hardwired circuit packs that plugged into a pre-built chassis that included cable connectors for attaching station equipment. Developed by the Bell System and first marketed in 1963, 1A2 systems were an improvement over earlier 1A and 1A1 systems, as they allowed the addition of a limited number of enhanced features through common control cards in the form of circuit packs. Electronic common control (ECC) systems made 1A2 systems obsolete long ago. See also *1A*, *1A1*, *ECC*, and *KTS.*

1Base5 (1 Mbps; Baseband; 500 meters) The IEEE standard (mid-1980s) for Ethernet transmission over Cat 3, Cat 4, or Cat 5 unshielded twisted pair (UTP) cable. 1Base5 translates to 1 Mbps (theoretical transmission rate), Baseband (one transmission at a time over a single, shared channel), and 500 meters maximum segment length. The predecessor to 10Base-T, 1Base5 is considered obsolete. See also *10Base-T*, *baseband*, *Cat 3*, *Cat 4*, *Cat 5*, *channel*, *Ethernet*, *IEEE*, and *transmission rate.*

1B+D Referring to a variation of the ITU-T specification for ISDN basic rate interface (BRI), also known as basic rate access (BRA). BRI supports two bearer (B) channels and one data (D) channel.

Intended for applications that require only one information-bearing channel, perhaps for voice or data transmission, 1B+D supports one bearer (B) channel and one data (D) channel. The D channel is designated for signalling and control purposes, but also can be used for low-speed packet data applications such as transaction processing, credit card verification, and telemetry. See also *0B+D*, *BRI*, and *ISDN*.

1+ Dialing Synonymous with *Direct Distance Dialing* (DDD). See *DDD*.

1-persistent carrier sense multiple access (1-Persistent CSMA) See *CSMA*.

1-Persistent CSMA (1-Persistent Carrier Sense Multiple Access) See *CSMA*.

1+ WATS See *Virtual WATS*.

1.5-way paging A variation of two-way paging, which supports guaranteed message delivery, as the network does not attempt to download messages until such time as the pager is within range, turned on, and has enough memory to support the download. The general location of the pager is communicated upstream, so the messages can be downloaded to the system antennas supporting that particular geographic area, rather than being broadcast across the entire paging network. Once downloaded successfully, the pager acknowledges to the network the receipt of the page. See also *antenna*, *download*, *two-way paging*, and *upstream*.

10 Gigabit Attachment Unit Interface (XAIU) See *XAUI*.

10Base2 (10 Mbps; Baseband; 200 meters) The IEEE standard (1986) for relatively thin coaxial cable in support of Ethernet transmission. Also known as *ThinNet* (Thin EtherNet), 10Base2 translates to 10 Mbps (theoretical transmission rate), Baseband (one transmission at a time over a single, shared channel), and 200 meters maximum segment length (actually 185 meters, rounded up). The thinner cable is less costly to acquire and deploy than earlier 10Base5, although its performance is less in terms of transmission distance due to increased attenuation. See also *10Base2*, *attenuation*, *baseband*, *channel*, *coaxial cable*, *Ethernet*, *IEEE*, and *transmission rate*.

10Base5 (10 Mbps; Baseband; 500 meters) The IEEE standard (1983) for coaxial cable assemblies in support of Ethernet transmission. Also known as *ThickNet* (Thick EtherNet), 10Base5 specifies traditional thick coaxial cable, often known affectionately as goldenrod, referring to its high cost, high value, and the yellow cable sheath used by some manufacturers. Other manufacturers used orange cable sheaths for thick coax, giving rise to the term orange hose. 10Base5 translates to 10 Mbps (theoretical transmission rate), Baseband (one transmission at a time over a single, shared channel), and 500 meters maximum segment length. See also *10Base2*, *baseband*, *channel*, *coaxial cable*, *Ethernet*, *IEEE*, and *transmission rate*.

10Base-T (10 Mbps; Baseband; Twisted pair) The IEEE standard (1990), and for Cat 3, Cat 4, or Cat 5 unshielded twisted pair (UTP) in support of Ethernet transmission. 10Base-T translates to 10 Mbps (theoretical transmission rate), Baseband (one transmission at a time over a single, shared channel) transmission, and over Twisted pair media. The maximum segment length between the 10Base-T hub and the attached device (e.g., workstation or printer) is specified at 100 meters or less, although good Cat 5 cable will perform well over somewhat longer distances. The 10Base-T hub is a wire hub that serves as a multiport repeater, as well as a central point of interconnection. See also *baseband*, *Cat 3*, *Cat 4*, *Cat 5*, *channel*, *Ethernet*, *IEEE*, *hub*, *repeater*, *standard*, *transmission rate*, and *UTP*.

10Broad36 (10 Mbps, Broadband, 3600 meters) The IEEE standard for a broadband LAN, 10Broad36 derives multiple channels through frequency division multiplexing (FDM). The aggregate bandwidth is 550 MHz and the FDM channels are 14 MHz wide, with 4 MHz guardbands. The modulation technique is differential phase shift keying (DPSK). The total span is a maximum of 3600 meters, and can be divided into multiple segments, each with a maximum distance of 1800 meters. 10Broad36 is considered obsolete. See also *bandwidth*, *broadband*, *channel*, *DPSK*, *FDM*, *guard band*, *IEEE*, *LAN*, and *modulation*.

1G (1st Generation) In cellular radio, referring to analog systems such as Advanced Mobile Phone System (AMPS), Narrowband AMPS (N-AMPS), Nordic Mobile Telephone (NMT), and Total Access Communications System (TACS). See also *2G*, *2.5G*, *3G*, *AMPS*, *analog*, *cellular radio*, *N-AMPS*, *NMT*, and *TACS*.

10GBase-CX4 (802.3ak) An IEEE standard for twinaxial cable patch cord assemblies in support of 10G Ethernet over distances of up to 50 feet. The standard specifies the same connectors used in 4X InfiniBand and the XAIU (10 Gigabit Attachment Unit Interface) specified in 802.3ae, spreading the 10 Gbps datastream over four paired transmitters and receivers, with each pair operating differentially over a thin twinaxial cable. As the 10 Gbps is spread over the twinax bundle, each cable supports a data rate of 2.5 Gbps over a 3.125 GHz channel with 8B10B coding. This approach requires four differential pairs in each direction, for a total of eight twinax cables per assembly. The tight operating tolerances require that the cable assemblies be factory-terminated. Also, and in consideration of the impact of resistance on signal attenuation, the diameter of the center conductors is sensitive to cable length. At a distance of 20 feet, the outside diameter a CX-4 cable is approximately the same as that of a Cat 5e cable. See also *8B/10B*, *InfiniBand*, *twinaxial cable*, and *XAUI*.

10GBase-SR, SW (10 Gbps; Baseband; Short Range, Short Wavelength) The IEEE 802.3ae media specification for 10 Gbps Ethernet (10GbE) transmission over multimode fiber (MMF) with a core diameter of 62.5 Ì, using a wavelength of 850 nm, with a modal bandwidth of 160 MHz/km, and a maximum distance of 160 meters. The letter R indicates that 64B/66B signal encoding is used. The W refers to the WAN Interface Sublayer (WIS) that enables compatibility between 10GbE equipment and SONET long haul equipment in a LAN-to-WAN interface scenario. See also *64B/66B*, *802.3ae*, *IEEE*, *MMF*, *modal bandwidth*, and *wavelength*.

10GBase-LR, LW (10 Gbps; Baseband; Long Range, Long Wavelength) The IEEE 802.3ae media specification for 10 Gbps Ethernet transmission over single-mode fiber (SMF) with a core diameter of 8.3μ, 9μ, or 10μ; using a wavelength of 1310 nm; and a maximum distance of 10 kilometers. The letter R indicates that 64B/66B signal encoding is used. See also *64B/66B*, *802.3ae*, *IEEE*, *SMF*, and *wavelength*.

10GBase-ER, EW (10 Gbps; Baseband; Extended Range, Long Wavelength) The IEEE 802.3ae media specification for 10 Gbps Ethernet (10GbE) transmission over single-mode fiber (SMF) with a core diameter of 8.3μ, 9μ, or 10μ; using a wavelength of 1550 nm; and a maximum distance of 40 kilometers. The letter R indicates that 64B/66B signal encoding is used. The W refers to the WAN Interface Sublayer (WIS) that enables compatibility between 10GbE equipment and SONET long haul equipment in a LAN-to-WAN interface scenario. See also *64B/66B*, *802.3ae*, *IEEE*, *SMF*, and *wavelength*.

10GBase-LX4 (10 Gbps; Baseband; Long range times 4 wavelengths) The IEEE 802.3ae media specifications for 10 Gbps Ethernet (10GbE) transmission over optical fiber, multiplexing four wavelengths through coarse wavelength division multiplexing (CWDM). There are two specifications, one of which is for multimode fiber (MMF) with a core diameter of 50μ or 62.5μ, using a wavelength of 1310 nm, with a modal bandwidth of 500 MHz/km; using a wavelength of 1550 nm; and a maximum distance of 40 kilometers. The second specification is for single-mode fiber (SMF) with a core diameter of 10μ, using a wavelength of 1310 nm, and with a maximum distance of 10 kilometers. See also *802.3ae*, *CWDM*, *IEEE*, *MMF*, *modal bandwidth*, *SMF*, and *wavelength*.

10GBase-T An IEEE standard (802.3an, 2006) for 10 Gigabit Ethernet (10GbE). 10GBase-T translates to 10 Gbps (theoretical transmission rate), Baseband (one transmission at a time over a single, shared channel), over Twisted pair media. Specifically, 10GBase-T specifies Cat 6 cable for distances up to at least 55 meters, although distances generally can be extended to 100 meters. Cat 7 cable is expected to extend those distances even farther. See also *baseband*, *Cat 6*, *Cat 7*, *channel*, *Ethernet*, *IEEE*, *standard*, *STP*, and *transmission rate*.

10GbE (10 Gbps Ethernet) See *10 Gigabit Ethernet*.

10 Gigabit Ethernet (10GbE) Standardized by the IEEE as 802.3ae, 10GbE uses the same frame format and medium access control (MAC) layer as predecessor Ethernets. 10GbE runs only in full-duplex (FDX) mode, which makes collision control unnecessary. The primary line coding technique used in both 10GbE and GbE is 8B/10B, which carries a 25 percent overhead penalty, thereby forcing the system to run at a signaling rate of 125 Gbps. Some 10GbE systems use the more recently developed 64B/66B line coding, which is similar but much more efficient. Although the signaling rates discourage the use of copper transmission media, 802.3ae currently specifies 10GBase-T for short distances using Cat 6 and Cat 7 twisted pair. Fiber optic systems are preferable, however, and 802.3ae specifies a number of options, including 10GBase-SR, SW; 10GBase-LR, LW; 10GBase-ER, EW; and 10GBase-LX4. 10GbE has application in the backbones of very bandwidth intensive local area networks (LANs) and metropolitan area networks (MANs). See also *10GBase-CX4*; *10GBase-SR, SW*; *10GBase-LR, LW*; *10GBase-ER, EW*; *10GBase-LX4*; *64B/66B*; *802.3ae*; *8B/10B*; *Cat 6*; *Cat 7*; *IEEE*; *LAN, line coding*; *MAC*; *MAN*; *overhead*; and *signaling rate*.

10GigE (10 Gbps Ethernet) See *10 Gbps Ethernet*.

10XXX See *1010XXX*.

128-QAM A variation on the quadrature amplitude modulation (QAM) signal modulation scheme. 128-QAM yields 128 possible signal combinations, with each symbol representing seven bits ($2^7 = 128$). The yield of this complex modulation scheme is that the transmission rate is seven times the signaling rate. See also *amplitude, bit, modulation, QAM, signal, signaling rate, symbol,* and *transmission rate*.

1/3 FEC A type of forward error correction (FEC) used in Bluetooth networks. The Bluetooth packet header is 16 bits in length, but is repeated three times to ensure that there are no errors in header transmission. As a result, the header consumes a total of 54 bits. This approach is overhead-intensive, but reliable. See also *Bluetooth, FEC, header, overhead,* and *packet*.

16-QAM A variation on the quadrature amplitude modulation (QAM) signal modulation scheme that splits the carrier into two waveforms that are 90° out of phase, and specifies two possible amplitude values for each of eight phase shifts separated by 45° (0°, 45°, 90°, 135°, 180°, 225°, 270°, and 315°). Thereby, each symbol carries one of 16 possible signal combinations and represents four bits ($2^4 = 16$). At a signaling rate of 2400 baud, for example, this quadbit modulation scheme yields a transmission rate of 9600 bps. 16-QAM is specified in the ITU-T V.29 Recommendation and the IEEE 802.11a (Wi-Fi5) standard. See also *802.11a, amplitude, carrier, ITU-T, modulation, QAM, quadbit, signal, signaling rate, symbol, transmission rate, V series,* and *waveform*.

16-QPSK A variation on the quadrature phase shift keying (QPSK) signal modulation scheme, 16-QPSK is a quadbit technique that impresses four bits on a baud by defining 16 phase shifts. See also *baud, bit, modulation, phase, QPSK,* and *signal*.

100Base-FX A 100Base-T extension that specifies optical fiber transmission media, and operating in full duplex (FDX). The standard specifies multimode fiber (MMF) over distances up to 2 kilometers and single-mode fiber (SMF) over distances up to 40 kilometers. See also *100Base-T, FDX, MMF,* and *SMF*.

100Base-T (100 Mbps; Baseband; Twisted pair) An IEEE standard (802.3u, 1995) similar to 10Base-T, 100Base-T translates to 100 Mbps (theoretical transmission rate), Baseband (one transmission at a time over a single, shared channel), over Twisted pair media. The maximum segment length between the 100Base-T hub and the attached device (e.g., workstation or printer) originally was specified at 100 meters or less over four pairs of Cat 3 unshielded twisted pair (UTP). The predominant version is 100Base-TX, which extends the distance to 350 meters over two pairs of Cat 5e. The 100Base-T hub is a wire hub that serves as a multiport repeater, as well as a central point of interconnection. 100Base-T uses the 4B5B line coding technique. See also *4B/5B, baseband, Cat 3, Cat 5e, channel, Ethernet, IEEE, hub, repeater, standard, transmission rate,* and *UTP*.

100Base-T4 A 100Base-T extension that specifies Cat 3 or better unshielded twisted pair (UTP), operating in either half-duplex (HDX) or full duplex (FDX) over distances up to 100 meters. See also *100Base-T, Cat 3, FDX, HDX,* and *UTP.*

100Base-TX A 100Base-T extension that specifies Cat 5 or better unshielded twisted pair (UTP), operating in either half-duplex (HDX) or full duplex (FDX) over distances up to 100 meters. See also *100Base-T, Cat 5, FDX, HDX,* and *UTP.*

100VG-AnyLAN (100 Mbps Voice Grade Any LAN) Standardized as IEEE 802.12, 100VG-AnyLAN supports Ethernet, Token Ring, and other LAN standards, incorporating a collisionless polling technique. The standard is considered obsolete, having been overwhelmed by Ethernet. See also *Ethernet, IEEE,* and *Token Ring.*

1000Base-LX (1000 Mbps; Baseband; Long range) The IEEE 802.3z media specifications for Gigabit Ethernet (GbE) transmission over multimode fiber (MMF) and single-mode fiber (SMF), using a wavelength of 1300 nm and 1310 nm, respectively. The modal bandwidth for the MMF options ranges from 400 MHz/km to 500 MHz/km, and maximum distance is 550 meters. The letter X indicates that 8B/10B signal encoding technique is used. See *802.3z* for comparisons. See also *8B/10B, IEEE, MMF, modal bandwidth,* and *wavelength.*

1000Base-SX (1000 Mbps; Baseband; Short range) The IEEE 802.3z media specifications for Gigabit Ethernet (GbE) transmission over multimode fiber (MMF) with a core diameter of either 50μ or 62.5μ, using a wavelength of 850 nm. The modal bandwidth ranges from 160 MHz/km to 500 MHz/km, and maximum distance ranges from 220 meters to 550 meters. The letter X indicates that 8B/10B signal encoding technique is used. See *802.3z* for comparisons. See also *8B/10B, IEEE, MMF, modal bandwidth,* and *wavelength.*

1000Base-T (1000 Mbps; Baseband; Twisted pair) An IEEE standard (802.3ab, 1999) similar in concept to the predecessor 10/100Base-T, 1000Base-T translates to 1000 Mbps (theoretical transmission rate) Baseband (one transmission at a time over a single, shared channel), over Twisted pair media. The original specifications called for Cat 5 cable to support Gigabit Ethernet (GbE) over four pairs and distances up to 100 meters. Cat 6 cabling specifications include unshielded twisted pair (UTP), shielded twisted pair (STP), and screened twisted pair (ScTP) rated at 250 MHz over distances up to 220 meters. See also *10Base-T, 100Base-T, baseband, Cat 5, Cat 6, channel, Ethernet, GbE, hub, IEEE, repeater, ScTP, standard, STP, transmission rate,* and *UTP.*

1010XXX The number format in the United States for accessing an interexchange carrier (IXC) for casual callers without a presubscription agreement with an IXC and for dial-around purposes in the event that the presubscribed carrier is suffering from network congestion or a failure. The format originally was 10XXX, but competition encouraged the formation of so many IXCs that the format had to be expanded. See also *IXC* and *presubscription.*

1394 The IEEE specification for a data transport bus between a host computer and peripherals, such as high-density storage devices and high-resolution still and video cameras, and is designed to eliminate the bottleneck at the serial port of a LAN-attached PC. 1394 increases the speed of the Small Computer System Interface (SCSI) to100, 200, and 400 Mbps to support take full advantage of high speed LANs. A single 1394 port can support up to 63 peripherals over a six-conductor cable up to 4.5 meters in length, and as many as 16 cables can be daisy-chained to extend the total length to as much as 72 meters. IEEE 1394 is known as *FireWire* in Apple Computer terminology. See also *bus, daisy chain, host, LAN, PC, peripheral,* and *SCSI.*

1xEV-DO (one carrier EVolution-Data Optimized) A high data rate (HDR) version of Code Division Multiple Access 2000 (CDMA2000) that employs 16-QPSK modulation in support of a peak data rate of 2.4 Mbps on the downlink and 153 kbps on the uplink. In a fully loaded cell, 1xEV-DO supports average aggregate throughput of 4.1 Mbps on the downlinks and 660 kbps on the uplinks.

1xEV-DO can run in any band and can coexist in any type of network. See also *16-QPSK*, *CDMA2000*, *downlink*, *modulation*, *throughput*, and *uplink*.

1xRTT (one times Radio Transmission Technology) The initial version of CDMA2000, with *one times* referring to standard channel width). See *CDMA2000*.

256-QAM A variation on the quadrature amplitude modulation (QAM) signal modulation scheme. 256-QAM yields 256 possible signal combinations, with each symbol representing eight bits ($2^8 = 256$). The yield of this complex modulation scheme is that the transmission rate is eight times the signaling rate. In the United States, 256-QAM is used in digital CATV applications. See also *amplitude*, *bit*, *modulation*, *QAM*, *signal*, *signaling rate*, *symbol*, and *transmission rate*.

2B+D Synonymous with *basic rate interface* (BRI) and *basic rate access* (BRA). Referring to the ITU-T specification for an ISDN interface supporting two bearer (B) channels and one data (D) channel. See also *ISDN* and *BRI*.

2B1Q (2 Binary 1 Quaternary) A line coding technique, with echo cancellation, used in North America for ISDN basic rate interface (BRI). 2B1Q is a form of pulse amplitude modulation (PAM) that uses four (i.e., quaternary) levels of amplitude (i.e., voltage), each of which represents two adjacent bits in a bit stream, and is accomplished by varying the voltage at nominal levels of ±1 (actually 0.833) and ±3 (actually 2.5) volts, as illustrated in the accompanying figure. As 2B1Q impresses two bits on each baud, the baud rate is halved, and a baud rate of 80 kilobaud will support a signaling rate of 160 kbps. Because statistics force the line voltage to be positive half the time and negative half the time, on average, the signal power lies at a frequency of 40 kHz, which is half the baud rate. 2B1Q also is the electrical line coding technique used in high-bit-rate digital subscriber line (HDSL). 2B1Q also is known as *4-level pulse amplitude modulation* (4 PAM). In European and many other countries, the line coding technique employed is 4 Binary 3 Ternary (4B3T). See also *4B3T*, *baud rate*, *bit rate*, *BRI*, *echo cancellation*, *HDSL*, *ISDN*, and *PAM*.

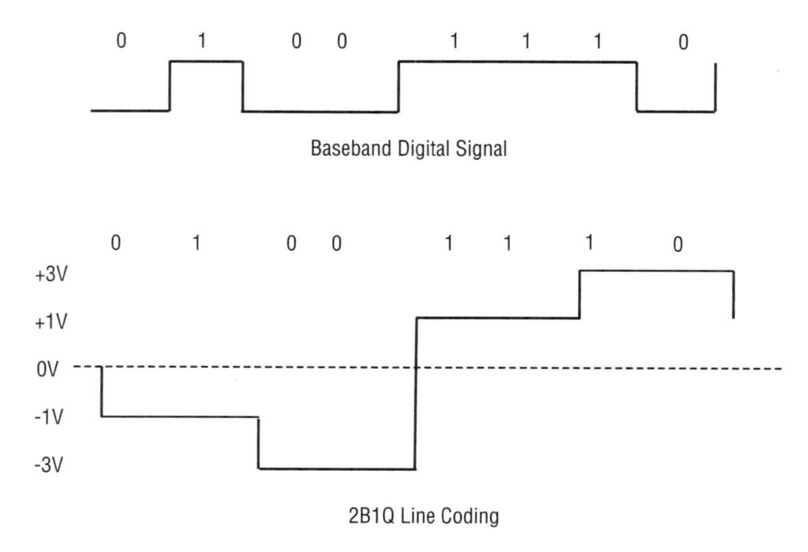

2 Binary 1 Quaternary (2B1Q) See *2B1Q*.

23B+D Synonymous with *primary rate interface* (PRI). Referring to the ITU-T specification for an ISDN interface supporting 23 bearer (B) channels and 1 data (D) channel. 23B+D is compatible with T1 and J-1, and is used in North America and Japan. See also *ISDN* and *PRI*.

30B+D Synonymous with *primary rate access* (PRA). Referring to the ITU-T specification for an ISDN interface supporting 30 bearer (B) channels and one data (D) channel. 30B+D is compatible with E-1, and is used outside of North America and Japan. See also *ISDN* and *PRA*.

2G (2nd Generation) In cellular radio, referring to systems based on the first digital standards, including Digital-AMPS (D-AMPS), Global System for Mobile Communications (GSM), Personal Digital Cellular (PDC), and Personal Communications System (PCS). See also *cellular radio*, *D-AMPS*, *digital*, *GSM*, *PDC*, and *PDH*.

2.5G (Two point Five Generation) In cellular radio, referring to digital systems and standards that are midway, or transitional, between 2G and 3G. Such systems include Enhanced Data rates for GSM Evolution (EDGE), General Packet Radio Service (GPRS), High-Speed Circuit Switched Data (HSCSD), and Universal Mobile Telecommunications System (UMTS). See also *2G*, *3G*, *cellular radio*, *digital*, *EDGE*, *GPRS*, and *HSCSD*.

218–219 MHz Service In the United States, a short distance, interactive licensed communications service for the transmission of information, product, and service offerings. The service is licensed in two blocks, with Block A operating in the 218.0–218.5 MHz band and Block B in the 218.5–219.0 MHz band. A system comprises one or more cell transmitter stations (CTSs) and response transmitter units (RTSs). Anticipated applications include ordering goods and services offered by television services, viewer polling, remote meter reading, vending machine inventory control, and cable television theft deterrence. The Federal Communications Commission (FCC) includes the service in the family of personal radio services. The service formerly was known as *Interactive Video and Data Services* (IVDS), although the bandwidth is insufficient for video transmission. See also *bandwidth*, *FCC*, and *personal radio services*.

3G (3rd Generation) Referring to digital cellular radio systems and standards that fit under the umbrella of International Mobile Telecommunications-2000 (IMT-2000), an ITU initiative for a twenty-first century wireless network architecture. Specifications include 128/144 kbps for high-mobility applications, 384 kbps for pedestrian speed (i.e., walking speed) applications, and 2.048 Mbps for both fixed WLL (Wireless Local Loop) and in-building applications such as WLANs (Wireless LANs). 3G systems include Universal Mobile Telecommunications System (UMTS), also known as Wideband CDMA (W-CDMA), Code Division Multiple Access 2000 (CDMA2000), and Time Division-Synchronous Code Division Multiple Access (TD-SCDMA). See also *CDMA2000*, *cellular radio*, *IMT-2000*, *ITU-T*, *TD-SCDMA*, *UMTS*, *W-CDMA*, *WLAN*, and *WLL*.

3GPP (3rd Generation Partnership Project) A collaboration that brings together a number of national standards parties to develop 3G mobile systems based on evolved Global System for Mobile Communication (GSM) core technologies and standards. The scope includes maintenance and development of General Packet Radio Service (GPRS), Enhanced Data rates for GSM Evolution (EDGE), and Universal Mobile Telecommunications System (UMTS). For contact information, see Appendix A. See also *3G*, *EDGE*, *GPRS*, *GSM*, *standards*, and *UMTS*.

3rd Generation Partnership Project (3GPP) See *3GPP*.

3x In cellular radio systems, also known as *IS-2000-A*. An enhancement to CDMA2000 that uses three cdmaOne carriers for total bandwidth of 3.75 MHz. See also *bandwidth*, *carrier*, *cellular radio*, *CDMA2000*, and *cdmaOne*.

419 Advance Fee Fraud Also known as the *Nigerian Connection* or *Nigerian Scam*. Named after a formerly relevant section of the Criminal Code of Nigeria, where the scam originated and where most activity still originates. The target receives an unsolicited e-mail (i.e., spam) indicating something along the lines that some member of a previous government or royal family of Nigeria or some other West African nation (or now some country in the Middle East or some other *exotic* place) has had substantial funds frozen by the current government and cannot access or expatriate those funds. The recipient of the e-mail can help by wiring $5,000 or so as an advance fee, transfer fee, or performance bond, for which the target will receive a much larger sum of money. Sometimes the offer is for the target to accept a cashier's check for $100,000 and return $75,000 of that via wire transfer, but keep the rest as a fee. The $100,000 cashier's check is bogus, of course, while the $75,000 wire transfer is real. Sometimes the scam involves alleged over-invoicing or double-invoicing, with the party receiving the fake cashier's check asked to cash it and wire the overage back to the scam artist, less a handsome fee for the inconvenience, of course. This sort of fraudulent activity has gone on at least since the 1970s through the postal service and via fax machines, but is much more prevalent these days due to the widespread reach and ease of use of Internet e-mail. There are endless variations on the scam, and not all originate in Nigeria, although many originate in West African countries. See also *e-mail*, *Internet*, and *spam*.

4B3T (4 Binary 3 Ternary) A line coding technique used in European and many other countries outside North America for ISDN basic rate access (BRA). 4B3T combines four bits to represent one ternary signal state. Therefore, the baud rate is three-fourths of the signaling rate, and ISDN BRA at 160 kbps requires a baud rate of 120 KBaud. 4B3T yields shorter ISDN transmission distances than 2B1Q, but distances in Europe and elsewhere often are much shorter between the central office exchange (COE) and the customer premises. See also *2B1Q*, *baud rate*, *BRI*, *COE*, *ISDN*, and *signaling rate*.

4B/5B (4 Bits/5 Bits) A line coding technique used in 100Base-TX, 100Base-FX, and Fiber Distributed Data Interface (FDDI) LANs. 4B/5B refers to the fact that every nibble of 4 Bits of data is encoded into 5 Bits of signal. Specified by the American National Standards Institute (ANSI) X3T9.5 committee recommendation for FDDI, 4B/5B is sometimes referred to as block coding, as a block of data bits is mapped into a block of signaling bits. This approach increases the number of bit patterns from 16 ($2^4 = 16$) to 32 ($2^5 = 32$), which means that each five-bit signal block includes enough clocking pulses and signal transitions to synchronize the network. 4B5B also provides some level of error detection. So, the signaling rate must be 125 MHz to support a unipolar code (such as that used with classic 10 Mbps Ethernet) with a signaling rate of 125 Mbps, which, in turn, supports a data rate of 100 Mbps. To address this issue, 100Base-TX uses an intermediate step known as multi-level transition (MLT) that reduces the carrier frequency to only 31.25 MHz. See also *100Base-TX*, *100Base-FX*, *8B/10B*, *ANSI*, *block coding*, *carrier*, *FDDI*, *LAN*, *MLT*, *quadbit*, *synchronize*, and *unipolar*.

4 Binary 3 Ternary (4B3T) See *4B3T*.

4-level pulse amplitude modulation (4 PAM) More commonly known as *2 Binary 1 Quaternary* (2B1Q). See *2B1Q*.

4 PAM (4-level Pulse Amplitude Modulation) More commonly known as *2 Binary 1 Quaternary* (2B1Q). See *2B1Q*.

500 Service In the North American Numbering Plan (NANP), 500 service is intended to support premium follow-me personal communications services, which are defined as a set of capabilities that allows some combination of personal mobility, terminal mobility, and service profile management. 500 service is designed to allow the subscriber to select from a menu of services and to initiate or receive calls at any terminal, fixed or mobile, across networks, regardless of geographic carrier. 500 numbers can be protected by the personal identification number (PIN) for network and feature access. As 500 numbers are not location-specific and are transportable across carriers and carrier domains, one number theoretically can be retained for life. 500 service numbers are in the format 500-NXX-XXXX. See also *NANP*.

512-QAM A variation on the quadrature amplitude modulation (QAM) signal modulation scheme. 512-QAM yields 512 possible signal combinations, with each symbol representing nine bits ($2^9 = 512$). The yield of this complex modulation scheme is that the transmission rate is nine times the signaling rate. See also *amplitude*, *bit*, *modulation*, *QAM*, *signal*, *signaling rate*, *symbol*, and *transmission rate*.

64B/66B A block line coding technique in which 8 data octets comprising 64 data bits is encoded into a block of 66 signaling bits prior to transmission. The data are scrambled in a self-synchronous scrambler function that is intended to even the distribution of 1s and 0s and thereby both achieve DC balance on the line and prevent intersymbol interference. See also *balance*, *bit*, *block coding*, *line coding*, *DC*, *intersymbol interference*, *octet*, and *synchronize*.

64-QAM A variation on the quadrature amplitude modulation (QAM) signal modulation scheme. 64-QAM yields 64 possible signal combinations, with each symbol representing six bits ($2^6 = 64$). The yield of this complex modulation scheme is that the transmission rate is six times the signaling rate. In the United States, 64-QAM is used in digital CATV applications and is specified in the IEEE 802.11a (Wi-Fi5) standard. See also *802.11a*, *amplitude*, *bit*, *CATV*, *modulation*, *QAM*, *signal*, *signaling rate*, *symbol*, and *transmission rate*.

80/20 Rule See *Pareto principle*.

800 Service The original area code prefix for toll free long distance services in the United States was 800. As the service gained in popularity during the 1990s, the numbering scheme was expanded to include 866, 877, and 888 area codes. See also *area code* and *toll free service*.

802 The IEEE 802 LAN/MAN Standards Committee develops local area network (LAN) standards and metropolitan area network (MAN) standards. The most widely used standards are for the Ethernet family, Token Ring (TR), wireless LAN (WLAN), bridging and virtual bridged LANs. An individual Working Group (WG) provides the focus for each area. See also *bridge*, *Ethernet*, *IEEE*, *LAN*, *MAN*, *standard*, *Token Ring*, and *WLAN*.

802.1 The IEEE Working Group that concerns itself with standards and recommendations in the areas of architecture and internetworking of local area networks (LANs) and metropolitan area networks (MANs), security, network management, and protocol issues above the Data Link Layer. See also *Data Link Layer*, *IEEE*, *LAN*, and *MAN*.

802.1p The IEEE specification (September 1998) that enables LAN switches and other devices (e.g., bridges and hubs) to prioritize traffic into one of eight classes. Class 7, the highest priority, is reserved for network control data such as Open Shortest Path First (OSPF) and Routing Information Protocol (RIP) table updates. Classes 6 and 5 can be used for voice, video, and other delay-sensitive traffic. Classes 4 through 1 address streaming data applications through loss-tolerant traffic such as File Transfer Protocol (FTP). Class 0, the default class, is a best effort class. In conjunction with the 802.1q specification for VLAN tagging, 802.1p paved the way for standards-based multivendor grade of service (GoS). See also *FTP*, *GoS*, *IEEE*, *LAN*, *OSPF*, *RIP*, and *VLAN*.

802.1q The IEEE specification that defines a virtual LAN (VLAN), which allows multiple logical LANs to share the same physical infrastructure. 802.1q defines a 32-bit tag that is added to the frame header to identify the specific VLAN and provide for priority indication. See also *frame*, *IEEE*, *header*, *LAN*, *logical*, *physical*, and *VLAN*.

802.2 The IEEE Working Group that develops standards for Logical Link Control (LLC), which corresponds to the upper sublayer of the Data Link Layer in the OSI Reference Model. The LLC sublayer is concerned with issues of multiplexing, flow control, and detection and retransmission of dropped frames. See also *Data Link Layer*, *flow control*, *IEEE*, *LLC*, *multiplexer*, and *OSI Reference Model*.

802.3 The IEEE standard that defines the carrier sense multiple access (CSMA) method of medium access control (MAC), and Physical Layer specifications. Although the term is commonly used interchangeably with Ethernet, 802.3 actually is a variation on the original Ethernet standard. (*Note:* Project 802 took its name from the fact that it was established in the year 1980, and the month 2, i.e., February.) See also, *CSMA*, *Ethernet*, *IEEE*, *LAN*, *MAC*, and *standard*.

802.3ab The IEEE standard for 1000Base-T. See *1000Base-T.*

802.3ae The IEEE media specifications for 10 Gigabit Ethernet (10GigE). Those specifications address both multimode (MMF) and single-mode (SMF) optical fiber, core diameter, wavelength, modal bandwidth (MMF), and distance limitations. The specifications are listed in the following table.

10 Gigabit Ethernet (IEEE 802.3ae) Media Specifications

Standard	*Fiber Type*	*Core Diameter*	*Wavelength*	*Modal Bandwidth (MHz/km)*	*Distance (Maximum)*
10GBase-SR, SW	MMF	62.5μ	850 nm	160	300 m
10GBase-LR, LW	SMF	8.3μ, 9μ, 10μ	1310 nm	Not Applicable	10 km
10GBase-ER, EW	SMF	8.3μ, 9μ, 10μ	1550 nm	Not Applicable	40 km
10GBase-LX4	MMF	50μ, 62.5μ	1310 nm	500	300 m
10GBase-LX4	SMF	10μ	1310 nm	Not Applicable	10 km

See each standard for more detail. See also *MMF, modal bandwidth, SMF,* and *wavelength.*

802.3af The IEEE standard (June 2003) for power over Ethernet (PoE). See *PoE.*

802.3ak The IEEE standard for 10GBase-CX4. See *10GBase-CX4.*

802.3an The IEEE standard for 10GBase-T. See *10GBase-T.*

802.3u The IEEE standard for 100Base-T. See *100Base-T.*

802.3z The IEEE media specifications for Gigabit Ethernet (GbE, or GigE). Those specifications address both multimode (MMF) and single-mode (SMF) optical fiber, core diameter, wavelength, modal bandwidth (MMF), and distance limitations. The specifications are listed in the following table.

Gigabit Ethernet (IEEE 802.3z) Media Specifications

Standard	*Fiber Type*	*Core Diameter*	*Wavelength*	*Modal Bandwidth (MHz/km)*	*Distance (Maximum)*
1000Base-SX	MMF	62.5 μ	850 nm	160	220 m
1000Base-SX	MMF	62.5 μ	850 nm	200	275 m
1000Base-SX	MMF	50.0 μ	850 nm	400	500 m
1000Base-SX	MMF	50.0 μ	850 nm	500	550 m
1000Base-LX	MMF	62.5 μ	1300 nm	500	550 m
1000Base-LX	MMF	50.0 μ	1300 nm	400	550 m
1000Base-LX	MMF	50.0 μ	1300 nm	500	550 m
1000Base-LX	SMF	9 μ	1310 nm	Not Applicable	5 km

See each standard for more detail. See also *MMF, modal bandwidth, SMF,* and *wavelength.*

802.4 The IEEE standard for Token Bus, a local area network (LAN) token passing protocol based on a physical bus topology. Token Bus is considered an orphaned standard, as the 802.4 committee disbanded in 2004 due to lack of interest. See also *bus, Ethernet, IEEE, LAN,* and *token passing.*

802.5 The IEEE standard for token-passing ring access method and Physical Layer specifications. The 802.5 recommendations include Token Ring. See also *IEEE*, *Physical Layer*, *token passing*, and *Token Ring*.

802.6 The IEEE standard for metropolitan area network (MAN) access method and Physical Layer specifications. Distributed Queue Dual Bus (DQDB) is defined here. Switched Multimegabit Data Service (SMDS) was derived from 802.6. This standard has been withdrawn. See also *IEEE*, *DQDB*, *MAN*, *Physical Layer*, and *SMDS*.

802.7 The IEEE Broadband Technical Advisory Group, chartered to develop standards for definition of a broadband cable plant design and establish guidelines for LAN construction within a physical facility such as a building. This standard has been withdrawn.

802.8 The IEEE Fiber Optic Technical Advisory Group, established to assess impact of fiber optics and to recommend standards.

802.9 The IEEE Integrated Services LAN (ISLAN) design for the integration of voice and data networks, both within the LAN domain and interfacing to publicly and privately administered networks running protocols such as FDDI and ISDN. The 802.9 Working Group developed the Isochronous Ethernet (IsoEthernet or IsoEnet) standard, which has been withdrawn.

802.10 The IEEE standards for Interoperable LAN/MAN Security (SILS). This standard was withdrawn in 2004, and the working group is currently inactive. Security for wireless networks is being addressed in 802.11i. VLAN security is addressed in 802.11q. See also *security* and *VLAN*.

802.11 The family of IEEE standards describing the over-the-air interfaces for a number of wireless local area networks (WLANs). Variously referred to in the vernacular as Wi-Fi (Wireless Fidelity) and Wireless Ethernet (the Ethernet CSMA/CA protocol is used in 802.11), 802.11 standards include infrared (IR) and radio frequency (RF) solutions, although there currently appear to be no practical applications for IR. The RF standards fall into the 2.4 GHz and 5 GHz ISM bands and offer theoretical bandwidth up to 54 Mbps. The original 802.11 standard (1997) operated in the 2.4 GHz band and supported theoretical data rates up to 2 Mbps. This early standard included a great number of options, which made interoperability of products difficult, or at least uncertain. As a result, 802.11 never gained any real traction in the market. Soon afterward, however, much improved extensions to 802.11 were finalized, and WLANs quickly gained in popularity. Current extensions include 802.11a, 802.11b, and 802.11g. Still under development is 802.11n. See also *802.11a*, *802.11b*, *802.11g*, *802.11n*, *CSMA/CA*, *Ethernet*, *IEEE*, *ISM*, *RF*, *Wi-Fi*, *Wi-Fi5*, and *WLAN*.

802.11a Also known as *Wi-Fi5* (*Wi*reless *Fi*delity *5* GHz). The IEEE standard for an 802.11 wireless local area network (WLAN) operating in the 5-GHz range. 802.11a uses coded orthogonal frequency division multiplexing (COFDM), which sends a stream of data symbols in a massively parallel fashion across multiple subcarriers, each of which is 20 MHz wide and is subdivided into 52 subcarrier channels, each of which is approximately 300 KHz wide. Of those subcarrier channels, 48 are used for data transmission, and the remaining 4 for error control. The specific modulation scheme depends on link quality, with the highest link quality employing the most complex technique, which yields the greatest signaling rate. As the link quality deteriorates, the modulation technique ratchets down to the least capable. In order of complexity and yield, those techniques are as follows:

- Binary phase-shift keying (BPSK) at 125 kbps per channel for a total of 6 Mbps (125 kbps × 48 channels = 6 Mbps), and 187.5 kbps for a total of 9 Mbps
- Quadrature phase-shift keying (QPSK) at 250 kbps per channel for a total of 12 Mbps, and 375 kbps per channel for a total of 18 Mbps

- 16-level quadrature amplitude modulation (16-QAM) at 500 kbps per channel for a total of 24 Mbps, and 750 kbps per channel for a total of 36 Mbps
- 64-level quadrature amplitude modulation (64-QAM) at 1 Mbps per channel for a total of 48 Mbps, and 1.125 Mbps per channel for a total of 54 Mbps

In Europe, 802.11a competes for spectrum with HiperLAN, the standard developed and promoted by the European Telecommunications Standards Institute (ETSI). ETSI requires that two additional protocols be used in conjunction with 802.11a in order to protect incumbent applications and systems running over previously allocated shared spectrum. Dynamic frequency selection (DFS) allows the 802.11a system to dynamically shift frequency channels, and transmission power control (TPC) reduces the power level. In combination, these protocols serve to eliminate interference issues with incumbent signals. See also *16-QAM, 64-QAM, 802.11, BPSK, channel, ETSI, IEEE, modulation, QAM, QPSK, protocol, signaling rate, sub-carrier, symbol,* and *WLAN.*

802.11b Also known as *Wi-Fi* (*Wi*reless *Fi*delity). The IEEE standard for an 802.11 wireless local area network (WLAN) operating in the 2.4-GHz band. 802.11b transmission options include infrared (IR), which is rarely used, and radio frequency (RF). 802.11b uses direct sequence spread spectrum (DSSS) modulation. DSSS involves the transmission of a bit stream that is modulated with the Barker code chipping sequence. Each bit is encoded into a redundant 11-bit Barker code (e.g., 10110111000), with each resulting data object forming a chip. The chip is put on a carrier frequency in the 2.4 GHz range (2.4–2.483 GHz) and the waveform is modulated using one of several modulation schemes, depending on link quality. The highest link quality employs the most complex technique, which yields the greatest signaling rate. As the link quality deteriorates, the modulation technique ratchets down to the least capable. In order of complexity and yield, those techniques are as follows:

- Binary phase-shift keying (BPSK) at 1 Mbps.
- Quadrature phase-shift keying (QPSK) at 2 Mbps.
- Complementary code keying (CCK) at 5.5 and 11 Mbps.

Radio link quality is always uncertain, especially when using unlicensed frequencies due to the potential for mutual interference with other systems. At the lowest rate, however, 802.11b link quality generally is acceptable at distances of up to 100 meters or so. 802.11b divides the available spectrum into 14 channels, each of which has a width of 25 MHz. In the United States, the FCC allows the use of 11 channels. Four channels are available in France, 13 in the rest of Europe, and only 1 in Japan. There also is overlap between adjacent channels as each has a width of 25 MHz and all share a band that is only 83 MHz (2.4–2.483 GHz) wide. See also *802.11, Barker code, BPSK, carrier, CCK, DSSS, frequency, IR, modulation, QPSK, RF,* and *WLAN.*

802.11e The IEEE standard (2005) for wireless LAN (WLAN) quality of service (QoS) based on access priority classes. 802.11e introduces a coordination function that provides a station with high priority traffic such as voice with more frequent network access than a station with low priority traffic such as e-mail. The station with the high priority traffic also is granted a longer transmit opportunity, or window, in which to transmit as many frames as possible. In all, 802.11e defines four access priority classes, which the Wi-Fi Alliance terms *Wi-Fi MultiMedia Extensions* (WMMs, or WMEs). Those classes are as follows:

- **Voice Priority**, the highest level, is defined in support of low latency voice.
- **Video Priority**, the second highest level, prioritizes video relative to other data traffic. One 802.11a/b channel can support three-to-four standard definition television (SDTV) datastreams or one high definition television (HDTV) datastream.

- **Best Effort Priority** is intended to support traffic from legacy devices and from applications or devices that lack QoS capabilities. Web browsing is an example of best effort traffic.
- **Background Priority** is defined in support of low priority traffic without strict latency and throughput requirements. Examples cited include file downloads and print jobs.

802.11e is critical to the successful implementation of voice over Wi-Fi (VoWiFi). See also *802.11a*, *802.11b*, *frame*, *HDTV*, *latency*, *SDTV*, *throughput*, *VoWiFi*, *Wi-Fi Alliance*, and *WLAN*.

802.11g Also known as *Wi-Fi* (*Wi*reless *Fi*delity). The IEEE standard (June 2003) for an 802.11 wireless local area network (WLAN) operating in the 2.4-GHz band at a signaling speed of up to 54 Mbps. Backward-compatible with 802.11b, 802.11g divides the available radio frequency (RF) spectrum into 14 channels, each of which has a width of 25 MHz. In the United States, the FCC allows the use of 11 channels, only three of which can be used in a confined area at any given time without overlap. Four channels are available in France, 13 in the rest of Europe, and only 1 in Japan. Like 802.11a, 802.11g uses orthogonal frequency division multiplexing (OFDM) at data rates of 6, 9, 12, 18, 24, 36, 48, and 54 Mbps, with the attainable speed being highly sensitive to distance and line of sight (LOS). At 5.5 Mbps and 11 Mbps, the modulation technique reverts to complementary code keying (CCK), which also is used in 802.11b. At 2 Mbps, it reverts to direct sequence spread spectrum (DSSS) and quadrature phase-shift keying (QPSK), and at 1 Mbps to DSSS and binary phase-shift keying (BPSK), again defaulting to the 802.11a specification. Tri-mode components allow 802.11a/b/g-equipped terminals and access points (APs) to interoperate, although supporting multiple simultaneous protocols affects performance negatively. See also *802.11*, *802.11a*, *802.11b*, *AP*, *BPSK*, *CCK*, *channel*, *DSSS*, *FCC*, *IEEE*, *LOS*, *modulation*, *OFDM*, *QPSK*, *RF*, *spectrum*, and *WLAN*.

802.11i The IEEE standard (June 2004) for 802.11 wireless LAN (WLAN) security mechanisms. The standard is more commonly known as *WPA2* (Wi-Fi Protected Access version 2), which is the fully compliant mechanism developed and promoted by the Wi-Fi Alliance. See also *802.11*, *IEEE*, *Wi-Fi Alliance*, *WLAN*, and *WPA*.

802.11n The developing IEEE standard (estimated March 2009) for an 802.11 wireless local area network (WLAN) operating in the 2.4-GHz band and operating at a signaling speed of up to 108 Mbps, with an option to increase speed to as much as 600 Mbps. 802.11n will be backward-compatible with 802.11a/b/g, building on them by introducing antenna technology known as multiple-input multiple-output (MIMO), which is based on the concept of spatial diversity. The transmitter splits the signal among multiple transmit antennas separated by some amount of space, but operating on the same frequency at the same time. The multiple receive antennas gather the signal, which has suffered from the effects of multipath propagation. Some signal elements will be stronger than others and will arrive ahead of others. Sophisticated signal processing software combines and correlates many signal elements arriving at different times into one linear combination of a stronger, synchronized, intelligible signal derived from each of the receive antennas and reconstitutes the original data stream. See also *802.11a*, *802.11b*, *802.11g*, *IEEE*, *frequency*, *MIMO*, *multipath propagation*, and *spatial diversity*.

802.12 The IEEE standards for 100+ Mbps LANs using demand priority access. The focus was on 100VG-AnyLAN, a standard that is considered obsolete.

802.14 The IEEE standards initiative for Broadband Cable Access Method and Physical Layer Specification in Local and Metropolitan Networks. The standard was withdrawn.

802.15 The IEEE medium access control (MAC) and Physical Layer (PHY) specifications for wireless personal area networks (WPANs). Bluetooth is formalized in 802.15.1 (2002). 802.15.4 is the specification for a low-data-rate WPAN comprising devices of very low complexity transmitting at very low power levels and very long battery life. ZigBee is based on 802.15.4. See also *Bluetooth*, *MAC*, *PAN*, and *ZigBee*.

802.16 The IEEE specification entitled Air Interface for Fixed Broadband Wireless Access (BWA) Systems. The 802.16 Working Group was formed to develop a set of specifications to standardize Local Multipoint Distribution Service (LMDS) and Multichannel Multipoint Distribution Services (MMDS), once promising BWA technologies that proved too costly and unreliable. The 802.16 specifications evolved over a number of years, as follows:

- 802.16 (2001) standardized LMDS, focusing on fixed wireless solutions in both point-to-point and point-to-multipoint configurations, operating at frequencies in the 10–66 GHz range, and requiring line of sight (LOS).
- 802.16a (2003) was based on MMDS and the European HiperMAN system. This extension includes both point-to-point and point-to-multipoint configurations, operates in the 2–11 GHz range, and does not require LOS in the lower bands.
- 802.16d, aka 802.16-2004 (2004), is a compilation and modification of previous versions and amendments. 802.16d specifies frequencies in the 2–11GHz range, and includes point-to-point, point-to-multipoint, and mesh topologies. This specification recommends, but does not require, LOS, and includes support for indoor CPE.
- 802.16e (October 2005), formally known as *Mobile WirelessMAN*, adds hand-off capability, thereby supporting portability and pedestrian-speed mobility for users of laptop, tablet, and handheld computers. Operating in the 2–6 GHz range, it is designed for point-to-multipoint applications and does not require LOS.

See *WiMAX* for more technical detail. See also *BWA*, *CPE*, *hand-off*, *HiperMAN*, *LMDS*, *LOS*, *mesh*, *MMDS*, *point-to-point*, *point-to-multipoint*, *portability*, and *WLL*.

802.17 The IEEE Working Group on Resilient Packet Ring (RPR) Access Method and Physical Layer Specifications. RPR standards address the resilient and efficient transfer of data packets over fiber optic local, metropolitan and wide area networks (LANs, MANs, and WANs) at rates scalable to many Gbps. See also *RPR*.

802.22 The IEEE Working Group on Wireless Regional Area Networks (WRANs). 802.22 is an initiative directed toward the development of a cognitive radio air interface for use by license-exempt radios on a non-interfering basis in spectrum currently allocated to television broadcast service. See also *cognitive radio*, *IEEE*, and *spectrum*.

850 Band The wavelength band defined by the ITU-T window as 810–890 nm. See also *window*.

866 An area code prefix for toll free services in the United States. See also *800 Service*, *area code*, and *toll free service*.

877 An area code prefix for toll free services in the United States. See also *800 Service*, *area code*, and *toll free service*.

888 An area code prefix for toll free services in the United States. See also *800 Service*, *area code*, and *toll free service*.

8B/10B (8 Bits/10 Bits) A line coding technique that maps eight data bits into a 10-bit symbol, or character. The eight-bit data octet is divided into two groups. The three most significant bits, or leftmost bits, are encoded into a four-bit group (3B/4B). The five least significant bits, or rightmost bits, are encoded into a six-bit group (5B/6B). The two groups are then concatenated, or joined together, and placed on the line. As eight bits yields 256 possible bit combinations ($2^8 = 256$) and 10 bits yields 1,024 ($2^{10} = 1024$) bit combinations, each eight-bit data octet can be phrased two different ways, with one being the bit-wise inverse of the other. For example, a data octet of 11001010 might be expressed the first time as 1000100111, and the second time as 0111011000. That encoding scheme yields direct current (DC) electrical balance on the line, as the number of 1s and 0s will be equal in the long term. This approach also ensures proper clocking as there is sufficient ones density. 8B/10B also provides an additional embedded error control mechanism similar to that discussed in 4B/5B, which is used in 100Base-TX. On the downside, 8B/10B adds 25 percent overhead (10/8 = 1.25) to the serial datastream. *Note:* The 10B format also provides for a number of control characters. 8B/10B is used in 10 Gigabit Ethernet (10GbE or 10GigE), Gigabit Ethernet (GbE or GigE), ESCON and Fibre Channel Storage Area Networks (SANs). See also *10 Gigabit Ethernet, 4B/5B, 64B/66B, balanced, DC, encode, ESCON, Fibre Channel, GigE, ones density*, and *line coding.*

8-level vestigial sideband (8-VSB) See *8-VSB.*

8-Phase Shift Keying (8-PSK) See *8-PSK.*

8-PSK (8-Phase Shift Keying) A signal modulation technique that involves eight levels of phase shift and, therefore, supports three bits per symbol. 8-PSK is specified in Enhanced General Packet Radio Service (EGPRS), the packet-switched transmission mode of Enhanced Data rates for GSM Evolution (EDGE), a 2.5G cellular radio standard. See also *2.5G, bit, cellular radio, EDGE, EGPRS, modulation, phase, PSK, signal*, and *symbol.*

8-VSB (8-level vestigial sideband) A radio frequency (RF) modulation technique specified by the Advanced Television Systems Committee (ATSC) for use in digital television (DTV), including both high definition television (HDTV) and standard definition television (SDTV). 8-VSB employs 8 levels of amplitude modulation (AM) to achieve a bit rate of 19.28 Mbps over a 6 MHz terrestrial broadcast channel. See also *AM, ATSC, bit rate, broadcast, channel, DTV, HDTV, RF, SDTV*, and *VSB.*

900 A dial prefix for toll free services in some countries. See also *800, area code*, and *toll free service.*

900 Service In the North American Numbering Plan (NANP), 900/976 numbers are used to access premium information services that carry either a flat cost per call or a cost per minute determined by the called party, or sponsoring party. The revenues are divided among the sponsoring party receiving the call and the various carriers involved. 900/976 service was originally intended for applications such as telethons and informational services, and was used widely by companies offering technical support on a per-call and per-minute basis. Subsequently, the service gained a bad reputation as many providers of telephone sex and other questionable services make extensive use of them. Fraud schemes actively use 900/976 numbers. In such a scheme, a caller will leave a message asking for a return call to a 900/976 number, with the call carrying an exorbitant charge. As a result, 900/976 services have fallen out of favor and are no longer widely used. 900 numbers are in the format 900-NXX-XXXX. See also *NANP.*

911 The dialing sequence used in the United States for calls for emergency assistance from police, fire, and ambulance services. Calls to 911 are routed to a public safety access point (PSAP). Enhanced 911 (E911) provides caller location information to the PSAP. See also *E911* and *PSAP.*

2001: A Space Odyssey The movie script was written by Arthur C. Clarke and Stanley Kubrick, and based on Clarke's short story "The Sentinel." In telecommunications circles, Clarke is better known as the inventor of communications satellites. He publicly unveiled the concept in an article entitled "Extra-Terrestrial Relays: Can Rocket Stations Give World-wide Radio Coverage," published in *Wireless World* in February 1945. See also *Clarke, Sir Arthur C.*

A (Ampere) See *ampere*.

AAL (ATM Adaptation Layer) In the ATM reference model, a layer that comprises two sublayers. The Convergence Sublayer (CS) functions are determined by the specifics of the service class supported by that particular AAL. The Segmentation and Reassembly (SAR) sublayer functions to segment the user data into payloads for insertion into cells, on the transmit side. On the receive side, the SAR extracts the payload from the cells and reassembles the data into the information stream as originally transmitted. There exist defined AAL Types 1, 2, 3/4, and 5, each of which supports a specific class of traffic. See also *ATM*, *ATM reference model*, *CS*, and *SAR*.

AAL1 (ATM Adaptation Layer Type 1) Supports Class A traffic, which is connection-oriented constant bit rate (CBR) traffic timed between the source and the sink. Such traffic is stream-oriented and intolerant of latency. Isochronous traffic such as digitally encoded, uncompressed voice is supported via AAL 1, which essentially supports the emulation of a T/E-carrier circuit. All such traffic is carefully timed and must depend on a guaranteed rate of network access, transport, and delivery. Such traffic is marked as high-priority in the cell header, as transmission delays could considerably impact presentation quality. See also *ATM*, *CBR*, *cell*, *compression*, *connection-oriented*, *E-carrier*, *emulation*, *header*, *isochronous*, *latency*, *sink*, *source*, *stream-oriented*, and *T-carrier*.

AAL2 (ATM Adaptation Layer Type 2) Supports Class B traffic, which is connection-oriented, real-time variable bit rate (rt-VBR), isochronous traffic timed between the source and the sink. Compressed audio and video are Class B. See also *ATM*, *cell*, *compression*, *connection-oriented*, *header*, *isochronous*, *rt-VBR*, *sink*, and *source*.

AAL3/4 (ATM Adaptation Layer Type 3/4) Supports Class C or Class D traffic, which is non real-time variable bit rate (nrt-VBR) data traffic with no timing relationship between the source and the sink. Class C traffic, such as X.25 and frame relay, is connection-oriented VBR traffic with no timing relationship between source and sink. Class D traffic, such as LAN and SMDS, is connectionless VBR traffic that is sensitive to loss, but not highly sensitive to delay. Message mode service is used for framed data in which only one interface data unit (IDU) is passed. Streaming mode service is used for framed data in which multiple IDUs are passed in a stream. See also *ATM*, *Class C ATM traffic*, *Class D*, *connectionless*, *connection-oriented*, *frame relay*, *IDU*, *message mode service*, *nrt-VBR*, *streaming mode service*, and *X.25*.

AAL5 (ATM Adaptation Layer Type 5) Supports Class C traffic in message mode, only. Such traffic is variable bit rate (VBR) traffic with no timing relationship between the source and the sink, and consists of only one interface data unit (IDU). AAL Type 5 also is known as *Simple and Efficient AAL* (SEAL), as some of the overhead has been stripped out of the Convergence Sublayer (CS). AAL 5 also supports Class X traffic, which is variable bit rate (VBR) and specifically either unspecified bit rate (UBR) or available bit rate (ABR), and is either connection-oriented or connectionless in nature. AAL 5 is used in support of a wide variety of data traffic, including LAN Emulation (LANE) and Internet Protocol (IP). See also *ABR*, *ATM*, *Class C ATM traffic*, *Class X ATM traffic*, *connectionless*, *connection-oriented*, *CS*, *IDU*, *IP*, *LANE*, *message mode service*, *sink*, *source*, *UBR*, and *VBR*.

AAV (Alternative Access Vendor) Synonymous with Competitive Access Provider (CAP). See *CAP*.

abbreviation A shortened form of a word or phrase. P-phone, for example, is an abbreviation of Proprietary phone. Acronyms, contractions, and initialisms are special forms of abbreviations comprising the initial letters or other parts of several words that constitute a term. See also *acronym*, *contraction*, and *initialism*.

Abilene Project A high-performance network by the University Consortium for Advanced Internet Development (UCAID) in support of Internet2. Abilene infrastructure comprises high-speed routers connected to several dozen GigaPOPs (Gbps Points of Presence) interconnected over fiber optic transmission systems (FOTS) operating at speeds up to 10 Gbps. See also *FOTS*, *GigaPOP*, *Internet2*, and *router*.

ABM (Asynchronous Balanced Mode) A peer-to-peer mode of asynchronous communications in which either of a pair of devices can initiate a transmission and send data over a point-to-point link at any time. ABM is a communication mode used in High-level Data Link Control (HDLC) and derivative protocols, such as Link Access Procedure-Balanced (LAP-B). See also *asynchronous*, *HDLC*, *LAP-B*, *link*, *master/slave*, and *point-to-point*.

Above 890 Decision In the United States, the Federal Communications Commission (FCC) decision (1959) that granted private microwave radio access to a dedicated portion of radio spectrum above 890 MHz. The decision also permitted construction of such networks, regardless of the economic impact on the established common carrier. See also *common carrier*, *FCC*, *microwave*, *radio*, and *spectrum*.

ABR (Available Bit Rate) Also known as *best-effort ATM*. In asynchronous transfer mode (ATM), a class of traffic in which the network attempts to pass the maximum number of cells, but with no absolute guarantees. Subsequent to the establishment of the connection, the network may change the transfer characteristics through a flow control mechanism that communicates to the originating end-user device. This flow control feedback mechanism is in the form of resource management cells (RM-Cells). During periods of congestion, the network can buffer cells and advise the sender to throttle back on the rate of transmission. ABR supports variable bit rate (VBR) traffic with flow control, a minimum transmission rate, and specified performance parameters. Traffic parameters include peak cell rate (PCR), cell delay variation tolerance (CDVT), and minimum cell rate (MCR). No quality of service (QoS) commitment is made. ABR traffic examples include bursty LAN traffic and e-mail, neither of which requires guarantees of network access, but rather can deal with time slot access on an as-available basis. ABR service is not intended to support real-time applications. ATM also defines constant bit rate (CBR), non real-time Variable Bit Rate (nrt-VBR), real-time Variable Bit Rate (rt-VBR), unspecified bit rate (UBR), and variable bit rate (VBR) traffic classes. See also *ATM*, *buffer*, *CDVT*, *cell*, *e-mail*, *flow control*, *LAN*, *MCR*, *nrt-VBR*, *PCR*, *QoS*, *RM-Cell*, *rt-VBR*, *time slot*, *UBR*, and *VBR*.

absorption The irreversible conversion of some or all of the energy of an electromagnetic wave to another form of energy as a result of its encounter and interaction with matter through which it is propagating or upon which it is incident. Generally, the sum of the electromagnetic energy converts to thermal energy, i.e., heat, which transfers to the matter, and which results in some amount of signal attenuation. An electrical signal propagating through a copper conductor, for example, attenuates as some electromagnetic energy is converted to thermal energy due to the vibration of free electrons in the copper. Similarly, an optical signal propagating through a glass optical fiber (GOF) suffers some attenuation as the photons interact with the crystalline silicon dioxide and dopants that comprise the fiber and convert to thermal energy. Radio waves also suffer considerably from absorption, which in fact is used to advantage in microwave ovens. See also *attenuation* and *propagation*.

abstruse Difficult to comprehend because of complexity and intellectual demands. See also *obtuse*.

AC (Alternating Current) Current is the flow of electrons through a metallic circuit, with the direction of flow being from positive (+) pole to negative ( ) pole outside of the source (generator or battery). Direct current (DC) travels in one direction, only, while alternating current (AC) travels in both directions across the circuit. A continuous flow of AC current travels first in one direction and then reverses polarity and flows in the opposite direction. See also *current*, *DC*, and *polarity*.

A Carrier See *non-wireline carrier*.

acceptance angle See *angle of acceptance.*

Access BPL (Access Broadband over Power Line) An access, or local loop, technology that runs over medium voltage (MV) power lines in the power utilities distribution networks. At utility substations, the high voltage (HV) lines, which run at 165,000 765,000 volts, are stepped down to medium voltage, which runs at up to 7,200 volts, for the distribution network (see Figure A-1). At the substation, the BPL provider typically terminates a fiber optic network connection in a device that accomplishes the opto-electric conversion process. Inductive couplers wrapped around the power lines, without touching them, serve both as injectors for downstream transmissions and as extractors for upstream signals. The injectors and extractors share a common frequency band on the MV power lines for both upstream and downstream communications through the use of a version of orthogonal frequency division multiplexing (OFDM) specially tailored for powerline environments, the radio frequency (RF) carrier supporting the communications signals can share the same line with the electrical signals. The BPL signal uses the frequency band between 2 MHz and 80 MHz, and repeaters must be spaced at intervals of 300 meters or so. Extractors remove downstream signals from the distribution power lines just ahead of the remote transformers that step the voltage down from MV to the low voltage (LV) level of 110/220 volts used within the premises. The connection to the premises can be over the LV lines, or via IEEE 802.11 wireless local area network (WLAN) technology, also known as Wi-Fi. Within the premises, communications can make use of In-House BPL or more traditional technologies. See also *FOTS, In-house BPL, local loop, OFDM, RF, transformer, volt, voltage, Wi-Fi,* and *WLAN.*

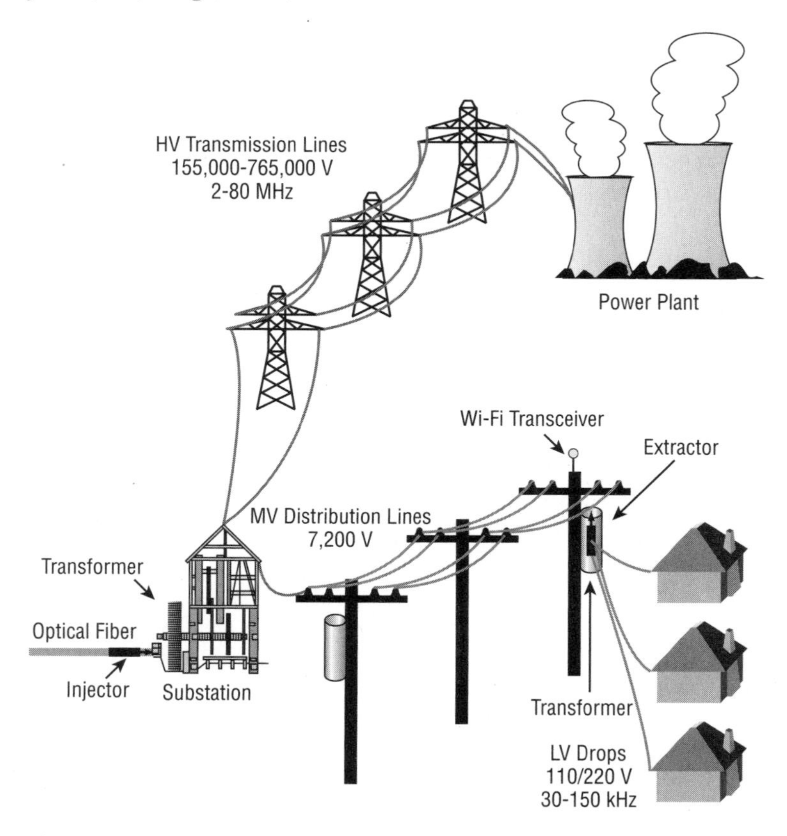

Figure A-1

access broadband over power line (Access BPL) See *Access BPL*.

access charges Charges imposed by the local exchange carrier (LEC) to compensate it for the cost of providing equal access capabilities, which enable a subscriber to access any competing interexchange carrier (IXC) with equal ease to place a long distance call. Although the structure of access charges varies from country to country, all include some combination of a Subscriber Line Charge (SLC) and Carrier Access Charge (CAC). In the United States, the LEC bills the monthly SLC to the subscriber. The LEC also bills the IXC for the CAC. See also *CAC*, *equal access*, *EUCL*, and *SLC*.

access circuit An access circuit is one used to gain access to, or entry to, a Wide Area Network (WAN) or Metropolitan Area Network (MAN). An access circuit usually is described as a local loop that connects a customer premises to a switch, router, multiplexer, or other device at the edge of the carrier or service provider network. See also *local loop* and *transport circuit*.

Access Manager An authorization mechanism that uses an application program interface (API) for application development, employing scripting. See also *API*, *authorization*, and *scripting*.

access node Synonymous with *service node* and *edge switch*. In packet networks, the outermost device on a carrier network, an access node is a switching point that comprises a point of end user access to the network. See also *edge switch* and *node*.

access point (AP) See *AP*.

access rate The maximum data rate of a channel between a user site and a network, as defined by the bandwidth of the access link available for data transmission.

access service A service that provides access to a network. Access services include residential lines, business lines, and PBX trunks for access to the public switched telephone network (PSTN). Broadband access services include broadband over power line (BPL), digital subscriber line (DSL), cable modem service, and passive optical network (PON). See also *BPL*, *cable modem*, *DSL*, *line*, *PON*, *PSTN*, and *trunk*.

access tandem switch A switch in the public switched telephone network (PSTN) that serves to connect the local exchange carriers (LECs), i.e., local telephone companies, to the interexchange carriers (IXCs), i.e., long distance carriers. See also *carrier*, *IXC*, *LEC*, and *PSTN*.

account code A code that a user enters via the telephone keypad when placing an outgoing call, particularly a toll call, through a PBX or KTS telephone system to track calling activity and to aid in client billing for time and expenses. Some users may be required to enter an account code, known as a *forced account code*, in order to place a call. Some systems allow a user to enter an account code for incoming calls, as well.

accounting management An element of network management, accounting management is the process of keeping and maintaining records. The call accounting module of a telemanagement system, for example, keeps track of network usage based on call detail recording (CDR) records output by the telephone system. On the basis of that data, the call accounting system can calculate calling costs, which it passes to a cost allocation module that creates reports of calling activity and associated costs by individual, station number, work group, department, project, division, etc. In another example, a RADIUS server gathers and maintains records of end user remote access to internal company resources and networks. See also *CDR*, *network management*, *RADIUS*, and *telemanagement software*.

ACD (Automatic Call Distributor) A system or application software program that serves to route incoming calls to the most available and appropriate agent. Incoming call centers make extensive use of such specialized software to enhance customer service. An ACD typically uses an automated attendant (i.e., front-end interactive voice processor) to answer incoming calls and provide callers with menu selections to guide the call through the system, perhaps sorting them into multiple queues associated with specialized agent groups according to the nature of the call and the caller. High priority calls (e.g., orders rather than returns) and calls from high priority callers (e.g., frequent customers) can be advanced to the head of the queue or placed in special queues. The routing of the call to an agent can be on the basis of next available, longest time since last call, least number of calls answered, or some other fairness routing algorithm. ACDs can be in the form of a specially equipped and partitioned PBX. Intensive call center applications typically make use of specialized ACDs that function as highly intelligent switches equipped primarily for the processing of incoming calls. An ACD is similar to, but much more sophisticated than a uniform call distributor (UCD). See also *call blending*, *PBX*, and *UCD*.

ACELP (Algebraic Code Excited Linear Prediction) A voice compression algorithm defined in ITU-T G.729, ACELP improves on CELP through the algebraic expression, rather than the numeric description, of each entry in the codebook. ACELP yields quality that is considered to be as good as ADPCM, but requiring bandwidth of only 8 kbps, which yields a compression ratio of 8:1. CS-ACELP is geared toward multi-channel operation. See also *ADPCM*, *algorithm*, *bandwidth*, *CELP*, *channel*, *compression*, *CS-ACELP*, *ITU-T*, and *LD-CELP*.

ACK (ACKnowledgement) **1.** A transmission control character sent by a station indicating that it is ready to receive a transmission. In ASCII, ACK is represented by the bit pattern 0110000. See also *ASCII*. **2.** A positive acknowledgement that a message, block, or frame has been received without error across a communications circuit, that the data set can be erased from buffer memory, and that the next data set can be sent. See also *block*, *circuit*, *data set*, *frame*, *message*, *NAK*, and *station*.

acknowledgement (ACK) See *ACK*.

ACL (Asynchronous Connectionless Link) A Bluetooth link option intended for packet data transmission. Bluetooth specifications also define a synchronous connection-oriented link (SCO) for real-time packet voice. See also *asynchronous*, *Bluetooth*, *connectionless*, *link*, and *packet*.

acoustics The branch of physics dealing with sound and its transmission.

ACR **1.** Attenuation-to-Crosstalk Ratio. Also known as *headroom*. The level of signal attenuation divided by the near-end crosstalk (NEXT), expressed in decibels (dB). If the level of crosstalk equals the level of the attenuated signal at any point in the cable, the signal is lost. See also *attenuation*, *crosstalk*, *dB*, *NEXT*, and *signal*. **2.** Anonymous Call Rejection. See *anonymous call rejection*.

acronym A pronounceable word formed of the initial letters or other parts of several words. An acronym generally comprises all upper case letters. SONET, for example, is the acronym for Synchronous Optical NETwork, a North American standard for fiber optic transmission systems. SONET became internationalized as SDH, an unpronounceable initialism for Synchronous Digital Hierarchy. Acronyms occasionally comprise all lower case letters. For example, bit is the acronym for binary digit, which is the basic unit of information in a binary numbering system. Bit also is a word unto itself, and with multiple meanings, including a small piece of something. Acronyms sometimes comprise both upper case and lower case letters. Sesame, for example, is the acronym for Secure European Systems for Applications in a Multivendor Environment. See also *abbreviation*, *anacronym*, *backronym*, *contraction*, *initialism*, and *portmanteau*.

acrylate coating Referring to a layer of polymer (a type of plastic) surrounding a glass optical fiber (GOF) to protect the glass from physical damage. See also *GOF*.

AC-3 (Adaptive Transform Coder 3) A specification from Dolby Digital and the Advanced Television Systems Committee (ATSC) for audio compression. The audio sampling rate is 48 kHz, and the system supports six channels in the Dolby Digital surround format. That format specifies multiple channel outputs, including center, left and right center, left and right surround, and low-frequency enhancement (LFE), also known as *subwoofer*. AC-3 is specified by the ATSC for use in high definition television (HDTV) and standard definition television (SDTV). See also *ATSC*, *audio*, *channel*, *compression*, *HDTV*, and *SDTV*.

active Energized, i.e., electrically powered, and in a state of readiness to perform a function such as amplifying a signal, detecting and correcting for errors in received data, retiming a signal, or retransmitting a signal. An amplifier or repeater, for example is an active device, but a reflector is passive. See also *passive*.

ACTS (Advanced Communications Technologies and Services) A European Union (EU) research and development program, ACTS addresses several hundred projects, including asynchronous transfer mode (ATM). ACTS is the successor to the Research for Advanced Communications in Europe (RACE) program. See also *ATM* and *RACE*.

adaptive differential pulse code modulation (ADPCM) See *ADPCM*.

Adaptive Transform Coder 3 (AC-3) See *AC-3*.

ADC (Analog-to-Digital Converter) A device in the form of a chipset that receives analog signals, measures the input at a regular sampling interval (or on command), and reports a digital output of the results. In a typical application, an ADC samples the analog signal at a fixed interval with enough resolution to accurately describe the analog waveform. In a typical voice application, for example, an ADC samples the audio stream 8,000 times per second at a precise interval of 125 microseconds (1/8,000 of a second) and reports a 14- or 16-bit value per sample. A digital signal processor (DSP) or other hardware then encodes the signal into pulse code modulation (PCM) format, employing an appropriate algorithm such as A-law or mu-law to produce a standard output like a DS-0 channel. At the receiving end of the connection, a matching DSP or other hardware and a digital-to-analog converter (DAC) reverses the process. See also *A-law*, *algorithm*, *analog*, *channel*, *digital*, *DSP*, *DS-0*, *encode*, *mu-law*, *PCM*, and *signal*.

ADCCP (Advanced Data Communications Control Procedures) An early synchronous data communications standard from the United States National Bureau of Standards (NBS), now the National Institute of Standards and Technology (NIST). ADCCP was a predecessor to the High-Level Data Link Control (HDLC) standard subsequently developed by the International Organization for Standardization (ISO). See also *HDLC* and *ISO*.

add/drop multiplexer (ADM) See *ADM*.

address The coded representation of the physical or logical location of a source or destination resource, such as a register, a memory partition, an application, or a node or station. An address may be contained in an address field associated with a data unit, such as a block, cell, frame, or packet, in order that switches, routers, and other devices can forward the data unit to the destination device across a network. Alternatively, an address might be used to set up a path between originating and destination devices, such as voice telephone sets, to connect a call. A PSTN telephone number, for example, is a logical address associated with a physical port on a physical central office (CO) switch connected to a physical copper circuit terminating in a physical device such as a PBX, key system, or telephone set at a fixed physical location. A cellular telephone number is a logical address associated with a physical station that typically is mobile, perhaps across networks. An Internet Protocol (IP) address is a logical address associated with a data terminal or other physical network element that may be either fixed in location or mobile, perhaps across networks. See also *E.164*, *IP address*, *logical*, and *physical*.

address book A typical feature of e-mail systems is the ability to create personal and corporate address books, perhaps importing them from other applications, for use in addressing outgoing mail. Address books typically provide links to a personal contact list that contains greater detail about each individual, including detailed contact information and free-form comments. E-mail systems also may support searches of Lightweight Directory Access Protocol (LDAP) Internet directories. See also *e-mail* and *LDAP*.

Address Resolution Protocol (ARP) See *ARP*.

ad hoc mode In wireless local area networks (WLANs), a temporary manner of operation that allows devices such as laptop computers to spontaneously discover each other and communicate directly, without the involvement of a centralized hub, or access point (AP). IEEE 802.11b and Bluetooth both include provisions for ad hoc mode. See also *802.11b*, *AP*, *Bluetooth*, *infrastructure mode*, and *WLAN*.

adjunct From the Latin *adjunctus*, meaning adjoin. Something added to something else, but secondary or not essential to it.

adjunct processor (AP) See *AP*.

ADM (Add/Drop Multiplexer) Multiplexers that have the capability to insert and extract individual lower speed channels (e.g., DS-1, DS-2, or DS-3) into a higher speed aggregate bit stream. ADM (see Figure A-2) offers considerable advantage over traditional time division multiplexers (TDM). The process of bit stuffing to adapt to slight clocking variations can require that a DS-3 frame be demultiplexed into its DS-2 and then DS-1 frames, which must be broken down into 24 DS-0 channels in order to extract and route an individual DS-0 channel. When that is accomplished, the process must be reversed to reconstitute the DS-3, minus the extracted DS-0. Contemporary T-carrier muxes are capable of add/drop multiplexing in the absence of stuff bits. Used extensively in SDH and SONET networks, ADMs perform the additional functions of dynamic bandwidth allocation, providing operation and protection channels, optical hubbing, and ring protection. In wavelength division multiplexing (WDM), an optical add/drop multiplexer (OADM) performs the same add/drop function on individual wavelengths. See also *channel*, *DS-1*, *DS-2*, *DS-3*, *hub*, *multiplexer*, *OADM*, *SDH*, *SONET*, *T-carrier*, *TDM*, *wavelength*, and *WDM*.

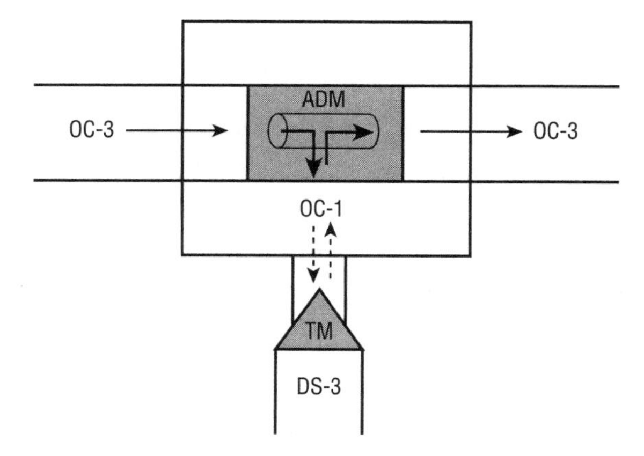

Figure A-2

ADPCM (Adaptive Differential Pulse Code Modulation) A voice encoding technique used to convert analog signals to digital format. ADPCM improves the quality of DCPM by adapting to the incoming signal, without increasing the bit requirement. ADPCM increases the range of signal changes that can be represented by a 4-bit value, thereby adapting to provide higher quality for voice transmission. ADPCM also overcomes the deficiency of DPCM with respect to support of modem transmissions. A modem transmission is characterized by abrupt shifts in amplitude and frequency levels that DPCM cannot accommodate. ADPCM can distinguish the presence of a modem tone, and can adapt by reverting to a channel width of 64 kbps, or by forcing the modem to adapt to a lower speed. ADPCM operates at 64, 56, 48, 40, 32, 24, and 16 kbps, with 32 kbps being the most commonly used. As central office (CO) exchanges are based on PCM rather than ADPCM, it is necessary to use a bit compression multiplexer (BCM) be used to insert two 32-kbps compressed voice conversations into a single PCM channel. See also *analog*, *BCM*, *digital*, *DPCM*, *encode*, *modem*, and *PCM*.

ADSL (Asymmetric Digital Subscriber Line) A broadband access technology designed to support voice, high speed Internet access, entertainment television over embedded telco unshielded twisted pair (UTP) local loops up to 18,000 ft in length. One of a family of xDSL standards, ADSL was developed by Bellcore (now Telcordia Technologies) at the request of the Regional Bell Operating Companies (RBOCs) in the United States, and was later standardized in 1999 by the American National Standards Institute (ANSI) as T1.413 and by the ITU-T as G.922.1. The term asymmetric refers to the fact that, in consideration of FEXT and NEXT crosstalk issues, ADSL offers considerably more bandwidth in the downstream direction than in the upstream direction. As ADSL employs discrete multitone (DMT) modulation, it sometimes is referred to as, G.dmt. Also known as *orthogonal frequency division multiplexing* (OFDM), DMT splits the signal over 256 narrowband subcarrier channels, within each of which quadrature amplitude modulation (QAM) is employed.

ADSL involves a pair of matching modems, with the ADSL Transmission Unit-Centralized (ATU-C) located at the central office (CO) or other headend location, and the ADSL Transmission Unit-Remote (ATU-R) located on the customer premises. The ATUs multiplex voice, data, and sometimes video signals over three separate frequency channels. A bi-directional voice grade analog channel at 0 4 kHz is provided for full duplex (FDX) voice and facsimile applications, that is, POTS service. Upstream data transmission is over what technically is a bidirectional channel provided in increments of 64 kbps, up to 640 kbps, in a frequency band from 26 140 kHz. Downstream transmission is in increments of 1.536 Mbps up to 6.144 Mbps, based on T1 specifications, in a frequency band that runs from approximately 140–552 kHz. (*Note:* The downstream increments are stated in maximum transmission rates. The available rates may be much lower, depending on carrier network design considerations and local loop characteristics.) In the context of the OSI Reference Model, ADSL is primarily a Layer 1 (Physical Layer) specification, although it includes Layer 2 (Data Link Layer) elements. The telcos currently avoid video over pure ADSL, although some provide video over hybrid fiber/ADSL loops in deployment scenarios known variously as fiber-to-the-curb (FTTC), fiber-to-the-neighborhood (FTTN), and fiber-to-the-node (FTTN). Table A-1 provides a comparative view of ADSL maximum downstream data rates and distance limitations based on acceptable options for wire gauges.

Table A-1: ADSL Maximum Data Rates

Maximum Data Rate	*American Wire Gauge (AWG)*	*Maximum Distance (ft.)*	*Metric Gauge*	*Maximum Distance (km.)*
1.544 Mbps (T1)				
2.048 Mbps (E-1)	24	18,000 ft.	0.5 mm	5.5 km
1.544 Mbps (T1)				
2.048 Mbps (E-1)	26	15,000 ft.	0.4 mm	4.6 km
6.144 Mbps (4 x T1)	24	12,000 ft.	0.5 mm	3.7 km
6.144 Mbps (4 x T1)	26	9,000 ft.	0.4 mm	2.7 km

Source: DSL Forum

Since the original ADSL standards were released in 1999, development efforts have continued and several enhanced versions have been released in the forms of ADSL2 and ADSL2+. See also *ANSI, asymmetric, AWG, bandwidth, Bellcore, broadband, channel, CO, crosstalk, Data Link Layer, DMT, downstream, DSL Forum, E-1, FEXT, frequency band, FTTC, FTTN, FTTP, G Series, headend, ITU-T, local loop, metric gauge, narrowband, NEXT, OFDM, OSI Reference Model, Physical Layer, POTS, QAM, RBOC, symmetric, T1, transmission rate, upstream, UTP, voice grade,* and *xDSL.*

ADSL2 (Asymmetric Digital Subscriber Line version 2) Specified by the ITU-T in Recommendations G.992.3 and G.992.4 (July 2002), ADSL2 supports increased data rates of as much as 12 Mbps downstream and 1 Mbps upstream, depending on local loop length and quality. ADSL2 achieves the higher downstream data rates by increasing the frequency band from 552 kHz (ADSL) to 1.1 MHz, and improving modulation efficiency through the introduction of trellis-coded modulation (TCM) quadrature amplitude modulation (QAM) constellations. In combination, these modulation techniques yield higher throughput on long loops with low signal-to-noise ratio (SNR). ADSL2 also uses receiver-determined tone reordering of the discrete multitone (DMT) channels to spread out the noise from AM radio interference and, thereby, to realize greater coding efficiency, which yields higher throughput. ADSL2 systems feature reduced framing overhead, enhanced power management, faster startup, seamless rate adaption, and improved diagnostics. ADSL2 also features an all-digital mode, in which the analog voice channel can be used for digital data transmission, thereby increasing aggregate upstream data transmission rates by as much as 256 kbps. ADSL2 adds a packet mode capability that enables packet-based services such as Ethernet. On long loops, ADSL2 can increase the data rate by as much as 50 kbps and extend the reach by about 600 feet (200 meters). ADSL2 supports bonding in asynchronous transfer mode (ATM) mode, based on the MFA Forum specification for Inverse Multiplexing over ATM (IMA). This allows two ADSL pairs to be bonded together to yield roughly double the single-pair rate. See also *ADSL*, *AM*, *analog*, *ATM*, *channel*, *digital*, *DMT*, *downstream*, *Ethernet*, *frequency band*, *IMA*, *interference*, *local loop*, *MFA Forum*, *modulation*, *overhead*, *QAM*, *rate adaption*, *SNR*, *TCM*, *throughput*, and *upstream*.

ADSL2+ (Asymmetric Digital Subscriber Line version 2 plus) Specified by the ITU-T in Recommendation G.992.5 and G.992.4 (January 2003), ADSL2+ doubles the downstream data rate, in comparison to ADSL2, to as much as 24.5 Mbps over local loops up to approximately 5,000 feet (1,500 meters) in length. The upstream rate remains at a maximum of 1 Mbps. In order to achieve this enhanced data rate, ADSL2+ increases the downstream frequency range to 2.2 MHz and increases the number of subcarriers to 512. The analog POTS channel remains at 4 kHz and the upstream data channel remains capped at 140 kHz. See also *ADSL*, *ADSL2*, *analog*, *downstream*, *frequency*, *local loop*, *POTS*, *subcarrier*, and *upstream*.

ADSL Lite (Asymmetric Digital Subscriber Line Lite) See *G.lite*.

ADSL transmission unit-centralized (ATU-C) See *ATU-C*.

ADSL transmission unit-remote (ATU-R) See *ATU-R*.

Advanced Communications Technologies and Services (ACTS) See *ACTS*.

Advanced Data Communications Control Procedures (ADCCP) See *ADCCP*.

Advanced Encryption Standard (AES) See *AES*.

advanced intelligent network (AIN) See *AIN*.

Advanced Mobile Phone System (AMPS) See *AMPS*.

Advanced Program-to-Program Communications (APPC) In the IBM Systems Network Architecture (SNA), Logical Unit (LU) 6.2. See *LU*.

Advanced Research Project Agency Network (ARPANET) See *ARPANET*.

advanced telecommunications capability The U.S. Federal Communications Commission (FCC) defines high-speed services as supporting a data rate of at least 200 kbps in at least one direction and advanced telecommunications capability as at least 200 kbps in both directions. See also *broadband* and *FCC*.

Advanced Television Systems Committee (ATSC) See *ATSC*.

adware **1.** A type of spyware that records search information and forwards it to an advertising agency or market research firm that later uses it to tailor pop-up ads for delivery to users without their knowledge or consent. See also *spyware.* **2.** Hardware, firmware, and software as it is advertised rather than as it exists. Unfortunately, truth in advertising is not a given. Always read the fine print and check references, especially for users of similar size, using the product in similar configurations with similar intensity in similar applications. Brochureware is a type of adware that you pick up at a trade show and take with you. See also *fine print.*

aerial cable An outside plant (OSP) communications cable designed to be suspended from poles or other overhead structures. As they are exposed to the elements, aerial cables must be well protected by cable jackets or sheathes, and from critters by armoring. As their hanging weight can be considerable, they also must incorporate load-bearing strength members. Aerial cables also are often pressurized to protect them from moisture in the event of failures in splice cases or insulation. Alternatively, a water-blocking gel or moisture-activated powdered gel can be used for moisture protection. See also *cable*, *icky-pic*, and *OSP.*

.aero (aeronautics) Pronounced *dot aero.* The generic Top Level Domain (gTLD) reserved exclusively for aeronautical interests. This domain was created in 2002 under the sponsorship of the Société Internationale de Télécommunications Aéronautiques (SITA). See also *gTLD*, *Internet*, and *sponsored domain.*

AES (Advanced Encryption Standard) A symmetric key encryption algorithm that supports key lengths of 128, 192, and 256 bits. AES superseded Triple DES (Data Encryption Standard), which has a 128-key length and is considered too slow and processor-intensive. Developed by the National Institute of Standards and Technology (NIST), AES has been adopted by the United States government for use with sensitive unclassified documents. See also *algorithm*, *DES*, *encryption*, *key*, *NIST*, and *Triple DES.*

AF (Assured Forwarding) The Differentiated Services (DiffServ) protocol identifies two primary types of per-hop behaviors (PHBs), representing two service levels, or forwarding classes. Expedited Forwarding (EF) provides minimal delay, jitter, and loss. Assured Forwarding (AF) comprises four classes, each of which contains three drop precedences and allocates certain amounts of buffer space and bandwidth. AF traffic exceeding the profile may be either dropped or demoted during periods of network congestion. See also *bandwidth*, *buffer*, *congestion*, *delay*, *DiffServ*, *EF*, *jitter*, *PHB*, and *protocol.*

African Network Information Center (AfriNIC) See *AfriNIC.*

AfriNIC (African Network Information Center) The Regional Internet Registry (RIR) responsible for assigning Internet Protocol (IP) addresses variously to National Internet Registries (NIRs) or directly to Local Internet Registries (LIRs) on the African continent. See also *IP*, *IP address*, *LIR*, *NIR*, and *RIR.*

AGC (Automatic Gain Control) Referring to the manner in which amplifiers adjust for amplitude variations of the input signal to ensure that the outgoing signal is of a constant strength. See also *amplifier*, *amplitude*, *gain*, and *signal.*

AIN (Advanced Intelligent Network) Bellcore built on earlier work done by Bell Labs on the intelligent network (IN) and defined AIN in the early 1980s (see Figure A-3). The initial AIN release was intended to provide a generic and modular set of tools to enable the creation, deployment, and management of services on a flexible basis within the public switched telephone network (PSTN). The software tools yielded a suite of service offerings accessible to all network switches, but operating independently from the switch logic. The services, therefore, can be defined, developed, and deployed quickly, and in a multivendor environment. Subsequent releases defined switching and database functions and the interactions between them. AINs include service creation toolkits, which enable the creation of centralized logic residing in centralized databases for the development and delivery of features across the network. AINs support all ISDN features and are intended to provide support for personal communications services (PCS), which permit subscribed features to be supported across networks of all types.

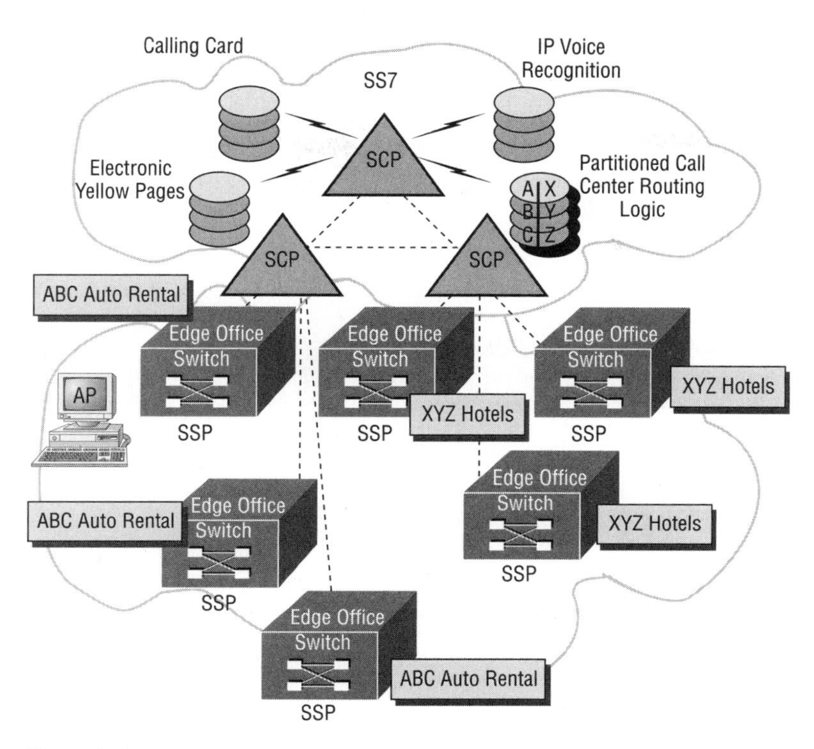

Figure A-3

The AIN architecture, as illustrated in Figure A-3, requires Common Channel Signaling System 7 (SS7). In addition to SS7, AIN comprises the following components:

- **Service Switching Points (SSPs):** PSTN switches that act on the instructions dictated by AIN centralized databases. SSPs can be end offices or tandem switches.
- **Signal Transfer Points (STPs):** Packet switches that route signaling and control messages between SSPs and SCPs, and between STPs.
- **Service Control Points (SCPs):** Nodes that contain all customer information in databases residing in centralized network servers. SCPs provide routing and other instructions to SSPs.
- **Service Management Systems (SMSs):** Network control interfaces that enable the service provider to vary the parameters of the AIN services. Under certain circumstances, the user organization may be provided access to a partition of the SMS.
- **Adjunct Processors (APs):** Decentralized SCPs that support service offerings limited either to a single SSP or to a regional subset of SSPs. APs might support routing tables or authorization schemes specific to a single switch or regional subset of switches.
- **Intelligent Peripherals (IPs):** Nodes that enhance the delivery of certain services by offloading processing demands from the SCPs and providing a basic set of services to the SCPs. The role of the IP typically includes collection of digits, collection and playing of voice prompts, collection of voice responses and their conversion to digits, menu services, and database lookups.

AIN services include find-me, follow-me, call pickup, store locator, multilocation extension dialing, call blocking, caller name, enhanced call return, enhanced call routing, call completion, and number portability.

air interface Referring to the radio portion of the link between wireless devices, such as a cellular telephone and a base station. In the context of the OSI Reference Model, the air interface operates at the Physical Layer and the Data Link Layer. See also *cellular radio, Data Link Layer, link, OSI Reference Model, Physical Layer, radio*, and *wireless*.

Airline Line Control (ALC) See *ALC*.

Airline Link Control (ALC) See *ALC*.

airwave transmission Synonymous with wireless transmission, as the term suggests open space rather than closed or confined space. Airwave is a bit of a misnomer, however, as the term also suggests transmission through the air, or atmosphere we breathe, which fortunately just happens to be there for us and the other living critters that share the Earths surface. Wireless transmission, however, definitely does not benefit from the atmosphere. Whether radio or optical in nature, wireless transmission suffers from any encounters with physical matter (e.g., atomic, molecular, and particulate matter) serving only to attenuate, impede, diffuse, distort, and otherwise interfere with the radio or optical signal. See also *free space* and *transmission medium*.

aka (also known as) Synonymous with, or having the same meaning as, another term. For example, software program is synonymous with software.

A-law A voice companding technique specified in the ITU-T G.711 Recommendation for pulse code modulation (PCM). This technique, which is used in the European digital hierarchy, converts 13-bit linear PCM samples into 8-bit compressed samples. Mu-law (μ-law) is a similar, but incompatible, companding technique used in the North American (T-carrier) and Japanese (J-carrier) digital hierarchies. In an international call, A-law is used if at least one of the national networks involved in the call uses A-law. See also *companding, E-carrier, G.711, ITU-T, J-carrier, mu-law, PCM*, and *T-carrier*.

ALC (Airline Line Control or Airline Line Control) An IBM protocol used in airline reservations systems such as American Airlines SABRE System and United Airlines APOLLO. Both ALC and P1024B, the Unisys version, employ a de facto standard six-bit coding scheme. See also *coding scheme, standard*, and *protocol*.

alchemy An early, unscientific form of chemistry practiced in the Middle Ages with aims including turning base metals into gold and discovering the elixir of perpetual youth, a universal cure for disease, and a universal solvent. Many alchemists were intelligent, well-meaning men and even distinguished scientists. Sir Isaac Newton, for example, was an alchemist. Pair-gain technologies such as ADSL do not involve alchemy, although sometimes they are characterized as turning copper into gold. See also *pair-gain*.

Algebraic Code Excited Linear Prediction (ACELP) See *ACELP*.

algorithm A logical, systematic, step-by-step procedure for solving a mathematical problem.

aliasing A phenomenon that occurs when different analog continuous signals overlap and become indistinguishable. If the sampling of the analog waveform is too infrequent (less than half the highest frequency present), the digitally encoded signal cannot reliably be decoded faithfully. Rather, it can be reconstructed as an alias of the true signal. Aliasing is a major concern in the digital encoding of analog audio and video signals. Aliasing in video signals results in artifacts in video images that can manifest as jagged blockings or a tiling effect. See also *analog, digital, encoding, PCM*, and *waveform*.

Alliance for Telecommunications Industry Solutions (ATIS) See *ATIS*.

Aloha From the Hawaiian *aloha*, meaning *hello* and *goodbye*. Also known as *pure Aloha*. A protocol developed at the University of Hawaii in the early 1970s as a contention management mechanism for use in inter-island wireless networks. Aloha is a very simple protocol in which the source just sends a frame of data whenever it desires. A target receiver confirms a frame whenever it receives one, and the source sends another whenever it desires. If the target receiver does not confirm the receipt of the frame within a specified time, the source resends it until it receives a confirmation. Pure Aloha is simple and inexpensive, but not useful in managing contention in large, complex networks. See also *AlohaNet* and *slotted Aloha*.

AlohaNet A packet radio system technology developed at the University of Hawaii in 1970. AlohaNet packet technology subsequently was incorporated into the first local area network (LAN) technology, which became known as *Ethernet*. See also *Aloha*, *Ethernet*, and *LAN*.

alpha Referring to a product, usually a software product, that is ready for initial testing, in-house and possibly under laboratory conditions. An alpha product generally is unstable and does not include all planned features or functionality. See also *beta* and *software*.

alternate mark inversion (AMI) See *AMI*.

alternating current (AC) See *AC*.

Alternative Access Vendor (AAV) Synonymous with Competitive Access Provider (CAP). See *CAP*.

Altos Aloha Network The original name for Ethernet. The experimental technology originally connected Altos computers through a network based on the AlohaNet packet radio system technology developed at the University of Hawaii. In 1973, it became known as Ethernet, from *luminiferous ether*. See also *ether* and *Ethernet*.

ALU (Arithmetic Logic Unit) The portion of the central processing unit (CPU) of a computer that performs mathematical calculations. See also *CPU*.

always on Referring to a continuous Internet connection. Such a connection is always available from the computer through the service providers central office (CO) or headend and to the Internet. Therefore, there are no dial-up delays such as those experienced when using a conventional modem to establish a circuit-switched connection to the Internet over the public switched telephone network (PSTN). The term generally is used to describe xDSL or cable modem services. See also *cable modem*, *circuit-switched*, *CO*, *headend*, *Internet*, *modem*, *PSTN*, and *xDSL*.

always on/dynamic ISDN (AODI) See *AODI*.

AM (Amplitude Modulation) Synonymous with ASK (Amplitude Shift Keying). A signal modulation technique in which the amplitude of the analog carrier sine wave is modulated to represent one or more 1 bits or 0 bits. The transmitting computer outputs a baseband electrical signal, defining 1 bits and 0 bits as discrete voltage levels. Using a unibit (single-bit) AM technique as illustrated in Figure A-4, each 1 bit entering the transmitting modem is expressed as one or more relatively high-amplitude sine waves, and each 0 bit is expressed as one or more low-amplitude sine waves. The high and low levels are defined in terms of a reference level or by the relative difference between the levels. At 2400 baud, this unibit technique yields a transmission rate of 2400 bps, with one bit transmitted per baud, that is to say that the bit rate equals the baud rate. In a dibit coding scheme, it is possible to express 2 bits with a single baud by defining 4 levels of amplitude, so the bit rate is double the baud rate. A tribit coding scheme expresses 3 bits per baud by defining 8 levels of amplitude, and a quadrabit coding scheme expresses 4 bits by defining 16 levels of amplitude. Amplitude modulation often is used in conjunction with frequency modulation (FM) and phase-shift keying (PSK) in electrically based networks. Digital fiber optics transmission systems (FOTS) use amplitude modulation. See also *AM*, *amplitude*, *analog*, *baseband*, *baud*, *baud rate*, *bit*, *bit rate*, *carrier*, *digital*, *FOTS*, *modem*, *PSK*, and *sine wave*.

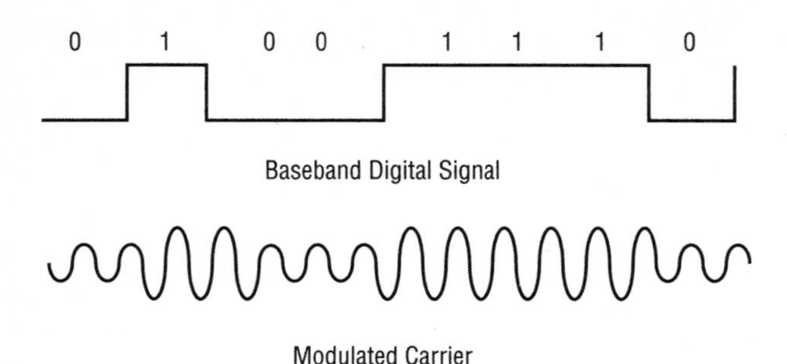

Figure A-4

amateur radio service Frequencies in the VHF, UHF, and microwave ranges above 30 MHz variously have been set aside in the United States and other countries for the use of radio hobbyists, or hams, for two-way voice and telegraphy communications. Some hams use inexpensive equipment operating in the unlicensed 2.4 GHz band for local communications, whereas others invest in more serious equipment for long distance, and even intercontinental, communications. Amateur radio service also is known as *ham radio*, in reference to the term *ham* that has long been applied to the hand of inept amateur telegraphers, and shortwave radio, in reference to the short wavelength (high frequency) of the permitted signals. See also *ham*, *UHF*, *VHF*, and *wavelength*.

American Registry for Internet Numbers (ARIN) See *ARIN*.

American Telephone and Telegraph (AT&T) See *AT&T*.

AMI (Alternate Mark Inversion) A line coding technique used in T1 networking. AMI is a bipolar transmission method that reverses the polarity of alternate marks, or 1 bits, expressing the first as a positive voltage of +3V, the second as a negative voltage of 3V, the third as +3V, and so on. Zero bits are coded as 0V. See Figure A-5. See also *bipolar*, *mark*, *polarity*, *signal*, and *voltage*.

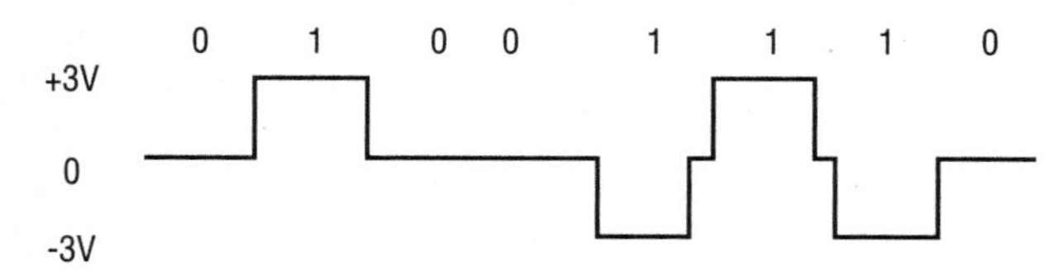

Figure A-5

AMIS (Audio Messaging Interchange Specification) A standard, published (1992) by the Industry Information Association, for networking voice mail systems, and specifying message file formats, addressing conventions, and message transmission.

amp (ampere) See *ampere*.

ampere (A) Abbreviated amp. **1.** The unit of electric current equivalent to the flow of one coulomb of charge per second past any cross section at any point in a circuit, with a coulomb being 6.24×10^{18} electrons. The ampere is named for AndrÈ-Marie AmpËre (1775 1836), who first distinguished the difference between electrical current and voltage. See also *current*, *sine wave*, and *voltage*. **2.** The steady flow of one volt (V) across a resistance of one ohm (Ω). **3.** The unit of constant electrical current that, when maintained in two straight parallel conductors of infinite length and of negligible circular cross section, and when placed one meter apart in a vacuum, would produce between the two conductors a force equal to 2×10^{-7} newtons per meter of length.

American National Standards Institute (ANSI) See *ANSI.*

American Standard Code for Information Interchange (ASCII) See *ASCII.*

American Wire Gauge (AWG) See *AWG.*

amplifier A device that actively boosts, or amplifies, a signal so that the output signal is a function of, and is of greater strength than, the input signal. An amplifier is a relatively simple device that transfers energy, at a controlled level, from an independent power source to an incoming signal in order to increase the strength of an outgoing signal. The increase, or gain, in signal strength usually is measured in positive decibels (+dB). Amplifiers are extensively used in analog networks to overcome the effects of signal attenuation, much as an amplifier in a radio receiver or TV set serves to boost a weak incoming signal to a level acceptable to the receiver. Amplifiers are spaced every 18,000 feet or so in a typical analog voice grade, twisted-pair telco local loop, for example. The exact spacing is sensitive to factors such as the transmission medium and the carrier frequency employed. Amplifiers are used in systems employing all transmission media, including not only twisted pair, but also coaxial cable, microwave radio, and optics.

An amplifier simply boosts the strength of a signal. So, whatever signal arrives at the amplifier leaves it with greater strength. In addition to attenuating, a signal accumulates noise as it transverses the network. The amplifier boosts the noise along with the signal. If there are multiple, cascading amplifiers in a long haul circuit, noise is compounded, thereby creating the potential for significant accumulated noise at the receiving end of the transmission. The resulting Signal-to-Noise Ratio (SNR) can be unacceptable.

Several types of optical amplifiers are employed in fiber optic systems. Erbium-Doped Fiber Amplifiers (EDFAs) amplify light signals falling in a narrow optical frequency range, performing much more cost-effectively than optical repeaters. Raman amplification makes use of pump lasers that send a high-energy light signal in the reverse direction (i.e., the direction opposite the signal transmission).

Digital transmission systems generally make use of regenerative repeaters, rather than amplifiers. A repeater not only amplifies, but also retimes and regenerates a signal. In combination, those processes serve to eliminate any accumulated noise, which improves signal quality considerably. See also *attenuation, dB, distributed amplification, EDFA, gain, lumped amplification, Raman amplifier, repeater,* and *SNR.*

amplitude The extreme range, or magnitude, of a fluctuating value such as an acoustic or electromagnetic signal, amplitude is measured perpendicular to the to the time axis of a time-plot, i.e., frequency, of a sine wave. Amplitude is a measure of the intensity, loudness, power, strength, or volume level of a signal. In an electrical circuit operating on alternating current (ac), amplitude is measured as the Voltage (V) level and is expressed as +V and V, depending on the direction of the current. See also *sine wave.*

amplitude distortion See *amplitude noise.*

amplitude modulation (AM) See *AM.*

amplitude noise A type of noise that occurs when the amplitude of the signal output by an amplifier or other device is not a linear function of the input amplitude. See also *noise.*

amplitude shift keying (ASK) Synonymous with amplitude modulation (AM). See *AM.*

AMPS (Advanced Mobile Phone System) A 1G analog cellular radio standard developed by Motorola and AT&T, and operating on 50 MHz of spectrum in the 800 MHz band. In the United States, 25 MHz and 333 (416 in some areas) channels each were provided to the A Carrier, or non-wireline carrier, and the B Carrier, or wireline carrier. Of the total number of channels awarded to each carrier, 21 channels are non-conversational channels dedicated to call setup, handoff, and teardown. The remaining communications channels are split into 30 kHz voice channels using frequency modulation (FM). As an analog system, AMPS derives channels using frequency division multiple access (FDMA) and bidirectional communications is achieved through frequency division duplex (FDD) with the downlink in the 869–894 MHz band and the uplink in the 824–849 MHz band. AMPS supports low speed modem transmission at rates generally limited to 6.8 kbps. Although once widely deployed in the United States, Australia, the Philippines, and other countries, AMPS has almost entirely been replaced by digital technology. Australian regulators mandated a cutover from analogue (Aussie for analog.) AMPS to digital GSM and CDMA, beginning December 31, 1999, in Melbourne, and gradually extending throughout the country during 2000. Motorola subsequently developed Narrowband AMPS (N-AMPS), which improved system capacity. In the United States, the Federal Communications Commission (FCC) authorized carriers to cease support for analog systems as of March 1, 2008. Digital AMPS (D-AMPS), standardized as IS-54 and IS-136, is essentially a digital version of AMPS. See also *1G*, *A Carrier*, *analog*, *B Carrier*, *carrier*, *CDMA*, *cellular radio*, *D-AMPS*, *digital*, *downlink*, *FCC*, *FDD*, *FDMA*, *FM*, *GSM*, *handoff*, *modem*, *N-AMPS*, *non-wireline carrier*, *uplink*, and *wireline carrier*.

anacronym (anachronism acronym) A portmanteau of anachronism and acronym that describes an acronym or initialism that has been used so long and become so ingrained in common language that its original spelled-out meaning is unknown to many, if not most. As examples, consider ASCII (American Standard Code for Information Interchange), radar (radio detecting and ranging), SCSI (Small Computer System Interface), scuba (self-contained underwater breathing apparatus), and sonar (sound navigation and ranging). See also *acronym*, *initialism*, and *portmanteau*.

analog **1.** A continuously present and continuously variable signal. In their native, or original, forms, audio and visual signals are analog. An active audio signal is a stream-oriented, i.e., continuously present, acoustic signal. A visual signal is a stream-oriented optical signal. Audio and visual signals travel in a waveform that can vary continuously and infinitely along two parameters amplitude and frequency. Amplitude refers to signal intensity or signal strength, which manifests as volume in audio signals and brightness in visual signals. Frequency refers to the number of waveforms per second, or cycles per second (cps), known in contemporary terminology as Hertz (Hz). Frequency manifests as pitch, or tone, in audio signals, and as color in image and video signals. All electromagnetic energy travels in continuous waveforms. The portions of the electromagnetic spectrum currently usable for telecommunications include electricity, radio, and infrared light. See also *digital*, *electromagnetic spectrum*, and *Hz*. **2.** Pertaining to the representation of data in the form of a continuous signal. Transmission of voice, image, and video information are relatively straightforward as they are analog in their native forms. Appropriate conversions to electrical, radio, and optical energy must be made, of course, and adjustments must be made in terms of amplitude and frequency levels, but the native signals and electromagnetic transmission signals are quite compatible in terms of their common analog nature. In order to accomplish the transmission of these native analog signals, the carrier signal (i.e., information-carrying signal) of the transmission system is modulated (i.e., varied, or changed) in order to create an analog of the original information stream. The transmission of digital computer data over an analog network is quite another matter, as a fundamental conversion in signal format must be made through a modem. To accomplish this conversion over an electrified analog network, the ones (1s) and zeroes (0s) of the digital bit stream must be translated into amplitude and frequency variations of the carrier signal. The electromagnetic sinusoidal waveform, or sine wave, can be varied in amplitude at a fixed frequency, using Amplitude Modulation (AM). Alternatively, the frequency of the sine wave can be varied at constant amplitude, using Frequency Modulation (FM). Additionally, both frequency and amplitude can be modulated simultaneously. Finally, the position of the sine wave can be manipulated (actually, can appear to be manipulated), adding the third technique of Phase Modulation, also known as *Phase Shift Keying* (PSK). See also *carrier*, *digital*, and *modem*. **3.** Something that is continuously present and continuously variable. For example, the shadow cast by the gnomon on the flat plate of a sundial is continuously present and continuously variable, at least during the daylight hours. Similarly, the hands of an analog clock, watch, or fuel gauge are continuously present and continuously variable, or essentially so. **4.** Something that is analogous to or similar to something else. The electrical waveform of a voice transmission over an analog network is analogous to the waveform of the native acoustic voice signal. The movement of the hour hand of an analog watch or clock is analogous to the movement of the shadow cast by a sundial.

analog-to-digital converter (ADC) See *ADC*.

analogue The British English spelling of analog. Analogue has two extra vowels, which seems like an awful waste. See *analog*.

angled physical contact (APC) See *APC*.

angle of acceptance The angle within which the core of an optical fiber will accept light injected by a light source, the boundary of the angle of acceptance is equal to the critical angle. Illustrated in three dimensions, the angle of acceptance becomes the cone of acceptance. Within that angle, a light source can inject an optical signal into the fiber core and the signal will remain in the core, reflecting off of the interface between the core and cladding, as illustrated in Figure A-6. At a more severe angle, i.e., outside the cone, the signal will penetrate the interface and enter, and perhaps be lost in, the cladding. The angle of acceptance and, therefore, the cone of acceptance are determined by the difference in index of refraction (IOR) between the core and cladding. The mathematical sine of the angle of acceptance is known as the *numerical aperture* (NA). See also *cone of acceptance*, *critical angle*, *IOR*, and *numerical aperture*.

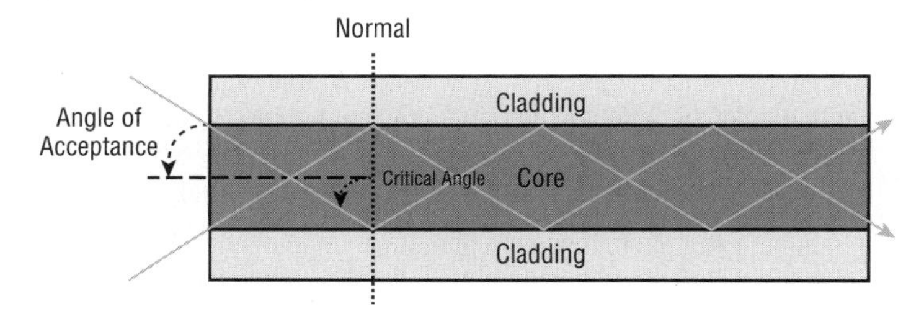

Figure A-6

angle of incidence The angle at which an incoming light ray strikes a reflecting or refracting surface. The angle is measured at the boundary, or interface, between two media, such as the core and cladding of an optical fiber. The angle is measured from the normal, which is a right angle to the surface. See also *cladding*, *core*, *critical angle*, *reflection*, and *refraction*.

ANI (Automatic Number Identification) A feature of the public switched telephone network (PSTN) that transmits the originating telephone number in advance of connecting the call. ANI is used by telephone companies for tracking and billing purposes. PBXs equipped with proper software and with special trunking arrangements (all at additional cost) can appear to the network as a central office (CO) switch and, therefore, can gain access to ANI information. This approach offers advantages compared to calling line identification (CLID), which the callers can block, and is often used by incoming call centers. ANI is available to interexchange carriers (IXCs) subscribing to Feature Group D (FGD) termination. See also *CLID*, *CO*, *FG*, *IXC*, and *PSTN*.

anonymous call rejection (ACR) A network-based CLASS service of the public switched telephone network (PSTN). The feature allows the subscriber to reject all calls from callers who have blocked the display of their calling identification information (calling number and calling name). Such calls are diverted to a recorded message advising the caller that the called party does not accept anonymous calls and to unblock the caller identification information before attempting the call again. See also *calling name and number blocking*, *CLASS*, and *PSTN*.

annotation A feature of voice mail systems that allows a message recipient to add critical or explanatory audio notes to a voice message before archiving it or forwarding it to another system user. See also *voice mail.*

ANSI (American National Standards Institute) ANSI coordinates the development and use of voluntary consensus standards in the United States across a wide variety of business sectors, and represents the needs and views of U.S. stakeholders in international standardization forums. Founded in 1918, ANSI membership comprises government agencies, private organizations, companies, academic and international bodies, and individuals. ANSI is the official U.S. representative to the International Organization for Standardization (ISO) and, via the U.S. National Committee, the International Electrotechnical Commission (IEC). See Appendix A for contact information. See also *EIA*, *IEC*, and *ISO*.

ANSI/EIA/TIA-568 A United States industry standard published jointly by ANSI and the EIA/TIA specifying a generic structured cabling system for commercial buildings. Intended to support a multivendor environment, the standard specifies cable physical attributes and performance characteristics for unshielded twisted pair (UTP), shielded twisted pair (STP), and screened twisted pair (ScTP) cabling categories. The initial standard was released in 1985, and has been modified and updated several times since. Category (Cat) 3, 4, 5, 6, and 7 cables have been defined. Cat 1 and Cat 2 are not formally defined under this particular standard. The most recent (as of Spring 2007) and most capable standard is Cat 7, which supports bandwidth of up to 600 MHz and data rates up to 10 Gbps. The standards also have been internationalized as ISO/IEC 11801. See also *ANSI*, *Cat*, and *EIA/TIA*.

ANSI/ICEA S-80-576 The standard published by the American National Standards Institute (ANSI) and the Insulated Cable Engineers Association (ICEA) for Category 1 (Cat 1) and Category 2 (Cat 2) individually unshielded twisted pair (UTP) indoor cables for use in communications wiring systems. See also *ANSI*, *Cat 1*, *Cat 2*, *ICEA*, and *UTP*.

ANSI/ICEA S-91-661 The standard published by the American National Standards Institute (ANSI) and the Insulated Cable Engineers Association (ICEA) for Category 3 (Cat 3), Category 5 (Cat 5), and Category 5e (Cat 5e) individually unshielded twisted pair (UTP) indoor cable for use in general purpose and LAN communication wiring systems. See also *ANSI*, *Cat 3*, *Cat 5*, *Cat 5e*, *ICEA*, *LAN*, and *UTP*.

ANSI/ICEA S-101-699 The standard published by the American National Standards Institute (ANSI) and the Insulated Cable Engineers Association (ICEA) for Category 3 (Cat 3) individually unshielded twisted pair (UTP) indoor cable for use in general purpose non-LAN telecommunication wiring systems. See also *ANSI*, *Cat 3*, *ICEA*, *LAN*, and *UTP*.

ANSI NFPA 70 American National Standards Institute (ANSI) National Fire Protection Association (NFPA) publication 70. Synonymous with National Electrical Code (NEC). See also *NEC*.

answering machine A relatively simple machine that answers incoming calls after a user programmable number of rings. Answering machines are customer premises equipment (CPE), owned or leased by the end user and located on the customers premises. Answering machines largely have been replaced by central office (CO) based voice processors. See also *voice processor*.

antenna A device comprising an arrangement of wires, metal rods, and so on for radiating and receiving radio signals. A transmitting antenna converts electrical current to electromagnetic radio waves projected into free space or a waveguide. A receiving antenna converts electromagnetic radio waves into electric current. An antenna commonly performs both transmit and receive functions. The transmitters and receivers used in free space optics (FSO) also can be characterized as antennas. See also *free space*, *FSO*, and *waveguide*.

Antheil, George (1900–1959) The self-proclaimed *bad boy of music*, Antheil was a serious composer of significance and, later in life, a successful composer of film scores. In telecommunications, he and Hedy Lamarr, a famous beauty and film star, co-invented spread spectrum (SS) radio, which is the basis for code division multiple access (CDMA). See also *CDMA*; *Lamarr, Hedy*; and *SS*.

anycast In IPv6, a transmission mode in which a packet is delivered to the closest (in the sense of routing cost) interface with that address. See also *anycast address*, *cost*, *interface*, *IPv6*, and *packet*.

anycast address In Internet Protocol version 6 (IPv6), an address that is assigned to multiple interfaces, typically on multiple nodes. A packet with an anycast address is delivered to the closest interface with that address, as determined by the routing by the routing protocols measure of distance. Anycast addresses are allocated from the unicast address space. See also *interface*, *IPv6*, *IPv6 address*, *node*, *packet*, *protocol*, *router*, *unicast*, and *unicast address*.

AODI (Always On/Dynamic ISDN) An ISDN basic rate interface (BRI) service that maintains an always-on connection to an ISP server, or perhaps a corporate intranet server, using only the D channel. The D channel maintains the always on logical link between the client and the server systems, enabling the transfer of packet data, such as e-mail, at rates of up to 9.6 kbps over an X.25 switched virtual circuit (SVC). The always-on nature of the connection avoids the call set-up time required for a circuit-switched connection. From the end user terminal adapter (TA) to the central office (CO), the multilink point-to-point protocol (MPPP) is employed over the D channel. As the signaling and control requirements of BRI are relatively light, there is sufficient capacity on the D channel to reliably support AODI and the packet data transfers it occasions, as long as they do not exceed 9.6 kbps. See also *always on*, *BRI*, *circuit switching*, *client*, *CO*, *D channel*, *intranet*, *ISDN*, *ISP*, *MPPP*, *server*, *SVC*, *TA*, and *X.25*.

AP **1.** Application Processor. A computer that processes data associated with an end user application, such as e-mail, navigation, payroll, or voice mail. This is in contrast to a computer that performs utility or control functions, such as a front-end processor (FEP) that manages traffic, or a processor dedicated to load balancing or storage management. **2.** Access Point. In wireless local area networks (WLANs), a centralized hub through which computers and peripherals interconnect and intercommunicate in infrastructure mode. Terminal devices generally connect to the APs using unlicensed RF bands, although infrared (IR) sometimes is used. APs can interconnect directly over RF links, but generally are hardwired to switches. In terms of functionality, APs can be fat or thin. A fat AP is sufficiently intelligent to act independently, while a thin AP must act under the supervision of a controller. See also ad hoc mode, fat AP, hardwired, IR, RF, thin AP, unlicensed, and WLAN. **3.** Adjunct Processor. In the advanced intelligent network (AIN) architecture, a decentralized service control point (SCP) that supports service offerings limited either to a single service switching point (SSP) or to a regional subset of SSPs. An AP might support routing tables or authorization schemes specific to a single switch or to a regional subset of switches. See also *adjunct*, *AIN*, *SCP*, and *SSP*.

APC (Angled Physical Contact) A type of optical fiber connector that joins two fiber endfaces at a slight angle to minimize attenuation and back reflection. See also *attenuation*, *back reflection*, *connector*, and *optical fiber*.

APD (Avalanche PhotoDiode) A type of diode used as a light detector used in high speed, long haul fiber optic transmission systems (FOTS) employing laser diode light sources and single-mode (SMF) glass optical fiber (GOF) media. APDs are especially sensitive detectors as they use a strong electric field to accelerate the electrons flowing in the semiconductor. As a result, an APD generates an avalanche of electrons with a multiplication factor that can be in the range of 70, i.e., an APD can generate 70 electrons from 1 photon. Therefore, a very weak incoming light pulse will create a much stronger electrical effect that can be interpreted more effectively and understood more clearly. So, an APD can be characterized as a very high gain photodiode receiver, i.e., a one-way photonic receiver with a high ratio of (electrical) output power to (optical) input power in the range of 70:1. Although more sensitive and more effective than PIN detectors used in lesser systems, APDs require more electrical power to operate, are more sensitive to extremes of ambient temperature and are more expensive. See also *diode*, *FOTS*, *gain*, *GOF*, *laser diode*, *light detector*, *light source*, *PIN*, and *SMF*.

API (Application Program Interface, aka Application Programming Interface) **1.** A set of routines by which an application program can call specific programs or services of a computer operating system (OS) or network operating system (NOS). See also *application*, *NOS*, *OS*, and *routine*. **2.** A set of routines by which an application program allows another application program to work directly with it. See also *application*, *program*, and *routine*.

APNIC (Asia Pacific Network Information Center) The Regional Internet Registry (RIR) responsible for assigning Internet Protocol (IP) addresses variously to National Internet Registries (NIRs) or directly to Local Internet Registries (LIRs) in the Asia-Pacific region. See also *IP*, *IP address*, *LIR*, *NIR*, and *RIR*.

APON (ATM-based Passive Optical Network) The original PON specifications set by FSAN and ratified by the ITU-T as G.983.1 (1998). APON specifies asynchronous transfer mode (ATM) as the Data Link Layer protocol. APON runs in asymmetric mode at a signaling rate of 622 Mbps downstream and 155 Mbps upstream, or in symmetric mode at 155 Mbps. The more contemporary broadband passive optical network (BPON) is an APON variant. See also *asymmetric*, *ATM*, *BPON*, *Data Link Layer*, *downstream*, *G Series*, *PON*, *protocol*, *signaling rate*, and *upstream*.

app (application) See *application*.

APP (Atom Publishing Protocol) An Application Layer protocol for publishing Web resources associated with periodically updated Web sites. APP builds on the RSS protocol. The Atom Syndication Format is documented in the IETF RFC 4287 (2005). See also *Application Layer*, *IETF*, *protocol*, *RSS*, and *WWW*.

APPC (Advanced Program-to-Program Communications) In the IBM Systems Network Architecture (SNA), Logical Unit (LU) 6.2. See also *LU*.

application A program designed to perform a function or suite of related functions of benefit to an end user, with examples being accounting, mathematical analysis, video editing, and word processing. Application software differs from utilities, which are devoted to system management tasks. See also *utility*.

Application Layer Layer 7, the highest layer, of the Open Systems Interconnection (OSI) Reference Model. Software at the Application Layer provides support services for user and application tasks such as file transfer, interpretation of graphic formats and documents, document processing, and user authentication in remote access applications. X.400 e-mail messaging, for example, takes place at Layer 7. TCP/IP application protocols such as Simple Mail Transfer Protocol (SMTP), Telnet, and File Transfer Protocol (FTP) also take place at this layer. See also *e-mail*, *FTP*, *layer*, *network architecture*, *OSI Reference Model*, *SMTP*, *Telnet*, and *X.400*.

application program interface (API) Synonymous with application programming interface. See *API*.

application programming interface (API) Synonymous with application program interface. See *API*.

application processor (AP) See *AP*.

application service provider (ASP) See *ASP*.

application software See *application*.

application-specific integrated circuit (ASIC) See *ASIC*.

appointment call An international long distance calling method that requires the caller to make an appointment for an international operator to seize an international trunk and place an international call at an appointed time. Appointment calling largely has been replaced by international direct distance dialing (IDDD). See also *IDDD*.

APS (Automatic Protection Switching) An automatic service restoration function by which a network senses a circuit or node failure and automatically switches traffic over an alternate path. SDH and SONET specifications require APS in order that the self-healing fiber optic network can recover from a fiber or node failure. The physical topologies specified in SDH/SONET standards are path-switched ring and line-switched ring. See also *circuit, line-switched ring, node, path, path-switched ring, SDH, self-healing, SONET,* and *topology.*

aramid (aromatic polyamide) Invented by Dupont, which markets it under the name KEVLAR™, aramid is a poly para-phenyleneterephthalamide, and is more properly known as a para-aramid. Belonging to the family of nylons, aramids are chemically and thermally stable, strong, lightweight, and resistant to impact and abrasion damage. Aramids also are dielectrics, or non-conductors of electric current. Uses for aramid include protective equipment (e.g., body armor such as bulletproof vests and helmets), fire-blocking fabrics, tire reinforcements, high-performance composites for aircraft, and strength members for telecommunications cables. See also *KEVLAR™* and *strength member.*

Archie A corruption of *archive.* An Internet browser based on the File Transfer Protocol (FTP). Archie enables the user to search for a file (exact name unknown) on a file server (name unknown) somewhere on the Internet. Archie servers contain directory listings of all such files, updated on a monthly basis through a process of polling file servers. Archie provides a user definable number of file hits, as well as file names, server names, and directory paths to access each listed file. Archie capabilities are limited to specific search strings, thereby providing little flexibility. Archie was first deployed in 1991 and currently is often integrated into Gopher or WWW clients and activated when the user accesses an Archie server. See also *browser, client, FTP, Gopher, Internet, polling, server, string,* and *WWW.*

architecture **1.** The organizational structure of an entity, such as a computer, data processing, or communications system. **2.** The organizational structure of a protocol suite or protocol stack, such as the OSI Reference Model. See also *OSI Reference Model, protocol stack,* and *protocol suite.*

archiving A feature of a messaging system (e.g., e-mail or voice mail) that allows the recipient to archive, or save, messages, usually for a limited amount of time unless they are resaved. Messages may be archived on an external storage medium for longer periods of time, perhaps to comply with document retention laws. For example, telephone companies in the United States must archive recorded user confirmation statements for long periods of time when effecting a change of carrier. Similarly, stockbrokers and certain telemarketers must retain confirmations associated with solicited trades or sales. See also *e-mail* and *voice mail.*

ARCNET (Attached Resource Computer NETwork) A popular LAN protocol developed by John Murphy at Datapoint Corporation in 1976, ARCNET was one of the first networking solutions for microcomputers. ARCNET employed coaxial cable to connect host computers, workstations, and peripherals through hubs in a star configuration. ARCNET employed a deterministic token passing bus medium access control (MAC) protocol, operated at signaling speeds up to 2.5 Mbps, and supported as many as 255 devices over link lengths up to 2,000 feet. More recent versions deliver 20 Mbps and 100 Mbps, although they have never been in great demand and are not widely available. ARCnet resembles, but does not adhere to, the IEEE 802.4 specification. ARCNET also is CamelCased as ARCnet. See also *802.4, CamelCase, coaxial cable, deterministic, IEEE, MAC, star,* and *token passing.*

area code Also known as the *Numbering Plan Area* (NPA). In the North American Numbering Plan (NANP), the three-digit number that corresponds to a geographic area within the area loosely defined as North America. The NPA follows the pattern NXX, with N indicating that only digits 2 9 are allowed, as 0 or 1 would confuse the network, and X indicating that any digit is allowed. The area code originally was used only when a call crossed an area code boundary. In such a case, the dialing sequence is 1.NNX.NNX.xxxx. the 1+ identified the call as long distance and triggered the involvement of an interexchange carrier. Where overlay area codes have been implemented, however, it is necessary to dial a full 10-digit number (NNX.NNX. xxxx) within a geographic area. Also, cellular networks treat area codes differently in terms of dialing pattern and rating. Cellular networks do not require 1+ dialing, and generally bill only for airtime, regardless of whether calls are local or long distance in nature. As a great many individuals subscribe exclusively to cellular service, area codes largely have lost their significance to those users in terms of calling costs, and many of them retain their old telephone numbers even when permanently moving their residences across area code boundaries. See also *NANP*, *NPA*, and *overlay area code*.

ARIN (American Registry for Internet Numbers) The Regional Internet Registry (RIR) responsible for assigning Internet Protocol (IP) addresses variously to National Internet Registries (NIRs) or directly to Local Internet Registries (LIRs) in Canada, many Caribbean and North Atlantic islands, and the United States. See also *IP*, *IP address*, *LIR*, *NIR*, and *RIR*.

arithmetic coding A technique used for lossless data compression that establishes a model of the entire data set and establishes probabilities of the occurrences of symbols and patterns or sequences of symbols that can then be expressed in the form of a single number. Arithmetic coding is much more efficient than a run-length encoding algorithm such as Huffman coding, which uses a discrete number of bits for each symbol, but is more processor-intensive. See also *algorithm*, *compression*, *Huffman coding*, *lossless compression*, *run-length encoding*, and *symbol*.

arithmetic logic unit (ALU) See *ALU*.

armored cable Cable armored to protect against cable-seeking backhoes, posthole diggers, cable-loving rodents, and other adverse forces of man and nature. The armor may be in the form of lead or lead alloy sheathing, or interlocking aluminum or galvanized steel cladding.

ARP (Address Resolution Protocol) A protocol that translates between network addresses, such as between Ethernet and Internet Protocol (IP) addresses or between asynchronous transfer mode (ATM) and Ethernet addresses. See also *ATM*, *Ethernet*, and *IP*.

.arpa (address routing and parameter area) Pronounced *dot arpa*. The generic Top Level Domain (gTLD) reserved exclusively for Internet infrastructure purposes. This is an unsponsored domain named for the Advanced Research Project Agency (ARPA). See also *ARPANET*, *gTLD*, *Internet*, and *unsponsored domain*.

ARPANET (Advanced Research Project Agency NETwork) Generally accepted as the first (1971) sophisticated packet network architecture, ARPANET was designed to link computers on a time-share basis in order to share computer resources more cost-effectively in support of various defense, higher education, and research and development organizations. In 1983, the majority of ARPANET users spun off to form the Defense Data Network (DDN), also called MILNET (Military Network), which included European and Pacific Rim continents. Locations in the United States and Europe that remained with ARPANET then merged with the Defense Advanced Research Project Agency Network to become DARPA Internet.

ARPA protocol suite See *TCP/IP protocol suite*.

ARQ (Automatic Repeat reQuest) An error control protocol that automatically initiates a request to repeat the transmission of any packet or frame not acknowledged as received correctly, in other words, to retransmit the last errored or lost frame or packet and any transmitted afterwards. Incremental redundancy (IR), also known as Hybrid ARQ II, is an enhanced ARQ technique employed in EGPRS (Enhanced General Packet Radio System), the packet-switched mode of Enhanced Data rates for GSM Evolution (EDGE) cellular radio networks. See also *EDGE*, *EGPRS*, *error control*, *frame*, *IR*, *packet*, and *protocol*.

ARS (Automatic Route Selection) Also known as *Least Cost Routing* (LCR). An optional, programmable PBX software feature that enables the system to route a call over the most appropriate carrier and service offering based on factors such as the type of call (e.g., local, local long distance, or long-haul long distance), the Class of Service (CoS) of the user, the time of day (e.g., prime time and non prime time), and the day of the year (e.g., weekday, weekend day, or holiday). In countries where there are lower rates for cellular-to-cellular calls than for calls between cellular phones and landlines, ARS sometimes is used to route the landline leg through a cellular interface to take advantage of the lower rates. ARS is of greatest value if the telecom environment is liberalized or deregulated and there are multiple competing carriers and rate plans from which to choose. In practice, ARS generally is on the basis of a table lookup rather than a hierarchical parsing of a dialed telephone number and calculation of a least cost route. See also *carrier, cellular radio, CoS, landline, local long distance, long distance, parse, PBX,* and *software.*

artifact Unintended and unwanted distortions or other aberrations in reproduced audio or video due to transmission errors or signal processing operations. Artifacts often result from the use of lossy compression algorithms at high compression ratios. Artifacts in video images can manifest as jagged blockings or a tiling effect known as aliasing, banding of colors, white spots, and even dropped frames. See also *aliasing, compression, distortion, lossy compression,* and *signal.*

AS (Autonomous System) Referring to a group of routers within the same administrative domain. The term is used in exterior protocols such as the Exterior Gateway Protocol (EGP) and the Border Gateway Protocol (BGP). See also *BGP, domain, EGP,* and *router.*

ASCII (American Standard Code for Information Interchange) A standard coding scheme specifically oriented toward data processing applications, ASCII was developed in 1963 and modified in 1967 by the American National Standards Institute (ANSI). ASCII employs a 7-bit coding scheme, supporting 128 (2^7) characters, which is quite satisfactory for both upper case and lower case letters of the English alphabet and similarly simple Roman alphabets, Arabic numerals, punctuation marks, a reasonable complement of special characters, and a modest number of control characters. As ASCII was designed for use in asynchronous communications (involving non-IBM computers, in those days), relatively few control characters were required, making a 7-bit scheme acceptable. IBM computers, which were relatively complex mainframes, required the 8-bit EBCDIC coding scheme to accommodate the necessary complement of control characters. Table A-2 shows the ASCII code.

Table A-2: ASCII Code

Bit positions 1, 2, Ï3, 4	*Bit positions 5, 6, 7*							
	000	*100*	*010*	*110*	*001*	*101*	*011*	*111*
0000	NUL[1]	DLE	SP	0	@	P		p
1000	SOH[2]	DC1	!	1	A	Q	a	q
0100	STX[3]	DC2		2	B	R	b	r
1100	ETX[4]	DC3	#	3	C	S	c	s
0010	EOT[5]	DC4	$	4	D	T	d	t
1010	ENQ[6]	NAK[7]	%	5	E	U	e	u
0110	ACK[8]	SYN	&	6	F	V	f	v
1110	BEL[9]	ETB[10]	`	7	G	W	g	w
0001	BS	CAN[11]	(	8	H	X	h	x
1001	HT	EM[12]	)	9	I	Y	i	y
0101	LF	SUB[13]	*	:	J	Z	j	z
1101	VT	ESC[14]	+	;	K	[	k	{
0011	FF	FS	,	<	L	\	l	|
1011	CR[15]	GS	-	=	M	]	m	}
0111	SO	RS	.	>	N	^	n	~
1111	SI	US	/	?	O	_	o	DEL

Although the full explanations of all control codes are outside the scope of this book, the following control characters are representative:

1. **NUL (NULl):** A transmission control character used to serve a media-fill or time-fill requirement, i.e., a stuff character or padding character.
2. **SOH (Start Of Header):** A transmission control character that indicates the start of a message heading.
3. **STX (Start of TeXt):** A transmission control character that alerts the receiving device to start the reading, transmission, reception, or recording of text.
4. **ETX (End of TeXt):** A transmission control character that alerts the receiving device to terminate the reading, transmission, reception, or recording of text.
5. **EOT (End Of Transmission):** A transmission control character that alerts the receiving device to terminate a transmission that may include one or more texts or messages.
6. **ENQ (ENQuiry):** A transmission control character to request a response from a station to which a connection has been established. The request may be for the station identification, type of equipment, and station status.
7. **NAK (Negative AcKnowledgement):** A transmission control character sent by the receiving device to the transmitting device to indicate that a received block of data contained one or more errors. A NAK will trigger the transmitting device to retransmit that errored block.
8. **ACK (ACKnowledgement):** A transmission control character sent by the receiving device to the transmitting device to indicate that a received block of data contained no errors.
9. **BEL (BELl):** A transmission control character that alerts the receiving device that causes a bell to ring or activates some other audio or visual device to gain the attention of the operator at the receiving station.
10. **ETB (End of Transmission Block):** A code-extension character used to indicate the end of the transmission of a block of data.
11. **CAN (CANcel):** A transmission control character indicating that the associated data is in error or is to be ignored.
12. **EM (End of Medium):** A control character indicating the physical end of a data storage medium, or the usable portion of the medium.
13. **SUB (SUBstitute):** Used in place of a character that is known to be invalid, i.e., in error. Also used to indicate a character used in place of one that cannot be represented on a given device, e.g., *e* may be used in place of ε (epsilon) or *d* may be used in place of Δ (delta).
14. **ESC (ESCape):** A code-extension character used to indicate a change in code interpretation to another character set, according to some convention or agreement. This is much like the use of the shift key in Baudot code to indicate a shift between figures and characters.
15. **CR (Carriage Return):** A format-control character that causes the print or display position to move to the first position, or left-hand margin, of the screen or print medium. Now often associated with an LF (Line Feed), which moves the print position down to the next line

In Unicode terms, ASCII is known as Unicode Transformation Format-7 (UTF-7). See also *asynchronous*, *code set*, *EBCDIC*, and *Unicode*.

Ashbacker Radio Corporation vs. the FCC The United States Supreme Court ruling (1945) that established that radio spectrum allocation is to be on the basis of comparative hearings. See also *spectrum management.*

Asia Pacific Network Information Center (APNIC) See *APNIC.*

ASIC (Application-Specific Integrated Circuit) A semiconductor integrated circuit designed for a specific application. An ASIC, for example, can be designed specifically a real-time processing task such as running a particular type of encryption, or running a cell phone or personal digital assistant (PDA). Contemporary ASICs often contain complete processors, and RAM, ROM, Flash, and other types of memory. See also *encryption, flash, memory, RAM, ROM,* and *semiconductor.*

ASK (Amplitude Shift Keying) Synonymous with AM (Amplitude Modulation). See *AM.*

ASP (Application Service Provider) A company that provides access to Internet-based software for a fee that generally is based on the number of users. See also *Internet* and *software.*

aspect ratio In video display, the relationship between the width and the height of the image. The NTSC standard, for example, specifies a 4:3 (4 wide to 3 high) aspect ratio. See also *NTSC* and *video.*

assured forwarding (AF) See *AF.*

asymmetric Lack of symmetry, i.e. lack of balance or proportion. **1.** In telecommunications, a link that supports more bandwidth in one direction than another. Asymmetric digital subscriber line (ADSL), for example, supports more bandwidth downstream than upstream. Bluetooth supports an asynchronous data channel that can operate in asymmetric mode at up to 721 kbps in either direction and 57.6 kbps in the reverse direction. Alternatively, the Bluetooth data channel can operate in symmetric mode at speeds of up to 432.6 kbps. See also *ADSL, asynchronous, bandwidth, Bluetooth, channel, downstream, symmetric,* and *upstream.* **2.** In compression, a process that is not equally time-consuming and processor-intensive in terms of compression and decompression. See also *compression.*

asymmetric digital subscriber line (ADSL) See *ADSL.*

asynchronous From Latin and Greek origins, asynchronous translates as not together with time. Referring to signals or events that bear no relationship to timing and, therefore, can be considered occurring at random instants and, for recurring events, at random intervals. See also *asynchronous transmission* and *synchronous.*

asynchronous balanced mode (ABM) See *ABM.*

asynchronous connectionless link (ACL) See *ACL.*

asynchronous transfer mode (ATM) See *ATM.*

asynchronous transmission Also known as start-stop transmission. Data transmission that is not synchronized between two or more computers across a circuit. The transmitting device sends data intermittently, rather than in a steady stream or at regular intervals. Such transmission is characterized as character-framed, as each character is preceded by a start bit that alerts the receiving computer of its arrival and succeeded by one or two stop bits that signal the end of the character. As illustrated in Figure A-7, an optional parity bit may be included for error control. Multiple characters commonly are organized into blocks, with an additional error control mechanism, such as a cyclic redundancy check (CRC), for improved error performance. Kermit, XMODEM, and ZMODEM are examples of asynchronous protocols. See also *asynchronous, CRC, error control, frame, Kermit, parity bit, synchronous, synchronous transmission, XMODEM,* and *ZMODEM.*

Start Bit	1	2	3	4	5	6	7	Parity Bit	Stop Bit

Figure A-7

AT&T (American Telephone and Telegraph) On July 9, 1877, the Bell Telephone Company was formed as a voluntary, unincorporated association. In 1878, the company split into the New England Telephone Company, charged with licensing telephone operating companies in New England, and the Bell Telephone Company, charged with licensing operating telephone companies elsewhere. In 1879, the two companies recombined to form the National Bell Telephone Company, which reorganized in 1880 and became known as American Bell Telephone Company, a Massachusetts corporation. Restrictive Massachusetts corporate laws forced American Bell to merge with its long distance subsidiary, the American Telephone and Telegraph Corporation (AT&T), a New York corporation. On December 30, 1899, the last business day of the nineteenth century, AT&T became the new parent company. AT&T grew to become the largest company in the world, employing over 1,000,000 people, and with a solid reputation for providing the best telephone service in the world. In 1984, the company was forced under the terms of the Second Computer Inquiry to spin its 22 wholly owned Bell Operating Companies (BOCs) into 7 Regional Bell Operating Companies (RBOCs).

AT&T reorganized into two business units. AT&T Long Lines became AT&T Communications, operating as an interexchange carrier (IXC). AT&T Technologies was formed of Western Electric, the manufacturing arm of AT&T, and AT&T Bell Telephone Laboratories (Bell Labs), the research and development organization. AT&T did very well over the next 13 years, focusing on its core businesses, although it did acquire and later divest NCR Corp. in a failed and costly attempt to get into the computer business. IBM previously experienced a similarly dismal failure with its acquisition of ROLM Corp., an almost legendary PBX manufacturer, which it subsequently sold to Stromberg-Carlson at a substantial loss.

On January 1, 1997, AT&T conducted the largest voluntary breakup in history. The US$75 billion company split into three market-focused companies, also selling AT&T Capital Corp., its captive financing business. Approximately 8,500 employees, all in the Global Information Solutions (GSI) computer business, lost their jobs fairly immediately. GSI resulted from the NCR acquisition, which did not live up to expectations. Hundreds of thousands of others lost their jobs over time. The post-divestiture AT&T boasted assets of US$79.2 billion, annual revenues of US$75.1 billion, and a total workforce of 303,000, which was down from over 1,000,000 prior to divestiture.

AT&T then went on a spending spree, variously acquiring and merging with a number of companies. In 1999, AT&T acquired MediaOne, which previously had been spun off from US West, in a bidding war against Comcast Corporation. The winning bid was in the form of AT&T stock worth US$58 billion at the time, plus the assumption of US$4.5 billion in debt. Together, these acquisitions formed AT&T Broadband, the largest CATV provider in the United States. Under extreme financial pressure due to the inflated cost of its acquisitions and the high costs of upgrading its CATV systems, AT&T Broadband agreed to merge with Comcast to form AT&T Comcast in a deal that initially valued AT&T Broadband at US$72 billion and later shrunk to US$53 billion, which did not compare favorably with the US$110.5 billion AT&T spent to form the company.

In 2006, the tattered remnants of AT&T were acquired by SBC for approximately US$16 billion, which named the combined entity AT&T. In just over 20 years, one of the oldest, largest and most respected companies in the world was reduced to a property for acquisition. On a personal note, I am so very glad that I was not there to see it up close. I left the Bell System of my own free will long, long before AT&T collapsed. Heck, I never did fit in, anyway.

AT&T Bell Telephone Laboratories (Bell Labs) See *Bell Labs.*

AT&T Technologies The company formed of Western Electric, the manufacturing arm of AT&T, and AT&T Bell Telephone Laboratories (Bell Labs), the research and development organization, as a result of the Modified Final Judgement (MFJ) that broke up the AT&T Bell System in 1984. AT&T Technologies later became Lucent Technologies, which was acquired by the French company Alcatel in 2006. The combined company is known as Alcatel-Lucent, as of Spring 2007. See also *Bell System* and *MFJ.*

ATIS (Alliance for Telecommunications Industry Solutions) Formerly the Exchange Carriers Standards Association (ECSA) A U.S. organization that develops and promotes technical and operations standards for the telecommunications and related information technology industries. ATIS standards activities address both wireless and wireline networks and include interconnection standards, number portability, improved data transmission, Internet telephony, toll-free access, telecom fraud, and order and billing issues. ATIS is accredited by the American National Standards Institute (ANSI). See also *ANSI.*

ATM (Asynchronous Transfer Mode) A fast-packet, connection-oriented, cell-switching technology for broadband signals. ATM was an outgrowth of the ITU-T development efforts towards broadband integrated services digital network (B-ISDN). Although B-ISDN faltered, ATM became the switching technology of choice in the broadband backbone of the public telephone network, at least for a time. ATM is designed to accommodate any form of data, including voice, facsimile, computer data, video, image, and multimedia, whether compressed or uncompressed, whether real-time or non-real-time in nature, and with guaranteed quality of service (QoS). ATM generally operates at minimum access speeds of DS-1 (e.g., T1 at 1.544 Mbps and E-1 at 2.048 Mbps) and DS-3 (e.g., E-3 at 34.368 Mbps and T1 at 44.736 Mbps). Designed to operate at very high speeds, ATM benefits from fiber optic transmission systems (FOTS) and commonly is provisioned over SDH/SONET networks. Access circuits operating at OC-3 (155 Mbps) are not unusual and backbone transmission rates generally are OC-3, at a minimum. ATM traffic consists of three basic types.

- **Constant Bit Rate** (CBR) traffic requires access to time slots at regular and precise intervals. Real-time, uncompressed voice and video, and circuit emulation are examples of CBR traffic.
- **Variable Bit Rate** (VBR) traffic, such as compressed voice and video and bursty data traffic, requires access to time slots at a rate that can vary dramatically from time to time but each logical connection is guaranteed a level of service defined by burst size, average bandwidth, etc.
- **Available Bit Rate** (ABR) traffic, also known as best-effort ATM, supports bursty LAN traffic and other traffic that can deal with time slot access on an as-available basis.

ATM organizes data into cells, as illustrated in Figure A-8. Each cell comprises a header of 5 octets and payload of 48 octets, with the payload including some amount of overhead attributable to Convergence Sublayer and Data Link Layer and Network Layer headers. Although the total overhead is in the range of 10 percent, the small cell size offers the advantage of effectively supporting any type of data. The fixed cell size offers the advantage of predictability, very much unlike the variable-length frames associated with services such as X.25, frame relay, and Ethernet, or the variable-length packets associated with the Internet Protocol (IP). This level of predictability yields much improved access control and congestion control. ATM multiplexes the cells, which contend for access to a broadband facility that ideally is SDH or SONET in nature. ATM also is used in some passive optical network (PON) local loops.

<table>
<tr><td>GFC</td><td colspan="2">VPI</td></tr>
<tr><td>VPI</td><td colspan="2">VPI</td></tr>
<tr><td colspan="3">VPI</td></tr>
<tr><td>VPI</td><td>PT</td><td>CLP</td></tr>
<tr><td colspan="3">HEC</td></tr>
<tr><td colspan="3">Information
Payload
(48 octets)</td></tr>
</table>

Figure A-8

The ATM cell header provides limited Data Link Layer functionality, managing the allocation of the resources of the underlying Physical Layer of the transmission facility. The ATM cell switches also perform Layer 1 functions such as clocking, bit encoding, and physical-medium connection. The header also is used for channel identification, thereby ensuring that all cells travel the same physical path and, therefore, arrive in sequence. The header is structured as follows:

- **Generic Flow Control (GFC):** 4 bits that provide local flow control.
- **Virtual Path Identifier (VPI):** 8 bits identifying the Virtual Path (VP).
- **Virtual Channel Identifier (VCI):** 16 bits identifying the Virtual Channel (VC). Together, VPI and VCI constitute the cell address, which has only local significance. That is, each switch maps the address on an incoming port to an address on an outbound port, so the local address changes on each hop.
- **Payload Type Indicator (PTI):** 3 bits distinguishing between cells carrying user information and cells carrying service information.
- **Cell Loss Priority (CLP):** 1 bit identifying one of two priority levels of the cell to determine the eligibility of that cell for discard in the event of network congestion.
- **Header Error Control (HEC):** 8 bits providing error checking of the header, but not the payload. Errored cells are discarded. There is no provision for error correction, which is handled at higher layers.

ATM standards largely are outgrowths of B-ISDN standards set by the ITU-T. The ATM Forum, now merged into the MFA Forum, developed interoperability specifications. The Frame Relay Forum (FRF), also now merged into the MFA Forum, worked with the ATM Forum in the development and publishing of joint Implementation Agreements (IAs) that specify the protocol interworking functions between frame relay and ATM networks. The Internet Engineering Task Force (IETF) also got involved in standards development as ATM has significant implications relative to the Internet backbone. ITU-T Standards Recommendations of significance include the following:

- I.113: B-ISDN Vocabulary
- I.121: Broadband Aspects of ISDN
- I.150: B-ISDN ATM Functional Characteristics
- I.211: B-ISDN Service Aspects
- I.311: B-ISDN General Network Aspects
- I.321: B-ISDN Protocol Reference Model
- I.327: B-ISDN Functional Architecture Aspects

- I.361: B-ISDN ATM Layer Specification
- I.362: B-ISDN ATM Adaptation Layer Functional Description
- I.363: B-ISDN ATM Adaptation Layer Specification
- I.413: B-ISDN User-Network Interface
- I.432: B-ISDN User-Network Interface-Physical Layer Specification
- I.555: Frame Relay and ATM Internetworking
- I.610: B-ISDN Operations and Maintenance Principles and Functions

See also *ABR*, *backbone*, *B-ISDN*, *broadband*, *CBR*, *cell*, *cell tax*, *channel*, *compression*, *congestion*, *connection-oriented*, *Data Link Layer*, *encode*, *Ethernet*, *FOTS*, *frame*, *frame relay*, *header*, *IETF*, *Internet*, *IP*, *ITU-T*, *MFA Forum*, *multiplex*, *Network Layer*, *non-real-time*, *packet*, *payload*, *PON*, *real-time*, *SDH*, *SONET*, *VBR*, and *X.25*.

ATM Adaptation Layer (AAL) See *AAL*.

ATM-based passive optical network (APON) See *APON*.

ATM Forum A not-for-profit special interest group of manufacturers, vendors, carriers and others with interests in the development and promotion of asynchronous transfer mode (ATM) technology. The ATM Forum merged with the Frame Relay Forum and MPLS Forum to form the MFA Forum. See also *ATM* and *MFA Forum*.

ATM reference model A multidimensional model, with three planes and four layers, as illustrated in Figure A-9. The lower two layers of this reference model loosely compare to the Physical Layer of the OSI Reference Model. As in the OSI model, each layer of the ATM model functions independently, yet all layers are tightly linked and the functions are highly coordinated. The layers of the ATM reference model are the Physical Layer, the ATM Layer, ATM Adaptation Layer, and higher layers and functions.

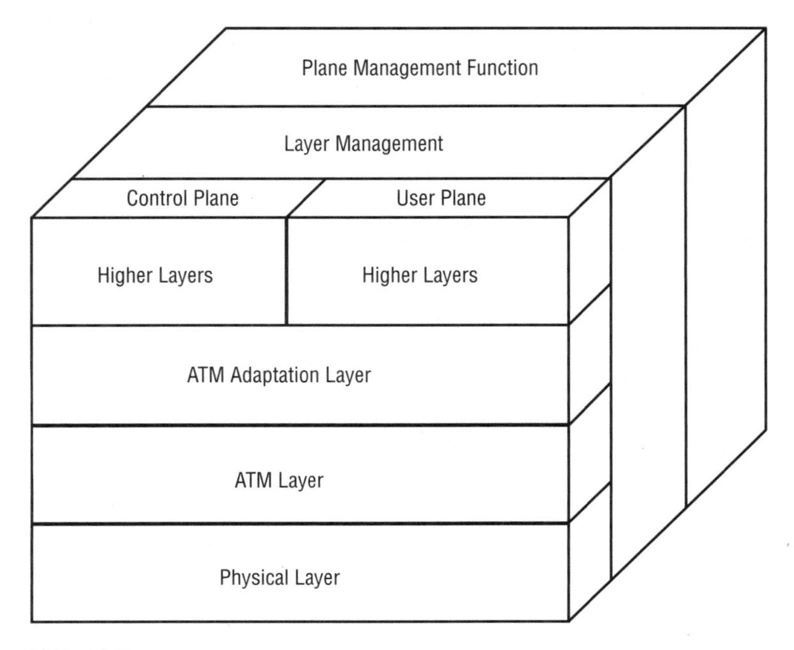

Figure A-9

- **Physical Layer (PHY)** functions are addressed through two sublayers: the Physical Medium and Transmission Convergence. The ATM Forum specifications for various User Network Interfaces (UNIs) address the implementation of the Physical Layer. The B-UNI, or Public UNI, is the specification for carrier internetworks. The UNI and DXI are Private UNIs, describing the implementation specifics for user access to the ATM network. Physical Medium (PM) sublayer specifies the physical and electro-optical interfaces with the transmission medium. The PM also provides timing functions. The Transmission Convergence (TC) sublayer handles frame generation, frame adaption, cell delineation, header error control (HEC), and cell rate decoupling.
- **ATM Layer (ATM)** functions include multiplexing of cells, selection of appropriate Virtual Path Identifiers (VPIs) and Virtual Channel Identifiers (VCIs), generation of headers, and flow control. At this layer, all multiplexing, switching, and routing takes place for presentation to the appropriate Virtual Paths (VPs) and Virtual Channels (VCs) of the SONET fiber optic transport system, which interfaces through the Physical Layer.
- **ATM Adaptation Layer (AAL)** functions are divided into sublayers. The Convergence Sublayer (CS) functions are determined by the specifics of the service class supported by that particular AAL. The Segmentation and Reassembly (SAR) sublayer functions to segment the user data into 48-byte payloads for insertion into cells, on the transmit side. On the receive side, the SAR extracts the payloads from the cells and reassembles the data into the information stream as originally transmitted, e.g. IP packets.

The planes include the Control Plane, User Plane, and Management Plane. See also *AAL*, *B-UNI*, *cell*, *CS*, *flow control*, *frame*, *header*, *HEC*, *OSI Reference Model*, *Physical Layer*, *PM*, *Private UNI*, *Public UNI*, *SAR*, *TC*, *VC*, *VCI*, *VP*, and *VPI*.

atmosphere The mixture of gases that surrounds and is retained by the gravity of a celestial body such as the Earth. The atmosphere is denser near the Earths surface, and becomes gradually thinner until it fades away into space. Particularly near the Earths surface, the physical matter in the atmosphere attenuates electromagnetic signals due to absorption, refraction and other phenomena. At the outer limits of the atmosphere are four layers of the ionosphere, which is useful for skywave radio propagation. See also *attenuation*, *ionosphere*, *propagation*, *refraction*, and *skywave*.

A-to-D (Analog-to-Digital) See *codec* and *modem*.

Atom Publishing Protocol (APP) See *APP*.

ATSC (Advanced Television Systems Committee) An ad hoc advisory group formed by the United States Federal Communications Commission (FCC) for the purpose of reviewing, testing, and documenting digital television (DTV) standards recommendations developed by the Grand Alliance. Specifically, standards recommendations were developed for standard definition television (SDTV) and high definition television (HDTV). The ATSC completed its work in the summer of 1995 and the standards were approved by the FCC in December 1996. See also *digital*, *DTV*, *FCC*, *Grand Alliance*, *HDTV*, and *SDTV*.

Attached Resource Computer Network (ARCNET) See *ARCNET*.

attachment unit interface (AUI) See *AUI*.

attendant access A feature of voice mail systems that allows a caller to reach a live human attendant or alternative answering point if the caller does not want to leave a message. Attendant access usually is presented as a menu option, at least by companies that place any value on customer satisfaction. Companies that do not care about customer satisfaction are happy to condemn the caller to voice mail jail. See also *human*, *voice mail*, and *voice mail jail*.

attenuation Loss in signal power. Electromagnetic signals tend to weaken, or attenuate, over a distance. Some of the signal is absorbed and converted to thermal energy as it interacts with the physical matter between the transmitter and receiver. Some of the signal is absorbed at the molecular level, and some of the signal is emitted and scattered in all directions, some of it at different frequencies. Twisted-pair copper wire systems attenuate electrical signals due to factors including the interaction of the signal with the copper in the conductors as the described by the level of resistance or impedance in the wire, and the tendency of the signal to radiate, or spread out, from the wire. Signal attenuation occurs in terrestrial radio systems due to interaction with the physical matter in the air and the tendency of the signal to disperse, or spread out.

Attenuation is a relatively minor issue with respect to satellite radio systems, at least with respect to signal propagation in the vacuum of space, where there is no physical matter to interact with the signal. The portion of the satellite link that travels through the atmosphere is very much subject to attenuation, however. Attenuation also affects fiber optic systems, as some optical energy is absorbed at the molecular level, some is converted to thermal energy, some is dispersed, and some suffers frequency shifts. In some fiber optic systems, some amount of optical energy can be lost in the cladding that surrounds the crystalline core. (*Note:* Glass actually is not crystalline, but is an extremely viscous fluid.)

Attenuation is sensitive to carrier frequency. In electrical and radio systems, for example, higher-frequency signals generally attenuate more than lower-frequency signals. The same phenomenon generally holds true in fiber optic systems, as well, although the measurement is in wavelengths, rather than frequencies, i.e., longer wavelength signals (lower frequency) signals attenuate less than shorter wavelength (higher frequency) signals. All else being equal, the impacts of attenuation increase with distance, and can become so severe over a long distance that the receiver cannot interpret the signals correctly. A variety of measures can be employed to overcome the effects of attenuation. Most commonly, amplifiers and regenerative repeaters are placed on circuits. The level of attenuation is described as insertion loss and is measured in decibels (dB) or decibels per kilometer (dB/km). See also *amplifier*, *dB*, *dB/km*, *frequency*, *gain*, *insertion loss*, *repeater*, and *wavelength*.

attenuation-to-crosstalk ratio (ACR) See *ACR*.

attenuator A passive optical component used to intentionally decrease the level of optical power propagating in an optical fiber.

ATIS (Alliance for Telecommunications Industry Solutions) Formerly the Exchange Carriers Standards Association (ECSA) A U.S. organization that develops and promotes technical and operations standards for the telecommunications and related information technology industries. ATIS standards activities address both wireless and wireline networks and include interconnection standards, number portability, improved data transmission, Internet telephony, toll-free access, telecom fraud, and order and billing issues. ATIS is accredited by the American National Standards Institute (ANSI). See also *ANSI*.

ATU-C (ADSL Transmission Unit-Centralized) An asymmetrical digital subscriber line (ADSL) modem located at the telco central office (CO) or other headend location. The ATU-C is the line side interface of a digital subscriber line access multiplexer (DSLAM). A matching modem, known as an ADSL transmission unit-remote (ATU-R) is located on the customer premises. See also *ADSL*, *ATU-R*, *CO*, *DSLAM*, *headend*, and *modem*.

ATU-R (ADSL Transmission Unit-Remote) An asymmetrical digital subscriber line (ADSL) modem located on the customer premises. A matching modem, known as an ADSL transmission unit-centralized (ATU-C) is located at the telco central office (CO) or other headend location. See also *ADSL*, *CO*, *headend*, and *modem*.

audio Sound. Generally referring to sound *recorded* and reproduced, including voice and music. Unwanted audio is noise. See also *noise*.

Audio Messaging Interchange Specification (AMIS) See *AMIS.*

audiotex Also known as audiotext. A simple voice processing technology that is essentially a voice bulletin board, audiotex allows callers to select prerecorded messages from a menu. Audiotex is used to provide information that seldom changes or that must be available to large numbers of callers. Examples of such messages include time and temperature, hours of operation, travel directions, facsimile (fax) numbers, web addresses, and school closings.

audiotext See *audiotex.*

AUI (Attachment Unit Interface) A standard that defines the manner in which an Ethernet cable, especially a coaxial cable, physically attaches to a network interface card (NIC). See also *coax* and *NIC.*

authentication Security measures designed to verify or validate the identity of a user or station prior to granting access to resources. Authentication mechanisms include passwords and intelligent tokens. See also *intelligent token NAS, password, RADIUS,* and *RAS.*

Autonomous System (AS) See *AS.*

authorization The process of granting approval or permission to a person or device seeking access to a resource, such as a database or network. Authorization involves complex software that resides on every secured computer on the network. Authorization systems include Access Manager, Kerberos, and Sesame. See also *Access Manager, Kerberos, security,* and *Sesame.*

authorization code A code that a user inputs to a system in order to gain access to resources such as applications, files, or networks.

auto dialer (automatic dialer) A peripheral device that connects to a telephone set and that automatically dials a telephone number.

automated attendant An application in which an interactive voice processor automates many of the functions of a human attendant, answering an incoming call and prompting the caller through a series of spoken menu options to directly access a department or station through touchtone or speech input. In the event that the caller does not know the desired station number, an automated directory can provide that information on the basis of a name search. When the station number is identified, the voice processor signals the telephone system (e.g., KTS, PBX, Centrex, or CO), instructing it to connect the call. See also *audiotex, human, voice mail,* and *voice processor.*

automatic callback Also known as call return. A network-based CLASS service of the public switched telephone network (PSTN). When activated by the caller who reaches a busy line, the central office (CO) monitors the target telephone number for a period of time, e.g., 30 minutes, and advises the caller with a (usually distinctive) callback ring when that line becomes available. When the caller answers the ringback call, the CO automatically redials the target number. See also *CLASS* and *PSTN.*

automatic call distributor (ACD) See *ACD.*

automatic line selection A key telephone system (KTS) feature that automatically selects an outside line when a station user picks up the telephone receiver.

automatic number identification (ANI) See *ANI.*

Automatic Protection Switching (APS) See *APS.*

automatic route selection (ARS) See *ARS.*

automatic repeat request (ARQ) See *ARQ.*

automatic set relocation Also known as customer rearrangement. A PBX administrative feature that allows the end user to accomplish set relocations without technical assistance. The user simply takes the phone from one location to another, plugs the set into the wall jack and dials a relocation code. The set identifies itself to the PBX, which changes the station port assignment and reassociates the station number and all assigned features to the new port. The feature considerably simplifies Move, Add, and Change (MAC) activity and lowers the associated costs.

available bit rate (ABR) See *ABR*.

avalanche photodiode (APD) See *APD*.

AWG (American Wire Gauge) The standard measurement of gauge in United States for all metals other than iron and steel. The gauge numbers are retrogressive; in other words, the larger the number, the thinner the conductor. The AWG number indicates the approximate number of wires that, laid side-by-side, span one inch. Historically, the AWG number indicated the number of times during the manufacturing process that the copper wire was cold drawn through the wire machine, with each draw involving a die of slightly smaller diameter in order to reduce the diameter of the wire a bit more. The contemporary process involves many fewer draws. A 24-gauge (AWG) wire, for example, has a diameter of 0.0201 in. (0.511mm), a weight of 1.22 lbs/kft (1.82 kg/km), maximum break strength of 12.69 lbs (5.756 kg), and DC resistance ohms of 25.7/kft (84.2/km). Twisted-pairs commonly employed in telco networks vary from 19 to 28 gauge, with the most common being 24 gauge. Table A-3 provides diameter, weight, and resistance comparisons of bare copper wire gauges. AWG originally was known as Brown and Sharp (B&S) Wire Gauge. See also *gauge*, *Imperial Standard Wire Gauge*, and *metric gauge*.

Table A-3: American Wire Gauge (AWG): Select Physical Attributes

AWG	*Nominal Diameter*	*Nominal Weight*	*Nominal Resistance*			
	Inches	*Millimeters*	*lb/kft*	*kg/km*	*Ohms/kft*	*Ohms/km*
10	.1019	2.60	9.55	46.78	0.9989	3.2763
11	.0907	2.30	7.57	37.09	1.2596	4.1328
12	.0808	2.05	6.00	29.42	1.5883	5.2086
13	.0720	1.83	4.76	23.33	2.0028	6.5698
14	.0641	1.63	3.78	18.50	2.5255	8.2820
15	.0571	1.45	2.99	14.67	3.1845	10.444
16	.0508	1.29	2.37	11.64	4.0156	13.172
17	.0453	1.15	1.88	9.219	5.0636	16.610
18	.0403	1.02	1.49	7.313	6.3851	20.942
19	.0359	0.912	1.18	5.807	8.0514	26.407
20	.0320	0.812	0.939	4.600	10.153	33.292
21	.0285	0.724	0.745	3.649	12.802	41.984
22	.0253	0.643	0.591	2.895	16.143	52.939
23	.0226	0.574	0.468	2.295	20.356	66.781
24	.0201	0.511	0.371	1.820	25.669	84.197
25	.0179	0.455	0.295	1.443	32.368	106.17

continued

Table A-3: American Wire Gauge (AWG): Select Physical Attributes *(continued)*

AWG	*Nominal Diameter*	*Nominal Weight*	*Nominal Resistance*			
27	.0142	0.361	0.185	0.9077	51.467	168.82
28	.0126	0.320	0.147	0.7198	64.898	212.87
29	.0113	0.287	0.117	0.5712	81.835	268.40
30	.0100	0.254	0.0924	0.4531	103.19	338.50
31	.0089	0.227	0.0733	0.3577	130.12	426.73
32	.0080	0.203	0.0581	0.2847	164.08	538.25
33	.0071	0.180	0.0461	0.2250	206.90	678.63
34	.0063	0.160	0.0365	0.1790	260.90	855.75
35	.0056	0.143	0.0290	0.1415	328.99	1079.1
36	.0050	0.127	0.0230	0.1126	414.85	1360.0
37	.0045	0.113	0.0182	0.0890	523.11	1715.0
38	.0040	0.102	0.0145	0.0708	659.63	2163.0

axis **1.** In geometry and optics, a straight line, either real or imaginary, around which a body or figure, or parts thereof, are symmetrically or evenly arranged or composed. In an optical fiber, for example, the axis is the centerpoint of a cross-section. **2.** In optics, an imaginary line perpendicular to the center of a lens or mirror.

B **1.** bel. In physics, the abbreviation for bel, a measure of relative loudness. See *bel.* **2.** B channel. Referring to an ISDN bearer (B) channel, which is an information-bearing channel designated for user payload. See *B channel.*

B&S (Brown and Sharp Wire Gauge) The original name for American Wire Gauge (AWG). See *AWG.*

B8ZS (Bipolar with Eight-Zeros Substitution) A technique used with alternate mark inversion (AMI) to support the transmission of long strings of zero bits, which are not unusual in data communications applications. When B8ZS encounters a string of eight zeros (00000000), it substitutes a specific bit pattern that intentionally violates AMI, i.e., includes an intentional bipolar violation (BPV). If the preceding mark (1 bit) was represented as a +3V (positive 3 volts), the substituted bit pattern is 0 0 0 + – 0 – +. If the preceding mark was represented as a –3V (negative 3 volts), the substituted bit pattern is 0 0 0 – + 0 + –. Since the bit pattern is known to both the transmitting and receiving multiplexer, the receiving multiplexer can restore the original 00000000 bit pattern (see Figure B-1). AMI and B8ZS are used in T1 networks. A similar technique known as High Density Bipolar order 3 (HDB3) is used in E-1 networks. See also *AMI, BPV, HDB3,* and *multiplexer.*

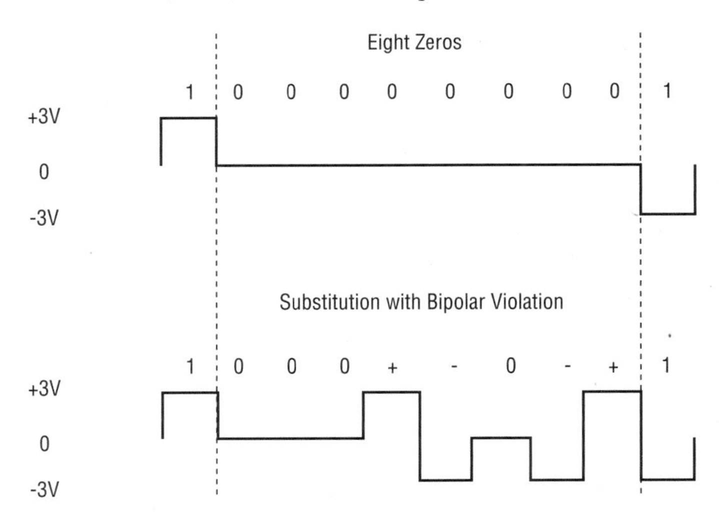

Figure B-1

backbone The central or essential part of a network is commonly known as the backbone, or core. The backbone comprises very high capacity elements and subsystems such as transmission systems, multiplexers, switches, and routers. The term is used in the context of a wide area network (WAN), metropolitan area network (MAN), and local area network (LAN). See also *core, LAN, MAN,* and *WAN.*

backbone switch Also known as a core switch and a tandem switch, a backbone switch is a high-capacity switch positioned in the physical core, or backbone, of a network. In the context of a public wide area network (WAN), a backbone switch serves to interconnect edge switches, which are positioned at the network edge, and does not connect to desktop machines or other end user terminals. In the context of a local area network (LAN), a backbone switch serves to interconnect relatively low capacity workgroup switches that serve the needs of groups of workers who are geographically clustered. See also *core switch*, *LAN*, *switch*, *tandem switch*, *WAN*, and *workgroup switch*.

back door A means of gaining access to a computer program or system by bypassing the normal authentication and other security procedures and mechanisms. Programmers often create back doors so that they can fix bugs and speed development work. If the back door code is left in place when the software goes into general release, it creates a considerable security risk. See also *authentication*, *bug*, and *Clipper Chip*.

backhaul **1.** In telecommunications, referring to a leased line network configuration in which traffic is transported to a point that is geographically beyond and then transported back (hauled back) to the destination site due to the lack of a direct path between the originating and destination sites. Such an indirect design is much like the indirect route one might be forced to take from New York City west to Seattle and then back east to get to Spokane, Washington. **2.** In telecommunications, and particularly wireless networks, to transport traffic from a distributed node, such as a cellular base station or Wi-Fi access point (AP), to a centralized node, such as a mobile telephone switching office (MTSO) or Internet service provider (ISP), respectively. See also *AP*, *base station*, *cellular radio*, *ISP*, *MTSO*, *node*, *Wi-Fi*, and *wireless*.

backhoe fade A circuit or network failure caused by a cable-seeking backhoe, posthole digger, auger, or other piece of earth-moving equipment.

back reflection In optical fiber installation, referring to light reflected back toward the source from the air gap at the point where two fiber endfaces meet in a connector. There is always a slight air gap, as the endfaces are never perfectly cleaved and never can be aligned in the connectors so that they mate perfectly. The considerable difference in index of refraction (IOR) between glass and air causes some amount of light to reflect. An angled physical contact (APC) connector, which joins two fiber endfaces at a slight angle, sometimes is used to minimize attenuation and back reflection. See also *APC*, *attenuation*, *backscatter*, *connector*, *IOR*, and *optical fiber*.

backronym (back acronym) The treatment of a word as an acronym even though it is not. For example, *ping* is a utility used to test a path from one host computer to another across an IP-based network in what is essentially a command to echo the packet from the remote host computer back to the originating host. *Ping* is a word, not an acronym. However, Dr. David L. Mills reverse-engineered *ping* into an acronym for *packet Internet groper* and a great many people believe that was the original meaning. It was not. See also *acronym*, *anacronym*, and *ping*.

backscatter In a fiber optic transmission system (FOTS), the portion of an optical signal that is deflected back towards the transmitter through interaction with the glass or plastic medium. See also *back reflection*, *FOTS*, *GOF*, and *POF*.

backward-compatible Referring to something (e.g., a device, machine, system, or program) that can be used with or is interoperable with an earlier generation, model, or version.

Backward Explicit Congestion Notification (BECN) See *BECN*.

bait rod The basic structure for creating a glass preform cylinder used in the mass production of glass optical fiber. The process begins with heating silica and germanium to the point that it vaporizes. The glass vapor cools and is deposited as layers of soot on the outside of a rotating hollow bait rod, also known as a seed rod. When the deposition process is complete, the bait rod is removed and the remaining glass cylinder is collapsed. See also *outside vapor deposition (OVD)*.

balanced **1.** Referring to electrical symmetry. A balanced line or balanced medium such as twisted pair, in which both twisted pair conductors serve for signal transmission and reception. Each conductor carries a similar electrical signal with identical direct and return current paths. At any given point in the cable, the signals are equal in voltage to ground but opposite in polarity, which has the effect of reducing radiated energy and, therefore, reducing attenuation. See also *unbalanced* and *UTP*. **2.** Referring to symmetrical relationship. For example, the X.25 protocol suite includes Link Access Procedure-Balanced (LAP-B), a balanced protocol that operates in Asynchronous Balanced Mode (ABM), which refers to the fact that the devices have a balanced, rather than a master/slave, relationship. Therefore, a device at either end of the link can initiate a dialogue at any time. See also *ABM*, *LAP-B*, *master/slave*, and *X.25*.

balun (balanced/unbalanced) A passive device, often a transformer, used to couple an electrically balanced device, medium, or system and an electrically unbalanced device, medium, or system. A balun commonly is used to connect an electrically balanced twisted pair to an electrically unbalanced coaxial cable. The term *balun* is a contraction of *balanced to unbalanced transformer*. See also *balanced*, *coaxial cable*, *passive*, *transformer*, *twisted pair*, and *unbalanced*.

band A continuous group, or range, with an upper limit and a lower limit. In analog terms, the width of a band or channel is defined as the upper and lower frequencies in a range of frequencies. The ITU-T defines standard optical transmission windows in bands of wavelengths. See also *bandwidth* and *window*.

band-pass filter A device that passes all signals in a designated frequency (electrical) or wavelength (optical) band, but absorbs, attenuates, blocks, rejects, or removes all other signals. See also *absorption*, *attenuation*, *band*, *electrical*, *frequency*, *high-pass filter*, *low-pass filter*, *optical*, *signal*, and *wavelength*.

bandwidth The measure of the capacity of a circuit or channel. More specifically, bandwidth refers (1) to the total frequency range on the available carrier in Hertz (Hz) for the transmission of data, or (2) the capacity of a circuit in bits per second (bps). There is a direct relationship between the bandwidth of an analog circuit or channel and both its frequency and the difference between the minimum and maximum frequencies supported. Although the information signal (bandwidth usable for data transmission) does not occupy the total capacity of a circuit, it generally and ideally occupies most of it. The balance of the capacity of the circuit may be used for various signaling and control (overhead) purposes. In other words, the total signaling rate of the circuit typically is greater than the effective transmission rate. In an analog transmission system, bandwidth is measured in Hertz (Hz). In a digital system, bandwidth is measured in bits per second (bps). See also *bps*, *carrier*, *Hz*, *overhead*, *signaling and control*, *signaling rate*, *throughput*, and *transmission rate*.

bandwidth-on-demand Referring to capacity available through a network as required by an application, perhaps during the course of a call. Asynchronous transfer mode (ATM) offers guaranteed bandwidth-on-demand, at least theoretically, adjusting the amount of bandwidth required to support a call once the call is established and guaranteeing that it will be available when required. As an example, real-time compressed voice over ATM may require no bandwidth during periods of prolonged silence, but requires guaranteed bandwidth at precise intervals during periods of speech activity. Frame relay offers bandwidth-on-demand within the limits of the committed information rate (CIR), on average, and within the limits of the port speed, as resources are available. ISDN and some other network services also offer bandwidth-on-demand, defined in various ways. See also *ATM*, *bandwidth*, *call*, *CIR*, *compression*, *frame relay*, *ISDN*, *port*, and *real-time*.

barge-in A feature of a key telephone system (KTS) or PBX, barge-in allows an authorized user from an authorized station to join, without invitation, an active call on a call in progress through the use of an authorization code the user enters via the telephone keypad. See also *KTS*, *PBX*, and *station*.

Barker code A coding scheme used in direct sequence spread spectrum (DSSS) radio systems. Barker code is a sequence of N values of +1 and –1, with N equaling 2, 3, 4, 5, 7, 11, or 13 bits. IEEE 802.11b wireless LANs (WLANs) operating at 1 Mbps and 2 Mbps use an 11-bit Barker code, 10110111000. The code has certain mathematical properties that make it ideal for modulating radio waves. The basic data stream is subjected to a swap algorithm with the Barker code to generate a series of data objects called chips. Each bit is encoded by the 11-bit Barker code, with each group of 11 chips encoding one bit of data. At 5.5 Mbps and 11 Mbps, 802.11b specifies the use of the more efficient complementary code keying (CCK). See also *802.11b*, *algorithm*, *bit*, *CCK*, *chip*, *DSSS*, and *WLAN*.

baseband **1.** Refering to a signal in its original form, without being altered in any way, whether by modulation or conversion. **2.** A single-channel transmission system, i.e., a transmission system that supports a single transmission at any given time. All contemporary wired local area networks (LANs) are baseband. See also *broadband*, *channel*, and *LAN*.

base station (BS) See *BS*.

basic input/output system (BIOS) See *BIOS*.

basic rate access (BRA) See *BRA*.

basic rate interface (BRI) See *BRI*.

basic service Pure and simple transmission capability over a communication path subject only to the technical parameters of fidelity and distortion criteria, or other conditioning. Basic service does not alter the form, content, or nature of the information. See also *enhanced service* and *POTS*.

battery A connected group of (one or more) electrochemical cells that store electric charges and generate direct current (DC) through the conversion of chemical energy into electrical energy. See also *common battery*, *DC*, *electricity*, *energy*, and *local battery*.

baud A signal event, signal change, or signal transition, such as a change from positive voltage to zero voltage, from zero voltage to negative voltage, or from positive voltage to negative voltage. The baud is named for Emile Baudot, inventor of the teletype. See also *Baudot code* and *baud rate*.

Baudot, Emile (1845–1903) Best known as the inventor of the teletypewriter, or teletype, an automatic printing telegraph machine that used a typewriter-style keyboard rather than a telegraph key. As the dot-and-dash Morse code system was not highly compatible with this automated approach, he invented and patented (1874) a five-bit coding scheme that became known as Baudot code. See also *Baudot code*, *Morse code*, *teletype*, and *typewriter*.

Baudot code A five-bit data coding scheme invented by Emile Baudot in the 1870s for use in the Baudot Distributor, a sort of automatic telegraph that supported higher speed transmission over a circuit between two synchronized electromechanical devices. The Baudot Distributor soon gave way to the teletype (TTY), which also employed the Baudot coding scheme, subsequently known as *International Telegraph Alphabet #2* (ITA #2). Updated in 1930, Baudot is limited to 32 (2^5) characters. Considering that each bit has two possible states (1 or 0), 5 bits in sequence yield 2^5 (32) possible combinations. Because 32 values is not sufficient to represent all 26 characters in the English alphabet, plus the 10 decimal digits, necessary punctuation marks and the space character, the shift key operates to shift between letters and other characters. Baudot employs asynchronous transmission, with start and stop bits separating characters. Telephone Devices for the Deaf (TDDs) and telex machines still use ITA #2. See also *asynchronous transmission*, *code set*, *TDD*, *telegraph*, *telex*, and *TTY*.

baud rate The number of signal events, signal changes, or signal transitions occurring per second over an analog circuit, such as changes from positive voltage to zero voltage, from zero voltage to negative voltage, or from positive voltage to negative voltage. The baud rate can never be higher than the raw bandwidth of the channel, as measured in Hz. Baud rate and bit rate, often and incorrectly, are used interchangeably. The relationship between baud rate and bit rate depends on the sophistication of the modulation scheme used to manipulate the carrier. The bit rate and baud rate can be the same, if each bit is represented by a signal transition in a unibit modulation scheme. The bit rate can be higher that the baud rate, as a single signal transition can, and generally does, represent multiple bits. See also *bit rate*, *carrier*, and *modulation*.

BBS (Bulletin Board System) A computer system running software that enables one to connect over the Internet to what is essentially an electronic bulletin board. BBSs generally are focused on specific topics such as a rock band, a software application, or unusual computer or network technical issues. Generally, anyone can access the BBS to post messages, reply to messages, post software applications for downloading by others, play games, and otherwise communicate and share with others. See also *Internet.*

B_c Maximum Burst Size (MBS) See *MBS.*

B Carrier See *wireline carrier.*

BCC (Block Check Character) **1.** The checksum comprising one or two bytes appended to a data block prior to transmission. See also byte, block, and checksum. **2.** The checksum comprising one or two bytes appended specifically to a Binary Synchronous Communications (BSC) data block prior to transmission. See also *byte, block, BSC,* and *checksum.*

BCH (Bose, Chaudhuri, and Hocquengham) A multi-level, variable-length, cyclic, error-correcting code used in forward error correction (FEC) applications. BCH has the ability to detect random error patterns involving up to approximately 25 percent of the total number of digits in a block. BCH is not limited to use with binary codes, but also can be used with multi-level phase-shift keying (PSK) modulation whenever the number of levels is a prime number or a power of a prime number. See also *binary, error control, FEC,* and *PSK.*

B channel (Bearer channel) In the integrated service digital network (ISDN), a 64-kbps channel that bears the end user data, or payload. Standard ISDN interfaces include multiple B channels and a D channel (Delta channel or Data channel) for signaling and control purposes. Basic rate interface (BRI) comprises two B channels and one D channel, and is often referred to as 2B+D. Primary rate interface (PRI) comprises 23 B channels, plus a D channel, is compatible with North American T1 and Japanese J-1 standards, and is often referred to as 23B+D. Primary rate access (PRA) comprises 30 B channels, plus a D channel, is compatible with European E-1 standards, and is often referred to as 30B+D. (*Note:* A 32nd channel is added for overhead and alarms.) See also *BRI, D channel, E-1, ISDN, J-1, payload, PRA, PRI,* and *T1.*

BCM (Bit Compression Multiplexer) A conversion device used to convert between voice signals encoded using adaptive differential pulse code modulation (ADPCM) and those encoded using pulse code modulation (PCM). As an ADPCM-encoded voice transmission generally is encoded at 32 kbps, two such signals can fit into a channel designed to support 64 kbps PCM-encoded voice. A BCM performs the necessary processing to pack two 4-bit ADPCM samples into a single 8-bit PCM time slot. Alternatively, a BCM can perform the necessary signal processing to convert an 8-bit PCM sample so that it will fit into a 4-bit ADPCM time slot. Such conversions are necessary when an E-carrier circuit supporting ADPCM channels connects to a central office (CO) exchange based on PCM, for example. A BCM generally is in the form of a printed circuit board (PCB) that fits into a standard time division multiplexer (TDM). See also *ADPCM, encode, PCM, TDM,* and *time slot.*

B_e Excess Burst Size See *Excess Burst Size.*

beaconing **1.** In wireless networks the transmission by a base station of precisely timed signals as a clear-to-send indicator, essentially advertising the presence of the base station and the availability of a time slot for use by a sender, or source. Beaconing is a contention method used in some wireless protocols, including slotted Aloha. See *slotted Aloha, clear to send, signal,* and *time slot.* **2.** In a Token Ring local area network (LAN), the continuous transmission of small frames if a network failure is detected. A beacon frame identifies the transmitting station, the nearest active upstream neighbor, and everything in between. This triggers a process of autoreconfiguration, in which nodes within the failure domain automatically initiate diagnostic measures in an attempt to identify, isolate, and bypass the point of failure. See also *LAN, node,* and *Token Ring.*

beamsplitter Also known as a *splitter*. In a fiber optic transmission system (FOTS), a passive device that divides an optical signal into two or more signals. See also *FOTS*, *passive*, and *splitter*.

bearer channel (B channel) See *B channel*.

BECN (Backward Explicit Congestion Notification) Pronounced *beckon*. In the frame relay LAPF frame, a 1-bit field used by the network to advise devices of congestion in the direction opposite of the primary traffic flow, i.e., opposite of the direction of the frame encountering the congestion. If the target frame relay access device (FRAD) responds to the originating FRAD in the backward direction, the BECN bit is set in a backward frame. If there is no data flowing in the backward direction, the frame relay network creates a frame in that direction, setting the BECN bit. The BECN bit essentially advises the originating FRAD to reduce the frame transmission rate, if it is capable of doing so, as the network may be forced to discard frames once the notification is posted. Forward explicit congestion notification (FECN) performs a congestion control function in the forward direction. See also *congestion*, *ECN*, *FECN*, *frame*, *frame relay*, and *LAPF*.

beeper Diminutive for pager, attributable to the beeping sound many use to alert the user to an incoming message. See *pager*.

bel (B) In physics, a measure of comparative power ratio, or relative loudness. In other words, a unit of power ratio. The number of bels is the decimal logarithm of the power ratio, which is expressed mathematically as follows:

$$B = \log_{10} (P_1/P_2)$$

where B = Bel, and P_1 and P_2 are power levels. One bel is equal to 10 decibels. The bel is named for Alexander Graham Bell, inventor of the telephone, among other things. See *Bell, Alexander Graham*; *decibel*; *logarithm*; and *power*.

Bell, Alexander Graham (1847–1922) The scientist and inventor of the telephone (1876), Bell was born and raised in Scotland and emigrated to Canada in 1870 and to the United States in 1871. Bell's grandfather and father were teachers of elocution, and Bell followed his father as a teacher of the deaf, expanding his work through the study of acoustics. Bell's work led to the development of various means of communicating with electricity. In addition to the telephone, Bell invented the photophone (1880), a system for transmitting voice utilizing mirrors to focus modulated sunlight onto a selenium cell. He was successful in transmitting voice over a distance of 700 feet on sunny days and was granted four patents for the invention. In 1881, Bell hurriedly invented the metal detector, which he used in an attempt to find an assassin's bullet in the body of President James A. Garfield. Although the device worked, it was confused by the metal bed frame on which Garfield was lying and could not locate the bullet. Bell also held patents for the phonograph, hydrofoil watercraft, aerial vehicles, and selenium cells. See also *bel*, *photophone*, and *telephone*.

Bell Communications Research (Bellcore) See *Bellcore*.

Bellcore (Bell Communications Research) The research and development arm of the Regional Bell Operating Companies (RBOCs), Bellcore was formed in 1984 under the terms of the Modified Final Judgement (MFJ), which forced AT&T to divest the Bell Operating Companies (BOCs). Bellcore originally focused on standards development, test procedures, and operations support system (OSS) development, rather than the physical sciences. Bellcore was privatized and acquired by SAIC in 1998, as the interests of the RBOCs were no longer common in a deregulated, competitive environment. The name was changed to Telcordia Technologies in April 1999, with the stated focus of emerging technologies. Telcordia is now a private, standalone organization involved in the development of OSSs and network management software, as well as consulting, testing services, and research services. See also *BOC*, *MFJ*, *network management*, *OSS*, and *RBOC*.

Bell Labs (AT&T Bell Telephone Laboratories) The research and development arm of the AT&T Bell System. As a result of the Modified Final Judgement (MFJ) that broke up the Bell System in 1984, Bell Labs and AT&T Technologies merged to form Lucent Technologies, which was acquired in 2006 by Alcatel, a French company to form Alcatel-Lucent. Undoubtedly, Bell Labs was once one of the greatest scientific laboratories the world has ever known. Bell Labs innovations include the transistor (1947), cellular telephone (1947), solar cells (1954), the laser (1958), digital transmission (1962), communications satellites (1962), the Unix operating system (1969), and the digital signal processor (DSP) (1979).

Bell Operating Company (BOC) See *BOC*.

Bell System The American Telephone and Telegraph Company (AT&T) organization as it existed prior to 1984, when the Modified Final Judgement (MFJ), also known as the Divestiture Decree, caused AT&T to divest itself of the 22 wholly owned operating companies and reorganize the remainder. The Bell System comprised AT&T, the Western Electric Company, Bell Telephone Laboratories (Bell Labs), and the operating companies. AT&T comprised the General Departments (e.g., Accounting, Finance, Legal, Engineering, Marketing, Human Resources, Public Relations, and Labor Relations) and the Long Lines Department. Long Lines owned and operated long distance transmission facilities and certain switching systems to interconnect the operating telephone companies and provide connectivity with foreign countries. Western Electric was the manufacturing and supply unit for Long Lines and the operating telephone companies. Bell Laboratories (Bell Labs) was funded by AT&T and Western Electric and operated as a non-profit corporation charged with research and development. Bell Labs was organized into 9 areas, including Research and Patents, Electronics Technology, Transmission Systems, Switching Systems, Military Systems, Computer Technology and Information Systems, and Business Information Systems Programs. The Bell System Operating Companies comprises 24 operating telephone companies, 22 of which were wholly owned (counting Bell Telephone Company of Nevada, which actually was wholly owned by Pacific Telephone and Telegraph Company. AT&T also owned a minority interest in Cincinnati Bell and Southern New England Telephone Company (SNET). The wholly owned Bell Operating Companies (BOCs) and their states of operation were as follows:

- Bell of Pennsylvania (Pennsylvania)
- The Chesapeake and Potomac Companies (District of Columbia, Maryland, Virginia, and West Virginia)
- Diamond State Telephone (Delaware)
- Illinois Bell (Illinois)
- Indiana Bell (Indiana)
- Michigan Bell (Michigan)
- Mountain Bell (Arizona, Colorado, Idaho, Montana, New Mexico, Utah, and Wyoming)
- Nevada Bell (Nevada)
- New England Telephone (Massachusetts, Maine, New Hampshire, Rhode Island, and Vermont)
- New Jersey Bell (New Jersey)
- New York Telephone (New York)
- Northwestern Bell (Iowa, Minnesota, North Dakota, Nebraska, and South Dakota)
- Ohio Bell (Ohio)
- Pacific Bell (California)

- Pacific Northwest Bell (Oregon and Washington)
- South Central Bell (Alabama, Kentucky, Louisiana, Mississippi, and Tennessee)
- Southern Bell (Florida, Georgia, North Carolina, and South Carolina)
- Southwestern Bell (Arkansas, Kansas, Missouri, Oklahoma, and Texas)
- Wisconsin Telephone (Wisconsin)

See also *Bell Labs*, *BOC*, and *MFJ*.

Bell Telephone Laboratories (Bell Labs) See *Bell Labs*.

bend diameter The diameter of the bend in a wire, fiber, or cable. Too severe a bend will cause a crimp, crack, or break in a wire or fiber, in the shielding or insulation surrounding it or the cable in which it resides, or otherwise will compromise the integrity of the physical medium or cabling system. Cable specifications include bend tolerances, generally stated in terms of minimum bend diameter. Absent those specifications, there are rules of thumb that guide in cable installation.

- **Fiber optic cable:** The dynamic bend, i.e., the bend in a cable under short-term physical load while being installed, should be no less than 20–15 times the outside diameter of a fiber optic cable at the point that the pulling load (i.e., tension in the cable sheath) approaches the maximum tensile strength of the cable. Dynamic bend minimums are intended to protect the cable from physical damage during installation. The static bend, i.e., the long-term bend in a cable at rest, should be no less than 10 times the outside diameter (OD) of a fiber optic cable. Static bend minimums are intended primarily to avoid bending loss, which is the loss of optical energy into the fiber cladding. If the bend is too severe, the angle of incidence, i.e., the angle at which the optical signal strikes the core/cladding interface, is too severe and the signal is not reflected back into the core. Rather, it penetrates the core/cladding interface and is lost in the cladding, or escapes the fiber altogether.
- **Unshielded twisted pair (UTP):** Manufacturers typically recommend a minimum bend radius of one (1) inch for a four-pair Category 5 UTP cable. The rule of thumb is four (4) times the outside diameter of the cable. A tighter bend can disturb the critical geometry of the twists, affecting the local impedance and balance, and reducing performance through increased sensitivity to external noise and increased near-end cross talk (NEXT) within the cable. Also, long-term damage to the cable jacketing and insulating material can result from bending stress.
- **Coaxial cable:** The rule of thumb is bend diameter no less than 6 times the outside diameter of the cable. A tighter bend can cause long-term damage to the cable jacketing, outer shield or conductor, and insulating material.

Some specifications state bend tolerance in terms of the bend radius, which is half the bend diameter. See also *angle of incidence*, *NEXT*, *rule of thumb*, and *tensile strength*.

bending loss Attenuation occurring as a result of either a bend in an optical fiber that exceeds the minimum bend radius or an abrupt discontinuity in the core/cladding interface. The incident light rays strike the boundary between the core and the cladding at an angle less than the critical angle and enter the cladding, where they are lost. See also *attenuation*, *cladding*, *core*, *critical angle*, *macrobend*, *microbend*, and *optical fiber*.

bend radius The radius of the bend in a wire, fiber, or cable. Too severe a bend will cause a crimp, crack, or break in a wire or fiber, in the shielding or insulation surrounding it or the cable in which it resides, or otherwise will compromise the integrity of the physical medium or cabling system. Some specifications state bend tolerance in terms of the bend diameter, which is twice the bend radius. See also *bend diameter.*

bent pipe In reference to a typical satellite configuration in which a satellite repeater, or transponder, accepts the weak incoming signals, boosts them, shifts them from the uplink to the downlink frequencies, and transmits them to the earth stations. The satellite performs no routing or switching function and there are no intersatellite links. See also *downlink, repeater, satellite, transponder,* and *uplink.*

BER (Bit Error Rate) The number of bits that are in error at one point in a circuit divided by the total number of bits transmitted. A BER of 1 bit in 1,000,000 typically is expressed as 1×10^{-6} or just 10^{-6}.

Berkeley Internet Name Daemon (BIND) See *BIND.*

Berners-Lee, Sir Timothy ("Tim") John (1955–) The inventor of the World Wide Web (WWW, aka the Web). Berners-Lee was employed at CERN (l'Conseil Européen pour la Recherche Nucléaire), the European Laboratory for Particle Physics in Geneva, Switzerland, when he developed the WWW in 1989 as a collaborative tool for high-energy physicists. He currently is a director of the World Wide Web Consortium (W3C), which oversees ongoing Web development. See also *W3C* and *WWW.*

best effort A quality-of-service (QoS) level that provides no guarantees in terms of metrics such as error performance, latency, and even loss. Best effort is the lowest QoS level. See also *best-effort ATM.*

best-effort ATM Also known as available bit rate (ABR). In asynchronous transfer mode (ATM), a class of traffic that does not require guarantees of network access, but rather can deal with time slot access on an as-available basis. Bursty LAN traffic and e-mail are examples of best-effort ATM traffic. ATM also defines constant bit rate (CBR) and variable bit rate (VBR) traffic classes. See also *ATM, CBR, e-mail, LAN, time slot,* and *VBR.*

beta Referring to a product (usually a software product) that is ready for pre-release testing by selected customers in real-world situations prior to general release. A beta product generally has completed alpha testing, which is conducted by in-house customers or under laboratory conditions. A beta product generally is considered to be stable and to include all features and functionality intended for the initial general release. See also *alpha* and *general release.*

BFT (Binary File Transfer) The transfer of a file containing bytes or words in binary format, which is computer-readable, but generally is neither viewable on screen nor printable. A binary file compares to a text file, which is a binary file that contains only printable characters. See also *binary.*

BGP (Border Gateway Protocol) An inter-Autonomous System (AS) protocol like the Exterior Gateway Protocol (EGP), BGP is concerned with conveying routing reachability information between groups of routers that fall within a single administrative domain. Although EGP runs on top of the Internet Protocol (IP), BGP runs on top of the Transmission Control Protocol (TCP), thereby ensuring a connection-oriented data flow and reliability of datastream transport. The IETF described the current version, BGP-4, in RFC 1771. BGP is assigned TCP well-known port number 179 and supports Classless Inter-Domain Routing (CIDR). See also *AS, CIDR, connection-oriented, domain, EGP, IETF, IP, port, protocol, routing, TCP,* and *well-known port.*

BHCAs (Busy Hour Call Attempts) The number of call attempts that a telephone system can support during the busy hour of the day. BHCAs is a measure of system processor capacity and a factor considered in traffic engineering. See also *BHCCs, busy hour,* and *traffic engineering.*

BHCCs (Busy Hour Call Completions) The number of calls that a telephone system can complete during the busy hour of the day. BHCCs is a measure of system processor capacity and a factor considered in traffic engineering. See also *BHCAs, busy hour,* and *traffic engineering.*

BHT (Busy Hour Traffic) The volume of calls variously attempted or completed curing the busy hour of the day. BHT is key to traffic engineering. See also *busy hour, Erlang, Poisson distribution, traffic,* and *traffic engineering.*

B-ICI (B-ISDN InterCarrier Interface) A specification from the ATM Forum (now MFA Forum) for a public network-to-network interface (PNNI) between two ATM-based network service providers or carriers, using permanent virtual circuits (PVCs). The B-ICI specification is based on Broadband ISDN User Part (B-ISUP) signaling messages and parameters. B-ICI includes service-specific functions above the ATM layer required to transport, operate and manage a variety of intercarrier services. See also *ATM*, *FUNI*, *B-ISDN*, *MFA Forum*, *PNNI*, and *PVC*.

bi-endian Referring to a system or network that can operate with either a big-endian or little-endian orientation. See also *endianess*, *big-endian*, and *little-endian*.

big-endian Referring to the orientation of a computer system, application, or network design with respect to the placement of most significant bit, digit, or byte in a coding scheme or with respect to storage in memory or order of transmission. Big-endian places the most significant bit, digit, or byte in the first, or leftmost, position, which is transmitted first. Little-endian places the most significant bit, digit, or byte in the last, or rightmost, position, which is transmitted last. Bi-endian systems can work either way. Telephone numbers, for example, are big-endian, beginning with a country code, followed by an area code, a central office prefix, and a line number. Motorola processors employ the big-endian approach, while Intel processors take the little-endian approach. The terms derive from Jonathan Swift's *Gulliver's Travels*, in which the Big-Endians were a faction of people on the islands of Lilliput and Blefuscu who defied the emperor's decree that soft-boiled eggs should be broken at the small end before being consumed. See also *bi-endian*, *bit*, *byte*, *digit*, *endianess*, and *little-endian*.

big iron See *heavy metal*.

bigit See *bit*.

binary See *binary notation*.

binary file transfer (BFT) See *BFT*.

binary notation (binary) **1.** A system with only two possible states, such as on or off, positive (+) or negative (–), or true or false. A simple light switch, for example, is in either the on position or the off position. **2.** The base-2 numbering system. A system of representing numbers characterized by a series of digits, each of which has only two possible states, one (1) or zero (0). See also *bit*, *decimal notation*, and *hexadecimal notation*.

binary phase-shift keying (BPSK) See *BPSK*.

Binary Synchronous Protocol (Bisync or BSC) See *BSC*.

BIND (Berkeley Internet Name Daemon) A domain name server (DNS) for UNIX operating systems (OSs), BIND was originally written for the BSD (Berkeley Software Distribution) version of UNIX written at the University of California at Berkeley. See also *daemon*, *DNS*, *OS*, and *UNIX*.

binder group A group of wire pairs bound together, usually by some sort of color-coded plastic tape or thread. In a large twisted pair cable, there may be many pairs combined into binder groups of 25 pairs for ease of connectivity management. Each pair within a binder group is uniquely color-coded for further ease of management. See also *cable* and *wire*.

binit See *bit*.

BIOS (Basic Input/Output System) On PC systems, a set of routines that tests the hardware (e.g., disk drives, keyboard, and monitor) at startup, starts the operating system (OS), and supports the transfer of data between hardware devices at startup. Until the early 1990s, BIOS was stored in firmware , i.e., read-only memory (ROM). In contemporary computers, BIOS is written to erasable programmable read-only memory (EPROM) or flash memory to facilitate updates.

BIP (Bit Interleaved Parity) In SDH and SONET networks, an error control mechanism comprising parity bytes associated with each frame. BIP is included in Line Overhead (LOH) and Section Overhead (SOH). See also *LOH*, *parity*, *SDH*, *SOH*, and *SONET*.

biphase coding See *Manchester coding*.

bipolar A digital signaling technique that makes use of a positive (+) and a negative (–) voltage to represent data in binary form, i.e., ones (1s) and zeroes (0s).

bipolar coding Synonymous with alternate mark inversion (AMI). See *AMI*.

bipolar violation (BPV) See *BPV*.

bipolar with eight-zeros substitution (B8ZS) See *B8ZS*.

bis From the Latin bis, meaning twice, or repeated. In standards terminology, bis refers to the second version, e.g., X.32bis.

B-ISDN (Broadband Integrated Services Digital Network) A set of specifications from the ITU-T for an integrated services digital network (ISDN) requiring transmission channels capable of supporting rates greater than the primary rate, which is defined in the North American primary rate interface (PRI) standard as 1.544 Mbps and in the European primary rate access (PRA) standard as 2.048 Mbps. There are three underlying sets of technologies and standards critical to B-ISDN:

- **Signaling and control**: Signaling System 7 (SS7) supports B-ISDN, just as it supports narrowband ISDN (N-ISDN).
- **Switching and multiplexing**: Asynchronous transfer mode (ATM).
- **Transmission**: SDH/SONET fiber optics transmission system (FOTS).

B-ISDN user access is specified at two SDH/SONET levels. User network interface A (UNI A) operates at OC-3 rates of 155 Mbps, whereas user network interface B (UNI B) operates at OC-12 rates of 622 Mbps. A network-to-network interface (NNI) is required for network access to B-ISDN from frame relay and N-ISDN networks. B-ISDN defines interactive services and distribution services.

- **Interactive services** involve bidirectional transmission and include three classes of service. Conversational services include voice, interactive data, and interactive video. Messaging services include compound document mail and video mail. Retrieval services include text retrieval, data retrieval, image retrieval, video retrieval, and compound document retrieval.
- **Distribution services** may or may not involve user presentation control. For example, interactive TV is a service requiring presentation control. Interactive TV actually enables the viewer to interact with the program, perhaps to select a product marketed over TV or to change the camera angle to view a sporting event from a different perspective. Conventional broadcast TV exemplifies a service requiring no presentation control.

B-ISDN user equipment is an extension of that described for N-ISDN. Broadband terminal equipment type 1 (B-TE1) is defined as B-ISDN compatible. Broadband terminal equipment type 2 (B-TE2) is defined as terminal equipment that supports a broadband interface other than B-ISDN and must interface with the network through a broadband terminal adapter (B-TA).

Cost and complexity issues forestalled the deployment of B-ISDN, in favor of broadband public networking based variously on ATM, Ethernet, frame relay, and Internet Protocol (IP). B-ISDN seems to have little future other than as an evolutionary dead end. See also *ATM*, *BRI*, *broadband*, *channel*, *Ethernet*, *FOTS*, *frame relay*, *IP*, *ISDN*, *ITU-T*, *narrowband*, *NNI*, *OC*, *SDH*, *SONET*, *SS7*, and *UNI*.

Bisync (Binary Synchronous Protocol) See *BSC*.

bit (binary digit) **1.** A small piece or quantity. **2.** A contraction of the term *binary digit*, a bit is an individual 1 or 0 in a binary numeration system, a base 2 numbering system. So, a bit is the smallest unit of digital data. The word first appeared in print in 1948 in a paper written by Claude Shannon, who credited John Tukey, an early computer scientist at Bell Telephone Laboratories with coining the term in 1947. Tukey later wrote that the term evolved as an alternative to bigit or binit. See also *binary* and *bit rate*. **3.** In coinage, originally a small silver coin worth one-eighth (⅛) of a Spanish peso. Later, a small British coin, a threepenny bit. Now commonly used to mean one-eighth (⅛) of a U.S. dollar, or twelve and a half (12½ cents), usually in the phrases two bits (¼ of a dollar, or 25 cents), four bits (½ of a dollar, or 50 cents), and six bits (¾ of a dollar, or 75 cents). As the story goes, coins, especially small coins, were scarce in colonial America, so it was common practice to cut a bit (or two bits) off of a dollar coin to make change.

bit compression multiplexer (BCM) See *BCM*.

bit density **1.** The number of bits transmitted per second (bps). See also *bps*. **2.** The number of bits per unit of length or area in a data storage medium such as a disk or magnetic tape.

bit error rate (BER) See *BER*.

bit interleaved parity (BIP) See *BIP*.

bit map A data structure that represents image information as a collection of bits organized into a grid of rows and columns that translate into pixels (i.e., color dots) for display or printing. See also *bit* and *pixel*.

bit-oriented protocol A synchronous communications protocol requiring only a single bit to communicate a command signal to the target station. Bit-oriented protocols transmit information without regard to character boundaries and thus handle all types of information images. Bit-oriented protocols are much less overhead-intensive, as compared to byte-oriented protocols, also known as character-oriented protocols. Bit-oriented protocols are usually full-duplex (FDX) and operate over dedicated, four-wire circuits. Examples include Synchronous Data Link Control (SDLC) and the High-Level Data Link Control (HDLC). See also *bit*, *byte-oriented protocol*, *FDX*, *four-wire circuit*, *HDLC*, *protocol*, *SDLC*, and *synchronous*.

bits per second (bps) See *bps*.

bit rate The number of bits per second (bps) transmitted. Bit rate and baud rate are often used interchangeably, and incorrectly so. Baud rate refers to the number of signal events, signal changes, or signal transitions occurring per second over an analog circuit, such as changes from positive voltage to zero voltage, from zero voltage to negative voltage, or from positive voltage to negative voltage. The relationship between baud rate and bit rate depends on the sophistication of the modulation scheme used to manipulate the carrier. The bit rate and baud rate can be the same, if each bit is represented by a signal transition in a unibit modulation scheme. The bit rate can be higher than the baud rate, as a single signal transition can, and often does, represent multiple bits. Further, multiple bits can be transmitted before a signal transition occurs. A modulation scheme that impresses multiple bits on a baud makes more effective use of analog bandwidth, which is always in limited supply.

A purely digital transmission system uses an entirely different approach. Rather than varying the signal state of an analog carrier, a purely digital system is a two-state system that involves simply turning a signal on and off. In an electrically based telegraph system, for example, a dot (.) is a short electrical pulse transmitted by holding a telegraph key down for a short time, thereby closing an electrical contact, and a dash (__) is a longer pulse. Fiber optic transmission systems (FOTS) can achieve bit rates of many Gbps through diode laser infrared light sources that pulse on and off billions of times per second. See also *analog*, *baud rate*, *bit*, *carrier*, *digital*, *FOTS*, and *modulation*.

bit robbing Referring to the process in a channel bank or T1 time division multiplexer (TDM) whereby the least significant bit (LSB) in a byte is robbed and a signaling bit is inserted, thereby truncating the eight-bit voice sample to seven bits. Bit robbing does not affect the quality of pulse-code modulated (PCM) voice, as seven bits are quite satisfactory for reconstructing a high-quality approximation of the analog voice input, and the robbing occurs in only every sixth frame. (The LSB really is just fine-tuning of the voice signal, so to speak.) Bit robbing does, however, seriously impact other data types. As a result, data transmission in T-carrier networks is restrained to 56 kbps per channel. See also *bit, byte, channel, channel bank, LSB, PCM, T1, T-carrier,* and *TDM.*

bit/s Synonymous with *bps*, at least to all but certain purists. See also *bps* and *purist.*

bit stream A continuous series of bits, as represented by electromagnetic pulses in a digital transmission system. See also *bit.*

bit stuffing A synchronization technique used in time division multiplexing (TDM) to adjust for slight timing discrepancies between incoming bit streams. As necessary, bit stuffing adds some number of pulses to incoming bits streams to synchronize them with the mux clock and position each of them properly in the outgoing aggregate bit stream. The mux also inserts a code into the outgoing bit stream to advise the receiving mux of the stuff bits in order that it can properly destuff the signal. See also *bit, bit stream, pulse, synchronize,* and *TDM.*

bit time The duration of an individual one (1) or zero (0) bit in a digital transmission. The transmitter ejects bits at a given rate (*x* bits per second), with each bit occupying the circuit for a given amount of time, which is the inverse of the bit rate (1/*x* seconds per bit). The receiver must monitor the circuit at precisely the same rate and at the same instants in time in order to distinguish the individual bits. See also *bit, circuit,* and *digital.*

bit transparent Referring to a network component that simply acts upon bits, with no knowledge of, or concern for, the higher level content. Components at the Physical Layer are bit transparent, with examples being E-carrier, Synchronous Digital Hierarchy (SDH), Synchronous Optical Network (SONET), and T-carrier. See also *bit, E-carrier, Physical Layer, SDH, SONET,* and *T-carrier.*

.biz (bizness) Pronounced *dot biz.* The Internet generic Top Level Domain (gTLD) reserved exclusively for business organizations. This is an unsponsored domain. See also *gTLD, Internet,* and *unsponsored domain.*

Bless his heart Someone once noted that a Texan can get away with the most awful kind of insult just as long as it's prefaced with the words, "Bless her heart" or "Bless his heart." As in, "Bless his heart, if they put his brain on the head of a pin, it'd roll around like a BB on a six-lane highway." My personal favorite is "Bless her heart, she cain't help bein' ugly, but she could'uh stayed home." Bless his heart saves a lot of energy trying to think of politically correct (PC) euphemisms. See also *euphemism* and *PC.*

block A group of bits, often comprising an integral number of bytes, encoded, processed, transmitted, or otherwise treated as a unit. See also *bit, byte,* and *encode.*

block check character (BCC) See *BCC.*

block cipher A private key encryption method that encrypts data in blocks of fixed size (usually 64 bits) that are the same size as the unencrypted data. See also *encryption* and *private key encryption.*

block coding A method in which a block of data bits is mapped into a block (often slightly larger) of signaling bits prior to transmission. Advantages of block coding include electrical balance, error detection, and synchronization See also *4B/5B, 64B/66B, 8B/10B, balanced, block, data, error control, signaling,* and *synchronization.*

blocking **1.** The intentional denying of access to a network, networked device such as a switch, router, circuit, or networked resource, such as a database. **2.** In reference to a voice switch (see Figure B-2) or other device intentionally engineered to block some call attempts during the busiest hour of the day. This approach involves the concept of Grade of Service (GoS), which is the probability of blockage during the busy hour, expressed as a decimal fraction of the calls anticipated to be blocked. See also *busy hour, GoS,* and *non-blocking.* **3.** A characteristic of a switch that is not able to handle simultaneous connections for all attached devices or all switch ports.

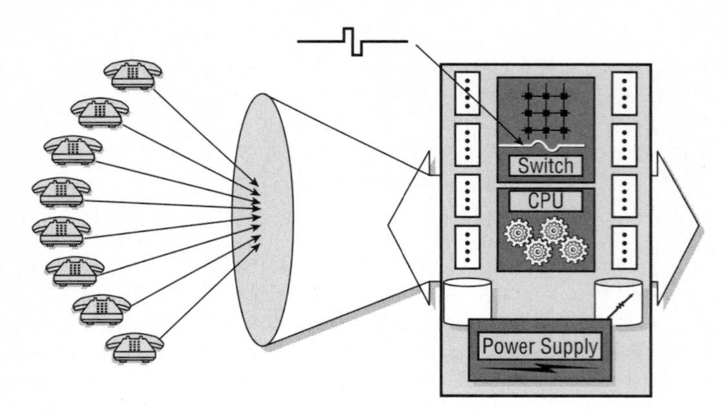

Figure B-2

blog (Web log) A Web site where an individual maintains a personal journal or even an interactive forum much like a personal newsgroup. The vast majority of blogs, or so it seems, are pretty silly electronic diaries posted by adolescents. Many blogs, however, are quite serious. Some companies maintain public blogs to foster dialogue amongst employees with respect to projects, strategies, and other matters of interest. Some blogs take the form of well-researched personal opinion columns on politics or other controversial and weighty subjects. See also *WWW*.

blue sky See *Rayleigh scattering*.

Blue Screen of Death (BSOD) See *BSOD*.

Bluetooth A specification to standardize wireless transmission and synchronize data among a wide variety of devices such as PCs, cordless telephones, headsets, printers, and PDAs. The initial effort (April, 1998) was in the form of a consortium of Intel, Microsoft, IBM, Toshiba, Nokia, Ericsson, and Puma Technology and was code-named Bluetooth after Harald Blaatand, the tenth-century Danish king who brought warring tribes together and unified Denmark. Bluetooth is now formalized in IEEE 802.15.1 (2002) as the specification for a wireless personal area network (WPAN) operating in the 2.45 GHz range of the ISM frequency band. Bluetooth employs frequency-hopping spread spectrum (FHSS), with devices stepping through a carefully choreographed pseudorandom hop sequence that makes data collisions highly unlikely even though large numbers of transmissions share the same frequency band. Devices hop through a set of 79 (United States and Europe) or 23 (Spain, France, and Japan) channels spaced 1 MHz apart at a rate of about 1600 hops per second, with each hop lasting 62.5 μs. The baseband signal is modulated using Gaussian frequency shift keying (GFSK). Bluetooth devices are organized into three classes based on transmit power and range:

- Class 1 radios (maximum 100 mW) have a nominal link range of 100 meters, or 300 feet. Class 1 radios generally are employed in industrial applications.
- Class 2 radios (maximum 2.5 mW) have a nominal range of 10 meters, or 3 feet, and are the most common.
- Class 3 radios (maximum 1 mW) have a nominal range of 1 meter, or 3 feet.

Bluetooth technology supports both synchronous connection-oriented (SCO) links for packet voice and asynchronous connectionless links (ACLs) for packet data. Bluetooth supports an asynchronous data channel in asymmetric mode of up to 721 Kbps in either direction and 57.6 Kbps in the reverse direction.

Alternatively, the data channel can be supported in symmetric mode of up to 432.6 Kbps. As yet another alternative, Bluetooth supports up to three simultaneous synchronous packet voice channels, or a channel that simultaneously supports both asynchronous data and synchronous voice. Bluetooth supports full-duplex (FDX) communications using time division duplex (TDD) as the access method. Voice coding employs the continuously variable slope delta modulation (CVSDM). Bluetooth specifies 16 packet types, including 4 types of control packets and 12 types of data packets. There are 3 types of voice packets, each running at a rate of 64 kbps, including overhead and, in some cases, a forward error control (FEC) mechanism. In the event that an error is detected in the user payload of a packet transmitted over the ACL, Bluetooth specifies the use of an automatic repeat request (ARQ) for retransmission of errored packets. The variable-length packet is limited to 366 bits, although the theoretical limit is 625 bits (62.5 μs × 1 Mbps = 625 bits). The limit of 366 provides the transmitters and receivers with enough time to hop to the next frequency and stabilize. As the access code and packet header consume 126 bits, the payload cannot exceed 240 bits, or 30 octets. There is a provision for multislot packets in support of larger payloads. The payload includes an error control mechanism in the form of a cyclic redundancy check (CRC).

Bluetooth operates in point-to-point, point-to-multipoint, and mesh configurations. As many as eight devices can link together in a piconet, or very small network, with as many as seven devices slaved to a single master. Multiple, overlapping piconets can form a scatternet. See also *ACL*, *ARQ*, *asymmetric*, *asynchronous*, *CVSDM*, *FDX*, *FEC*, *FHSS*, *ISM*, *master/slave*, *piconet*, *point-to-multipoint*, *point-to-point*, *scatternet*, *SCO*, *symmetric*, *synchronous*, *TDD*, and *WPAN*.

BOC (Bell Operating Company) An operating telephone company owned by the AT&T Bell System and reporting directly to AT&T general headquarters. The 22 BOCs were spun off from AT&T as a result of the Modified Final Judgement (MFJ), also known as the Divestiture Decree, which took effect January 1, 1984. At that point, the BOCs were formed into seven Regional Bell Operating Companies (RBOCs). See also *Bell System*, *MFJ*, and *RBOC*.

Boolean logic A branch of algebra in which all operations are either true or false, i.e., yes or no, and all relationships between the operations can be expressed with logical operators such as AND, OR, or NOT. Invented by English mathematician George Boole (1815–1864), Boolean logic was obscure until the rise of digital computing, which is based on two values: 1 and 0, which essentially translate into yes or no, on and off, and so on.

Border Gateway Protocol (BGP) See *BGP*.

bonding **1.** The process or method of permanently joining the metallic shields, screens, or armor of multiple wire and cable segments in order to establish electrical continuity between them, to a ground strap or wire that connects to a ground rod, and eventually to ground. Bonding serves to ensure that electrical noise will be conducted to ground, rather than coupling with and, therefore, interfering with the desired signal. **2.** Synonymous with *channel aggregation*, dynamic bandwidth allocation, multirate ISDN, and Nx64. A feature of ISDN-compatible terminal adapters (TAs), PBXs, and routers that enables the system to dynamically allocate, or bond, multiple contiguous 64-kbps bearer (B) channels to serve an application that requires more than a narrowband channel. From the transmitter, through the network, and to the receiver, the narrowband channels are bonded and treated as a single superrate channel known as a high-speed (H) channel. As an example, a videoconference might require 128 kbps (2 channels) or 384 kbps (6 channels). Dial-up Internet access typically benefits from bonding two channels for a connection at 128 kbps. Also, multiple IP, ATM, or frame relay links can be joined to emulate a faster channel. See also *B channel*, *H channel*, *ISDN*, *narrowband*, *PBX*, *router*, *superrate*, and *TA*.

BOOTP (BOOTstrap Protocol) A protocol employed by a workstation on a local area network (LAN) to find its Internet Protocol (IP) address. BOOTP originally was intended to allow a diskless client machine to discover its own IP address, the address of a server host computer, and the name of a file to be loaded into memory and executed. First described in IETF RFC 951 (1985), BOOTP runs on top of the User Datagram Protocol (UDP). The Dynamic Host Configuration Protocol (DHCP) is based on

BOOTP, but is far more complex. See also *client*, *DHCP*, *host*, *IETF*, *IP*, *IP address*, *LAN*, *protocol*, *server*, *UDP*, and *workstation*,

bootstrap **1.** A leather strap looped and sewn on the side or back of a boot to assist the wearer in pulling it on. **2.** Self-reliant and self-sustaining, i.e., not needing anyone else's help. Such a person is capable of "pulling himself up by his own bootstraps," as the saying goes. I've been a bootstrapper all my life, but I have had a helping hand now and then, and always appreciated it. I occasionally need help getting my boots off. I use a bootjack for that.

Bootstrap Protocol (BOOTP) See *BOOTP*.

bounded medium A transmission medium that binds the signal within (i.e., constrains the signal to stay within) a conductor or other waveguide. See also *transmission medium*.

BPL (Broadband over Power Line) A set of specifications for broadband communications over the existing electrical power grid. BPL is based on power line carrier (PLC) technology developed in 1928 by AT&T Bell Telephone Laboratories, and which has been used since that time by the electric power utilities in select internal low speed data communications applications. Access BPL is a local loop technology that runs over medium voltage (MV) power lines in the power distribution networks. In-house BPL is a data communications transmission technology that allows a device to connect to a local area network (LAN) directly through the low voltage (LV) electric grid inside the premises. See also *Access BPL*, *broadband*, *In-house BPL*, *LAN*, *local loop*, *LV*, *MV*, and *PLC*.

BPON (Broadband Passive Optical Network) An ATM-based PON specification described in ITU-T G.983.3 (2001) and based on the original APON (ATM-base PON) specifications ratified by the ITU-T in 1998. BPON runs in asymmetric mode at 622 Mbps downstream and 155 Mbps upstream, or in symmetric mode at 155 Mbps over a distance of as much as 20 kilometers (12 miles). BPON supports as many as 32 splits, that is, splitters can divide the signal to serve as many as 32 premises from a single optical fiber. BPON employs wavelength division multiplexing (WDM) for downstream transmission, with as many as 16 wavelengths with 200 GHz spacing and 32 wavelengths with 100 GHz spacing between channels. BPON provides for enhanced security through a technique known as churning in which the encryption key is changed at least once a second between the Optical Line Terminal (OLT) at the headend and the Optical Network Terminal (ONT) at the customer premises. PON variants also include Ethernet-based PON (EPON), and gigabit PON (GPON). See also *APON*, *asymmetric*, *ATM*, *churning*, *downstream*, *EPON*, *GPON*, *ITU-T*, *PON*, *splitters*, *symmetric*, *upstream*, and *WDM*.

bps (bits per second) The measurement of bandwidth of digital transmission. The scale is bps, kbps kilobits per second (thousands of bits per second), Mbps or Megabits per second (millions of bits per second), Gbps or Gigabits per second (billions of bits per second), and Tbps or Terabits per second (trillions of bits per second). *Note:* ISO purists prefer bits/s, as in Mbit/s. See also *bandwidth*.

Bps (Bytes per second) The measurement of the transmission rate in select storage networking systems. Storage technologies such as Fibre Channel and Enterprise Systems Connection (ESCON) measure the speed of information transfer in Bps. Although storage is byte-oriented and transmission is bit-oriented, the transmission system in such an application is purely storage-oriented, so the rate is stated in storage terms.

BPSK (Binary Phase Shift Keying) A unibit PSK modulation technique in which the continuous analog waveform is interrupted and restarted at the baseline with a 180° phase shift to indicate a change in value from a 1 bit to a 0 bit or from a 0 bit to a 1 bit. As examples, Wi-Fi5 (802.11a) wireless LAN (WLAN) standards specify BPSK at 6 Mbps, Personal Communications System (PCS) cellular radio specifies BPSK in the 800–900 MHz band, and ZigBee specifies BPSK in the 868 and 915 MHz bands. See Figure B-3. See also *802.11a*, *analog*, *cellular radio*, *Manchester coding*, *modulation*, *PCS*, *phase*, *PSK*, *unibit*, *waveform*, *Wi-Fi5*, *WLAN*, and *ZigBee*.

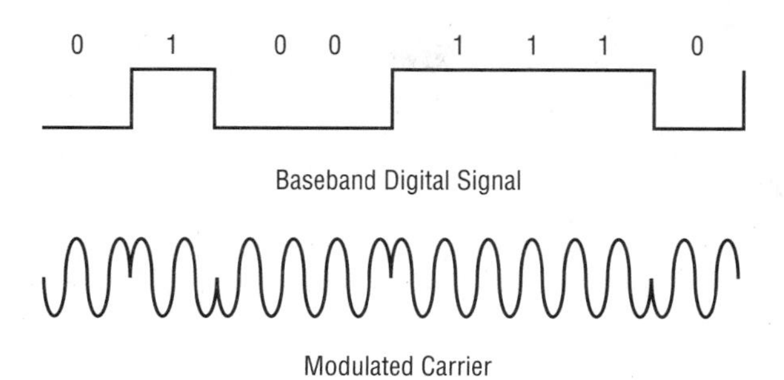

Figure B-3

BPV (BiPolar Violation) A violation of the line coding convention required by alternate mark inversion (AMI) in T-carrier networks or high density bipolar order 3 (HDB3) in E-carrier networks. AMI reverses the polarity of alternate marks, or 1 bits, expressing the first as a positive voltage of +3V, the second as a negative voltage of –3V, the third as +3V, and so on. Most deviations from this pattern indicate an error in transmission. If bipolar with eight-zeros substitution (B8ZS) is employed in conjunction with AMI, specific intentional bipolar violations can be used to indicate the legitimate transmission includes a long string (8 or more) of zero (0) bits. Since the violating bit patterns contain pulses, they prevent the receiver from losing track of the count of bit intervals that should represent 0s, which are sent in AMI as 0V on the line. B8ZS patterns must be known and configured into both the transmitting and receiving multiplexers so that the receiving multiplexer can restore the original 00000000 bit pattern. HDB3 is similar to AMI, but imposes a limit of three successive 0 bits. See also *AMI*, *bipolar*, *HDB3*, and *voltage*.

BRA (Basic Rate Access) Synonymous with *BRI* (Basic Rate Interface). The term BRA generally is used outside North America. See *BRI*.

BRAN (Broadband Radio Access Networks) A project chartered by the European Telecommunications Standards Institute (ETSI) for the development of standards for wireless access to wireline networks at broadband speeds of 25 Mbps or more. BRAN includes HiperLAN2, a mobile short-range access network, and two fixed wireless broadband access technologies. HiperACCESS is intended to operate above 11 GHz and HiperMAN below 11 GHz. See also *broadband*, *ETSI*, *HiperACCESS*, *HiperLAN2*, *HiperMAN*, *wireless*, and *WiMAX*.

brand spoofing See *phishing*.

break and make See *hard handoff*.

break strength The ability to withstand stress without breaking, break strength is measured as the tensile or compressive load required to fracture something. Break strength is an important consideration in the design of telecommunications conductors and cables, for example. See also *flex strength* and *tensile strength*.

BRI (Basic Rate Interface) The user interface to an integrated services digital network (ISDN) intended for residential and small business applications. Also known as basic rate access (*BRA*) outside North America, BRI supports two bearer (B), or information-bearing, channels, each operating at the clear-channel rate of 64 kbps. Each B channel can support an independent data transmission, a PCM-encoded voice conversation, or a number of statistically multiplexed subrate (low-speed) data transmissions. The B channels also can be aggregated, or bonded, to provide up to 128 kbps to a given conversation, such as a videoconference or Internet session. BRI also provides a data (D) channel at 16 kbps, which is intended primarily for purposes of signaling and control, messaging, and network management. The D channel also generally is made available for packet data transmission and low-speed telemetry when not in use for signaling purposes. Also known as 2B+D, the combination of two B channels at 64 kbps each and the D channel at 16 kbps yields a three-channel configuration at a total signaling rate of 144 kbps over a conditioned, four-wire, twisted-pair local loop within 18,000 feet of the service central office (CO). A single BRI line can support up to 16 devices (e.g., telephones, facsimile machines, computers, and video cameras) that contend for access to the BRI channels through a terminal adapter (TA), also known as an ISDN modem. BRI also can support as many as 64 individual service profile identifiers (SPIDs), which are equivalent to directory numbers, one per terminal device. Many carriers offer BRI variations such as 0B+D and 1B+D targeted at specific applications such as credit card verification and telemetry, but all share the same BRI technology and the same signaling rate of 160 bps.

The BRI interface between the CO and the customer premises is known as the U interface or Reference Point U, which runs at a signaling rate of 160 kbps, carrying two 64 kbps B channels, one 16 kbps D channel, and 16 kbps of overhead for framing, echo cancellation, and an embedded operations channel (EOC) for line testing and monitoring. In order to support this transmission rate in full duplex (FDX) mode over a physical two-wire, logical four-wire, twisted-pair local loop, the 2B1Q (2 Binary 1 Quaternary) encoding technique, with echo cancellation, is used in North America. In European and many other countries, the line coding technique employed is 4B3T (4 Binary 1 Ternary), a block code that combines four bits to represent one ternary signal state. See also *0B+D*, *1B+D*, *2B1Q*, *4B3T*, *bonding*, *EOC*, *FDX*, *four wire*, *ISDN*, *local loop*, *Reference Point U*, *signaling and control*, *signaling rate*, *SPID*, *TA*, and *two-wire*.

bridge **1.** A simple, protocol-specific device that interconnects two or more links in a circuit, reading the destination address of an incoming data frame and forwarding to the next link in the direction of the target device. A bridge also acts as a repeater, amplifying, reshaping, and retiming the input signal. A bridge does not perform complex processes on the data frames, and neither does it attempt to evaluate the network as a whole to make end-to-end routing decisions. **2.** A simple, protocol-specific device that interconnects two or more segments in a local area network (LAN), or two or more LANs of the same architecture (e.g., Ethernet-to-Ethernet). A bridge reads the destination address of an incoming data frame and forwards it to the next segment in the direction of the target device. A bridge also acts as a repeater, amplifying, reshaping and retiming the input signal to extend the physical reach of the LAN. Bridges operate at the lower two layers of the OSI Reference Model, providing Physical Layer and Data Link Layer connectivity. Specific bridge protocols include source routing protocol (SRP), source routing transparent (SRT), spanning tree protocol (STP). See also *architecture*, *circuit*, *Data Link Layer*, *encapsulating bridge*, *filtering bridge*, *frame*, *LAN*, *link*, *OSI Reference Model*, *Physical Layer*, *protocol*, *repeater*, *router*, *self-learning bridge*, *SRP*, *SRT*, and *STP*.

bridged tap A section of cable that is not on the direct physical path between the user premises and the central office (CO), but is bridged, i.e., temporarily spliced, onto the path. A bridged tap results in multiple appearances of the same cable pair, usually as a result of old drops to premises where previous subscribers had telephone service, but that were not removed when they disconnected service. Bridged taps cause signal distortion due to reflection and are best removed, especially if high frequency services such as ADSL are to be provisioned over the cable pair. See also *ADSL*, *cable*, *CO*, *distortion*, *drop*, *pair*, *premises*, and *reflection*.

British Standard Gauge (BSG) See *BSG*.

broadband **1.** In the Wide Area Network (WAN) domain, broadband is an imprecise term referring to a circuit or channel providing a relatively large amount of bandwidth. The ITU-T defines broadband in Recommendation I.113 as a transmission rate faster than the primary rate (referring to ISDN), which translates into 1.544 Mbps in North America and 2.048 Mbps in most of the rest of the world. The U.S. Federal Communications Commission (FCC) does not define broadband, but defines high-speed services as supporting a data rate of at least 200 kbps in at least one direction and advanced telecommunications capability as at least 200 kbps in both directions. Asymmetric digital subscriber line (ADSL) generally is described as a broadband access technology, even though many ADSL services operate at less than T1 and E-1 rates. In this context, ADSL certainly operates at much higher rates than the predecessor modem technology, which operates at narrowband rates of less than 64 kbps. Relatively speaking, ADSL is broadband in nature, even at very low operating rates. See also *bandwidth*, *FCC*, *ISDN*, *ITU-T*, *narrowband*, and *wideband*. **2.** In the Local Area Network (LAN) domain, broadband refers to a multichannel RF-based (Radio Frequency-based) LAN, with the channels derived through frequency division multiplexing (FDM). The workstations and other attached digital devices access analog channels through radio frequency (RF) modems that accomplish the digital-to-analog conversion process. Broadband LANs commonly use 75-ohm CATV-type coax, and use CATV-style connectors, taps, filters, and amplifiers in a tree and branch topology, which essentially is a variation of the bus with multiple branches off of a main root bus, much as there are branches off of the main trunk of a tree. The only broadband LAN to gain any significant following was 10Broad36, which has long been considered obsolete. All other LANs are baseband in nature. See also *10Broad36*, *baseband*, *B-ISDN*, *bus topology*, *CATV*, *channel*, *FDM*, *LAN*, and *tree topology*.

broadband integrated services digital network (B-ISDN) See *B-ISDN*.

broadband ISDN (B-ISDN) See *B-ISDN*.

broadband ISDN intercarrier interface (B-ICI) See *B-ICI*.

broadband over power line (BPL) See *BPL*.

Broadband Passive Optical Network (BPON) See *BPON*.

Broadband Radio Access Networks (BRAN) See *BRAN*.

broadband switching system (BSS) See *BSS*.

Broadband User Network Interface (B-UNI) See *B-UNI*.

broadband wireless access (BWA) See *BWA*.

broadcast A transmission mode in which a station sends a message to all stations on a network. A bridge or switch, for example, might broadcast a ping to all stations in order to update its routing tables. An authorized user of a voice mail system might broadcast a message to all stations to advise users of scheduled downtime for system maintenance. See also *anycast*, *unicast* and *multicast*.

broadcast and unknown server (BUS) See *BUS*.

broadcast radio An over-the-air radio communication service intended for direct reception by the general public.

broadcast television Television programming sent over the air to all receivers. The initial standards were set in the United States, where broadcast TV originated. In 1945, the FCC set the initial transmission standards at 4.5 MHz, in the very high frequency (VHF) band. The National Television Standards Committee (NTSC) was formed in 1948 to standardize the characteristics of the broadcast signal. Ultimately, the Radio Corporation of America (RCA), which was owned by AT&T, lobbied the Electronics Institute of America (EIA) and set the initial black-and-white TV standards. Color TV was commercialized some years later. See also *narrowcast*, *NTSC*, *television*, and *VHF*.

broadcast quality Referring to television video signal that has the strength, color balance, and lack of artifacts good enough to be used as a source for a TV station transmission of the best quality. The frame rate is 30 frames per second (fps) or more. At that rate, the perception is one of complete fluidity of motion. See also *artifact*, *frame rate*, and *motion picture quality*.

brochureware Hardware, firmware, and software as it is advertised in a brochure rather than as it exists. Unfortunately, "truth in advertising" is not a given. Always read the fine print and check references, especially for users of similar size, using the product in similar configurations with similar intensity in similar applications. Brochureware is a type of adware that you pick up at a trade show and take with you. See also *fine print*.

broomsticking A technique used in drawing a glass optical fiber. The tip of a preform cylinder is heated to a temperature of 2,500 degrees in a drawing tower. The resulting gob of molten glass is carefully drawn by gravity, in a process known as broomsticking, into a fiber as long as 20 kilometers. Common fiber manufacturing techniques include inside vapor deposition (IVD) and outside vapor deposition (OVD). See also *IVD*, *OVD*, and *Wicked Witch of the West*.

Brown and Sharp (B&S) Wire Gauge The original name for American Wire Gauge (AWG). See *AWG*.

brownfield **1.** In the building industry, referring to a parcel of land previously built upon and, especially, polluted. **2.** In telecommunications, referring to network deployments in territories where there is an existing telecommunications infrastructure that is removed or abandoned in place. For example a passive optical network (PON) fiber-to-the-premises (FTTP) configuration may require the removal or involve the abandonment of existing wireline infrastructure. See also *FTTP*, *greenfield*, *overlay*, *PON*, and *wireline*.

browser Also known as *Web browser*. A client software program that runs against a Web server or other Internet server and enables a user to navigate the World Wide Web (WWW) to access and display data. Web browsers are built on the concept of hyperlinks on which a user can click with a mouse to jump from page to page, document to document, or even site to site. Browsers can download files, display graphics, play audio and video files, and execute programs. The first primitive text browser was developed by CERN in 1991. In 1993, the first graphical browsers appeared: Viola for X Windows, Mac browser from CERN, and Mosaic for X Windows. Early graphical browsers that also support hyperlinks include Archie, Gopher, Jughead, Veronica, and WAIS. Currently popular browsers include Internet Explorer (IE), Mozilla, and Netscape Navigator. See also *Archie*, *CERN*, *client*, *Gopher*, *hyperlink*, *Internet*, *Internet Explorer*, *JUGHEAD*, *Mozilla*, *Netscape Navigator*, *server*, *software*, *Veronica*, *WAIS*, and *WWW*.

BS (Base Station) In a wireless network, the fixed central transmit/receive antenna and transceiver to which end user transceivers or mobile stations (MSs) connect. The base station also generally connects to the public switched telephone network (PSTN) or other public network, usually via microwave or wireline facilities. The base station may connect to a wireless network backbone or directly to other base stations, usually on a wireless basis. See also *antenna*, *backbone*, *microwave*, *PSTN*, *transceiver*, *wireless*, and *wireline*.

BSC (Binary Synchronous Protocol) Also known as Bisync. A byte-oriented communications protocol that organizes data into blocks of up to 512 characters framed with control codes that apply to the entire data set. Bisync was developed by IBM in 1966. BSC operates in half-duplex (HDX) mode, with the sending station transmitting one block at a time and the receiving device returning an acknowledgement following the receipt of each block. A positive acknowledgement (ACK) indicates that the data were received without error and that the sending station can transmit the next block. A negative acknowledgement (NAK) indicates that the data block was errored in transmission and that the sending station should retransmit it. Error control is based on a block checking character (BCC) that is transmitted along with the data. The receiving device independently calculates the BCC and compares the calculated byte with the received byte. There are six basic Bisync block formats, all of which comprise synchronizing bits, data, and control characters sent in a continuous data stream, block-by-block. See the illustration of a generic Bisync block in Figure B-4.

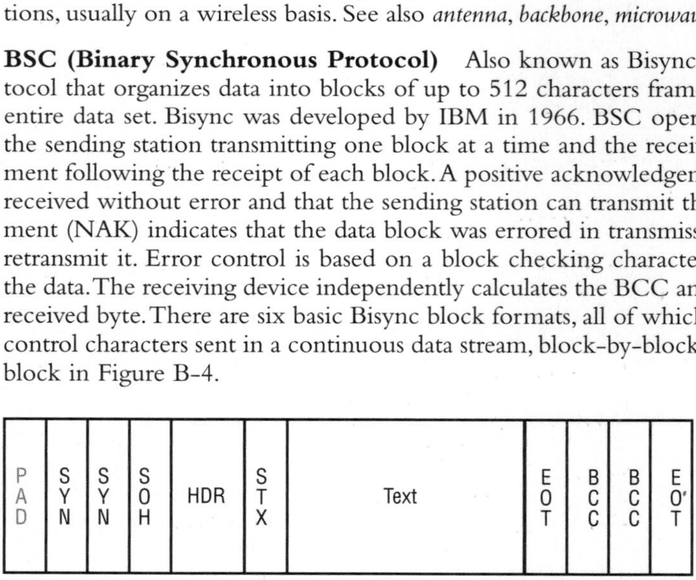

PAD	SYN	SYN	SOH	HDR	STX	Text	EOT	BCC	BCC	EOT

Figure B-4

The fields in the BSC block are as follows:

- **PAD:** An optional PADding character to alert the receiving device of the transmission of a block of data and to ensure that the receiving device is in sync with the data bits.
- **SYN:** SYNchronizing characters (usually two) establish character synchronization between the transmission and the receiving devices.
- **SOH:** A Start-Of-Header control character preceding the optional header.
- **Header:** An optional field of one or more octets of auxiliary data such as the addresses of the transmitting and receiving devices.
- **STX:** A Start-of-TeXt control character indicates the beginning of the data field.
- **Text:** Also known as the payload or data field. The text field can be up to 512 octets in length.
- **ETX:** An End-of-TeXt control character signaling the end of the data field.
- **BCC:** One of two Block Check Characters for error detection.
- **EOT** or **PAD:** An End-Of-Transmission character or PADding character trails the transmission to ensure the receipt of all previous characters and to indicate the end of the block.

See also *ACK*, *BCC*, *block*, *byte-oriented protocol*, *data set*, *HDX*, *hexadecimal*, *NAK*, *octet*, *payload*, and *text field*.

BSG (British Standard Gauge) Synonymous with *Imperial Standard Wire Gauge*. The measure of the diameter, or thickness, of a conductor in England. See also *gauge*.

BSOD (Blue Screen of Death) Slang phrase referring to the error message when Microsoft Windows experiences a fatal error condition. The monitor screen turns blue and fills with a typically cryptic description of the reason that your computer is going to crash and any data not saved to the hard drive will be forever lost. If you are very lucky, you can reboot your computer and restore the system. If you are not very lucky, you had better hope that you had the good sense to back up your system recently. Earlier MS-DOS systems lacked color and suffered the Black Screen of Death.

BSS (Broadband Switching System (BSS) A carrier exchange switch capable of switching and transport at broadband speeds. A BSS typically supports multiple broadband ports with fiber optic interfaces and a highly redundant internal switching matrix. See also *broadband*, *carrier*, *exchange*, *fiber optics*, *port*, *switch*, and *switch matrix*.

BT (Burst Tolerance) In asynchronous transfer mode (ATM), an expression of the ability of a network to tolerate bursts of traffic. See also *ATM* and *MBS*.

buddy list A closed correspondent list of friends, associates, and others that one invites to exchange instant messages (IM). Anyone not on the buddy list cannot determine your presence or communicate with you via IM. See also *IM*.

buffer **1.** A region of memory that temporarily stores data in a networking device, commonly to compensate for congestion at an incoming or outgoing port on a concentrator, multiplexer, switch, or router. If, for example, the level of incoming traffic exceeds the resources of a switch, a buffer at the incoming switch port can temporarily store the excess traffic until the switch has sufficient resources to process the traffic. If the level of outgoing traffic exceeds the capacity of the circuit, a buffer at the outgoing switch port can temporarily store the excess traffic until the circuit can accommodate it. A buffer also can serve to store packet data temporarily to allow retransmission in the event that a downstream device does not receive the packet without error within an acceptable period of time. **2.** A standalone storage device specifically designed to store data until such time as it is downloaded by another device for processing. A PBX system, for example, commonly outputs call detail records to a buffer, which stores them in temporary memory until a centralized poller accesses the buffer and downloads the data for subsequent processing into various call detail reports for purposes of cost allocation and management, traffic analysis, and security analysis. **3.** Electrically powered mechanical device that rotates abrasive pads for cleaning floors. Buffers are known to cause network problems when plugged into shared outlets by injecting electrical noise or blowing fuses. *Note:* It is always a good idea to make sure that a computer system has access to clean power, i.e., a dedicated circuit that, by definition, will be free of power dips caused by buffers and other devices. See also *CDR* and *register.* **4.** A protective material sometimes extruded directly on the acrylate coating of an optical fiber to further allow individual fibers to be handled easily during installation, while protecting them from physical damage. A tight buffered cable is used for short jumper cables and many other indoor applications where the temperature is controlled and the differences in thermal expansion and contraction are not so great between the buffer and fiber as to cause bending, which ultimately can lead to cracking and breaking of the fiber. See also *loose-tube cable.*

bug **1.** An error in coding or logic that causes a computer software program to malfunction (i.e., perform erratically, suffer catastrophic failure (crash), or produce invalid results. Although the origin of the term bug is somewhat in dispute, folklore dates it the mid-to-late 1940s, when a moth attracted to the light given off by the vacuum tubes of one of the first digital computers, either the Mark I or the ENIAC, died and shorted one of the circuits, causing a system failure. Others attribute the term to Thomas Edison (1878). So much for the etymology of bug. As for entomology, a moth is not a bug, strictly speaking. A true bug has thickened forewings and mouth parts adapted for piercing and sucking, and is of the insect order Hemiptera, which includes aphids, cicadas, and squash bugs. Moths suck, but do not pierce, and belong to the order Lepidoptera, as do butterflies. (Note: There are two kinds of people in this world — those who just love this arcane trivia, and those who hate it. I love it, in moderation. "Arcane trivia" is somewhat tautological, which is a pedantic word for redundant. There are two types of people in this world — those who love pedants and those who hate them. I hate them.) By the way, a bug fix is in the form of a patch, update, upgrade, or insecticide, depending on the type of bug involved. See also *bug fix*; *Edison, Thomas Alva*; *entomology*; *etymology*; and *pedant* **2.** An undocumented feature. (That's a joke.)

bug fix A solution for a bug. See *bug*, *patch*, *update*, and *upgrade.*

bulletin board system (BBS) See *BBS.*

bulletproof **1.** Impervious to bullets or other high velocity projectiles. On a personal note, when I worked for Continental Telephone Company (CONTEL) in the 1980s, the central office (CO) in Espanola, New Mexico, was a regular target for disgruntled customers or folks just out for a little target practice. (We never were 100 percent sure. Espanola was a pretty wild place.) The CO technicians regularly had to dig rifle slugs out of the analog step-by-step (SxS) CO switch. With relatively minor repairs, that old electromechanical switch could be restored to full operational status, which attests to the ruggedness of the technology. The CO wasn't bulletproof, but was close to it. When we installed a new, and much more sensitive, electronic common control (ECC) digital switch, we had to bulletproof the CO with an armored door. See also *aramid.* **2.** Impervious to attack. Totally secure, generally in reference to software that is impervious to attack from hackers, crackers, and other miscreants. Interestingly, however, Microsoft defines bulletproof as [software] capable of overcoming hardware problems. (I suppose a definition is a matter of perspective to some extent.)

bundled service The Telecommunications Act of 1996 specifies three ways that competitive local exchange carriers (CLECs) in the United States can provide competing local telephone service: build and interconnect, bundled wholesale purchase, and unbundled service. Securing bundled service from the incumbent LECs (ILECs) was at government-controlled wholesale prices, which typically were 15–25 percent below retail prices, and included local loops, local exchange and tandem switches (including software features), interoffice transmission facilities, signaling and call-related database facilities, operations support systems (OSSs) and information, and operator and directory assistance facilities. See also *CLEC, CO, database, ILEC, local loop, OSS, signaling and control, tandem switch,* and *Telecommunications Act of 1996.*

B-UNI (Broadband User Network Interface) Also known as the *Public UNI*. In asynchronous transfer mode (ATM), the specifications for the interface between an ATM endpoint or private ATM switch and an ATM switch in a public network. See also *ATM, ATM reference model, Private UNI,* and *UNI.*

burglar alarm circuit Also known as a *dry copper pair*. A twisted pair that is not electrified, that is, has no associated electronics. Providers of digital subscriber line (DSL) services and burglar alarm services order dry copper pairs from the telco and place their own electronics on them. All they want is a pair of wires from one point to another. As a point of interest, the first central telephone exchange was invented by E.T. Holmes, a young man whose father, in 1858, had originated the idea of protecting property by electric wires connected to a central alarm office. Holmes obtained telephone numbers 6 and 7 and attached them to a wire in his office. He then placed six box telephones on a shelf in his office. Any of these telephones could be switched into connection with the burglar alarm wires and any two of the six wires could be joined by a wire cord. At night, when the telephone operator was off duty, the telephone network reverted to a burglar alarm network. See also *dark fiber, twisted pair,* and *xDSL.*

burst tolerance (BT) See *BT.*

bursty transmission Referring to asynchronous communications, which is characterized by long periods of inactivity interspersed by sudden short periods of intense activity, such as keystrokes, data file transfers, or packet traffic. See also *asynchronous.*

bus A common electrical conductor shared by multiple related devices. There may be a number of buses in a given system, such as a switch or router. A single bus may consist of multiple (e.g., 16 or 32) conductors.

BUS (Broadcast and Unknown Server) In LAN Emulation (LANE), a network-based server that handles broadcast and multicast traffic, as well as initial unicast frames before address resolution. See also *broadcast, LANE, multicast, server,* and *unicast.*

bus topology A network structure in the form of a multipoint electrical circuit. The original 802.3 local area networks (LANs), commonly referred to as Ethernets, employed thick coaxial cable specified in 10Base5. The network was in the form of a physical bus topology, as illustrated in Figure B-5. All devices connected to the cable, and communicated over a single, shared channel on a shared electrical circuit. Each coaxial cable segment was limited to 500 meters due to issues of signal attenuation at the relatively high carrier frequency. Each segment supported as many as 1,024 (2^{10} = 1024) network addresses, each of which was associated with an attached device, such as a workstation or peripheral device. Ethernet segments could connect through bridges, which function as signal repeaters. The total route length of the entire Ethernet was limited to 2.5 kilometers, which is a function of both signal propagation time and medium access control (MAC) mechanisms. Subsequently, the 10Base-T specification allowed workstations and peripheral devices to interconnect through a hub, with each device connecting directly to a hub port over unshielded twisted pair (UTP). The physical topology is that of a star, but the logical topology is that of a bus. That is to say that, although the devices connect to the hub over circuits that emanate from the hub like the rays of a star, they interconnect through a collapsed bus housed within the hub. Bus networks employ a decentralized MAC method known as carrier sense multiple access (CSMA). A tree topology is a variation on the bus theme, with multiple branches off the trunk of the central bus. See also *10Base-T, 802.3, attenuation, bridge, bus, carrier, CSMA, MAC, star topology, topology,* and *UTP.*

busy The condition that exists when a central office (CO), PBX, destination telephone circuit, or other voice telecommunications component is in use at its full capacity and not available to accept additional calls. If a caller attempts to place a call to a device that is in a busy condition, the device returns a busy signal in the backward direction. For a telephone, busy is the same as off hook. See also *busy signal.*

busy/DA call forwarding (busy/Don't Answer) call forwarding) A voice telephone system feature method that allows a station user to program the system to automatically forward calls from the primary, or target, station to another station or perhaps a voice mail system after if the station either is engaged (i.e., busy, or off-hook) or does not answer after a selectable number of rings. See also *call forward.*

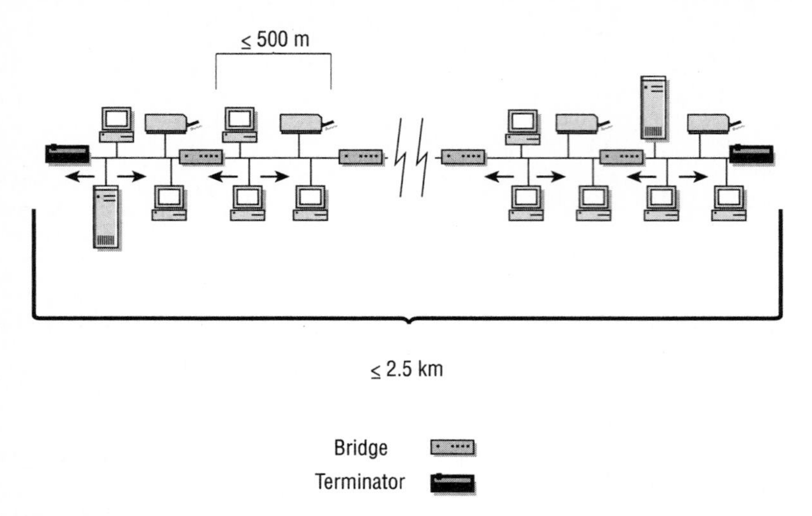

Figure B-5

busy hour In traffic engineering, the busiest 60-minute period of the day, ideally as determined by empirical traffic studies. See also *traffic engineering*.

busy hour call attempts (BHCAs) See *BHCAs*.

busy hour call completions (BHCCs) See *BHCCs*.

busy hour traffic (BHT) See *BHT*.

busy lamp field A modular component of a special telephone set typically used by the receptionist or operator of a key telephone system (KTS), a busy lamp field is an array of line status lamps. A busy lamp field allows the operator to view the status, e.g., busy or available, of the stations.

busy out To invoke a feature of a telecommunications system that rejects incoming calls. See also *do not disturb*.

busy signal An audible (or visual) signal indicating to the calling party that the called circuit is engaged and, therefore, unavailable. In the public switched telephone network (PSTN), a central office (CO) sends the busy signal in the backward direction, i.e., back to the caller. A fast busy signal, or reorder tone, is an indication to the calling party that network resources are not available to process the call. See also *PSTN*.

BWA (Broadband Wireless Access) Wireless local loop (WLL) technologies supporting broadband performance to individual end users. BWA technologies include both radio frequency (RF) and infrared (IR) solutions. RF-based solutions include Local Multipoint Distribution Service (LMDS), Multichannel Multipoint Distribution Services (MMDS), and Wireless Interoperability for Microwave Access (WiMAX). Free space optics (FSO) systems are wireless optical transmissions using wavelengths in the infrared range. See also *broadband*, *FSO*, *IR*, *LMDS*, *local loop*, *MMDS*, *RF*, *WiMAX*, and *WLL*.

byte (B) A character in a computer coding scheme. A byte is a unique set of adjacent bits with a unique meaning in a computer coding scheme. A byte generally comprises eight (8) bits that represent a letter in an alphabet (e.g., a, A, z, or Z), a diacritical mark (e.g., ~ or `), a single digit number (e.g., 0, 1, 2, or 3), a punctuation mark (e.g., ,, ., or !), or a control character (e.g., paragraph break, page break, or carriage return). The size of a byte is specific to the coding scheme involved. For example, ACSII employs a coding scheme of seven (7) bits, so a byte is actually seven information bits, although the addition of a parity bit for error control results in a byte of eight (8) bits for storage and transmission purposes. An EBCDIC byte is truly eight (8) information bits. Unicode variously employs eight (8) and 16 bits. Computers create, store, manage, and output information in bytes. The origin of the term is uncertain, although it is certain that Dr. Werner Buchholz originated the term byte in 1956 when working for IBM on the STRETCH computer. Some suggest that byte is an alteration and contraction of bit, referring to the basic unit of information, and bite, referring to a morsel or chunk of data consumable by the early computer eight-bit processors. Others suggest byte is an acronym formed from binary digit eight. Still others suggest that byte is short for binary term. Computing and storage systems measure memory and storage capacities in bytes. For example, a kB (kiloByte) is actually 1,024 (2^{10}) bytes, since the measurement is based on a base 2, or binary, number system. The term kB comes from the fact that 1,024 is nominally, or approximately, 1,000. So, 64 kB of memory is actually $1{,}024 \times 64 = 65{,}536$ (2^{16}) bytes. Transmission rates are measured in bits per second (bps), with the exception of some storage technologies that measure transmission rates in bytes per second (Bps). In transmission measurements, a kbps is 1000 bps, not 1024 bps. See also *MB*, *nibble*, *octet*, and *TB*.

byte interleaving The transmission of multiple independent data streams over a single circuit where one byte from each input stream is taken at one time and framed in a time-division sequence. In a typical voice application, a T1 transmission system, for example, supports a periodic frame of 24 information bytes, one from each channel, interleaved through a time division multiplexer (TDM). As long as the same 24 conversations remain active, byte samples of each conversation are interleaved in the same order 8,000 times per second. See also *byte*, *T1*, and *TDM*.

byte-oriented protocol A text-oriented synchronous communications protocol that handles only full bytes or characters of text, thereby requiring an entire byte to communicate a command signal to the target station. Control characters are embedded in the header and trailer of each data byte or block. As byte-oriented protocols are overhead-intensive, they are used exclusively in older computer protocols at the Data Link Layer. Byte-oriented protocols generally are synchronous and half-duplex (HDX) in nature, and operate over dial-up, two-wire circuits. Bisynchronous Communications (BSC) is an example of a byte-oriented protocol. See also *bit-oriented protocol*, *block*, *BSC*, *Data Link Layer*, *dial-up access*, *overhead*, *synchronous*, *synchronous transmission*, and *two-wire circuit*.

bytes per second (Bps) See *Bps*.

c **1.** Symbol for the velocity of light in a vacuum, which is exactly 299,792.458 kilometers per second, or 186,282.397 miles per second. The nominal figures, used for ease of reference and approximate calculations, are 300,000 km/s and 186,000 miles per second. The velocity of propagation (Vp) is sensitive to the medium. See also *Vp*. **2.** In 1900 the *Ladies Home Journal* magazine predicted that by the year 2000 the letters *c*, *x*, and *q* would be banished from the English alphabet. The expectation was that spelling would be phonetic and that those letters would be unnecessary. It was a perfectly logical prediction, but no more correct than the predictions that the use of jetpacks, teleportation, videophones, and smell-o-vision would be widespread. (*Note:* I make no predictions in this dictionary.) See also *Smell-O-Vision*, *teleportation*, and *videophone*.

C **1.** The symbol for capacitance, or capacity. See *capacitance*. **2.** The symbol for coulomb. See *coulomb*.

CA (Certificate Authority) In a public key infrastructure (PKI), an entity that issues and verifies digital certificates that contain an encryption key and attest to the authenticity of the transaction party. See also *authentication*, *digital certificate*, *encryption*, *key*, and *PKI*.

cable A group or bundle of conductors, fibers, or wires, or bound together, sharing a common protective sheath or jacket, and perhaps strength members and shielding. See also aerial cable, direct bury cable, submarine cable, and underground cable. See also *conductor*, *fiber*, and *wire*.

Cable Communications Policy Act (CCPA) In the United States, an act of Congress (1984) that amended the Communications Act of 1934 and established a national policy with respect to cable television (CATV) communications. The Act sought to facilitate cable system and services deployment by regulating what were considered the excessive demands that some local governments attempted to impose on CATV providers. In exchange for what was widely considered a curtailment of the authority of local governments to regulate cable systems, local governments were granted the right to collect franchise fees not to exceed five percent of the cable operator's gross revenues. The Act, among other things, regulates the local franchise renewal and transfer process as well as protects the personal information of customers of CATV operators by restricting the collection, maintenance, and dissemination of subscriber data by cable operators. The CCPA was later amended by the Cable Television Protection and Competition Act of 1992 and the Telecommunications Act of 1996. See also *Cable Television Protection and Competition Act*, *CATV*, *Communications Act of 1934*, *franchise*, and *Telecommunications Act of 1996*.

cable jacket Synonymous with *cable sheath*. See *cable sheath*.

CableLabs (Cable Television Laboratories) A not-for-profit research and development organization for the benefit of its member companies in the cable television (CATV) business. CableLabs' stated mission is that of pursuing new cable telecommunications technologies and helping its cable operator members integrate them into their business objectives. Of particular significance is the Data over Cable Service Interface Specification (DOCSIS) suite of specifications for which CableLabs assumed administrative responsibility from Multimedia Cable Network Systems Partners Ltd. (MCNS). DOCSIS defines the interface requirements for cable modems in support of high speed data communications and Internet access over CATV networks. The CableHome initiative is geared toward developing the interface specifications necessary to extend cable-based services to devices within the home. PacketCable is a CableLabs initiative aimed at developing interoperable interface specifications of real-time multimedia over two-way cable plant. For contact information, see Appendix A. See also *cable modem*, *CATV*, *DOCSIS*, and *modem*.

cable modem (CM) A modem designed to support high speed data communications over hybrid fiber/coax (HFC) CATV networks. Cable modems are positioned at the customers' premises and in a cable modem termination system (CMTS) at the service provider's headend. The initial Data over Cable Service Interface Specification (DOCSIS), released in 1997, standardized cable modems. Continuing research and development efforts directed at cable modems and related CATV network standards are largely the responsibility of CableLabs. See also *CableLabs*, *CATV*, *CMTS*, *DOCSIS*, *HFC*, and *modem*.

Cable Modem Termination System (CMTS) See *CMTS*.

cable sheath A close-fitting tubular covering that protects a cable. See also *cable*.

cable television Formally known as *community antenna television* (CATV), cable TV is transmitted from a service provider's headend to the subscribers' premises via a cable system, as opposed to over-the-air broadcast via ultra high frequency (UHF) or very high frequency (VHF) radio, or direct broadcast satellite (DBS). See also *cable*, *CATV*, *DBS*, *headend*, *UHF*, and *VHF*.

Cable Television Protection and Competition Act In the United States, an Act of Congress (1992) that amended the Communications Act of 1934 to provide increased consumer protection and to promote increased competition in the cable television (CATV) and related markets. The Act included provisions for rate regulation, customer service guidelines, and broadcast radio retransmission. The Act further included provisions with respect to local signal carriage regulation, specifically, must-carry and may-carry, i.e., retransmission consent. The Act was amended in important respects by the Telecommunications Act of 1996, particularly with respect to regulation of cable rates. See also *broadcast radio*, *CATV*, *Communications Act of 1934*, *Telecommunications Act of 1996*.

Cable Television Laboratories (CableLabs) See *CableLabs*.

cable TV (cable TeleVision) Synonymous with *community antenna television* (CATV). See *CATV*.

CAC **1.** Carrier Access Code. In the United States, a seven-digit code (101 XXXX) that a caller can use to dial around the presubscribed interexchange carrier (IXC) to reach another IXC. The first three digits (101) of the CAC signal the network of the caller's intent. The last four digits (XXXX) of the CAC are the Carrier Identification Code (CIC), which is used for call routing purposes. See also CIC, dial around, and IXC. **2.** Carrier Access Charge. In the United States, a set of access charges billed by the local exchange carrier (LEC) to the interexchange carrier (IXC) to compensate the LEC for the costs associated with connecting interexchange long distance calls over LEC local loop facilities. First, the CAC includes flat-rate, recurring charges that apply for tandem exchange termination. Second, the Carrier Common Line Charge (CCLC) is a minutes-of-use charge that applies to each call connected to the IXC. See also *access charges*, *IXC*, and *LEC*.

caching In the context of the Internet, the process of temporarily storing frequently accessed Web content in a local or regional server in order to reduce long haul Internet traffic. The cached content must be refreshed frequently in order to maintain currency with the source content on the distant server. See also *Internet*, *server*, and *Web*.

CAD (Computer-Aided Design) The use of computers and software in design work. CAD is used extensively in architecture, aeronautical engineering, civil engineering, electrical engineering, genetic engineering, mechanical engineering, facilities planning, and interior design.

CAD/CAM (Computer-Aided Design/Computer-Aided Manufacturing) The use of computers and software in both the design and manufacture of a product. For example, a machine part, aircraft component, or boat hull can be designed with a CAD program that can provide the final design instructions for use in a computer-driven, automated fabrication and assembly process. See also *CAD*.

CAI (Common Air Interface) See *air interface*.

CALC (Customer Access Line Charge) Synonymous with *subscriber line charge* (SLC). See *SLC*.

CALEA (Communications Assistance for Law Enforcement Act) In the United States, a federal law (1994) passed to protect public safety and ensure national security by preserving law enforcement's ability to conduct lawfully authorized electronic surveillance while preserving public safety, protecting the public's right to privacy, and the telecommunications industry's competitiveness. In 2006, the Federal Communications Commission (FCC) adopted an order extending the act to facilities-based broadband Internet access providers and connected VoIP providers. See also *broadband*, *FCC*, *Internet*, *VoIP*, and *wiretap*.

calibrate To measure the accuracy of a device, instrument, system, or process relative to a known standard and to adjust it, as necessary, in order that it conform to that standard over the range of its operation.

call **1.** In telecommunications, a connection through a telephone network in support of a communication between two or more stations. A call comprises a sequence of events that begins when an end user at an originating station initiates a call request to a switch that may work in conjunction with other switches to establish a connection to an end user at a destination station, and concludes when one party (user) terminates the connection. In other words, a call encompasses the operations required to set up or establish, maintain, and terminate or release a connection. **2.** In switching systems, a demand to set up a connection between two or more stations. **3.** A telephone conversation between two or more simultaneously present end users. In other words, a real-time telephone conversation between two or more live people. **4.** The attempt to reach someone by telephone, whether successful or not, in other words, a call attempt. **5.** In computing, the act of causing a program, routine, or subroutine to execute.

call agent Sometimes referred to as a *call controller*. In voice over Internet Protocol (VoIP) networks, a softswitch often is decomposed into a media gateway (MG) and a call agent. The call agent is the logic responsible for the registration and management of resources at the media gateway (MG) and is responsible for functions such as billing and routing. In H.323-compliant networks, the call agent is known as a *gatekeeper*. In Megaco-compliant networks, the term is media gateway controller (MGC). In the Session Initiation Protocol (SIP), the Call Agent registers and proxies for all endpoints in a domain, including phones as well as MGs. See also *endpoint*, *gatekeeper*, *H.323*, *media gateway*, *Megaco*, *proxy*, *SIP*, *softswitch*, and *VoIP*.

call-back messaging A feature of automatic call distributors (ACDs) that enables an impatient caller to register the desire to be called back by a call center agent, rather than wait in a seemingly interminable queue. When the queue has been satisfied and an agent becomes available, the system will call the customer back and connect him to an agent automatically, perhaps using a predictive dialer in a call blending application. See also *ACD*, *call blending*, *call center*, and *predictive dialer*.

call blending A feature of standalone automatic call distributors (ACDs) that supports intensive outgoing, as well as incoming, calling activity. Call blending employs a predictive dialer that monitors the status of incoming calling activity and the level of availability of the agent pool. When the level of incoming calling activity drops to a level such that agents are idle, the system introduces outgoing calls. See also *ACD* and *predictive dialer*.

call block Also known as *selective call block* and *selective call rejection*. **1.** A CLASS service feature of the public switched telephone network (PSTN). Selective call rejection screens incoming calls against a user-defined list of acceptable telephone numbers. Calls from those numbers are diverted to a recorded message. Calls from all other numbers ring through as usual. See also *CLASS* and *PSTN*. **2.** As defined in the advanced intelligent network (AIN) specifications, call block can work on incoming or outgoing basis. On an incoming basis, the feature allows the subscriber to program some number of telephone numbers from which incoming calls are denied. Incoming call blocking is widely available. On an outgoing basis, this feature typically supports the blocking of calls to international destinations, either in total or to specific country codes. Content blocking supports the blocking of calls to specific numbers, such as 900/976 numbers. This capability is deployed fairly commonly in many foreign networks, but is not widely available in the United States. See also *AIN* and *selective call rejection*.

call center A central office staffed by agents dedicated to handling large volumes of customer contacts by telephone. Call centers generally focus on processing incoming calls, but may handle outgoing calls, as well. Incoming calls generally are processed by a standalone automatic call distributor (ACD), although a small call center might make use of a PBX with ACD software or perhaps more primitive uniform call distributor (UCD) software. A call center that handles both incoming and outgoing calls generally does so through the use of a predictive dialer that facilitates call blending. Call centers increasingly are located off-shore, in consideration of lower labor costs. Call centers also increasingly are virtual in nature, as ISDN and especially IP networks have enabled agents to telework, i.e., work from home. See also *ACD*, *call blending*, *IP*, *ISDN*, *PBX*, *predictive dialer*, and *UCD*.

call centre The British English spelling of the American English *call center.* As an American, I prefer call center, although call centre certainly has the advantage of being quaint. See *call center.*

call completion service In the advanced intelligent network (AIN) specifications, a feature that enables the directory assistance operator to extend the call automatically, perhaps at an additional charge. This capability is widely offered by providers of cellular service to prevent accidents caused by *driving and dialing*. See also *AIN*.

call controller See *call agent*.

call detail record (CDR) Synonymous with *station message detail record* (SMDR). See *SMDR*.

caller ID (caller IDentification) Synonymous with *calling line identification* (CLID) and *calling number delivery*. See *CLID*.

call extender A device that allows a call center to extend an incoming call to a remote agent, perhaps a telecommuter in a home office. Such devices support integrated voice and data communications between the automatic call distributor (ACD) and the remote agent and can operate over narrowband analog or ISDN connections through the public switched telephone network (PSTN), or broadband ADSL or cable modem networks through the Internet. See also *ACD*, *ADSL*, *cable modem*, and *ISDN*.

call forking Also known as *call splitting*. A feature of the Session Initiation Protocol (SIP) that enables an incoming call to ring several extensions, i.e., other telephones, at once. The first telephone to answer takes control of the incoming call. See also *SIP*.

call forward A feature of a key telephone system (KTS), PBX, or central office (CO) that enables a station user to forward calls to another station by one of several methods. One method allows the user to dial a feature code to forward calls from the primary, or target, station to a specific secondary station on an ad hoc basis. Another method allows the user to program the system to automatically forward calls from the primary station to the attendant station, a colleague's station, or perhaps a voice mail system after a selectable number of rings or if the station is engaged. The latter feature is commonly known as Busy/DA (Busy/Don't Answer) call forwarding.

call notification service Synonymous with *call pick-up service*, a feature of the advanced intelligent network (AIN). See also *AIN* and *call pick-up*.

call park A voice telephone system feature that enables a station user to park a call in system orbit. The call then can be retrieved by any other user by dialing the associated call park code.

call pick-up A voice telephone system feature that enables users in a programmed call pick-up group (e.g., workgroup) to answer calls for each other. In a key telephone system (KTS) environment, a user can pick up a call at any phone where the line appears by dialing a feature code, usually of three characters. In a PBX environment, an authorized user can enter a code on the set keypad or by depressing a designated feature button on an electronic set, with no line appearance on the set necessary. Call pick-up service, also known as *call notification service*, is defined in the advanced intelligent network (AIN) specifications as a service of the public switched telephone network (PSTN) that provides for calls to be answered automatically by a voice processor. The called party can be notified of a deposited voice message by pager, fax, e-mail, or other means. The caller can enter a privilege code provided by the called party to distinguish the priority of the calling party. To pick up the call, the called party dials a direct inward system access (DISA) port on the network switch and enters password codes in a manner similar to that used to access contemporary voice mail systems. See also *AIN*, *DISA*, *KTS*, *PBX*, and *PSTN*.

call return Also known as *automatic callback*. See *automatic callback*.

call splitting See *call forking*.

call supervision See *supervision*.

call trace Also known as *customer-originated trace*. See *customer-originated trace*.

call transfer A voice telephone system feature that enables a station user to transfer an incoming or outgoing call to another station by dialing a feature code, usually of three characters.

call vectoring Also known as *custom control routing*. The process of customer-programmable call handling and routing through an automatic call distributor (ACD) in a call center environment. Once a call successfully completes to a call center, the caller's identity can be established through several means, including Calling Line Identification (CLID), Automatic Number Identification (ANI), or an account number or some other Personal Identification Number (PIN). The system can then search a computer database in order to establish the caller's profile, analyze the profiles of the available agents to identify those most capable and available to handle the call, select an agent, and present the caller's profile to the agent in advance of the connection of the call through a screen pop. Thereby, the most available and capable agent has access to full account information and theoretically can provide the highest possible level of service. Skills-based routing also considers special agent skills, such as language skills. See also *ACD*, *ANI*, *call center*, *CLID*, *PIN*, and *screen pop*.

call waiting **1.** A CLASS service of the public switched telephone network (PSTN). Call waiting allows an incoming call to wait in queue if the target telephone line is engaged. A call waiting signal alerts the user of the target telephone line to the fact that a call is waiting. See also *CLASS* and *PSTN*. **2.** A voice telephone system (e.g., PBX or Centrex) feature that allows an incoming call to wait in queue if the target station is engaged. A call waiting signal alerts the user to the fact that a call is waiting. See also *camp-on*.

calling line identification (CLID) See *CLID*.

calling name and number blocking A feature of CLASS service that allows the calling party to block calling name delivery and calling number delivery, also known as calling line identification (CLID), on a permanent basis, or on an ad hoc basis by dialing a code prior to dialing the destination telephone number. See also *CLASS* and *CLID*.

calling name delivery Also known as *caller name*. A CLASS service of the public switched telephone network (PSTN). The service delivers the name of the calling party, as listed in the directory, number to the called line, where it can appear on a telephone set equipped with a display or on a peripheral display unit. Because this feature works in conjunction with CLID, it automatically is blocked if CLID is blocked. See also *calling name and number blocking* and *CLID*.

calling number delivery Synonymous with *Calling Line Identification* (CLID) and *Caller ID*. See *CLID*.

CamelCase The convention in which a compound word, acronym, or phrase is spelled using medial capitals. CamelCase is used in a number of protocols, such as Extensible Markup Language (XML). ARCnet and WikiWiki are examples of CamelCase. The term is so named as the bumps of the medial capital letters are reminiscent of the humps of a camel. See also *acronym* and *initialism*.

camp-on A voice telephone system feature that enables the system attendant to forward or extend a call to a station, even if the station is engaged in a call. Camp-on essentially is much like call waiting, although the camped-on call will be routed back to the attendant if not answered within a programmable time interval. See also *call waiting.*

Canadian Radio-television and Telecommunications Commission (CRTC) See *CRTC.*

candela (cd) The basic SI unit of luminous intensity. The standard originally was based on the light emission of a candle flame, then as the glow from molten platinum, but then things got more complicated. In contemporary SI terms, the candela is the luminous intensity, in the perpendicular direction, of an area of $\frac{1}{600{,}000}$ square meter of a blackbody radiator at the temperature of freezing platinum (2,045 Kelvin), under a pressure of 101,325 newtons per square meter. One candela emits 4π lumens of light flux. For more information, get a degree in electrical engineering or physics, just like the person who came up with this way too complicated, but delightfully precise, measurement. See also *luminance* and *SI.*

candle See *candela.*

CAN-SPAM Act (Controlling the Assault of Non-Solicited Pornography And Marketing Act) A law passed by the United States Congress in 2003, extending the ban on certain unsolicited marketing to e-mail. In 2005, that ban was extended to text messaging directed towards cellular phones and other mobile devices. See also *do-not-call registry, Telephone Consumer Protection Act,* and *text messaging.*

CAP **1.** Competitive Access Provider. Also known as *alternative access vendor* (AAV). A company that provides access circuits between an end user enterprise and an interexchange carrier (IXC), or long distance carrier, in competition with the incumbent local exchange carrier (ILEC). A CAP is facilities-based and generally offers access services at attractive rates compared to those of the ILEC. See also *carrier, ILEC,* and *IXC.* **2.** Carrierless Amplitude Phase. A non-standard variation of quadrature amplitude modulation (QAM), CAP was the early de facto standard modulation technique for asymmetric digital subscriber line (ADSL) modems. CAP was replaced by discrete multitone (DMT) prior to the release of ADSL standards recommendations by the ITU-T. See also *ADSL, DMT, ITU-T, modulation, QAM,* and *standard.*

capacitance (C) Capacitance, or capacity, is the property of a device or material medium to store an electrostatic charge. A capacitor is a device specifically designed to do so. See also *capacitor.*

capacitive reactance (X_C) The opposition to the flow of alternating electric current (AC) in a capacitor, Capacitive reactance is an inertial reaction to changes in the electromagnetic field created when an alternating voltage is applied. When AC passes through a component that contains reactance, energy is alternately stored in and released from a magnetic field or electric field. In the case of electric energy, the reactance is capacitive. As voltage is applied to the capacitor, charges accumulate on it and the voltage increases until it reaches the level of the applied voltage. At that point, no more current will flow. As the current reverses again, the capacitor discharges, reaching zero charge and zero voltage, and the cycle begins again. The greater the amount of capacitance, the greater the inertial opposition. The faster the reversal of current, the greater the inertial opposition. Capacitive reactance is measured in Ohms (Ω). See also *capacitance, inductive reactance,* and *reactance.*

capacitor An electrical device specifically designed to store an electrostatic charge, a capacitor is a system of conductors and dielectrics. A capacitor opposes changes in voltage, whereas an inductor opposes changes in current. See also *capacitive reactance, conductor, current, dielectric, inductor,* and *voltage.*

capacity See *capacitance.*

Captioned Telephone Service A Telecommunications Relay Service (TRS) that involves a special telephone with a text display. Rather than using TTY technology, the called party's speech is re-voiced by the communications assistant (CA), converted into text by a voice recognition system, and transmitted directly to the callers display telephone. See also *text, TRS, TTY,* and *voice recognition.*

carding See *phishing.*

carrier **1.** A continuous signal, or waveform, at a certain frequency on a circuit, or within a certain frequency range, and that can be modulated to support an information-bearing signal. In other words, the carrier carries the information signal, which the transmitter impresses on the carrier by varying the signal in some fashion. The carrier also can support signaling and control information used to coordinate and manage various aspects of network operations, although signaling and control can also occur over a subcarrier frequency. A carrier also is known as a *carrier wave.* **2.** A company that provides information transport services. For example, a Local Exchange Carrier (LEC) provides local information transport services, and an Interexchange Carrier (IEC or IXC) provides transport services between LECs. A heavy carrier is facilities-based, i.e., owns the switching and transmission systems that comprise the network it uses to provide services to its customers. A light carrier is not facilities-based, i.e., leases rather than owns the network it uses to provide services. A common carrier provides message transport services to the general public and generally is regulated to a considerable extent, at least with respect to fundamental aspects of service such as availability and basic rates. See also *common carrier, IXC,* and *LEC.*

Carrier Access Code (CAC) See *CAC.*

Carrier Access Charge (CAC) See *CAC.*

carrier class Synonymous with *carrier grade.* Referring to hardware or software that is durable and reliable enough to satisfy the demands of a carrier, i.e., incumbent telephone company, competing telecommunications service provider, or Internet service provider (ISP). See also *carrier.*

carrier frequency **1.** The frequency of an unmodulated carrier wave. **2.** A frequency capable of being modulated by an information-bearing signal of another frequency or frequencies. The carrier frequency is referred to as the center frequency in Frequency Modulation (FM), as the carrier frequency is used as a centerpoint, or reference point, with higher and lower frequency signals used to represent 1 bits and 0 bits. See also *carrier* and *frequency.*

carrier grade Synonymous with *carrier class.* See *carrier class.*

Carrier Identification Code (CIC) See *CIC.*

carrierless amplitude phase modulation (CAP) See *CAP.*

carrier sense multiple access (CSMA) See *CSMA.*

carrier sense multiple access with collision detection (CSMA/CD) See *CSMA/CD.*

carrier sense multiple access/collision avoidance (CSMA/CA) See *CSMA/CA.*

carrier serving area (CSA) See *CSA.*

carrier system A system that derives multiple logical channels from a single physical communications path, thereby supporting multiple communications over a single circuit. The first practical analog carrier systems were invented in the 1890s for telegraphy, based on what is now termed frequency division multiplexing (FDM). Early work on analog carrier multiplexing for telephony networks was done in the laboratories at American Bell Telephone Company, the predecessor to AT&T, as early as 1894. The first commercial carrier system was installed between Baltimore, Maryland and Pittsburgh, Pennsylvania in the United States, in 1918. That Type A system provided four full duplex (FDX) carrier channels above the 4-kHz voice band on short haul open wire circuits in the frequency range from 5 kHz to 25 kHz. The last commercial analog carrier system was L5E (1978), which used 22 pairs of coaxial cables to support a total of 132,000 simultaneous voice grade transmissions. The first commercial digital carrier system was T-carrier, specifically T1, which Bell System activated in 1962 in Chicago, Illinois. Digital carrier systems use the technique of time division multiplexing (TDM). See also *channel, circuit, FDM, FDX, multiplexer, T1, T-carrier,* and *TDM.*

Carterfone Decision In the United States, the Federal Communications Commission (FCC) decision (1968) that allowed the interconnection of foreign (i.e., provided by an entity other than the telco) equipment through standard protective coupling device provided by the telco. The Carterfone Decision effectively countered the Hush-a-Phone Decision. See also *FCC* and *Hush-a-Phone Decision.*

Cartesian coordinates In mathematics, any of the coordinates that locate a point on a plane and measure its distance from an axis at which two or more planes intersect. Such a point is identified by its position relative to the x-axis (horizontal) and y-axis (vertical) in a two-dimensional grid, chart, or graph, or by its position relative to the x-axis, y-axis, and z-axis (depth) in a three-dimensional grid, chart, or graph. The underlying mathematical concepts were developed by René Descartes, the Latin form of whose name is Cartesius. See also *x-axis*, *y-axis*, and *z-axis.*

Carty, John Joseph (1861–1932) As a young American Bell technician and one of the original telephone operators (1879), Mr. Carty suggested in 1881 the use of a second copper wire to complete the telephone circuit and, thereby, to avoid the emanation of electrical noise from the ground. Previously, copper telephone circuits were provisioned over a single wire. Carty later introduced the concept of a phantom circuit, through which two telephone lines can operate in tandem to support three simultaneous conversations. He was the first to install loading coils on a circuit (circa 1900) and was involved in the engineering of the first transcontinental telephone circuit. Carty became AT&T chief engineer in 1907, vice president in charge of the Department of Development and Research in 1919, and Chairman of the Board of Directors of AT&T Bell Telephone Laboratories in 1923. See also *loading coil*, *phantom circuit*, and *twisted pair.*

CAS **1.** Centralized Attendant Service. A PBX feature that allows the consolidation of multiple console operators for a number of PBXs. Thereby, a single group of operators can perform all attendant functions, including placing calls, answering and extending calls, and establishing conference calls. CAS will work whether the PBXs are distributed across a campus, a metropolitan area, or even a wide area, as long as the PBXs are networked over dedicated circuits. **2.** Channel-Associated Signaling. A signaling and control technique by which the signaling and control information associated with a communications channel is carried within the channel, itself, or in a separate channel permanently associated with that communications channel. See also *CCS*, *NFAS*, and *signaling and control.*

.cat (.Catalan) Pronounced *dot cat.* The generic Top Level Domain (gTLD) intended to support the Catalan linguistic and cultural community. This agreement is most unusual, as Catalan is spoken by less than 16 million people in the world and understood by less than 21 million. Catalan is the language of Catalonia in Spain, the city of Valencia, the Principality of Andorra, and other isolated cities, regions, and islands in Spain, France, Italy, and that general area of Europe. This domain was created in 2005 under the sponsorship of Fundació puntCat, a non-profit special interest organization. See also *gTLD*, *Internet*, and *sponsored domain.*

Cat (Category) A set of structured cabling standards have been developed over time by standards bodies acting in various collaborations. The standards bodies are the American National Standards Institute (ANSI), Electronic Industries Alliance (EIA), Insulated Cable Engineers Association (ICEA), International Electrotechnical Commission (IEC), International Organization for Standardization (ISO), and Telecommunications Industry Association (TIA). EIA/TIA 568A, released in 1985, built on earlier work at AT&T, IBM, and other companies, and set the tone for formally standardization structured wiring standards. That standard subsequently was supplemented and improved as EIA/TIA 568B. It also was internationalized in 1995 as ISO/IEC 11801. The standards are known as Cat 1, Cat 2, Cat 3, Cat 4, Cat 5, Cat 6, and Cat 7, and are progressively more capable due to tighter tolerances on dimensions and different twist pitches. Table C-1 compares the various categories with gauge measurements stated in American Wire Gauge (AWG). See also *ANSI*, *AWG*, *Cat 1-7*, *EIA*, *EIA/TIA-568*, *ICEA*, *IEC*, *ISO*, *ISO/IEC 11801*, *TIA*, and *twist pitch.*

Table C-1: Twisted-Pair Categories of Performance

Cat	AWG	Rating	Typical Applications	Standards
Cat 1	Various	Unspecified; < 1 MHz	POTS, ISDN BRI, RS-232 & RS-422, low-speed data, speaker wire, alarm cable	ANSI/ICEA S-80-576 ANSI/ICEA S-91-661
Cat 2	24 gauge	1 MHz	4 Mbps Token Ring LANs	ANSI/ICEA S-80-576
Cat 3	24 gauge	16 MHz	POTS, ISDN, T1, 10Base-T LAN	ANSI/ICEA S-91-661 ANSI/ICEA S-101-699 ANSI/TIA/EIA 568 ISO/IEC 11801
Cat 4	24 gauge	20 MHz	16 Mbps Token Ring LAN	ANSI/TIA/EIA 568 ISO/IEC 11801
Cat 5	24 gauge	100 MHz	10/100Base-T LAN	ANSI/ICEA S-91-661 ANSI/TIA/EIA 568 ISO/IEC 11801
Cat 5e	24 gauge	100+ MHz	10/100Base-T LAN, 155 Mbps ATM, 1000Base-T (Gigabit Ethernet, or GbE)	ANSI/ICEA S-91-661 ANSI/TIA/EIA 568 ISO/IEC 11801
Cat 6	22–24 gauge	250 MHz	1000Base-T	ANSI/TIA/EIA 568 ISO/IEC 11801
Cat 7	22–24 gauge	600 MHz	10 GbE	ANSI/TIA/EIA 568 ISO/IEC 11801

Cat 1 (Category 1) A type of cabling that is specified in ANSI/ICEA S-80-576 and ANSI/ICEA S-91-661, but is not recognized in either the EIA/TIA or ISO/IEC standards. Cat 1 is obsolete in telecommunications applications, but much remains in place in applications such as POTS and ISDN BRI telco local loops, analog PBX and key telephone systems, inside wire and cable systems, low speed data cables (e.g., RS-232 and RS-422), alarm cabling, and audio speaker wire. Cat 1 runs over unshielded twisted pair (UTP) of unspecified gauge and is rated at less than 1 MHz, but typically supports bandwidth less than 100 kHz. In ISO/IEC 11801 terms, Cat 1 variously falls into Class A (≤ 100 kHz) and Class B (≤ 1 MHz). See also *bandwidth, gauge, ISDN BRI, POTS,* and *UTP.*

Cat 2 (Category 2) A type of cabling that is specified in ANSI/ICEA S-80-576, but is not recognized in either the EIA/TIA or ISO/IEC standards. Now considered obsolete, Cat 2 was developed as IBM Type 3, in support of 4 Mbps IBM Token Ring LANs. Cat 2 runs over 24 AWG unshielded twisted pair (UTP) and is rated at ≤ 1 MHz. In ISO/IEC 11801 terms, Cat 2 falls into Class B (≤ 1 MHz). See also *AWG, LAN, Token Ring,* and *UTP.*

Cat 3 (Category 3) A type of cabling that is specified in ANSI/ICEA S-91-661, ANSI/ICEA S-101-699-2001, ANSI/EIA/TIA 568, and ISO/IEC 11801. Originally developed in support of 10Base-T Ethernet LANs, Cat 3 is commonly used in telco local loops in support of POTS, ISDN, and T1 and E-1 services. Cat 3 runs over 24 AWG unshielded twisted pair (UTP), with 3–4 twists per foot, and is rated at 16 MHz. In ISO/IEC 11801 terms, Cat 3 falls into Class C (≤ 16 MHz). See also *10Base-T, AWG, E-1, ISDN, LAN, POTS, T1,* and *UTP.*

Cat 4 (Category 4) A type of cabling that is specified in ANSI/EIA/TIA 568, and ISO/IEC 11801. Originally developed in support of 16 Mbps IBM Token Ring LANs, Cat 4 is an orphaned category. Cat 4 runs over 24 AWG unshielded twisted pair (UTP) and is rated at 20 MHz. In ISO/IEC 11801 terms, Cat 4 falls into Class D (≤ 100 MHz). See also *AWG, LAN, Token Ring,* and *UTP.*

Cat 5 (Category 5) A type of cabling that is specified in ANSI/ICEA S-91-661, ANSI/EIA/TIA 568, and ISO/IEC 11801. Originally developed in support of 10/100Base-T Ethernet LANs, Cat 5 quickly became preferred for all inside wire and cable applications. Cat 5 runs over 24 AWG unshielded twisted pair (UTP) and is rated at 100 MHz. In ISO/IEC 11801 terms, Cat 5 falls into Class D (≤ 100 MHz). See also *10/100Base-T*, *AWG*, *Ethernet*, *LAN*, and *UTP*.

Cat 5e (Category 5 enhanced) A type of cabling that is specified in ANSI/ICEA S-91-661, SI/EIA/TIA 568, and ISO/IEC 11801. Originally developed as a performance-enhanced version of Cat 5 in support of 10/100Base-T Ethernet LANs, Cat 5e quickly became, and remains, preferred for all inside wire and cable applications. Cat 5e performance is achieved through a tighter twist of 3–4 twists per inch, electrical balancing between the pairs, and fewer cable anomalies, such as inconsistencies in both conductor diameter and thickness of the dielectric insulation. Cat 5e typically runs over 24 AWG unshielded twisted pair (UTP), and is rated at 100 MHz over distances up to 350 meters. Cat 5e also may be shielded twisted pair (STP). Some manufacturers have increased the signaling speed up to 250 MHz in support of 1000Base-T, although over shorter distances of 100 meters or so. In ISO/IEC 11801 terms, Cat 5e falls into Class D (≤ 100 MHz). See also *10/100BaseT*, *1000Base-T*, *AWG*, *Ethernet*, *LAN*, *STP*, and *UTP*.

Cat 6 (Category 6) A type of cabling that is specified in ANSI/ICEA S-91-661, ANSI/EIA/TIA 568, and ISO/IEC 11801. Cat 6 was developed specifically in support of 1000Base-T, also known as *GbE* or *GigE*, spreading the signals over each of four pairs. Cat 6 runs over 22-24 AWG unshielded twisted pair (UTP), shielded twisted pair (STP), and screened twisted pair (ScTP) and is rated at 250 MHz at distances up to 220 meters. Some manufacturers claim performance up to 400 MHz. In ISO/IEC 11801 terms, Cat 6 falls into Class E (≤ 250 MHz). See also *1000Base-T*, *AWG*, *ScTP*, *STP*, and *UTP*.

Cat 7 (Category 7) A type of cabling that is specified in ANSI/EIA/TIA 568, and ISO/IEC 11801. Cat 7 was developed specifically in support of 10 GbE, spreading the signals over each of four pairs. Cat 7 runs over 22–24 AWG shielded twisted pair (STP) with a combination foil and braided screen construction. Cat 7 is rated at 600 MHz, although the usable spectrum can be up to 750 MHz. In ISO/IEC 11801 terms, Cat 7 falls into Class F (≤ 600 MHz). See also *10GbE*, *AWG*, and *STP*.

Category (Cat) See *Cat*.

CATV (Community Antenna TeleVision) More commonly known as *cable television* or *cable TV*. In a traditional CATV network, the community antenna comprises multiple satellite and microwave antennas located at a headend, which is the point of signal origin. At the headend, multiple analog broadcast TV signals are interleaved by frequency division multiplexing (FDM) and transmitted downstream over an analog coaxial cable trunk system to the community or neighborhood to be served. In a tree and branch architecture, the trunk system branches off into distribution cables to the neighborhoods, where taps and drops serve to connect individual subscribers, as illustrated in Figure C-1.

CATV originated in the mountains of Pennsylvania (United States) in the late 1940s, when there were only a few TV stations, all located in major cities. All TV transmission was broadcast over the air and reception was poor in the cities. In remote rural areas and particularly in mountainous areas where line of sight (LOS) was not possible, reception was awful. John and Margaret Walson, owners of the Service Electric Company, a retail appliance store in Mahanoy City, were having a difficult time selling TV sets as reception was so poor in the valley where the town was situated. In order to demonstrate TV sets to their best advantage, Mr. Walson placed an antenna on top of a tall utility pole on a nearby mountaintop and ran antenna wire to the store. In June 1948, he built some amplifiers to bring the signal to customers who had bought his TV sets, thereby creating the first CATV network. Walson also was the first to use microwave to import TV signals from distant stations and the first to use coaxial cable to improve reception. In 1972, Walson also was the first to distribute Home Box Office (HBO), which marked the beginning of pay TV and the explosive growth of the CATV industry. CATV remains largely a North American phenomenon, although CATV is widely available in England and a few other countries.

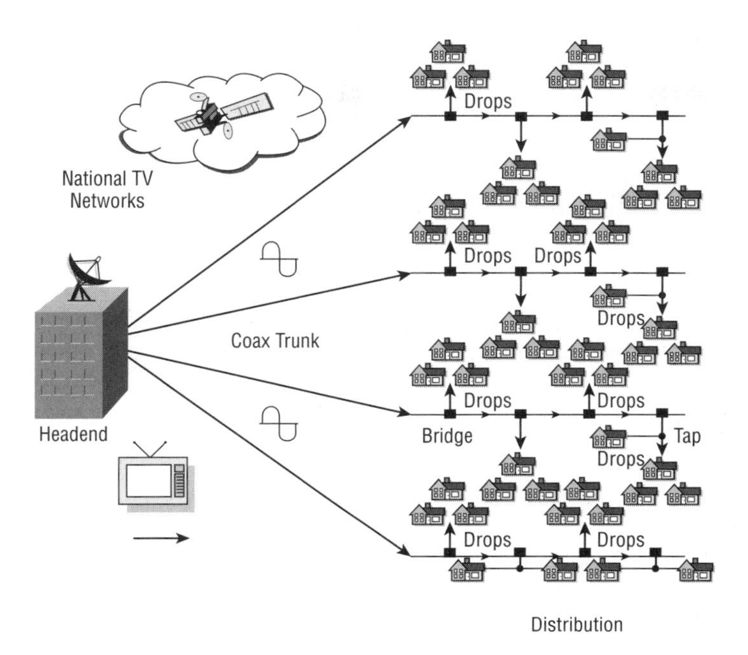

Figure C-1

Contemporary coax-based CATV systems commonly support aggregate raw bandwidth of 500–850 MHz. In the United States, each 6 MHz video channel carries a signal in the National Television Standards Committee (NTSC) format. (*Note:* In European and many other countries, the phase alternate line (PAL) format requires an 8 MHz channel.) Within each 6 MHz channel, approximately 250 kHz is transition bands, or guard bands, 4.2 MHz is required for the video signal, 300 kHz is required for the associated audio signal, and the balance is due to the vestigial side-band amplitude modulation (VSB-AM) technique utilized as the signal is placed on the radio frequency (RF) carrier.

In the mid-to-late 1990s, CATV networks began to merge and were acquired by large multiple system operators (MSOs) in the United States. Many coax-based systems were upgraded to support not only downstream television, but also high-speed full-duplex (FDX) Internet access and, more recently, voice. The process began of converting many systems from analog to digital format and deploying optical fiber, at least in the trunk network. The Multimedia Cable Network Systems Partners Ltd. (MCNS) developed the Data over Cable Service Interface Specification (DOCSIS) standards for cable modems, for which CableLabs assumed administrative responsibilities. DOCSIS supports high speed data, full duplex (FDX) data communications, primarily targeted at Internet access.

Contemporary digital CATV networks also support voice communications. The most common approach currently employs pulse code modulation (PCM) encoding and time division multiplexing (TDM), although some CATV networks use adaptive differential pulse code modulation (ADPCM) with silence suppression. These are precisely the same techniques traditionally employed in the public switched telephone network (PSTN). The emphasis, going forward, is on voice over Internet Protocol (VoIP). See also *ADPCM*, *amplifier*, *antenna*, *cable modem*, *coaxial cable*, *DOCSIS*, *downstream*, *drop*, *FDM*, *FDX*, *headend*, *HFC*, *Internet*, *LOS*, *microwave*, *MSO*, *NTSC*, *optical fiber*, *over the air*, *PAL*, *PCM*, *PSTN*, *RF*, *satellite*, *silence suppression*, *tap*, *TDM*, *transition band*, *trunk*, and *VSB*.

CB (Citizens Band) Radio Service Also known in Canada as *General Radio Service*. A private, two-way voice communication radio service for use by the general public on an unlicensed basis for personal and business activities. CB radio operates over a short-range of 1–5 miles in the 26.965–27.405 MHz band. In the United States, the Federal Communications Commission (FCC) regulates CB radio as one of a family of personal radio services. The FCC also considers Family Radio Service (FRS) to be in the general category of CB Radio Service. Similar services exist in Australia, Indonesia, Germany, the United Kingdom, and other countries. CB radio was invented by Al Gross, who also invented the cordless telephone, pager, and walkie talkie. See *Gross, Al*. See also *cordless telephone*, *FCC*, *FRS*, *pager*, *personal radio services*, *radio*, *shortwave*, and *walkie talkie*.

C band The portion of the microwave radio spectrum in the ranges of 3.7–4.2 GHz and 5.924–6.425 GHz, as specified by the ITU-T. Applications include satellite and terrestrial microwave systems. See also *electromagnetic spectrum*, *ITU-T*, *microwave*, and *satellite*.

C-Band (Conventional Band) The ITU-T standard optical transmission window in the wavelength range of 1,530–1,565 nm. See also *wavelength* and *window*.

CBDS (Connectionless Broadband Data Service) The name by which *Switched Multimegabit Data Service* (SMDS) is known in Europe. See *SMDS*.

CBR (Constant Bit Rate) In asynchronous transfer mode (ATM), a class of traffic that requires access to time slots at regular and precise intervals. Traffic parameters include peak cell rate (PCR) and cell delay variation tolerance (CDVT). Quality of Service (QoS) parameters include cell delay variation (CDV), cell transfer delay (CTD), and cell loss ratio (CLR). Real-time, uncompressed voice and video are examples of CBR traffic. ATM also defines available bit rate (ABR), non real-time Variable Bit Rate (nrt-VBR), real-time Variable Bit Rate (rt-VBR), unspecified bit rate (UBR), and variable bit rate (VBR) traffic classes. See also *ABR*, *ATM*, *compression*, *CDVT*, *CTD*, *nrt-VBR*, *PCR*, *QoS*, *rt-VBR*, *time slot*, *traffic*, *UBR*, and *VBR*.

CCIR (Comité consultatif international des Radiocommunications) The International Radio Consultative Committee, as translated from French. Now the International Telecommunication Union-Telecommunication Standardization Sector (ITU-T). See *ITU* and *ITU-T*.

CCITT (Comité consultatif international télégraphique et téléphonique) The International Telephone and Telegraph Consultative Committee, as translated from French. Now the International Telecommunication Union-Radiocommunication Sector (ITU-R). See *ITU* and *ITU-R*.

CCK (Complementary Code Keying) In IEEE 802.11b wireless LANs (WLANs), a process that organizes data bits into a series of complementary sequences that form 64 unique six-bit code words. Differential quaternary phase-shift keying (DQPSK) then modulates the code bits to form complex bit pairs known as chips, eight of which form a symbol, which is the smallest CCK transmission unit. In 802.11b WLANs operating at 5.5 and 11 Mbps, the symbol rate is 1.375 MSps and the chip rate is 11 Mchips per second. At 5.5 Mbps, 4 data bits are mapped into the 8 chips of each symbol. At 11 Mbps, 8 data bits are mapped into the 8 chips of each symbol. See also *802.11b*, *chip*, *DQPSK*, *encode*, *IEEE*, *modulate*, *symbol*, and *WLAN*.

CCLC (Common Carrier Line Charge) In the United States, a minutes-of-use charge billed by the local exchange carrier (LEC) to the interexchange carrier (IXC) to recover costs associated with providing subscriber access to the IXC over LEC local loop plant. The CCLC is one element of the Carrier Access Charge (CAC). See also *CAC* and *equal access*.

CCPA (Cable Communications Policy Act of 1984) See *Cable Communications Policy Act of 1984*.

CCS **1.** Centum Call Seconds. In traffic engineering, hundred (centum) call seconds. One hour contains 3600 call seconds (60 seconds times 60 minutes), or 36 CCS. **2.** Common Channel Signaling. A technique that uses a highly robust subnetwork, or separate network, to support the signaling and control and the network management requirements of the primary communications network. A CCS link is digital in nature, based on packet switching, and often in the form of a dedicated T1 or E-1 channel over a broadband optical fiber. A CCS subnetwork connects the various network switches and routers to centralized computer systems and databases in order to monitor and control the operations of an entire communications network, from end to end. Signaling System 7 (SS7), also known as Common Channel Signaling System 7 (CCS7) is a CCS network. See also *network management*, *packet switching*, *signaling and control*, and *SS7*.

CCS7 (Common Channel Signaling System 7) Also known as *SS7*. See *CCS* and *SS7*.

ccTLD (country code Top Level Domain) In the Internet Domain Name System (DNS), the rightmost portion of the address — the domain — identifies the type of entity owning or sponsoring the address. The several types of Top Level Domains (TLDs) include generic Top Level Domains (gTLDs) and country codes. If a country code is used in the address, the gTLD becomes a secondary domain. Country codes are neutral two-character codes established and maintained by the ISO 3166 Maintenance Agency. The government of each nation manages the use of country codes, which are appended to the standard address, and are necessary only if the target country domain differs from the country domain of origin. Example country codes include .am (Armenia), .ca (Canada), .fm (Federated States of Micronesia), .jp (Japan), .tv (Tuvalu), .us (United States), and .za (South Africa). In March 2005, ICANN approved .eu as a regional country code for the European Union. This first regional TLD is administered by EURid, a consortium of the ccTLD registry operators of Belgium, the Czech Republic, Sweden, and Italy. See also *DNS*, *domain*, *gTLD*, *Internet*, and *ISO*.

CCTV (Closed Circuit TeleVision) A private television system involving one or more cameras connected to one or more monitors for security, surveillance, law enforcement, and general purpose monitoring applications. Unlike public broadcast TV, CCTV is a closed system in intended for private use.

cd The symbol for candela. See *candela*.

CDDI (Copper Distributed Data Interface) Also known as *TPDDI* (Twisted Pair Distributed Data Interface). A standard for extending Fiber Distributed Data Interface (FDDI) connections to workstations via unshielded twisted pair cable (UTP) over distances of 100 meters or less. See also *FDDI* and *UTP*.

CDMA (Code Division Multiple Access) A multiplexing technique used in radio networks, CDMA is rooted in spread spectrum (SS) technology developed in the 1940s. Spread spectrum is a wideband radio transmission technology that spreads of the transmitted signal over a spectrum of radio frequencies that is much wider than that required to support the native narrowband transmission. Thereby, multiple transmissions can simultaneously use the entire system wideband, rather than just individual time slots or frequency channels. CDMA employs a variant known as frequency hopping spread spectrum (FHSS), which transmits short bursts of data over a range of frequency channels within the wideband carrier. Each transmission is assigned a 10-bit pseudorandom binary code sequence, which comprises a series of ones and zeros in a seemingly random pattern known to both the transmitter and receiver. The original code sequence is mathematically self-correlated to yield a code that stands out from all others, at least on average. The paired transmitters and receivers recognize their assigned and correlated code sequences, which look to all others as pseudorandom noise (PN). FHSS phase-modulates the carrier wave with a continuous string of PN code symbols, or chips. So, the chip rate is much faster than the bit rate. Thereby, the noise signal occurs with much greater frequency than the original data signal and spreads the signal energy over a much wider band. The transmitter and receiver hop from one frequency to another in a carefully choreographed hop sequence under the control of the centralized base station transceiver. Each transmission dwells on a particular frequency for a very short period of time (no more than 400 milliseconds for FCC-controlled applications), which may be less than the time interval required to transmit a single data packet, or symbol, or even a single bit. So, the chip rate can be faster than the bit rate. A large number of other transmissions also may share the same range of frequencies simultaneously, with each using a different hop sequence. The potential remains, however, for the overlapping of packets. The receiving device can distinguish each packet in a packet stream by reading the various codes prepended to the packet data transmissions, and treating competing signals as noise. See also *bandwidth*, *carrier*, *channel*, *chip*, *chip rate*, *DSSS*, *FHSS*, *frequency*, *hop sequence*, *modulation*, *narrowband*, *PN*, *signal*, *SS*, *symbol*, *transceiver*, and *wideband*.

CDMA Digital Cellular (Code Division Multiple Access Digital Cellular) A U.S. term for cellular radio systems based on EIA/TIA IS-95a, and also known as *cdmaOne and Personal Communications System* (PCS). See also *cellular radio* and *PCS*.

cdmaOne (code division multiple access One) A U.S. term for cellular radio systems based on EIA/TIA IS-95a, and also known as *CDMA Digital Cellular and Personal Communications System* (PCS). See also *cellular radio* and *PCS*.

CDMA2000 (Code Division Multiple Access 2000) A 3G cellular radio system based on earlier CDMA versions TIA/EIA IS-95a and IS-95b, CDMA2000 has been standardized by the EIA/TIA as IS-856 and approved by the ITU-R as part of the IMT-2000 family. The initial version, known as CDMA2000 1xRTT offers 2.5G capabilities within a single standard 1.25 MHz channel, effectively doubling the voice capacity of the predecessor 2G cdmaOne systems and offering theoretical data speeds up to 153 kbps through the use of quadrature phase shift keying (QPSK) modulation. An enhanced 3G version known as 1xEV-DO is a high data rate (HDR) version that employs 16-QPSK modulation in support of a peak data rate of 2.4 Mbps on the downlink and 153 kbps on the uplink. 3x, also known as IS-2000-A, is an enhancement that uses three cdmaOne carriers for total bandwidth of 3.75 MHz. See also *16-QPSK, 1XEV-DO, 1xRTT, 2G, 2.5G, 3G, 3x, bandwidth, carrier, CDMA, cdmaOne, cellular radio, downlink, IMT-2000, ITU-R, modulation, QPSK*, and *uplink*.

CDPD (Cellular Digital Packet Data) A method for data transmission over analog Advanced Mobile Phone System (AMPS) cellular radio networks. CDPD takes advantage of the natural idleness between disconnections and connections during the break and make process to transmit packetized data at rates up to 19.2 kbps, using either the Internet Protocol (IP) or the ISO Connectionless Network Protocol (CLNP). The frequency-agile CDPD modems search for available channels over which to send encrypted packets during the periods of channel idleness. Ultimately, CDPD proved too complex, too bandwidth-limited, and too expensive, especially as data-ready 2.5G and 3G networks made their appearances and analog cellular networks were phased out. The last of the CDPD networks were decommissioned around the end of 2005. See also *AMPS, analog, break and make, cellular radio, CLNP, encryption, IP, ISO*, and *modem*.

CDR (Call Detail Record). Synonymous with *station message detail record* (SMDR). See *SMDR*.

CDSU (Channel Data Service Unit) Also known as a *CSU/DSU*. A combined channel service unit (CSU) and data service unit (DSU). See also *CSU* and *DSU*.

CDV (Cell Delay Variation) In asynchronous transfer mode (ATM), the variation in actual cell transfer delay (CTD) and the expected transfer delay of an individual cell. CTV is a form of jitter, which can seriously degrade the quality of voice and video payloads. If cells arrive sooner than expected, the clumping can cause the peak cell rate (PCR) to be exceeded, and the excess cells to be discarded. If some cells arrive too late, the result may be gaps in the received information stream. Cell delay variation tolerance (CDVT) is a measurement of the maximum allowable CDV tolerance between two end stations. Peak-to-peak CDV is negotiated between the end station and the network, with peak-to-peak referring to the best case compared with the worst case, i.e., the difference between the earliest and the latest arriving cells on a connection. See also *ATM, cell, CTD, jitter, payload*, and *PCR*.

CDVT (Cell Delay Variation Tolerance) In asynchronous transfer mode (ATM), a measurement of the maximum allowable cell delay variation (CDV) between two end stations. See also *ATM, CDV*, and *cell*.

cell **1.** In asynchronous transfer mode (ATM) and Switched Multimegabit Data Service (SMDS) networks, a small protocol data unit (PDU) comprising 53 octets. The ATM cell consists of a header of 5 octets and payload of 48 octets. The small cell size offers the advantage of effectively supporting any type of data, including voice, fax, text, image, video, and multimedia, whether compressed or uncompressed and whether real-time or non-real-time. The fixed cell size offers the advantage of predictability, unlike the variable-length frames associated with services such as X.25, frame relay, and Ethernet, or the variable-length packets associated with the Internet Protocol (IP). This level of predictability yields much improved access control and congestion control. See also *ATM, compression, Ethernet, frame relay, header, IP, non-real-time, packet, payload, PDU, real-time, SMDS*, and *X.25*. **2.** In radio systems, a relatively small geographical area of coverage determined by factors such as frequency band, power level, and line of sight (LOS). The formal concept of radio cells dates back to 1947, when Bell Telephone engineers developed a radio system concept that included numerous, low-power transmit/receive antennas positioned throughout a metropolitan area. This sort of architecture served to increase the effective subscriber capacity of radio systems by breaking the area of coverage into cells, or smaller areas of coverage. Thereby, each frequency band could be reused in nonadjacent cells. Additionally, the cells can be split, or subdivided, further as the traffic demands of the system increase. In the context of radio telephony, including cellular telephony, cells can be characterized as falling into three broad descriptive categories, as illustrated in Figure C-2.

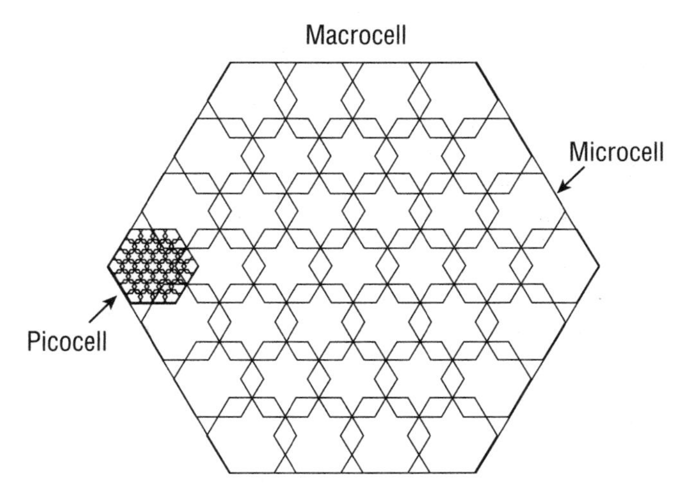

Figure C-2

- **Macrocells** cover relatively large areas, perhaps an entire metropolitan area of 50–60 miles in diameter.
- **Microcells** cover a smaller area of perhaps 6–8 miles in diameter.
- **Picocells** cover a very small area of perhaps only a few city blocks, a tunnel, walkway, or single floor of a parking garage.
- **Femtocells** are very small radio cells associated with cellular radio base stations located in homes or small offices.

See also *antenna*, *architecture*, *cellular radio*, *frequency*, *LOS*, and *power*.

cell delay variation (CDV) See *CDV*.

cell delay variation tolerance (CDVT) See *CDVT*.

cell error ratio (CER) See *CER*.

cell loss ratio (CLR) See *CLR*.

cell loss priority (CLP) See *CLP*.

cell misinsertion rate (CMR) See *CMR*.

cell tax Referring to the overhead penalty imposed by the small size of the asynchronous transfer mode (ATM) cell. Although the ATM cell imposes only 5 octets of overhead, there are only 48 octets of payload, so the overhead accounts for approximately 10 percent of the cell. Additionally, there is overhead embedded in the payload attributable to Convergence Sublayer and Data Link Layer and Network Layer headers. See also *ATM, Convergence Sublayer, Data Link Layer, header, Network Layer, octet, overhead,* and *payload.*

cell transfer delay (CTD) See *CTD.*

Cellular Digital Packet Data (CDPD) See *CDPD.*

cellular radio A radio system that positions numerous, low-power transmit/receive antennas throughout a metropolitan area, thereby dividing a large area of coverage, or macrocell, into smaller microcells, or even smaller picocells. Allocated spectrum is divided into a number of bands and further subdivided into a number of voice grade channels. The bands are distributed among the cells and each frequency band can be reused in nonadjacent cells. The basic concept of cellular radio dates to 1947, when numerous, low-power transmit/receive antennas were scattered throughout a metropolitan area to increase the effective subscriber capacity of specialized mobile radio (SMR) systems. The first prototype cellular radio system was developed by AT&T and Bell Labs in 1977. The first commercial cellular system was activated in Tokyo, Japan by Nippon Telephone and Telegraph (NTT) in 1979. The first commercial system in the United States was activated in Chicago in 1983.

The coverage area of each individual cell overlaps those of neighboring cells, with the cell diameter generally a minimum of about one mile and a maximum of about five miles, sensitive to factors such as topography and traffic density. Approximately in the center of each cell is a fixed antenna known as a base station (BS) that establishes and maintains connections with mobile stations (MSs). As a mobile station moves out of the effective range of one cell, the call switches from one base station to another through a process known as *handoff,* in order to maintain connectivity at acceptable signal strength. The handoff is controlled through a mobile telephone switching office (MTSO), which is the functional equivalent of a central office in the public switched telephone network (PSTN). The MTSOs generally are interconnected and also provide connections to the PSTN. Cellular radio initially was analog in nature, although contemporary systems are largely digital. Analog standards include Advanced Mobile Phone System (AMPS), Narrowband AMPS (N-AMPS), Total Access Communications System (TACS), and Nordic Mobile Telephone (NMT). Digital standards include Digital AMPS (D-AMPS), Global System for Global Communications (GSM), Personal Communications System (PCS), and Personal Digital Cellular (PDC). See also *AMPS, analog, antenna, band, BS, channel, D-AMPS, digital, GSM, handoff, macrocell, microcell, MS, MTSO, N-AMPS, NMT, PCS, PDC, picocell, PSTN, SMR, TACS,* and *voice grade.*

CELP (Code Excited Linear Prediction) A family of voice compression algorithms used in voice over frame relay (VoFR), voice over Internet Protocol (VoIP), and in digital cellular radio systems based on cdmaOne and PDC standards. A key element of CELP and its variants is the construction and maintenance of a codebook, which comprises binary descriptions of sets of voice samples. CELP gathers a set of 80 8-bit PCM voice samples, representing 10 milliseconds (ms) of a voice stream, in a buffer. CELP employs silence suppression to remove periods of silence and redundancy from the data set, normalizes the volume level, compares the resulting data set to a set of candidate shapes in the codebook, and selects the shapes that most closely match the actual data. (*Note:* CELP uses a stochastically overlapped codebook, with each entry sharing all but two samples with its neighboring entries.) CELP then transmits the index number of the selected code description and the average loudness level of the set of samples. Every 10 ms, the code is sent across the network in a block of 160 bits, yielding a data rate of 16 kbps, which is a compression ratio of 4:1, compared with toll quality PCM voice over the circuit-switched public switched telephone network (PSTN) at 64 kbps. At the receiving end of the transmission, the transmitted code is compared to the codebook, the PCM signal is reconstructed, and, eventually, the analog signal is reconstructed. See also *algorithm, analog, binary, buffer, cdmaOne, circuit switching, compression, CS-ACELP, LD-CELP, PCM,* μ*/4 QPSK, PDC, PSTN, silence suppression,* and *toll quality.*

centimeter (cm) See *cm.*

centralized attendant service (CAS) See *CAS.*

central office (CO) See *CO.*

central office exchange (COE) The international term for *central office* (CO). See *CO.*

central office exchange code See *central office prefix.*

central office prefix Synonymous with *central office exchange* (COE) *code.* As defined by the North American Numbering Plan (NANP), the three-digit part of a telephone number that identifies the central office and the associated geographic carrier serving area (CSA). The central office (CO) prefix specifies the dialing pattern NXX, with N indicating that only digits 2–9 are allowed, as 0 or 1 would confuse the network, and X indicating that any digit is allowed. The complete telephone number takes the form NXX.xxxx. See also *CSA* and *NANP.*

central processing unit (CPU) See *CPU.*

Centrex (CTX) A contraction of *Central exchange.* A service that provides PBX-like features from a central office (CO) via a special software load. Centrex service generally is provided from the CO via a dedicated voice grade local loop connected to each voice telephone set or other terminal located on the customer premises. Although it is unusual, Centrex service also can be provided by locating a remote CO partition or a remote line shelf on the customer premises, and connecting it to the main CO with one or more high capacity circuits. Centrex first appeared in the early 1960s in the United States and Canada, were it was de-emphasized it in the 1970s in favor of PBXs, but regained popularity in the mid-1980s. As a CO-based service, Centrex is primarily offered by the incumbent local exchange carriers (ILECs), also known as telephone companies. The emergence of IP Centrex in the mid-1990s has expanded Centrex offerings to competitive service providers providing IP-based services over the Internet. Proprietary voice terminals (P-phones) are required for ease of feature access, although most systems support generic sets, as well. P-Phones are switch-specific. See also *IP Centrex* and *PBX.*

centum call seconds (CCS) See *CCS.*

CEPT (Conférence Européenne des administrations des Postes et des Télécommunications) In English, the *European Conference of Postal and Telecommunications Administrations* (or Confederation of European Posts and Telecommunications) was established in 1959 by 19 countries to coordinate the activities of the monopoly postal and telecommunications administrations with respect to commercial, operational, regulatory, and technical standardization issues. In 1988, CEPT established the European Telecommunications Standards Institute (ETSI), which assumed responsibility for all matters of technical standardization. In 1992, the postal and telecommunications operators created their own organizations, Post Europe and the European Telecommunications Network Operators' Association (ETNO). CEPT now act largely as a forum for coordination and a force for building relationships among the member nations, now numbering nearly 50 throughout Europe. See also *ETNO* and *ETSI.*

CEPT-1 The original European name for *Cordless Telephony generation 1* (CT1), which was developed in Europe. See also *CEPT* and *CT1.*

CER (Cell Error Ratio) In asynchronous transfer mode (ATM), a dependability parameter expressed as the ratio of the number of errored cells to the total number of transmitted cells sent over a measurement interval. See also *ATM* and *cell.*

Cerberus Also known as *Kerberos.* In Greek mythology, the three-headed dog that guarded the gates of Hades. Cerberus also had a snake for a tail and a serpentine mane. See also *Kerberos.*

CERN (l'Conseil Européen pour la Recherche Nucléaire) Translates from French as *The European Council for Nuclear Research*, and is generally known as the *European Laboratory for Particle Physics* in Geneva, Switzerland. Tim Berners-Lee developed the World Wide Web (WWW) at CERN in 1989 as a collaborative tool for high-energy physicists. See Appendix A for contact information. See also *Berners-Lee, Tim* and *WWW.*

CERT (Computer Emergency Response Team) A group of experts at Carnegie-Mellon University's Software Engineering Institute responsible for overseeing security issues on the Internet. CERT was formed in 1988 by the Defense Advanced Research Projects Agency (DARPA) as an incident response team, specifically in response to the Internet Worm. See Appendix A for contact information. See also *DARPA*, *Internet*, and *US-CERT*.

certificate authority (CA) See *CA*.

Challenge Handshake Authentication Protocol (CHAP) See *CHAP*.

channel **1.** In communications terminology, a channel is a means of supporting a connection between two devices, such as a transmitter and a receiver, in support of a single communication. More specifically, a channel is a logical connection over a physical circuit in support of a single conversation. A physical circuit may support only a single channel in support of a single conversation. A multichannel circuit can support many channels through some form of multiplexing that supports many conversations. **2.** In data processing terminology, particularly IBM, a channel is a high-speed two-way connection between mainframe host computer and a peripheral. See also *multiplexer*.

channel-associated signaling (CAS) See *CAS*.

channel bank An early device designed to interface analog PBX or central office (CO) voice circuits to a DS-1 circuit. Channel banks perform two functions, in sequence. First, they multiplex up to 24 analog signals on a common pulse amplitude modulation (PAM) electrical bus. Second, they encode the individual PAM channels into a digital format, using pulse code modulation (PCM), for transmission over a DS-1 circuit. Channel banks place each voice conversation on a separate channel. A given channel can support a digital data transmission, rather than a voice transmission. For example, a data transmission at 9.6 kbps or 19.2 kbps originally occupied a full DS-0 channel of 64 kbps, just as does a 56 kbps data transmission or a digitized voice conversation. Later, sub-rate multiplexing allowed as many as 5 channels of 9600 bps to share a DS-0 channel. See also *bus*, *DS-0*, *DS-1*, *multiplexer*, *PAM*, and *PCM*.

channelized voice over DSL (CVoDSL) See *CVoDSL*.

channel service unit (CSU) See *CSU*.

channel width The bandwidth of a channel, defined as a range of frequencies. See also *bandwidth* and *channel*.

CHAP (Challenge Handshake Authentication Protocol) An authentication scheme used in Point-to-Point Protocol (PPP) remote access servers (RASs) to validate the identity of a remote user. CHAP employs a challenge-response mechanism that challenges the remote user with a random number. The user responds with a digest, which is an encrypted password based on the random number challenge. The RAS then decrypts the password using that same random number key to verify the identity of the remote user. This approach is much more secure that the predecessor Password Authentication Protocol (PAP). See also *authentication*, *encryption*, *PAP*, *password*, *PPP*, and *RAS*.

character A letter (e.g., a, A, b, and B), number (0, 1, 2, or 3), punctuation mark (e.g., ;, :, ', or "), or other symbol (e.g., @, #, $, or %) or control code (e.g., carriage return (CR), line feed (LF), acknowledgement (ACK), or negative acknowledgement (NAK)) that can be represented by a single data byte. See also *byte*.

character-oriented protocol See *byte-oriented protocol*.

chat room Referring to a teleconferencing channel that allows participants to converse in real-time text mode. Chat rooms typically are topical in nature, being dedicated to the discussion of specific subject matter.

check bit A bit added to an array of information bits at the originating device and used by the receiving device to check for errors that might have occurred during transmission. See also *parity bit*.

checksum (summation check) A manner of checking the integrity of a set of data by summing all of the bytes of data, or otherwise combining through a series of arithmetic or other logical operations. The originating device appends the result of the calculation to the data set prior to storing or transmitting the data. The device that retrieves the stored data or receives the transmitted data repeats the calculation and compares the two checksums. If the two do not match exactly, it is assumed that the dataset is errored. This process detects many, but not all, errors, and includes no mechanism for error correction. Checksum is employed in an error control mode known as recognition and retransmission. See also *cyclic checksum*, *error control*, *LRC*, and *recognition and retransmission*.

chip **1.** In computer hardware, a miniaturized integrated electronic circuit etched on a tiny wafer of silicon. See also *electronic*, *hardware*, *integrated circuit*, and *silicon*. **2.** In spread spectrum (SS) radio, a random pseudonoise (PN) code symbol. A sequence of chips, each of which has a much shorter duration than an information bit, are used to modulate the bits. The IEEE 802.11b standard for wireless LANs (WLANs), for example specifies Barker code at transmission rates of 1 Mbps and 2 Mbps, and complementary code keying (CCK) at 5.5 Mbps and 11 Mbps. Both Barker code and CCK code data bits into chips to form symbols prior to transmission. See also *Barker code*, *CCK*, *code*, *DSSS*, *FHSS*, *SS*, and *symbol*.

chip rate In spread spectrum (SS) radio, the rate at which chips, or code symbols, are used to modulate the data bits. The chip rate is expressed in chips per second (cps). See also *bit*, *chip*, *code*, *modulate*, *SS*, and *symbol*.

chromatic dispersion The spreading of a pulse in an optical fiber caused by differences in wave velocity in the medium, chromatic dispersion is measured in picoseconds of pulse spreading per nanometer of spectral width per kilometer of fiber length. Chromatic dispersion is the sum of waveguide dispersion and material dispersion. Material dispersion is caused by the fact that the speed of light in a medium is sensitive to the wavelength, i.e., the velocity of light in a medium depends on its wavelength. Waveguide dispersion is caused by the fact that a given wavelength travels at different speeds in the core and cladding of a single-mode fiber (SMF). Material dispersion, waveguide dispersion, and, therefore, chromatic dispersion, are issues in long haul fiber optic transmission systems (FOTS) employing single-mode fiber (SMF) of step-index construction. Multimode fiber (MMF) and graded-index fiber suffer so much from modal dispersion over short distances that material dispersion and chromatic dispersion never become factors. Chromatic dispersion is so named as different wavelengths of light in the visible spectrum display as different colors. See also *graded-index fiber*, *material dispersion*, *MMF*, *SMF*, *step-index fiber*, and *waveguide dispersion*.

chrominance Referring to color. Different video standards permit varying levels of color depth. A video image is more pleasing and lifelike when the range of color is as broad as possible.

C-HTML (Compact HyperText Markup Language) A simplified version of HTML developed for use in Web-enabled microbrowsers employed in Japanese i-Mode compatible cellular networks. C-HTML is similar to Wireless Markup Language (WML) used in WAP networks. See also *browser*, *cellular*, *HTML*, *i-Mode*, *WAP*, *Web*, and *WML*.

churning A security mechanism employed in broadband passive optical network (BPON). Downstream transmissions are encrypted through the use of a byte-oriented churn key exchanged between the optical line terminal (OLT) at the network headend and the optical network unit (ONU). The ONU or optical network terminal (ONT) at the customer premises generates the key and sends it to the OLT, which uses it to encrypt downstream transmissions. The key is changed at least once a second, hence the term churning. See also *BPON*, *byte*, *downstream*, *encryption*, *headend*, *OLT*, *ONT*, *ONU*, and *PON*.

CIC (Carrier Identification Code) In the United States, a four-digit code (XXXX) that identifies the interexchange carrier (IXC) the subscriber has pre-selected for interstate long distance calling purposes. The CIC allows the originating local exchange carrier (LEC) to identify and hand off the call to the correct interstate IXC. The CIC is part of the Carrier Access Code (CAC) that subscribers can use to dial around pre-subscribed carriers. See also *CAC*, *IXC*, and *LEC*.

CIDR (Classless Inter-Domain Routing) An address assignment and aggregation strategy that allows multiple Internet Protocol version 4 (IPv4) Class C subnet address blocks to be grouped under a single address. Thereby, all addresses in those blocks can be routed to the same host. CIDR uses shorthand to specify the subnet mask, which is written in dotted decimal notation, as are the IPv4 addresses. CIDR reduces the number of routes to be learned and, therefore, reduces the size and complexity of the routing tables that the Internet switches and routers must support. CIDR is specified in RFCs 1518 and 1519. See also *dotted decimal notation*, *host*, *Internet*, *IPv4*, *IPv4 address*, *router*, *subnet*, *subnet mask*, *supernetting*, and *switch*.

CIF (Common Intermediate Format) In the ITU-T H.320 umbrella standard for videoconferencing and multimedia communications over narrowband ISDN (N-ISDN) an optional video format that supports resolution of 352 × 288 pixels. See also *H.320*, *ITU-T*, *multimedia*, *N-ISDN*, *pixel*, *QCIF*, *resolution*, *video*, and *videoconference*.

CI II (Computer Inquiry II) The Second Computer Inquiry. See *Second Computer Inquiry*.

CIP (Classical IP over ATM) See *Classical IP over ATM*.

cipher A system of secret writing based on a set of rules for encrypting or encoding data, converting plain text into symbolic code that can be interpreted only with the key to the code. A cipher is used to secure information against unauthorized use.

CIR (Committed Information Rate) In packet networks, the average rate, measured in bit per second (bps), at which a public carrier network commits to support data transfer over a virtual circuit (VC) during normal operations. During periods of low network traffic volumes, the network may allow a given user or traffic flow to exceed the CIR by a rate known as the Excess Information Rate (EIR). During periods of congestion, the network restricts the user or traffic flow to a level below CIR according to some fairness algorithm that apportions available bandwidth to all. Frame relay and Resilient Packet Ring (RPR) employ CIR mechanisms. In frame relay, the calculation of the CIR is affected by the access rate and the maximum burst size (B_c). See also *access rate*, *bandwidth*, *carrier*, *EIR*, *frame relay*, *maximum burst size*, *packet*, *RPR*, and *VC*.

circuit **1.** A closed loop comprising a number of elements such as resistors, capacitors, transistors, and power sources connected together for the purpose of carrying an electrical current. **2.** An end-to-end communications path between a transmitter and a receiver. Circuit generally implies a logical connection, or session, over a physical medium. Circuit often is used interchangeably with path, link, line, and channel, although the usage can be specific to the underlying technology, the overall context, and other factors. Circuits can be characterized in many ways. For example, a circuit can be either two-wire or four-wire in nature, for either access or transport applications. A circuit can be simplex, half-duplex, or full-duplex in nature. A circuit can be dedicated, switched, or virtual in nature. A circuit sometimes is referred to in the vernacular as a pipe, as in a broadband pipe. See also *channel*, *four-wire circuit*, *line*, *link*, *path*, *two-wire circuit*, and *virtual circuit*.

circuit switch A circuit switch establishes connections between links, on demand and as available, in order to establish an end-to-end circuit between devices. The connections are temporary, continuous, and exclusive in nature. The connections are temporary as they are established and maintained only for the duration of the logical session, or call. They are continuous as they provide a specific amount of bandwidth, or capacity, continuously for the duration of the call. They are exclusive as the connection and the associated bandwidth are committed to only that call, i.e., are not shared with other transmissions. Circuit switches were developed for uncompressed, realtime voice communications, but will support any type of information transfer. Common examples of circuit switches include Private Branch Exchanges (PBXs) and Central Office Exchanges (COs or COEs). See also *circuit*, *CO*, *cordboard*, *ECC*, *packet switch*, *PBX*, *switch*, *switchboard*, *SxS*, and *Xbar*.

circuit-switched data (CSD) See *CSD*.

circuit-switched network Also known as *switched circuit network* (CSN). A network based on circuit switching, rather than packet switching. The traditional public switched telephone network (PSTN) is a circuit-switched network, although it rapidly is transitioning to packet switching based on the Internet Protocol (IP), the fundamental protocol of the Internet. See also *circuit switch*, *Internet*, *IP*, *packet switch*, and *PSTN*.

circuit terminating equipment Also known as *data circuit terminating equipment* or *data communications equipment* (DCE). See *DCE*.

Citizens Band (CB) See *CB*.

cladding **1.** A metal coating bonded onto another metal for protective purposes or to alter its conductive properties. Copper-clad aluminum coaxial cables sometimes are used in long-haul CATV distribution networks, for example. Coax cables sometimes are tinned or silvered, as well. See also *coaxial cable*. **2.** The layer or layers of glass that surround the core of a glass optical fiber (GOF). The cladding serves as a waveguide that variously reflects or steers the light signal back into the core, which is the central and primary light conducting portion of a glass fiber. Step-index fiber is characterized by a sharp decrease in the index of refraction (IOR) between the core and cladding, i.e., the cladding is sharply clearer or purer than the core material. This sharp step of approximately one percent in IOR at the core/cladding interface causes any errant light rays to reflect back into the core in a phenomenon known as total internal reflection. Graded-index fiber is characterized by a gradual decrease in the refractive index of the cladding through a great many layers of successively clearer glass. This approach causes the errant light rays to gradually gain in velocity and bend, or refract, back towards the core. Step-index construction is used largely in single-mode fiber (SMF) and graded-index construction in multimode fiber (MMF). See also *core*, *GOF*, *laser diode*, *LED*, *MMF*, *reflection*, *refraction*, *SMF*, and *total internal reflection*.

Clarke, Sir Arthur C. (1917–) The inventor of the communications satellite. Clarke was a physicist at the British Interplanetary Society in February 1945 when he published the concept in an article entitled "Extra-Terrestrial Relays: Can Rocket Stations Give World-wide Radio Coverage" in *Wireless World* magazine. Clarke is better known in popular culture as the author of "The Sentinel," a short story that served as the basis for the movie "2001: A Space Odyssey," which he co-wrote with Stanley Kubrick. In a subsequent article entitled "A Short Pre-History of Comsats, Or: How I Lost a Billion Dollars in My Spare Time," Clarke lamented his failure to patent the concept. (*Note:* Clarke reportedly received the princely equivalent of US$40.00 for the article.) While Clarke did not achieve great wealth directly from his invention, he was knighted on May 26, 2000.

Clarke orbit So named in honor of Arthur C. Clarke, the inventor of the communications satellite, Clarke orbit is synonymous with *geosynchronous earth orbit* (GEO). See also *GEO*.

CLASS (Custom Local Access Signaling Services) Also known as *custom calling services*. A group of network-based services offered by local exchange carriers (LECs) through the public switched telephone network (PSTN). CLASS services include anonymous call rejection, automatic callback, calling name delivery, calling number delivery, calling number blocking, call waiting, distinctive ringing, call trace, ring again, selective call acceptance, selective call forwarding, selective call rejection, and selective call screening. For more detail, see also the preceding terms.

Class 1 In the traditional public switched telephone network (PSTN) hierarchy, a regional toll center. The 10 Class 1 offices that existed prior to the breakup of the Bell System served to interconnect Class 2 offices, or sectional toll centers. At the end of 2001, seven remained in the United States, and two in Canada. See Figure C-3. See also *Bell System*, *Class 2*, and *PSTN*.

Class 2 In the traditional public switched telephone network (PSTN) hierarchy, a sectional toll center. Class 2 offices served to interconnect Class 3 offices, or primary toll centers, largely for interstate calling within a geographic region of North America. Approximately 67 sectional toll centers existed prior to the breakup of the Bell System. It is doubtful that any remain. See also *Bell System*, *Class 3*, and *PSTN*.

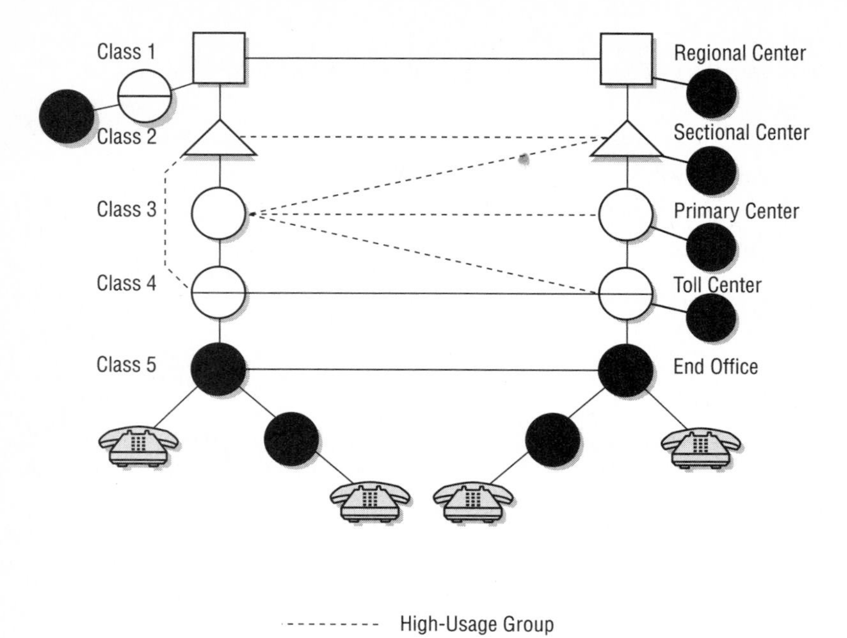

Figure C-3

Class 3 In the traditional public switched telephone network (PSTN) hierarchy, a primary toll center. Class 3 offices served to connect Class 4 offices for intrastate toll calling, and to interconnect independent telcos and the Bell operating companies (BOCs). Approximately 200 Class 3 offices existed prior to the breakup of the Bell System. It is doubtful that any remain. See also *Bell System*, *BOC*, *Class 4*, *independent telephone company*, and *PSTN*.

Class 4 In the traditional public switched telephone network (PSTN) hierarchy, a tandem toll center. Class 4 offices serve to interconnect Class 5 offices, or central offices (COs) not interconnected directly. As the lowest class of toll center, Class 4 offices interconnect within a relatively local toll network and provide access to higher-order toll centers. In the contemporary PSTN, a Class 4 office commonly serves as a Class 5 office, as well, with the separate functions provided through logical and physical partitioning within the switch. See also *Class 4/5*, *Class 5*, *CO*, and *PSTN*.

Class 4/5 In the public switched telephone network (PSTN) hierarchy, a Class 4 office that also serves as a Class 5 office. The separate functions are provided through logical and physical partitioning within the switch. See also *Class 4*, *Class 5*, and *PSTN*.

Class 5 A local central office (CO), which is the lowest class switching office in the public switched telephone network (PSTN) hierarchy. A Class 5 office is the point at which subscriber local loops and network trunks terminate and interconnect. Synonymous with *central office* (CO), *central office exchange* (COE), and *end office*. See also *CO*, *end office*, and *PSTN*.

Class A cable The ISO/IEC 11801 standard for copper wire cable rated at up to 100 kHz. See also *ISO/IEC 11801*.

Class A ATM traffic In ITU-T standards for asynchronous transfer mode (ATM), a class of traffic supported by ATM Adaptation Layer 1 (AAL1). Such traffic is connection-oriented constant bit rate (CBR) traffic that must be timed between the source and the sink. Class A traffic is stream-oriented and intolerant of latency. See also *AAL, AAL1, ATM, CBR, connection-oriented, ITU-T, latency, sink, source,* and *stream-oriented.*

Class A IP address In Internet Protocol version 4 (IPv4), a unicast address in the range 1.0.0.1 to 126.255.255.254. Class A addresses are identified by a beginning 0 bit. The next 7 bits identify the specific network, with 128 (2^7) theoretically possible. As addresses 0 and 127 are reserved, 126 network addresses remain available for assignment. As the specific host on the network is identified in bits 8 through 31, as many as 16,777,214 (2^{24} -2) hosts can be supported per network. Class A addresses are intended for very large networks supporting a great number of host computers. See also *binary notation, bit, host, IPv4, network,* and *unicast.*

Class B cable The ISO/IEC 11801 standard for copper wire cable rated at up to 1 MHz. See also *ISO/IEC 11801.*

Class B ATM traffic In ITU-T standards for asynchronous transfer mode (ATM), a class of traffic supported by ATM Adaptation Layer 2 (AAL2). Such traffic is connection-oriented, real-time variable bit rate (rt-VBR), isochronous traffic timed between the source and the sink. Compressed audio and video are Class B. See also *AAL, AAL2, ATM, cell, compression, connection-oriented, header, isochronous, ITU-T, rt-VBR, sink,* and *source.*

Class B IP address In Internet Protocol version 4 (IPv4), a unicast address in the range 128.0.0.1 to 191.255.255.254. Class B addresses are identified by a beginning set of 2 bits in a 10 sequence. The next 14 bits identify the specific network, with 16,384 (2^{14}) theoretically possible. As addresses 0 and 16,383 are reserved, 16,382 network addresses remain available for assignment. As the specific host on the network is identified in bits 16 through 31, as many as 65,634 (2^{16} –2) hosts can be supported per network. See also *binary notation, bit, host, IPv4, network,* and *unicast.*

Class C cable The ISO/IEC 11801 standard for copper wire cable rated at up to 16 MHz. See also *ISO/IEC 11801.*

Class C ATM traffic In ITU-T standards for asynchronous transfer mode (ATM), a class of traffic supported by ATM Adaptation Layer 3/4 (AAL3/4). Such traffic is connection-oriented variable bit rate (VBR) traffic with no timing relationship between the source and the sink. Examples of Class C traffic include X.25 and frame relay. See also *AAL, AAL3/4, ATM, connection-oriented, frame relay, ITU-T, sink, source, VBR,* and *X.25.*

Class C IP address In Internet Protocol version 4 (IPv4), a unicast address in the range 192.0.0.1 to 233.255.255.254. Class C addresses are identified by a beginning set of 3 bits in the binary sequence 110. The next 21 bits identify the network, with 2,097,154 networks (2^{21}) theoretically possible. As addresses 0 and 2,097,151 are reserved, 2,097,152 network addresses remain available for assignment. As the specific host on the network is identified in bits 24 through 31, as many as 254 (2^8 – 2) hosts can be supported per network. (Host addresses 0 and 255 are reserved.) Class C addresses are reserved for smaller networks such as LANs. The vast majority of end users make use of Class C addresses. See also *binary notation, bit, host, IPv4, LAN, network,* and *unicast.*

Class D cable The ISO/IEC 11801 standard for copper wire cable rated at up to 100 MHz. See also *ISO/IEC 11801.*

Class D ATM traffic In ITU-T standards for asynchronous transfer mode (ATM), a class of traffic supported by ATM Adaptation Layer 3/4 (AAL3/4). Such traffic is connectionless variable bit rate (VBR) traffic that is sensitive to loss, but not highly sensitive to delay. Examples of Class D traffic include LAN and SMDS. With the demise of SMDS, Class D has all but disappeared in favor of AAL5 and Class C. See also *AAL, AAL3/4, AAL5, ATM, Class C, connectionless, ITU-T, LAN, sink, SMDS, source,* and *VBR.*

Class D IP address In Internet Protocol version 4 (IPv4), an address beginning with a binary 1110. Class D addresses are reserved for multicast applications. See also *binary notation*, *IPv4*, and *multicast*.

Class E cable The ISO/IEC 11801 standard for copper wire cable rated at up to 250 MHz. See also *ISO/IEC 11801.*

Class E IP address In Internet Protocol version 4 (IPv4), an address beginning with a binary 1111. Class E addresses are reserved for future use. See also *binary notation* and *IPv4.*

Class F cable The ISO/IEC 11801 standard for copper wire cable rated at up to 600 MHz. See also *ISO/IEC 11801.*

Classical IP over ATM (Classical Internet Protocol over Asynchronous Transfer Mode) Also informally known as *CIP*. An IETF specification for transmitting IP datagrams and ATM Address Resolution Protocol (ATMARP) requests and replies over ATM Adaptation Layer 5 (AAL5) where ATM is configured as to include multiple logical IP subnetworks (LISs).In Classical IP over ATM, ATM replaces a legacy local area network (LAN) such as Ethernet. The term classical derives from the fact that IP packets between two logical subnets on the same ATM network must go through an intervening router. Subsequently, alternative methods were defined for transmitting IP datagrams over ATM networks, with those methods including LAN Emulation (LANE), and Multiprotocol over ATM (MPOA). Classical IP over ATM was originally defined in IETF RFC 1577 (1994) and most recently in IETF RFC 2225 (1998). See also *AAL5*, *ARP*, *ATM*, *datagram*, *IETF*, *IP*, *LANE*, *MPOA*, and *subnet*.

Classless Inter-Domain Routing (CIDR) See *CIDR*.

class of service (CoS) See *CoS*.

Class X ATM traffic In ITU-T standards for asynchronous transfer mode (ATM), a class of traffic supported by ATM Adaptation Layer 5 (AAL5). Class X traffic is variable bit rate (VBR) and specifically either unspecified bit rate (UBR) or available bit rate (ABR) in nature. Class X traffic can be characterized as either connection-oriented or connectionless traffic with no timing relationship between the source and the sink. Class X traffic examples include LAN Emulation (LANE) and Internet Protocol (IP). See also *AAL*, *AAL5*, *ABR*, *ATM*, *connectionless*, *connection-oriented*, *IP*, *ITU-T*, *LANE*, *sink*, *source*, *UBR*, and *VBR*.

clear channel A DS-0 channel of 64 kbps, all of which can be used for user payload, network management, or other applications, as the requisite signaling and control functions for the circuit are accomplished out-of-band. As the signaling and control functions are performed in separate channels or even in a separate network designed specifically for that purpose, there is no bit robbing or otherwise intrusive technique that consumes channel capacity intended for payload. As examples, E-carrier and ISDN support clear channel communications. See also *bit robbing*, *E-carrier*, *DS-0*, *ISDN*, *out-of-band signaling and control*, and *payload*.

clear to send (CTS) See *CTS*.

CLEC (Competitive Local Exchange Carrier) A local exchange carrier (LEC), i.e., carrier providing local telephone service, in competition with the incumbent local exchange carrier (ILEC). See also *carrier*, *ILEC*, and *LEC*.

CLID (Calling Line IDentification) **1.** A network-based CLASS service of the public switched telephone network (PSTN). The feature delivers the calling number to the called line, where it can appear on a telephone set equipped with a display or on a peripheral display unit. In a call center environment, the calling number also can be linked to a database and used to access a customer profile in order to route the incoming call through an automatic call distributor (ACD) to an agent who can provide the caller with improved customer service. Calling number blocking is a feature that allows the calling party to block the transmission of CLID information on a permanent basis, or on an ad hoc basis by dialing a code prior to dialing the destination telephone number. Synonymous with *caller ID*. See also *ACD*, *ANI*, *call center*, *CLASS*, *LEC*, *PSTN*, and *screen pop*. **2.** A voice telephone system feature that supports the CLID network service and offers a similar capability for station-to-station PBX calls. See also *CLASS*.

client In a client/server architecture, a complete, standalone computer that optimizes the user interface, relying on servers to handle the more mundane tasks associated with application and file storage, network administration, security, and other critical functions. See also *architecture*, *client/server*, and *server*.

client mesh See *pure mesh*.

client/server A network architecture that distributes intelligence and responsibilities at several levels, with some machines designated as servers to serve the needs of client machines. A server can be a mainframe, minicomputer, or personal computer that operates in a time-sharing mode to provide for the needs of many clients. Client machines are complete, standalone computers that optimize the user interface, relying on servers to handle the more mundane tasks associated with application and file storage, network administration, security, and other critical functions. See also *peer-to-peer*.

Clipper Chip An integrated circuit that uses the Skipjack voice encryption algorithm developed by the United States National Security Agency (NSA) for the National Institute of Science and Technology (NIST). Skipjack is a block coding algorithm that encrypts 64-bit data blocks with an 80-bit key. Data encrypted by the Skipjack algorithm can be provided not only to the intended recipient through the use of a key, but also by the U.S. government through the use of a back door into a Law Enforcement Access Field (LEAF). The Clipper Chip is manufactured by the U.S. government, which has tried unsuccessfully to make it, and similar technologies, mandatory for voice encryption in the United States. Privacy advocates feared that government authorities would abuse the back door. Law enforcement authorities fear that the widespread use of other voice encryption technologies will make it impossible to place legal wiretaps. See also *algorithm*, *back door*, *encryption*, *integrated circuit*, and *wiretap*.

CLNP (ConnectionLess Network Protocol) A Network Layer datagram protocol from the International Organization for Standardization (ISO) for use over OSI (Open Systems Integration) networks and specified in ISO 8473. CLNP is very similar to Internet Protocol (IP). The datagram size is the same as IP, and there are similar mechanisms for fragmentation, error control, and lifetime control. CLNP, however, has an address space of 20 octets compared the IPv4 address space of only 4 octets. OSI networks have not been well accepted, however, and the OSI protocol stack has been relegated to the status of OSI Reference Model. See also *datagram*, *error control*, *fragmentation*, *IP*, *ISO*, *lifetime control*, *Network Layer*, *OSI*, *OSI Reference Model*, *protocol*, and *protocol stack*.

clocking pulse Periodic signals generated by a timing source for purposes of synchronizing the flow of data within a computer or between computers across a circuit. See also *synchronous transmission*.

closed circuit television (CCTV) See *CCTV*.

closed-loop algorithm In frame relay, a congestion control mechanism that prevents the frame relay network device (FRND) from accepting incoming frames unless there is an extremely high probability of the network's being able to deliver them without discard. A closed-loop algorithm fairly allocates backbone bandwidth among all the permanent virtual circuits (PVCs) configured on a particular trunk, and in proportion to the Committed Information Rate (CIR) of each PVC. See also *backbone*, *bandwidth*, *CIR*, *congestion*, *frame relay*, *FRND*, *PVC*, and *trunk*.

closed user group (CUG) See *CUG*.

cloud A wide area network (WAN) commonly is depicted as a cloud, which serves to obscure its complex inner workings from view. Data just pops in on one side of the cloud and pops out on the other side, so to speak.

CLP (Cell Loss Priority) In asynchronous transfer mode (ATM), one bit in the cell header that identifies the priority level of the cell to determine the eligibility of that cell for discard in the event of network congestion. Applications such as LAN-to-LAN traffic and e-mail are tolerant of loss. Applications such as real-time voice and video are highly intolerant of loss. See also *ATM*, *cell*, *congestion*, *e-mail*, *header*, *LAN*, *real-time*, *traffic*, *video*, and *voice*.

CLR (Cell Loss Ratio) In asynchronous transfer mode (ATM), a dependability parameter expressed as the ratio of the number of lost cells to the number of transmitted cells. Cell loss can occur for reasons that include misdirection of cells by a switch, a congestion problem causing a discard in consideration of buffer capacity, a station exceeding its peak cell rate (PCR) resulting in cell discard, or a cell that exceeds the maximum cell transfer delay (CTD) and arrives too late for processing. CLR applies to all service categories except unspecified bit rate (UBR). See also *ATM*, *buffer*, *cell*, *congestion*, *CTD*, *PCR*, and *UBR*.

cm (centimeter) One one-hundredth (10^{-2}, or 1/100) of a meter. See also *meter*.

CM (Cable Modem) See *cable modem*.

CMR (Cell Misinsertion Rate) In asynchronous transfer mode (ATM), a dependability parameter expressed as the number of cells received over a time interval at a destination endpoint that were not transmitted originally by the source endpoint of the virtual circuit (VC). CMR is expressed as a rate, rather than as a ratio, because the number of misinserted cells is beyond the control of the originating and destination endpoints. Although the header checksum is designed to prevent misinsertion, CMR can result from the corruption of a cell header, which would cause a cell to be misinserted into the cell stream of another source-destination pair of end points. See also *ATM*, *cell*, *checksum*, *endpoint*, *header*, and *VC*.

CMTS (Cable Modem Termination System) The head-end portion of a CATV network designed to support high speed data, as described in the Data over Cable Service Interface Specification (DOCSIS). Matching DOCSIS cable modems (CMs) in the CMTS and the customer premises support high speed, full duplex (FDX) data communications over a hybrid fiber/coax (HFC) system. The CMTS supports a packet data connection to an IEEE 802.3 10/100-Mbps Ethernet port on a router. In terms of the OSI Reference Model, the system runs the Internet Protocol (IP) at the Network Layer in Ethernet frames at the Data Link Layer. Associated with the CMTS are various servers for security, address translation, data caching, and video caching. A CMTS can support as many as 2,000 cable modem users on a single 6-MHz channel (8 MHz in Europe), with issues of congestion for shared bandwidth becoming more severe as the number of active users increases. The modem on the customer premises is in the form of a set-top box, which supports traditional coax connections to multiple TV sets and a 10/100BaseT Ethernet connection to a PC or to a hub serving multiple PCs. See also *10BaseT*, *100BaseT*, *802.3*, *bandwidth*, *caching*, *coaxial cable*, *Data Link Layer*, *DOCSIS*, *Ethernet*, *HFC*, *IP*, *Network Layer*, *optical fiber*, *OSI Reference Model*, *server*, and *set-top box*.

CO (Central Office) **1.** A local telephone company office that provides a central point for the termination of lines and trunks, and where they can be interconnected, i.e., connections can be exchanged. An integral part of the public switched telephone network (PSTN), a CO traditionally houses one or more voice-optimized circuit switches to interconnect subscriber lines within a local area known as the carrier serving area (CSA) and to connect subscriber local loops to network trunks. A contemporary CO may also house a variety of voice and data switches, multiplexers, concentrators, and so on. Synonymous with *central office exchange* (COE), *Class 5 office*, *end office*, and *local exchange*. See also *CSA* and *PSTN*. **2.** The CO switch, rather than the building that houses it. Synonymous with *Class 5 switch*, *edge switch*, *end office*, and *local exchange*.

Coarse Wavelength Division Multiplexing (CWDM) See *CWDM*.

coax (coaxial cable) See *coaxial cable*.

coaxial cable (coax) A very robust shielded copper cable. All components are symmetrically arranged around a common axis, or center point, hence the term coaxial. A coax cable has a relatively thick center conductor (in comparison to a twisted-pair conductor), generally solid, although stranded wire sometimes is used in applications requiring greater flex strength. The metal used for the inner conductor may be bare copper, silvered copper, tinned copper, copper-clad aluminum or copper-covered steel. A layer of dielectric material, either foam or solid, generally surrounds the inner conductor, serving to separate it from the single outer conductor, or sometimes two outer conductors. The conductor(s) comprising the outer shield generally consists of a solid aluminum foil, although a braided or stranded metal screen of aluminum, bare copper, silvered copper, copper-clad aluminum, or tinned copper may be used. The entire cable is then protected by a sheath of dielectric material such as PVC or Teflon®. Coaxial cable types are identified by RG (Radio Guide) number. Invented by AT&T Bell Telephone Laboratories in 1934, the first coaxial cables were hollow tubes about one-quarter inch in diameter. A single copper wire ran down the center of each pipe and was held in place by insulating discs. The pipes were in pairs, with one for transmission in each direction. The first coaxial system was placed into service in New York City in 1936. See also *flex strength*, *ScTP*, *shield*, *STP*, and *twinaxial cable*. See also *RG* for a listing of example coaxial cable types.

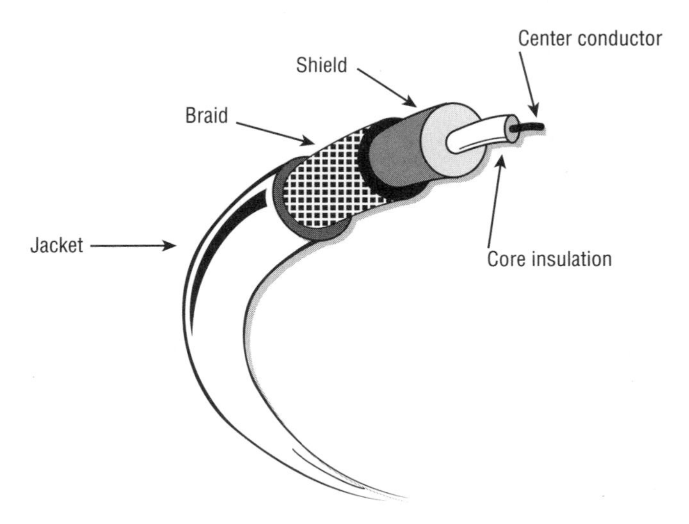

Figure C-4

cobweb From the Middle English *coppeweb*, meaning *spider web*. *Coppe* is an abbreviation of the Old English *attercoppe*, meaning *poison head*. In contemporary usage, an abandoned spider web. This definition has absolutely nothing to do with telecommunications, except for the fact that I noticed a cobweb in my office while I was writing this book, and I was compelled to research the term. (It was one more diversionary tactic of mine. The alternatives were load coil, SMDS, and ytterbium.) See also *World Wide Web (WWW)*.

CO Centrex (Central Office Centrex) Centrex service provisioned from a CO, rather than from a premises-based switch. CO Centrex is the most typical method for delivering Centrex service, as few organizations are large enough to justify a CO switch on premises. See also *Centrex*.

COCOT (Customer-Owned Coin-Operated Telephone) A payphone that is owned by the end user who owns or occupies the premises in which it is located. See also *pay telephone*.

code **1.** Program instructions, i.e., instructions that comprise programs that computers execute in order to perform processes. Source code comprises human readable instructions written in a programming language. Source code is compiled or converted into machine code, i.e., machine language, which is a set of numerical instructions that a computer can read and execute. **2.** A set of rules or conventions that clearly specifies the manner for representing data in symbolic form. A code that intentionally conceals the information for security purposes is known as a cipher. **3.** A system of symbols that provides information about something, like a postal code, a telephone country code or area code, or an Internet Protocol (IP) country code. **4.** A system by which some combination of bits is used within a computer and between computers to represent a character or symbol, such as a letter, number, punctuation mark, or control character. See also *code set*.

codec (coder/decoder) A device that interfaces an analog device to a digital circuit or channel. Codecs operate in balanced and symmetrical pairs, with one at each end of the communications circuit and with both having the same capabilities, at least at a minimum level. On the transmit side of the connection, a codec accepts an incoming analog signal, encodes it (i.e., converts it into digital form), and places it on a digital circuit. On the receive side of the connection, a codec with matching capabilities accepts the digital signal and decodes it to (i.e., recreates) an approximation of the original analog signal. Many codecs are capable of operating in full duplex (FDX), simultaneously encoding signals as they transmit them and decoding signals as they receive them. See also *analog*, *channel*, *circuit*, *digital*, *encode*, and *FDX*.

code division multiple access (CDMA) See *CDMA*.

Code Division Multiple Access 2000 (CDMA2000) See *CDMA2000*.

code excited linear prediction (CELP) See *CELP*.

coded orthogonal frequency division multiplexing (COFDM) See *COFDM*.

code set Also known as *coding scheme*. A set of binary codes used by a computer system to create, store, and exchange information. A code set establishes a specific combination of 1s and 0s of a specific total length in order to represent a character, such as a letter, number, punctuation mark, or control character (e.g., carriage return, line feed, space, blank, and delete). Contemporary standard coding schemes include Baudot, ASCII, EBCDIC, and Unicode. See also *ASCII*, *Baudot code*, *EBCDIC*, and *Unicode*.

coding scheme See *code set*.

COE (Central Office Exchange) Synonymous with *central office* (CO). See *CO*.

COFDM (Coded Orthogonal Frequency Division Multiplexing) A signal modulation scheme that sends a stream of data symbols in a massively parallel fashion, with multiple independent subcarriers, that is, small slices of spectrum within the designated carrier frequency band. Each subcarrier carries a small of the total data stream. In the case of 802.11a, aka Wi-Fi5, for example, each carrier channel is 20 MHz wide, and is subdivided into 52 subcarrier channels, each of which is approximately 300 kHz wide. See also *802.11a*, *carrier*, *channel*, *DMT*, *frequency band*, *modulation*, *orthogonal*, *signal*, *symbol*, and *Wi-Fi*.

cognitive radio A radio that is able to acquire knowledge of the condition of the spectrum to which it has access, determine which channels and services are in use and the intensity of the usage patterns, and avoid those channels in order to optimize performance. Cognitive radio is especially advantageous in applications using unlicensed spectrum such as the ISM band, as that spectrum is often and unpredictably subject to congestion. See also *channel*, *ISM*, *radio*, and *spectrum*.

coherence From the Latin *co-* (together) and *haerere* (to stick), translating as *sticking together*. **1.** The property of a set of electromagnetic waves, consistently similar in terms of a feature such as polarization or phase. Signals consistently synchronized in phase are characterized by oscillations that maintain a fixed relationship, with the sine waves rising and falling in unison. See also *phase*, *polarization*, and *sine wave*. **2.** The property of a light source that fires within a narrow range of wavelengths, ideally only one, so that all photons act identically. See also *wavelength* and *window*.

collaborative computing An interactive multimedia conferencing application that enables multiple parties to collaborate on textual and graphic documents. Through special software, each party to the call can contribute to such documents, working together with the other parties. During such a collaborative session, the original text document is saved, while each party contributes changes that are identifiable as such, by contributor. When the parties agree to the collaborative edits and enhancements, the entire text file is refreshed and saved. Similarly, a design or a concept can be developed graphically and on a collaborative basis through whiteboarding, much as the parties would do on a physical whiteboard in a face-to-face meeting. Typically, each party to the conference has access to a special whiteboard pad and stylus, which is used to draw. Each party can modify the initial drawing, with each individual's contribution identified by separate color. Again, and once the group has agreed on the final graphic rendition, the graphic is saved and all screens are refreshed.

collimation The process by which a beam of radiant electromagnetic energy is lined up to minimize divergence or convergence. Ideally, a collimated beam is a bundle of parallel rays perfectly lined up along an optical line-of-sight (LOS) between a transmitter and receiver, perhaps through, and in perfect parallel with, a waveguide. In a fiber optic transmission system (FOTS), a perfectly collimated optical beam would be perfectly lined up with the fiber core. See also *LOS*.

collision domain A physical region of a local area network (LAN) in which data collisions can occur. Collisions are most likely in LANs, such as Ethernets, that use non-deterministic medium access control (MAC) protocols such as carrier sense multiple access with collision detection (CSMA/CD). See also *CSMA/CD*, *Ethernet*, *LAN*, *MAC*, and *non-deterministic*.

collocation **1.** A physical arrangement in which things are placed close together. **2.** In telecommunications, referring to the placement of the equipment of a competitive local exchange carrier (CLEC) or Internet service provider (ISP) in the incumbent LEC's (ILEC's) central office (CO). A collocation arrangement generally requires that the ILEC provide a separately area, such as a cage, for the CLEC or ISP to secure its termination equipment, switches, routers, and other equipment. See also *CLEC*, *CO*, *ILEC*, and *ISP*.

colocation See *collocation*.

color sampling See *color-space conversion*.

color-space conversion Also known as *color sampling*. A step in the video compression process that involves the reduction of color information in the image. As the human eye is not highly sensitive to slight color variations, the impact is not noticeable. Black and white are prioritized, as the human eye is very sensitive at that level to differences in total brightness. See also *compression*.

.com (commercial) Pronounced *dot com*. The Internet generic Top Level Domain (gTLD) reserved exclusively for commercial organizations. This is an unsponsored domain. See also *gTLD*, *Internet*, and *unsponsored domain*.

combination trunk A PBX trunk that supports both incoming and outgoing calls. See also *PBX* and *trunk*.

comfort noise See *white noise*.

Committed Burst Size (B_c) In frame relay, the maximum amount of data that the carrier agrees to handle without discard under normal conditions. The B_c and access rate affect the calculation of the Committed Information Rate (CIR) for a virtual circuit (VC). See also *access rate*, *CIR*, *frame relay*, and *VC*.

Committed Information Rate (CIR) See *CIR*.

common air interface (CAI) See *air interface*.

common battery A battery that serves as a single source of electrical energy, in the form of direct current (DC), for more than one circuit and perhaps for more than one connected device. The common battery may supply energy for an entire system, such as a central office (CO) or PBX, and the circuits connecting that system to terminal devices. In a telephone company application, the common battery provides loop current for the CO and a great many local loops. In many telecommunications applications, the common battery is 48 volts (V). See also *battery*, *circuit*, *DC*, *electricity*, *energy*, *local battery*, *local loop*, *loop current*, and *V*.

common carrier **1.** A company transporting goods, persons, or messages for a fee, at uniform rates available to the public. **2.** In telecommunications, a company that is licensed to provide message transport services to the general public and generally is regulated to a considerable extent, at least with respect to fundamental aspects of service such as availability and basic rates. Such a license grants the holder certain rights, such as the right to control and assign globally unique telephone numbers (i.e., E.164 numbers), the right to collect certain fees from other carriers when handling calls jointly, and status under certain laws and regulations requiring interconnection. Common carrier status also imposes certain responsibilities, including collecting taxes from users, publishing tariffs, providing interconnection arrangements to other carriers, and paying certain fees to other carriers. In the United States, the Federal Communications Commission (FCC) and the state public utilities commissions (PUCs) regulate incumbent local exchange carriers (ILECs), i.e., telephone companies or telcos, and interexchange carriers (IXCs) to various extents. See also *FCC*, *ILEC*, *IXC*, and *PUC*.

Common Carrier Line Charge (CCLC) See *CCLC*.

common channel signaling (CCS) See *CCS*.

Common Channel Signaling System 7 (CCS7) See *CCS* and *SS7*.

common control A common set of stored program logic that controls the activities of a system and all of its various elements. A common control unit generally consists of multiple microprocessors operating under a stored program, and is synonymous with stored program control (SPC).

Common Intermediate Format (CIF) See *CIF*.

Common Profile for Instant Messaging (CPIM) See *CPIM*.

communication manager See *media gateway*.

Communications Act of 1934 In the United States, the act of Congress that established the Federal Communications Commission (FCC) to regulate interstate, international, and maritime communications, with universal service stated as the goal. See also *FCC* and *universal service*.

Communications Act of 1962 In the United States, the act of Congress that placed authority with Federal Communications Commission (FCC) to assign commercial satellite frequencies. The act also established the Communications Satellite Corporation (Comsat) to act as a carriers' carrier (wholesaler) for international satellite service and in conjunction with the International Telecommunications Satellite Organization (Intelsat). Intelsat was established as an international financial cooperative that owns and operates satellites for international communications. See also *carrier*, *FCC*, and *satellite*.

Communications Assistance for Law Enforcement Act (CALEA) See *CALEA*.

Communications Decency Act (CDA) Title V of the Telecommunications Act of 1996. In the United States, an act of Congress (1996) enacted to hold both creators of content and service providers responsible for access of minors to indecent or offensive material over the Internet. Portions of the act subsequently were ruled unconstitutional, in violation of free speech guaranteed by the First Amendment. See also *Internet* and *Telecommunications Act of 1996*.

communications software Software that assists a computer operating system (OS) in managing local and remote terminal access to host resources, managing security, and performing certain checkpoint activities. Communications software generally is embedded in the OS, although it can take the form of a systems task under the control of the OS. Communications software, for example, is used to control a modem, performing terminal emulation and file transfer tasks. See also *host*, *modem*, *OS*, *terminal*, and *terminal emulation*.

community antenna television (CATV) Synonymous with cable television. See *CATV*.

Compact HTML (C-HTML) See *C-HTML*.

compaction See *compression*.

companding (compressing/expanding) Referring to the twin processes of compression and decompression as used in the conversion of a voice signal from analog to digital format and then converting the signal back from digital to analog. The ITU-T G.711 Recommendation for pulse code modulation (PCM) specifies both μ-law (mu-law) and A-law companding techniques. See also *A-law*, *codec*, *compression*, *decompression*, *G.711*, *ITU-T*, *μ-law*, and *PCM*.

compatible **1.** Referring to the fact that a device, program, or system can interface with another without interfering with each other and without requiring the intervention of another device or program, such as a gateway or middleware. Fully compatible devices are even interchangeable. See also *gateway* and *middleware*. **2.** Referring to a device or system that fully conforms to a standard. (*Note:* meaning 2 does not guarantee meaning 1.) See also *standard*.

Competitive Access Provider (CAP) See *CAP*.

competitive local exchange carrier (CLEC) See *CLEC*.

complementary code keying (CCK) See *CCK*.

Compressed SLIP (CSLIP) See *CSLIP*.

compression A means of reducing the amount of data to be transmitted or stored. Compression is possible since there always is some amount of data redundancy or there may be a predictable flow to the data. These characteristics of a set of data or a stream of data allow the use of a sort of mathematical algorithm to represent or describe the original data in fewer bits. A matching decompression process reverses the compression process and restores the data to its original form, or an approximation thereof. Compression serves to improve the efficiency of data transmission and storage, and is especially valuable if bandwidth and memory resources are limited. Data compression techniques can include the following:

- **Formatting:** A technique that removes formatting from a commonly used form prior to transmission or storage. The receiving device reformats the data, placing the various fields of data in the appropriate places on the form, which it maintains in primary memory.

- **Redundant data:** Also known as *string coding*. A technique that identifies and deletes redundant data prior to transmission or storage. See also *run-length encoding*.

- **Commonly used characters:** A technique that involves the identification and abbreviation of commonly used characters, similar to the technique used by Samuel Morse in the development of Morse code. See also *Huffman coding* and *Morse code*.

- **Commonly used strings of characters:** A technique that relies on the probability of character occurrence following a specific character. For example, the letter q generally is followed by the letter u. See also *Markov source*.

A number of steps are involved in video compression, including filtering, color-space conversion, scaling, transforms, quantization and compaction, and interframe compression. Lossless compression enables faithful reproduction of the signal, with no data loss, although compression rates tend to be relatively low. Lossy compression tends to produce artifacts, which are unintended and unwanted distortions or aberrations that result in a degraded signal, but supports very high compression rates. Additional compression techniques include Modified Huffman (MH), Modified Read (MR), and Modified Modified Read (MMR). See also *color-space conversion*, *filtering*, *interframe compression*, *lossless compression*, *lossy compression*, *Markov source*, *MH*, *MMR*, *Morse code*, *MR*, *quantization and compaction*, *run-length encoding*, *scaling*, *string coding*, *suppression*, and *transform*.

compromise The negotiated settlement to a dispute in which at least some of the parties agree to accept less than they originally wanted. Typically, none of the parties that make concessions in the spirit of compromise is ecstatic about the settlement, but all can accept it. The standards-making process is characterized by compromise, with multiple manufacturers, governments, and other interested parties lobbying to enhance their individual positions and ultimately compromising on a specification that often is not the optimum technical solution, but is acceptable to a majority.

CompTIA (Computing Technology Industry Association) A not-for-profit association that represents the computing industry on public policy issues and offers vendor-neutral certification exams. See Appendix A for contact information.

computer A machine that computes. Specifically, a modern computer is a digital electronic system that performs complex calculations or compiles, correlates, or otherwise processes data based on instructions in the form of stored programs and input data. A device that can receive, store, retrieve, process, and output data.

Computer-Aided Design (CAD) See *CAD*.

Computer-Aided Design/Computer-Aided Manufacturing (CAD/CAM) See *CAD/CAM*.

Computer Emergency Response Team (CERT) See *CERT*.

Computer Inquiry II (CI II) The Second Computer Inquiry. See *Second Computer Inquiry*.

Computer Supported Telephony Applications (CSTA) See *CSTA*.

computer telephony (CT) The blending of telecommunications switching with computer processing power and programmable logic. CT was an intermediate step in the evolution from the third generation of digital electronic common control (ECC) telephone systems to the fourth generation of IP-based systems. The term *computer telephony* was coined by Howard Bubb of Dialogic Corporation (subsequently acquired by Intel and later sold to another company that renamed itself Dialogic Corporation).

Computing Technology Industry Association (CompTIA) See *CompTIA*.

concatenation To sequentially link together two or more information units, such as character strings, packets, or frames, into a single unit. Concatenation allows the component units to transverse a network or subnetwork as a single entity, which not only ensures that they all work their way through the network together, but also reduces network processing time and, thereby, decreases latency. See also *frame*, *network*, *packet*, *SONET*, *string*, and *subnetwork*, and *T1C*.

concentrator A simple form of data multiplexer that concentrates traffic from multiple low speed asynchronous devices onto a single high speed synchronous circuit, usually by time division multiplexing (TDM). In a simple concentrator, the total speed of the low speed incoming channels is equal to or less than the speed of the high speed outgoing circuit, so the maximum incoming load placed on the concentrator never exceeds the capacity of the outgoing circuit. In a more sophisticated concentrator, the incoming traffic load may exceed the capacity of the outgoing circuit, with buffers serving to store the excess data traffic for short periods of time until capacity is available on the outgoing circuit. At this more sophisticated level, a concentrator is a relatively unsophisticated statistical time division multiplexer (STDM). See also *buffer*, *multiplexer*, *STDM*, and *TDM*.

conditioned circuit See *conditioning*.

conditioning The addition, or removal, of certain equipment in order for a circuit to achieve the performance characteristics required by analog or, more typically, certain types of data transmission. For example, an especially long copper local loop might require the addition of load coils or amplifiers to achieve the proper level of performance for voice grade analog applications, or regenerative repeaters for digital data applications. A copper local loop might require the removal of load coils to achieve the proper level of performance for an ADSL application. See also *ADSL*, *amplifier*, *load coil*, and *repeater*.

conductance A measure of the ability of a substance to allow electric current to pass through in relation to the applied voltage (E). Conductance is the reciprocal of resistance (R), as measured in ohms (Ω). Conductance is measured in mhos, which is ohms spelled backwards. (Isn't that clever?)

conductor **1.** A substance that allows electricity, light, heat, sound, or other forms of energy to pass through. In the context of telecommunications transmission systems, a conductor allows electric current or photonic energy to pass through. **2.** In fiber optics transmission systems, various highly specialized types of glass or plastic are used to conduct photonic energy. As dielectrics, glass and plastic are not conductors of electric current; therefore, optical fibers are immune to electromagnetic interference (EMI). See also *EMI*. **3.** In electric circuits, such as twisted pair and coaxial cable systems, a conductor is a material that readily permits the flow of electrons through itself in response to an electric field. Metals are excellent electrical conductors because of the high concentration of free electrons. Telecommunications networks most commonly use copper in electrified circuits, although copper-covered steel, copper alloy, nickel- or gold-plated copper, and even aluminum metallic conductor are used. Gold, silver, and platinum are sometimes used in short circuits internal to various devices such as switches. CATV networks make use of aluminum and copper-clad aluminum in coaxial cable trunks. See also *dielectric* and *insulator*.

conduit A protective tube, pipe, or trough for wires, fibers, and cables. Early conduits for telecommunications cables were made of vitrified clay pipe, creosoted lumber, and even hollowed-out logs. Contemporary conduits commonly are made of aluminum, steel, polyethylene, and polyvinyl chloride (PVC).

cone of acceptance The maximum angle, represented in three-dimensional view as a cone, at which an optical fiber will accept incident light. Within that cone, as defined by those angles, a light source can inject an optical signal into the fiber core and the signal will remain in the core, reflecting off of the interface between the core and cladding, as illustrated in Figure C-5. At a more severe angle, i.e., outside the cone, the signal will penetrate the interface and enter, and perhaps be lost in, the cladding. The angle of acceptance and, therefore, the cone of acceptance are determined by the difference in index of refraction (IOR) between the core and cladding. See also *angle of acceptance*, *cladding*, *core*, and *IOR*.

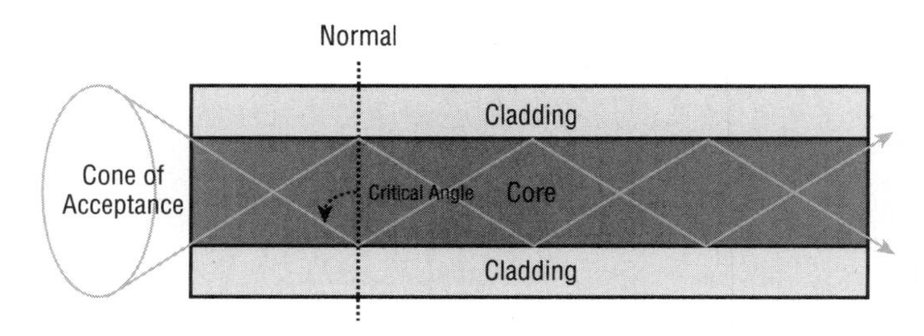

Figure C-5

Confederation of European Posts and Telecommunications (CEPT) See *CEPT.*

conference bridge A device, usually in the form of a printed circuit board (PCB), that bridges, or connects, multiple circuits or channels in order to effect a conference call. Conference bridges are available for Centrex systems, key telephone systems (KTSs), and PBXs. See also *bridge* and *conference call.*

conference call A voice telephone call involving more than two parties. Conferencing is a typical feature of voice telephone systems, commonly allowing the attendant or an authorized user to bridge as many as 2, 4, 8, or 16 parties on a given conference call. See also *bridge.*

Conférence Européenne des administrations des Postes et des Télécommunications (CEPT) Translated from French as *European Conference of Postal and Telecommunications Administrations.* See *CEPT.*

configuration management An element of network management, configuration management comprises the management of the logical and physical arrangement and interconnection of the components and elements that comprise a given system and network, including software, firmware, hardware, circuits, and channels. At a more detailed level, configuration management includes the processes of load balancing, network optimization, and traffic analysis. See also *load balancing, network management, network optimization,* and *traffic analysis.*

confirmation A typical feature of e-mail systems, allowing the sender to request that the recipient send a receipt to confirm that the message has been received and, perhaps, send another receipt to indicate that the message has been read. See also *e-mail.*

congestion The condition that arises when a system or network experiences a level of offered calling activity or message traffic that exceeds its capacity.

congestion management The process of managing a congestion condition. Congestion management mechanisms include the use of buffers that can be used to temporarily store data in one or more queues until the data can be forwarded through the internal bus or switching matrix of a switch or router, or through an outgoing port and across a communications link. As the buffers fill to capacity, data can be discarded, perhaps selectively based on a priority or quality of service (QoS) mechanism. If a network is configured as a mesh, a component router may have the ability to identify and exercise alternate paths if the primary path is suffering congestion levels that exceed definable parameters established in consideration of QoS objectives. Some network protocols provide for a router, for example, to advise its peers of congestion conditions and to instruct them to throttle back their transmission rates to avoid compounding the situation. Similarly, public network-based routers can advise customer edge routers to throttle back, or even temporarily suspend transmission of offered traffic until the congestion condition relaxes. Finally, a network-based router or switch can simply reject a call or message transmission. In a voice network, a PBX or central office (CO) provides the rejected caller with a fast busy signal.

conjugate structure algebraic code excited linear prediction (CS-ACELP) See *CS-ACELP.*

connection From the Latin *connectere,* meaning to *bind together.* **1.** A physical joining of two or more things. **2.** In telecommunications, a physical joining of two stations or nodes by a circuit, perhaps comprising multiple links. See also *circuit, link, node, physical,* and *station.* **3.** A physical joining of two conductors accomplished with a mechanical splice or fusion splice. See also *conductor, fusion splice, mechanical splice,* and *physical.* **4.** A logical relationship or association of two or more things.

connectionless In packet data transmission, a mode in which there is no call setup phase before transmission begins, no predetermined path set up for frames or packets to travel through a network, and no call teardown phase after transmission ceases. As each packet header contains enough information to enable its independent delivery without the aid of any additional instructions, each packet can take an entirely different route from originating host to destination host. In X.25 networks, for example, datagram service is connectionless. Local area networks (LANs) are connectionless. Internet Protocol (IP) is a connectionless datagram service. See also *always on*, *connection-oriented*, *datagram service*, *frame*, *LAN*, *packet*, and *X.25*.

Connectionless Broadband Data Service (CBDS) The name by which *Switched Multimegabit Data Service* (SMDS) was known in Europe. See *SMDS*.

Connectionless Network Protocol (CLNP) See *CLNP*.

connection-oriented A transmission mode in which a network establishes a logical relationship, and often a predetermined path, for all frames or packets associated with an originating and destination address pair to travel. The path can be permanent or can exist for a single session. The public switched telephone network (PSTN) is connection-oriented. Transmission Control Protocol (TCP) is a connection-oriented protocol, where Internet Protocol (IP) is a connectionless datagram service. See also *connectionless*, *frame*, *IP*, *packet*, *protocol*, *PSTN*, *session*, and *TCP*.

connector A simple device that physically links, couples, or connects, two things together. A male connector has pins that fit into the sockets, or receptacles, of a female connector, as the connectors mate. A male connector sometimes is referred to as a plug, and a female connector as a jack.

l'Conseil Européen pour la Recherche Nucléaire (CERN) See *CERN*.

Consent Decree The 1956 negotiated settlement between AT&T and the United States Department of Justice (DOJ) that allowed AT&T to retain ownership of Western Electric if it manufactured only equipment of a type to be used for the provision of telephone service and only for Bell companies. The decree also prevented the Bell System from offering data processing services and other services not related to functions of a common carrier and required that Bell System patents be licensed to others on basis of reasonable fees. As a result, AT&T was forced to license its transistor technology to any company for $25,000. See also *AT&T*, *common carrier*, *patent*, and *transistor*.

console **1.** A control unit or terminal, including a display and keyboard, that an operator or administrator uses to communicate with and control a computer system such as a central office (CO), PBX, or general-purpose computer. **2.** The keyboard and display components of a personal computer (PC), or microcomputer.

constant bit rate (CBR) See *CBR*.

constraint-based routing In Multiprotocol Label Switching (MPLS), a technique that considers factors such as bandwidth, hop count, and performance requirements of the traffic flow in selecting the end-to-end Label Switched Path (LSP). See also *bandwidth*, *flow*, *hop*, *LSP*, *MPLS*, and *traffic*.

Consultative Committee for International Telephone and Telegraph (CCITT) See *ITU-T*.

content blocking In the advanced intelligent network (AIN), a feature that supports the blocking of calls to specific numbers, such as 900/976 numbers. Content blocking is a variation of call blocking. See also *AIN* and *call block*.

continuity algorithm A mathematical mechanism integral to predictive voice compression employed in packet voice technologies such as voice over frame relay (VoFR) and voice over Internet Protocol (VoIP). Continuity algorithms intelligently fill the void of missing or errored voice frames and packets by stretching the previous voice frames or packets and blending several together. See also *compression*, *VoFR*, and *VoIP*.

continuous redial Also known as *repeat dial*. A CLASS programmable service feature of the public switched telephone network (PSTN). Continuous redial enables the caller encountering a busy signal or no answer to request that the network continuously redial the telephone number for a period of time or until the call is successfully completed. See also *CLASS* and *PSTN*.

continuously variable slope delta modulation (CVSDM) See *CVSDM.*

contraction The shortening of a word or phrase by omitting letters or syllables. In the English language, a contraction generally, but not always, marks the omitted letters or syllables with an apostrophe or a period. *Telco*, for example, is a contraction of *telephone company*. Doesn't is a contraction of does not. *Mr.* is a contraction of *mister,* a title of courtesy for a man. *Mrs.* is a contraction of *mistress*, a title of courtesy for a married or widowed woman. Once upon a time, people used the title of courtesy Miss to denote a girl or unmarried woman. The feminist movement of the 1970s forced a change to Ms., which makes no distinction in a woman's marital status. Now we frequently delete such titles, altogether, leaving those who don't know the person to guess at both his or her gender and marital status, which is all quite silly, even if it is PC (politically correct). (*Note:* PC is an initialism.) See also *initialism* and *portmanteau.*

control plane In the ATM reference model, and other architectures, the functions defining all aspects of network signaling and control, such as call control and connection control. See also *ATM reference model, management plane, signaling and control,* and *user plane.*

control unit The portion of the central processing unit (CPU) of a computer that retrieves instructions from memory, accepts calculations from the arithmetic logic unit (ALU), and executes the instructions. See also *CPU.*

Controlling the Assault of Non-Solicited Pornography and Marketing Act (CAN-SPAM Act) See *CAN-SPAM Act.*

conventional band (C-Band) See *C-Band.*

convergence **1.** The moving from different directions towards union or one another, especially referring to entities that were very different or even opposed. **2.** In telecommunications, the coming together of voice, facsimile, data, video, and image applications, systems, and networks, both wireline and wireless. The developing IP Multimedia Subsystem (IMS) currently comes closest to full-on convergence. IMS is an architectural concept built around a packet core and providing an environment in which a user can access a wide range of multimedia services using any device and any type of network connection. See also *IMS.*

Convergence Sublayer (CS) The term applied to the top portion of a protocol, typically at the Data Link Layer (Layer 2), that functions to format data originating in higher layers for processing by the lower layers. The CS adds a header or wraps the data in a header and trailer that contain information necessary to provide the necessary services. Typically, error control and priority information are added at this layer. In asynchronous transfer mode (ATM), for example, the CS functions established in the header (there is no trailer) are determined by the specifics of the service supported by a given ATM Adaptation Layer (AAL). Service classes are designated as Class A (AAL1), Class B (AAL2), Class C (AAL3/4 and AAL5), and Class D (AAL3/4). See also *AAL1, AAL2, AAL3/4, AAL5, Class A ATM traffic, Class B ATM traffic, Class C ATM traffic, Class D ATM traffic, Data Link Layer, error control, header, protocol,* and *trailer.*

converter box See *set-top box.*

.coop (cooperative) Pronounced *dot co-op.* The generic Top Level Domain (gTLD) reserved exclusively for cooperative associations. This domain was created in 2002 under the sponsorship of Dot Cooperation LLC. See also *gTLD, Internet,* and *sponsored domain.*

coordinated dialing plan Synonymous with *uniform dialing plan* (UDP). See *UDP.*

copper (Cu) A reddish-brown metallic element that is highly malleable, ductile, corrosion-resistant, and is an excellent conductor of electricity and heat. Only silver is a better conductor of electricity at room temperature. Copper is extensively used in electrical cables. Copper has an atomic number of 29.

Copper Distributed Data Interface (CDDI) See *CDDI.*

copyright (©) The exclusive legal right of an author or publisher to publication, production, or sale of the rights to an original literary, dramatic, musical, or artistic work that has been tangibly expressed. I, Ray Horak, am the author of this original work, to which Wiley owns the copyright. Please enjoy the book, but don't even think about violating the copyright. Thank you. Have a nice day. See also *intellectual property.*

cordboard A manual switching technology that requires the operator to establish connections on a plug and jack basis, with the plugs on cords and the jacks mounted on a board. As was the case with the earlier switchboard technology, the cordboard operator establishes a unique physical and electrical connection that remains in place for the duration of the call. When either party disconnects, the operator is alerted and manually disconnects the circuit, which then becomes available for use in support of another call. The size of such switches, the complexity of interconnecting long distance calls across multiple switches, and the labor intensity of this approach all contributed to their functional obsolescence many years ago. Although cordboards were rendered technically obsolete by automatic step-by-step (SxS) switches in 1891, many thousands remained in service for many years, and many remain in service to this day. See also *switchboard, SxS,* and *tip and ring.*

cordless telephone A system comprising one or more telephone handsets that connect on a wireless radio frequency (RF) basis to a base station that connects via a standard plug and jack for access to the public switched telephone network (PSTN). The original cordless telephones (circa 1980) in the United States were assigned one of 10 channels in the 27 MHz range. In 1986, the Federal Communications Commission (FCC) changed the cordless frequency range to the 46 and 49 MHz bands and reduced the allowable power levels. Contemporary digital versions operate in the 900 MHz, 2.4 GHz, and 5.8 GHz bands, which are in the unlicensed industrial/scientific/medical (ISM) band. Contemporary standards include Cordless Telephony generation 1 (CT1), Cordless Telephony generation 1 plus (CT1+), Cordless Telephony generation 2 (CT2), Cordless Telephony generation2 plus (CT2+), Cordless Telephony generation 3 (CT3), Digital Enhanced (nee European) Cordless Telecommunications (DECT), Personal Handyphone System (PHS), Personal Access Communications Services (PACS), Personal Communications Services (PCS), and Personal Wireless Telecommunications (PWT). The cordless telephone was invented by Al Gross, who also invented the CB radio, paging system, and walkie talkie. See *Gross, Al.* See also *CB radio service, CT1, CT1+, CT2, CT2+, CT3, DECT, FCC, ISM, jack, PACS, paging system, PCS, PHS, plug, PSTN, PWT, RF, walkie talkie,* and *wireless.*

Cordless Telephony generation 0 (CT0) See *CT0.*

Cordless Telephony generation 1 (CT1) See *CT1.*

Cordless Telephony generation 1 plus (CT1+) See *CT1+.*

Cordless Telephony generation 2 (CT2) See *CT2.*

Cordless Telephony generation 2 plus (CT2+) See *CT2+.*

Cordless Telephony generation 3 (CT3) See *CT3.*

core **1.** The central or essential part of a Wide Area Network (WAN) or Metropolitan Area Network (MAN) is commonly known as the core, or backbone. The network core comprises very high capacity elements and subsystems such as transmission systems, multiplexers, switches, and routers. See also *MAN* and *WAN.* **2.** The central and primary light-conducting portion of a glass optical fiber (GOF). The core is the inner portion of the fiber into which the optical signal is injected by either a light-emitting diode (LED) or one of many types of laser diodes. A single-mode fiber (SMF) used in a high speed, long haul fiber optic transmission system (FOTS) has a very narrow inner core, 5–10 microns in diameter. A multimode fiber (MMF) used in a relatively low speed, short haul system has a relatively broad core that typically is either 50 microns or 62.5 microns in diameter. See also *cladding, GOF, laser diode, LED, MMF,* and *SMF.*

core switch A core switch, also known as a *tandem switch* and a *backbone switch*, is a high-capacity switch positioned in the physical core, or backbone, of a network. In a public Wide Area Network (WAN) a core switch serves to interconnect edge switches, which are positioned at the network edge. In a Local Area Network (LAN), a core switch serves to interconnect workgroup switches, relatively low capacity switches that serve groups of workers in geographic clusters. See also *switch*.

CoS (Class of Service) **1.** The level of privilege afforded, or level of restriction imposed upon, a system user. Each key system (KTS), PBX, or Centrex user is assigned a CoS that defines that individual's level of access privileges to internal and network resources, with examples being feature assignments and priority levels for access to circuits. See also *Centrex*, *KTS*, and *PBX*. **2.** A priority level assigned to a particular traffic type in a packet data network. Real-time, uncompressed voice and video traffic, for example, typically are assigned the highest priority level, as they are not tolerant of latency and loss. E-mail and certain types of signaling and control messages typically are assigned the lowest priority level, as they are highly tolerant of latency and loss. CoS is a highly effective means of managing traffic on a best effort basis, but does not offer the performance assurances or guarantees of a Quality of Service (QoS) mechanism. As examples, CoS mechanisms are employed in local area networks (LANs), frame relay networks, Internet Protocol (IP) networks, and Multiprotocol Label Switching (MPLS) networks. QoS mechanisms are employed in networks based on asynchronous transfer mode (ATM). See also *ATM*, *Frame Relay*, *IP*, *latency*, and *QoS*.

cost **1.** The amount of money paid to acquire something, or spent in producing something. **2.** The amount of time, effort, or other resources expended in accomplishing something. **3.** In telecommunications, the cost of transmitting data along a given path or route can be measured in terms of bandwidth consumption and quality of service (QoS) parameters such as number of hops, total latency, bit error rate (BER), and packet loss, or any number of considerations other than the direct monetary cost of passing traffic to another carrier or service provider. See also *bandwidth*, *BER*, *carrier*, *hop*, *latency*, *path*, *packet*, *QoS*, *route*, and *traffic*.

coulomb (C) The unit of electric charge equal to the quantity of electricity transferred by one ampere (A) in one second, a coulomb is the flow of 6.24×10^{18} electrons. The coulomb is named for Charles Augustin de Coulomb (1736–1806), a French physicist who worked in the field of electrostatics. See also *ampere*.

counterintuitive Contrary to intuition, instinct, or commonsense expectations. See also *normal*.

country code In the context of the public switched telephone network (PSTN), the leading one-, two-, or three-digit number associated with an international call. Numbering plan administration (NPA) is the responsibility of the ITU-T and is standardized in E.164. The ITU-T assigns the country codes, for example, 1 for the United States, 27 for South Africa, and 352 for Luxembourg. See also *E.164*, *ITU-T*, *NPA*, and *PSTN*.

country code Top Level Domain (ccTLD) See *ccTLD*.

coupler **1.** A passive device that combines or divides signals. See also *splitter*. **2.** A passive device that joins three or more optical fiber ends. In one direction, a coupler splits an incoming signal into two or more outgoing signals. In the other direction, a coupler combines two or more incoming signals into one outgoing signal. See also *splitter*.

coupling efficiency The efficiency with which a light source physically connects to an optical fiber. The more precisely the light source can inject a tightly focused signal directly into the inner core of a fiber, the stronger the resulting signal and the better the signal performs over a distance. Coupling efficiency is a key advantage of pairing a diode laser with a single-mode fiber (SMF), which has an inner core of only 5–10 microns. A less sophisticated light-emitting diode (LED) is incompatible with SMF as the emitted optical signal is too crudely focused and, therefore, would overfill the fiber core. LEDs are designed to mate with multimode fiber (MMF), which has a thicker inner core. The level of inefficiency, or coupling loss, is described as insertion loss, as measured in decibels (dB). See also *core*, *coupling loss*, *dB*, *insertion loss*, *laser diode*, *LED*, *MMF*, *signal*, and *SMF*.

coupling loss Referring to the extent to which a signal attenuates, or loses power across an interface between two components, e.g., from the input side of the interface between a circuit and a switch or router to the output side of the interface, or from the output endface of one optical fiber to the input interface of another across a pair of connectors. Coupling loss, or coupling inefficiency, is generally described as insertion loss, as measured in decibels (dB). See also *coupling efficiency*, *dB*, *insertion loss*, and *signal*.

Coy, George W. The inventor of the first practical telephone exchange switch, which was placed into service on January 28, 1878, in New Haven, Connecticut. This manual exchange, or cordboard, allowed the flexible interconnection of 21 subscribers. See also *central office* and *cordboard*.

CPE (Customer Premises Equipment) All communications equipment located on the customer's premises; owned, leased, or rented by the customer; connected to a public or private network through a network interface of some sort; and on the customer side of the demarcation point (demarc). CPE primarily refers to voice equipment, including telephone sets, key equipment, PBXs, ACDs, and peripheral equipment such as answering machines. The term data terminal equipment (DTE) generally applies to data terminals, hubs, switches, routers and multiplexers, all of which also are considered CPE in the broader context. Inside wire and cable systems are not considered CPE. Equipment owned and operated by a telephone company or third party is not considered CPE, with examples being public pay stations and protectors.

CPIM (Common Profile for Instant Messaging) An Internet Engineering Task Force (IETF) specification (RFC 3860) that defines common semantics and data formats for instant messaging (IM) to facilitate the development of gateways between services. See also *gateway*, *IETF*, *IM*, and *SIMPLE*.

cps (chips per second) See *chip rate*.

CPU (Central Processing Unit) The central computational and control unit of a computer system, the CPU controls the interpretation and execution of instructions. The CPU contains the arithmetic logic unit (ALU), which performs mathematical calculations, and the control unit, which retrieves instructions from memory, accepts calculations from the ALU, and executes the instructions. The CPU often is contained on a single silicon chip, known as a *microprocessor*. CPU commonly is interchangeable with *microprocessor* and *processor*.

cracker A computer enthusiast, or computerphile, who gains, or attempts to gain, unauthorized access to computers or computer networks and tamper with operating systems, application programs, and databases. See also *hacker*.

CRC (Cyclic Redundancy Check) A commonly used error detection mechanism that validates the integrity of a data set, formatted in a block or frame, through the use of a statistical sampling process and a unique mathematical polynomial. In a data communications application, the transmitting device statistically samples the data in the block or frame and applies a 17-bit generator polynomial based on a Euclidean algorithm to generate a description of the text field, or cyclic checksum, which is appended to the block or frame or text as either a 16- or 32-bit value prior to transmission. The receiving device executes the identical process and compares the results of its process to the CRC value appended to the data block. If the two values match, the data block almost certainly was unerrored in transmission. The integrity factor is 10^{-14}, which means that the possibility of an undetected error is 1 in 100 trillion. If the receiving device determines that the block or frame is unerrored, it returns a positive acknowledgement (ACK). If, however, it determines that the block or frame is errored, it returns a negative acknowledgement (NAK), which prompts the transmitting device to retransmit that specific block or frame, which has been stored in a buffer. When that block or frame has been positively acknowledged, the sending device erases it from buffer memory and transmits the next. In this example, CRC is part of an error control mode known as recognition and retransmission, and is used by communications protocols such as Kermit and XMODEM. CRC also is used by MS-DOS when writing data to a hard drive or floppy disk, and by file compression utilities such as PKZIP. See also *block*, *buffer*, *cyclic checksum*, *error control*, *frame*, *Kermit*, *MS-DOS*, *recognition and retransmission*, *text field*, and *XMODEM*.

critical angle Light striking the interface between two substances can either reflect off of the substance it encounters or enter it, with the difference depending on the nature of the substances and the angle at which the incident light ray strikes the interface. A glass optical fiber (GOF) comprises an inner core of glass of a given refractive index, or index of refraction (IOR), surrounded by one or more layers of cladding of lower refractive index. The critical angle is measured from the normal, which is at 90 degrees from (i.e., perpendicular to) the surface of the core/cladding interface, or boundary. If, as illustrated in Figure C-6, the incident light rays strike the interface at an angle greater than the critical angle, they reflect off the interface, with the angle of reflection being the same as the angle of incidence. The light rays glance off of the interface, so to speak. If, on the other hand, the incident light rays strike the boundary at an angle less than the critical angle, they enter the cladding, where they either are lost or refracted back into the core, depending on the type of fiber and the angle of incidence. See also *angle of acceptance*, *angle of incidence*, *cladding*, *cone of acceptance*, *core*, *GOF*, *graded-index fiber*, *IOR*, *numerical aperture*, *step-index fiber*, and *total internal reflection*.

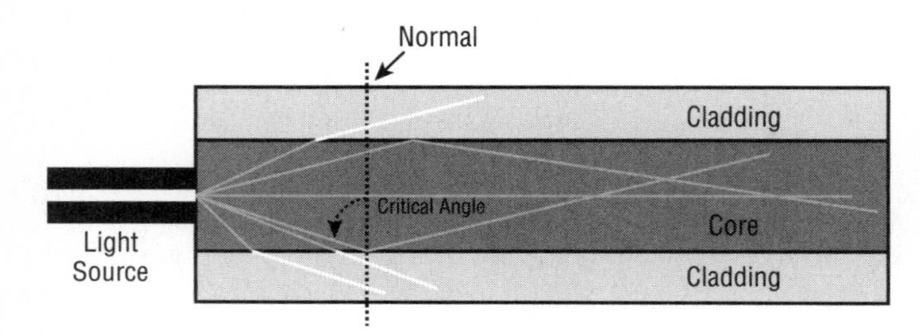

Figure C-6

crossbar See *Xbar*.

cross-connect A device that allows conductors or channels to be interconnected, either physically or electronically, on a semi-permanent basis. A cross-connect can be in the form of a main distribution frame (MDF), intermediate distribution frame (IDF), or terminal block where wire pairs from cables are mechanically terminated on a punch-down block, with short wire jumpers interconnecting the cable pairs on each side of the block. A digital cross-connect (DXC), also known as a digital access cross-connect system (DACS), is a device that allows digital circuits (e.g., T1s or E-1s), or even individual channels, to be cross-connected via an electronic cross-connect matrix. An optical cross-connect (OXC) is used in fiber optical transmission systems (FOTS). A cross-connect is much like a static switch in that connections can be changed to alter physical paths, but not on a call-by-call basis. See also *DACS*, *DXC*, *FOTS*, *IDF*, *MDF*, *OXC*, and *switch*.

crossreed An electromagnetic circuit-switching technology for central office (CO) and PBX voice applications, developed by Stromberg-Carlson and similar to crossbar (Xbar) technology. See also *circuit switching*, *CO*, *PBX*, and *Xbar*.

cross-talk See *crosstalk*.

crosstalk (XT) The unwanted coupling of energy between two circuits or channels, crosstalk is a form of co-channel interference that superimposes a transmission occurring on one circuit or channel onto another transmission occurring on another circuit or channel. In a voice scenario, the parties talking on the disturbed channel can hear one or more of the parties talking on the disturbing channel. Near-End CrossTalk (NEXT) occurs at or near the transmitting end of the connection and Far-End CrossTalk (FEXT) occurs at or near the far end. See also *FEXT* and *NEXT*.

CRTC (Canadian Radio-television and Telecommunications Commission) An independent agency responsible for regulating broadcasting and telecommunications systems in Canada. The CRTC reports to Parliament through the Minister of Canadian Heritage.

cryptology The scientific study of cryptography. See *cryptography*.

cryptography From the Greek *kryptos*, meaning *hidden*, and *graphos*, meaning *written*. The art or science, or system, of writing messages in code, or cipher, to disguise, and thereby secure, the content. When encrypted, a plain text message can be revealed only through the use of the key to the code. Cryptography does not mask the existence of the message, but does disguise its content. See also *steganography*.

crystal oscillator (XO) An oscillator in which the frequency is controlled by a piezoelectric crystal. See also *frequency*, *oscillator*, and *piezoelectric*.

CS (Convergence Sublayer) See *Convergence Sublayer*.

CSA (Carrier Serving Area) The geographical area served by a Central Office (CO), a CSA is generally limited to a radius of approximately 18,000 feet, which is the maximum reach of a typical copper Unshielded Twisted Pair (UTP) local loop. An amplifier or regenerative repeater is required to extend a UTP loop beyond that distance. See also *CO* and *local loop*.

CS-ACELP (Conjugate Structure-Algebraic Code Excited Linear Prediction) A voice compression algorithm defined in ITU-T G.729, ACELP improves on CELP through the algebraic expression, rather than the numeric description, of each entry in the codebook. ACELP yields quality that is considered to be as good as ADPCM, but requiring bandwidth of only 8 kbps, which yields a compression ration of 8:1. CS-ACELP is geared toward multi-channel operation. See also *ADPCM*, *algorithm*, *bandwidth*, *CELP*, *channel*, *compression*, *ITU-T*, and *LD-CELP*.

CSD (Circuit-Switched Data) Referring to data communications over a circuit-switched network, particularly a cellular radio network such as Digital Advanced Mobile Phone Service (D-AMPS). See also *cellular radio*, *circuit switch*, and *D-AMPS*.

CSLIP (Compressed Serial Line Internet Protocol) A method for improving TCP/IP performance over low-speed (300 bps to 19.2 kbps) serial lines by compressing the TCP/IP headers. See *SLIP*.

CSMA (Carrier Sense Multiple Access) A decentralized, contentious medium access control (MAC) protocol used in Ethernet and other bus-oriented local area networks (LANs). The carrier frequency is sensed by each of multiple stations, or nodes, to determine network availability before accessing the medium to transmit data. Each station must monitor the network to determine if a collision has occurred and the data require retransmission. CSMA variations include Nonpersistent CSMA, 1-Persistent CSMA and P-Persistent CSMA.

- **Nonpersistent CSMA:** A machine may transmit data whenever it senses an idle channel. If the channel is busy, the machine backs off the network, calculates a random time interval, and again monitors the channel when that interval expires. This approach mathematically distributes the temporal monitoring of the network, thereby reducing the likelihood that multiple stations will sense its availability at approximately the same time and transmit simultaneously.
- **1-Persistent CSMA:** A machine may transmit data whenever it senses an idle channel. If the channel is in use, the machine will continuously sense it until the channel becomes free. The protocol is so named as the machine is persistent in its monitoring of the channel, and transmits with a probability of 1.0, i.e., 100 percent certainty of access success, whenever the channel is idle. If the network includes a large number of stations persistently monitoring the network, a great many of them might sense the availability of the network and begin to transmit simultaneously, virtually guaranteeing a collision.

- **P-Persistent CSMA:** A machine may transmit a frame during an idle time with probability p or lower, based on the length of the idle time as measured by a time slot. A time slot is the maximum packet transmission time for a station at one extreme end of the network to send a packet to a station at the opposite extreme end of the network, and is based on the physical length of the cable, the physical size of the frames, and the speed of signal propagation through the wire or fiber. If a machine senses an idle condition on the channel, it transmits with probability p for one time slot. The machine then delays for worst-case propagation delay for one packet with probability 1–p. If the channel is busy, the machine listens persistently until the channel becomes idle, and starts over. For example, p.01 means that there is a probability of 1 percent that the transmission will be unsuccessful. If p is set very low (e.g., .01), throughput is nearly 100 percent, but transmission delays will be very long, as the machine will wait a very long time between idle periods.

CSMA is implemented in two standard means: CSMA/CD provides collision detection, and CSMA/CA provides collision avoidance. See also *carrier*, *channel*, *CSMA/CA*, *CSMA/CD*, *Ethernet*, *LAN*, *MAC*, *node*, *propagation*, *propagation delay*, *station*, and *time slot*.

CSMA/CA (Carrier Sense Multiple Access/Collision Avoidance) A medium access control (MAC) protocol used in some bus networks, CSMA/CA includes a priority scheme to guarantee the transmission privileges of high-priority stations. CSMA/CA requires a delay in network activity after each completed transmission. That delay is proportionate to the priority level of each device, with high-priority nodes programmed for short delays and low-priority nodes programmed for relatively long delays. As collisions still may occur, they are managed either through collision detection or through retransmission after receipt of a negative acknowledgment (NAK). CSMA/CA is more expensive to implement than CSMA/CD (CSMA with Collision Detection) because it requires that additional programmed logic be embedded in each device or network interface card (NIC). CSMA/CA does, however, offer the advantage of improved access control, which serves to reduce collisions and, thereby, improve the overall performance of the network. CSMA/CA is half-duplex (HDX) in nature.

Wireless LANs (WLANs), as standardized in IEEE 802.11, employ CSMA/CA. The 802.11 standard uses a positive acknowledgement (*ACK*) mechanism, which requires that the transmitting station first check the medium to determine its availability. The transmitter sends a short request to send (RTS) packet that contains the source and destination network addresses, as well as the duration of the subject transmission. If the shared medium is available, the destination station responds with a clear to send (CTS) packet. All devices on the network recognize and honor this acknowledged claim to the shared network resources. If the source station does not receive an ACK packet from the destination station, it retransmits RTS packets until access is granted. See also *802.11*, *bus*, *CSMA/CD*, *HDX*, and *MAC*.

CSMA/CD (Carrier Sense Multiple Access with Collision Detection) The most common medium access control (MAC) protocol used in bus networks, including 802.3 (Ethernet). The transmitting Ethernet station sends a data frame in both directions of the bus. Each transceiver of each station in the path of the frame reads the address in the frame header. If the address matches, the transceiver provides the frame to the target device. If the address does not match, the transceiver forwards the frame to the next transceiver. If any node detects a data collision, that station sends a brief jamming signal over a subcarrier frequency to advise all stations of the collision. All devices then back off the network. If the network is running the Nonpersistent CSMA protocol, each station then calculates a random time interval before monitoring the network again, and attempting a retransmission. See also *802.3*, *bus*, *CSMA*, *CSMA/CA*, *Ethernet*, *frame*, *MAC*, *Nonpersistent CSMA*, *subcarrier*, and *transceiver*.

CSTA (Computer Supported Telephony Applications) The first truly open computer telephony (CT) development standard for link-level protocols. CSTA was developed by the European Computer Manufacturers Association (ECMA) and subsequently was improved and formally standardized by the ITU-T, incorporating the Switch-to-Computer Applications Interface (SCAI). CSTA is a full protocol stack that requires an open system interface to a PBX, automatic call distributor (ACD), or Centrex central office (CO). See also *ACD*, *Centrex*, *computer telephony*, *ECMA*, *ITU-T*, *PBX*, and *SCAI*.

CSU (Channel Service Unit) Data circuit terminating equipment, i.e., data communications equipment (DCE), that provides the customer interface to a digital circuit. A CSU performs a number of functions, including isolation of the data terminal equipment (DTE) from the circuit for purposes of network testing, and electrical isolation from the circuit for protection from aberrant voltages. Many contemporary CSUs also have the ability to perform various line analyses, including monitoring the signal level. The CSU also serves to resolve issues of electrical coding between DTE and the circuit, and to ensure that ones density is achieved. Depending on the carrier network, 15–80 zeros can be transmitted in a row as long as the density of ones is at least 12.5 percent (1 in 8) over a specified interval of time. Also, a CSU inserts, or stuffs, 1 bits on a periodic basis in order to ensure that the various network elements maintain synchronization. A CSU also serves to provide signal regeneration and generates keep-alive signals to maintain the circuit in the event of a DTE transmission failure. Finally, the CSU stores various performance data in temporary memory for analysis by an upstream element management system (EMS). A CSU and data service unit (DSU) commonly are combined in a CSU/DSU, also known as a *CDSU.* See also *DCE*, *DSU*, *EMS*, *ones density*, and *stuff bit.*

CSU/DSU (Channel Service Unit/Data Service Unit) Also known as a *CDSU.* A combined channel service unit (CSU) and data service unit (DSU). See also *CSU* and *DSU.*

CT (Computer Telephony) See *computer telephony.*

CT0 (Cordless Telephony generation 1) A variation of CT1 that was primarily used in the United Kingdom. CT0 is an early standard for analog cordless telephony that specified eight paired channels, with the base station transmission in the 1.642 GHz-1.782 GHz range, and portable station transmission in the 47 MHz range. A number of parochial CT0 versions were developed in other countries. See also *analog*, *channel*, *cordless telephone*, and *CT1.*

CT1 (Cordless Telephony generation 1) An early standard for analog cordless telephony developed by the European Conference of Postal and Telecommunications Administrations (CEPT) in Europe, where it was known as *CEPT-1.* CT1 operates in the 915 MHz and 960 MHz bands over 40 paired channels 25 kHz wide and employs frequency modulation (FM). Frequency division multiple access (FDMA) and frequency division duplex (FDD) are used to derive two separate channels, one downstream and one upstream, each of which is 12.5 kHz wide. CT1 is limited in range to approximately 150 meters. A variation on this standard is CT0, which was primarily used in the United Kingdom. See also *analog*, *CEPT*, *channel*, *cordless telephone*, *CT0*, *downstream*, *FDD*, *FDMA*, *FM*, and *upstream.*

CT1+ (Cordless Telephony generation 1 plus) A variation on the early CT1 standard for analog cordless telephony, CT1+ was developed jointly by Belgium, Germany, and Switzerland. CT1+ was intended as the basis for a public wireless service, along the lines of Telepoint. CT1+ operated in the 887 MHz and 932 MHz bands over 80 channels 25 kHz wide, with one channel per carrier. CT1+ employed frequency division multiple access (FDMA) and frequency division duplex (FDD) to derive two separate channels per conversation, one for downstream and one for upstream transmission, each of which is 12.5 kHz wide. Although CT1+ was not successful, it originated the concept of a common air interface (CAI), which enables multiple manufacturers to develop products in support of a public cordless telephony service offering. See also *air interface*, *analog*, *carrier*, *channel*, *cordless telephone*, *CT1*, *downstream*, *FDD*, *FDMA*, *Telepoint*, *upstream*, and *wireless.*

CT2 (Cordless Telephony generation 2) An early standard for digital cordless telephony, CT2 was developed in the United Kingdom, where it formed the basis for the Telepoint public cordless service. CT2 employs time division multiple access (TDMA) and time division duplex (TDD), and is deployed on a limited basis in Europe, Canada, and the Asia-Pacific. Although it originally supported only outgoing calling, contemporary CT2 implementations support two-way calling. As CT2 does not support handoff, the user must remain within range of the antenna used to set up the call. CT2 operates in the 864–868 MHz range, supports 40 channels spaced at 100 kHz with one channel per carrier, and uses Gaussian frequency shift keying (GFSK) modulation. Dynamic channel allocation requires a frequency-agile handset. CT2 was the first international standard providing a common air interface (CAI) for systems operating in the 800 MHz and 900 MHz bands. CT2 supports digital speech at 32 kbps and data communications at rates up to 72 kbps. See also *air interface, antenna, carrier, channel, cordless telephone, digital, GFSK, handoff, modulation, TDD, TDMA*, and *Telepoint*.

CT2+ (Cordless Telephony generation 2 plus) An early standard for digital cordless telephony, CT2+ was an improvement on CT2, supporting two-way calling and call handoff. CT2+ uses 8 MHz of bandwidth in the 944–948 MHz range supports 40 channels spaced at 100 kHz with one channel per carrier, and uses Gaussian frequency shift keying (GFSK) modulation. CT2+ is based on dynamic channel allocation, requiring frequency-agile handsets. Encryption is supported for improved security. A common signaling and control channel offers improved call set-up times, increased traffic capacity, and longer battery life because the handset must monitor only the signaling channel. CT2+ has been used in applications such as the Walkabout public cordless telephony market test in Canberra, Australia. See also *carrier, channel, cordless telephone, CT2, digital, encryption, GFSK, handoff, modulation*, and *signaling and control*.

CT3 (Cordless Telephony generation 3) A proprietary digital cordless telephony system developed by Ericsson in 1990 as a wireless office telecommunications system (WOTS) for application in high-density office environments. CT3 is based on time division multiple access (TDMA) and time division duplex (TDD), runs in the 944–948 MHz range, and supports roaming and call handoff. See also *cordless telephone, digital, handoff, TDD, TDMA*, and *WOTS*.

CTD (Cell Transfer Delay) In asynchronous transfer mode (ATM), the average time it takes a cell to transverse the network, from source to destination between a user network interface (UNI) at each end. CTD is the sum of all delays imposed by coding and decoding, segmentation and reassembly (SAR), propagation across transmission media, cell processing at the nodes, queuing of the cell in input and output buffers, and loss and recovery. If a cell arrives too late at the receiving station, it may be considered lost or late, and may be disregarded. If the subject cell is a segment of a larger data packet, the entire packet must be discarded and forgotten, or retransmitted. Maximum CTD (maxCTD) is negotiated between the end stations and the network. See also *ATM, buffer, cell, code, network, node, packet, propagation, SAR*, and *UNI*.

CTS (Clear To Send) **1.** A message sent to a device clearing it for access to a wireless network. Demand Aassigned Multiple Access (DAMA), for example, is a protocol that assigns available channel capacity to an Earth station from a pool of bandwidth, on demand and as available. The Earth station transmits request to send (RTS) messages to the satellite until it responds with a clear to send (CTS), at which time the message transmission ensues. See also *DAMA*. **2.** A conductor (pin) used on serial data interfaces, as between a terminal and modem, to indicate one device is ready to accept data from the other. See also *conductor, modem, serial communication*, and *terminal*.

CTX (CenTreX) See *Centrex*.

Cu Symbol for copper. See *copper*.

CUG (Closed User Group) A group of users on the same public network who communicate with each other by mutual agreement, and who exclude others. A CUG prevents unwanted correspondence in an instant messaging (IM) system, for example, and provides a significant level of security. See also *IM* and *security*.

current (I) The flow of electrons through a metallic circuit, like the flow of water down a riverbed. The direction of flow is from positive (+) pole to negative (–) pole at opposite ends of the circuit. Direct current (DC) travels in one direction, only. Alternating current (AC) travels first in one direction and then in the other as the polarity changes at the ends of the circuit. Current is measured in amperes (A), or amps.

custom calling services The popular conversational term for customer local access signaling services (CLASS). See *CLASS*.

custom control routing Also known as *call vectoring*. See *call vectoring*.

customer access line charge (CALC) Synonymous with subscriber line charge (SLC). See *SLC*.

customer contact center See *call center*.

customer-originated trace Also known as *call trace*. A CLASS service of the public switched telephone network (PSTN). The feature enables the subscriber to initiate a trace on the last call received. The service is intended as a countermeasure for obscene or harassing calls. The customer initiates the trace by depressing the telephone switchhook and dialing a code. The central office (CO) records the calling number, which can be provided to law enforcement authorities on request, when the called party has filed a proper complaint. See also *CLASS* and *PSTN*.

customer-owned coin-operated telephone (COCOT) See *COCOT*.

customer premises equipment (CPE) See *CPE*.

customer rearrangement Also known as *automatic set relocation*. See *automatic set relocation*.

Custom Local Access Signaling Services (CLASS) See *CLASS*.

custom ringing Also known as *distinctive ringing*. See *distinctive ringing*.

cutoff frequency Referring to the frequency above or below which a band-pass filter absorbs, attenuates, blocks, rejects, or removes signals, allowing only the signals within the designated band to pass through. Generally, the filter introduces insertion loss of 3 dB or more, which results in signal attenuation of 50 percent or more. See also *absorption*, *attenuation*, *band*, *band-pass filter*, *dB*, *frequency*, *insertion loss*, and *signal*.

cutoff wavelength In fiber optic transmission, the wavelength beyond which a single-mode fiber (SMF) supports only a single mode of propagation. See also *fiber optics*, *mode*, *propagation*, *SMF*, and *wavelength*.

cut-through switch A LAN matrix switch that quickly reads the address of a data frame and quickly flows it through the switching matrix, bit by bit, starting to flow the front of the frame forward before the last of the frame arrives. See also *fragment-free switch*, *LAN switch*, *matrix switch*, and *switch*.

CVoDSL (Channelized Voice over Digital Subscriber Line) A technique that enables multiple derived time-division multiplexed (TDM) voice conversations to be transported simultaneously over DSL in 64 kbps channels. The term largely is applied to voice over ADSL2 and ADSL2+. See also *ADSL2*, *ADSL2+*, *channel*, *TDM*, and *voice*.

CVSD (Continuously Variable Slope Delta) A voice compression technique that encodes the changes (i.e., deltas) in the slope (i.e., rate of change), rather than the amplitude, of analog voice signals. CVSD yield compression ratios of 4:1 (16 kbps) or 8:1 (9.6 kbps), as compared to pulse code modulation (PCM) at 64 kbps. Although CVSDM has given way to techniques such as adaptive differential pulse code modulation (ADPCM) in contemporary voice networks, it is used in Bluetooth voice applications. See also *ADPCM*, *Bluetooth*, *compression*, *encode*, and *PCM*.

CWDM (Coarse Wavelength Division Multiplexing) An optical multiplexing technique specified by the ITU-T as 18 wavelengths in the 1270–1610 nm range, with spacing of 20 nm (2500 GHz at 1550 nm). Targeted at networks with a reach of 50 kilometers or less, CWDM offers the advantage of using uncooled laser sources and filters, which are not only less expensive, but also consume less power and possess smaller footprints that the cooled lasers used in Dense WDM (DWDM). While CWDM does not allow channels to be spaced as tightly as DWDM and, therefore, does not offer the same spectral efficiency, it is a cost-effective alternative for short haul metropolitan and local rings supporting applications such as GbE. See also *DWDM*, *filter*, *GbE*, *ITU-T*, *laser*, *multiplexer*, *spectral efficiency*, *wavelength*, and *WDM*.

cybercafé **1.** An establishment that provides patrons with Internet access, usually high-speed access, on a fee basis. A cybercafé may also sell coffee and light fare. See also *Internet.* **2.** A virtual meeting place in cyberspace where people can engage in a chat session on a bulletin board system (BBS) or instant messaging (IM) system. See also *BBS*, *cyberspace*, *IM*, and *virtual.*

cybercast A streaming audio or video broadcast in cyberspace, i.e., the Internet. See also *broadcast*, *cyberspace*, and *Internet.*

cyberspace The virtual space created by interconnected computers and computer networks on the Internet. Cyberspace is a conceptual electronic space unbounded by distance or other physical limitations. William Gibson coined the term in his novel *Neuromancer* (1982) to describe an advanced virtual reality network. See also *Internet* and *virtual.*

cycle time Synonymous with *rise and fall time*, cycle time is the time it takes for a fiber optic light source to cycle through a rise to its peak and a fall to its trough level of signal intensity, or power. Digital fiber optic systems use Amplitude Modulation (AM), so the faster the light source can cycle, the higher the bit rate. (*Note:* Most fiber optic system are digital, although a few are analog in nature.) Light sources never completely turn off, as that would limit their speed, so there is always some amount of residual signal present. Diode lasers are the fastest light sources, followed by vertical cavity surface-emitting lasers (VCSELs) and light-emitting diodes (LEDs). See also *AM*, *diode laser*, *LED*, and *VCSEL.*

cyclic checksum A number of any length, but often either 6, 8, or 32 characters long, that is calculated from a much larger data block (typically a frame, file, or image) in a way that is sensitive to any change in the original data. The checksum is calculated by the sender and either appended to the data block or delivered separately, as are MD5 checksums for software files. The receiver independently calculates a checksum and compares the two values as a means of detecting errors created in transit, or perhaps deliberate alterations of the data. The term cyclic applies to the calculation method, which is binary long division of the original data by a derived number until there is only a remainder, which is the checksum. See also *checksum* and *cyclic redundancy check.*

cyclic redundancy check (CRC) See *CRC.*

D **1.** In physics, the symbol for dispersion. See *dispersion.* **2.** Referring to an ISDN data (D) channel, also known as a delta (D) channel (from the phonetic alphabet), which is a channel designated for out-of-band signaling and control functions. See also *D channel.*

D1 An early (1962) T1 framing convention that robbed the least significant bit (LSB) in each channel of each frame in order to insert a signaling bit. D1 is obsolete. See also *bit robbing, D2, D3, D4, ESF, frame, LSB,* and *T1.*

D2 An early T-carrier framing convention used to create a 12-frame sequence or superframe. D2 framing is considered obsolete. See also *D2, D3, D4, ESF, frame,* and *T-carrier.*

D3 A T-carrier framing convention that assumes that all inputs are analog. D3 uses a superframe format. See also *D1, D2, D4, ESF, frame,* and *T-carrier.*

D4 Also known as M24 Superframe, with M meaning *Multiplex.* A T-carrier framing convention that enables bit robbing of the least significant bit (LSB) of the sixth and twelfth frames, only, of a 12-frame sequence or superframe. Voice and data both are accommodated, with data treated as a digital input. Ones density must be maintained through the insertion of stuff bits. See also *bit robbing, D1, D2, D3, ESF, frame, LSB, ones density, superframe,* and *T-carrier.*

DAC **1.** Dual-Attached Concentrator. A concentrator or LAN switch attached to both rings of a Fiber-Distributed Data Interface (FDDI) LAN. See also *concentrator, FDDI,* and *LAN.* **2.** Digital-to-Analog Converter. A device in the form of a chipset that receives digital signals and changes them into analog format. In a typical voice application, for example, an ADC (Analog-to-Digital Converter) receives analog signals, measures the input at a regular sampling interval (or on command), and reports a digital output of the results. At the receiving end of the connection, a DAC reverses the process. For a full explanation of the process, see *ADC.* See also *analog, digital,* and *signal.*

DACS (Digital Access Cross-connect System) Synonymous with *digital cross-connect system* (DCCS or DXC). A non-blocking, electronic common control (ECC) switch that serves to cross-connect digital carrier bit streams on a buffered basis by redirecting individual channels or frames from one circuit to another through an electronic cross-connect matrix. A DACS is much like a static switch in that connections can be changed to alter physical paths, on a semi-permanent basis, rather than a call-by-call or packet-by-packet basis. A DACS is akin to a digital, electronic version of a manual distribution frame, which can take the form of a main distribution frame (MDF) or intermediate distribution frame (IDF). See also *ECC, IDF, MDF, non-blocking,* and *switch.*

daemon From the Greek *daimon,* meaning *divine power.* A utility that resides in RAM, waiting in the background until an event triggers it to take action. Print spoolers, e-mail handlers, and automatic backup utilities are examples of daemons. In mythology, a daemon was variously a guardian spirit or secondary divinity in the form of a demigod, i.e., half-man and half-god, that was tasked with duties deemed too insignificant for the gods' attention. See also *RAM* and *utility.*

daisy chain A method of connecting devices together with a series of cables, one plugged into another, much as one would intertwine flower stems to create a chain of daisies. See also *home run.*

DAMA (Demand Assigned Multiple Access) A wireless access technique used extensively in satellite systems. DAMA is a variation of frequency division multiple access (FDMA), which assigns Earth stations specific uplink and downlink frequencies within an allotted range licensed for use. DAMA assigns available channel capacity to an Earth station from a pool of bandwidth, on demand and as available in a first come, first served manner. The Earth station transmits request to send (RTS) messages to the satellite until it responds with a clear to send (CTS), at which time the message transmission ensues. DAMA allows several carriers, and perhaps a great many Earth stations, to share a given frequency band, as needed. See also *downlink*, *FDMA*, and *uplink*.

D-AMPS (Digital Advanced Mobile Phone Service) Also known as US Digital Cellular (USDC), US TDMA and NA-TDMA (North American TDMA), and specified in IS-54, which evolved into IS-136. D-AMPS is a North American 2G digital cellular radio standard that essentially is a digital version of the earlier analog AMPS specification. D-AMPS operates in the same 800 MHz band as AMPS and the two can coexist in the same network. D-AMPS uses the same 30 kHz bands as AMPS, and supports up to 416 frequency channels per carrier. D-AMPS employs time division duplex (TDD) to subdivide each frequency channel into six time slots, each of which operates at 8 kbps. Each call initially uses two time slots in each direction, for a total of 16 kbps. The standard recommends speech compression using vector-sum excited linear predictive coding (VSELP) at an average rate of 7.95 kbps that can burst up to 48 kbps. IS-136 is known as a dual-mode standard because both D-AMPS and AMPS can coexist on the same network, with both using the same 21 control channels for call setup, call handoff, and call teardown. Thereby, IS-136 offers carriers the advantage of a graceful transition from analog to digital. IS-136 also includes a non-intrusive digital control channel (DCCH), which adds features such as short message service (SMS) and caller ID. D-AMPS supports symmetric data communications at up to 9.6 kbps per channel through paired downlink and uplink time slots, and as many as three channels can be aggregated for speeds up to 28.8 kbps. D-AMPS is considered to be at the end of its technological life cycle and is being replaced by GSM/GPRS and CDMA2000. The RF modulation technique is π/4 differential quaternary phase shift keying (π/4 DQPSK). See also *2G*, *AMPS*, *analog*, *caller ID*, *carrier*, *CDMA2000*, *cellular radio*, *channel*, *DCCH*, *digital*, *downlink*, *GPRS*, *GSM*, *handoff*, *IS-54*, *IS-136*, *modulation*, *π/4 DQPSK*, *RF*, *SMS*, *symmetric*, *TDD*, *time slot*, *uplink*, and *VSELP*.

dark fiber Optical fiber left unlit in a fiber optic cable. Fiber optic cables often contain a great many fibers, some of which are lit, and others of which are left unlit, or dark. The dark fibers can be spares for backup purposes, can be held in reserve to accommodate future demand, or can be available for lease or sale to other carriers or user organizations with private line requirements. See also *dim fiber* and *dry copper pair*.

DARPA (Defense Advanced Research Projects Agency) The central research and development organization of the United States Department of Defense (DoD). DARPA manages and directs selected basis and applied science projects and pursues research and technology considered high in risk and payoff and that hold the potential for dramatic advances for traditional military roles and missions. DARPA was formed in 1958 in response to the Soviet launching of the Sputnik satellite. See also *DARPANET*.

DARPA Internet See *DARPANET*.

DARPANET (Defense Advanced Research Projects Agency Network) An early United States government network that merged with the ARPANET (Advanced Research Projects Agency Network) to form the DARPA Internet, which eventually gave rise to the Internet.

DAS (Dual Attached Station) A station attached to both rings of a Fiber Distributed Data Interface (FDDI) LAN. The cost of dual attachment generally reserves this approach for servers and routers. Lesser devices such as workstations are single attached. See also *FDDI*, *LAN*, and *station*.

dash **1.** The longer of the two signal elements in Morse code, created by closing an electrical circuit with a mechanical key for a relatively long period of time. An audible dash is a long click or buzz, known as a *dah* to radio telegraph operators, and is graphically represented as a short horizontal line —. See also *dot*, *Morse code*, and *telegraph*. **2.** Thomas Alva Edison Jr. (1876–1935) was nicknamed "Dash" by his father, Thomas Alva Edison (1847–1931), the "Wizard of Menlo Park," who invented such devices as the phonograph, electric light bulb, carbon microphone, and electric chair. Much of his early work was in telegraphy, and the two-way telegraph and quadraplex telegraph were among his early financial successes. See also *dot* and *telegraph*.

data From the Latin datum, meaning what is given. What is known or assumed, and upon which conclusions can be drawn. Factual information in a form that can be input to, created by, processed by, stored in, and output by a computer. Data can take the form of characters such as letters, numbers, punctuation marks, mathematical operators, and control characters. Data also can take the form of photographic display elements, such as pixels. *Note:* Data is the plural form of the Latin datum, although data is used conversationally to represent both singular and plural.

database A file comprising a collection of related information organized in a manner that enables operations such as searching and sorting the data.

data circuit terminating equipment (DCTE or DCE) Synonymous with data communications equipment (DCE). See *DCE*.

data communications equipment (DCE) See *DCE*.

data compression See *compression*.

Data Encryption Standard (DES) See *DES*.

Data Exchange Interface (DXI) See *DXI*.

data field See *field*.

data file An electronic file containing the work created with a computer program. The contents of a data file are information in text or graphic format. See also *graphics*, *program file*, and *text*.

data format A critical part of a communications protocol that enables the receiving device to logically determine what is to be done with the data and how to go about doing it. As illustrated in Figure D-1, a data format generally involves a header, text field, and a trailer. Although the header and trailer are overhead, they serve critical functions in support of the successful transfer of the data content. In total, the header, text, and trailer compose what is known variously as a packet, block, frame, or cell, with the specific terminology being sensitive to the specific protocol involved. *Note:* Some protocols, such as Asynchronous Transfer Mode (ATM), do not involve a trailer at the cell level, but do have trailers at higher (convergence) layers. See also *header*, *text field*, and *trailer*.

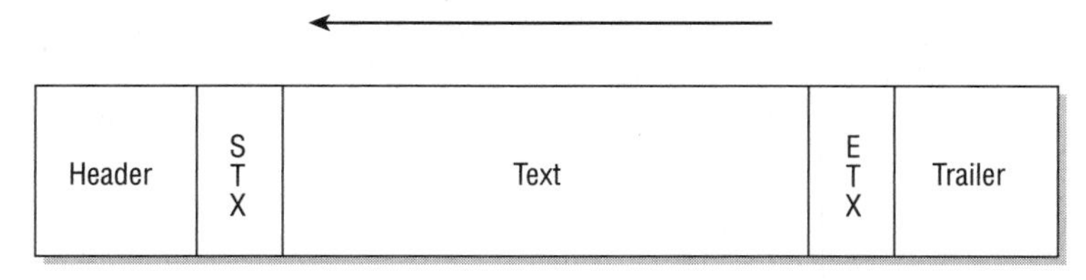

Figure D-1

datagram A fully addressed and self-contained packet that can be switched through a packet-switched network and delivered to the intended target station independent of any other packets. That is, a datagram contains enough information to enable the network routers to deliver it to the correct destination. A datagram is a standalone, totally independent packet that is not considered by the routing protocol to be part of a datastream. See also *datagram service*, *packet*, *packet switch*, and *protocol*.

datagram service A service mode in which a packet-switched network routes each datagram without regard to any other datagrams that might have preceded it or that might follow it. Datagram service is connectionless, meaning that there is no need to set up a predetermined path for the packets to travel through the network. Rather, each packet in a stream of datagrams can take an entirely different route from originating host to destination host. As a result, the datagrams can arrive at the destination host out of sequence. *Note:* In the wild, the routing protocol commonly determines the path in advance, anyway. See also *connectionless*, *datagram*, *packet*, and *packet switch*.

Data Interchange Standards Association (DISA) See *DISA*.

Data Link Connection Identifier (DLCI) See *DLCI*.

Data Link Layer (DLL) Layer 2 of the seven-layer Open Systems Interconnection (OSI) Reference Model. Software at the Data Link Layer is responsible for addressing, signal formatting (i.e., line encoding and decoding), framing, frame synchronization and sequencing, flow control, error control, and otherwise managing traffic between devices across a link. HDLC, LAP-B, and LAP-D are Data Link Layer protocols, as is frame relay. Local area networks (LANs) operate at the Data Link and Physical Layers. See also *error control*, *flow control*, *frame relay*, *HDLC*, *LAP-B*, *LAP-D*, *layer*, *link*, *network architecture*, *OSI Reference Model*, *protocol*, *software*, and *synchronization*.

Data over Cable Service Interface Specification (DOCSIS) See *DOCSIS*.

DataPhone In obsolete (1959) AT&T terminology, modem. See also *modem*.

data service unit (DSU) See *DSU*.

data set or dataset **1.** A group of related information comprising separate data elements stored, retrieved, or otherwise organized and treated as a unit, i.e., file. The term largely is used in the mainframe community; others use the term *file*. **2.** In obsolete (1957) AT&T terminology, modem. See also *modem*.

data terminal equipment (DTE) See *DTE*.

Data Under Voice (DUV) See *DUV*.

dB (decibel) In physics, one-tenth (1/10) of a bel (B). Bel is a measure of comparative power, or loudness, i.e., a unit of relative power ratio. The number of decibels is the decimal logarithm of the relative power ratio, which is expressed mathematically as follows:

$$dB = 10 \log_{10} (P_1/P_2)$$

where dB = deciBel, P_1 is the power level to be measured, and P_2 is the reference power level. If the result of the calculation is positive (+), there is a gain in signal power. If the result of the calculation is negative (–), there is a loss of signal power, i.e., attenuation. If, for example, the ratio of power at the output port of an optical amplifier to the power at the input port is 2.000, i.e., twice the reference level, the gain in signal power is expressed as:

$$2.000 = +3.0103 \text{ dB}$$

If the ratio of power at the point of signal origin where the optical signal from an LED or laser enters an optical fiber to the power at the end of the fiber is 0.500, i.e., one-half the reference level, the attenuation, or insertion loss, is expressed as:

$$0.500 = -3.0103 \text{ dB}$$

Decibel is the commonly used relative power ratio, as the unit of measurement is smaller and more convenient than the bel. The bel is named for Alexander Graham Bell, inventor of the telephone, among other things. See also *amplifier*; *attenuation*; *Bell, Alexander Graham*; *decibel*; *gain*; *insertion loss*; *logarithm*; and *power*.

dB/km (deciBels per kilometer) A measure of the attenuation of a signal over a distance. See also *attenuation*, *dB*, and *signal*.

dBm (deciBels referenced to a milliWatt) A unit of measure used to reference signal strength to electrical power level. Unlike dB, which is a relative measure, dBm is an absolute measure, and can be used to express very small values (dBm) and very large values (dBW). The baseline reference relationship is 1mW = 0 dBm. A doubling of signal power yields an increase of approximately 3 dBm (2mW = 3 dBm) and halving of signal power yields a loss of approximately 3 dBm (500 μW = -3 dBm. Similarly, a tenfold increase in signal power yields an increase of approximately 10 dBm (10mW = 10 dBm) and a decrease of 90 percent of signal power yields a loss of approximately 10 dBm (100 μW = -10 dBm). The dBm measurement is used extensively in radio and fiber optics networking. Bluetooth Class 1 radios operate at a maximum 100 mW (20 dBm) and have a nominal link range of 100 meters (300 feet). Bluetooth Class 3 radios operate at a maximum of 1 mW (0 dBm) and have a nominal range of 1 meter (3 feet). A typical 100Base-FX Ethernet LAN might use a fiber optic laser with an operating wavelength of 1550 nm and a launch power range from a maximum of 0 dBm and minimum –5 dBm and a receiver with a sensitivity of –34 dBm. See also *100Base-FX*, *Bluetooth*, *dB*, *Ethernet*, *laser*, *signal*, and *wavelength*.

DBS (Direct Broadcast Satellite) A geosynchronous earth-orbiting (GEO) satellite system that broadcasts signals directly to subscribers' premises, thereby bypassing traditional terrestrial broadcast television stations and CATV providers. DBS systems that offer two-way Internet services also receive signals directly from subscriber's premises, thereby bypassing traditional wireline Internet service providers (ISPs). See also *broadcast*, *CATV*, *GEO*, *Internet*, *ISP*, and *wireline*.

DC (Direct Current) The flow of electrons through a metallic circuit, with the direction of flow being from positive (+) pole to negative (–) pole. Direct current (DC) travels in one direction only, while alternating current (AC) travels in both directions across the circuit. A continuous flow of AC current travels first in one direction and then reverses polarity and flows in the opposite direction. See also *AC*, *electromagnetic spectrum*, *Hz*, and *wavelength*.

DCCH (Digital Control CHannel) In North American digital cellular radio systems based on IS-136 and generally known as Digital Advanced Mobile Phone System (D-AMPS), a channel that supports residential and in-building coverage, increased battery standby time, text messaging via short message service (SMS), over-the-air activation, circuit-switched data (CSD), and caller ID. See also *cellular radio*, *channel*, *CSD*, *D-AMPS*, *digital*, *IS-136*, *SMS*, and *text messaging*.

DCCS (Digital Cross-Connect System) See *DXC*.

DCE (Data Communications Equipment) Also known as data circuit terminating equipment (DCTE). Equipment that interfaces data terminal equipment (DTE) to a wide area network (WAN), resolving any issues of incompatibility. Incompatibility issues can include signal format (digital versus analog), voltage level, signaling speed, and bit density. DCE includes modems, digital service units (DSUs), channel service units (CSUs), and front-end processors (FEPs). See also *analog*, *bit density*, *CSU*, *digital*, *DSU*, *DTE*, *FEP*, *modem*, *signaling speed*, *voltage*, and *WAN*.

D channel (Delta channel or Data channel) In the integrated service digital network (ISDN), a channel designated for out-of-band signaling and control functions. Standard ISDN interfaces include multiple B (Bearer) channels that carry user payload and a D channel for signaling and control. Basic rate interface (BRI) comprises two B channels at 64 kbps and one D channel at 16 kbps, and is often referred to as 2B+D. Primary rate interface (PRI) comprises 23 B channels at 64 kbps, plus a D channel at 64 kbps, is compatible with North American T1 and Japanese J-1 standards, and is often referred to as 23B+D. Primary rate access (PRA) comprises 30 B channels at 64 kbps, plus a D channel at 64 kbps, is compatible with European E-1 standards, and is often referred to as 30B+D. A D channel also can support low-speed end user packet data and telemetry applications. See also *D*, *B channel*, *BRI*, *E-1*, *ISDN*, *J-1*, *packet*, *payload*, *PRA*, *PRI*, and *T1*.

DCS 1800 (Digital Cellular System 1800) Also known as *Personal Communications Network* (PCN). An upbanded version of GSM, the pan-European digital cellular radio system. DCS 1800 operates in the 1800 MHz (1.8 GHz) range, rather than the 900 MHz range of GSM. See also *GSM.*

DCT (Discrete Cosine Transform) A technique used in signal and image processing and particularly in lossy compression techniques. DCT separates an image into discrete blocks of pixels of differing importance with respect to the overall image. DCT expresses a function or signal in terms of a sum of sinusoidal waveforms that vary in amplitude and frequency, essentially transforming the image from the spatial domain into the frequency domain. In the process, the average luminance of each block is evaluated using the DC coefficient. Transform compression is based on the premise that the low-frequency components of a signal are more important than the high-frequency components. Therefore, a substantial reduction in the number of bits used to represent a high-frequency component will degrade the quality of the image only slightly. JPEG and MPEG both specify DCT, as do the ITU-T H.261 and H.263 videoconferencing standards. See also *amplitude, DC, frequency, H.261, H.263, ITU-T, JPEG, lossy compression, MPEG, pixel, signal*, and *sine wave.*

DCTE (Data Circuit Terminating Equipment) Synonymous with data communications equipment (DCE). See *DCE.*

DDD (Direct Distance Dialing) A feature of the public switched telephone network (PSTN) that allows a subscriber to dial a long distance telephone number directly, i.e., without the intervention of an operator. DDD was introduced by the Bell System in 1951, and is virtually ubiquitous today. See also *long distance* and *PSTN.*

DDN (Defense Data Network) Also known as MILNET (Military Network). A packet network formed in 1983 of ARPANET users in European and Pacific Rim continents. See also *ARPANET.*

DDS (Digital Data Service) Also known as Digital Data Service and Subrate Digital Loop (SRDL). An end-to-end, fully digital, dedicated data service available in point-to-point and point-to-multipoint configurations. In either configuration, a designated headend system (Front-End Processor [FEP] or communications server) controls all access to the network through a process of polling. The DDS generally supports line rates of 2400 bps, 4800 bps, 9600 bps, 19.2 kbps, 56 kbps or 64 kbps, and digital carrier rates of 1.544 Mbps (T1) and 2.048 Mbps (E-1). The DDS signals actually are carried inside T-carrier or E-carrier channels in the backbone carrier networks. DDS was introduced by AT&T in 1974 and is now provided by most incumbent carriers, worldwide, under a variety of names. Fractional E-1 and T1 local loops are often tariffed as DDS circuits. See also *backbone, carrier, digital, DUV, E-1, E-carrier, FDX, four-wire circuit, HDX, headend, point-to-multipoint, point-to-point, simplex, T1*, and *T-carrier.*

DE (Discard Eligible) In the frame relay LAPF frame, a 1-bit field indicating the eligibility of the frame for discard under conditions of network congestion. Theoretically, the user equipment sets the DE in consideration of the acceptability of the application to packet loss. Should the user equipment not set the DE, the network routers and switches may do so on a random basis or to police traffic that exceeds the CIR. SDLC, voice, and video traffic, for example, do not tolerate loss gracefully, and so typically are engineered to avoid DE status. File transfer and e-mail applications are not time-sensitive and, therefore, are highly insensitive to latency and loss. Therefore, the user may offer them to the network marked DE and pay a lower cost. See also *congestion, frame relay, LAPF, latency*, and *SDLC.*

decibel (dB) See *dB.*

decibels per kilometer (dB/km) See *dB/km.*

decimal notation The base-10 numbering system. A computational system based on the number 10. The 10 digits in decimal notation are 0, 1, 2, 3, 4, 5, 6, 7, 8, and 9. See also *binary notation, digit*, and *hexadecimal notation.*

DECT (Digital Enhanced Cordless Telecommunications) Originally known as Digital European Cordless Telecommunications. The pan-European standard for digital cordless telephony, DECT was ratified by the European Telecommunications Standards Institute (ETSI) in 1992, and is intended primarily for indoor applications. Through frequency division multiplexing (FDM), DECT provides 10 carriers in the 1880–1990 MHz band, with channel spacing is at 1.728 MHz. Each channel will support 1.152 Mbps with Gaussian frequency shift keying (GFSK) as the modulation technique. Each channel supports 12 users through time division multiple access (TDMA) and time division duplex (TDD), for a total system load of 120 users. Voice encoding is adaptive differential pulse code modulation (ADPCM) at 32 kbps. DECT supports call handoff, so users can roam from cell to cell at pedestrian speeds as long as they remain within range of the system. DECT antennas can be equipped with optional spatial diversity to deal with multipath fading. Security is provided through authentication and encryption mechanisms. In North America, DECT is the basis for the Personal Wireless Telecommunications (PWT) standard, which operates in the unlicensed 1910–1920 MHz band. PWT-E is an extension into the licensed bands of 1850–1910 MHz and 1930–1990 MHz. See also *ADPCM*, *antenna*, *authentication*, *carrier*, *channel*, *cordless telephone*, *digital*, *encode*, *encryption*, *ETSI*, *FDM*, *GFSK*, *handoff*, *modulation*, *multipath fading*, *PWT*, *PWT-E*, *spatial diversity*, *TDD*, and *TDMA*.

dedicated circuit A distinct physical circuit dedicated to directly connecting devices, such as multiplexers, PBXs, and host computers. A dedicated circuit, also known as a *leased line*, can be provisioned over a private network comprising facilities owned by the end user organization, although it more typically is in the form of a leased line provisioned over a public network. In the latter case, the circuit includes an access circuit, or local loop, that connects the originating device at the customer premises to the service provider's point of presence (POP) at the edge of the carrier network. In the case of an incumbent local exchange carrier (ILEC), the POP typically is housed in a central office (CO). At the POP, the access circuit terminates in a wire center, where it is cross-connected directly to a long-haul transport circuit, bypassing any switching devices. The long haul portion of the dedicated circuit typically comprises multiple interconnected links and terminates in a POP at the egress edge of the network, where it is cross-connected to another access circuit that connects to the premises housing the destination device. A dedicated circuit offers the advantages of dedicated availability, dedicated bandwidth, and excellent performance overall. Because a dedicated circuit is not in shared public use but is dedicated to the requirements of a specific customer, it tends to be expensive, with the cost sensitive to bandwidth and distance. The cost, however, is typically a flat rate, with no usage-sensitive component; therefore, the end user organization can use the circuit constantly to maximum capacity at the same cost as if it were to not use the circuit at all. The nature of the traffic over a dedicated circuit generally is not restricted, and can include voice, computer data, facsimile, image, video, and multimedia traffic. The ability to integrate such a wide variety of traffic over a single facility offers considerable efficiencies.

A multi-site user organization might consider a private, leased line network to interconnect the sites. However, the process of designing such a network can be difficult, as it is necessary to determine the points of termination in the optimal topology, the correct number of circuits, and the bandwidth requirements of each. When the design is established, the provisioning time required by the carrier can be quite lengthy. As a dedicated circuit involves a specific set of network elements, the circuit is susceptible to disruption. Therefore, backup circuits or services are required to ensure connectivity in the event of either a catastrophic failure or serious performance degradation.

Organizations with intense communications requirements commonly consider dedicated circuits to be viable alternatives to switched circuits. Large data centers that communicate intensively in support of applications such as data backup traditionally have opted for dedicated circuits. Large multi-site end user organizations often use dedicated circuits known as tie trunks to tie together multiple PBXs. In such applications, the advantages of assured availability, capacity, and performance in support of mission-critical, time-sensitive applications, particularly when coupled with low comparative cost, can outweigh considerations of configuration difficulty and risk of circuit failure. Dedicated circuits sometimes are referred to as nailed-up circuits because, in bygone days, the twisted-pair copper physical circuits were hung from nails driven in the walls of the carrier's wire centers. See also *switched circuit* and *tie trunk*.

de facto From Latin, literally *from what is done*, meaning *in fact*. See also *standard*.

Defense Advanced Research Project Agency Network (DARPANET) See *DARPANET*.

Defense Data Network (DDN) See *DDN*.

de jure From Latin, literally meaning *from the law*. See also *standard*.

delay The total time required for a signal to travel from one point to another, generally from a transmitter through a network to a receiver. See also *delay skew* and *latency*.

delay skew The difference in time required by a signal to propagate through conductors in the same cable due to differences in the physical lengths of the pairs caused by different twist ratios. For example, each of the four pairs in a Cat 5 or Cat 5e cable has a slightly different twist ratio, in order to minimize crosstalk. The difference in twist ratios results in a slightly different lay length, i.e., physical length if the cable were to be untwisted and laid flat, for each pair. This causes built-in propagation delay skew simply because it takes more time for a signal to travel a longer physical path, as illustrated in Figure D-2. As too much delay skew will cause transmission errors because of timing differences between the signals spread across the various pairs, some cable manufacturers take advantage of the skin effect and use foamed insulation, rather than solid insulation, on the conductors. Other manufacturers simply stagger the twists. In fiber optics and wireless systems, the comparable term is *delay spread*, which is due, respectively, to modal dispersion and multipath propagation. See also *crosstalk*, *delay spread*, *lay length*, *modal dispersion*, *multipath propagation*, *skin effect*, and *twist ratio*.

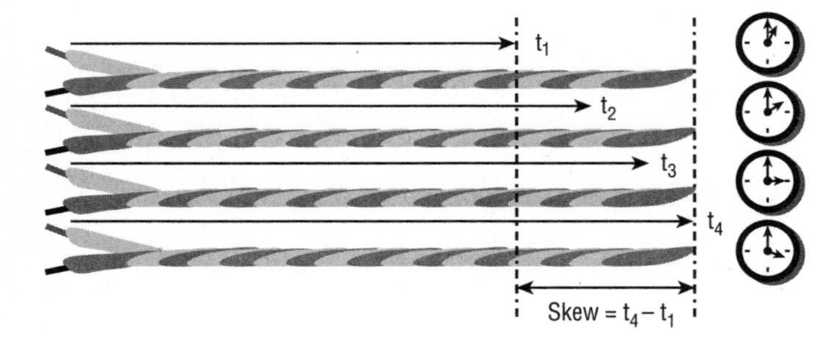

Figure D-2

delay spread See *delay skew*.

delta (Δ) A change (increase or decrease) in the value of a variable.

delta modulation A modulation technique for encoding an analog-to-digital signal conversion. Delta modulation samples the analog signal at precise points in time, converting the signal into a series of segments, each of which is compared to the original signal to determine if the amplitude changes. If there is a change in relative amplitude from one sample to the next, a series of bits is sent to indicate the extent of the increase or decrease in signal amplitude. If there is no change in amplitude, no bits are transmitted. Variations on the theme of delta modulation include continuously variable-slope delta (CVSD) modulation. See also *amplitude*, *CVSD*, *delta*, *encode*, and *modulation*.

demand assigned multiple access (DAMA) See *DAMA*.

demarc (demarcation point) The point of demarcation between the public carrier network and the customer premises. At the demarc there typically is a physical interface between the local loop, on the network side, and the subscriber's inside wire and cable system and customer premises equipment (CPE). The demarc, whether physical or logical in nature, constitutes a boundary of responsibility, with the carrier being solely responsible for the network side and the customer being primarily, if not solely, responsible for the subscriber side. In a public network connecting to a residential or small business subscriber via an electrical local loop, the physical interface commonly is described as a network interface unit (NIU) or network interface device (NID), and includes a surge protector. If the subscriber is a larger business, the demarc is positioned at the Minimum Point of Entry (MPOE), defined as the closest logical and practical point within the customer domain. In a high-rise office building, for example, it typically is defined as a point of the entrance cable 12 inches from the inside wall. See also *CPE*, *local loop*, and *protector*.

demarcation point See *demarc*.

demod/remod (demodulation/remodulation) Also known as *fax relay*. See *fax relay*.

demodulate The opposite of modulate. Demodulation reverses the process of modulation in order to extract a signal from a circuit. See also *modulate*.

dense wavelength division multiplexing (DWDM) See *DWDM*.

DES (Data Encryption Standard) An encryption standard developed by IBM and adopted by the United States government in 1976 for non-classified applications. DES also was widely used by financial institutions for electronic financial transfers. DES uses a 56-bit key, which subsequently proved to be vulnerable to attack as computer processors increased in speed and cryptanalysis tools became more sophisticated. DES has since been superseded by standards such as Advanced Encryption Standard (AES) and Triple DES. See also *AES*, *encryption*, *security*, and *Triple DES*.

deterministic **1.** Predictive. Referring to the ability to objectively predict an outcome or result of a process due to knowledge of a cause and effect relationship. **2.** In telecommunications switching and routing, the programmed predetermination of a path between nodes. See also *node*, *path*, *route*, and *switch*. **3.** In local area networks (LANs), a non-contentious medium access control (MAC) convention that allows both the centralized master station, which commonly is in the form of a server, and each slaved station to determine the maximum length of time that passes before a station gains access to the network, perhaps on the basis of programmed access priorities. Deterministic access employs token passing, a protocol in which a token, which consists of a specific bit pattern, indicates the status of the network — available or unavailable. The token is generated by a centralized master control station and transmitted across the network. The station in possession of the token (the last station to receive the message containing the token) controls the access to the network. That station may either transmit or require other stations to respond. Transmission is in the form of a data packet of a predetermined maximum size, determined by the number of nodes on the ring and the traffic to be supported. After transmitting, the station passes the token to a successor station, in a predetermined sequence. Token passing is complex and overhead-intensive, but the associated high level of network control avoids data collisions. Token Ring is a deterministic LAN protocol. See also *LAN*, *MAC*, *master/slave*, *node*, *non-deterministic*, *token passing*, and *Token Ring*.

DFB laser (Distributed FeedBack laser) A diode laser that operates at high output power levels over very short cycle times, and emits a signal at long wavelengths with very narrow spectral width. A DFB uses diffraction gratings, which act as distributed reflectors, to create the photonic resonance and oscillation in the cavity. This approach improves on the mirrors used in Fabry-Perot and vertical cavity surface-emitting laser (VCSEL) cavities. The high output power level means that the signal can suffer attenuation over longer distances and survive. The short cycle time translates into a very high bit rate. The diffusion grating supports signal emissions at spectral widths under 1 nm, which effectively means that a DFB laser emits a single wavelength. This narrow spectral width is significant in the context of wavelength division multiplexing (WDM). DFB lasers run in the 1300 and 1550 nm regions. As signals at these wavelengths attenuate relatively little, they are preferred for long haul applications. Also, WDM is centered on the 1550 nm window. In total, these characteristics currently make DFB lasers the overwhelming choice of telecommunications carriers and CATV providers, in high speed local loop and long haul wide area network (WAN) applications, where their relatively high cost is justifiable. See also *attenuation, bandwidth, diffraction grating, Fabry-Perot laser, laser, laser diode, LED, local loop, spectral width, VCSEL, WAN, wavelength, WDM,* and *window.*

DFS (Dynamic Frequency Selection) A protocol that allows an IEEE 802.11a wireless LAN (WLAN) to actively shift frequency channels. As 802.11a competes for spectrum with HiperLAN, the standard developed and promoted by the European Telecommunications Standards Institute (ETSI), 802.11a implementations in Europe must use DFS and transmission power control (TPC), which reduces the power level to no more than needed. In combination, these protocols serve to eliminate interference issues with incumbent signals. See also *802.11a, ETSI, frequency, IEEE, HiperLAN/1, HiperLAN2, interference, protocol,* and *WLAN.*

DG XIII (Directorate General XIII) The directorate within the European Union (EU) that is responsible for dealing with issues of telecommunications regulation and standardization.

DHCP (Dynamic Host Configuration Protocol) A method to automate the assignment of Internet Protocol (IP) addresses, subnet masks, and other parameters. DHCP allows a router connecting a LAN to the Internet to assign public IP addresses on a temporary basis. When a workstation logs onto the network, the DHCP client requests an IP address, which is provided from a pool of available addresses stored on a server, which may be integrated into the router. That address is assigned to that client for the duration of the session, or for a specific time period, after which the address is returned to the pool for reassignment to another client station. DHCP commonly works in conjunction with Network Address Translation (NAT), which translates between the private IP address the host workstation uses within the LAN domain and the public IP address (and specific TCP or UDP port number) it uses in the public domain. DHCP is specified in RFC 2131. In combination, these protocols relieve pressure on the IPv4 address scheme, which otherwise would be in risk of depletion. See also *client, domain, Internet, IP, IP address, IPv4, IPv4 address, NAT, private IP address, protocol, public IP address, router, server, session, subnet mask, TCP,* and *UDP.*

dial **1.** A face, usually in the form of a disk, upon which some measurement is registered in graduations, and to which a pointer indicates a specific value. An analog clock or watch has a dial with 12 major graduations (hours) and 60 minor graduations (minutes and seconds). The second, minute, and hour hands sweep around the dial in a carefully synchronized manner. **2.** A device, usually in the form of a disk, that is manually rotated to make electrical connections or to control the operation of a machine. **3.** In telephone systems, a disk with holes in it that correspond to numbers 0 through 9 and that is mounted on a telephone set. The user places a call by addressing another telephone set identified by a sequence of numbers. In a simplified example, the caller sticks his finger in the hole associated with the first number in the sequence and rotates the dial until it reaches a hard stop and then releases it. As the dial returns to the starting position, it makes and breaks electrical contacts and sends a sequence of electrical pulses across a link to a switch. The switch counts the pulses and stores the number. When all numbers in the sequence have been dialed, the switch sets up the connection between the telephone sets and the conversation ensues. Dial pulse telephone sets are considered primitive today. Most telephone sets have keypads that generate either tones or digital signals. However, people still talk about dialing a telephone number.

dial-up access Pertaining to establishing a connection to the Internet over the public switched telephone network (PSTN) by dialing the telephone number of an Internet service provider (ISP). The term implies the use of an analog local loop and a modem or ISDN service and a terminal adapter (TA). *Note:* A dial-up connection usually involves dialing the number using a telephone equipped with a keypad, rather than a dial. See also *analog, dial, Internet, ISDN, ISP, keypad, local loop, modem, PSTN,* and *TA.*

dial around In the United States, a caller can dial a seven-digit Carrier Access Code (CAC) in the format 101 XXXX to dial around the presubscribed interexchange carrier (IXC) for that line or trunk to reach another IXC. The first three digits (101) of the CAC signal the network of the caller's intent. The last four digits (XXXX) of the CAC are the Carrier Identification Code (CIC), which is used for call routing purposes. The full dialing sequence is 101 XXXX + 1 + NXX (area code) + NXX (central office prefix) XXXX (line number). See also *area code*, *CAC*, *CO prefix*, *CIC*, and *IXC*, and *line number*.

dialed number identification service (DNIS) See *DNIS*.

dial tone **1.** An audible signal indicating that a telephone set is connected to a telephone switching system that is available to process an outgoing call. **2.** Central office (CO) dial tone, sometimes referred to as hard dial tone, is provided by the CO switch to a terminal device. When the dial tone is seized, the user is free to dial a telephone number. See also *CO*. **3.** PBX dial tone, sometimes referred to as internal dial tone or soft dial tone, is provided to a PBX station indicating that the PBX switch is available. The user is then free to dial an internal PBX station number. If the target telephone number is an external number, the user must dial an access code in order to gain access to an external trunk connected to the public switched telephone network (PSTN). See also PBX and PSTN. The conventional access code is nine (9) in the United States and Canada, and zero (0) in most other countries. **4.** Stutter, or stuttered, dial tone is dial tone interrupted by short, regular periods of silence, and is used by some centrex and PBX systems as a message indicator, typically indicating that a voice message has been deposited in a voice mailbox either integrated with or interfaced directly to the system. Stuttered dial tone also is often used to confirm that a feature, such as call forwarding, has been activated or deactivated. See also *call forwarding*, *centrex*, and *PBX*. **5.** Video dial tone, or visual dial tone, refers to the notion of a broadband network that provides videoconferencing capability on demand. See also *broadband*.

dial-up circuit Referring to a circuit established by dialing the number of the remote telephone or other device over a public switched telephone network (PSTN), rather than over a dedicated circuit. There also are switched data services (e.g., Switched 56), many of which operate over the PSTN, as well. See also *circuit*, *dedicated circuit*, *dial*, *PSTN*, and *Switched 56*.

dibit Referring to a modulation technique that impresses two bits on a baud, so that the bit rate is double the baud rate. Such a technique employs four signal states. Quadrature phase-shift keying (QPSK) is a dibit technique achieved by defining four phase shifts separated by 90 degrees. Quadrature amplitude modulation (QAM) is an amplitude modulation (AM) scheme in ISDN BRI that yields the same result. There are similar frequency modulation (FM) schemes, as well. See also *AM*, *baud*, *baud rate*, *bit*, *bit rate*, *BRI*, *FM*, *QAM*, *QPSK*, *quadbit*, *signal*, *tribit*, and *unibit*.

DID (Direct Inward Dial) A PBX feature that allows incoming calls to connect directly to the station, without operator assistance. To accomplish this, each station is assigned a DID telephone number drawn from a bank of such numbers so designated by the local exchange carrier (LEC). (The last three or four digits of the DID number correspond to the internal station number, so the PBX dialing plan must be flexible enough to accommodate the DID numbering scheme.) When an outside caller dials that number, the terminating CO recognizes that fact and connects the call over a special DID trunk. The CO passes the DID number to the PBX in advance of the call, thereby enabling the PBX to automatically route the call directly to the station, without the intervention of an attendant. The service provider rents DID numbers to user organizations in groups or blocks of 50, 100, or 250, typically.

dielectric A substance that is not a conductor of direct electric current, a dielectric is an insulator, rather than a conductor. A dielectric permits the passage of the lines of force associated with an electromagnetic field, but does not conduct the current. As dielectrics, however, can sustain an electromagnetic field, they are commonly used in capacitors and between wires in a cable. Dielectrics include rubber, gutta percha, wood pulp, polyethylene, polyvinyl chloride, flouropolymer resin, and Teflon®, all of which have been used at various times as insulation in telecommunications cable and wire applications. The dielectric properties of plastic and glass make them ideal optical conductors in fiber optic cables, which are immune from electromagnetic interference (EMI). The dielectric nature of aramid fiber and fiberglass make them excellent strength members for improving the tensile strength of fiber optic and copper cables, respectively. See also *conductor*, *EMI*, *insulator*, and *strength member*.

differential phase shift keying (DPSK) See *DPSK*.

differential pulse code modulation (DPCM) See *DPCM*.

Differential Quadrature Phase Shift Keying (DQPSK) See *DQPSK*.

Differential Quaternary Phase Shift Keying (DQPSK) See *DQPSK*.

Differentiated Services (DiffServ) See *DiffServ*.

diffraction The process by which the propagation of radiant waveforms or light waves is modified as those waves encounter an obstacle or discontinuity, or exit one medium and enter another. Some of the waves deviate from their paths, i.e., bend, by diffraction. See also *knife-edge diffraction*, *reflection*, and *refraction*.

diffraction grating An optical device comprising a set of fine parallel slits or grooves etched onto a material in order to product interference patterns that separate components of a spectrum of incoming light. The fundamental principle is that different wavelengths diffract to different extents, so they can be separated much as a prism separates a composite polychromatic incident light beam into its constituent wavelengths of light. Human beings see this as the separation of white light into a rainbow of colored light. There are two types of diffraction gratings:

- A transmission grating comprises grooves etched into a transparent material such as glass. As the elements of light in the incident spectrum strike the grooves at a certain angle, they are diffracted and, therefore, separated to various degrees, with blue light diffracted the least and red light the most. This approach is used effectively in fiber optics transmission systems (FOTS) employing wavelength division multiplexing (WDM).
- A reflection grating comprises grooves ruled into a surface that can be either plane or concave. In a distributed feedback laser (DFB laser), a diffraction grating with a concave surface serves to focus light without affecting the spectra. This approach is much more effective than the mirror technique employed with a Fabry-Perot laser and vertical cavity surface-emitting laser (VCSEL).

See also *DFB laser*, *diffraction*, *Fabry-Perot laser*, *FOTS*, *laser*, *VCSEL*, and *WDM*.

DiffServ (Differentiated Services) Defined by the IETF in RFC 2474 (1998) as a framework for enabling the deployment of scalable service discrimination in the Internet. DiffServ operates at the Network Layer (Layer 3) to assign relative priorities to packets on the basis of an 8-bit code point in the Differentiated Services (DS) field in the IP header. The DS field occupies the same position as the IPv4 Type of Service (ToS) octet or the IPv6 Traffic Class field. At the ingress to each node, the DS field is analyzed and a routing table is consulted in order to determine queuing considerations at the packet's output interface on that node, which considerations reflect the differential level of treatment to be afforded that packet in accordance with a policy that may be based on application, customer, traffic type, or as expressed in a Service Level Agreement (SLA). Such policy criteria might include time of day, source and destination address pair, and port number (i.e., application identifier). There are two primary types of per-hop behaviors (PHBs), representing two service levels, or forwarding classes. Expedited Forwarding (EF) provides minimal delay, jitter, and loss. Assured Forwarding (AF) comprises four classes, each of which contains three drop precedences and allocates certain amounts of buffer space and bandwidth. DiffServ operates on a packet-by-packet and hop-by-hop basis, which explains why it ultimately did not scale up well to the size of the Internet. See also *AF*, *bandwidth*, *buffer*, *EF*, *header*, *hop*, *IETF*, *Internet*, *IPv4*, *IPv6*, *jitter*, *node*, *packet*, *routing*, and *SLA*.

diffused propagation A technique used in short-range transmission using infrared (IR) light and free space optics (FSO). Within a room, for example, it is possible to bounce the light signal off of a wall, ceiling, or other surface between the transmitter and receiver as long as the signal retains sufficient strength. The signal naturally diffuses, or scatters, to some extent. Line-of-sight (LOS) is always preferable. See also *diffusion*, *IR*, *FSO*, and *LOS*.

diffusion The scattering or deviation in the path of an electromagnetic waveform as it strikes an obstacle in its path. In transmission systems, diffusion generally refers to the scattering of light as it strikes an impurity or interacts with molecular matter in an optical fiber, or to the scattering of a radio signal as it strikes solid matter in the path between transmitter and receiver.

digerati A portmanteau of digital and literati. Synonymous with *digiterati*. The digitally literate, or those who claim to be, i.e., those who are or claim to be knowledgeable about, and perhaps influential in, the digital world of Cyberspace, the Internet, and related subjects. See also *guru*, *nerd*, *portmanteau*, and *techie*.

digit From Latin digitus, translating as finger or toe. Any of the Arabic numerals 1–9, and usually 0, used to represent numbers in the decimal system, and so-called because they originally were counted on the fingers. *Note:* Romans counted with letters, rather than numbers, partially because they hadn't invented 0 (zero). The theory is that the Roman system derived from the Etruscan system for counting sheep by making notches on tally sticks. I don't know what the Romans were doing with their fingers during this period in history. See also *decimal notation*.

digital Pertaining to the representation of data by means of digits, or discrete quantities such as numbers or signals that can be interpreted as numbers. By contrast, analog signals have meaning at all intermediate levels. In telecommunications, digital transmission systems make use of pulses or varying levels of electromagnetic energy, such as electricity, radio waves, or light. Digital communications originates in telegraphy, in which a mechanical key is used to close an electrical circuit for varying lengths of time to send a series of short pulses (dots) and long pulses (dashes) that, in specific combinations, represent specific characters or series of characters. Early mechanical computers used a similar concept for input and output. Contemporary computer systems communicate in binary mode through variations in electrical voltage.

Digital signaling in a contemporary electrical transmission system involves a signal that varies in voltage to represent one of two discrete and well-defined states. Two of the simplest approaches are unipolar and bipolar signaling. Unipolar signaling makes use of a positive (+) voltage and a null, or zero (0), voltage. Bipolar signaling, makes use of a positive (+) or a negative (–) voltage. Using a more complex signaling scheme, ISDN Basic Rate Interface (BRI) uses four voltage levels to signal two digital bits per baud. The transmitter creates the signal at a specific carrier frequency and for a specific duration (bit time), and the receiver monitors the signal to determine its state (+ or –). Various data transmission protocols employ different physical signal states and sequences, such as voltage level, voltage transition, or direction of the transition. Because of the discrete nature of each bit transmitted, the bit form is often referred to as a square wave.

Digital signaling in an optical transmission system can involve either the pulsing on and off of a light source, or a discrete variation in the intensity of the light signal. Digital transmission over a radio transmission system (e.g., microwave, cellular, or satellite) can be accomplished by discretely varying the amplitude, frequency or phase of the signal. Digital signaling over optical and radio systems also requires careful synchronization of the transmitter and receiver in order to coordinate bit timing. Bandwidth, in digital terms, is measured in bits per second (bps). See also *analog*, *bandwidth*, *binary*, and *sine wave*.

digital access cross-connect system (DACS) See *DXC*.

Digital Advanced Mobile Phone Service (D-AMPS) See *D-AMPS*.

Digital-AMPS (D-AMPS) See *D-AMPS*.

Digital Cellular System 1800 (DCS 1800) See *DCS 1800*.

digital certificate An encrypted and digitally signed attachment that authenticates a user on the Internet or an intranet. A digital certificate is issued by a certificate authority (CA), and attests to the legitimacy of an online transfer of information, funds, or other sensitive materials through the use of encryption. A digital certificate includes the sender's name, a serial number, expiration dates, a copy of the certificate holder's public key, and the digital signature of the issuing CA. A digital certificate holder has both a private key and a public key. The private key is held only by the user and is for signing outgoing messages and decrypting incoming messages. The public key is available to anyone for encrypting data to send to the holder of that public key, who then uses the private key to decrypt the message. Many digital certificates conform to the X.509 standard. See also *authentication*, *CA*, *encryption*, *Internet*, *intranet*, *PKI*, *private key encryption*, *public key encryption*, and *X.509*.

digital control channel (DCCH) See *DCCH*.

digital cross-connect (DXC) See *DXC*.

digital cross-connect system (DCCS) See *DXC*.

Digital Data Service (DDS) A synonym for *Dataphone Digital Service* (DDS). See *DDS*.

Digital Enhanced Cordless Telecommunications (DECT) See *DECT*.

Digital European Cordless Telecommunications (DECT) See *DECT*.

digital loop carrier (DLC) See *DLC*.

Digital Port Line Charge (DPLC) See *DPLC*.

digital signal hierarchy The hierarchy of digital signal levels in the European E-carrier, Japanese J-carrier, and North American T-carrier systems. In the North American hierarchy, the term digital signal (DS) level is specifically used. See Table D-1 for a side-by-side comparison. See *E-carrier*, *J-carrier*, and *T-carrier*.

Table D-1: Digital Hierarchy: T-Carrier, E-Carrier, and J-Carrier

Signal Level	*Number of Data Channels*	*Total Signaling Rate (Mbps)*		
		T-Carrier (North America)	E-Carrier (Europe)	J-Carrier (Japan)
0	1	0.064	0.064	0.064
1	24	1.544		1.544
	30		2.048	
1C	48	3.152		
2	96	6.312		6.312
	120		8.448	7.786
3	480		34.368	32.064
	672	44.736		
3C	1,344	91.053		
4	1,440			97.728
	1,920		139.268	
	4,032	274.176		
5	5,760	400.352		
	7,680		565.148	565.148
6	30,720		2200.00	

digital signal processor (DSP) See *DSP.*

digital signature A security mechanism issued by a certificate authority (CA) and appended to a digital certificate in order to allow a receiver to verify that a message has not been altered since its creation by a sender. See also *CA* and *digital certificate.*

digital speech interpolation (DSI) See *DSI.*

digital subscriber line (DSL) See *xDSL.*

digital subscriber line access multiplexer (DSLAM) See *DSLAM.*

Digital Switched Access (DSA) See *Switched 56.*

digital television (DTV) See *DTV.*

digital-to-analog converter (DAC) See *DAC.*

digital TV (DTV) See *DTV.*

digital wrapper Described in the ITU-T G.709 Recommendation (2003) "Interface for the Optical Transport Network (OTN)," digital wrapper is a method for encapsulating an existing frame of data, regardless of the native protocol, to create an optical data unit (ODU) similar to that used in SDH/SONET. A digital wrapper, however, allows multiple existing frames of data to be wrapped together into a single entity that can be more efficiently managed through a lesser amount of overhead. A digital wrapper includes a Reed-Solomon forward error correction (FEC) mechanism that improves error performance on noisy links. Digital wrappers have been defined for 2.5-, 10-, and 40-Gbps systems. See also *digital, error control, FEC, ITU-T, link, OTN, protocol, Reed-Solomon,* and *wavelength.*

digiterati See *digerati.*

digroup (digital group) A basic multiplexing group. In the various digital carrier hierarchies, two digroups form a level-one signal. In the North American T-carrier system and the Japanese J-carrier system, each digroup is 12 channels, or the equivalent amount of bandwidth (12 × 64 kbps = 768 kbps). Two digroups plus overhead bits equal a T1 or J-1. (2 × 768 kbps = 1.536 Mbps; 1.536 Mbps + 8 kbps = 1.544 Mbps). In the European E-carrier system, each digroup is 15 channels. See *E-1, E-carrier, J-1, J-carrier, T1,* and *T-carrier.*

dikes A corruption of diagonal cutters. Side-cutting pliers, wire clippers, or wire cutters. See also *PC.*

dim fiber Optical fiber only partially lit in a fiber optic transmission system (FOTS) employing wavelength division multiplexing (WDM). WDM technology can support a considerable number of wavelengths running simultaneously over a single optical fiber within a cable comprising perhaps a great number of fibers. A dim fiber is one over which not all available wavelengths have been lit and which, therefore, has excess capacity. See also *dark fiber, FOTS, wavelength,* and *WDM.*

diode An electronic device with two (di) electrodes, originally in the form of a needle on a natural quartz crystal, and later an electron tube with an anode (positive terminal) and a cathode (negative terminal), but now usually in the form of a transistor with a p-n (positive-negative) junction between a positive and a negative layer of semiconducting materials. The negative layer is doped with impurities to create extra electrons, which are negatively charged. The positive layer is doped to create extra holes into which electrons can migrate when a charge is applied, which has the effect of adding extra positive particles. When current is applied and the electrons move across the junction, from the n semiconductor layer to the p semiconductor layer, and settle into the holes, they release energy in the form of photons, i.e., light. (Note: A diode conducts current in only one direction, like a one-way gate.) The composition of the semiconductor material determines the color of light, how much of it is absorbed, and how much of it is released. In electrical applications, a diode is primarily used as a rectifier, to convert alternating current

(AC) to direct current (DC). In free space optics (FSO) and fiber optic transmission systems (FOTS), light sources variously use light-emitting diodes (LEDs) and diode lasers to convert electrical signals to optical signals. FSO and FOTS optical receivers variously use positive-intrinsic-negative (PIN) diodes and avalanche photodiodes (APDs) to convert light to electrical current. See also *AC*, *APD*, *DC*, *diode laser*, *FOTS*, *FSO*, *LED*, *PIN*, and *rectifier*.

diode laser See *laser diode*.

diplex A circuit or device that supports simultaneous transmission or reception of two independent signals. Diplex communications technology is a simple form of multiplexing that was considered quite revolutionary in the early days of telegraphy, when a diplex circuit would support transmission of two independent signals in one direction only. The term continues to be used in radio communications. See also *quadraplex* and *multiplexer*.

direct broadcast satellite (DBS) See *DBS*.

direct bury cable An underground cable designed to be buried in direct contact with the Earth, rather than housed in a protective conduit. Direct bury cable must be well protected by a watertight jacket, or sheath, and usually by a water blocking gel or a moisture-activated powder that turns into a gel. Direct bury cables also commonly are armored to protect against cable-seeking backhoes and posthole diggers, cable-loving rodents, and other adverse forces of man and nature. The armor may be in the form of lead or lead alloy sheathing, or interlocking aluminum or galvanized steel cladding. See also *cladding*, *icky-pic*, and *sheath*.

direct current (DC) See *DC*.

direct distance dialing (DDD) See *DDD*.

direct inward dial (DID) See *DID*.

direct inward system access (DISA) See *DISA*.

directionality **1.** The extent to which an optical beam is lined up with the core of an optical fiber. Collimated light beams are lined up in perfect parallel. Divergent light beams spread out as they exit the source. Light-emitting diodes (LEDs) emit the most divergent light beams, and laser diodes emit the least divergent. See also collimation, laser diode, and LED. **2.** The property of a radio antenna that transmits or receives more RF signal in one direction than another. See also *antenna*, *radio*, *RF*, and *signal*.

directivity In the context of the radiation pattern of an antenna, the ratio of the power radiated in a given direction to the average of the power radiated in all directions. See also *isotropic* and *radiation*.

Directorate General XIII (DG XIII) See *DG XIII*.

directory number (DN) Telephone number. See *telephone number*.

direct outward dial (DOD) See *DOD*.

direct sequence spread spectrum (DSSS) See *DSSS*.

direct-to-home (DTH) See *DTH*.

DISA **1.** Direct Inward System Access. A PBX feature that allows remote access to the system, generally on a toll-free basis. The caller dials the telephone number associated with a special DISA trunk. After the PBX answers the call, the caller can gain access to connected resources by entering the proper authorization codes. DISA is useful to authorized users requiring remote access to e-mail servers, voice processors, computer systems, and outgoing toll trunks. However, DISA may open the system and all connected resources to hackers, who often target them in order to gain unauthorized access to the same resources.

2. Data Interchange Standards Association. A not-for-profit organization created for the development of cross-industry business interchange standards. DISA serves as the Secretariat for the ANSI Standards Committee (ASC) X12 and their X12 Electronic Data Interchange (EDI) and Extensible Markup Language (XML) standards development processes. See also *ANSI*, *EDI*, *standards*, and *XML*.

discard eligible (DE) See *DE*.

discrete amplification A type of amplification that occurs in a single, discrete location. Also known as *lumped amplification*. Erbium-doped fiber amplifiers (EDFAs), and most amplifiers, perform the process of amplification on a discrete, or lumped, basis. See also *amplifier*, *distributed amplification*, *EDFA*, and *Raman amplifier*.

discrete cosine transform (DCT) See *DCT*.

discrete multitone (DMT) See *DMT*.

disk operating system (DOS) See *DOS*.

dispersion (D) **1.** The process by which light rays are distorted, scattered, or redirected differently depending on their wavelength. **2.** Specifically, the effect of different propagation speeds for different wavelengths of light. See also *chromatic dispersion*, *material dispersion*, *modal dispersion*, *polarization mode dispersion*, *propagation*, *pulse dispersion*, *waveguide dispersion*, and *wavelength*.

dispersion-shifted fiber (DSF) See *DSF*.

distance-vector routing protocol A protocol used in packet-switched networks by which routers calculate the direction and distance between any two points and route packets based on their calculation of the fewest number of hops. The calculations are based on topology information that each is required to share with neighboring routers when changes occur. Those routers then share the information with their neighbors, and so on, with the updates spreading like ripples in a pond. Distance-vector routing is sometimes referred to as *routing by rumor*, as most routers depend on hearsay information, rather that on their personal knowledge of network topology. Routing Information Protocol (RIP) is a distance-vector routing protocol.

distinctive ringing Also known as *custom ringing and priority ringing*. **1.** A network feature that allows the subscriber to (a) enter a list of telephone numbers, calls from which are announced with a special ringing tone, or (b) associate a ringing pattern with different directory numbers, all of which ring on the same physical line. Distinctive ringing is one of a suite of services of the public switched telephone network (PSTN) known as custom local area signaling services (CLASS). See also *CLASS* and *PSTN*. **2.** A feature of voice telecommunications systems that uses distinctive tones to distinguish between internal and outside incoming calls. Distinctive ringing is a common feature of key telephone systems (KTS), PBX, and Centrex systems. See also *Centrex*, *KTS*, and *PBX*.

distortion The presence of undesired changes in some attribute of a waveform. Distortion is a type of noise that typically occurs in an amplifier, repeater, or transducer. Distortion commonly manifests as a nonlinear relationship between an input waveform and an output waveform, a non-uniform frequency shift, or a phase shift not proportional to the frequency. See also *noise*.

distributed amplification A type of amplification that is spread over an entire link or portion of a link, rather than occurring in a single, discrete location. Raman amplification is a distributed approach, generally occurring over the entire length of an optical fiber link. See also *discrete amplification* and *Raman amplifier*.

distributed feedback laser (DFB laser) See *DFB laser*.

Distributed Queue Dual Bus (DQDB) See *DQDB*.

distribution frame A structure in the form of a frame of metal uprights and cross pieces with termination points on each side into which conductors can be mechanically connected and semi-permanent cross-connections can be made. The term generally is applied to distribution frames for twisted pair cables, with jumpers used to connect the pairs on each side of the frame. See also *frame*, *intermediate distribution frame (IDF)*, *main distribution frame (MDF)*, and *patch panel*.

diurnal wander Telecommunications wires, fibers, and cables exposed to the elements expand and contract daily due to variations in ambient temperature from day to night. Aerial cables suspended from poles stretch during the warmth of the day, with the weight of the cable magnifying the effect, and shrink in the cool of the night. Copper cables are more susceptible than glass fiber optic cables, but both stretch and shrink to some extent. As they do so, the length of the path increases and decreases, and signal propagation delay increases and decreases slightly. As a result, high speed digital systems can suffer timing problems that can cause losses of synchronization and, ultimately, temporary system failures. See also *propagation delay*.

Divestiture Decree Synonymous with Modified Final Judgement (MFJ). See *MFJ*.

DLC (Digital Loop Carrier) A digital carrier system that serves to increase the efficiency of twisted pair local loop facilities by multiplexing voice grade channels over a single high speed facility. The Subscriber Line Carrier-96 (SLC-96) system introduced by Western Electric (now Alcatel-Lucent), for example, comprises up to four T1s from the central office (CO) to a remote neighborhood node. The node is a remote line shelf and time division multiplexer (TDM mux) that terminates the T1s, or one T2, from the CO on one side, and 96 voice grade analog circuits from individual customer premises on the other side. The SLC-96 DLC thereby supports 96 voice grade, two-wire local loops over four four-wire (i.e., four two-pair) circuits, which results in a savings of 88 pairs (96 – 8 = 88) from the CO to the node. DLC is an example of a pair-gain technology. See also *carrier*, *circuit*, *channel*, *four-wire circuit*, *local loop*, *multiplexer*, *pair-gain*, *T1*, *T2*, *TDM*, *two-wire circuit*, and *voice grade*.

DLCI (Data Link Connection Identifier) **1.** In ISDN, A two-octet address field in a LAPD frame, the DLCI is used for addressing at the Data Link Layer, which is Layer 2 of the OSI Reference Model. The first octet is the service access point identifier (SAPI), which identifies the destination service access point, each of which can support multiple terminal devices. The second octet is the terminal endpoint identifier (TEI), which is the address of the destination terminal device. See also *Data Link Layer*, *ISDN*, *LAPD*, and *OSI Reference Model*. **2.** In frame relay, 10 bits in a LAPF frame that identify the data link, the virtual circuit (VC). A switch may associate a DLCI with a set of service parameters, which include frame size, Committed Information Rate (CIR), Committed Burst Size (B_c), Excess Burst Size (B_e), and Committed Rate Measurement Interval (T_c). See also *CIR*, *frame relay*, *LAPF*, and *VC*.

DLL (Data Link Layer) See *Data Link Layer*.

DMT (Discrete MultiTone) Also known as orthogonal frequency division multiplexing (OFDM). A modulation technique that splits the signal into a stream of data symbols for massively parallel simultaneous transmission over 256 or more narrowband, low data rate subcarrier frequencies, each of which is a voice grade channel with a width of 4 kHz. DMT is used in IEEE 802.11a (Wi-Fi5), 802.11g (Wi-Fi), and 802.16 (WiMax) wireless systems, as well as asymmetric digital subscriber line (ADSL) and broadband over power line (BPL) technologies. See also *802.11a*, *802.11g*, *802.16*, *ADSL*, *BPL*, *COFDM*, *modulation*, *narrowband*, *signal*, *subcarrier*, *symbol*, *voice grade*, *Wi-Fi*, *Wi-Fi5*, and *WiMAX*.

DN (Directory Number) Telephone number. See *telephone number*.

DND (Do Not Disturb) See *do not disturb*.

DNIS (Dialed Number Identification Service) A service offered by PSTN carriers that passes the dialed digits to the destination PBX in advance of connecting the call. DNIS operates much like direct inward dial (DID) service, but over toll-free lines, enabling the PBX to determine which toll-free number was dialed and to route the call to the proper agent or agent group. DNIS is heavily used in call centers that service multiple clients or client groups, each of which has one or more unique toll-free numbers. See also *call center*, *DID*, and *toll free*.

DNR **1.** Domain Name Resolver. A program that looks up an Internet domain name or attribute of a domain (e.g., host, alias, network, protocol, or service) name on a Domain Name Server (DNS) to translate domain names and Uniform Resource Locators (URLs) into Internet Protocol (IP) addresses (and vice versa) in order that the routers can route e-mail and other traffic correctly. See also *DNS*, *e-mail*, *Internet*, *IP*, *IP address*, *router*, and *URL*. **2.** Do Not Resuscitate. A doctor's order that a patient should not be resuscitated in the event of cardiac or respiratory arrest. A DNR can be instituted on the basis of an advance directive from the person involved or from someone authorized to make medical decisions on their behalf. A DNR prevents a hospital from resuscitating someone and placing that person on life support systems. Many people issue advance DNR directives in order to ensure that they will die a natural death, without extraordinary, invasive, and painful measures being taken to prolong their lives. I have issued a DNR directive and so has my wife. Think about it. Also, please donate blood regularly. It doesn't hurt enough to even talk about, and it could save a life.

DNS **1.** Domain Name System. The naming convention that divides the Internet into logical domains identified in Internet Protocol version 4 (IPv4) as a 32-bit portion of the total address. The standard convention is *user@organization.domain*, where user is the name of the end user, organization is the name of the enterprise or other organization owning or sponsoring the address, and domain is the type of entity owning or sponsoring the address. The administration of the DNS is the responsibility of Internet Corporation for Assigned Names and Numbers (ICANN). See also *domain*, *ICANN*, *Internet*, and *IPv4*. **2.** Domain Name Server. All Internet domain names are maintained in mirrored databases on root servers distributed around the world. These servers are updated by the domain registrars. Thousands of Domain Name Resolvers (DNRs) located strategically with Internet Service Providers (ISPs) and institutional networks periodically download database updates from the root servers. See also *database*, *DNR*, *Internet*, *ISP*, *root*, and *server*.

DOCSIS (Data Over Cable Service Interface Specification) A set of specifications for high speed, full duplex (FDX) data communications over CATV networks, DOCSIS was developed by the Multimedia Cable Network Systems Partners Ltd. (MCNS), and currently is administered by Cable Television Laboratories (CableLabs) an industry research and development organization. The original specifications were for set-top boxes and cable modems (CMs). Matching cable modems positioned at the service provider's headend and the customer's premises communicate over a hybrid fiber/coax (HFC) network comprising optical fiber, at least in the trunk portion, and embedded coaxial cable from the neighborhood node to the premises. DOCSIS specifications are for always-on Internet access, which avoids the dial-up delays characteristic of circuit-switched modem access via the public switched telephone network (PSTN). DOCSIS also specifies asymmetric data transmission, which is typical of high speed Internet access services.

Downstream transmission takes place over one or more 6-MHz channels in the range between 50 MHz and 750-850 MHz. DOCSIS 2.0 specifies several variations of Quadrature Amplitude Modulation (QAM). At the lowest level, 64-QAM yields six bits per symbol and a signaling rate of 36 Mbps per 6-MHz channel. Alternatively, 128-QAM yields seven bits per symbol and increases the signaling rate to 42 Mbps, although it is more sensitive to noise. The standards also provide for the use of 256-QAM, which increases the raw signaling rate to 48 Mbps. These variations on QAM are compatible with Moving Picture Experts Group-2 (MPEG-2), the compression technique specified for digital video transmission in CATV networks. Downstream data is encapsulated into MPEG-2 packets of 188 bytes.

Upstream transmission, in early DOCSIS versions, is supported over 6-MHz channels in the range between 5 MHz and 42 MHz for United States systems, and 8-MHz channels in the range between 5 MHz and 65 MHz for European systems. The channels each support a signaling rate of 12 Mbps through use of the quadrature phase shift keying (QPSK) modulation technique. DOCSIS 1.1 specifies 16-QAM, which roughly doubles the data rate, and doubles the channel width. DOSCIS 2.0 further increases the upstream rate by again doubling the channel width, and using either time division multiple access (TDMA) in combination with 64-QAM or synchronous code division multiple access (S-CDMA) in combination with 128-QAM trellis-coded modulation (TCM). See also *128-QAM, 16-QAM, 256-QAM, 64-QAM, always on, asymmetric, byte, CableLabs, cable modem, CMTS, coaxial cable, compression, downstream, encapsulate, FDX, headend, HFC, modulation, MPEG-2, noise, packet, PSTN, QAM, QPSK, S-CDMA, set-top box, signaling rate, symbol*, and *upstream*.

DOD (Direct Outward Dial) A PBX feature that allows a station user to dial an outside number without the assistance of an operator. The user generally first dials an access code, which by convention is the number 9 (nine) in North America, and 0 (zero) in most of the rest of the world.

domain **1.** Sphere of influence. **2.** In the Internet Domain Name System (DNS), the rightmost portion of the address, the domain identifies the type of entity owning or sponsoring the address. The several types of Top Level Domains (TLDs) include generic Top Level Domains (gTLDs) and country codes. If a country code is used in the address, the gTLD becomes a secondary domain. See also *country code, DNS, gTLD, Internet, secondary domain*, and *TLD*.

domain name In the context of the World Wide Web (WWW), the name under which an organization or individual registers a Web address. The domain name follows a convention known as the *Domain Name System* (DNS). See also *DNS* and *WWW*.

Domain Name Resolver (DNR) See *DNR*.

Domain Name System (DNS) See *DNS*.

Domsat Decision (Domestic satellite Decision) In the United States, the Federal Communications Commission (FCC) decision (1972) that permitted the development of the domestic satellite market. AT&T was excluded from the market for three years. See also *FCC* and *satellite*.

dongle A small hardware device that dangles from a cable that plugs into a port to add some sort of capability. In asymmetric digital subscriber line (ADSL) installations, for example, the customer generally is provided with a number of dongles that are microfilters that plug into telephone sets or fax machines that share the ADSL line. The filters prevent the telephones and fax machines from causing interference between the analog voice channel and the digital data channel when they go off hook or on hook, or when the telephone or fax receives a ringing signal. Dongles often are used for smart cards and other security devices that provide copyright protection for very expensive software programs or that protect access to virtual private networks (VPNs). See also *ADSL, cable, channel, hardware, interference, microfilter, port, software*, and *VPN*.

do-not-call registry The United States Federal Communications Commission (FCC) and Federal Trade Commission (FTC) established the National Do-Not-Call Registry in 1993 as a proactive means of addressing the problem of telephone solicitation calls. Commercial telemarketers are not allowed to call a residence if the subscriber's number is on the registry, subject to certain exceptions. As a result, consumers can, if they choose, reduce the number of unwanted phone calls to their homes. The registry applies to all telemarketers (with the exception of certain non-profit organizations), and covers both interstate and intrastate telemarketing calls. A number of state governments already had such registries in place. The National Do-Not-Call Registry was extended to cellular phones in late 2004. The Controlling the Assault of Non-Solicited Pornography and Marketing (CAN-SPAM) Act was passed in 2003, extending the ban on certain unsolicited marketing to e-mail. See also *CAN-SPAM Act, FCC, FTC, telemarketing*, and *Telephone Consumer Protection Act*.

do not disturb (DND) A feature of voice telecommunications systems that enables end users to busy out a station and reject incoming calls by depressing a feature button or soft key, or entering a code via the telephone keypad. Once the station is busied out, incoming calls typically are automatically forwarded to another station, the system attendant, or perhaps voice mail.

dopant A material added to another material during the manufacturing process to alter its composition and create a material with more desirable characteristics. The manufacture of glass optical fiber (GOF) involves adding dopants such as erbium and germanium to pure silicon dioxide in order to variously alter the light conducting properties of the core and cladding. See also *cladding*, *core*, *erbium*, *germanium*, *GOF*, and *silicon dioxide*.

doping See *dopant*.

Doppler effect The phenomenon by which the observed frequency of a wave changes as a result of a time change in the effective length of the path of propagation between the source of the wave and the point of observation. If there is a source of wave energy and an observer of the wave energy, the frequency of the waveform increases as the observer moves closer to the source, the source moves closer to the observer, or both. The frequency of the waveform decreases as the observer and source move farther apart. The phenomenon applies to all waveforms, including acoustical and electromagnetic waveforms. In acoustics, the pitch of the sound increases as the observer and source move closer together, and decreases as they move farther apart, as you may have noticed when listening to a train whistle as the train comes closer and then goes farther away. The combination of the Doppler effect and that of multipath fading causes the wooooo-wooooo sound. In telecommunications, the Doppler effect creates difficulties when a mobile device, such as a cellular telephone moves towards or away from a fixed base station at a high rate of speed. The Doppler effect is used in some forms of radar to determine the speed and direction of a moving object. The Doppler effect was first hypothesized by Johann Christian Andreas Doppler (1803–1853), an Austrian mathematician and physicist. See also *acoustics*, *cellular radio*, *electromagnetic*, *frequency*, *propagation*, and *waveform*.

DOS (Disk Operating System) **1.** A generic term describing an operating system (OS) that loads from a disk device when the computer system is turned on. Earlier, more primitive OSs loaded from magnetic tape or even paper tape, or were resident in memory. See also *OS*. **2.** Referring to MS-DOS (MicroSoft Disk Operating System), a single-user, single-tasking OS released in 1981 for use in IBM personal computers (PCs) and compatibles. See also *OS* and *PC*.

dot **1.** The shorter of the two signal elements in Morse code telegraphy, created by closing an electrical circuit with a mechanical key for a short period of time. A dot is audible as a brief click or buzz, called a dit by radiotelegraph operators, and is graphically represented as a small round mark. See also *dash*, *Morse code*, and *telegraph*. **2.** Marion Estelle Edison (1873–1965) was nicknamed "Dot" by her father, Thomas Alva Edison (1847–1931), the "Wizard of Menlo Park," who invented such devices as the phonograph, electric light bulb, carbon microphone, and electric chair. Much of his early work was in telegraphy, and the two-way telegraph and quadraplex telegraph were among his early financial successes. See also *dash* and *telegraph*.

dots per inch (dpi) A measure of the resolution of a printed image referring to the number of dots of color per linear inch. The more dpi, the better the resolution, at least up to a point. See also *lpi*.

dotted decimal notation Referring to the format of the Internet Protocol version 4 (IPv4) address. An IPv4 address comprises four fields separated by dots and expressed as xxx.xxx.xxx.xxx, with each field given a value in decimal notation of 0–255, the range expressed by a single octet in binary notation (2^8 = 256, or 0–255). This format sometimes is referred to as *dotted quad*. IP addresses are always stored in memory in network byte order, which is big-endian byte order. See also *big-endian*, *binary notation*, *decimal notation*, *dotted quad*, *hexadecimal notation*, *IP*, *IPv4*, *IPv4 address*, *IPv6*, *IPv6 address*, and *octet*.

dotted quad Referring to the format of the Internet Protocol version 4 (IPv4) address. All IP addresses are written in dotted decimal notation. An IPv4 address comprises four fields separated by dots and expressed as xxx.xxx.xxx.xxx, with each field given a value in decimal notation of 0–255, the range expressed by a single octet in binary notation ($2^8 = 256$, or 0–255). See also *binary notation, decimal notation, dotted decimal notation, IPv4,* and *octet.*

double sideband modulation (DSB) See *DSB.*

downlink **1.** The microwave radio link from a satellite to a terrestrial receive antenna. See also *link, microwave, radio,* and *uplink.* **2.** In a cellular network, the radio link from the base station (BS) to the mobile station (MS). See also *cellular radio, link, radio,* and *uplink.*

download To transfer a file copy from a remote computer to a local computer over a network. See also *upload.*

downstream The signal direction from the network edge to the customer premises. See also *upstream.*

downtime The time during which a functional machine or system is not functioning properly or is unavailable to users for other reasons. Scheduled downtime is planned in advance for reasons including scheduled maintenance, system updates and patches, and system upgrades. Unscheduled downtime is unplanned downtime due to system or environmental (e.g., power) failures. See also *patch, update, upgrade,* and *uptime.*

DPCM (Differential Pulse Code Modulation) A voice encoding technique used to express analog signals in digital format. DPCM is more efficient than PCM in terms of bandwidth utilization, as only the quantized differences in signal values are encoded and transmitted. DPCM compares two successive analog amplitude values, quantizes and encodes the difference, and transmits the differential value. As it can be assumed that the change, or differential, in the voice signal occurs relatively gradually, relatively few bits can be used to represent each sample. DPCM will work with various numbers of bits, but a 4-bit approach generally is used in this technique, which yields a 2:1 (2-to-1) compression ratio, when compared to 8-bit PCM. At 32 kbps, DPCM generally provides voice quality comparable to that of PCM at 64 kbps, although distortion can result if the signal varies significantly from one sample to another. A modem transmission, for example, is characterized by amplitude (i.e., volume) and frequency (i.e., pitch, or tone) levels that vary abruptly. Although DPCM is unusual, a variation known as adaptive differential pulse code modulation (ADPCM) is commonly used. See also *ADPCM, analog, bandwidth, compression, digital, encode, modem, modulation, PCM,* and *quantize.*

DPLC (Digital Port Line Charge) A Subscriber Line Charge (SLC) that the Federal Communications Commission (FCC) authorized the local exchange carriers (LECs) to collect as a monthly surcharge on digital local loops. See *SLC* and *EUCL.*

DPSK (Differential Phase Shift Keying) A modulation technique in which each 1 bit triggers a phase shift of 180 degrees, but 0 bits have no effect. A number of applications specify various PSK techniques. For example, Wi-Fi5 (802.11a) wireless local area network (WLAN) standards specify binary phase shift keying (BPSK) at 6 Mbps and quaternary phase shift keying (QPSK) at 12 Mbps. See also *BPSK, modulation, phase, PSK, QPSK,* and *WLAN.*

DQDB (Distributed Queue Dual Bus) A technology defined by the IEEE 802.6 standard for metropolitan area networks (MANs), as a means of extending the reach of a local area network (LAN) across a metropolitan area. The original work on the DQDB concept was done at the University of Western Australia where it was known as *Queued Packet Synchronous Exchange* (QPSX). Subsequently, Switched Multimegabit Data Service (SMDS) evolved from these efforts. This technology is considered obsolete. See also *802.6, IEEE, LAN, MAN,* and *SMDS.*

DQPSK (Differential Quaternary Phase Shift Keying) Also known as *Differential Quadrature Phase Shift Keying*. A variation on the QPSK modulation technique, DQPSK relies on the difference between successive phases of a signal rather than the absolute phase position. DQPSK is commonly used in cellular radio systems and other radio systems, such as LMDS. See also *cellular radio*, *LMDS*, *π/4 DQPSK*, *PSK*, and *QPSK*.

Draft-Martini VPN In Multiprotocol Label Switching (MPLS), referring to a Data Link Layer (Layer 2) Virtual Private Network (VPN) so called because Luca Martini, a senior architect at Level 3 Communications, was a major contributor to the draft standard. A Draft-Martini VPN emulates a point-to-point virtual circuit connection, or pseudowire, between two routers or switches. Also commonly referred to as Pseudowire Emulation (PWE) VPN. See also *connection*, *Data Link Layer*, *MPLS*, *point-to-point*, *pseudowire*, *PWE*, *virtual circuit*, and *VPN*.

drain wire A wire wrapped around or part of a shield within a cable that reduces the resistance from any point on the shield to ground. A drain wire serves to complete an electrical circuit from the shield, thereby carrying extraneous electrical noise to ground and away from the circuit or system the shield is intended to protect. See also *ground*, *ground wire*, and *shield*.

driver A hardware device or software program that operates, controls, or regulates another device or network element. A line driver, for example, converts a digital electrical signal into a low voltage, low impedance signal that can travel longer distances over a dedicated copper circuit. A software driver is a device-specific program that controls the operation of a peripheral device, such as a monitor or printer. As a separate software program, a driver relieves the computer operating system (OS) of the burden of dealing with mundane, device-specific tasks. See also *hardware*, *OS*, *program*, and *software*.

drop **1.** The portion of a local loop that extends from a distribution cable terminal to a customer's building. A drop, or drop wire, generally is in the form of a cable containing one to five twisted pairs, or perhaps one or two optical fibers. The term refers to the fact that the small cable often drops from a larger aerial cable suspended above the ground from poles. See also *cable* and *local loop*. **2.** The portion of an inside wire and cable system that drops from a false ceiling, also known as drop ceiling, perhaps connecting to a pre-wired cubicle or electrical pole, rather than directly to an electrically powered device.

drop wire See *drop*.

dry Referring to a current-carrying conductor or circuit that is not electrified or that carries current of such low voltage that it will not arc. Dry is a reference to the inability of the conductor or circuit to produce any result. See also *circuit*, *conductor*, *current*, *dry copper pair*, and *voltage*.

dry copper pair Also known as a burglar alarm circuit. A twisted pair that is not electrified, that is, has no associated electronics. Providers of digital subscriber line (DSL) services and burglar alarm services order dry copper pairs from the telco and place their own electronics on them. All they want is a pair of wires from one point to another. See also *dark fiber*, *dry*, *twisted pair*, and *xDSL*.

DS (Digital Signal) Referring to a digital signaling level in the North American digital carrier hierarchy. See *digital signal hierarchy*, *DS-0*, *DS-1*, *DS-1C*, *DS-2*, *DS-3*, and *DS-4*.

DS-0 (Digital Signal Level Zero) A 64 kbps channel, which is the fundamental building block of the generic digital signal hierarchy. DS-0 is a voice grade channel designed to support voice digitized using pulse code modulation (PCM), or to support other data at a rate of up to 64 kbps. A DS-0 channel comprises eight-bit binary values transmitted at a rate of 8,000 per second (8 bits × 8,000 per second = 64,000 bps, or 64 kbps). In a typical voice application, 24 PCM-encoded voice conversations are interleaved to form a T1 frame. See also *digital signal hierarchy*, *DS-1*, *DS-1C*, *DS-2*, *DS-3*, *DS-4*, *PCM*, *T1*, and *voice grade*.

DS-1 (Digital Signal Level One) The signal level for a digital carrier system operating at level one of the North American digital signal hierarchy. A DS-1 corresponds with a T1 in the North American hierarchy. See also *digital signal hierarchy* and *T1.*

DS-1C (Digital Signal Level One Concatenated) The signal level for a digital carrier system that contains two level one signals concatenated. A DS-1C corresponds to a T1C in the North American hierarchy. There is no equivalent in the European hierarchy or the Japanese hierarchy. See also *concatenated digital signal hierarchy* and *T1C.*

DS-2 (Digital Signal Level Two) The signal level for a digital carrier system operating at level two of the North American digital signal hierarchy. A DS-2 corresponds to a T2 in the North American hierarchy. See also *digital signal hierarchy* and *T2.*

DS-3 (Digital Signal Level Three) The signal level for a digital carrier system operating at level three of the North American digital signal hierarchy. A DS-3 corresponds to a T3 in the North American hierarchy. See also *digital signal hierarchy* and *T3.*

DS-4 (Digital Signal Level Four) The signal level for a digital carrier system operating at level four of the North American digital signal hierarchy. A DS-4 corresponds to a T4 in the North American hierarchy. See also *digital signal hierarchy* and *T4.*

DS-5 (Digital Signal Level Five) The signal level for a digital carrier system operating at level five of the North American digital signal hierarchy. A DS-5 corresponds a T-5 in the North American hierarchy. See also *digital signal hierarchy* and *T-5.*

DSA (Digital Switched Access) See *Switched 56.*

DSB (Double Sideband) The process of amplitude modulation (AM) results in the creation of two sidebands. An upper sideband is above the carrier frequency and a lower sideband is below the carrier frequency. DSB transmission uses both sidebands at full power. Single sideband (SSB) transmission suppresses one of the sidebands. See also *AM, amplitude, carrier, frequency, modulation, sideband, SSB,* and *VSB.*

DSF (Dispersion-Shifted Fiber) A type of single-mode fiber (SMF) that improves on earlier non–dispersion-shifted fiber (NDSF), shifting the optimal dispersion point by adjusting the refractive index profile of the core and the cladding. There are two types of DSF. Zero Dispersion-Shifted Fiber (ZDSF) shifts the point of zero dispersion by increasing material dispersion to the point that it cancels out chromatic dispersion at 1550 nm, rather than 1310 nm. Dense Wavelength Division Multiplexing (DWDM) and Erbium-Doped Fiber Amplifiers (EDFAs) both work in this higher window, which can create yet another noise problem in the form of four-wave mixing (FWM), a phenomenon by which wavelengths interact to create additional wavelengths. The EDFAs amplify those signals, and superimpose them on the DWDM channels. Non-Zero Dispersion-Shifted Fiber (NZDF) addresses this issue by shifting the optimal dispersion point slightly above the range in which EDFAs operate. See also *chromatic dispersion, cladding, core, dispersion, NDSF, NZDF, material dispersion, refractive index, SMF, window,* and *ZDSF.*

DSI (Digital Speech Interpolation) Also known as *voice activity detection* (VAD). A voice compression mechanism based on the fact that there are predictable periods of inactivity in normal human speech. During those pauses and other periods of silence, the pulse code modulation (PCM) signals are halted through a technique known as silence suppression. On the receiving end of the transmission, a matching algorithm reconstitutes the PCM signals, reinserting the periods of silence. As DSI works on the basis of statistical probabilities, it is employed effectively only when there are a significant number of voice conversations supported. For example, DSI applied to 96 channels of PCM-encoded voice yields compression of 2:1, which reduces the PCM bandwidth requirement of 64 kbps to only 32 kbps, on average. DSI builds on the earlier analog technique of time-assignment speech interpolation (TASI). See also *bandwidth, compression, encode, PCM,* and *TASI.*

DSL (Digital Subscriber Line) See *xDSL.*

DSL access multiplexer (Digital Subscriber Line access multiplexer or DSLAM) See *DSLAM.*

DSLAM (Digital Subscriber Line Access Multiplexer) In digital subscriber line (DSL) networks, a packet multiplexer that provides the interface between the DSL local loop and the service provider's point of presence (POP). Most DSLAMs are ATM-based, although some earlier models are based on frame relay, with the specific technology depending on what the carrier has in place. In an ADSL scenario the DSLAM receives upstream DSL traffic and splits, or demultiplexes, the voice and data traffic. The voice traffic then is encoded into pulse code modulation (PCM) format and time division multiplexed over a channelized T1, T3, or, perhaps, SONET link to the public switched telephone network (PSTN). The data traffic is multiplexed or concentrated in asynchronous transfer mode (ATM) cells over an unchannelized T1, T3, or SONET circuit directly to the Internet backbone, or to perhaps to an independent Internet service provider (ISP). The DSLAM generally represents the first potential point of contention and congestion that affects upstream end user traffic, as the local loop is a dedicated circuit. In asymmetric DSL (ADSL) implementations, the line side of the DSLAM in known as an ADSL transmission unit-centralized (ATU-C). See also *ADSL, ATM, ATU-C, carrier, DSL, ISP, line side, PCM, POP, SONET, T1, T3,* and *upstream.*

DSL Forum A consortium of companies dedicated to the development of the full potential of digital subscriber line (DSL) through defining the core technology, establishing advanced architecture standards, and maximizing effectiveness in deployment, reach, and application support. The DSL has been heavily involved in the global standardization of ADSL, ADSL2+, SHDSL, VDSL, and VDSL2. See Appendix A for contact information. See also *ADSL, ADSL2+, SHDSL, VDSL,* and *VDSL2.*

DSP (Digital Signal Processor) An integrated circuit, usually in the form of a microprocessor, designed specifically for high-speed signal processing (i.e., data manipulation) tasks, typically in real time, and built into another device. In a typical telecommunications application, an analog-to-digital converter (ADC) samples an analog signal and converts it into digital format. A DSP then encodes the signal into pulse code modulation (PCM) format, employing an appropriate algorithm such as A-law or mu-law to produce a standard output like a DS-0 channel. A DSP might then filter noise and remove interference from the signal, and transcode, compress, and perhaps encrypt the signal prior to placing it on a carrier of an electrically based digital circuit for transmission. A matching DSP reverses the process on the receiving side of the communication, sending the signal to a digital-to-analog converter (ADC) for decoding back into real world analog form, perhaps with improved clarity, a shifted frequency, or demodulated. DSPs are used in audio, communications, image manipulation, and video applications. DSPs are built into cellular telephones, fax machines, modems, and many other devices. See also *ADC, A-law, analog, carrier, channel, codec, compression, DAC, digital, DS-0, encode, encryption, interference, mu-law, noise, PCM,* and *transcode.*

DSSS (Direct Sequence Spread Spectrum) A radio transmission technique that spreads a narrowband signal across a wider carrier frequency band. Each transmission is assigned a 10-bit pseudorandom binary code sequence, which comprises a series of ones and zeros in a seemingly random pattern known to both the transmitter and receiver. The original code sequence is mathematically self-correlated to yield a code that stands out from all others, at least on average. The paired transmitters and receivers recognize their assigned and correlated code sequences, which look to all others as pseudonoise (PN). DSSS phase-modulates the carrier wave with a continuous string of PN code symbols, or chips, each of which has a much shorter duration than a data bit. So, the chip rate is much faster than the bit rate. Thereby, the noise signal occurs with much greater frequency than the original data signal and spreads the signal energy over a much wider band. DSSS is used in the 802.11b (Wi-Fi) and 802.11g specifications for wireless local area networks (WLANs). See also *802.11b, 802.11g, band, carrier, FHSS, frequency, narrowband, SS, symbol, wideband, Wi-Fi,* and *WLAN.*

DSU (Data Service Unit) A device that converts the unipolar signal received from data terminal equipment (DTE) into a bipolar signal required by a digital network. DSU functions variously include regeneration of digital signals, insertion of control signals, signal timing, and signal reformatting, although some of these functions can be ceded to either the channel service unit (CSU) or to the DTE. A CSU and DSU commonly are combined in a CSU/DSU, also known as a CDSU. See also *CSU* and *DTE*.

DTE (Data Terminal Equipment) **1.** Equipment comprising a keyboard, a video adapter, and a monitor. A data terminal is a dumb terminal, i.e., it does no independent processing, but relies on the computational resources of a computer to which it is connected over a dedicated circuit or through a network. A data terminal essentially is an input/output (I/O) device. See also terminal and terminal emulation. **2.** Referring to the subset of customer premises equipment (CPE) used in computer data, rather than voice, applications. DTE generally applies to data hubs, multiplexers, routers, switches, and terminals. See also *CPE*.

DTH (Direct-To-Home) Referring to a satellite television signal transmitted directly to the home, rather than to a broadcast television station or to a cable television (CATV) provider for retransmission to the subscriber. Synonymous with *direct broadcast satellite* (*DBS*). See also *CATV* and *DBS*.

DTMF (Dual Tone MultiFrequency) A signaling method that employs fixed pairs of frequencies. Most analog touchtone telephone sets are equipped with DTMF keypads in a grid three keys on the vertical axis and four keys on the horizontal axis, although some special purpose sets have a 4 × 4 grid. As the user depresses each key on the pad, the telephone set transmits a paired set of tones, comprising one high frequency tone and one low frequency tone, across a link to a switch that receives and stores them in a register for use in call setup. See Figure D-3 for an illustration of the DTMF grid.

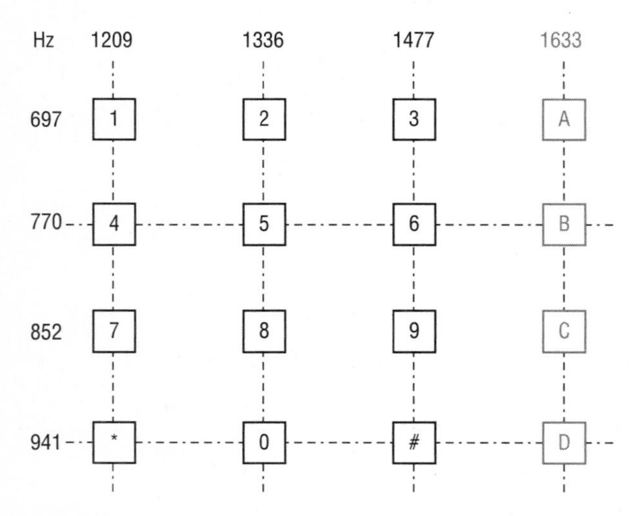

Figure D-3

D-to-A (Digital-to-Analog) See also *codec* and *modem*.

DTV (Digital TeleVision) Referring to television in digital, rather than analog, form. As is the case with digital communications, in general, DTV offers the advantages of enhanced bandwidth efficiency through compression, improved signal quality due to reduced noise, and enhanced overall management and control. DTV does not suffer from the ghosting, snowy images, and generally poor audio quality associated with analog TV. Issues of signal quality in DTV transmission manifest in artifacts such as blocking, or tiling, and stuttering. Digital video content benefits from enhanced processing, storage, and manipulation.

More specifically, digital content advantages include editing, alteration (e.g., morphing), reproduction, compression, and store-and-forward capability. DTV standards include high definition television (HDTV) and standard definition television (SDTV). The ATSC standards specify MPEG-2 compression, and the transport subsystem as ISO/IEC 13818. Packet transport involves a serial data stream of packets of 188 octets, one octet of which is a synchronization byte and 187 octets of which are payload. This packet approach is suitable for ATM switching, as each 188-octet MPEG-2 packet maps into the payload of four ATM cells, with only 4 octets of padding required. SDTV employs Reed-Solomon forward error correction (FEC) and 8-level vestigial sideband (8 VSB) RF modulation to support a bit rate of 19.28 Mbps over a 6 MHz terrestrial broadcast channel. Audio compression is based on the AC-3 specification from Dolby Digital and the ATSC. SDTV standards were developed by the Grand Alliance and reviewed, tested, and documented by the Advanced Television Systems Committee (ATSC) at the request of the United States Federal Communications Commission (FCC). See also *8-VSB, AC-3, analog, artifact, ATM, ATSC, bandwidth, broadcast, byte, channel, compression, digital, DTV, FCC, FEC, ghosting, Grand Alliance, HDTV, modulation, MPEG-2, NTSC, octet, packet, padding, PAL, payload, Reed-Solomon, RF, SDTV, SECAM, signal, store and forward, synchronize, TV*, and *video*.

dual attached concentrator (DAC) See *DAC*.

dual attached station (DAS) See *DAS*.

dual counter-rotating ring A network structure in which the nodes are laid out in a physical ring topology, or closed loop configuration, and are interconnected by two optical fibers, with one fiber transmitting in one direction and the other in the other direction. Such a configuration is highly resilient, not only because there is a redundant physical ring, but also because the counter-rotating transmission ensures that any node can communicate with any other node in one direction or the other in the event of the failure of only a single node or link. Fiber Distributed Data Interface (FDDI), Resilient Packet Ring (RPR), Synchronous Digital Hierarchy (SDH), and Synchronous Optical Network (SONET) all specify dual counter-rotating ring configurations. See also *FDDI, ring topology, RPR, SDH*, and *SONET*.

dual tone multifrequency (DTMF) See *DTMF*.

duct A tube, pipe, channel, or conduit through which a gas or liquid flows or through which cables or wires are run. See also *duct tape*.

duct tape Arguably one of the two greatest inventions of the twentieth century (WD-40 is the other), duct tape is a strong, pressure-sensitive, fabric-based adhesive tape invented in 1942 by scientists at Permacel, then a division of Johnson & Johnson, as a waterproof tape for sealing ammunication cases during World War II. In addition to its use in sealing air conditioning and heating ducts, duct tape is used for emergency repairs in NASCAR events, military field maneuvers, and manned space exploration, and is generally considered indispensable for sticking almost any thing together. As the saying goes, "If it is stuck and shouldn't be, WD-40 it. It if is unstuck and shouldn't be, duct tape it." See also *WD-40*.

du jour From French, meaning of the day. See also *standard*.

dumb terminal Also known as a *thin client*. A data terminal that does no independent processing, but relies on the computational resources of a computer to which it is connected over a dedicated circuit or through a network. Essentially an input/output (I/O) device with no internal processing power, a dumb terminal is a simple input-output device comprising a keyboard, monitor, network interface, and buffer memory that has only enough intelligence to respond to simple control codes from a computer. See also *computer, terminal*, and *thin client*.

duplex **1.** Two-way, or duplex, transmission over a single physical circuit. Duplex transmission can be either half-duplex or full duplex in nature, depending on the communications protocol. Half-duplex transmission is in both directions, but only one direction at a time. Full duplex transmission is simultaneous transmission in both directions. See also half-duplex, full-duplex, and simplex. **2.** A fiber optic system physical configuration that employs two fibers, one for transmission in each direction. Passive optical network (PON) standards, for example, include a duplex configuration option. See also *PON* and *simplex*.

duty cycle **1.** The proportion of time that a system, device, or component on intermittent duty is operating, rather than remaining idle. The duty cycle includes the time starting, running, stopping, and idling and generally is expressed as a percentage or fraction. A compressor, for example, that operates for 1 minute and then shuts off for 99 minutes, is said to have a duty cycle of 1⁄100, or 1 percent. **2.** In pulse systems, such as digital fiber optic transmission systems (FOTS) or baseband digital electrical transmission systems, the ratio of the sum of all pulse durations to the total period, during a specified period of continuous operation, with the pulse duration, including the rise to peak power and the fall to zero power. See also *baseband*, *digital*, *FOTS*, and *pulse*.

DUV (Data Under Voice) A technique developed by Bell Labs in 1971 that supported the transmission of high speed data signals on the lower end of frequencies in existing microwave radio systems. DUV became the basis for Dataphone Digital Service (DDS). See also *DDS*.

dweeb An unpleasant, tiresome, and socially inept person. A techie dweeb is a nerd or geek. See also *geek*, *nerd*, and *techie*.

DWDM (Dense Wavelength Division Multiplexing) An optical multiplexing technique defined in ITU-T Recommendations G.692 and G.959.1 as supporting eight or more wavelengths in the 1530–1565 nm C-band (Conventional band) and the 1565–1625 nm L-band (Long band). DWDM currently supports channel spacings of 200 GHz (1.6 nm at 1550 nm) and 100 GHz, with spacings of 50 GHz and even 25 GHz expected in the future. DWDM is used extensively in long haul networks. Coarse wavelength division multiplexing (CWDM) is defined by the ITU-T as supporting as many as 18 wavelengths in the 1270–1610 nm range, with spacing of 20 nm (2500 GHz at 1550 nm). See also *CWDM*, *ITU-T*, *multiplexer*, *wavelength*, *WDM*, and *window*.

DXC (Digital CROSS-Connect) A cross-connect device that enables digital circuits (e.g., T1s or E-1s), or even individual channels, to be cross-connected via an electronic cross-connect matrix. A cross-connect is much like a static switch in that connections can be changed to alter physical paths, on a semi-permanent basis, rather than a call-by-call basis. Synonymous with digital access cross-connect system (DACS) and digital cross-connect system (DCCS). A DXC can be used in an optical system, although the signal must be converted from optical format to electrical format and back to optical format. An optical cross-connect (OXC) can perform cross-connect functions in optical format. See also *DACS*, *IDF*, *MDF*, *OXC*, and *switch*.

DXI (Data eXchange Interface) A private user network interface (UNI) for end-user access to an ATM network from equipment such as a bridge, router, or DSU. At the Data Link Layer, DXI encapsulates the user payload in a frame, employing a variation of the HDLC protocol. The data communications equipment (DCE) converts the frame to the appropriate, class-specific ATM protocol through segmentation and reassembly (SAR) and presents the data to the ATM network switch in ATM cell format. DXI modes correspond to AAL 3/4 and AAL 5. See also *AAL 3/4*, *AAL 5*, *ATM*, *bridge*, *cell*, *Data Link Layer*, *DCE*, *DSU*, *encapsulate*, *HDLC*, *payload*, *private UNI*, *router*, and *SAR*.

Dy (Dysprosium) See *dysprosium*.

dynamic address Referring to an Internet Protocol (IP) address assigned to a host each time a connection is established, rather than being permanently assigned. Dynamic IP address assignment usually is accomplished through a router running the Dynamic Host Configuration Protocol (DHCP) protocol, as specified in RFC 1541 and the Network Address Translation (NAT) protocol specified in RFC 3022. See also *DHCP*, *host*, *IP address*, *NAT*, *router*, and *static address*.

dynamic bandwidth allocation Synonymous with *bonding*, *channel aggregation*, and *Nx64*. See *bonding*.

dynamic bend The bend in a cable under short-term physical load while being installed. See also *bend diameter.*

dynamic load The short-term load placed on a cable during installation. See also *load.*

dynamic frequency selection (DFS) See *DFS.*

Dynamic Host Configuration Protocol (DHCP) See *DHCP.*

dynamic port Synonymous with *private port.* A port that can be used by any computer application program to communicate with any other application program running Transmission Control Protocol (TCP) or User Datagram Protocol (UDP), with no registration requirements. Dynamic ports are numbered from 49,152 through 65,535. See also *port, registered port,* and *well-known port.*

dynamic rate adaptation Referring to the ability of a device, such as a modem, to dynamically adjust the speed of data transfer to varying line conditions in order to ensure the integrity of the datastream. The actual transmission speeds that modems can realize in either direction depends on the attributes of the analog local loop and various transient interference issues. Modem speeds are stated in terms of their maximums, based on the assumption that conditions are optimal. When modems initially connect, they negotiate a transmission rate based on their individual capabilities and their common evaluation of the quality of the connection. Should the quality of the connection deteriorate, dynamic rate adaptation enables fallback modems to negotiate a lower rate of transmission, using a less sophisticated modulation technique and perhaps adjusting the baud rate, and to ratchet up the transmission rate when conditions improve. See also *baud rate, modem, modulation, signaling rate,* and *transmission rate.*

dynamic rate adaption In ISDN, referring to the ability of a device, such as a terminal adapter (TA), to dynamically adjust the number of 64-kbps bearer channels (B channels) to the demands of the call or calls. In one scenario, rate adaption effectively throttles down the transmission rate from 64 kbps to the rate at which the non-ISDN device is capable. Rate adaption also serves to bond multiple B channels into H channels for more bandwidth-intensive applications. Rate adaption is accomplished in North America through the ITU-T V.120 protocol; the European standard is V.110. See also *B channel, bonding, H channel, TA, V.110,* and *V.120.*

dysprosium (Dy) A silvery white rare earth metallic element with an atomic number of 66. Dysprosium is from the Greek *dysprositos,* meaning *hard to get at,* which undoubtedly was inspired by the difficulty in isolating it. After all, dysprosium was first discovered in 1866 by Paul-Émil Lecoq de Boisbaudran, a French chemist, as an impurity in erbia, the oxide of erbium. It was not until 1906 that Georges Urbain, another French chemist, isolated it. (*Note:* Rumor has it that the French are not in a hurry to isolate elements.) There are no commercial applications for dysprosium. When combined with certain other rare earth elements, dysprosium is used in some lasers. See also *erbium, iridium,* and *laser.*

e The symbol for electron. See *electron*.

E **1.** Exa. From the Greek *hexa*, meaning *six*, translates to *quintillion*, referring to the fact that, in terms of order of magnitude in base 1,000, exa is $1{,}000^6$. In order, that puts it right behind kilo (thousand), Mega (million), Giga (billion), Tera (trillion), and Peta (quadrillion). **2.** In terms of the electromagnetic spectrum, EHz (ExaHertz) is a quintillion (10^{18}) Hertz, which is in the range of X-rays, gamma rays and cosmic rays, none of which are currently have any application in telecommunications. An Ebps would be a quintillion (10^{18}) bits per second (bps). In transmission systems, therefore, a quintillion would be exactly 1,000,000,000,000,000,000 since the measurement is based on a base 10, or decimal, number system. That definitely would be broadband, if it were possible, but it is difficult to imagine an application for that level of bandwidth. See also *bandwidth, bps, broadband*, and *electromagnetic spectrum*. **3.** Voltage. See *voltage*. **4.** In physics, the symbol for energy. See *energy*. **5.** In physics, the symbol for electromotive force. See *electromotive force*.

E-1 The first level of the European E-carrier digital hierarchy. An E-1 system comprises circuit terminating equipment in the form of a combination of a channel service unit (CSU) and data service unit (DSU) that jointly serve to interface a device to a full duplex (FDX) four-wire digital circuit and to perform various signal formatting, signal timing, monitoring, and diagnostic functions. E-1 operates at a signaling rate of 2.048 Mbps, which supports a frame rate of 8,000 frames per second (fps), with each frame comprising 32 eight-bit time slots, 30 of which are for user payload and 2 of which are set aside for signaling and control data. As illustrated in Figure E-1, time slot 0 (zero) is used for synchronization and alarming. Time slot 16 is used by the equipment for all signaling functions, such as circuit monitoring and diagnostics. At a rate of 8,000 fps, each time slot is repeated 8,000 times per second, which translates into a DS-O channel at 64 kbps (8 bits × 8,000 per second = 64,000 bps, or 64 kbps). Taken together, the 32 8-bit time division multiplexed (TDM) channels at 8,000 fps yields an aggregate payload signaling rate of 2.048 Mbps (32 channels × 64 kbps = 2.048 Mbps). Of that total, the two signaling and control channels consume 128 kbps (2 × 64 kbps = 128 kbps), leaving the user payload transmission rate at 1.920 Mbps.

E-1 was designed to operate over an unshielded twisted-pair (UTP) circuit comprising two two-wire pairs, each of which operates in simplex mode. One pair supports transmission in one direction and the other pair, in the opposite direction. In the aggregate, the physical four-wire circuit supports full-duplex (FDX) transmission. The line coding technique employed in traditional E-1 is alternate mark inversion (AMI), which yields 2.048 Mbps at a nominal carrier frequency of 1.168 MHz, which is exactly half the

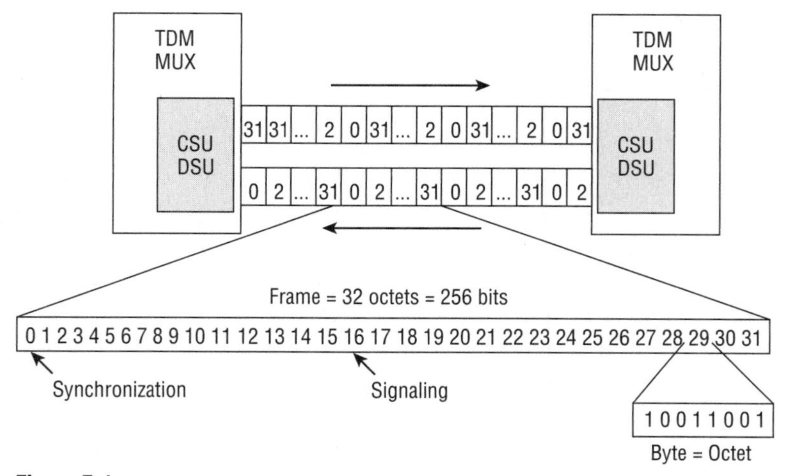

Figure E-1

E-1 bit rate, plus some overhead for error control. At such a high frequency, issues of attenuation are significant and mutual interference between cable pairs must be considered, so repeaters must be placed every 6,000 feet. Contemporary E-1 circuits typically are provisioned using high-bit-rate digital subscriber line (HDSL) technology, which mitigates these issues.

E-1 is considered medium independent and will run over coaxial cable, optical fiber, microwave, satellite, and free space optics (FSO), as well as twisted pair. E-1 generally is used in local loops and other short haul applications. In long haul applications, E-3 and other, higher speed, standards generally are employed. See also *AMI, attenuation, carrier, CSU, DS-0, DSU, E-3, E-carrier, FDX, four wire, frame, framing bit, frequency, FSO, HDSL, line coding, payload, signaling rate, simplex, synchronization, TDM, time slot, transmission rate, two-wire circuit,* and *UTP.*

E.164 The ITU-T recommendation entitled *The International Public Telecommunication Numbering Plan.* E.164 specifies the current international Numbering Plan Administration (NPA) convention for the global switched telephone network (GSTN) and the national public switched telephone networks (PSTNs) that comprise it. E.164 specifies a maximum of 15 digits, although the number of digits required for calling within a nation varies. See also *GSTN, NPA,* and *PSTN.*

E-2 The second level in the European E-carrier digital hierarchy. E-2 runs at 8.448 Mbps and comprises four E-1s at 2.048 Mbps each, plus 256 kbps of overhead and justification, or bit stuffing, to adjust for variations in the clocking rates of the incoming E-1s. Multiplexing is performed by M12 (Multiplex T1-to-T2) terminals that multiplex four E-1 signals, which yield 120 DS-0 channels at 64 kbps per channel for user payload. See also *bit stuffing, digital hierarchy, DS-0, E-1,* and *E-carrier.*

E-3 The third level in the European E-carrier digital hierarchy. Designed for long haul transmission in support of interoffice trunking in the public switched telephone network (PSTN), E-3 runs at a signaling rate of 34.368 Mbps. An E-3 actually begins as four E-1s multiplexed into an E-2, by an M13 (Multiplex E-1 to E-3) mux, which then multiplexes four E-2s to yield a signaling rate of 33.792 Mbps. Stuff bits are added as necessary to adjust for variations in the clocking rates of the incoming E-2s, bringing the signaling rate up to 34.368 Mbps, comprising 480 DS-0 channels at 64 kbps for payload, plus 16 embedded channels for signaling and control. See also *bit stuffing, channel, digital hierarchy, DS-0, E-carrier, long haul, multiplexer, payload, PSTN, signaling rate,* and *trunk.*

E-4 The fourth level in the European E-carrier digital hierarchy. E-4 operates at a signaling rate of 139.268 Mbps, which supports 1,920 DS-0 channels at 64 kbps for payload. E-4 has been superseded by the SDH fiber optic transmission system (FOTS). See also *channel, digital hierarchy, DS-0, E-carrier, FOTS, payload, signaling rate,* and *SDH.*

E-5 The fifth level in the European E-carrier digital hierarchy. E-5 operates at a signaling rate of 565.148 Mbps, which supports 7,680 DS-0 channels at 64 kbps for payload. E-5 has been superseded by the SDH fiber optic transmission system (FOTS). See also *channel, digital hierarchy, DS-0, E-carrier, FOTS, payload, signaling rate,* and *SDH.*

E911 (Enhanced 911) An enhanced version of the 911 system used in the United States for calls for emergency assistance from police, fire, and ambulance services. Calls to 911 are routed to a public safety access point (PSAP). E911 identifies the location of the caller, routes the call to the appropriate local PSAP, and provides the PSAP with location information in order to speed response. Wireline and cellular service providers both are required by the Federal Communications Commission (FCC) to support E911. VoIP over the Internet presents considerable E911 difficulties as the location of the caller can be difficult, if not impossible, to determine. See also *911, FCC, Internet, PSAP,* and *VoIP.*

easement In law, a legal interest in real property that affords one the right to make limited use of the property of another for a specified purpose, which often is a right of way across the property. See also *right of way.*

eavesdrop To secretly listen to or overhear a conversation without physically wiretapping a circuit. In medieval times, eavesdrop was rainwater that dropped to the ground from the eaves of a building. An eavesdropper was one who secretly hid in the area of the eavesdrop to overhear a private conversation. Contemporary electronic eavesdropping undoubtedly is a much drier and more comfortable endeavor. See also *Echelon* and *wiretap*.

EB (ExaByte) A quintillion (10^{18}) bytes. An EB is actually 1,152,921,504,606,846,976 bytes ($1{,}024^6$, or 2^{60}) bytes, as the measurement of internal computer memory is based on a base 2, or binary, number system. The term EB comes from the fact that 1,152,921,504,606,846,976 is nominally, or approximately, 1,000,000,000,000,000,000. In reality, an exabyte is rarely mentioned and never encountered. To put an EB in perspective, some sources suggest that the total volume of information generated annually, worldwide, is in the range of 2 EB, and that the total of all words ever spoken by human beings is in the range of 5 EB. See also *byte*.

E-Band (Extended Band) The ITU-T standard optical transmission window in the wavelength range of 1,360–1,460 nm. See also *wavelength* and *window*.

EBCDIC (Extended Binary Coded Decimal Interchange Code) An improvement over earlier (1950) Binary Coded Decimal (BCD) and (1951) Extended Binary Coded Decimal (Extended BCD), EBCDIC was developed by IBM in 1962 to enable different IBM computer systems to communicate based on a standard coding scheme, which users have the ability to modify. EBCDIC is an 8-bit coding scheme, yielding 2^8 (256) possible combinations. As a result, English and similarly complex alphabets can be supported, as can upper- and lowercase letters, a full range of numbers (0–9), and all necessary punctuation marks. EBCDIC also supports a large number of control characters, which is critical in the coordination of communications between the complex mainframe and midrange computers that were the core of IBM's business. Table E-1 of the EBCDIC code is based on a 0–255 scale in decimal notation (dec), and Table E-2 is based on 00–FF in the hexadecimal notation (hex).

Table E-1: EBCDIC Code — 0–255 Decimal Notation

<table>
<tr><th>Dec</th><th>Hex</th><th>Code</th><th>Dec</th><th>Hex</th><th>Code</th><th>Dec</th><th>Hex</th><th>Code</th><th>Dec</th><th>Hex</th><th>Code</th></tr>
<tr><td>0</td><td>00</td><td>NUL</td><td>32</td><td>20</td><td></td><td>64</td><td>40</td><td>space</td><td>96</td><td>60</td><td>-</td></tr>
<tr><td>1</td><td>01</td><td>SOH</td><td>33</td><td>21</td><td></td><td>65</td><td>41</td><td></td><td>97</td><td>61</td><td>/</td></tr>
<tr><td>2</td><td>02</td><td>STX</td><td>34</td><td>22</td><td></td><td>66</td><td>42</td><td></td><td>98</td><td>62</td><td></td></tr>
<tr><td>3</td><td>03</td><td>ETX</td><td>35</td><td>23</td><td></td><td>67</td><td>43</td><td></td><td>99</td><td>63</td><td></td></tr>
<tr><td>4</td><td>04</td><td></td><td>36</td><td>24</td><td></td><td>68</td><td>44</td><td></td><td>100</td><td>64</td><td></td></tr>
<tr><td>5</td><td>05</td><td>HT</td><td>37</td><td>25</td><td>LF</td><td>69</td><td>45</td><td></td><td>101</td><td>65</td><td></td></tr>
<tr><td>6</td><td>06</td><td></td><td>38</td><td>26</td><td>ETB</td><td>70</td><td>46</td><td></td><td>102</td><td>66</td><td></td></tr>
<tr><td>7</td><td>07</td><td>DEL</td><td>39</td><td>27</td><td>ESC</td><td>71</td><td>47</td><td></td><td>103</td><td>67</td><td></td></tr>
<tr><td>8</td><td>08</td><td></td><td>40</td><td>28</td><td></td><td>72</td><td>48</td><td></td><td>104</td><td>68</td><td></td></tr>
<tr><td>9</td><td>09</td><td></td><td>41</td><td>29</td><td></td><td>73</td><td>49</td><td></td><td>105</td><td>69</td><td></td></tr>
<tr><td>10</td><td>0A</td><td></td><td>42</td><td>2A</td><td></td><td>74</td><td>4A</td><td>[</td><td>106</td><td>6A</td><td>|</td></tr>
<tr><td>11</td><td>0B</td><td>VT</td><td>43</td><td>2B</td><td></td><td>75</td><td>4B</td><td>.</td><td>107</td><td>6B</td><td>,</td></tr>
<tr><td>12</td><td>0C</td><td>FF</td><td>44</td><td>2C</td><td></td><td>76</td><td>4C</td><td><</td><td>108</td><td>6C</td><td>%</td></tr>
<tr><td>13</td><td>0D</td><td>CR</td><td>45</td><td>2D</td><td>ENQ</td><td>77</td><td>4D</td><td>(</td><td>109</td><td>6D</td><td>_</td></tr>
<tr><td>14</td><td>0E</td><td>SO</td><td>46</td><td>2E</td><td>ACK</td><td>78</td><td>4E</td><td>+</td><td>110</td><td>6E</td><td>></td></tr>
<tr><td>15</td><td>0F</td><td>SI</td><td>47</td><td>2F</td><td>BEL</td><td>79</td><td>4F</td><td>| !</td><td>111</td><td>6F</td><td>?</td></tr>
</table>

continued

Table E-1: EBCDIC Code — 0–255 Decimal Notation *(continued)*

Dec	*Hex*	*Code*	*Dec*	*Hex*	*Code*	*Dec*	*Hex*	*Code*	*Dec*	*Hex*	*Code*
16	10	DLE	48	30		80	50	&	112	70	
17	11		49	31		81	51		113	71	
18	12		50	32	SYN	82	52		114	72	
19	13		51	33		83	53		115	73	
20	14		52	34		84	54		116	74	
21	15		53	35		85	55		117	75	
22	16	BS	54	36		86	56		118	76	
23	17		55	37	EOT	87	57		119	77	
24	18	CAN	56	38		88	58		120	78	
25	19	EM	57	39		89	59		121	79	‘
26	1A		58	3A		90	5A	!]	122	7A	:
27	1B		59	3B		91	5B	$	123	7B	#
28	1C	IFS	60	3C		92	5C	*	124	7C	@
29	1D	IGS	61	3D	NAK	93	5D	)	125	7D	‘
30	1E	IRS	62	3E		94	5E	;	126	7E	=
31	1F	IUS	63	3F	SUB	95	5F	^	127	7F	“

Table E-2: EBCDIC Code — 00–FF Hexadecimal Notation

Dec	*Hex*	*Code*	*Dec*	*Hex*	*Code*	*Dec*	*Hex*	*Code*	*Dec*	*Hex*	*Code*
128	80		160	A0		192	C0	{	224	E0	\
129	81	a	161	A1	~	193	C1	A	225	E1	
130	82	b	162	A2	s	194	C2	B	226	E2	S
131	83	c	163	A3	t	195	C3	C	227	E3	T
132	84	d	164	A4	u	196	C4	D	228	E4	U
133	85	e	165	A5	v	197	C5	E	229	E5	V
134	86	f	166	A6	w	198	C6	F	230	E6	W
135	87	g	167	A7	x	199	C7	G	231	E7	X
136	88	h	168	A8	y	200	C8	H	232	E8	Y
137	89	i	169	A9	z	201	C9	I	233	E9	Z
138	8A		170	AA		202	CA		234	EA	
139	8B		171	AB		203	CB		235	EB	
140	8C		172	AC		204	CC		236	EC	
141	8D		173	AD		205	CD		237	ED	
142	8E		174	AE		206	CE		238	EE	
143	8F		175	AF		207	CF		239	EF	

Table E-2: EBCDIC Code — 00–FF Hexadecimal Notation *(continued)*

Dec	*Hex*	*Code*	*Dec*	*Hex*	*Code*	*Dec*	*Hex*	*Code*	*Dec*	*Hex*	*Code*
144	90		176	B0		208	D0	}	240	F0	0
145	91	j	177	B1		209	D1	J	241	F1	1
146	92	k	178	B2		210	D2	K	242	F2	2
147	93	l	179	B3		211	D3	L	243	F3	3
148	94	m	180	B4		212	D4	M	244	F4	4
149	95	n	181	B5		213	D5	N	245	F5	5
150	96	o	182	B6		214	D6	O	246	F6	6
151	97	p	183	B7		215	D7	P	247	F7	7
152	98	q	184	B8		216	D8	Q	248	F8	8
153	99	r	185	B9		217	D9	R	249	F9	9
154	9A		186	BA		218	DA		250	FA	
155	9B		187	BB		219	DB		251	FB	
156	9C		188	BC		220	DC		252	FC	
157	9D		189	BD		221	DD		253	FD	
158	9E		190	BE		222	DE		254	FE	
159	9F		191	BF		223	DF		255	FF	

Although the full explanations of all control codes are outside the scope of this book, the following is a representative list:

- **NUL (NULl):** A transmission control character used to serve a media-fill or time-fill requirement, i.e., a stuff character or padding character.
- **SOH (Start Of Header):** A transmission-control character indicating the start of a message heading.
- **STX (Start of TeXt):** A transmission-control character to start the reading, transmission, reception, or recording of text.
- **ETX (End of TeXt):** A transmission-control character to terminate the reading, transmission, reception, or recording of text.
- **EOT (End Of Transmission):** A transmission-control character to terminate a transmission that may have included one or more texts or messages.
- **ENQ (ENQuiry):** A transmission-control character used to request a response from a station to which a connection has been established. The request may be for the station identification, type of equipment, and station status.
- **NAK (Negative AcKnowledgement):** A transmission-control character sent by the receiving device to the transmitting device to indicate that a received block of data contained one or more errors. A NAK will trigger the transmitting device to retransmit that errored block.
- **ACK (ACKnowledgement):** A transmission-control character sent by the receiving device to the transmitting device to indicate that a received block of data contained no errors.

- **BEL (BELl):** A transmission-control character that causes a bell to ring or activates some other audio or visual device to gain the attention of the operator at the receiving station.
- **ETB (End of Transmission Block):** A code-extension character used to indicate the end of the transmission of a block of data.
- **CAN (CANcel):** A transmission-control character indicating that the associated data is in error or is to be ignored.
- **EM (End of Medium):** The physical end of a data storage medium, or the usable portion of the medium.
- **SUB (SUBstitute):** Used in place of a character that is known to be invalid, i.e., in error. Also used to indicate a character used in place of one that cannot be represented on a given device, e.g., *e* may be used in place of ε (epsilon) or *d* may be used in place of Δ (delta).
- **ESC (ESCape):** A code-extension character used to indicate a change in code interpretation to another character set, according to some convention or agreement. This is much like the use of the shift key in Baudot code to indicate a shift between figures and characters.
- **CR (Carriage Return):** A format-control character that causes the print or display position to move to the first position, or left-hand margin, of the screen or print medium.
- **LF (Line Feed):** A format-control character that moves the print position down to the next line.

In Unicode terms, EBCDIC is known as *Unicode Transformation Format-EBCDIC* (UTF- EBCDIC). See also *code set*, *decimal system*, *hexadecimal notation*, and *Unicode*.

EBPP (Electronic Bill Presentation and Payment) A vendor service that involves rendering an invoice on a Web site and providing for electronic payment in the form of an authorization for a wire transfer or a credit card charge. See *WWW*.

E-carrier (European carrier) A hierarchy of standards for digital transmission, E-carrier is based on the original North American T-carrier digital carrier system, although the specifics are quite different with respect to signaling rates, framing conventions, line coding technique, and PCM companding technique (A-law rather than μ-law). In many respects E-carrier is a considerable improvement over T-carrier. For example, E-1 supports 30 DS-0 payload channels, compared with T1 at 24 channels, and the higher E-carrier levels build on that difference. E-carrier also supports non-intrusive signaling and control through two channels reserved for such purposes. As a result, E-carrier supports clear channel communications of a full 64 kbps per DS-0, compared to 56 kbps data with T-carrier. The DS-0 (Digital Signal level Zero) is the fundamental building block of E-carrier, as it is with T-carrier and J-carrier, the Japanese version. Through time division multiplexing (TDM), E-carrier interleaves DS-0 channels at various signaling rates to create the services that comprise the European digital hierarchy, as detailed in Table E-3.

Table E-3: European Digital Hierarchy: E-Carrier

E-carrier Level	*Data Rate (Mbps)*	*Number of 64 kbps Channels (DS-0s)*	*Number of E-1s*
E-1	2.048	30	1
E-2	8.448	120	4
E-3	34.368	480	16
E-4	139.268	1920	64
E-5	565.148	7680	256

See *digital signal hierarchy* for a side-by-side comparison of the North American, European, and Japanese digital hierarchies. See also *carrier, channel, companding, digital, DS-0, E-1, E-2, E-3, E-4, E-5, J-carrier, PCM, signaling rate, T-carrier,* and *TDM.*

ECC (Electronic Common Control) A specialized microprocessor controlled circuit switch. The first ECC switch was the Electronic Switching System (ESS), developed by AT&T Bell Telephone Laboratories (Bell Labs) with the assistance of Western Electric. Based on the transistor, invented at Bell Labs in 1948, the ESS involved a development effort that began in earnest in the early 1950s. The first ESS central office (CO) began service in Succasunna, New Jersey, on May 30, 1965, connecting 200 subscribers. By 1974, there were 475 such offices in service, serving 5.6 million subscribers. The development effort was estimated to involve 4,000 man-years and a total cost of $500 million. See also *circuit switch* and *electronic.*

eccentric **1.** Elliptical or off-center, rather than perfectly circular with a precisely centered axis. Eccentricities in the core of an optical fiber can cause signal attenuation and distortion. **2.** A euphemism for someone who is crazy and rich, as opposed to being just plain crazy like the rest of us.

Echelon A system operated by the United States National Security Agency (NSA), Echelon reportedly eavesdrops on approximately three billion conversations a day in defense of national security. Echelon apparently can tap any electromagnetic transmission system, including fiber optics, anywhere on the globe. See also *wiretap.*

echo **1.** Also known as the *rain-barrel effect*, echo is signal reflection. At any point in a circuit where an electromagnetic wave meets a discontinuity, a portion of the wave is reflected back in the direction of the transmitter. Such discontinuities can be caused by impedance mismatches, mismatches between line and balancing networks, and irregular spacing of loading coils. Echo is imperceptible in human-to-human conversations as long as the echo return is weak and the total roundtrip delay is not longer than 30–40 milliseconds (ms). Echo generally is not an issue, except in very long haul copper circuits or over satellite circuits. Contemporary networks are designed with echo cancellers, which remove a portion of the delayed transmitted signal from the received signal. There are also devices known as echo suppressors, which often convert full duplex (FDX) phone connections into half duplex (HDX). See also *echo canceller, echo suppressor, FDX, HDX, impedance,* and *loading coil.* **2.** A signal intentionally returned to the transmitter by the receiver for purposes of primitive error control. See also *echo checking.* **3.** A packet intentionally returned to the transmitter by the receiver for purposes of testing an end-to-end path. The ping utility is an application of the Internet Control Message Protocol (ICMP) used to test a path from one host computer to another across an IP-based network in what is essentially a command to echo the packet from the remote host back to the originating host. See also *host, ICMP, IP, ping,* and *utility.*

echo canceller Transmission equipment designed to suppress echo in a two-way circuit by attenuating the signals propagating in one direction caused by reflected (i.e., echoed) signal currents in the other direction. See also *attenuation, current, echo, echo suppressor,* and *propagation.*

echo checking Synonymous with *echoplex.* A primitive error control method in which the receiving device echoes the received data back to the transmitting device, character by character. The transmitting operator can view the data as received and echoed, and make corrections as appropriate, assuming that he hasn't lost his sight or mind due to the ddoouubbllee vviissiioonn effect. As errors also can occur in the transmission of the echoed data, this approach is highly unreliable. See also *echo* and *error control.*

echoplex Synonymous with *echo checking.* A primitive error control protocol in which the receiving station retransmits each received character back to the transmitting station. See also *echo, echo checking, error control, full duplex, half duplex, protocol,* and *simplex.*

echo suppressor A voice-operated device designed to suppress echo in a two-way circuit by shutting off the return path to prevent echo signals propagating back to the speaker. See also *attenuation, circuit, echo, echo canceller,* and *propagation.*

ECM (Error Control Mode) A communication mode that invokes error control, i.e., error detection and correction. Some fax machines, for example, allow the user to toggle error control on and off. See also *error control.*

ECMA (European Computer Manufacturers Association) See *Ecma International.*

Ecma International Née ECMA (European Computer Manufacturers Association). An industry association with the stated purpose of the standardization of information and communication technology (ICT) and consumer electronics (CE) in cooperation with appropriate national, European, and international organizations.

ECN Explicit Congestion Notification The means by which the frame relay network advises devices of network congestion. Forward Explicit Congestion Notification (FECN) is a one-bit field in the LAPF frame used to advise the target (i.e., receiving) frame relay access device (FRAD) that the frame experienced congestion on the network so the FRAD can adjust its expectations. Backward Explicit Congestion Notification (BECN) is a one-bit field used to advise the transmitting FRAD that it is transmitting into a congested network so that the FRAD can reduce its rate of transmission. See also *BECN*, *FECN*, *FRAD*, *frame*, *frame relay*, *Implicit Congestion Notification*, and *LAPF.*

e-commerce (electronic commerce) The use of the Internet for business transactions. See also *Internet.*

ECSA (Exchange Carriers Standards Association) Now the Alliance for Telecommunications Industry Solutions (ATIS). See *ATIS.*

ECSD (Enhanced Circuit Switched Data) In Enhanced Data rates for GSM Evolution (EDGE) cellular radio networks, an enhancement of the native Global System for Mobile Communications (GSM) circuit-switching protocol. ECSD adds 8-Phase Shift Keying (8-PSK) as a modulation option, thereby increasing the efficiency of data transmission and yielding greater throughput. See also *8-PSK*, *EDGE*, *GSM*, *modulation*, *protocol*, and *throughput.*

ECTF (Enterprise Computer Telephony Forum) Under the umbrella of the Computing Technology Industry Association (CompTIA), the ECTF promotes interoperability and standard approaches to computer telephony (CT). See *CompTIA* and *computer telephony.*

EDFA (Erbium-Doped Fiber Amplifier) A type of amplifier used in fiber optic transmission systems (FOTS) and comprising a short length of fiber that has been doped with erbium and spliced into the operating single-mode fiber (SMF) in a configuration known as discrete amplification, or lumped amplification. A three-port wavelength division multiplexer (WDM) is used, with one incoming port connected to the operating fiber carrying the primary signal in the 1550 nm window, one incoming port attached to a pump laser operating at 980 nm or 1480 nm, and the one outgoing port connected to the operating fiber (see Figure E-2). The pump laser excites the erbium atoms. Weak incoming light from the operating system causes the erbium atoms to drop from their excited state. As they do so, they release the extra energy, which transfers to the primary signal and amplifies it. An EDFA can simultaneously amplify a number of wavelengths in an operating range around 1550 nm, which is in the optical C-Band. A single-pump EDFA involves a pump laser on the upstream side of the erbium-doped fiber section and provides a gain varying from +10 dB (1,000%, or 10:1), to as much as approximately +17 dB (approximately 8,000 percent, or 80:1). A double-pump EDFA involves one pump laser on the upstream side and another on the downstream side of the erbium-doped fiber section, and provides a gain of close to 30 dB (100,000 percent, or 1,000:1). *Note:* The pump lasers can operate in either direction. Optical isolators, placed on both sides of the EDFA, act like diodes, serving to prevent optical signals from traveling in more that one direction. EDFAs are highly effective and less costly than optical repeaters, but generally are limited to no more than 10 spans over a total distance of 800 km or so. At that point a repeater must be applied to retime and reshape the signal, thereby filtering out the accumulated noise caused by various forms of dispersion.

EDFAs are further limited by their inability to amplify wavelengths shorter than 1525 nm. An EDFA performs a type of amplification known as lumped amplification, as it is concentrated in a single point. See also *amplifier, C-Band, diode, dispersion, dopant, erbium, FOTS, lumped amplification, noise, repeater, SMF, WDM,* and *window.*

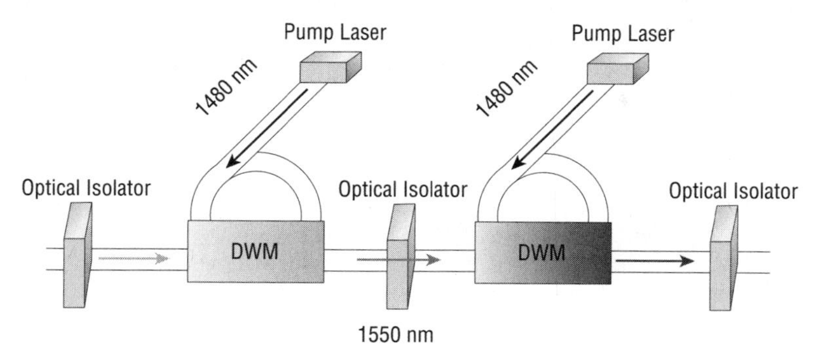

Figure E-2

EDGE (Enhanced Data rates for GSM Evolution) Originally, Enhanced Data rates for Global Evolution. A 2.5G standard (1999) developed by the European Telecommunications Standards Institute (ETSI) as the final stage in the evolution of data communications within the Global System for Mobile Communications (GSM) standards. The only IMT-2000 specification based on time division multiple access (TDMA) rather than code division multiple access (CDMA), EDGE supports data transmission rates up to 473.6 kbps over GSM channels 200 kHz wide through an improved modulation technique known as 8-Phase Shift Keying (8-PSK), which involves eight levels of phase shift and, therefore, supports three bits per symbol. EDGE employs frequency division duplex (FDD) to support bidirectional communications over 124 channels, each of which supports 8 time slots. EDGE supports two modes of operation:

- Enhanced GPRS (EGPRS) is a packet switching transmission mode that supports transmission rates as high as 473.6 kbps..
- Enhanced Circuit Switched Data (ECSD) is an enhancement of the native GSM circuit switching protocol.

EDGE also runs over IS-136 D-AMPS TDMA networks in the United States. In either case, EDGE is an intermediate step between 2G TDMA and 3G W-CDMA. See also *2.5G, 8-PSK, CDMA, channel, code, D-AMPS, circuit switch, ETSI, FDD, GPRS, GSM, IMT-2000, modulation, packet switch, symbol, TDMA,* and *transmission rate.*

edge switch An edge switch is positioned at the physical edge of a public network. The user organization gains access to an edge switch via an access link, or local loop. A *central office* (CO) is an example of an edge switch in the context of the circuit-switched Public Switched Telephone Network (PSTN). In the context of asynchronous transfer mode (ATM), an edge switch may be referred to as an access node or service node. See also *ATM, CO, local loop, switch, tandem switch,* and *PSTN.*

EDI (Electronic Data Interchange) Standards for the computer-to-computer electronic exchange of business data, such as invoices and purchase orders, in standard formats. The parties engaged in an EDI transaction agree to a format that allows data transfer requiring no human intervention or re-keying on either end. The American National Standards Institute (ANSI) Accredited Standards Committee (ASC) developed the ANSI ASC X12 standard, which is popular in North America and is used widely throughout the world. The UN/EDIFACT international standard, a United Nations recommendation, is predominant outside of North America. EDI standards specify data formats, character sets, and data elements. See also *ANSI* and *standard*.

EDIFACT (Electronic Data Interchange For Administration, Commerce, and Transport) See *EDI*.

Edison, Thomas Alva (1847–1931) Inventor of such devices as the phonograph, electric light bulb, carbon microphone, and electric chair. Not only was much of his early work was in telegraphy, but the full-duplex quadraplex telegraph was among his early financial successes. Always fascinated with telegraphy, he even nicknamed his first two children "Dot" and "Dash." Edison was known popularly as the "Wizard of Menlo Park." See also *dash*, *dot*, and *telegraph*.

.edu (education) Pronounced *dot e-d-u*. The Internet generic Top Level Domain (gTLD) reserved exclusively for accredited degree-granting educational institutions. This is an unsponsored domain. See also *gTLD*, *Internet*, and *unsponsored domain*.

EEB (Extended Erlang B) See *Extended Erlang B*.

EF (Expedited Forwarding) The Differentiated Services (DiffServ) protocol identifies two primary types of per-hop behaviors (PHBs), representing two service levels, or forwarding classes. Expedited Forwarding (EF) provides minimal delay, jitter, and loss. EF traffic exceeding the traffic profile, as defined by the Service Level Agreement (SLA), is discarded. Assured Forwarding (AF) comprises four classes, each of which contains three drop precedences and allocates certain amounts of buffer space and bandwidth. See also *AF*, *bandwidth*, *buffer*, *delay*, *DiffServ*, *jitter*, *PHB*, *protocol*, and *SLA*.

EFF (Electronic Frontier Foundation) A donor-funded, not-for-profit organization dedicated to defending free speech, privacy, innovation, and consumer rights in the context of the digital age and, particularly, the Internet. For contact information, see Appendix A.

e.g. (exempli gratia) Translates from Latin as *for example*, or literally *as example by favor*. *Note:* It strikes me as odd that those Latins seem to have a different word for just about everything. <grin> However, they come in handy when you want to explain the difference between two things, which would be hard if they had the same name. <big grin> See also *emotag*.

egosurf To surf the Internet for one's own name, or for links to one's own website.

EGP (Exterior Gateway Protocol) An inter-Autonomous System (AS) protocol concerned with conveying routing reachability information between groups of routers that fall within a single administrative domain. EGP runs on top of the connectionless Internet Protocol (IP) and is assigned well-known port number 8. The Border Gateway Protocol (BGP), which builds on and enhances EGP, runs on top of the Transmission Control Protocol (TCP), thereby ensuring a connection-oriented data flow and reliability of datastream transport. EGP was described in IETF RFC 827 (1982). See also *AS*, *BGP*, *connectionless*, *connection-oriented*, *domain*, *IETF*, *IP*, *port*, *protocol*, *routing*, and *TCP*.

EGPRS (Enhanced General Packet Radio System) In Enhanced Data rates for GSM Evolution (EDGE) cellular radio networks, a packet-switched transmission mode that supports transmission rates as high as 473.6 kbps. EGPRS estimates link quality in order to adapt the modulation and coding scheme (MCS), of which there are nine levels. Four levels employ Gaussian minimum-shift keying (GMSK) and yield transmission rates up to 140.8 kbps. Five levels employ 8-phase shift keying (8-PSK) and yield transmission rates up to 473.6 kbps. If the system senses that the link quality is good, it will elect to employ the more efficient 8-PSK and, therefore, realize higher signaling rates per time slot and higher data throughput. If the link quality is estimated to be poor, the system will ratchet down to the less capable GMSK. Incremental Redundancy (IR) is an enhanced automatic repeat request (ARQ) technique that forward error correction (FEC) overhead in an attempt to maximize throughput. See also *8-PSK*, *ARQ*, *cellular radio*, *coding*, *EDGE*, *FEC*, *GMSK*, *GPRS*, *link*, *modulation*, *overhead*, *packet switch*, *signaling rate*, *throughput*, and *time slot*.

EHF (Extremely High Frequency) EHF radio is in the frequency range of 30 GHz – 300 GHz and has a wavelength of 1 cm – 1 mm. EHF radio has applications in microwave and satellite radio, and radiolocation systems. EHF radio is at the very upper limit of the radio spectrum. Higher frequency signals fall into the infrared light spectrum. See also *electromagnetic spectrum*, *frequency*, *Hz*, *Ir*, and *wavelength*.

EIA (Electronic Industries Alliance) A national United States trade organization that is a partnership of electronic and high-technology associations and companies. Founded in 1924 as the Radio Manufacturers Association, the EIA is accredited by the American National Standards Institute (ANSI) and provides a forum for standards development in the areas of electronic components, consumer electronics, electronic information, telecommunications, and Internet security. See Appendix A for contact information. See also *ANSI* and *TIA*.

EIA-232 A standard interface for data terminal equipment (DTE) first published by the Electronics Industry Alliance (EIA) in the early 1960s and originally known as *RS-232* (Recommended Standard 232), EIA-232 addresses signal voltages, signal timing, signal function, a protocol for information exchange, and either 25-pin or 9-pin mechanical connectors. Most personal computers have an RS-232 serial port for connecting external modems, printers, scanners, and other peripheral devices. See also *1394*, *EIA*, *modem*, *protocol*, *short haul modem*, *USB*, and *voltage*.

EIA/TIA See *ANSI/EIA/TIA-568*

EIA/TIA (Electronic Industry Alliance/Telecommunications Industry Alliance) See *EIA* and *TIA*.

EIEIO The chorus of "Old MacDonald Had a Farm," which has nothing to do with telecommunications, unless perhaps Old MacDonald had a telephone on the farm. The song mentions only a chick (cluck-cluck), cow (moo-moo), duck (quack-quack), pig (oink-oink), horse (neigh-neigh), and various other barnyard animals. It would be easy enough, however, to equip Old MacDonald with a telephone with a ring-ring here and a ring-ring there. Those who use cell phones and downloadable ring tones are advised to sing another song.

EIR (Excess Information Rate) In packet networks, the data rate, measured in bit per second (bps), in excess of the Committed Information Rate (CIR) to which a public carrier network will allow a virtual circuit (VC) to burst during periods of no congestion. If the EIR is set to zero (i.e., disabled), the VC can burst up to the full port speed. If the EIR is set to a non-zero (i.e., enabled) the VC can burst up to a rate equal to CIR+EIR, but no more than the full port speed. During periods of congestion, the VC is throttled back to the CIR speed. Frames in excess of the CIR are marked discard eligible (DE), which means that they may be discarded in the event of congestion within the network core. Frame relay and Resilient Packet Ring (RPR) employ CIR and EIR mechanisms. ATM services offer similar features based on cell counts. See also *bandwidth*, *carrier*, *CIR*, *DE*, *frame relay*, *packet*, *RPR*, and *VC*.

EKTS (Electronic Key Telephone System) A semiconductor-based, software controlled KTS. EKTS systems appeared in the 1970s and soon obsoleted electromechanical KTS systems. See also *electronic* and *KTS*.

ELAN (Emulated LAN) See *emulation* and *LANE*.

electric telegraph See *telegraph*.

electricity From the Greek *elektor*, meaning *shining* or the *sun*. A fundamental form of energy created by the movement of electrons (negative charges), protons, or positrons (positive charges) and generating current. See also *current*.

electromagnetic A device that operates on the basis of electromagnetic fields and that contains few, if any, mechanical components. A solenoid is an example of an electromagnetic device. See also *electromagnetism*, *electromechanical*, and *electronic*.

electromagnetic interference (EMI). See *EMI*.

electromagnetic spectrum The full range of electromagnetic energy that can be radiated, as defined by frequency (f), or wavelength (λ), which is the inverse of frequency. In terms of frequency, the spectrum begins at almost zero (0) and extends to infinity. In terms of wavelength, the spectrum begins at almost zero and extends to infinity, but in reverse. The portion of the spectrum currently usable for telecommunications includes electricity, radio, and infrared light. Table E-4 includes frequency band designations, nominal frequency ranges, nominal wavelengths, and example telecommunications applications.

Table E-4: Frequency Spectrum: Band Designations, Nominal Frequency Ranges, Nominal Wavelengths, and Example Communications Applications

Band Designation	*Frequency (Hz)*[1]	*Wavelength (Λ)*[2]	*Applications*
Audible	20 Hz – 20 kHz	>100 km	Acoustics
Extremely Low Frequency (ELF) Radio	30 Hz – 300 Hz	10,000 km – 1,000 km	Submarine Communications
Infralow Frequency (ILF)	300 Hz – 3 kHz	1,000 km – 100 km	Not Applicable
Very Low Frequency (VLF) Radio	3 kHz – 30 kHz	100 km – 10 km	Navigation, Weather
Low Frequency (LF) Radio	30 kHz – 300 kHz	10 km – 1 km	Navigation, Maritime Communications, Information and Weather Systems, Time Systems
Medium Frequency (MF) Radio	300 kHz – 3 MHz	1 km – 100 m	Navigation, AM Radio, Mobile Radio
High Frequency (HF) Radio	3 MHz – 30 MHz	100 – 10 m	Citizens Band (CB) Radio (aka Shortwave Radio), Mobile Radio, Maritime Radio
Very High Frequency (VHF) Radio	30 MHz – 300 MHz	10 m – 1 m	Amateur (Ham) Radio, VHF TV, FM Radio, Mobile Satellite, Mobile Radio, Fixed Radio
Ultra High Frequency (UHF) Radio	300 MHz – 3 GHz	1 m – 10 cm	Microwave, Satellite, UHF TV, Paging, Cordless Telephony, Cellular and PCS Telephony, Wireless LAN
Super High Frequency (SHF) Radio	3 GHz – 30 GHz	10 cm – 1 cm	Microwave, Satellite, Wireless LAN
Extremely High Frequency (EHF) Radio	30 GHz – 300 GHz	1 cm – 1 mm	Microwave, Satellite, Radiolocation
Infrared Light (IR)	300 GHz – 400 THz	1 mm – 750 nm	Wireless LAN Bridges, Wireless LANs, Fiber Optics
Visible Light	400 THz – 1 PHz	750 nm – 380 nm	Not Applicable

Table E-4: Frequency Spectrum: Band Designations, Nominal Frequency Ranges, Nominal Wavelengths, and Example Communications Applications *(continued)*

Band Designation	*Frequency (Hz)*[1]	*Wavelength (Λ)*[2]	*Applications*
Ultraviolet Light (UV)	1 PHz – 30 PHz	380 nm – 10 nm	Not Applicable
X-Rays	30 PHz – 30 EHz	10 nm – .01 nm	Not Applicable
Gamma and Cosmic Rays	>3 EHz	<.1 nm	Not Applicable

[1]k = kilo = 1,000 (1 thousand)
[1]M = Mega = 1,000,000 (1 million)
[1]G = Giga = 1,000,000,000 (1 billion)
[1]T = Tera = 1,000,000,000,000 (1 trillion)
[1]P = Peta = 1,000,000,000,000,000 (1 quadrillion)
[1]E = Exa = 1,000,000,000,000,000,000 (1 quintillion)
[2]km = kilometer (1,000 meters)
[2]m = meter
[2]cm = centimeter (1/100 meter)
[2]mm = millimeter (1/1,000 meter)
[2]μ = micron (1/1,000,000 meter)
[2]nm = nanometer (1/1,000,000,000 meter)

The wavelength figures assume transmission in a vacuum. Wavelength in a medium will be shorter due to the fact that the frequency remains the same while the signal propagates at speeds less than 300 km/s. For example, in glass the speed of light is reduced by the index of refraction, which is about 1.5 in practice, so the velocity of propagation (Vp) is approximately 200 km/s (300/1.5 = 200). Index of refraction (IOR) is the ratio of speed in a vacuum divided by speed in the medium. See also *IOR*, *medium*, *vacuum*, and *Vp*.

electromagnetism **1.** Magnetism produced by an electric current, and electric current produced by a changing magnetic field. **2.** The branch of physics that deals with the interaction of electric and magnetic fields.

electromechanical A device that comprises electrically operated components that move mechanically. See also *electromagnetic* and *electronic*.

electromotive force (emf) See *voltage*.

electron An elementary particle of matter that carries a negative charge of approximately 1.6021×10^{-19} coulomb (C) and having a mass, when at rest, of approximately 9.109534×10^{-28} grams, which is $\frac{1}{1836}$ the mass of a proton. Ordinarily, an atom has the same number of negatively charged electrons orbiting the nucleus as there are positively charged protons within the nucleus. Electrons are the moving matter that contributes the most to electric currents and voltages. Metals are conductors of electricity as they contain free electrons, also known as conduction electrons. Copper is a particularly good conductor as each copper atom contains one free electron, i.e., one electron that is free to detach from an atom and to flow through the conductor when voltage is applied to create current. Coined by George Johnstone Stoney in 1891, the word is from the Greek *elektron*, meaning *amber*. (*Note:* When rubbed against wool, amber, which is fossilized tree sap, attracts free electrons from the wool and becomes negatively charged and will attract small objects that are positively charged through a process known as electrostatic induction. If rubbed long enough and electrostatically charged enough, amber will generate sparks of static electricity.) See also *coulomb*, *current*, *inductance*, and *voltage*.

electronic A device that operates on the basis of the controlled flow of electrons through semiconductors. See also *electromagnetic* and *electromechanical*.

Electronic Bill Presentation and Payment (EBPP) See *EBPP*.

electronic common control (ECC) See *ECC.*

electronic data interchange (EDI) See *EDI.*

Electronic Frontier Foundation (EFF) See *EFF.*

Electronic Industries Alliance (EIA) See *EIA.*

electronic key telephone system (EKTS) See *EKTS.*

electronic mail (e-mail or email) See *e-mail.*

Electronic Messaging Association (EMA) See *The Open Group.*

electronic number (ENUM) See *ENUM.*

electronic private automatic branch exchange (EPABX) See *PBX.*

Electronic Switching System (ESS) See *ESS.*

electrophotography Also known as *xerography*. See *xerography*.

electrothermochemical A printing technology that varies the temperature of a print head to cause the image to be reproduced on chemically treated paper. This technology is used in older facsimile (fax) machines. See also *facsimile.*

element management system (EMS) See *EMS.*

ELF (Extremely Low Frequency) ELF radio has a frequency of 30–300 Hz and a wavelength of 10,000–1,000 km. ELF radio has application in submarine radio communications. See also *electromagnetic spectrum*, *frequency*, *Hz*, and *wavelength.*

ELV (Extra low voltage) According to the International Electrotechnical Commission (IEC), alternating current (AC) voltage less than 50V, or direct current (DC) voltage less than 120V. Unlicensed personnel can safely install ELV wiring. See also *AC*, *DC*, and *IEC.*

EMA (Electronic Messaging Association) See *The Open Group.*

e-mail (electronic mail) Application software system originally developed for store-and-forward text messaging over a packet-based computer network. E-mail originated in the mid-1960s for communications between time-share computer users. E-mail quickly became popular for government and military communications in the late 1960s and early 1970s, especially as an application on the Advanced Research Projects Agency Network (ARPANET), which was the predecessor to the Internet. E-mail was popularized in the late 1970s and early 1980s, as part of the office automation concept designed to lead us toward the paperless office. E-mail relies on a client/server architecture can be implemented over local area networks (LANs) or wide area networks (WANs) such as the public Internet. Some e-mail systems, such as Microsoft Outlook, support not only plain text, but also rich text and Hypertext Markup Language (HTML) formatting. Unfortunately, communication with e-mail clients not supporting rich text or HTML creates considerable formatting incompatibilities. E-mail now permits the attachment of other forms of information, including binary files, images, graphics, and even digitized voice and video. E-mail system features typically include address book, confirmation, and formatting. See also *address book*, *ARPANET*, *client/server*, *confirmation*, *e-mail address*, *format*, *HTML*, *IMAP*, *Internet*, *MIME*, *plain text*, *POP*, *rich text*, *SMTP*, *spam*, *store-and-forward*, and *time-sharing.*

embedded operations channel (EOC) See *EOC.*

emf (electromotive force) See *voltage.*

EMI (ElectroMagnetic Interference) Interference with a desired signal caused by the coupling of an undesired signal due to electromagnetic radiation. The source of the electromagnetic interference may be natural, such as solar radiation, or artificial, such as a generator, compressor, fluorescent light, or electrified copper circuit. The radiation may be in many forms, including radio waves, light waves, and gamma rays. Unshielded twisted pair (UTP) and radio circuits are particularly susceptible to EMI. EMI that is in the radio frequency range is known as radio frequency interference (RFI). See also *noise* and *RFI.*

emotag A pseudo-HTML tag used in chat rooms, e-mail messages, or newsgroup postings to convey the sort of emotion or feeling that plain text does not otherwise support. An emotag mimics the format of an actual HTML tag. An emotag typically follows a sentence, with an example being <grin>. See also *chat room, e-mail, emoticon, HTML, newsgroup,* and *plain text.*

emoticon (emotion icon) A string of ASCII text characters used after a sentence in e-mails and newsgroup postings intended to represent a facial expression and to convey the sort of emotion that plain text does not otherwise support. Common examples of emoticons (meant to be viewed sideways) include those shown in the following table.

Emoticon	*Meaning*
:-) or :) or =)	Smile or happy
;-) or ;)	Winking and smiling
:-(	Sad
:D or :-D or =D	Big smile
:-0 or :0 or =0	Surprise or shock
>:-(	Angry
>:-)	Evil smile
0:-)	Innocence (Halo over the head)
:-x	No comment or My lips are sealed or I shouldn't have said that

Common examples of emoticons meant to be viewed without rotation include those shown in the following table.

Emoticon	*Meaning*
^_^	Smile or happy
;_;	Sad and crying
-_-	Annoyed
\V/	Peace or live long and prosper (Mr. Spock of Star Trek)
\@^@/ or \O^O/	Look closer (glasses)
("\(^_^)/")	Big hug

See also *e-mail, emotag,* and *newsgroup.*

empty suit A derisive term for an anonymous business executive or bureaucrat lacking in both individuality and substance, i.e., a phony. Such a person is little more than a suit of clothes. As we used to say when

I was a young man in Texas, such a person is all hat and no horse, i.e., a drugstore cowboy. A suit, especially an empty suit, is in sharp contrast to a techie. See also *suit* and *techie*.

EMS (Element Management System) A network management system (NMS) that manages one or more network elements (NEs) of a specific type, e.g., modems or multiplexers, and manufacturer. Multiple EMSs may be managed by a higher level NMS commonly known as a manager of managers (MOM) and as described in the layered telecommunications management network (TMN) model. See also *MOM*, *network management*, *NMS*, and *TMN*.

emulated LAN (ELAN) See *LANE*.

emulation **1.** The process of imitating a computer or computer software program. Terminal emulation, for example, is the process by which a microcomputer imitates a dumb terminal in order to communicate with a mainframe computer. **2.** Circuit emulation is the process by which a broadband circuit can support many virtual circuits (VCs), with each performing as a distinct legacy physical circuit. See also *broadband*, *circuit*, *dumb terminal*, *mainframe*, *microcomputer*, and *VC*.

EN (Enterprise Network) An imprecise term referring to a private network for the exclusive use of a commercial enterprise or government, educational, or other organization. An enterprise may own the infrastructure or lease it. An EN may be metropolitan or wide area in nature, but is not a public metropolitan area network (MAN) or wide area network (WAN).

encapsulate To frame or enclose a unit of information with control data in order that a network can process it properly. An encapsulating bridge, for example, can interconnect an Ethernet LAN and a Token Ring LAN, surrounding the native LAN frame with control information appropriate to the LAN on which the target device is attached. In other words, an encapsulating bridge places an Ethernet frame inside a Token Ring frame, or vice versa. Similarly, a frame relay access device (FRAD) encapsulates an Internet Protocol (IP) packet before presenting it to a frame relay network. See also *bridge*, *Ethernet*, *FRAD*, *frame*, *frame relay*, *IP*, *LAN*, *packet*, and *Token Ring*.

encapsulating bridge Also known as a *Medium Access Control* (MAC) bridge. A bridge that can interconnect two or more unlike networks, such as local area networks (LANs). In order to bridge an Ethernet LAN and a Token Ring LAN, an encapsulating bridge surrounds the native LAN frame with control information appropriate to the LAN on which the target device is attached, placing an Ethernet frame inside a Token Ring frame, or vice versa. See also *encapsulate*, *Ethernet*, *frame*, *LAN*, and *Token Ring*.

encode The process of coding data into symbolic form. See also *code*.

encrypt The process of coding or ciphering data into symbolic form. See also *code* and *scramble*.

encryption The art or science, or system, of coding or ciphering data into symbolic form to disguise, and thereby secure, the contents of a message. Generally in the form of firmware, rather than software, encryption logic commonly both scrambles and compresses message units (e.g., blocks or packets) prior to transmission. The receiving device is equipped with the necessary logic to decompress and decrypt the data. Private key is a symmetric encryption method that uses the secret same key to encrypt and decrypt data. Public key is an asymmetric encryption method with an encryption (encoding) key that can be used by all authorized network users and a decryption (decoding) key that is kept secret. Encryption algorithms and mechanisms used in telecommunications include Advanced Encryption Standard (AES), Data Encryption Standard (DES), RSA, and Triple DES. See also *AES*, *algorithm*, *block*, *code*, *cryptography*, *DES*, *message unit*, *packet*, *RSA*, *scramble*, *security*, *steganography*, and *Triple DES*.

endianess Referring to the orientation of a computer system, application, or network design with respect to the placement of the most significant bit, digit, or byte in a coding scheme. Big-endian places the most significant bit, digit, or byte in the first, or leftmost, position. Little-endian places the most significant bit, digit, or byte in the last, or rightmost, position. Bi-endian systems can work either way. Motorola processors employ the big-endian approach, whereas Intel processors take the little-endian approach. Telephone numbers, for example, are big-endian, beginning with a country code, followed by an area code, a central office prefix, and a line number. Table E-5 illustrates how the decimal value 47,572 would be expressed in hexadecimal and binary notation (two octets) and how it would be stored using these two methods.

Table E-5: Endianess

Number	*Big-Endian*	*Little-Endian*
Hexadecimal		
B9D4	B9D4	4D9B
Binary		
10111001	10111001	11010100
11010100		10111001
	11010100	

Endianess is a pointless matter of philosophical orientation, with no right or wrong, but often with intense feelings on both sides of the argument. The dispute is a classic example of a holy war, or jihad, in which the various impossibly rigid positions are based on (or at least justified on) the basis of irreducible pseudoreligious principles rather than reason. (Fortunately, there are no documented casualties resulting from the Endian Wars.) The terms derive from Jonathan Swift's *Gulliver's Travels*, in which the Big-Endians were a faction of people on the islands of Lilliput and Blefuscu who defied the emperor's decree that soft-boiled eggs should be broken at the small end before being consumed. See also *bi-endian*, *big-endian*, *bit*, *byte*, and *little-endian*.

end office A local central office (CO), which is at the end, or edge, of the public switched telephone network (PSTN). An end office is the point at which subscriber local loops and network trunks terminate and interconnect. Synonymous with *central office* (CO), *central office exchange* (COE), and *Class 5 office*. See also *Class 5*, *CO*, and *PSTN*.

endpoint **1.** In asynchronous transfer mode (ATM) a switch or other device at the end of the ATM network. An endpoint serves as the source (transmitter) and sink (receiver) of data in ATM cell format. See also *ATM*, *cell*, *network*, *sink*, *source*, and *switch*. **2.** In H.323-compliant multimedia networks, a terminal device on a local area network (LAN). See also *H.323*, *LAN*, *multimedia*, *network*, and *terminal*.

end user The ultimate user of a product or service, especially of a computer system, application, or network. The end user is at the bottom of the hierarchy, yet is (or should be) the focus of all attention, for it is the end user (or end user organization) that makes purchase decisions and ultimately pays the bills, although vendors sometimes seem to forget that detail.

End User Common Line Charge (EUCL) See *EUCL*.

energy **1.** In physics, the capacity of a system for doing work. It took a lot of energy to write this book. (Take my word for it, so to speak.) See also *physics*. **2.** In physics, referring to a source of energy, electrical, mechanical, or otherwise.

engineer A person skilled in the science of putting scientific knowledge to practical use, specifically in the design, planning, construction, or maintenance of manufactured things. Engineering is divided into branches such as chemical, civil, electrical, mechanical, and software.

enhanced call routing In the advanced intelligent network (AIN) specifications, a network-based enhancement to toll-free calling in the public switched telephone network (PSTN). The callers are presented with options that enable them to specify their needs and then be connected with the offices or individuals best able to satisfy them. See also *AIN* and *PSTN*.

Enhanced Circuit Switched Data (ECSD) See *ECSD.*

Enhanced Data rates for Global Evolution (EDGE) See *EDGE.*

Enhanced Data rates for GSM Evolution (EDGE) See *EDGE.*

Enhanced General Packet Radio System (EGPRS) See *EGPRS.*

Enhanced GPRS (EGPRS) See *EGPRS.*

enhanced service See *value-added service.*

Enhanced Switched Mobile Radio (ESMR) See *SMR.*

Enhanced TDMA (E-TDMA) See *E-TDMA.*

Enhanced Variable Rate Vocoder (EVRC) See *EVRC.*

ensure **1.** Make certain. **2.** Safeguard. See also *insure.*

Enterprise Computer Telephony Forum (ECTF) See *ECTF.*

Enterprise Network (EN) See *EN.*

Enterprise Systems Connection (ESCON) See *ESCON.*

entomology The branch of zoology concerned with the study of insects. Entomology is not to be confused with etymology, the study of the origin of words. See also *bug.*

entropy (S) **1.** In physics, and particularly in the area of thermodynamics, a measure of the amount of energy unavailable to do work in a closed system. **2.** The degradation of the matter and energy in the universe to the point of inert uniformity. The dispersal of energy. **3.** In information theory, a measure of the content of a message evaluated with respect to its probability of occurrence, or uncertainty of occurrence, depending on your perspective. **4.** In communications, a measure of the randomness of signal noise occurring in transmission.

ENUM (Electronic NUMber) A standard (RFC 2916) issued by the Internet Engineering Task Force (IETF) for translating between PSTN and Internet addresses. ENUM translates between PSTN telephone numbers, as specified by the ITU-T in E.164, and Internet Protocol (IP) addresses, as specified for IPv4 in RFC 791 and IPv6 in RFC 2460. ENUM requires that both E.164 and IP addresses be registered with the ENUM Domain Name Service (DNS), which can be consulted by gateways that interconnect the two disparate networks. Thereby, a given call can traverse both the PSTN and the Internet or other IP-based packet network. See also *DNS*, *E.164*, *gateway*, *IETF*, *Internet*, *IP*, *IP address*, *IPv4*, *IPv6*, *ITU-T*, *PSTN*, *RFC*, and *telephone number.*

EOC (Embedded Operations Channel) **1.** A control channel integral to the T-carrier extended superframe (ESF) frame format for network management purposes and embedded in the framing bits (one bit per frame). **2.** In ISDN, a basic rate interface (BRI) facility depends on the EOC from the central office (CO) to command the NT1 device for purposes of network management and testing. See also *channel*, *CO*, *ESF*, *framing bit*, *ISDN*, *network management*, and *NT1.*

EPABX (Electronic Private Automatic Branch Exchange) See *PBX.*

EPON (Ethernet-based Passive Optical Network) A PON specified by the IEEE in 802.3ah (2004) as employing 802.3 (aka Ethernet) at the Data Link Layer. EPON runs at a signaling rate of 1.244 Gbps in symmetric mode and the maximum logical reach is approximately 20 kilometers (12 miles). EPON supports as many as 16 splits, that is, splitters can divide the signal to serve as many as 16 premises from a single optical fiber. PON variants also include ATM-based PON (APON), broadband PON (BPON), and gigabit PON (GPON). See also *802.3*, *APON*, *BPON*, *Data Link Layer*, *Ethernet*, *GPON*, *IEEE*, *logical reach*, *optical fiber*, *PON*, *signaling rate*, and *splitter*.

equal access Referring to the ability of a telephone subscriber to place a long distance call through any competing interexchange carrier (IXC) with equal ease, i.e., simply by dialing the telephone number. The implementation of equal access requires that subscribers be surveyed and afforded the right to pre-select an IXC. Users who do not respond are assigned a default carrier. All user choices and default selections are compiled in a centralized database that is queried as each call is placed. Based on the originating circuit number, the database is consulted, the Carrier Identification Code (CIC) associated with the pre-selected carrier is determined, and the call is routed through the designated IXC. Equal access rules were established in the United States in 1982 as part of the Modified Final Judgement (MFJ). See also *access charges*, *carrier*, *CIC*, *IXC*, and *MFJ*.

equal access trunk Referring to a trunk side, or Feature Group D (FGD) termination between a local exchange carrier (LEC) and an interexchange carrier (IXC). Such a trunk supports equal access. See also *equal access*, *FG*, *IXC*, *LEC*, *trunk*, and *trunk side*.

EQEEB (Equivalent Queue Extended Erlang B) See *Equivalent Queue Extended Erlang B*.

equilibrium mode distribution Also known as *modal equilibrium*. Referring to a state in which optical power is evenly distributed across all modes, i.e., physical paths in an optical fiber. See also *mode* and *optical fiber*.

Equivalent Queue Extended Erlang B (EQEEB) A traffic engineering model that assumes that calls encountering blockage are queued, but only for a predetermined period of time. If a circuit in the primary group does not become available during that time, either the call is routed over a more expensive circuit or the caller is given the option of trying to place the call again at a later time. A percentage of callers retry their calls until they are successfully completed. Developed by Jim Jewitt of Telco Research, EQEEB was used in incoming call centers in the applications where circuits are very expensive and poor GoS levels (e.g., P.10) are acceptable. See also *Erlang*, *Erlang B*, *Erlang C*, *Extended Erlang B*, *GoS*, *Poisson distribution*, *traffic*, and *traffic engineering*.

E-rate Program See *Schools and Libraries Program*.

erbium (Er) A soft, malleable, silvery rare-earth element used in various alloys. Erbium-doped fiber amplifiers (EDFAs) are used extensively in long haul fiber optic transmission systems (FOTS). Erbium is number 68 in the Periodic Table of Elements. Erbium is named for the village of Ytterby, Sweden, where it was discovered. So were ytterbium, yttrium, and terbium. See also *EDFA*, *FOTS*, and *Ytterby*.

erbium-dope fiber amplifier (EDFA) See *EDFA*.

Erlang A measure of the traffic intensity of a transmission facility, such as a circuit or channel. One Erlang is the maximum traffic that a facility can support during an hour, and is equivalent to 36 CCS. The Erlang measurement is named for A.K. Erlang, the Danish mathematician and traffic engineer who developed the various Erlang traffic engineering models. See also *Erlang, H. K.*; *Erlang B*; *Extended Erlang B*; *Erlang C*; *Equivalent Queue Extended Erlang B*; *GoS*; *Poisson distribution*; *traffic*; and *traffic engineering*.

Erlang, A. K. (1878–1929) The Danish mathematician and traffic engineer for the Copenhagen Telephone Company who developed the various Erlang traffic engineering models. These formulas calculate grade of service (GoS) based on Busy Hour Traffic (BHT) expressed in hours of traffic, or Erlangs, presented to circuits. See also *Erlang*, *GoS*, and *traffic engineering*.

Erlang B A traffic engineering model that assumes that an offered call is cleared immediately, with no queuing. In other words, Erlang B assumes that a call encountering blockage will not appear again. Either the caller will hang up and not attempt to place the call again, or the call will automatically be routed over another circuit if one exists, even if the use of that circuit is more expensive. See also *Erlang*, *Erlang C*, *Equivalent Queue Extended Erlang B*, *Extended Erlang B*, *GoS*, *Poisson distribution*, *traffic*, and *traffic engineering*.

Erlang C A traffic engineering model that assumes that calls encountering blockage are queued indefinitely until a circuit is available, with no overflow to more expensive circuits. Erlang C commonly is used to engineer circuit requirements for automatic call distributors (ACDs) in incoming call centers. See also *Erlang*, *Erlang C*, *Equivalent Queue Extended Erlang B*, *Extended Erlang B*, *GoS*, *Poisson distribution*, *traffic*, and *traffic engineering*.

ERMES (European Radio MEssage System) A digital paging system supported by the European Telecommunications Standards Institute (ETSI) and the European Union (EU). ERMES operates at 6,250 bps in the 169.4–169.8 MHz band and uses frequency shift keying (FSK) modulation. See also *band*, *digital*, *ETSI*, *FSK*, *modulation*, and *paging system*.

error **1.** The discrepancy between a computed, estimated, or measured value or condition and that which is true, specified, expected, or theoretically correct. **2.** In a computer, a discrepancy in a calculation, in a file, or in the execution of a program. **3.** In telecommunications, the discrepancy between data as transmitted and data as received. See also *error control*.

error control The process of improving communications through techniques designed variously to detect, flag, and correct errors created in transmission. There are three specific error control modes. Recognition and flagging involves simply flagging detected errors, with no mechanism for automatic error correction. Recognition and retransmission provides for retransmission of errored data. Forward error correction (FEC) triggers an automatic error correction process in the receiver when it detects an error in a data packet. See also *FEC*, *parity checking*, *recognition and flagging*, and *recognition and retransmission*.

error control mode (ECM) See *ECM*.

ESCON (Enterprise Systems CONnection) A proprietary storage area network (SAN) developed by IBM for a high speed serial interface between mainframe computers and peripherals such as external disk drives. ESCON supports data transfer rates up to 17 MBps in half-duplex (HDX) over distances up to 43 km. ESCON is yielding to Fibre Connections (FICON), a faster technology that runs over Fibre Channel. See also *Bps*, *Fibre Channel*, *FICON*, *HDX*, *mainframe*, *peripheral*, *SAN*, and *serial*.

ESF (Extended SuperFrame) A T-carrier framing convention that extends the superframe sequence from 12 to 24 frames; with signaling performed in frames 6, 12, 18, and 24. ESF offers the advantages of non-disruptive error detection through a six-bit cyclic redundancy check (CRC), and an embedded operations channel (EOC) for network management. See also *CRC*, *D1*, *D2*, *D4*, *D4*, *frame*, *network management*, *superframe*, and *T-carrier*.

ESMR (Enhanced Switched Mobile Radio) See *SMR*.

ESS (Electronic Switching System) The ESS was the first electronic common control (ECC) circuit switch. Developed by AT&T Bell Telephone Laboratories with the assistance of Western Electric, and based on the transistor, invented at Bell Labs in 1948, the ESS involved a development effort that began in earnest in the early 1950s. The first ESS central office (CO) began service in Succasunna, New Jersey, on May 30, 1965, connecting 200 subscribers. By 1974, there were 475 such offices in service, serving 5.6 million subscribers. The development effort was estimated to involve 4,000 man-years and a total cost of $500 million. See also *circuit switching*, *CO*, and *ECC*.

ESSID (Extended Service Set IDentifier) See *SSID*.

etalon A spectroscopic instrument used to measure and control optical wavelengths. Also known as an *interferometer*, an etalon comprises two parallel reflecting plane surfaces. Etalons are widely used in lasers. See also *Fabry-Perot laser*.

etymology The study of the origin of words. Etymology is not to be confused with entomology, the branch of zoology concerned with the study of insects. See also *bug.*

equalization A form of conditioning that reduces the frequency distortion or phase distortion, or both, in an electrical signal on a metallic conductor or in a radio signal traveling through the atmosphere. Equalization compensates for the differences in signal attenuation and delay associated with different frequency components. Around a center frequency, relatively high frequency signals attenuate more than relatively low frequency signals over a distance, so an equalizer may reduce (cut) the amplitude of the low frequency signals and increase (boost) the amplitude of the high frequency signals in order that the signals at the receiver are in the same relative balance as they were at the transmitter. Adaptive equalizers automatically adjust to levels of distortion that vary as the signal path or its characteristics change over time. See also *amplitude, attenuation, delay, distortion, phase,* and *signal.*

ETACS (Extended Total Access Communications System) A version of the TACS 1G analog cellular radio technology developed for use in the United Kingdom. ETACS operates in the 900 MHz band, employs frequency modulation (FM), and supports 1,000 channels of 25 kHz. As an analog system, TACS derives channels using frequency division multiple access (FDMA) and bidirectional communications is achieved through frequency division duplex (FDD) with the downlink in the 916–949 MHz band and the uplink in the 871–904 MHz band. See also *1G, analog, cellular radio, downlink, FDD, FDMA, FM, narrowband, TACS,* and *uplink.*

E-TDMA (Enhanced Time Division Multiple Access) A multiplexing technique developed by Hughes Network Systems as an improvement over TDMA, which is employed in many digital cellular networks. E-TDMA employs digital speech interpolation (DSI) compression, also known as voice activity detection (VAD), and half-rate vocoders (voice coders) operating at 4.8 kbps to enhance bandwidth utilization. See also *bandwidth, cellular, compression, digital, DSI, TDMA,* and *vocoder.*

ether Luminiferous ether. The omnipresent passive medium once thought to pervade all space and to support the propagation of electromagnetic energy, even through a vacuum. The existence of the ether was disproved around 1900 by a number of scientists, including Albert Einstein, Albert A. Michaelson, and Edward W. Morley. In 1973, Robert Metcalfe chose the name Ethernet to describe the local area network (LAN) technology he and his associates invented at the Xerox Palo Alto Research Center (Xerox PARC). See also *Ethernet; LAN; Metcalfe, Robert M.;* and *Xerox PARC.*

Ethernet Robert M. Metcalfe and his associates at the Xerox Palo Alto Research Center (Xerox PARC) first developed both the concept of a local area network (LAN) and the enabling technology. That first network originally was known as the Altos Aloha Network, because it connected Altos computers through a network based on the University of Hawaii's AlohaNet packet radio system technology. Subsequently (1973), it was renamed Ethernet, from luminiferous ether, the omnipresent passive medium once theorized to pervade all space and to support the propagation of electromagnetic energy. The original Ethernet supported a transmission rate of 2.94 Mbps over coaxial cable. Xerox commercialized the technology, renaming it the Xerox Wire. Gordon Bell, vice president of engineering at Digital Equipment Corporation (DEC, subsequently acquired by Compaq, which later merged with Hewlett-Packard), hired Metcalfe as a consultant in 1979 specifically to develop a LAN network technology that would not conflict with the Xerox patent. Metcalfe brought DEC, Intel, and Xerox together to form into a joint venture known as DIX, which improved the technology, increasing the bandwidth to 10 Mbps and reverting to the name Ethernet. The technology quickly became a de facto standard. In February 1980, the IEEE established Project 802 to develop a set of LAN standards. In December 1982, the first standard was published and circulated as IEEE 802.3, which actually is a variation on the now obsolete Ethernet standard. Although the two do not interoperate, the terms 802.3 and Ethernet are used interchangeably in informal conversation.

Ethernet has evolved considerably since 1980. The signaling rate has increased from 10 Mbps to 100 Mbps, 1 Gbps, and 10 Gbps. The original 10Base5 specification for coaxial cable has given way to 10/100/1000Base-T specifications for twisted pair, and various 10GBase-XX specifications for optical fiber. Relatively unchanged have been the frame format and the protocols for medium access control (MAC), which include carrier sense multiple access with collision detection (CSMA/CD) and carrier sense multiple access/collision avoidance (CSMA/CA). The Ethernet frame, as illustrated in Figure E-3, is formatted as follows:

- **Preamble:** A field of seven octets in an alternating pattern of 1s and 0s that advises the receiving stations that a frame of data is arriving.
- **Start of Frame (SOF):** A delimiter of one octet that ends with two consecutive 1 bits that serve to synchronize the receiving stations on the rate of transmission. If multiple stations start sending to an idle network at nearly the same time, the preamble is long enough to ensure that a collision occurs before user data is sent by either station.
- **Destination and Source Addresses:** The addresses of the target station and the originating station, respectively. Each address comprises each six octets, the first three of which are specified by the by the IEEE on a vendor-dependent basis and the last three of which are assigned by the vendor.
- **Length:** A field of two octets that indicates the number of octets in the data field.
- **Data:** A minimum of 64 octets and a maximum of 1518 octets. In consideration of the fact that 18 octets are consumed with Layer 1 and Layer 2 processing, the Data field, or payload, must comprise 46–1500 octets. In the event that the payload is less than 46 octets, padding bytes are inserted.
- **Frame Check Sequence (FCS):** A 32-bit Cyclic Redundancy Check (CRC) comprising the frame trailer for purposes of error control.

	Preamble	SOF	Destination Address	Source Address	Length	Data	FCS
Octets	7	1	6	6	2	46-1500	4

Figure E-3

See also *10Base5*, *10/100/1000Base-T*, *10GBase-XX*, *802.3*, *bandwidth*, *coaxial cable*, *CRC*, *CSMA/CA*, *CSMA/CD*, *ether*, *Gigabit Ethernet*, *LAN*, *Metcalfe*, *payload*, *standard*, *transmission rate*, and *Xerox PARC*.

Ethernet-based passive optical network (EPON) See *EPON*.

ethics The study of the effects of moral principles and standards on human conduct. Business ethics deal with ethics in business, and with the constant process of optimizing profitability in the context of what is right and what is wrong.

ETNO (European Telecommunications Network Operators' Association) An organization spawned in 1992 from the European Conference of Postal and Telecommunications Administrations (CEPT) to become the principal policy group for European electronic communications network operators. See also *CEPT*. For contact information, see Appendix A.

ETSI (European Telecommunications Standards Institute) An independent, not-for-profit organization officially responsible for the standardization of information and communications technologies (ICT) within the European Union (EU). ICT technologies include telecommunications, broadcasting, and related areas such as intelligent transportation and medical electronics. The membership includes administrations, manufacturers, network operators, research bodies, service providers, and end-user organizations. For more information see the contact information in Appendix A.

EUCL (End User Common Line Charge) Previously known variously as Customer Access Line Charge (CALC), Service Line Charge (SLC), and Subscriber Line Charge (SLC). In the United States, an access charge approved by the Federal Communications Commission (FCC) and billed by the incumbent local exchange carrier (ILEC), the EUCL is intended to compensate the ILEC for the costs of connecting a call to competitive local exchange carrier (CLEC) or an interexchange carrier (IXC) through local exchange facilities (the local loop, central office, and associated equipment), maintaining the equal access database, and other related costs. The EUCL applies to all ILEC local loops, but varies by type of facility (e.g., residence line, business line, and CO trunk). If the access circuit is digital in nature, a Digital Port Line Charge (DPLC) also applies. See also *CLEC*, *DPLC*, *equal access*, *FCC*, *ILEC*, and *IXC*.

euphemism An agreeable, inoffensive, less offensive, or politically correct (PC) synonym for a word or phrase that is harsh, unpleasant, or offensive. For example, eccentric is a euphemism for someone who is crazy and rich, as opposed to being just plain crazy like the rest of us poor folks. See also *Bless his heart*, *leverage*, and *PC*.

European Computer Manufacturers Association (ECMA) See *Ecma International*.

European Conference of Postal and Telecommunications Administrations (Conférence Européenne des administrations des Postes et des Télécommunications or CEPT) See *CEPT*.

European Radio Message System (ERMES) See *ERMES*.

European Telecommunications Network Operators' Association (ETNO) See *ETNO*.

European Telecommunications Standards Institute (ETSI) See *ETSI*.

EV (EVolution-Data Optimized) See *EV-DO*.

EV-DO (EVolution-Data Optimized) Also known as *1xEV-DO* (one carrier EV-DO). A high data rate (HDR) version of Code Division Multiple Access 2000 (CDMA2000). Revision 0 (Rev 0) employs 16-QPSK modulation in support of a peak data rate of 2.4 Mbps on the downlink and 153 kbps on the uplink. In a fully loaded cell, 1xEV-DO supports average aggregate throughput of 4.1 Mbps on the downlinks and 660 kbps on the uplinks. 1xEV-DO can run in any band and can coexist in any type of network. Rev A supports peak speeds of 3.1 Mbps on the downlink and 1.8 Mbps on the uplink, and average speeds of 450-800 Mbps and 300-400 Mbps, respectively. Rev B, still under development, is anticipated to yield peak downlink speeds up to 4.9 Mbps per carrier, with as many as 3 carriers linked for aggregate peak downlink speed of 14.7 Mbps. See also *16-QPSK*, *CDMA2000*, *downlink*, *modulation*, *throughput*, and *uplink*.

Evolution-Data Optimized (EV-DO or EV) See *EV-DO*.

EVRC (Enhanced Variable Rate Vocoder) A speech encoding mechanism specified in Personal Communications System (PSC) digital cellular radio standard IS-95a. EVRC runs at 13 kbps at maximum speech activity and varies the rate downward to as low as one-eighth rate if the level of speech activity permits. See also *cellular radio*, *encode*, and *PCS*.

Excess Burst Size (B_e) In frame relay, the maximum amount of data that the network will accept in a block from a user and will attempt to deliver without discard, if bandwidth is available, and over a specified time interval (T). In recognition of the bursty nature of LAN-to-LAN communications, the transmitting device may burst above the Committed Information Rate (CIR) and Committed Burst Size (Bc) for a brief period of time; and the network will attempt to accommodate those bursts. The network reserves the option to mark the excess data above B_c as discard eligible (DE) should the user CPE not have done so already. See also *bandwidth*, *CIR*, *CPE*, *DE*, *frame relay*, *LAN*, and *Maximum Burst Size*.

Excess Information Rate (EIR) See *EIR*.

exchange **1.** A central office exchange (CO or COE) of the public switched telephone network (PSTN) and all of the equipment contained therein for the purpose of interconnecting (i.e., exchanging connections between) the lines and trunks terminating there. See also CO, line, line side, PSTN, trunk, and trunk side. **2.** The area served by a central office exchange (CO or COE). Synonymous with *carrier serving area* (CSA). See also *CO* and *CSA*.

Exchange Carriers Standards Association (ECSA) Now the Alliance for Telecommunications Industry Solutions (ATIS) See *ATIS*.

exclusive hold See *hold*.

Expedited Forwarding (EF) See *EF*.

Explicit Congestion Notification (ECN) See *ECN*.

extended band (E-Band) See *E-Band*.

extremely high frequency (EHF) See *EHF*.

extremely low frequency (ELF) See *ELF*.

Extended Binary Coded Decimal Interchange Code (EBCDIC) See *EBCDIC*.

Extended Erlang B (EEB) A traffic engineering model that, like Erlang B, assumes that an offered call is cleared immediately, with no queuing. However, Extended Erlang B assumes that the caller encountering blockage (e.g., busy signal or no dial tone) will hang up and immediately attempt the call again, with no overflowing of calls to more expensive routes. EEB was developed by Jim Jewitt and Jaqueline Shrago of Telco Research. See also *Erlang*, *Erlang B*, *Erlang C*, *Equivalent Queue Extended Erlang B*, *GoS*, *Poisson distribution*, *traffic*, and *traffic engineering*.

extended service set identifier (ESSID, or SSID) See *SSID*.

extended superframe (ESF) See *ESF*.

Extended TACS (ETACS) See *ETACS*.

Extended Total Access Communications System (ETACS) See *ETACS*.

Extensible Markup Language (XML) See *XML*.

Extensible Messaging and Presence Protocol (XMPP) See *XMPP*.

Exterior Gateway Protocol (EGP) See *EGP*.

exterior protocol A protocol concerned with routing between Autonomous Systems (ASs), which are groups of routers within the same administrative domain. Exterior protocols include the Exterior Gateway Protocol (EGP) and the Border Gateway Protocol (BGP). See also *BGP*, *domain*, *EGP*, *interior protocol*, and *routing*.

Extra low voltage (ELV) See *ELV*.

extranet An intranet opened to select groups of users outside of a company. Access generally is provided to groups of vendors, suppliers, customers, and others who have a requirement to access select databases and processes. Extranets, for example, can enable customers to place orders electronically and to track them to fulfillment, and vendors can track retail sales of their products, perhaps store-by-store. See also *intranet*.

extrinsic loss In fiber optics transmission, signal attenuation due to external forces or factors, including imperfect joining of optical fibers through splices or connectors. Bending loss also is a form of extrinsic loss as macrobends are caused by excessive bending of the fiber as a result of its physical manipulation and microbends can be caused by excessively tight crimping. See also *attenuation*, *bending loss*, *connector*, *intrinsic loss*, *microbend*, *macrobend*, *optical fiber*, *signal*, and *splice*.

f **1.** Symbol for frequency. **2.** femto. From Danish or Norwegian *femten,* translating as *fifteen* and referring to one quadrillionth (10^{-15}). See also *femtocell.*

Fabry-Perot laser A general purpose laser light source with a relatively narrow spectral width in the range of 3–6 nm. As a Fabry-Perot laser oscillates at several wavelengths, it emits a narrow range of less intense wavelengths around the center wavelength in which the power is concentrated. For example, a Fabry-Perot laser operating at a nominal wavelength of 1310 nm might also emit weaker signals at wavelengths ranging from 1307–1313 nm. This spectral width causes some amount of chromatic dispersion, which limits bandwidth in single-mode fiber (SMF) systems due to pulse spreading, which limits the bit rate. Fabry-Perot lasers Perot lasers are more precise than LEDs, are moderately fast at 1 Gbps or less, and are moderately priced. The Fabry-Perot laser is named for Charles Fabry and Alfred Pérot, who together invented the Fabry-Perot interferometer, or etalon, which formed the basis for this type of laser. See also *DFB laser, etalon, laser, LED,* and *wavelength.*

face time Time spent with people, face-to-face. Face time is highly productive. Telephone conversations are less so, but still very productive. E-mail and instant messaging are not good substitutes for looking people in the eye or at least hearing their voices. People will rediscover that some day.

facilities bypass Referring to the technologies that a service provider or user organization uses to bypass the local loop facilities of the incumbent local exchange carrier (ILEC) in order to gain access to an interexchange carrier (IXC) or Internet service provider (ISP). Facilities bypass generally involves wireless technologies. See also *ILEC, ISP,* and *IXC.*

facsimile (fax) From the Latin *facere simile*, which translates to *make similar.* Technology that enables the transmission of images between paired transmitters and receivers. The transmitting fax scans the image document from top to bottom and from left to right, looking for dots of color — most machines support only black and white, some systems will also support 256 levels of grayscale, and some will support a large color palette — at various levels of resolution, as measured in lines per inch (lpi) and dots per inch (dpi). The fax machine translates the dots into data bits, which it compresses in order to reduce transmission time, and transmits through a network to the receiving machine. If the local loop is analog in nature, as is usually the case, an embedded modem transmits the digital data by modulating the analog waveforms.

Just in case you were wondering about the origin and evolution of facsimile technology, Edward Davy invented the first practical facsimile machine in 1837, but abandoned the invention soon thereafter. Alexander Bain (1811–1877), a Scottish clockmaker and inventor, revived the concept and patented the recording telegraph in 1843. The first commercial facsimile service was established in 1865 by Giovanni Casselli over a circuit between Paris and Lyon, France. Circuits were added to other cities, and Casselli sent 5,000 faxes in the first year using his patented Pantelegraph machine, which was based on the Bain recording telegraph. A number of other inventors developed various wireline facsimile devices over the next 50 years or so, but all soon failed in the face of competition from the much more functional and practical electric telegraph (1844), invented by Samuel F.B. Morse (1791–1872) and Alfred Vail (1807–1859). Fax development began anew in the 1920s, but it was not until the 1970s that the ITU-T set international interoperability standards and the technology found some level of market acceptance. The ITU-T developed standards for fax machines in stages, designated as Group I, Group II, Group III, and Group IV. The Group III specification (1980) succeeded in making facsimile truly affordable for mass business markets. The Group IV specification (1984) addressed digital technology, which led to the development of high-capacity, networked fax servers and made fax broadcasting and fax-on-demand (FOD) possible. It is possible for a personal computer (PC) to emulate a fax machine through the use of a fax board and fax software, although this approach never achieved general market acceptance. See also *compression, fax emulation, grayscale, Group I, Group II, Group III, Group IV, ITU-T, modem, modulation*, and *resolution*.

Facsimile over Internet Protocol (FoIP) See *FoIP*.

fair use policy A policy of some Internet service providers (ISPs) that imposes bandwidth restrictions on users who exhibit patterns of system usage that exceed certain thresholds for extended periods of time. Direct broadcast satellite (DBS) providers offering two-way Internet access have fair use policies in place as bandwidth is so highly limited and a small number of highly active users can consume large amounts of bandwidth, leaving little for consumption by others. More traditional terrestrial ISPs in various countries (e.g., Australia and South Africa) have similar policies with respect to international usage such as web surfing and file transfers. See also *bandwidth, DBS, Internet*, and *ISP*.

fallback modem **1.** A modem that has the ability to adjust its transmission rate downward in the event that the connection quality degrades. For example, V.90 modems also are V.34bis modems. Assuming that the terminating modem is V.34bis, the V.90 modem adjusts its maximum downstream rate of 56 kbps downward to the V.34bis maximum transmission rate of 33.6 kbps. See also *downstream, dynamic rate adaptation, modem, transmission rate, V.34bis*, and *V.90*. **2.** A modem that dials a connection only when a primary, usually leased, line has failed.

Family Radio Service (FRS) See *FRS*.

fantail circuit Also known as a *multi-drop circuit* and, more formally, as a *point-to-multipoint circuit*. A dedicated circuit that connects a single device (i.e., point) to multiple devices, with the circuit fanning out like a tail from the headend. The drops, or tail circuits, connect to the main circuit through a simple bridge. Fantail circuits generally are phrased in the context of a wide area network (WAN), and generally are provided as a carrier service. See Figure F-1. See also *bridge, drop, headend, point-to-multipoint circuit*, and *WAN*.

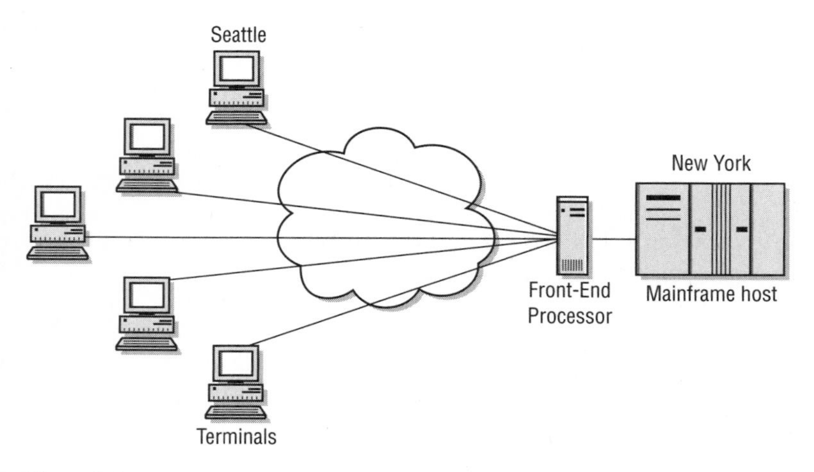

Figure F-1

farther **1.** More distant in space or time, particularly where there is a notion of physical distance. **2.** Erroneously used to mean to a greater extent. See also *further*.

far-end cross talk (FEXT) See *FEXT*.

FAST (Framed ATM over SONET/SDH Transport) A specification from the ATM Forum (July 2000) that defines the mechanisms and procedures required to support the transport of variable-length datagrams, known as ATM frames, over an ATM infrastructure using SONET/SDH facilities. FAST is similar to Data Exchange Interface (DXI) and Frame User Network Interface (FUNI), which are designed for access to an ATM network over relatively low-speed plesiochronous transmission facilities. FAST, however, is designed for access and/or inter-switch trunking over very high speed SONET/SDH transmission facilities. See also *ATM, ATM Forum, datagram, DXI, frame, FUNI, plesiochronous, SDH*, and *SONET*.

fast busy signal A signal indicating to the calling party that network resources are not available to process the call. Synonymous with *reorder tone*. See also *busy signal*.

fast packet services Referring to a group of packet services operating at broadband speeds and including asynchronous transfer mode (ATM), frame relay, and Switched Multimegabit Data Service (SMDS). See also *ATM, broadband, frame relay*, and *SMDS*.

fast retrain A feature of asymmetric digital subscriber line (ADSL) modems that allows the transmission rate to resume normal levels after having been reduced by the power back-off feature in order to avoid interference between voice and data channels. See *ADSL, modem, power back-off*, and *transmission rate*.

fat access point See *fat AP*.

fat AP (fat Access Point) In wireless local area networks (WLANs), an AP with sufficient program logic and processing power to allow it to enforce policies relating to access and usage, rather than working under the supervision of a centralized controller. (A fat AP may use information from a RADIUS server, for example.) A network based on fat APs is more costly and complex, but offers the advantage of faster access as they can act independently rather than having to consult a centralized controller for authentication and other security purposes. In a mobile application, users moving between AP zones of coverage realize faster handoffs with fat APs. See also *authentication, RADIUS, server, thin AP*, and *WLAN*.

fat client In contrast to a thin client, a fat client possesses considerable resources (e.g., memory, hard drive storage, and processing power) and functionality independent of a server. See also *client, client/server, server*, and *thin client*.

FATE (Frame-based ATM Transport over Ethernet) A specification from the ATM Forum (February 2000 and July 2002) that allows ATM Adaptation Layer Type 5 (AAL5) services to be provided over Ethernet by transporting ATM data within an Ethernet frame. FATE has particular application in the context of an ATM-based ADSL environment interfacing to an Ethernet local area network (LAN) through a switch or hub on the customer premises. See also *AAL5, ADSL, ATM, ATM Forum, Ethernet, hub, LAN*, and *switch*.

fatware Software that is so rich in feature content or so bloated with inefficient design or poorly written code that it consumes excessive resources, such as RAM, hard disk storage, and processing power. See also *RAM* and *software*.

fault management An element of network management, fault management includes the detection of alarms and alerts, test and acceptance, and network recovery. Network elements (NEs) generate alarms and alerts are to indicate catastrophic failures or severe performance degradations. A network management system receives and correlates alarms and alerts from multiple NEs, and perhaps disables a failed port and enables another, or perhaps reroutes traffic around a failed switch or router after testing the alternate route. See also *NE* and *network management*.

fax (facsimile) See *facsimile*.

fax emulation Application software that enables a personal computer or fax server to behave like (i.e., function as) a fax machine.

fax-on-demand (FOD) See *FOD*.

Fax over Internet Protocol (FoIP) See *FoIP*.

fax relay Also known as *demod/remod*, fax relay is one of the implementation methods described by the ITU-T Recommendation T.38 specification for Fax over Internet Protocol (FoIP). Fax relay addresses the demodulation of standard analog fax transmissions from originating machines equipped with modems, and their remodulation for presentation to matching destination devices. Fax relay depends on a low latency IP network in order to avoid session time-out. See also *facsimile*, *fax spoofing*, *FoIP*, *latency*, *modulation*, and *T.38*.

fax spoofing An implementation methods described by the ITU-T Recommendation T.38 specification for Fax over Internet Protocol (FoIP). Fax spoofing is used for facsimile transmissions over IP networks characterized by relatively long and unpredictable levels of packet latency that could cause a session between conventional fax machines to time out. Fax spoofing compensates for both increased latency and jitter by padding the line with occasional keep-alive packets to keep the session active, rather than allowing it to time out. Thereby, T.38 spoofs, or fools, the receiving device into thinking that the incoming transmission is over a real-time, synchronous voice network. See also *facsimile*, *FoIP*, *jitter*, *latency*, *session*, *spoofing*, and *T.37*.

FC (Fibre Channel) See *Fibre Channel*.

FCC (Federal Communications Commission) An independent United States government agency, directly responsible to Congress and charged with regulating interstate and international communications by radio, television, wire, satellite, and cable. The FCC's jurisdiction covers the 50 states, the District of Columbia, and U.S. possessions. The FCC was established by the Communications Act of 1934. See also *Communications Act of 1934*.

FC/IP (Fibre Channel over Internet Protocol) A specification that extends Fibre Channel (FC) to operate through secure tunnels over long haul public Internet Protocol (IP) networks. See also *Fibre Channel*, *IP*, and *tunnel*.

FCS (Frame Check Sequence) A 16- or 32-bit field containing the cyclic redundancy check (CRC) character sequence used to check the integrity of both the payload and control fields of a frame, such as a Synchronous Data Link Control (SDLC) frame. See also *CRC*, *frame*, and *SDLC*.

FDD (Frequency Division Duplex) A means of providing duplex (bidirectional) communications in wireless networks, FDD makes use of separate frequencies for forward and backward channels. FDD is used with both analog and digital wireless technologies, including cordless telephony and cellular. See also *analog*, *cellular*, *channel*, *cordless telephony*, *digital*, *duplex*, *frequency*, and *wireless*.

FDDI (Fiber Distributed Data Interface) The ANSI standard (X3T9-5) for a dual, counter-rotating, fiber optic, token-passing ring LAN. The specification pegs the signaling rate at 125 Mbps and the transmission rate (i.e., data rate) at 100 Mbps due to the 4B/5B line coding technique. FDDI is intended for backbone applications, interconnecting major computing resources such as high speed switches, routers, and servers. As the FDDI maximum frame size is 9000 symbols (1 symbol = 4 bytes), Ethernet and Token Ring frames can easily be encapsulated within FDDI frames for backbone transport. FDDI specifies devices separations of as much as 1.2 miles (2 kilometers) over multimode fiber (MMF) and 37.2 miles (62 kilometers) over single-mode fiber (SMF), with excellent error performance. The dual counter-rotating ring provides considerable redundancy, but requires that all directly connected devices be dual-attached, which adds to the cost and complexity. In consideration of the high cost and fragility of optical fiber, standards were developed to extend connectivity to workstations via unshielded twisted pairs. Those standards are known variously as *CDDI* (Copper Distributed Data Interface) and *TPDDI* (Twisted Pair Distributed Data Interface). FDDI is considered obsolete, having been overwhelmed by simpler, higher speed switched Ethernet technologies such as 1000Base-LX, 1000Base-SX, and 10GBase-LR, LW. See also *10GBase-LR*, *1000Base-LX*, *1000Base-SX*, *4B/5B*, *ANSI*, *backbone*, *CDDI*, *Ethernet*, *fiber optics*, *frame*, *LAN*, *LW*, *MMF*, *signaling rate*, *SMF*, *symbol*, *token passing*, *Token Ring*, and *TPDDI*.

FDM (Frequency Division Multiplexing) A multiplexing method by which multiple low speed incoming transmissions can share a single high speed outgoing analog circuit. An analog voice conversation requires bandwidth of 4 kHz. A voice grade analog local loop, therefore, provides an analog channel of 0–4,000 Hz. Such a loop is a two-wire circuit, comprising two physical conductors in a pair configuration. It is possible to equip a four-wire circuit to support multiple 4 kHz channels. In order to do so, a FDM multiplexer, or mux, must be placed on each end of the circuit. The muxes subdivide the bandwidth of the circuit into 4 kHz channels, each of which can support a voice grade transmission. So, an FDM mux might multiplex 24 voice grade channels of 4 kHz onto a four-wire circuit with total bandwidth of 96 kHz. All 24 channels coexist on the same physical circuit, separated only by frequency, as illustrated in Figure F-2. Within each channel, the voice conversation occupies the 300–3,300 Hz band and signaling and control functions take place over the 3,300–3,700 Hz band. The 0–300 Hz and 3,700–4,000 Hz bands are guard bands that provide channel separation in order to minimize the likelihood of mutual interference should the frequency channels overlap due to equipment malfunction or electromagnetic interference (EMI) from and external source.

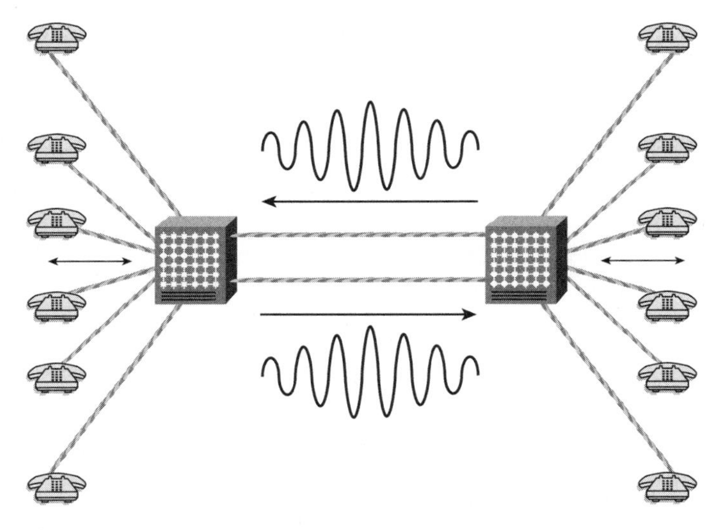

Figure F-2

FDMA (Frequency Division Multiple Access) A multiplexing technique used in radio networks, FDMA derives multiple narrowband frequency channels from a wider band of assigned radio spectrum, much as frequency division multiplexing (FDM) operates in the electrical wireline domain. Using a technique known as frequency division duplex (FDD), a given call takes place on one pair of frequencies, with one for transmission in the forward direction and another for transmission in the reverse direction. At the same time, another call takes place on another pair of frequencies. The forward and reverse channels in each frequency pair are separated in frequency in order to avoid crosstalk and other forms of co-channel interference. Analog cellular systems, such as Advanced Mobile Phone System (AMPS), employ FDMA. Alternative multiplexing techniques employed in various cellular radio networks are code division multiple access (CDMA) and time division multiple access (TDMA). See also *AMPS*, *channel*, *crosstalk*, *FDD*, *FDM*, *FDMA*, *frequency*, *multiplexer*, *narrowband*, *radio*, *spectrum*, and *TDMA*.

FDX (Full DupleX) A duplex transmission path, circuit, or channel designed to support information transfer in both directions, simultaneously. An FDX circuit can be a single physical circuit, such as a voice grade local loop. Alternatively, an FDX circuit can comprise two simplex circuits, one operating in each direction. Traditionally, T1 and E-1 circuits were provisioned over two simplex twisted pair circuits. A half-duplex (HDX) circuit differs in that it supports information transfer in both directions, but only one direction at a time. Most circuits are FDX in nature. Joseph B. Stearns of Boston, Massachusetts (United States) invented the first working FDX communications circuit, which was installed in 1872 on a one-wire telegraph system using a ground return. This system effectively doubled the traffic capacity of the circuit, and at much lower cost than stringing another wire. See also *HDX* and *simplex*.

Feature Group (FG) See *FG*.

FEC **1.** Forward Error Correction. An error control mode in which a detected error triggers an automatic error correction process in the receiver. FEC involves adherence to a set of specific rules of data construction and the addition of sufficient redundant data in order that the receiving device can identify, isolate, and correct a certain number (depending on the method) of errors without requiring retransmission. FEC often is used in networks where link quality is poor and bandwidth is limited, or where latency is high. FEC is used, for example, in cellular and other wireless networks in support of e-mail, short message service (SMS), and Internet access. Two commonly employed techniques are Hamming code and BCH (Bose, Chaudhuri, and Hocquengham). FEC also is used extensively in satellite communications. See also *1/3 FEC*, *BCH*, *error control*, *Hamming code*, *recognition and flagging*, and *recognition and retransmission*. **2.** Forwarding Equivalence Class In Multiprotocol Label Switching (MPLS), a class of packets, all of which are treated the same in terms of destination, priority level, and so on. See also *MPLS*.

FECN (Forward Explicit Congestion Notification) Pronounced *feckon*. In the frame relay LAPF frame, a 1-bit field available to the network to advise devices in the forward direction, that is, in the direction of the data flow, that the frame has experienced congestion in transit. FECN thereby alerts the receiving frame relay access device (FRAD) that subsequent frames might be delayed in transit or even discarded if the congestion condition worsens. In the event the receiving FRAD detects a frame loss, it recovers by requesting a retransmission. Backward explicit congestion notification (BECN) performs a congestion control function in the reverse direction, that is, in the direction opposite the congested data flow. See also *BECN*, *congestion*, *ECN*, *FRAD*, *frame*, *frame relay*, and *LAPF*.

Federal Communications Commission (FCC) See *FCC*.

Federal Trade Commission (FTC) See *FTC*.

femto- (f) See *f* and *femtocell*.

femtocell An imprecise term referring to a radio cell smaller than a picocell and used to describe a very small radio cell associated with a cellular radio base station located in a home or small office. Femtocells are proposed for use as extensions of public cellular radio service into the customer premises to counteract perceived competitive threats from Generic Access Network (GAN), Wi-Fi, WiMAX, and other wireless network technologies. See also *cellular radio*, *femto-*, *GAN*, *picocell*, *Wi-Fi*, and *WiMAX*.

FEP (Front End Processor) Synonymous with *communications processor*. An auxiliary processor, usually in the form of a dedicated computer, that assumes responsibilities for managing the interface between a host computer and networks, terminals, and peripherals. The FEP, thereby, relieves the host main processing unit of those responsibilities, allowing it to concentrate on running applications software. The FEP can assume responsibility for such tasks as authentication, access privileges, code translation, compression, encryption, and priority management. The FEP traditionally takes the form of a mid-range computer that manages access to a mainframe computer, although those distinctions are less significant in a contemporary data processing context. In more contemporary terms, FEP and host responsibilities both tend to be distributed across multiple servers in one or more clusters. See also *authentication*, *compression*, and *encryption*.

FEX (Foreign EXchange) See *FX*.

FEXT (Far-End CROSS Talk) The unwanted coupling of energy between two circuits or channels occurring at the far end of a link, i.e., far away from the point of signal origin. It is at the far end that the attenuated downstream signal from the network, for can experience crosstalk from the strong upstream signal emanating from equipment at the customer premises. FEXT is not a large issue for V.90 modems or asymmetric digital subscriber line (ADSL) services at the customer premises, as the cables are successively smaller, containing fewer and fewer twisted pairs, from the central office (CO) to the premises, so there are fewer opportunities for signals to experience co-interference. Near-end crosstalk (NEXT) occurs at the near end of the link. It is in consideration of the phenomenon of crosstalk and the differences between NEXT and FEXT, that V.90 modems and most DSL services are asymmetric, with the higher frequencies on the downstream side (i.e., from the edge of the telco network to the customer premises) in support of greater bandwidth in that direction. See also *ADSL, asymmetric, bandwidth, channel, circuit, CO, crosstalk, downstream, frequency, interference, link, NEXT, signal, upstream,* and *V.90.*

FF (Firefox) See *Firefox.*

FG (Feature Group) A trunking arrangement between a local exchange carrier (LEC) and an interexchange carrier (IXC) that enables end users to make long distance telephone calls using the IXC network. Feature groups are designated A, B, C, and D.

- **FGA:** A line side access method in which the caller must dial a local telephone number for IXC access, then dial the long distance telephone number and a PIN or password. The IXC gains access to the customer through a subscriber line, rather than an interoffice trunk.
- **FGB:** The caller must dial a 950-XXXX telephone number and then dial the long distance telephone number and a PIN or password. In the 950-XXXX telephone number, the last four digits correspond to the Carrier Identification Code (CIC) associated with each IXC. The 950-XXXX number, which can be dialed as a local call from any geographic location, directs the call setup request to a centralized database that is used to direct the call to the nearest carrier retail outlet. An FGB connection is a trunk side connection commonly provided through an access tandem (AT) switch, although it also can be provided through a direct trunking arrangement to a central office (CO) switch. An FGB connection is superior to an FGA connection.
- **FGC:** A traditional trunk side access service used prior to the implementation of equal access in 1984. Where it is available, which is almost universally, FGD has replaced FGC.
- **FGD:** An FGD trunk is sometimes referred to as an equal access trunk, as all carriers with FGD trunks in a local exchange network are afforded equal access (1+ dialing) to the subscriber at designated central office exchanges (COs), whether via a direct trunking arrangement or through an access tandem (AT) switch. FGD provides the IXC with trunk side access, connecting the IXC as an integral part of the exchange carrier network. FGD includes presubscription to a subscriber-specific IXC and a 10XXX access code for use by end users in originating and terminating connections. FGD also includes call supervision and the IXC receives calling party identification as part of the call setup, automatic number identification (ANI) data for billing purposes.

See also *10XXX, access tandem, ANI, call supervision, carrier, CIC, CO, equal access, IXC, LEC, line side, password, PIN, presubscription, trunk,* and *trunk side.*

FGE (FiberGlass Epoxy) FGE is a combination of fiberglass and epoxy resins used in products such as cable strength members. See also *fiberglass* and *strength member.*

FHSS (Frequency-Hopping Spread Spectrum) A signal modulation technique employed in radio communications, FHSS transmits short bursts of data over a range of frequency channels within the wideband carrier. Each transmission is assigned a 10-bit pseudorandom binary code sequence, which comprises a series of ones and zeros in a seemingly random pattern known to both the transmitter and receiver. The original code sequence is mathematically self-correlated to yield a code that stands out from all others, at least on average. The paired transmitters and receivers recognize their assigned and correlated code sequences, which look to all others as pseudorandom noise (PN). FHSS phase-modulates the carrier wave with a continuous string of PN code symbols, or chips, resulting in a chip rate that can be much faster than the bit rate. Thereby, the noise signal occurs with much greater frequency than the original data signal and spreads the signal energy over a much wider band. The transmitter and receiver hop from one frequency to another in a carefully choreographed hop sequence under the control of the centralized base station antenna. Each transmission dwells on a particular frequency for a very short period of time (no more than 400 milliseconds for FCC-controlled applications), which may be less than the time interval required to transmit a single data packet, or symbol, or even a single bit. So, the chip rate can be faster than the bit rate. A large number of other transmissions also may share the same range of frequencies simultaneously, with each using a different hop sequence. The potential remains, however, for the overlapping of packets. The receiving device can distinguish each packet in a packet stream by reading the various codes prepended to the packet data transmissions, and treating competing signals as noise. Code division multiple access (CDMA) and Bluetooth employ FHSS, which is much like the original SS technology patented by Hedy Lamarr in 1942. See also Bluetooth, CDMA, carrier, channel, chip, chip rate, frequency, hop sequence, modulation, PN, signal, symbol, and wideband. See also *Lamarr, Hedy*.

fiber See *optical fiber.*

Fiber Channel See *Fibre Channel.*

Fiber Connection (FICON) See *FICON.*

Fiber Distributed Data Interface (FDDI) See *FDDI.*

fiberglass Also known as *spun glass*, fiberglass is a composite of extremely fine fibers of glass. Invented in 1938 by Russell Games Slayter of Owens-Corning for use as insulating material, the raw spun glass is used as a reinforcing agent and combined with polymers and epoxies to create what is known popularly as fiberglass, which can be drawn, shaped, and molded for a wide variety of uses. The low weight, great tensile strength, and dielectric properties of fiberglass contribute to its wide use in both rigid and non-rigid applications, including boat hulls, swimming pools, hot tubs, surfboards, thermal insulation, automobile bodies, and cable strength members. Owens-Corning remains the largest manufacturer of fiberglass, which it markets as Fiberglas®. See also *strength member.*

fiberglass epoxy (FGE) See *FGE.*

fiber optics Referring variously to optical fiber as a transmission medium or a fiber optic transmission system (FOTS). See also *FOTS* and *optical fiber.*

fiber optic transmission system (FOTS). See *FOTS.*

fiber-to-the-curb (FTTC) See *FTTC.*

fiber-to-the-neighborhood (FTTN) See *FTTN.*

fiber-to-the-node (FTTN) See *FTTN.*

fiber-to-the-premises (FTTP) See *FTTP.*

Fibre Channel (FC) An American National Standards Institute (ANSI) specification (X.3230, 1994) for a high-speed link between computers and peripherals, primarily high speed external storage devices. Developed to replace High Performance Parallel Interface (HIPPI) and as a high-speed alternative to the distance-limited Small Computer System Interface (SCSI), Fibre Channel is intended to support applications such as data backup and mirroring, and is the predominant data link technology employed in storage area networks (SANs). Fibre Channel is connected at Layer 1, the Physical Layer, by fibre, a term the Fibre Channel industry coined to refer to a network comprising a close-knit fabric of access including both optical fiber and copper twisted pair for large data transfers with low overhead, low-latency switching, and minimal interruptions to the flow of data. The preferred physical medium is optical fiber, which can be multimode fiber (MMF) of either 62.5μ (300 meters) or 50μ (500 meters, maximum distance), or single-mode fiber (SMF) (50+ km). Fibre Channel over IP (FC/IP) technology extends Fibre Channel to operate through secure tunnels over long-haul public IP networks. The line coding technique is 8B/10B, which encodes each 8-bit byte into a 10-bit symbol, which adds a 25% overhead factor. Fibre Channel operates in full duplex (FDX) at 1 Gbps (200 MBps), 2 Gbps (400 MBps), 4 Gbps (800 MBps), and 10 Gbps (2400 MBps or 2.4 GBps). Gateways are responsible for protocol conversion to support interconnection to telecom networks such as asynchronous transfer mode (ATM) and SONET, as well as ESCON, FICON and SCSI SANs and Ethernet LANs. See also *8B/10B, ANSI, ATM, bit, byte, ESCON, Ethernet, FDX, FICON, gateway, HIPPI, IP, LAN, latency, line coding, MMF, optical fiber, overhead, Physical Layer, protocol, SAN, SCSI, SMF, SONET, symbol, tunnel,* and *twisted pair.*

Fibre Channel over IP (FC/IP) See *FC/IP* and *Fibre Channel.*

FICON (FIbre CONnection) A proprietary specification for storage area network (SAN) developed by IBM for a high-speed serial interface between mainframe computers and peripherals such as external disk drives. FICON supports data transfer rates up to 400 MBps in full duplex (FDX) over distances up to 100 km. FICON is replacing the earlier and slower Enterprise System Connection (ESCON). See also *Bps, ESCON, FDX, Fibre Channel, mainframe, peripheral, SAN,* and *serial.*

fidelity The extent to which an electronic device or process faithfully reproduces audio or visual information. Hi-fi, for example, is high fidelity.

field Synonymous with *data field.* **1.** A location or area in which certain data is located within a block or frame of transmitted data. See also *block* and *frame.* **2.** A location or area in which certain data is located on a storage medium, particularly in a database record.

FIFO (First-In-First-Out) A buffering or temporary storage method in which the entity that first exits is the one that first entered. Thereby, the entity served (e.g., processed or switched) is the one that waited the longest period of time. FIFO is commonly used in message switches such as automatic call distributors (ACDs), PBXs, switches, and routers in the absence of a priority mechanism employed to establish quality-of-service (QoS) differentiation between different types of calls, packets, or other message entities. See also *LIFO* and *queue.*

file **1.** Program file. An electronic file containing commands and instructions for execution by a computer. See also *program file.* **2.** Data file. An electronic file containing the work created with a program. See also *data file* and *data set.*

File Transfer Protocol (FTP) See *FTP.*

filter A device that allows some signals to pass through but absorbs, attenuates, blocks, rejects, or removes all other signals, depending on their frequency (electrical) or wavelength (optical). Active filters require electrical power to operate, while passive filters do not. A low-pass filter passes all frequencies below a certain value, but blocks all others. A high-pass filter passes all frequencies above a certain value, but blocks all others. A band-pass filter passes all frequencies in a designated band, but blocks all others. See also *absorption, active, attenuation, electrical, frequency, optical, passive, signal,* and *wavelength.*

filtering Also known as *image decimation.* In video compression, a step that reduces the total frequency of the analog signal through a process of averaging the values of neighboring pixels or lines. For example, adjoining black and white pixels become gray pixels. Taps are the number of lines or pixels considered in this process. MPEG, for example, uses a seven-tap filter. See also *analog, compression, frequency, MPEG, pixel, signal,* and *video.*

filtering bridge A bridge that examines the destination address of an incoming frame, consults an address table, and forwards the frame only over the link toward the target device. If the frame is intended for a station on the same LAN segment, the bridge simply ignores it, rather than passing it on. Since the frames are not forwarded across other links, a filtering bridge does a great deal to relieve overall congestion on a segment-by-segment basis. Most bridges are self-learning, filtering bridges and are standardized in IEEE 802.1D, which describes the spanning tree protocol (STP). See also *bridge*, *self-learning bridge*, and *STP*.

find-me A voice system (e.g., Centrex or PBX) feature that enables the user to preprogram telephone numbers (e.g., home office phone, cell phone, and home phone) that the system will attempt in sequence in order to complete an incoming call. If the system is unable to find the target user, it will so advise the caller and offer the opportunity to leave a message. Find-me service is defined in the advanced intelligent network (AIN) specifications as a service of the public switched telephone network (PSTN). See also *AIN* and *PSTN*.

fine print The really small print at the bottom of an advertisement that details the conditions under which a product or service offering is made, conditions on its use, limitations on associated liability, and other matters imposed or provided for by law or regulation. Fine print is not only really tiny, but also in legal or otherwise obscure language in hopes that you will not read it or not understand it and, therefore, will make your purchasing decision in ignorance.

finger In a rake receiver antenna system, an individual receiver that works with other receivers in a coordinated way to gather signal elements much like the tines of a garden rake work together to gather leaves. Each finger gathers a faded, or attenuated, signal element at a separate moment in time. The rake receiver employs spatial diversity and time diversity, combining and correlating the results of all four fingers to optimize the signal, thereby countering the effects of multipath fading and delay spread. Code-division multiple access (CDMA) systems employ rake receivers comprising four fingers to deal with issues of multipath interference (MPI). See also *antenna*, *attenuation*, *CDMA*, *delay spread*, *MPI*, *multipath fading*, *spatial diversity*, and *time diversity*.

Firefox (Fx, fx, or FF) A graphical Web browser developed by the not-for-profit Mozilla Corporation and made available free of charge for Windows, Mac, and Linux computers. Firefox began as a fork of the Navigator component of the Mozilla application suite. See also *Mozilla*.

firewall Security software that can actively block unauthorized entities from gaining access to internal resources such as systems, servers, databases, and networks. A firewall may also act to prevent internal users from accessing unauthorized external resources. A firewall is installed in a communications router, server, or some other device that physically and/or logically is a first point of access into a networked system. A packet-filtering firewall examines all data packets, forwarding or dropping individual packets based on predefined rules that specify where a packet is permitted to go, in consideration of both the authenticated identification of the user and the originating address of the request. A proxy firewall acts as an intermediary for user access requests by setting up a second connection to the resource. The proxy then decides if the message or file is safe. A stateful inspection firewall examines packets, notes the port numbers that they use for each connection, and shuts down those ports once the connection is terminated. See also *authentication*, *authorization*, *proxy firewall*, and *security*.

FireWire Apple Computer terminology for IEEE 1394. See *1394*.

firmware Software programs that are stored in a computer's read-only memory (ROM), where they are available for instantaneous use. Firmware is hard-coded and stored on a silicon chip and, therefore, is not affected by loss of electrical power, hence the term *firm*. See also *grayware*, *hardware*, and *software*.

First Computer Inquiry In the United States, a Federal Communications Commission (FCC) inquiry (1971) that drew a firm line between data processing services and data communications services. The FCC determined that it would continue to regulate data communications services in order to avoid the possibility of AT&T's subsidizing profit-making competitive activities with revenues from regulated telephone company activities. In the 1956 Consent Decree, AT&T had agreed to refrain from offering data processing services even through a separate subsidiary. See also *Consent Decree*, *FCC*, and *Second Computer Inquiry*.

first-in-first-out (FIFO) See *FIFO*.

first mile More commonly referred to as the last mile and generally referring to the telco local loop, which is the link between the central office (CO) at the edge of the telco network and the user premises. In a broader contemporary context, the term applies to the physical connection between the edge of any service provider's network and the end user's premises. In practice, the first mile is often much longer than a mile. In the United States, UTP local loops are generally 12,000 feet or less, but often are as long as 18,000 feet. Passive optical network (PON) standards allow for local loops as long as 12 miles (20 km). Whether the first mile or the last mile, which is a matter of perspective, it is seldom exactly a mile. See also *central office*, *local loop*, and *PON*.

Fixed Mobile Convergence A term coined by the 3rd Generation Partnership Project (3GPP) for the seamless melding of fixed IP–based fixed wireless (e.g., Wi-Fi and WiMAX) and cellular radio networks. Fixed Mobile Convergence is the ultimate goal of the 3GPP GAN (Generic Access Network). See also *3GPP*, *cellular radio*, *fixed wireless*, *GAN*, *IP*, *Wi-Fi*, and *WiMAX*.

fixed satellite system (FSS) See *FSS*.

fixed wireless Referring to a group of wireless local loop (WLL) transmission systems that involve antennas permanently or semi-permanently located at the edge of a public network and at the customers' premises, rather than being mobile. A number of 2.5G and 3G cellular standards also include fixed wireless options. See also *antenna*, *local loop*, and *WLL*.

flag **1.** A marker or indicator of a condition, such as an error condition, in a program or file. See also *file* and *program*. **2.** In Synchronous Data Link Control (SDLC), High-level Data Link Control (HDLC), and other frame-based communications protocols, a specific eight-bit pattern that alerts the receiving device to the beginning or end of a frame, i.e., message unit. The most commonly used flag character is 01111110 in binary code (7E in hexadecimal). Flags also fill all idle time on the line between frames. Only one flag is needed between frames. See also *frame*, *HDLC*, *hexadecimal notation*, *protocol*, and *SDLC*.

flash memory A type of non-volatile read-only memory (ROM) that can store data or programs, be erased, and be used again. Flash memory must be erased in blocks, rather than a byte at a time, which limits its use to applications such as a supplement to or replacement for a mechanical hard disk drive. Flash memory is unsuitable for use as main memory, or random access memory (RAM). See also *RAM* and *ROM*.

FLEX A set of proprietary (Motorola) protocols for radio paging systems, FLEX largely has replaced POCSAG in the United States, and has become the de facto standard throughout most of the world, excepting Western Europe, where the ERMES standard is favored. FLEX solutions support duplex messaging and data transmission. FLEX supports as many as 5 billion addresses, with up to 600,000 supported per channel. The FLEX family of protocols includes the following:

- **FLEX:** 1600 bps; 25 kHz channels; simplex downstream
- **ReFLEX:** 1600, 3200, 6400, or 9600 bps; 25 or 50 kHz channels downstream and 12.5 kHz channel upstream; duplex
- **InFLEXion:** up to 112 kbps, 50 kHz channels in the narrowband PCS (N-PCS) range; duplex; supports compressed voice downstream

See also *channel*, *compression*, *downstream*, *duplex*, *ERMES*, *narrowband*, *N-PCS*, *paging system*, *PCS*, *POCSAG*, *proprietary*, *protocol*, *simplex*, *standard*, and *upstream*.

flex strength The ability to withstand the stress of twisting and bending, flex strength is important in wire and cable applications that involve frequent bending and twisting. Illustrative applications include elevator telephone cables, telephone handset and headset cords, and microphone cords. Wires and cables designed for high flex strength commonly involve small diameter conductors and cables, stranded rather than solid core conductors, stranded or braided rather than solid metal shields, conductors and shields made of metal alloys that are flexible, and insulation made of materials that are not only flexible, but also abrasion resistant. See also *break strength* and *tensile strength*.

flow **1.** Movement in a manner suggestive of a liquid. Movement in a smooth and gentle manner, like water in a stream. See also *stream-oriented.* **2.** In telecommunications, a sequence of bits, bytes, datagrams, or packets between common endpoints identified by features such as network addresses and port numbers. See also *bit, byte, datagram, endpoint, packet,* and *port.*

flow control The process of controlling the rate of data transfer in order to prevent data loss due to congestion, flow control is an element of many data communications protocols. Between a PC and a local printer, for example, there is a simple flow control mechanism that throttles back the data transfer rate from the PC to ensure that the printer is not overwhelmed and that the printer buffer memory is not exceeded. ATM, frame relay IPv6, and TCP include more complex flow control mechanisms to ensure that switches and routers are not overwhelmed. A rate-based mechanism is an end-to-end flow-control scheme that considers resources edge to edge, communicating the level of available resources through a feedback loop. This approach requires that the transmitting end-user device adjust its rate of transmission downward in consideration of congestion. A credit-based approach either allows or disallows the end-user device to transmit data, based on end-to-end consideration of whether sufficient buffer space is available on each link of the network. See also *ATM, buffer, frame relay, IPv6,* and *TCP.*

flux **1.** The rate of flow of something such as energy, particles, or fluid volumes across or through a surface. Radiant flux is the time rate of energy flow of radiant energy as measures in watts or joules per second. Luminous flux is radiant flux evaluated with respect to its luminous (brightness) efficiency. See also *joule, luminance,* and *watt.* **2.** The strength of an energy field acting on a specific area.

FM (Frequency Modulation) Also known as *frequency-shift keying* (FSK). A signal modulation technique in which the frequency of the analog carrier sine wave is varied to distinguish between a 1 bit and a 0 bit. For example, FM is the sole technique used in low-speed, Hayes-compatible modems. When no bits are transmitted, the carrier is at a reference frequency of 1700 Hz. A unibit FM technique impresses one bit on each baud by shifting the carrier to 2200 Hz when transmitting a 1 bit, and to 1200 Hz when transmitting a 0 bit. At 2400 baud, therefore, the transmission rate is 2400 bps. The benefits of dibit transmission can be realized by defining four frequencies, with each sine wave or set of sine waves representing a 2-bit pattern (00, 01, 10, and 11). Thereby, at 2400 baud, the transmission rate is 4800 bps. Figure F-3 illustrates unibit frequency modulation, with the carrier waveform modulated to a relatively high frequency to represent a 1 bit and to a relatively low frequency to represent a 0 bit. FM often is used in conjunction with amplitude modulation (AM) and phase-shift keying (PSK). A variation on the FSK theme is Gaussian frequency-shift keying (GFSK), which is used in DECT and Bluetooth wireless networks. See also *AM, analog, Bluetooth, carrier, DECT, GFSK, modulation, PSK, sine wave,* and *transmission rate.*

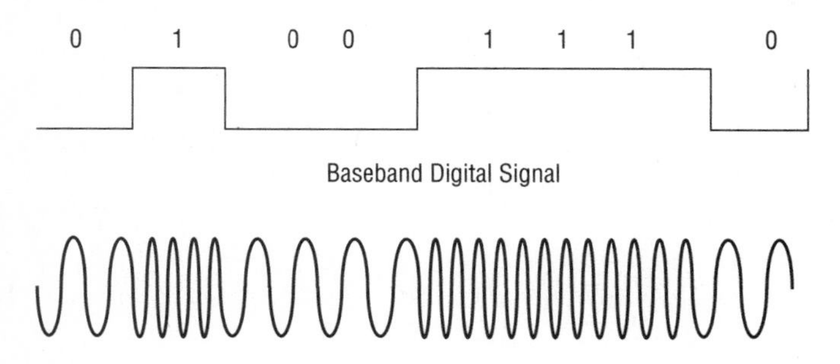

Figure F-3

FMNP (Full Mobile Number Portability) Synonymous with *wireless number portability* (WNP). See *WNP.*

FOD (Fax-On-Demand) A system that enables a customer to request that a document be faxed back to him or her. Traditional FOD involves a voice processing system that answers a telephone call and prompts the caller to select a document from a menu of options, enter a return fax number, and perhaps enter a credit card number for billing purposes. FOD also has been integrated with Web sites for access over the Internet, allowing customer to select documents from a menu, and enter a return fax number and billing information. FOD largely has been obsoleted by the Internet and World Wide Web (WWW), as documents easily can be attached to e-mail or printed directly from a Web page.

foil twisted pair (FTP) Synonymous with *shielded twisted pair* (STP). See *STP*.

FoIP (Fax over IP) A technique for facsimile (fax) transmission over the Internet or other IP-based packet network, rather than over the traditional public switched telephone network (PSTN). FoIP typically involves a fax gateway, which not only serves as a physical gate between the circuit-switched and the packet-switched networks, but also runs gateway protocols that convert from the PSTN to the IP-based packet network. Relevant IP fax standards include T.37, the specification for store-and-forward fax, and T.38, the specification for fax relay and fax spoofing. See also *circuit switch*, *facsimile*, *fax relay*, *fax spoofing*, *Internet*, *IP*, *packet switch*, *PSTN*, and *store-and-forward*.

follow-me A voice service of the public switched telephone network (PSTN) that provides for call forwarding on a predetermined schedule. A telecommuter, for example, might have the network forward calls to the home office three days a week during normal business hours. Calls would be directed to the traditional office two days a week. Calls clearly outside of normal business hours automatically would be directed to a voice mail system. Follow-me service is defined in the advanced intelligent network (AIN) specifications. See also *AIN*, *call forwarding*, *PSTN*, *telecommuting*, and *voice mail*.

FOMA (Freedom of Mobile Multimedia Access) The Japanese term for *Universal Mobile Telecommunications System* (UMTS), a 3G cellular radio standard. See *UMTS*.

footprint **1.** The physical space something occupies, as in the footprint of a computer or a PBX. **2.** The coverage area of a satellite or other radio transmitter system.

forced account code An account code that a KTS or PBX system requires an end user to enter prior to placing an outgoing call. See *account code*.

foreign exchange (FEX or FX) See *FX*.

forked ringing An IPBX feature that allows a call processor to ring multiple phones at once, rather than in turn, and serve the call to the first phone that answers. Forked ringing is enabled by the Session Initiation Protocol (SIP) signaling and control protocol. See also *IPBX* and *SIP*.

forklift upgrade Referring to a complete system replacement. The term originated in days of yore, when mainframe computers and PBXs were so big and heavy that it literally took a forklift to remove an obsolete system from the computer room or switch room and replace it with another, usually of at least equal size and weight. See also *heavy metal*, *mainframe computer*, and *PBX*.

format **1.** The structure, organization, presentation, or appearance of a set of data in a document. The format of a set of data can include the coding scheme (e.g., ASCII, EBCDIC, or Unicode) and any compression technique (e.g., GIF, JPEG, MPEG-2, or MPEG-4) that might have been employed. In an e-mail application, textual data can be formatted as plain text or rich text, with the latter supporting **bold**, *italics*, and underline. The format of the data must be compatible with the application software that attempts to read it. See also *ASCII*, *coding scheme*, *compression*, *EBCDIC*, *GIF*, *JPEG*, *MPEG-2*, *MPEG-4*, *plain text*, *rich text*, and *Unicode*. **2.** The arrangement of fields of data in a block, frame, or cell. See also *data format*.

forwarding equivalence class (FEC) See *FEC*.

FOTS (Fiber Optic Transmission System) An optical transmission system comprising, at the most basic level, a light source, an optical fiber, and a light detector. The light source can be in the form of a light-emitting diode (LED), vertical cavity surface-emitting laser (VCSEL), or laser diode. The optical fiber generally is one of many types of glass optical fiber (GOF), although plastic optical fiber (POF) is sometimes used. The light detector can be a positive-intrinsic-negative (PIN) diode or avalanche photodiode (APD). As is true of any transmission system, attenuation can be an issue over a long haul, so some form of amplification can be introduced. As FOTS systems generally are digital in nature, the typical approach is to apply a regenerative repeater, which detects the weak incoming signal, which it amplifies electrically, reshapes, retimes, and retransmits as an improved outgoing signal. Raman amplifiers increasingly are used, often in conjunction with repeaters. Long haul systems also typically conform to a set of ITU-T standards, with Synchronous Optical Network (SONET) being the North American version and Synchronous Digital Hierarchy (SDH) being preferred elsewhere. The optical signal is in the infrared (IR) range, within one of a number of windows, or wavelength bands, specified by the ITU-T. A single wavelength may be involved, or multiple wavelengths may coexist in a single fiber through a process known as wavelength division multiplexing (WDM), of which there are several levels, coarse (CWDM) and dense (DWDM). As a conducted system, FOTS systems are unparalleled in terms of bandwidth, error performance, signal attenuation, and security. See also *APD, bandwidth, diode, diode laser, DWDM, GOF, IR, laser, LED, PIN, POF, repeater, SDH, SONET, VCSEL, wavelength, WDM,* and *window*.

fourth estate A term first used by historian Thomas Carlyle (1795–1881) in his book, *On Heroes, Hero Worship, and the Heroic in History* (1841), to describe the press. Novelist Jeffrey Archer, in his book *The Fourth Estate* (1996), observed: "In May 1789, Louis XVI summoned to Versailles a full meeting of the 'Estate General'. The First Estate consisted of three hundred nobles. The Second Estate, three hundred clergy. The Third Estate, six hundred commoners. Some years later, after the French Revolution, Edmund Burke, looking up at the Press Gallery of the House of Commons, said, 'Yonder sits the Fourth Estate, and they are more important than them all.'" *Note:* The Estate General refers to the British Parliament. The First Estate refers to the Lords Temporal and the Second Estate to the Lords Spiritual, the two of which later combined to form the House of Lords. The Third Estate refers to the House of Commons. The Fourth Estate refers to the public press, which in those days was the newspapers. The contemporary reference is to the mass media, including both print and electronic media.

four-wave mixing (FWM) See *FWM*.

four-wire circuit A circuit that supports transmission in both directions over separate physical links or paths in support of full duplex (FDX), i.e., simultaneous two-way, transmission. The distinguishing characteristic of a four-wire circuit, as opposed to a two-wire circuit, is its ability to support multichannel communications and out-of-band signaling and control. A physical four-wire circuit, the traditional means of provisioning, is a circuit comprising four physical twisted-pair copper wires in a two-pair configuration, with one pair supporting transmission in the forward direction and the other pair supporting transmission in the reverse direction. A logical four-wire circuit comprises two-wires in a single-pair configuration. A four-wire circuit also can be provisioned as a radio circuit, which is wireless, of course. Four-wire circuits are used in bandwidth-intensive local loops, particularly multichannel loops, and backbone circuits. Specific examples include DDS, ISDN, T/E-carrier, and SDH/SONET. See Figure F-4. See also *channel, circuit, DDS, E-carrier, FDX, ISDN, out-of-band signaling and control, SDH, SONET, T-carrier, twisted pair,* and *two-wire circuit*.

forward error correction (FEC) See *FEC*.

forward explicit congestion notification (FECN) See *FECN*.

FPLMTS (Future Public Land Mobile Telecommunications System) An initiative of the ITU-R that defined a vision for a single global standard for digital wireless networks. FPLMTS was subsequently replaced by International Mobile Telecommunications-2000 (IMT-2000). See also *IMT-2000*.

fps (frames per second) The number of frames transmitted per second. See *frame, frame rate,* and *second*.

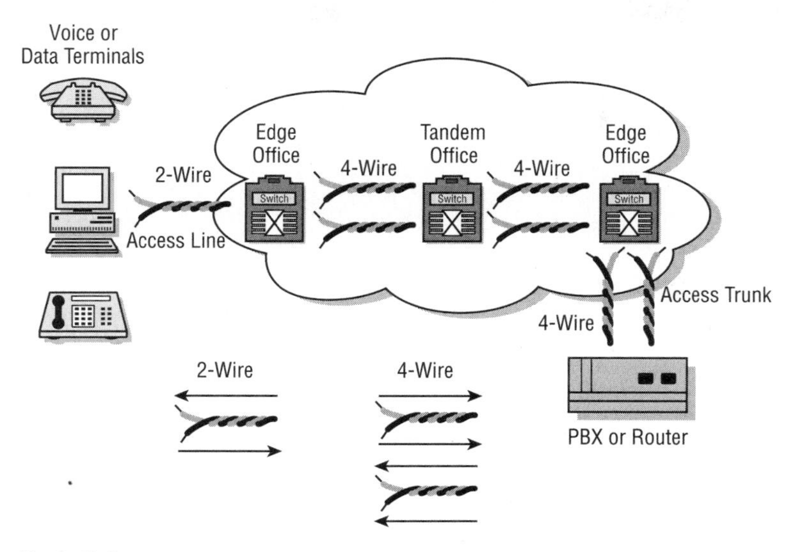

Figure F-4

fractal From the Latin *fractus*, translating as *broken* or *fractured*. An irregular or fragmented geometric shape that can be repeatedly subdivided into parts, each of which is a smaller copy of the whole. In words, a complex irregular object that is self-similar. Examples of fractal objects include mountain ranges, clouds, and lightening bolts. See also *fractal transform*.

fractal transform A technique for video compression. Fractal compression reduces an image into extremely small independent blocks which it translates into mathematical equations that are used to form a codebook that creates a mathematical model of the image. So, the image, itself, becomes the basis for the ad hoc codebook. Vector quantization is similar, but uses a standard codebook. Fractal compression is extremely efficient, but highly processor-intensive. Fractal compression can be either lossless or lossy in nature. See also *compression, fractal, lossless compression, lossy compression, transforms, vector quantization*, and *video*.

fractional T1 (FT1) See *FT1*.

FRAD (Frame Relay Access Device or **Frame Relay Assembler/Disassembler)** A device that assembles and disassembles frame relay frames. A FRAD is data communications equipment (DCE) that can be in the form of a standalone device, although it is generally embedded under the skin of a router or other device. The FRAD connects from the customer premises across a digital local loop, such as an E-1 or T1, to the frame relay network device (FRND) at the edge of the carrier network. The FRAD can be located at the network edge in support of dial-up users. See also *carrier, DCE, E-1, frame, frame relay, FRND, local loop, router*, and *T1*.

fragment-free switch A type of LAN matrix switch that quickly stores the first 64 octets of the frame before it reads the address and quickly flows it through the switching matrix, bit by bit. As most errors occur at the beginning of a frame, this approach eliminates the possibility that runt frames, i.e., truncated frames, will be transmitted. See also *cut-through switch, LAN, LAN switch, matrix switch*, and *switch*.

fragmentation Referring to the process by which a switch or router breaks up or divides a large datagram. If the receiving network cannot accommodate a datagram of a given total length, it must be fragmented. There must be some form of fragmentation control to ensure that the fragments can be re-associated when they exit the network and that the datagram can be reconstituted. In Internet Protocol version 4 (IPv4), for example, fragmentation control requires that each fragment contain a copy of an identification field and certain other fields in the header. The IPv4 header contains a fragment offset field that identifies where a fragment fits in the complete set of fragments that comprise the original datagram. This field is used to sequence the fragments correctly, as they may arrive at the destination device out of sequence. See also *datagram*, *IPv4*, *IPv6*, *network*, *router*, and *switch*.

frame **1.** A structure in the form of a structure of metal uprights and cross pieces with termination points on each side into which components can be mounted and conductors can be mechanically connected. The term is applied to distribution frames for cables and is the origin of the term *mainframe computer*. **2.** In asynchronous serial data communications, a transmission unit comprising a character of data and one or two parity bits, preceded by a start bit, succeeded by a stop bit. Asynchronous communications are said to be character-framed. See Figure F-5. See also *asynchronous transmission*. **3.** In synchronous data communications protocols such as HDLC and SDLC, a message unit. A frame comprises control data, address data, user data, and an error control mechanism. The frame is preceded by a beginning flag and succeeded by an ending flag. The data field of an SDLC frame, for example, can comprise as many as 4,096 octets and the various control fields add another four or six octets. See Figure F-6. See also *SDLC* and *synchronous transmission*. **4.** In digital carrier systems E-carrier, J-carrier, and T-carrier, a collection of time slots that repeats every 125 microseconds. In a channelized application, each time slot constitutes a channel. In an unchannelized application, the entire collection of time slots constitutes a channel. Figure F-7 is an illustration of a channelized T1 frame. *Note:* A T1 and J1 frame is always preceded by a framing bit, which is used for synchronization and other control purposes. An E-1 frame does not require a framing bit, as time slots 0 and 16 serve those functions. See also *carrier*, *channel*, *E-carrier*, *framing bit*, *J-carrier*, *synchronous*, *T-carrier*, and *time slot*. **5.** In video communications, a single photographic image that is one of many in a motion picture. See also *frame rate*.

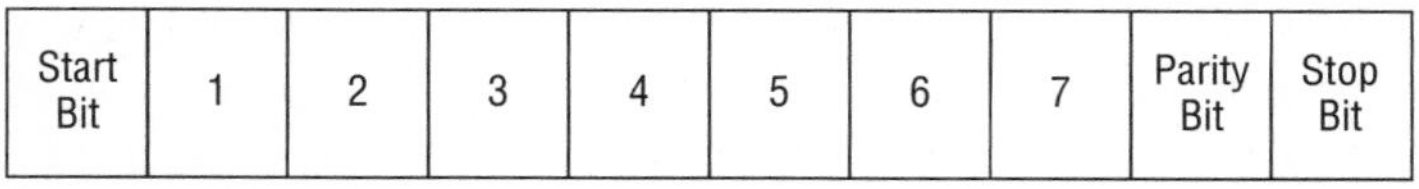

Start Bit	1	2	3	4	5	6	7	Parity Bit	Stop Bit

Figure F-5

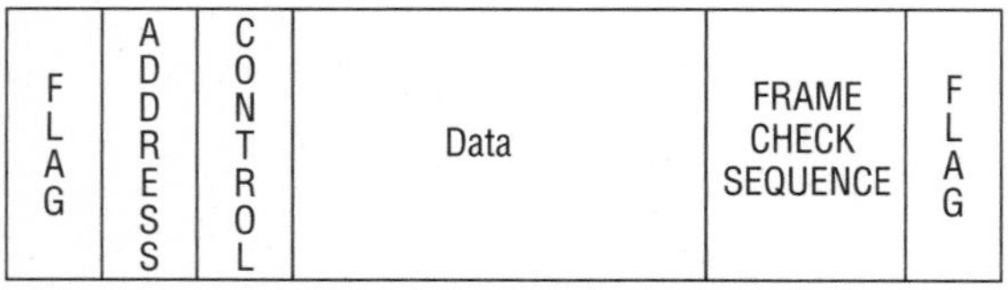

FLAG	ADDRESS	CONTROL	Data	FRAME CHECK SEQUENCE	FLAG

Figure F-6

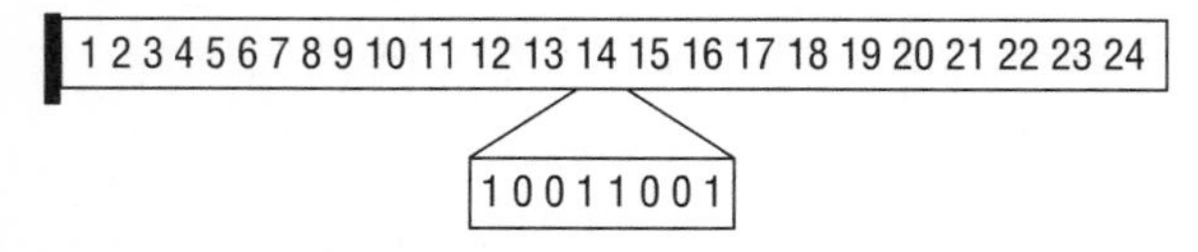

Figure F-7

Frame-based ATM Transport over Ethernet (FATE) See *FATE*.

frame check sequence (FCS) See *FCS*.

Framed ATM over SONET/SDH Transport (FAST) See *FAST*.

frame rate In video communications, the rate at which frames of still images are transmitted. Video is a series of still images transmitted in succession to create the perception of fluidity of motion. Motion picture quality is considered to be a frame rate of 24 frames per second (fps) and broadcast television quality is considered to be 30 fps. If the frames are transmitted at a slow rate, the result is a poor quality, herky-jerky video that creates a strobe-light effect. Particularly below 15 fps, quality suffers noticeably, as the fluidity of motion is lost even though the image quality may be high. See also *scanning*.

frame relay The ITU-T I.122 Recommendation (1988), Framework for Providing Additional Packet Mode Bearer Services established the basic framework for a packet communications mode over an integrated services digital network (ISDN). The access protocol specified in ITU-T Q.922 (1992) is Link Access Procedure for Frame Mode Services (LAPF), which is an adaptation of the Link Access Procedure Data channel (LAPD) signaling protocol developed for ISDN. Frame relay was originally intended as an ISDN framing convention for a bearer service, i.e., information-bearing service, anticipated for the ISDN D channel. The D channel is intended primarily for signaling and control purposes, in support of Signaling System 7 (SS7). The D channel runs at 16 kbps for ISDN basic rate interface (BRI) and at 64 kbps for primary rate interface (PRI). Certainly during the early stages of frame relay development, 16 kbps was not a particularly limiting signaling rate, particularly in the context of X.25 packet switching, which often was limited to 9.6 kbps. Over time, however, it became clear that ISDN was far too slow for data communications, even at B channel rates of 64 kbps, and certainly at D channel rates of 16 kbps. So, frame relay became a distinct service, independent of ISDN. In the context of the OSI Reference Model, frame relay standards address the Physical Layer and Data Link Layer, and do not specify internal network operations.

Frame relay is analogous to a streamlined and supercharged version of X.25. Although both are designed to support bursty data traffic, frame relay is intended specifically for LAN-to-LAN traffic, but also is used in support of SDLC and many other legacy protocols. Access to a frame relay network is generally over a dedicated digital circuit in the form of a DDS, a Fractional T1, an E-1 (2.048 Mbps), or a T1 (1.544 Mbps). Access via E-3 (34 Mbps) or T3 (45 Mbps) circuits is also generally available. Frame relay statistically multiplexes frames of data over virtual circuits (VCs), with specifications providing for both permanent virtual circuits (PVCs) and switched virtual circuits (SVCs). (*Note:* SVCs are virtually non-existent because of their additional complexity and cost and the fear of carriers that such a service would cannibalize the PSTN.) The user interface is in a frame relay access device (FRAD) that can be implemented on the customer premises and is analogous to an X.25 packet assembler/disassembler (PAD). Like X.25, frame relay is intended for bursty data traffic, although it works well with fixed bit rate applications, for which it offers assured bandwidth. Although both X.25 and frame relay can support voice, video, and audio, the inherently unpredictable levels of latency and loss over such a highly shared network translate into quality of service (QoS) issues. As frame relay specifies a completely digital network, error performance is excellent. Therefore, frame relay does not attempt to correct any errors created in transit, but simply discards errored frames. It is the responsibility of the receiving user equipment to discover and recover from such an action. As frame relay guarantees frame delivery in the order sent, there is no frame sequence numbering, and there are no acknowledgements of any sort provided. As a result, the load on the computational and bandwidth resources of the network is reduced, frame processing and forwarding are speeded up considerably, and latency is reduced significantly. There are, however, a number of congestion control mechanisms that variously work to provide some assurances of acceptable performance. Whereas frame relay specifies a variable size payload up to 4,096 octets, the Frame Relay Forum (now MFA Forum) developed an Implementation Agreement (IA) that sets the maximum size at 1,600 octets for purposes of interconnectivity and interoperability. This frame size easily supports the largest standard 802.3 Ethernet frame of 1,518 octets. See also *B channel, BRI, Data Link Layer, D channel, DDS, E-1, E-3, Fractional T1, FRAD, frame, IA, ISDN, ITU-T, latency, LAPD, LAPF, MFA Forum, OSI Reference Model, packet switch, PAD, Physical Layer, PRI, PVC, QoS, SDLC, signaling and control, signaling rate, SS7, statistical time division mulltiplex, SVC, T1, T3,* and *X.25.*

frame relay access device (FRAD) See *FRAD.*

frame relay assembler/disassembler (FRAD) See *FRAD.*

Frame Relay Forum (FRF) A not-for-profit special interest group of manufacturers, vendors, carriers, and others with interests in the development and promotion of frame relay technology. The Frame Relay Forum developed a number of Implementation Agreements (IAs) that address interoperability issues. The Frame Relay Forum merged with the ATM Forum and MPLS Forum to form the MFA Forum. See also *frame relay*, *Implementation Agreement*, and *MFA Forum*.

frame relay network device (FRND) See *FRND*.

frames per second (fps) See *fps*.

Frame UNI (FUNI) See *FUNI*.

framing bit A bit that precedes a T1 or J1 frame and is used for synchronization and other control purposes. See also *bit*, *frame*, *J1*, and *T1*.

Freedom of Mobile Multimedia Access (FOMA) The Japanese term for Universal Mobile Telecommunications System (UMTS), a 3G cellular radio standard. See *UMTS*.

Freephone The term used in some countries to refer to toll free service. See also *toll free service*.

free space **1.** A theoretical region of space utterly devoid of physical matter, gravitational fields, and electromagnetic fields. **2.** A region devoid of physical obstructions that might hinder the propagation of electromagnetic signals. See also *free space transmission*.

free space optics (FSO) See *FSO*.

free space transmission The transmission of radio or optical signals in free space, i.e., space devoid of physical obstructions that might hinder signal propagation. In this context, the term physical obstruction suggests trees, buildings, hills, mountains, and other significant material objects. The term does not suggest atomic, molecular, or particulate matter that commonly is present in the atmosphere. Neither does it suggest water vapor, rain, snow, sleet, or hail. Free space transmission does not include radio or optical transmission through waveguides. See also *free space*, *FSO*, *transmission medium*, and *waveguide*.

freeware Software given away free of charge, typically over the Internet or through a user group. Individual software developers often make beta versions of copyrighted software available as freeware for user trials, but charge for subsequent versions refined as a result of end user input. Some developers never charge for its use and even forbid others to do so. Some freeware is available in commercial versions, with charges for support, but not for licensing. Freeware is generally not public domain software, as there usually are restrictions on its use and modification. See also *beta*, *copyright*, *open source*, *public domain*, and *software*.

frequency (f) The Institute of Electrical and Electronics Engineers (IEEE) defines frequency as the number of complete cycles of sinusoidal variation per unit time, with the unit of time generally being that of one second. Plotting $y = \sin x$, where x is expressed in radians, yields a sine wave as illustrated in Figure F-8. (From the Latin *radius*, a radian is a unit of plane angular measurement equivalent to the angle between two radii that enclose a section of a circle's circumference [arc] equal in length to the length of a radius. There are 2π radians in a circle.) A complete sine wave entails a cycle as measured from a point of zero (0) amplitude to a point of maximum positive amplitude (+A), through zero to a point of maximum negative amplitude (–A), and back to a point of zero (0) amplitude. The frequency of an alternating current, for example, is the half number of times the polarity of a continuous waveform reverses direction, or polarity, per second. The frequency of the waveform is the number of times per second it makes a complete cycle from zero voltage (0V) to its maximum positive voltage (+V) or amplitude, back to zero (0 V), to its maximum negative voltage (–V) or amplitude, and back to zero (0V). Frequency traditionally is measured in cycles per second (cps) or, in more contemporary terms, Hertz (Hz). See also *amplitude*, *electromagnetic spectrum*, *Hz*, and *sine wave*.

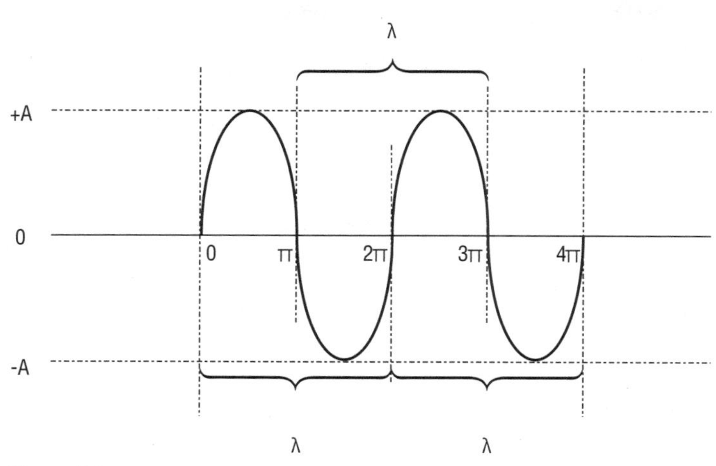

Figure F-8

frequency band A continuous group, or range, of frequencies with an upper limit and a lower limit. In analog terms, bandwidth and channel width are defined as a range of frequencies. See also *analog*, *bandwidth*, *channel*, and *frequency*.

frequency diversity The use of multiple paired transmit and receive antennas operating at different frequencies. As the likelihood is that the signals will not suffer the same level of attenuation at different frequencies, the receiver with the strongest signal assumes control of the transmission. Frequency diversity is sometimes employed in microwave systems. See also *attenuation*, *frequency*, and *microwave*.

frequency division duplex (FDD) See *FDD*.

frequency division multiple access (FDMA) See *FDMA*.

frequency division multiplexer (FDM) A device that performs Frequency Division Multiplexing (FDM). See *FDM*.

frequency division multiplexing (FDM) See *FDM*.

frequency-hopping spread spectrum (FHSS) See *FHSS*.

frequency modulation (FM) See *FM*.

frequency-shift keying (FSK) Also known as *Frequency Modulation* (FM). See *FM*.

frequency spectrum See *electromagnetic spectrum*.

Fresnel, Augustin-Jean (1788–1827) A French physicist who made significant contributions to the field of wave optics, particularly with respect to explaining the phenomenon of diffraction. See also *diffraction*, *Fresnel reflection*, *Fresnel refraction*, *optics*, and *wave*.

Fresnel reflection The reflection of a portion of the light incident on a planar surface (i.e., the interface) between two media having different refractive indexes. The reflected portion of the light remains in the same plane as the incident light and reflects at the same angle to the normal as the incident ray. The polarization is known as parallel. The extent to which the incident is reflected, rather than refracted, depends on the difference in the index of refraction (IOR) of the individual media (e.g., air and glass), the angle of incidence, the light polarization, and the direction in which the light is passing (e.g., from air to glass or from glass to air). See also *angle of incidence*; *Fresnel, Augustin-Jean*; *Fresnel refraction*; *IOR*; *polarization*; *reflection*; and *refraction*.

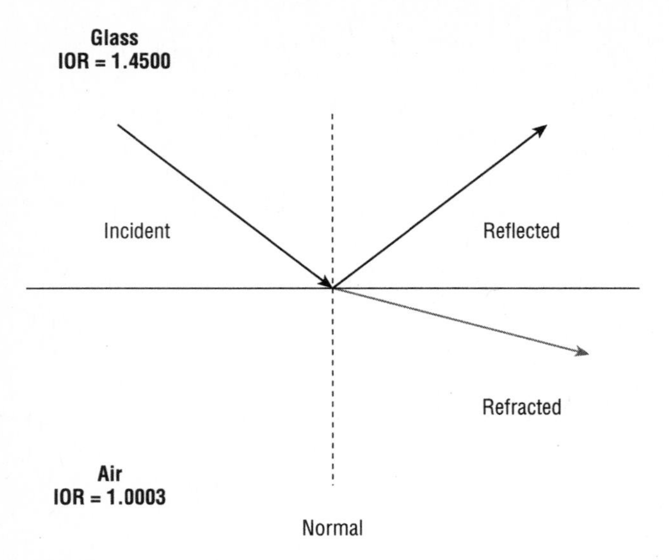

Figure F-9

Fresnel refraction The bending of light as it crosses the planar surface between two media having different refractive indexes. The refracted portion of the light crosses the interface and enters the plane of the incident medium, changes velocity, and bends while the reflected portion of the light remains in the same plane as the incident light, remains at the same velocity, and reflects at the same angle to the normal as the incident ray. The polarization is known as perpendicular. The extent to which the incident is refracted, rather than reflected, the difference in the index of refraction (IOR) of the individual media (e.g., air and glass), the angle of incidence, the light polarization, and the direction in which the light is passing (e.g., from air to glass or from glass to air). See also *angle of incidence*; *Fresnel, Augustin-Jean*; *Fresnel reflection*; *IOR*; *polarization*; *reflection*; and *refraction*.

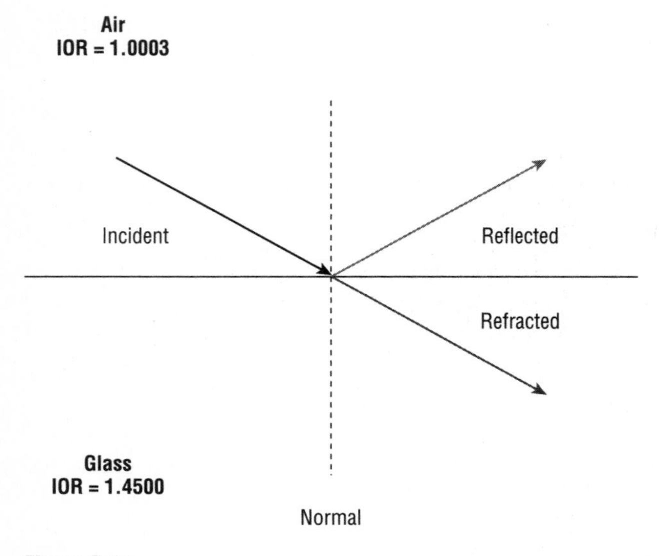

Figure F-10

Friis, Harald T. (1893–1976) An electrical engineer with a focus on radio systems, Friis was director of a research team at Bell Telephone Laboratories that developed the first microwave transmission systems. The first primitive systems, used in military applications in the European and Pacific theaters of World War II, could handle up to 2,400 voice conversations over five channels. The first public demonstration was conducted between the West Street lab and Neshanic, New Jersey in October 1945, and construction began on the first experimental microwave telephone network in 1947. Friis also was instrumental in the development of several radar systems and various millimeter-wave systems, and made fundamental contributions to the understanding of the role of noise in radio receivers. See also *Bell Labs*, *microwave*, *noise*, and *radar*.

FRND (Frame Relay Network Device) Pronounced *friend*. A device at the edge of the frame relay carrier network that provides the user network interface (UNI). The FRND connects across a digital local loop, such as an E-1 or T1, to the frame relay access device (FRAD) on the customer premises. See also *carrier*, *E-1*, *FRAD*, *frame relay*, *local loop*, *T1*, and *UNI*.

front end processor (FEP) See *FEP*.

FRS (Family Radio Service) In the United States, the Federal Communications Commission (FCC) regulates FRS as one of the family of CB (Citizens Band) Radio Services. FRS uses mobile handsets that operate in the 462.5625–467.7125 MHz band at a maximum power level of 0.5 watts (W) and have a range of less than 1 mile. FRS is intended for personal and business use by family, friends, and associates. Dual mode sets are available that also operate as General Mobile Radio Service (GMRS) radios. See also *CB Radio Service*, *FCC*, *GMRS*, *personal radio services*, and *watt*.

FSAN (Full Service Access Network) Pronounced *ef san*. An initiative of a consortium of vendors for a ATM-based passive optical network (APON). The specification subsequently was ratified by the ITU-T and incorporated within the G.983.1 standard. See also *APON*, *ATM*, *ITU-T*, and *PON*.

FSK (Frequency-Shift Keying) Also known as *frequency modulation* (FM). See *FM*.

FSO (Free Space Optics) Point-to-point airwave transmission systems that use focused infrared (IR) light beams between transmitters and receivers, much as microwave systems use focused radio beams. FSO requires optical line of sight (LOS) and will work through clear windows, depending on their chemical composition, but not through walls or other opaque objects. FSO transmitters use laser light sources in the form of either vertical cavity surface-emitting lasers (VCSELs) or laser diodes. FSO systems suffer from environmental interference, particularly fog, which absorbs, diffuses, and reflects the light beam, much as fog affects the beam from the headlights of an automobile. Under optimum conditions, transceiver separation is limited to about two to five kilometers, which limits the technology to short haul applications. Unlike radio systems, FSO systems are immune to electromagnetic interference (EMI) and radio frequency interference (RFI), require no licensing, and are inherently secure. FSO systems commonly operate at rates of 1.544 Mbps (T1), 2.048 Mbps (E-1), 34 Mbps (E-3) and 45 Mbps (T3), and are capable of rates up to 10 Gbps (OC-192). Systems running at up to 160 Gbps have been demonstrated in the labs. See also *absorption*, *airwave transmissions*, *diffusion*, *EMI*, *infrared*, *laser diode*, *LOS*, *reflection*, *RFI*, *short haul circuit*, and *VCSEL*.

FSS (Fixed Satellite System) A satellite that is at an altitude and in an orbit such that it maintains a fixed position relative to the Earth's surface. An FSS is in contrast to an MSS (Mobile Satellite System). FSS is synonymous with *geosynchronous earth orbiting* (GEO) satellite. See also *GEO*, *MSS*, and *satellite*.

FT1 (Fractional T1) A T1 circuit with some DS-0 channels disabled to provide a fraction of T1 capacity. FT1 generally is tariffed to provide 1, 2, 4, 6, 8, or 12 DS-0 channels. Subrate transmission also is available at speeds of 9.6 kbps. See also *DS-0*, *subrate*, and *T1*.

FTC (Federal Trade Commission) An independent United States government agency, reporting directly to Congress, the FTC was created in 1914 to prevent unfair methods of competition in commerce. In 1938, Congress passed the Wheeler-Lea Amendment, which included a broad prohibition against unfair and deceptive acts or practices. Since then, the Commission also has been directed to administer a wide variety of other consumer protection laws, including the Telemarketing Sales Rule, the Pay-Per-Call Rule and the Equal Credit Opportunity Act. In 1975, Congress passed the Magnuson-Moss Act, which gave the FTC the authority to adopt trade regulation rules that define unfair or deceptive acts in particular industries, and that have the force of law. The FTC is organized into the Bureaus of Consumer Protection, Competition, and Economics.

FTP **1.** Foil Twisted Pair. Synonymous with *shielded twisted pair* (STP). See *STP*. **2.** File Transfer Protocol (FTP). An Application Layer protocol of the TCP/IP protocol suite, FTP supports the exchange of files between two host computers across the Internet. FTP also supports interactive user interface in which humans must interact with a remote host. The specifics of the file type and data format (e.g., ASCII, EBCDIC, or binary notation; and compressed or uncompressed) can be determined from client to server. FTP also requires clients to satisfy security authorization measures in the form of login and password. FTP makes use of TELNET for control messages between the hosts, and relies on connection-oriented Transmission Control Protocol (TCP) for data transfer. FTP is defined in IETF RFC 959 (1985). See also *Application Layer, ASCII, binary notation, compression, connection-oriented, EBCDIC, host, IETF, Internet, login, password, protocol, protocol suite, TCP, TCP/IP,* and *TELNET.*

FTTC (Fiber-To-The-Curb) Also known as *FTTN* (Fiber-To-The-Node) A hybrid broadband local loop configuration that involves optical fiber from the edge of the telco or CATV network to an intelligent node at the curb, with one node serving perhaps many residences or small businesses. The few hundred meters or so of the local loop from the node to the premises generally is either unshielded twisted pair (UTP) in a telco application or coaxial cable (coax) in a CATV application, although some form of wireless technology is also possible. The fiber portion of a contemporary FTTC configuration generally conforms to one of the passive optical network (PON) standards. See also *broadband, CATV, local loop, node, optical fiber, PON, telco,* and *UTP.*

FTTN (Fiber-To-The-Neighborhood or **Fiber-To-The-Node)** Also known as *Fiber-To-The-Curb* (FTTC). A hybrid broadband local loop configuration that involves optical fiber from the edge of the telco or CATV network to an intelligent node in the neighborhood, with one node serving perhaps many residences or small businesses. The few hundred meters or so of the local loop from the node to the premises generally is either unshielded twisted pair (UTP) in a telco application or coaxial cable (coax) in a CATV application, although some form of wireless technology is also possible. The fiber portion of a contemporary FTTN configuration generally conforms to one of the passive optical network (PON) standards. See also *broadband, CATV, local loop, node, optical fiber, telco,* and *UTP.*

FTTP (Fiber-To-The-Premises) A broadband local loop configuration involving optical fiber from the edge of the telco or CATV network to the customer premises. A contemporary FTTP loop generally conforms to one of the passive optical network (PON) standards. See also *broadband, CATV, local loop, optical fiber, PON,* and *telco.*

FUBAR (Fouled Up Beyond All Recognition) Referring to something so mangled as to be unrecognizable. The term is generally attributed to U.S. military slang circa WWII. In contemporary usage, the spelled-out phrase is a bit more colorful and much less polite. (*Note:* This is a polite book, so you'll just have to use your imagination or search elsewhere for the more colorful version.) See also *SNAFU.*

FUD (Fear, Uncertainty, and Doubt) factor A marketing strategy used by a dominant company to confuse and freeze the competition by either maintaining secrecy with respect to future product or service plans or by changing them frequently and, thereby, creating concerns that competitors' products might not be compatible with those of the dominant vendor. The term was coined by Gene Amdahl after he left IBM to found Amdahl Corporation. Amdahl was quoted as saying "FUD is the fear, uncertainty, and doubt that IBM salespeople instill in the minds of potential customers who might be considering Amdahl products."

full duplex (FDX) See *FDX.*

full mobile number portability (FMNP) Synonymous with *wireless number portability* (WNP). See *WNP.*

full mode A mode defined by the ITU-T T.37 standard for Fax over Internet Protocol (FoIP). Full-mode extensions include mechanisms for ensuring call completion through negotiation of capabilities between transmit and receive devices. Full-mode also provides for delivery confirmation. See also *facsimile, FoIP, resolution, simple mode, T.37,* and *TIFF-F.*

full rate ADSL (full rate Asymmetric Digital Subscriber Line) See *ADSL.*

Full Service Access Network (FSAN) See *FSAN.*

FUNI (Frame User Network Interface) A derivative of the DXI standard, FUNI extends asynchronous transfer mode (ATM) access to smaller sites with relatively low speed access at rates from 56 kbps to 1.544 Mbps (T1). Data enters a router, which encapsulates the data in frames similar to those employed in frame relay. The router then forwards the data to an ATM switch, which converts them to cell format. See also *ATM, cell, DXI, frame, frame relay, router, switch,* and *UNI.*

further **1.** Additional; to a greater extent. **2.** More distant in place or time, especially where there is no notion of physical distance. I don't know about you, but I was always taught to use farther when referring to physical distances, and to use further otherwise. Anyway, it drives me nuts when people, especially weathermen, point to the map and talk about further north or further east. It sounds affected, like they are trying to impress us ignorant viewers at home. See also *farther* and *utilize.*

fusion splice In optical fiber installation and repair, the permanent joining of two fibers by melting the glass and allowing the molten ends to meld. *Note:* The melting point of glass is approximately 2000 degrees Celsius. See also *mechanical splice, optical fiber,* and *splice.*

Future Public Land Mobile Telecommunications System (FPLMTS) An initiative of the ITU-R that defined a vision for a single global standard for digital wireless networks. FPLMTS was subsequently replaced by International Mobile Telecommunications-2000 (IMT-2000). See also *IMT-2000.*

FWM (Four-Wave Mixing) A type of noise created in fiber optic transmission systems (FOTS) when wavelengths interact to create additional wavelengths. FWM occurs in optical fibers engineered with a zero dispersion point at or near the wavelengths being transmitted. The number of additional wavelengths that can result is determined by the following formula:

$$FWM = N^2 (N - 1)/2$$

Based on this formula, 2 wavelengths will produce 2 additional wavelengths, 4 wavelengths will produce 24 additional wavelengths, and 8 wavelengths will produce 224 additional wavelengths. FWM can particularly be a problem at 1550 nm in systems using dense wavelength division multiplexing (DWDM), and especially when erbium-doped fiber amplifiers (EDFAs) are employed. The EDFAs amplify the additional wavelengths and superimpose them on the original wavelengths. Non Zero Dispersion-Shifted Fiber (NZDF) addresses this issue by shifting the optimal dispersion point slightly away from the 1550 nm range where DWDM operates. Fiber fabricated to have a zero dispersion point at 1540 nm or 1560 nm, for example, will reduce dispersion significantly and eliminate the issue of four-wave mixing. See also *amplification, dispersion, DWDM, EDFA, NZDF,* and *wavelength.*

Fx (Firefox) See *Firefox.*

FX (Foreign eXchange) circuit Also abbreviated as FEX circuit. A dedicated circuit that draws dial tone from a foreign exchange, i.e., a central office exchange (COE) other than the one in which the system resides. An FX line (KTS) or trunk (PBX) allows a user organization to avoid toll charges for long distance calls, as the leased line or trunk is billed on a flat rate basis, sensitive to bandwidth and distance, rather than a usage sensitive basis. If the user organization places a great many calls to a foreign exchange area, the costs savings can be considerable, particularly if the foreign exchange is in reasonable proximity and the FX circuit, therefore, is a relatively short haul. The interface typically found at each end of the FX circuit is of two types. FXO faces the office or switch, receives battery and ringing, and seizes the line by closing the loop. FXS faces the subscriber, gives battery and ringing voltage, and responds to a phone going off hook.

G **1.** (Giga). From the Greek *gigas*, meaning *giant*, translates to *billion*. **2.** In transmission systems and internal computer busses, GHz (GigaHertz) is a billion (10^9) Hertz and Gbps (Gigabit per second) is a billion (10^9) bits per second. In transmission systems, therefore, a billion is exactly 1,000,000,000 since the measurement is based on a base 10, or decimal, number system. **3.** Hard disks and flash drives measure computer storage in SI units, which also are based on the base 10, or decimal, system, so GB is one billion (10^9) bytes, or 1,000,000,000 bytes. **4.** Internal computer memory is based on a base 2, or binary, number system. A GB of internal memory, therefore is 1,073,741,824 (2^{30}) bytes. The term GB comes from the fact that 1,073,741,824 is nominally, or approximately, 1,000,000,000. See also *byte*, *Hertz*, and *SI*. **5.** G. The G series of ITU-T Recommendations. The G series addresses transmission systems and media, digital systems and networks. See *G series*.

G.711 The ITU-T Recommendation for pulse code modulation (PCM) voice encoding. See *G series*, *ITU-T*, and *PCM*.

G.721 The ITU-T Recommendation for adaptive differential pulse code modulation (ADPCM) voice coding and compression of high-fidelity 7 kHz voice at 64/56/48 kbps. See *G series* and *ITU-T*.

G.722 The ITU-T Recommendation for adaptive differential pulse code modulation (ADPCM) voice coding and compression of high-fidelity 7 kHz voice at 64/56/48 kbps. See also *ADPCM*.

G.726 The ITU-T Recommendation for adaptive differential pulse code modulation (ADPCM) voice encoding at 40, 32, 24, and 16 kbps. See also *ADPCM*, *G series*, and *ITU-T*.

G.728 The ITU-T Recommendation for low-delay code excited linear prediction (LD-CELP) coding and compression of 3.3 kHz voice at 16 kbps. See also *compression*, *G Series*, *ITU-T*, and *LD-CELP*.

G.729 The ITU-T Recommendation for conjugate-structure algebraic-code-excited linear-prediction (CS-CELP) voice coding and compression at 8 kbps. See also *compression*, *CS-CELP*, *G Series*, and *ITU-T*.

gain An increase in signal power between two points, achieved by an active device or system such as an amplifier, which receives an attenuated input signal, applies controlled power to that signal, and outputs a signal that is a function of the input signal, but at a higher power level. Gain is the opposite of attenuation. The gain, or increase, in signal power is typically described in positive decibels (+dB). See also *amplifier*, *attenuation*, and *dB*.

GAN (Generic Access Network) Previously known as *Unlicensed Mobile Access* (UMA). The 3rd Generation Partnership Project (3GPP) global standard for the use of Wi-Fi (802.11) unlicensed spectrum to provide mobile handsets with seamless connectivity between wireless local area networks (WLANs) and both cellular and fixed mobile wide area networks (WANs). The ultimate GAN goal is Fixed Mobile Convergence, which is the seamless melding of fixed IP-based fixed wireless (e.g., Wi-Fi and WiMAX) and cellular radio networks. See also *3GPP*, *802.11*, *cellular radio*, *fixed wireless*, *WAN*, *Wi-Fi*, *WiMAX*, and *WLAN*.

gatekeeper In H.323-compliant networks, a central point of control in a zone. Endpoints may communicate directly, in either a unicast or a multicast environment, if no gatekeeper is present. If a gatekeeper is present, all endpoints in its zone must register with it. The gatekeeper performs the function of admission control, determining if devices are authorized to connect and if there is sufficient bandwidth to support the call. Gatekeepers serve to translate LAN addresses into IP or IPX addresses, as defined in the Registration/Admission/Status (RAS) specification. Gatekeepers also can act to route H.323 calls through gateways, if necessary, and monitor the network bit rate capacity, with the ability to deny access to a session if programmable bandwidth thresholds have been reached or exceeded. Gatekeepers also can perform certain administrative functions, such as accounting, billing, directory, and collecting network usage data. Gatekeepers may be distinct network elements (NEs), or gatekeeper functionality can be incorporated into multipoint control units (MCUs). See also *bit rate*, *endpoint*, *H.323*, *IP*, *IPX*, *LAN*, *MCU*, *multicast*, *NE*, *RAS*, and *unicast*.

gateway **1.** A node that interconnects two or more disparate networks, both physically and logically, serving as a protocol converter (e.g., PSTN to IP) and media converter (e.g., electrical twisted pair to optical fiber) as necessary. See also *IGF*, *node*, and *protocol converter*. **2.** The collection of hardware and software required to interconnect two or more disparate networks, including performing protocol conversion. See also *hardware*, *protocol converter*, and *software*. **3.** In H.323-compliant multimedia networks, a gateway is an optional element used for various levels of protocol conversion. The gateway serves as a protocol converter between devices and networks that have native H.323 capability and those that do not. The gateway also may translate between audio, video, and data formats, and may perform signaling conversions between the H.225 packet protocol and external protocols such as SS7 and Q.931. Alternatively, signaling conversions may be performed by gatekeepers, call processors, or session border controllers. See also *H.225*, *H.323*, *multimedia*, *network*, *packet*, *protocol*, *protocol converter*, *Q.931*, *session*, and *SS7*.

gauge The measure of the diameter, or thickness, of a conductor. The thicker the wire, the less the resistance, the stronger the signal over a given distance, and the better the overall performance of the medium. Thicker wires also offer the advantage of greater break strength. Thicker wires, however, also require more metal, which makes them heavier and more difficult to bend, which ultimately increases both acquisition and deployment costs. By way of example, the first long-line copper wire telephone circuits were strung between New York and Chicago. Consisting of uninsulated hard drawn copper conductors about as thick as a pencil, the two-wire circuit weighed 870,000 pounds, filled a twenty-two car freight train and cost US$130,000 for the copper alone. The most commonly used measurements of gauge are American Wire Gauge (AWG), Imperial Standard Wire Gauge, and metric gauge. See also *AWG*, *break strength*, *Imperial Standard Wire Gauge*, *metric gauge*, and *resistance*.

Gaussian frequency-shift keying (GFSK) See *GFSK*.

Gaussian minimum-shift keying (GMSK) See *GMSK*.

Gaussian noise White noise that has a probability density graphed as a normal distribution, or Gaussian distribution, also known as a bell curve because of its bell-like shape. The Gaussian distribution is named for Carl Friedrich Gauss (1777–1855), a great German mathematician, physicist, and scientist in fields including number theory, differential geometry, magnetism, and optics. See also *noise* and *white noise*.

GB (GigaByte) A billion bytes, often shortened to *gig* in conversation. **1.** Hard disks and flash drives measure computer storage in SI units, which are based on the base 10, or decimal, system, so GB is one billion (10^9) bytes, or 1,000,000,000 bytes. See also *byte*, *decimal*, and *SI*. **2.** Internal computer memory is based on a base 2, or binary, number system. A GB of internal memory, therefore is 1,073,741,824 (2^{30}) bytes. The term GB comes from the fact that 1,073,741,824 is nominally 1,000,000,000. See also *byte*, *G*, and *SI*.

GbE (Gigabit Ethernet) Also known as GigE and GigEnet. See *Gigabit Ethernet*.

Gbps (Gigabit per second) A billion (10^9) bits per second. A measure of bandwidth in a digital transmission system. See also *bps* and *G*.

GBps (GigaByte per second) A billion (10^9) bytes per second. A measure of bandwidth in select digital transmission systems (e.g., Fibre Channel) oriented towards storage area networks (SANs). See also *Bps*, *byte*, *Fibre Channel*, *G*, *GB*, and *SAN*.

G.dmt A term sometimes used for asymmetric digital subscriber line (ADSL), the term G.dmt derives from the fact that ADSL is specified in the ITU-T G Series Recommendations as G.922.1 and employs discrete multitone (DMT) modulation. See also *ADSL*, *DMT*, *G Series*, *ITU-T*, and *modulation*.

G.dmt.bis The term applied to asymmetric digital subscriber line version 2 (ADSL2) while it was under development. The term G.dmt.bis derives from the fact that ADSL is specified in the ITU-T G Series Recommendations as G.992.3 and G.992.4, employs discrete multitone (DMT) modulation, and is the second (bis) version. See also *ADSL, DMT, G Series, ITU-T,* and *modulation.*

Géant2 (Translated from French as Giant2.) An ultra fast European fiber optic backbone connecting some 5,000 institutions across Europe, including many high schools, in the European Union, plus Russia, Switzerland, Turkey, and Israel. Géant2 was lit in Milan in June, 2005, with the €186 million funded half by the European Union and half by member nations. Géant2 is the European equivalent of the U.S. Internet2. See also *Internet2.*

geek **1.** A performer in carnival sideshows whose act consists of grotesque or depraved acts, such as biting the heads off live chickens or snakes. The origin of the term is uncertain. Some suggest it has roots in the Low German *gek*, meaning *fool.* Others suggest it is echoic of unintelligible cries, much like eek. **2.** Anyone considered to be socially awkward, especially those with excessive interests in computers or related technologies. See also *nerd* and *techie.*

gender bender See *gender changer.*

gender changer Synonymous with gender bender. A device that converts a male connector, i.e., a connector with pins, to a female connector, i.e., a connector with sockets, in order that they can be mated. There are at least two quite graphic analogies that would work here, but this dictionary is rated for children of all ages.

Generalized MPLS (GMPLS) See *GMPLS.*

General Mobile Radio Service (GMRS) See *GMRS.*

General Packet Radio Service (GPRS) See *GPRS.*

General Radio Service In Canada, a private, two-way voice communication radio service for use by the general public on an unlicensed basis for personal and business activities. The identical service is known in the United States as Citizens Band (CB) Radio. See also *CB Radio Service.*

general release Referring to a product, usually a software product, that is ready for unconditional release to the general public. Such a product generally has completed pre-release beta testing by selected customers in real-world situations. See also *beta.*

Generic Access Network (GAN) See *GAN.*

generic flow control (GFC) See *GFC.*

Generic Routing Encapsulation (GRE) See *GRE.*

generic Top Level Domain (gTLD) See *gTLD.*

Geneva Convention In 1864, a number of world leaders, statesmen, and diplomats representing 16 nations convened in Geneva, Switzerland to devise and document a set of rules of engagement designed to outlaw the atrocities of war and, thereby, lessen the suffering of both combatants and noncombatants. Commonly known as the Geneva Convention, the conference was formally titled Convention for The Amelioration of The Condition of The Wounded in Armies in The Field. The resulting document has since been ratified, clarified, and expanded, most recently in 1977. Actually, there are four Geneva Conventions, signed on August 12, 1949. There are two additional Protocols dated June 8, 1977. The Geneva Conventions are referenced during times of war, particularly when organized armies of civilized nations are engaged in military conflict. The additional protocols apply to domestic conflicts such as civil wars. Those who fail to follow the rules must be held accountable by an international court or tribunal. Unfortunately, any rules of engagement or warriors' codes are unilateral when armies of civilized nations are in military conflict with barbarians, even when they claim to be engaged in jihad. Civilized peoples establish rules and live by them, while barbarians do not. See also *rules of engagement*, *voice mail jail*, and *warrior's code*.

GEO (Geosynchronous Earth Orbit) At a position approximately 22,235 statute miles (35,784 km) above the equator, a satellite in orbit is in synchronization with the revolution of the earth. In other words, the satellite rotates around the earth at the same speed as the Earth rotates on its axis. As a result, the satellite maintains its relative position over the same spot of the earth's surface and is kept from wandering by tidal forces. At that altitude, the centrifugal force acting to fling the satellite away from the Earth is equal to, and therefore cancels out, the centripetal force of gravity acting to pull the satellite into the Earth. Therefore, the satellite maintains its vertical distance, or altitude. As a GEO maintains a stable position in the heavens, transmit and receive earth stations can be pointed to fixed coordinates to establish a communications link, secure in the knowledge that the satellite will be there. A Geosynchronous Earth Orbiting (GEO) satellite is also known as a Fixed Satellite System (FSS) because of its fixed positions relative to the Earth's surface. Neither a Low Earth Orbit (LEO) nor a Medium Earth Orbit (MEO) offers this level of stability. See also *FSS*, *LEO*, *MEO*, and *satellite*.

geostatic earth orbit (GEO) Synonymous with geosynchronous earth orbit (GEO). See *GEO*.

geostationary earth orbit (GEO) Synonymous with geosynchronous earth orbit (GEO). See *GEO*.

geosynchronous earth orbit (GEO) See *GEO*.

germanium (Ge) A grayish-white semi-metallic element (No. 32 in the Periodic Table of Elements) used extensively in semiconductors transistors, diodes, and rectifiers. Fiber optic diodes commonly are made of germanium due to its efficient response to infrared light. Germanium commonly is used as a dopant in the manufacture of silica-based optical fibers.

GFC (Generic Flow Control) In the asynchronous transfer mode (ATM) cell header, 4 bits that provide local flow control. As intermediate ATM switches overwrite the GFC field with virtual path identifier (VPI) data, the GFC data has no significance on an end-to-end basis. See also *ATM*, *cell*, *flow control*, *header*, *switch*, and *VPI*.

GFR (Guaranteed Frame Rate) In asynchronous transfer mode (ATM), a class of traffic intended to support non real-time (nrt) applications that may require a minimum rate guarantee and can benefit from accessing additional bandwidth dynamically, as it becomes available. GFR does not require adherence to a flow control protocol. The GFR service guarantee is based on AAL5 protocol data units (PDUs), also known as *frames*. During periods of network congestion, GFR attempts to discard entire frames, rather than cells that are segments of frames. Traffic descriptors in both the forward and backward directions include maximum frame size (MFS), burst cell tolerance (BCT), minimum cell rate (MCR), and peak cell rate (PCR). Specific applications have yet to be identified. ATM also defines available bit rate (ABR), constant bit rate (CBR), non real-time Variable Bit Rate (nrt-VBR), real-time Variable Bit Rate (rt-VBR), unspecified bit rate (UBR), and variable bit rate (VBR) traffic classes. See also *AAL5*, *ABR*, *ATM*, *BCT*, *CBR*, *cell*, *frame*, *MCR*, *MFS*, *nrt*, *nrt-VBR*, *PCR*, *PDU*, *QoS*, *real-time*, *rt-VBR*, *UBR*, and *VBR*.

GFSK (Gaussian Frequency-Shift Keying) A type of frequency-shift keying (FSK) that uses a Gaussian filter to smooth out frequency deviations by limiting their spectral width in a process generically known as pulse shaping. GSFK is used by Bluetooth and DECT systems. See also *Bluetooth*, *DECT*, *FSK*, *Gaussian noise*, *pulse shaping*, and *spectral width*.

ghosting In broadcast television and poorly installed cable television, the effect of seeing faint duplicate images rather than a single strong image. Ghosting is the result of multipath fading as the signal breaks into multiple elements that travel different routes from transmitter to receiver and, therefore, arrive at different times. See also *broadcast television*, *cable television*, and *multipath fading*.

GHz (GigaHertz) A billion (1,000,000,000) Hertz. A measure of bandwidth in an analog transmission system. See also *Hertz* and *G*.

.gif The file extension that identifies files in the Graphics Interchange Format (GIF). See also *GIF*.

GIF (Graphics Interchange Format) Pronounced *jif*, like the peanut butter. A graphics file format for encoding and exchanging graphic files on the Internet. Originally developed by CompuServe, GIF includes a patented lossless compression technique known as LZW, but is limited to 8-bit format, which makes it suitable for grayscale, but unsuitable for full color images. See also *compression*, *grayscale*, *JPEG*, *lossless compression*, *LZW*, and *PNG*.

Giga- (G) See *G*.

Gigabit Ethernet (GbE, GigE, and GigEnet) Standardized by the IEEE (1998) as IEEE 802.3z, GigE is a backbone Ethernet solution that operates at 1 Gbps and is fully compatible with 10/100-Mbps Ethernet. GigE is available in both shared bus and switched versions, both of which support multiple ports that can run at 1 Gbps in full-duplex (FDX). Specifications include shared bus hubs, matrix switches, and transmission media. A GigE shared bus solution is a high-speed hub that uses CSMA/CD for medium access control (MAC) and can run at a speed of several Gbps. Switched GigE buffers incoming Ethernet frames, passing them to the output port when the shared bus becomes available. More substantial switched GigE products offer nonblocking matrix switching. Transmission media specifications include twisted pair (1000Base-T) and fiber optics (1000Base-LX and 1000Base-SX), and employs the same 8B/10B line coding technique used in ESCON and Fibre Channel.

Although GigE is much like the predecessor 10/100-Mbps Ethernet versions, differences include frame size. As the clock speed of GbE is one or two orders of magnitude greater, issues of roundtrip propagation delay affect error detection. To avoid potentially disastrous collision rates, the minimum frame size is increased from 64 octets to 512 octets, which generally is equivalent in duration to transmitting a 64-byte frame at 100 Mbps. The larger minimum frame size provides the same time for the transmitting device to receive and interpret a collision notification. Although nonstandard, some manufacturers have increased the maximum frame size from 1,518 bytes to a jumbo frame size of 9,000 bytes, which improves throughput as there are fewer frame headers to be analyzed and fewer inter-frame intervals. GigE is found not only in local area network (LAN) backbones, but also is used by service providers in metropolitan area networks (MANs). See also *1000Base-LX*, *1000Base-SX*, *1000Base-T*, *8B/10B*, *backbone*, *CSMA/CD*, *ESCON*, *Ethernet*, *FDX*, *Fibre Channel*, *LAN*, *LAN switch*, *MAC*, and *MAN*.

gigabit passive optical network (GPON) See *GPON*.

Gigabit per second (Gbps) See *Gbps*.

gigabyte (GB) See *GB*.

GigaHertz (GHz) See *GHz*.

GigaPOP (Gigabit-per-second Point Of Presence) The point at which a carrier establishes a physical presence in a geographic area with a switch, router, or other device offering access to a fiber optic transmission system (FOTS) supporting bandwidth in the Gbps range. See also *bandwidth*, *carrier*, *FOTS*, *Gbps*, *POP*, *router*, and *switch*.

GigE (Gigabit Ethernet) See *Gigabit Ethernet*.

GigEnet (Gigabit Ethernet) See *Gigabit Ethernet*.

GII (Global Information Infrastructure) Synonymous with International Information Infrastructure (III). The international version of the National Information Infrastructure (NII), or Information Superhighway. See also *Information Superhighway* and *NII*.

glare The condition that arises when a telephone line or trunk is seized at both ends for different reasons, perhaps causing the collision between an incoming call and an outgoing call, for example. Glare is a phenomenon associated with loop start signaling used to support single-line telephones, multi-line telephones, and key telephone systems (KTSs). When the handset of the telephone is lifted, the electrical loop is completed and current flows across the circuit. The central office switch detects that fact and returns dial tone for an outgoing call, or connects an incoming call, as appropriate. If the user picks up the handset to place an outgoing call at the same time that the central office switch is attempting to connect an incoming call, a collision, or glare condition, occurs. See also *ground start*, *loop*, and *loop start*.

glass optical fiber (GOF) See *GOF*.

G.lite Also known as ADSL Lite, universal ADSL, and splitterless ADSL. Specified by the American National Standards Institute (ANSI) in T1.413 and standardized by the ITU-T in Recommendation G.992.2, G.lite is an interoperable extension of asymmetric digital subscriber line (ADSL). Three deployment options exist, all of which support simultaneous voice and data communications over a single unshielded twisted pair (UTP) local loop comprising one physical pair. One option involves a centralized splitter and another, distributed splitters. The third, and most significant option, is that of splitterless DSL, supporting high-frequency data communications through an ADSL modem not requiring professional installation. G.lite operates on an asymmetric basis over local loops up to 18,000 feet in length, at signaling rates of up to 1.544 Mbps (T1) downstream and up to 512 kbps upstream, sensitive to loop characteristics. Upstream and downstream speeds both are selectable in increments of 32 kbps. Because of bandwidth limitations, G.lite never gained much traction in the market, but introduced the splitterless concept. See also *ADSL*, *ANSI*, *asymmetric*, *bandwidth*, *downstream*, *G Series*, *ITU-T*, *local loop*, *modem*, *upstream*, and *UTP*.

Global Information Infrastructure (GII) Synonymous with International Information Infrastructure (III). The international version of the National Information Infrastructure (NII), or Information Superhighway. See also *Information Superhighway* and *NII*.

global positioning system (GPS) See *GPS*.

global switched telephone network (GSTN) See *GSTN*.

Global System for Mobile Communications (GSM) See *GSM*.

global unicast address In Internet Protocol version 6 (IPv6), a conventional, publicly routable address that can be used in the Internet or any public domain that is associated with a single node, and can, in effect, identify the node. See also *domain*, *Internet*, *IPv6*, *link-local address*, *router*, *unicast*, and *unicast address*.

GMPLS (Generalized MultiProtocol Label Switching) A protocol that extends MPLS beyond packet switched interfaces to include Time Division Multiplexing (TDM), Wavelength Division Multiplexing (WDM), and Add/Drop Multiplexing (ADM). GMPLS adds the concept of label switching to photonics at the lambda level in a Wavelength Division Multiplexing (WDM) system, to time slots in a SONET/SDH system, and to physical optical fibers in an optical cross-connect (OXC). GMPLS is designed to speed the provisioning of end-to-end traffic-engineered paths in the TDM and optical domains, much as MPLS has done in the Internet Protocol (IP) domain. GMPLS was known as Multiprotocol Lambda Switching before it was extended to other media. See also *ADM*, *domain*, *lambda*, *optical fiber*, *OXC*, *packet switch*, *photonics*, *physical*, *PXC*, *SDH*, *SONET*, *TDM*, *time slot*, *traffic engineering*, *wavelength*, and *WDM*.

GMRS (General Mobile Radio Service) In the United States, a land-mobile licensed radio service designed for short distance two-way communications to facilitate the activities of an adult individual and his or her family members. GMRS radios operate at in the 462–467 MHz band at a power level of 1–5 watts and may have detachable antennas. Dual mode sets are available that also operate as Family Radio Service (FRS) radios. The Federal Communications Commission (FCC) regulates FRS and GMRS, both of which are in the family of personal radio services. See also *FCC*, *FRS*, and *personal radio services*.

GMSK (Gaussian Minimum-Shift Keying) A modulation technique that is a variant of frequency shift keying (FSK) in which the signal is smoothed with a low-pass Gaussian filter before being placed on the carrier. This process reduces the spectral width and minimizes co-channel interference. GMSK is specified for use in GSM cellular radio networks. See also *channel, filter, FSK, Gaussian noise, GSM, interference, modulation, signal,* and *spectral width.*

GMT (Greenwich Mean Time) Civilian terminology for Universal Coordinated Time (UTC), which is known as Zulu Time (Z) for military and aviation purposes. See also *UTC* and *Zulu Time.*

GOF (Glass Optical Fiber) A slender strand of extremely pure glass, specially constructed to serve as a conductor, or waveguide, for infrared (IR) light signals in a fiber optic transmission system (FOTS). There are a large number of GOF types, all of which support very high signaling rates over considerable distances with very low attenuation and, therefore, excellent error performance. The fiber comprises a core of pure silica doped with germanium or some other substance to alter its index of refraction (IOR) to slow the velocity of propagation (Vp), i.e., to slow down the speed of light, by approximately one-third to about 200,000 kilometers per second, or 124,000 miles per second. Surrounding the inner core is the cladding, which consists of pure silica doped in such a way as to have a lower refractive index, i.e., index of refraction (IOR), which translates into increased purity and, therefore, enhanced propagation speed, as compared to the core. A step-index fiber is characterized by a sharp step in the IOR at the core/cladding interface, which serves to reflect errant signal components back into the core. A graded-index fiber is characterized by cladding comprising many layers of doped silica that are gradually and successively lower in IOR, which construction serves to gradually refract, or bend, any errant signal components back into the core. In either case, the ultimate effect is that the light signal is essentially confined to propagate through the core through a process known as *total internal reflection.* Graded-index multimode fiber (MMF) generally is used in short haul applications requiring bandwidth of 1 Gbps or less, such as local area networks (LANs). Step-index single-mode fiber (SMF) generally is used in long haul, high-bandwidth applications such as wide area networks (WANs). Plastic optical fiber (POF) sometimes is used over short distances in low bandwidth applications where its flexibility, general durability, and low cost are advantageous. See also *diffraction, graded index fiber, IOR, IR, MMF, POF, propagation, SMF, step-index fiber,* and *Vp.*

goldenrod A term used to describe the thick coaxial cable specified by the IEEE as 10Base5, for use in early Ethernet networks. The term was in reference to the high cost and high value of the cable, as well as the yellow cable sheath used by some manufacturers. The cable also was about as thick and inflexible as a rod. See also *10Base5, coaxial cable, Ethernet,* and *IEEE.*

goodput The amount of useful data, user data, or payload that can be processed by, passed through, or otherwise put through a system when operating at maximum capacity and received at the correct destination address, minus any packet headers or other overhead, minus any information lost or errored in transit, and minus any duplicate transmissions or retransmissions. Goodput can be thought of as throughput seen by the receiver. Goodput is usually less than throughput, which is always less than bandwidth. See also *bandwidth* and *throughput.*

Gopher An early text-based browser developed at the University of Minnesota, where the Golden Gopher is the school mascot. As Gopher is designed to go for information as it tunnels through the Internet to dig for data, the name is a multi-layer pun. Gopher was developed as a user interface to ease access to server resources in educational institutions and quickly became a de facto standard. Gopher servers enable the user to access a directory of Gopher server sites, click the name of the server, browse its file resources on the basis of nested menus, and download files using the File Transfer Protocol (FTP). Gopher has been obsoleted by the World Wide Web (WWW), although a number of Gopher servers remain in service. See also *browser, de facto, FTP, Internet, server, standard,* and *WWW.*

Gopherspace The virtual space comprising the resources contained on Gopher-accessible servers. A search tool known as *Veronica* is used to develop an index of searchable titles in Gopherspace. See also *Gopher, server, Veronica,* and *virtual.*

GoS (Grade of Service) **1.** In the PSTN or with respect to a PBX, the probability of a call being blocked or queued for some period of time due to limited system resources during the busy hour of the day. GoS is expressed as a decimal fraction. For example, P.03 indicates that there is a probability (P) that three (3) in every one hundred (100) call attempts will encounter a blockage condition (i.e., will fail to gain access to the system) during the busy hour of the day. See also *traffic* and *traffic engineering*. **2.** In a packet network, a generic priority or similar preference given to a user, a location, or a type of traffic.

GOSIP (Government Open Systems Interconnection Profile) A United States government specification, published in 1990, that essentially required all government networking products to be compliant with the Open Systems Interconnection (OSI) protocols. In 1995, a new directive modified that position by allowing federal government agencies to acquire products in compliance with IETF, ITU-T, or ISO protocols. The GOSIP initiative fostered a significant movement towards standards-based, rather than proprietary, solutions. See also *IETF*, *ISO*, *ITU-T*, *OSI*, *protocol*, and *standards*.

.gov (government) Pronounced *dot gov*. The Internet generic Top Level Domain (gTLD) reserved exclusively for United States government agencies. This domain is unsponsored. See also *gTLD*, *Internet*, and *unsponsored domain*.

Government Open Systems Interconnection Profile (GOSIP) See *GOSIP*.

GPON (Gigabit Passive Optical Network) The PON specification described by the ITU-T in G.984 (2004), GPON operates in both asymmetric and symmetric configurations, and currently supports signaling rates as high as 2.488 Gbps with a maximum logical reach of approximately 60 kilometers (37 miles). GPON supports as many as 32 or 64 splits, that is, splitters can divide the signal to serve as many as 32 or 64 premises from a single optical fiber, and expectations are that as many as 128 splits will be supported in the future. At full speed of 2.488 Gbps with the current maximum of 64 splits, each premises has access to sustained bandwidth of more than 35 Mbps, which is far beyond that offered by other access technologies. GPON supports voice, data, and video in ATM format. GPON also supports voice in native PCM/TDM format and data in Ethernet format, and employs wavelength division multiplexing (WDM) for downstream transmission. GPON specifies advanced encryption standard (AES) to secure downstream transmissions. PON variants also include ATM-based PON (APON), broadband PON (BPON), and Ethernet-based PON (EPON). See also *AES*, *APON*, *asymmetric*, *ATM*, *bandwidth*, *BPON*, *EPON*, *Ethernet*, *ITU-T*, *logical reach*, *optical fiber*, *PCM*, *PON*, *splitter*, *symmetric*, and *TDM*.

GPRS (General Packet Radio Service) The 2.5G data service enhancement for GSM host networks. GPRS is a packet-switched service that takes advantage of available GSM time slots for data communications and supports both X.25 and TCP/IP packet protocols, with quality of service (QoS) mechanisms. GPRS enables high-speed mobile datacom usage and is considered most useful for bursty data applications such as mobile Internet browsing, e-mail, and various push technologies. Through concatenating as many as eight GSM channels, GPRS yields a theoretical transmission rate as high as 171.2 kbps. GPRS maintains the same Gaussian minimum shift keying (GMSK) modulation scheme used by GSM and provides always-on access. GPRS can run in either symmetric or asymmetric mode, with the speed in either direction sensitive to which of the 12 multislot service classes is selected. The multislot service class determines the number of time slots in each direction, with each time slot supporting a theoretical data rate of 21.4 kbps. The simplest is service class 1, which supports one time slot in each direction. The most capable is service class 12, which supports four time slots in each direction. GPRS specifications were developed in 1997 by the European Telecommunications Standards Institute (ETSI), which subsequently passed that responsibility to the 3rd Generation Partnership Project (3GPP). In Enhanced Data rates for GSM Evolution (EDGE) cellular radio networks, Enhanced GPRS (EGPRS) is a packet-switched transmission mode that supports transmission rates as high as 473.6 kbps. See also *2.5G*, *3GPP*, *always on*, *channel*, *concatenation*, *EDGE*, *EGPRS*, *ETSI*, *GMSK*, *GSM*, *modulation*, *packet switch*, *protocol*, *QoS*, *time slot*, and *transmission rate*.

GPS (Global Positioning System) A satellite-based navigation system comprising a constellation of 24 Navstar satellites launched by the United States Department of Defense from 1978 to 1994. The satellites are in multiple medium-earth orbital (MEO) paths at altitudes of approximately 11,000 miles, and are positioned such that signals from six of them can be received by a GPS terminal at virtually any point on the Earth's surface at any time. The satellites constantly broadcast timing signals based on atomic clocks that are accurate to within three nanoseconds. The signals are broadcast on two frequencies — the L1 civilian signal and the L2 military signal. The civilian signal is transmitted in the UHF band at 1575.42 MHz, which requires line-of-sight (LOS). The terminals receive those signals and correlate them based on their knowledge of the satellites' positions, adjusting for propagation delay. Assuming that three signals are received, the terminal can determine its two-dimensional position within a few meters of the exact longitude (x) and latitude (y) and. Signals from four satellites allow determination of its three-dimensional position, including altitude (z). See also *LOS*, *MEO*, *propagation delay*, and *satellite*.

graded-index fiber A type of glass optical fiber (GOF) characterized by many layers of cladding surrounding the inner core, as illustrated in Figure G-1. From the core outward, the layers of cladding are of doped silica that are gradually and successively lower in index of refraction (IOR). Multi-mode fiber (MMF) commonly is of graded-index fiber construction and couples to a light-emitting diode (LED) light source. Unlike more sophisticated laser diodes, LEDs do not tightly focus a collimated beam of light. Rather, they emit a poorly focused and physically broad beam, which they inject into the relatively thick inner core of a MMF. Some light rays strike the core/cladding interface at sharp angles that exceed the critical angle. Rather than reflecting off the interface, they penetrate it and enter the cladding. As the fiber is not perfectly constructed, there may be numerous physical anomalies in the core/cladding interface over the length of a cable run, numerous points at which the light rays encounter extreme angles in the interface, and, therefore, numerous opportunities for the light rays to enter the cladding. Also, a typical cable installation is not perfectly straight, but takes numerous twists and turns that create odd angles, once again causing light rays to enter the cladding. Some light rays enter the cladding at extreme angles and are simply lost in it. Those errant light rays that enter at more modest angles gain in velocity and refract, or bend, as they travel through it as the cladding has a slightly lower IOR than the core. As they do so through many very thin layers of glass with slightly and successively lower refractive indexes, they gradually gain in speed and gradually bend back into the core, which is the primary light conducting medium. So, the graded-index fiber variously reflects or refracts light rays, guiding them to propagate through the core through a process known as *total internal reflection*. Graded-index multimode fiber (MMF) generally is used in relatively short haul, low-bandwidth applications such as local area networks (LANs) running at 1 Gbps or less. Step-index single-mode fiber (SMF) is used in long haul, high-bandwidth applications such as carrier backbones. See also *collimation*, *critical angle*, *GOF*, *laser diode*, *LED*, *MMF*, *reflection*, *refraction*, *SMF*, *step-index fiber*, *total internal reflection*, and *VCSEL*.

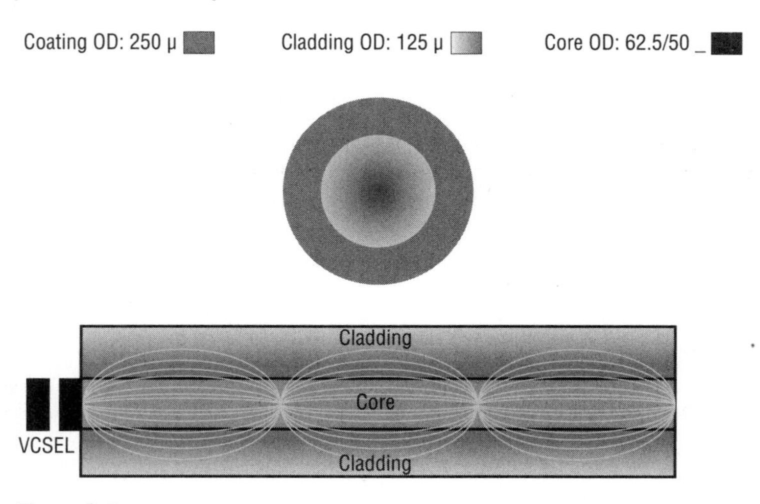

Figure G-1

grade of service (GoS) See *GoS.*

Graham-Willis Act The United States act (1921) that established telephone companies as natural monopolies. See also *natural monopoly.*

Grand Alliance An ad hoc advisory group formed by the United States Federal Communications Commission (FCC) in May 1993, comprising AT&T, General Instruments, Zenith, The Massachusetts Institute of Technology (MIT), Thompson Consumer Electronics, Philips Consumer Electronics, and the David Sarnoff Research Center. The efforts of the Grand Alliance led to a set of recommended Digital TV (DTV) standards for Standard Definition TV (SDTV) and High Definition TV (HDTV). Those recommended standards were tested and documented by the Advanced Television Systems Committee (ATSC) in the summer of 1995 and approved by the FCC in December 1996. See also *ATSC, digital, DTV, FCC, HDTV,* and *SDTV.*

graphical user interface (GUI) See *GUI.*

graphics Referring to diagrams and illustrations, as opposed to graphical representations, of letters, numbers, and symbols.

Graphics Interchange Format (GIF) See *GIF.*

grayscale A scale of shades from black to white used in digital image technology, such as facsimile machines. Grayscale is a variation on the theme of black and white. Grayscale measures the intensity of the light reflected from an area (dot) of a plane surface and defines each pixel (picture element) as a byte, generally of eight bits, which yields 2^8, or 256, shades of gray. Applications such as medical imaging require greater definition and, therefore, use 16-bit pixel descriptions, which yields 2^{16}, or 65,536, shades of gray. See also *facsimile* and *pixel.*

grayware The logic contained in the gray matter comprising the human brain. Some things are best done by using the grayware between your ears, rather than the software programs of a computer system. For example, mankind will never invent a computer that will eliminate man's need to know multiplication tables and fractions. See also *firmware, hardware,* and *software.*

GRE (Generic Routing Encapsulation) A tunneling protocol designed by Cisco to allow the encapsulation of a wide variety of packet types in Internet Protocol (IP) packets, thereby creating a virtual point-to-point link. GRE is described in IETF RFC 1701, 1702, and 2784. See also *encapsulate, IETF, IP, link, packet, point-to-point, protocol,* and *tunneling.*

greenfield **1.** In the building industry, referring to a parcel of undeveloped and unpolluted land ripe for exploitation. **2.** In telecommunications, referring to network deployments in virgin territories where there is no existing telecommunications infrastructure. See also *brownfield* and *overlay.*

greenphone The term used in some countries to refer to toll free service. See also *toll free service.*

Greenwich Mean Time (GMT) See GMT.

Gross, Al As a high school student in Cleveland, Ohio (United States) in 1938, Al Gross invented the walkie talkie, a portable handheld radio transmitter-receiver. Gross also is credited with inventing the pager, the citizens band (CB) radio, and the cordless telephone. Gross also lobbied the FCC to create the Personal Radio license spectrum, which later became Citizens Radio Service Frequency Band, or Citizens Band (CB) radio. Gross formed the Citizens Radio Corporation, which manufactured and sold personal two-way radios, mostly to farmers and the United States Coast Guard. See also *CB Radio Service, cordless telephone, pager, radio,* and *walkie talkie.*

Group I (G1) A group of specifications for facsimile (fax) machines conforming to a standard published by the Electronic Industries Alliance (EIA) in 1966 as EIA RS-328 and accepted by the ITU-T as the T.2 mode of operation. Now considered obsolete, Group I standards specified analog transmission with modems using double sideband modulation (DSB). Group I machines conforming to international standards use Frequency Modulation (FM), transmitting at two frequencies, with 1500 Hz pegged as the white frequency and 2300 Hz as the black frequency. The North American standards peg 1500 Hz as white and either 2300 Hz or 2400 Hz as black. As Group I machines used no compression mechanism, transmission was slow at about 4–6 minutes per page, even at the relatively poor resolution of about 100 scan lines per inch (lpi). Group I machines used an obsolete electrochemical printing process and are, themselves, considered obsolete. See also *compression*, *DSB*, *facsimile*, *FM*, and *resolution*.

Group II (G2) A set of standards for facsimile (fax) machines published by the ITU-T in 1978 as T.3. Group II specifies black-and-white mode of operation and compression through the use of encoding and vestigial sideband (VSB) transmission. Group II machines use amplitude modulation (AM) and phase modulation (PM) at a carrier frequency of 2100 Hz. Group II machines use compression to improve transmission speed to approximately two-three minutes per page, but resolution is relatively poor at 100 lines per inch (lpi). Group II machines use the same obsolete electrochemical printing process as Group I machines and are, themselves, considered obsolete. See also *AM*, *carrier*, *compression*, *facsimile*, *Group I*, *ITU-T*, *PM*, *resolution*, and *VSB*.

Group III (G3) A set of standards for facsimile (fax) machines published by the ITU-T in 1980 as T.4. Group III devices convert a document to digital form and employ a run-length encoding algorithm, as described below, that compresses the document prior to transmission. Group III machines are backward-compatible with Group I and Group II devices. See also *backward-compatible*, *compression*, *facsimile*, *Group I*, *Group II*, *ITU-T*, *run-length encoding*, *T.30*, and *T.4*.

Group IV (G4) A set of standards for facsimile (fax) machines published by the ITU-T in 1984 as T.6. Group IV machines are highly specialized and relatively expensive fax computer systems designed to make use of digital circuits to improve quality and improve transmission speed at rates up to 64 kbps. Group IV fax machines also are backward-compatible with Group III and can connect to analog circuits. See also *facsimile*, *Group I*, *Group II*, *Group III*, *T.30*, and *T.6*.

Groupe Spéciale Mobile (GSM) Translates from French into *Special Mobile Group*. Now *Global System for Mobile Communications* (GSM), the original name derives from the Groupe Spéciale Mobile (GSM) formed in 1982 by the Confederation of European Posts and Telecommunications (CEPT) to design a pan-European cellular radio technology. See also *cellular radio*, *CEPT*, and *GSM*.

ground A conducting connection, whether intentional or accidental, by which an electric circuit is connected at some point to the earth, or to some other large conducting body that can serve in place of the earth. The point can be a single point common to a great many circuits, such as an equipment frame, chassis, or cabinet. A ground serves as a reference point, a return path for an electrical signal, and to carry current safely away from a circuit in the event of a fault. In the event that earth is not available as a ground, the conducting frame of an aircraft, spacecraft, or land vehicle not conductively connected to the earth can serve as ground. See also *circuit*, *current*, *electricity*, and *signal*.

ground loop A complete circuit made up of the earth, represented by two different grounding points, and another conductor such as the shield of a cable grounded to devices at both ends. A surge of electricity results in the cable when the grounding points of different devices are at different voltage levels, and the data line connects the devices' ground together. See also *ground*, *surge*, and *voltage*.

ground rod A rod that is driven into the ground to provide electrical connection to the ground to carry current safely away from a circuit in the event of an electrical surge. A ground rod generally is solid copper, although it may be a copper clad metal rod or pipe, or a galvanized iron rod or pipe. See also *ground*.

ground start A signaling technique typically used between PBXs and central office switches. Ground start signaling momentarily grounds one side (usually the tip rather than the ring side) of the circuit, sending an immediate signal to the central office switch in order to start, i.e., seize, the trunk and get dial tone. Ground start is superior to loop start as it avoids glare, or collisions between incoming and outgoing calls.

Loop start is used in POTS applications between central offices and telephone sets or key telephone systems. See also *glare*, *loop start*, and *tip and ring*.

ground wave A radio wave that hugs the ground, following the curvature of the Earth rather than traveling in a straight line. The Earth is an electromagnetic conductor, although certainly not a perfect one. At the low end of the electromagnetic spectrum, signals propagate close to, are coupled to, interact with, and even travel through the Earth. A ground wave also is attenuated by the Earth, with the effect increasing as the frequency of the signal increases. Signals in the extremely low frequency (ELF), very low frequency (VLF), low frequency (LF), and medium frequency (MF) ranges are characterized as ground waves. See also *conductor*, *electromagnetic spectrum*, *ELF*, *LF*, *MF*, *propagation*, *skywave*, *surface wave*, and *VLF*.

ground wire A conductor that leads from a circuit, shield, or system to an electrical connection to the ground, generally through a ground rod driven into the earth. See also *ground* and *ground rod*.

grounding The process of connecting a circuit or system to an electrical ground to carry current safely away in the event of an electrical surge or an internal fault that connects a high voltage source to the exterior of the device. Grounding also is used to complete an electrical circuit from an electrical shield, thereby carrying electromagnetic interference (EMI) away from a wire or cable. The shield connects to a ground wire, or drain wire, that connect to a ground rod or some other conducting rod or pipe that connects to the ground. See also *current*, *EMI*, and *ground*.

G series The ITU-T Recommendations addressing transmission systems and media, digital systems and networks. See Table G-1 for selected G-series Recommendations. For a full listing of ITU-T Recommendations, see the contact information in Appendix A.

Table G-1: G-Series Recommendations

Recommendation	*Description*
G.168	Digital network echo cancellers
G.169	Automatic level control devices
G.651	Characteristics of a 50/125 μm multimode graded index optical fibre cable
G.652	Characteristics of a single-mode optical fibre and cable
G.653	Characteristics of a dispersion-shifted single-mode optical fibre and cable
G.654	Characteristics of a cut-off shifted single-mode optical fibre and cable
G.655	Characteristics of a non-zero dispersion-shifted single-mode optical fibre and cable
G.665	Generic characteristics of Raman amplifiers and Raman amplified subsystems
G.694.1	Spectral grids for WDM applications: DWDM frequency grid
G.694.2	Spectral grids for WDM applications: CWDM wavelength grid
G.702	Digital hierarchy bit rates
G.703	Physical/electrical characteristics of hierarchical digital interfaces
G.704	Synchronous frame structures used at 1544, 6312, 2048, 8448 and 44,736 kbit/s hierarchical levels
G.705	Characteristics of PDH equipment functional blocks
G.711	Pulse code modulation (PCM) of voice frequencies
G.722	7 kHz audio-coding within 64 kbit/s
G.722.1	Low-complexity coding at 24 and 32 kbit/s for hands-free operation in systems with low frame loss

Table G-1: G-Series Recommendations *(continued)*

Recommendation	*Description*
G.722.2	Wideband coding of speech at around 16 kbit/s using Adaptive Multi-Rate Wideband (AMR-WB)
G.723.1	Dual rate speech coder for multimedia communications transmitting at 5.3 and 6.3 kbit/s
G.726	40, 32, 24, 16 kbit/s ADPCM
G.727	5-, 4-, 3-, and 2-bit/sample embedded ADPCM
G.728	Coding of speech at 16 kbit/s using LD-CELP
G.729	Coding of speech at 8 kbit/s using CS-CELP
G.729.1	G.729 based Embedded Variable bit-rate coder: An 8-32 kbit/s scalable wideband coder bitstream interoperable with G.729
G.983.1	Broadband optical access systems based on Passive Optical Networks (PON)
G.991.1	High bit rate Digital Subscriber Line (HDSL) transceivers
G.991.2	Single-pair high-speed digital subscriber line (SHDSL) transceivers
G.992.1	Asymmetric digital subscriber line (ADSL) transceivers
G.992.2	Splitterless asymmetric digital subscriber line (ADSL) transceivers
G.992.3	Asymmetric digital subscriber line transceivers 2 (ADSL2)
G.992.4	Splitterless asymmetric digital subscriber line transceivers 2 (splitterless ADSL2)
G.992.5	Asymmetric Digital Subscriber Line (ADSL) transceivers - Extended bandwidth ADSL2 (ADSL2+)
G.993.1	Very high speed digital subscriber line transceivers
G.993.2	Very high speed digital subscriber line transceivers 2 (VDSL2)

G.shdsl A term sometimes used for symmetric high-bit-rate digital subscriber line (SHDSL), the term G.shdsl derives from the fact that SHDSL is specified in the ITU-T G Series Recommendations as G.991.2. See also *G Series*, *ITU-T*, and *SHDSL*.

GSM (Global System for Mobile Communications née Groupe Spéciale Mobile) The pan-European digital cellular radio standard developed by the Groupe Spéciale Mobile (Special Mobile Group) formed in 1982 by the Confederation of European Posts and Telecommunications (CEPT). GSM was adopted by the CEPT in 1987 and was commenced commercial operations in 1991. GSM operates in the 800 MHz and 900 MHz frequency bands and is ISDN-compatible. GSM derives 124 carriers of 200 kHz. Frequency division duplex (FDD) is employed to support bidirectional communications, with the 935–960 MHz band supporting downlink transmission and the 890–915 MHz band supporting uplink transmission. Time division multiple access (TDMA) is employed to derive 8 time slots from each carrier. Gaussian minimum-shift keying (GMSK) modulation yields a signaling rate of 270.833 kbps per carrier and a maximum transmission rate of 33.8 kbps per channel. Regular pulse excitation linear predictive coding (RPELPC) supports voice at 13 kbps and vector sum excited linear predictive (VSELP) coding at 8 kbps. Data throughput generally is limited to 9.6 kbps, due to overhead associated with forward error correction (FEC) and encryption. GSM commonly employs a four-cell reuse plan, rather than the seven-cell plan used in AMPS, and divides each cell into 12 sectors. GSM commonly uses frequency hopping and time-slot hopping, which also is used in CDMA systems.

GSM offers additional security in the form of a subscriber identification module (SIM) that plugs into a card slot in the handset. The SIM contains user-profile data, a description of access privileges and features, and identification of the cellular carrier that hosts the home registry. The SIM can be used with any GSM set, thereby providing complete mobility across nations and carriers supporting GSM, assuming that cross-billing relationships are in place. GSM is predominant in Europe, Africa, and much of Asia. GSM is the basis for DCS 1800, also known as *Personal Communications Network* (PCN), in Europe. DCS 1800, in large part, is an upbanded version of GSM, operating in the 1800 MHz (1.8 GHz) range. Also with minor modifications, it is the basis for PCS 1900 in the United States, where it also is known as *GSM*. PCS 1900 is the ANSI standard (J-STD-007, 1995) for PCS at 1900 MHz (1.9 GHz). See also *band, carrier, CDMA, cellular radio, CEPT, channel, digital, downlink, encode, encryption, FDD, FEC, frequency, frequency hopping spread spectrum, GMSK, ISDN, PCS, RPELPC, signaling rate, SIM, TDMA, time slot, throughput, transmission rate, uplink*, and *VSELP*.

GSTN (Global Switched Telephone Network) The generic ITU-T term for the international public telephone network that comprises the national Public Switched Telephone Networks (PSTNs) and the facilities that interconnect them. See also *ITU-T* and *PSTN*.

gTLD (generic Top Level Domain) In the Internet Domain Name System (DNS), the rightmost portion of the address — the domain — identifies the type of entity owning or sponsoring the address. The several types of Top Level Domains (TLDs) include generic Top Level Domains (gTLDs) and country codes (ccTLDs). If a country code is used in the address, the gTLD becomes a secondary domain. The original gTLDs are unsponsored, meaning that they operate under policies established by the global Internet community, directly through the administration process of the Internet Corporation for Assigned Names and Numbers (ICANN). The original gTLDs and their intended use are as follows:

- **.arpa:** Address Routing and Parameter Area, exclusively for Internet infrastructure purposes
- **.com:** commercial organizations
- **.edu:** accredited degree-granting educational institutions
- **.gov:** United States government agencies
- **.int:** organizations formed under international treaties between governments
- **.mil:** United States military
- **.net:** network access providers, originally; now unrestricted
- **.org:** noncommercial organizations, originally; now unrestricted

Another seven TLDs became active in 2002,and another four in 2005. The majority of these gTLDs are unsponsored, although some have a sponsor that represents the narrower community that is most affected. The new gTLDs and their intended uses are as follows:

- **.aero:** aeronautical interests (2002), sponsored by Societe Internationale de Telecommunications Aeronautiques (SITA)
- **.biz:** biznesses (2002), unsponsored
- **.cat:** Catalan linguistic and cultural community (2005), sponsored by Fundació puntCat.
- **.coop:** cooperative associations (2002), sponsored by Dot Cooperation LLC
- **.info:** informational sites, unrestricted (2002), unsponsored

- **.jobs:** job-related sites and job seekers (2005), sponsored by an alliance between Employ Media, the Society for Human Resource Management (SHRM), Verisign, and ICANN
- **.museum:** museum community (2002), sponsored by the Museum Domain Management Association
- **.mobi:** mobile products and services (2005), sponsored by mTLD Top Level Domain, Ltd.
- **.name:** individuals (2002), unsponsored
- **.pro:** certified professionals (e.g., doctors, .med.pro; lawyers, .law.pro; and accountants, .cpa.pro) and professional companies and associations (2002), unsponsored
- **.travel:** travel industry (2005), sponsored by Tralliance Corporation

In May 2006, ICANN's Board of Directors voted against a proposed agreement for an .xxx domain, which would have been a TLD for pornography. Some argued that .xxx would serve as a positive, if voluntary, means of segmenting the Internet. ICANN tentatively approved the new TLD before receiving an unprecedented level of correspondence in opposition. See also *ccTLD*, *DNS*, *ICANN*, and *Internet*.

guaranteed frame rate (GFR) See *GFR*.

guard band A frequency band that exists between two information-bearing channels, a guard band provides channel separation in order to avoid mutual interference. Analog voice grade bandwidth, for example, is defined as 4,000 Hz (4 kHz). As illustrated in Figure G-2, the voice signal runs in the 300–3,300 Hz range, and signaling and control functions take place in the 3,300–3,700 Hz range. The 0–300 Hz range and the 3,700–4,000 Hz range are used as guard bands. These guard bands prevent mutual interference between voice channels when multiplexed using frequency division multiplexing (FDM). Guard bands are equally important in video communications. An analog community antenna television (CATV, aka cable TV) video channel has a width of approximately 6,000,000 Hz, or 6 MHz, of which approximately 4.5 MHz is used for transmission of the video signal. The balance of the bandwidth is used for guard bands, also known as *transition bands* in this context, to separate adjacent channels riding the common, analog coaxial cable system. See also *bandwidth*, *CATV*, *channel*, *FDM*, *frequency band*, *interference*, *MUX*, and *voice grade*.

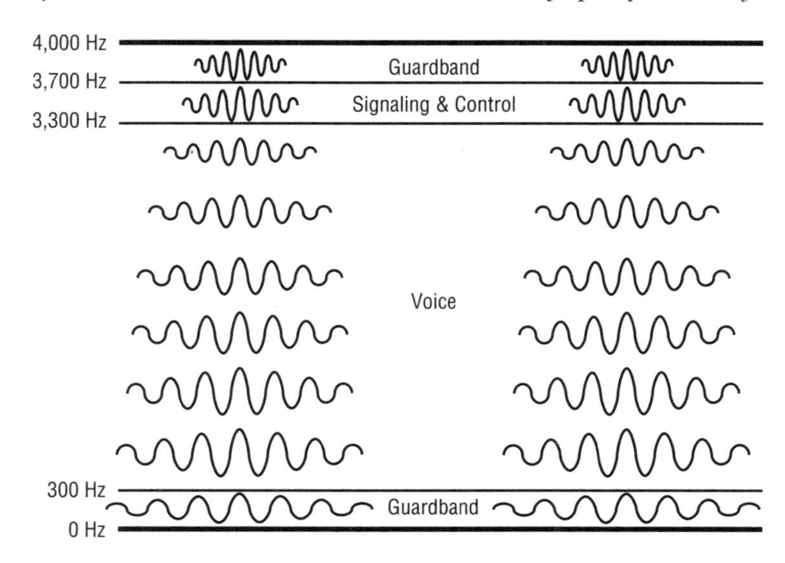

Figure G-2

GUI (Graphical User Interface) Pronounced *gooey*. A visual computer environment that uses graphical images such as icons, menus, and dialog boxes to represent files, file folders, programs, and options to ease the interface between the user and the computer. The user can use a mouse, keypad, thumbpad, or other device to manipulate an icon, button, scroll bar, check box, or other graphical item to activate a routine to activate a program feature, such as opening or closing a file. The first GUI was invented by a research team at the Xerox Palo Alto Research Center (Xerox PARC). GUI has largely superseded the Command Line Interface (CLI) in computer and communications gear except for routers and switches.

guided media See *transmission medium*.

guru **1.** In Hinduism or Sikhism, a religious teacher or spiritual guide. **2.** In the context of technology, an intellectual leader, teacher, or guide who is highly knowledgeable or skilled and who can express himself in an intelligible manner.

H **1.** henry. See *henry.* **2.** Horizontal. In line with the horizon. The x-axis on a two- or three-dimensional grid, chart, or graph in a Cartesian coordinate system. See also *Cartesian coordinates*, *V&H*, and *x-axis.* **3.** H. The H series of ITU-T Recommendations. The H series addresses audiovisual and multimedia systems. See *H series.*

H.221 The ITU-T Recommendation for frame structure for a channel of 64–1920 kbps in audiovisual teleservices. H.221 is included in the ITU-T H.320 umbrella recommendation for videoconferencing and multimedia communications over narrowband ISDN (N-ISDN). See also *H.320.*

H.225 The ITU-T Recommendation for call signaling protocols and media stream packetization for packet-based multimedia systems. See also *H.320, H.323, H Series, multimedia, packet,* and *protocol.*

H.230 The ITU-T Recommendation for frame structure for frame synchronous control and indication signals for audiovisual systems. H.230 is included in the ITU-T H.320 umbrella recommendation for videoconferencing and multimedia communications over narrowband ISDN (N-ISDN). See also *H.320.*

H.242 The ITU-T Recommendation for a system for establishing communications between audiovisual terminals using digital channels up to 2 Mbps. H.424 addresses call setup and teardown, in-band signaling and control, and channel management. H.242 is included in the ITU-T H.320 umbrella recommendation for videoconferencing and multimedia communications over narrowband ISDN (N-ISDN). See also *H.320.*

H.245 The ITU-T Recommendation for call control procedures for multimedia communications. See also *H.320, H.323, H Series,* and *multimedia.*

H.248 The ITU-T Recommendation for Gateway control protocol, better known as Media Gateway Control (Megaco). The matching standard in the Internet community is IETF RFC 3525. See also *IETF* and *Megaco.*

H.261 The ITU-T Recommendation for video codecs for use in audiovisual telephony over circuit-switched ISDN at Px64 kbps. See also *circuit switch, codec, H.320, H.323, ISDN, ITU-T,* and *video.*

H.263 The ITU-T Recommendation for video coding for low–bit rate communications. See also *ITU-T* and *H Series.*

H.310 The ITU-T Recommendation for broadband audiovisual communications systems and terminals. See also *broadband, H Series,* and *ITU-T.*

H.320 Also known as *Px64.* The ITU-T umbrella standard for videoconferencing and multimedia communications over narrowband ISDN (N-ISDN) bearer (B) channels at bit rates from 64 kbps to 1.920 Mbps, in increments of 64 kbps. Video compression makes use of codecs defined in H.261. Video formats include Common Intermediate Format (CIF), which is optional, and Quarter-CIF (QCIF), which is mandatory in compliant codecs. CIF supports resolution of 352 × 288 pixels, and QCIF supports resolution of 176 × 144 pixels, which is exactly ¼ the resolution of CIF. Frame rates are 30 frames per second (fps), or lower. Audio coding and compression recommendations include G.711, which is pulse code modulation (PCM) at 64 kbps. G.722 is high fidelity adaptive differential pulse code modulation (ADPCM) transmitting up to a 7 kHz range and compressing voice at 48/56/64 kbps. G.728 specifies low delay code-excited linear prediction (LD-CELP) running in the 3 kHz range and compressing voice at 16 kbps. The balance of the specification (H.221, H.230, and H.242) addresses techniques for call setup and teardown, data framing and multiplexing, and various other operational and administrative functions. See also

ADPCM, B channel, CIF, codec, frame, frame rate, fps, G.711, G.722, G.728, H.221, H.230, H.242, H Series, ITU-T, LD-CELP, multimedia, N-ISDN, PCM, pixel, QCIF, resolution, video, and *videoconference.* See Table H-1 for a listing of ITU-T Recommendations related to H.320.

Table H-1: H.320 Related Standards Recommendations

ITU-T Recommendation	*Description*
G.711	Pulse Code Modulation (PCM) voice coding at 64 kbps.
G.722	Adaptive Differential Pulse Code Modulation (ADPCM) voice coding and compression of high-fidelity 7 kHz voice at 64/56/48 kbps.
G.723	Dual-rate speech coder at 5.3 and 6.3 kbps for multimedia communications.
G.728	Low-Delay Code Excited Linear Prediction (LD-CELP) coding and compression of 3.3 kHz voice at 16 kbps.
G.729	Conjugate-Structure Algebraic-Code-Excited Linear-Prediction (CS-CELP) voice coding and compression at 8 kbps.
H.221	Frame Structure for channel of 64–1920 kbps in audiovisual teleservices.
H.223	Multiplexing protocol for low bit-rate multimedia communication. Annexes address mobile communications over low, moderate, and highly error-prone channels.
H.225	Call signaling protocols and media stream packetization for packet-based multimedia systems.
H.230	Frame synchronous control and indication signals for audiovisual systems.
H.242	System for establishing communications between audiovisual terminals using digital channels up to 2 Mbps. Addresses call setup and teardown, in-band signaling and control, and channel management.
H.245	Call control procedures for multimedia communications.
H.261	Video codec for audiovisual services at px64 kbps.
H.263	Video coding for low bit-rate communication at rates less than 64 kbps.
T.120	Multipoint transport of multimedia data.

H.321 The ITU-T Recommendation for the adaptation of H.320 visual telephone terminals to broadband ISDN (B-ISDN) environments. See also *B-ISDN, H.320, H Series,* and *ITU-T.*

H.323 The ITU-T Recommendation for multimedia communications over packet networks. The recommendation addresses service over local area networks (LANs), but extends to the Internet and other IP-based networks. H.323 is not linked to any specific hardware device or operating system (OS) and, therefore, can be deployed in a wide variety of devices, including PCs, telephone sets, cable modems, and set-top boxes. H.323 supports multicast communications, thereby avoiding the requirement for specialized multipoint control units (MCUs) in a network where routers assume the responsibility for packet replicating. Version 2 1998) provides a means for encryption, includes mechanisms for call transfer and call forward, supports URL-style addresses, and provides the ability for endpoints to set quality of service (QoS) levels through Resource Reservation Protocol (RSVP). The four major components specified for H.323 include terminals, gateways, gatekeepers, and MCUs.

- **Terminals** are the client endpoint devices on the LAN. All terminals must support voice, but data and video are optional. H.245 must be supported for negotiation of channel usage and capability. Q.931 is required for signaling and control. The Registration/Admission/Status (RAS) protocol communicates with the gatekeeper. Sequencing of audio and video packets is supported through Real-Time Protocol/Real-Time Control Protocol (RTP/RTCP). Endpoints can set quality of service (QoS) levels through Resource Reservation Protocol (RSVP). Terminals optionally may include video codecs, T.120 data conferencing capabilities, and MCU functionality.

- **Gateways** are optional elements used for various levels of protocol conversion. The gateway serves as a protocol converter between devices and networks that have native H.323 capability and those that do not. The gateway also may translate between audio, video, and data formats, and may perform signaling conversions between the H.225 packet protocol and external protocols such as SS7 and Q.931. Alternatively, signaling conversions may be performed by Gatekeepers, call processors, or session border controllers.
- **Gatekeepers** are optional elements that act as the central points in H.323 zones. Endpoints may communicate directly, in either a unicast or a multicast environment, if no gatekeeper is present. If a gatekeeper is present, all endpoints in its zone must register with it. The gatekeeper performs the function of admission control, determining if devices are authorized to connect and if there is sufficient bandwidth to support the call. Gatekeepers serve to translate LAN addresses into IP or IPX addresses, as defined in the RAS specification. Gatekeepers also can act to route H.323 calls through gateways, if necessary, and monitor the network bit rate capacity, with the ability to deny access to a session if programmable bandwidth thresholds have been reached or exceeded. Gatekeepers also can perform certain administrative functions, such as accounting, billing, directory, and collecting network usage data. Gatekeepers may be distinct network elements (NEs), or gatekeeper functionality can be incorporated into MCUs.
- **Multipoint Control Units (MCUs)** support conferencing among three or more participating terminals. The MCU comprises a Multipoint Controller (MC) and optional Multipoint Processors (MPs). The MC is responsible for call control negotiation to achieve common levels of communication. The MP may process either a single media stream or multiple media streams, depending on the nature of the conference.

See also *bandwidth*, *bit rate*, *cable modem*, *encryption*, *Internet*, *IP*, *IPX*, *H.225*, *H.245*, *ITU-T*, *LAN*, *MCU*, *multicast*, *NE*, *Q.931*, *QoS*, *RAS*, *RSVP*, *RTCP*, *RTP*, *set-top box*, *signaling and control*, *T.120*, and *unicast*.

H.324 The ITU-T Recommendation for low bit-rate multimedia communication over the analog public switched telephone network (PSTN) through V.34 modems. V.34 modems are limited to a maximum transmission rate of 28.8 kbps and V.34bis modems to 33.6 kbps. See also *analog*, *ITU-T*, *modem*, *multimedia*, *PSTN*, *transmission rate*, *V.34*, and *V.34bis*.

hacker **1.** A computer enthusiast, or computerphile, who enjoys computer technology and programming to the point of examining the code of operating systems to figure out how they work. *Note:* I promise that I do not make this stuff up. **2.** Synonymous with *cracker*. A person who gains, or attempts to gain, unauthorized access to computers or computer networks and tamper with operating systems, application programs, and databases.

half duplex (HDX) See *HDX*.

ham **1.** A ham-fisted person, i.e., someone with big, clumsy hands. **2.** An amateur radio operator. The exact origin and meaning of the term is vague. Some suggest that ham is a shortened and corrupted version of amateur. The most reliable origin seems to be in reference to a ham-fisted amateur telegrapher, as amateur radio operators traditionally were required to demonstrate a reasonable level of skill in Morse code telegraphy. Skilled telegraphers referred to someone lacking in proficiency using the pejorative terms plug or ham. This use of the term dates at least to 1899, when G.M. Dodge first included it in his book, *The Telegraph Instructor*, under the heading Definitions of Technical Terms Used in Railroad and Telegraphic Work. See also *amateur radio service*, *Morse code*, and *telegraph*.

ham radio See *amateur radio service.*

Hamming code A family of linear error-correcting codes used for forward error correction (FEC), Hamming code can detect and correct single-bit errors by adding multiple parity bits to a data set. As an example, one of the simplest Hamming codes is the 7,4 code, which uses each group of four bits to compute a three-bit value, which it appends to the original four bits prior to transmission. If any of the seven bits is altered in transit, the receiving device can easily identify, isolate, and correct the errored bit. The 7,4 code is generally considered impractical, as it involves a non-standard character length. More complex Hamming codes based on standard character lengths (e.g., 11,7 for ASCII and 12,8 for EBCDIC) can also detect and distinguish two-bit and three-bit errors, but not correct them. Hamming code was invented in the 1940s by Richard W. Hamming of Bell Labs. See also *ASCII, data set, EBCDIC, error control, FEC,* and *parity bit.*

handoff The process by which a cellular radio network transfers a call as the mobile station (MS) moves out of the range of one base station (BS) in one cell and into the range of another base station in another cell. A *hard handoff* is one in which the connection is briefly broken by the first base station before being re-established by the second. This technique is known as break and make. A *soft handoff* is one in which the connection is established by the second base station before being broken by the first. This technique is known as make and break. See also *BS, cellular radio, MS,* and *radio.*

handshaking In the context of a protocol, the sequence of events that occurs between devices over a circuit as they set up a session. The handshaking process establishes the fact that the circuit is available and operational, establishes the level of device compatibility, and determines the speed of transmission by mutual agreement. The process of handshaking occurs as the devices pass tones or frames of data back and forth in order to negotiate the basis on which they will communicate, in consideration of the performance characteristics of the circuit. Once the handshaking process is complete, the devices move to the next stage, which is that of line discipline. See also *line discipline* and *protocol.*

hard copy A computer output printed on paper, film, or other permanent, tangible medium, as distinguished from information on a computer disk or in computer memory. Hard copy is not exactly carved in stone, but it cannot easily be changed. See also *soft copy.*

hard handoff In cellular radio networks, a handoff process in which the connection is briefly broken by one base station (BS) before being re-established by another as a mobile station (MS) moves out of the range of the first and into the range of the second. This technique is also known as *break and make.* See also *BS, cellular radio, MS, radio,* and *soft handoff.*

hardphone A conventional telephone set, which is a single function terminal, hardwired to support voice communications. A hardphone is in sharp contrast to a softphone, which is a software-based telephone comprising a desktop, laptop, or tablet computer equipped with a microphone, a speaker, and software that allows it to emulate a hardphone. See also *softphone.*

hardware The physical components, peripherals, and equipment that comprise a computer system, as compared to the logical system software programs and routines that run the computer and the application programs that support the tasks of end users. If you can break it with a hammer, it's hardware. Otherwise, it's software. See also *firmware, grayware,* and *software.*

hardwire **1.** To physically wire components together to form a system or subsystem rather than using switches, plugs, or connectors. Early electromechanical key telephone systems (KTSs), for example, were hardwired and, therefore, both limited in feature content and highly inflexible. In contrast, contemporary electronic common control (ECC) systems are microprocessor-based, software-controlled, and, therefore, easily upgradeable and highly flexible. See also *ECC, KTS,* and *software.* **2.** To build a function into a system with hardware, rather than software. Hardwired logic is fixed, i.e., inflexible, and cannot be reprogrammed. See also *hardware* and *software.*

hang up and call back See *Huh?*.

harmonic A sinusoidal component of a waveform that is an integral multiple of a fundamental frequency. The signal waveform is known as the first harmonic. A waveform that has a component that is twice the frequency of the fundamental frequency, or signal waveform, is known as the *second harmonic.* An unwanted harmonic causes harmonic distortion. See *harmonic distortion.*

harmonic distortion Nonlinear distortion characterized by the output of harmonics in a signal waveform that do not correspond with the input signal waveform. Harmonic distortion is caused by an amplifier, transducer, or other element that malfunctions. See also *distortion, harmonic,* and *noise.*

hash See #.

H channel (High-speed channel) In the integrated service digital network (ISDN), a high-speed channel comprising multiple aggregated low-speed channels to accommodate bandwidth-intensive applications such as file transfer, videoconferencing, and high-quality audio. An H channel is formed of multiple bearer (B) channels bonded together in a primary rate access (PRA) or primary rate interface (PRI) frame in support of applications with bandwidth requirements that exceed the B channel rate of 64 kbps. The channels, once bonded, remain so end-to-end, from transmitter to receiver, through the ISDN network. The feature is known variously as multirate ISDN, Nx64, channel aggregation, and bonding. ISDN standards define H channels at the following levels:

- **H_0**: An aggregate bit rate of 384 kbps, which is the equivalent of six B channels (6×64 kbps = 384 kbps).
- **H_1**: A full DS-1, with no framing overhead. This channel is sensitive to the specifics of the DS-1 implementation. In a North American PRI implementation where non-facility associated signaling (NFAS) is in place, the aggregate bit rate is 1.536 Mbps, which is the equivalent of 24 B channels. In a European PRA implementation, H_1 supports an aggregate bit rate of 1.920 Mbps, which is the equivalent of 30 B channels.
- **H_{10}**: An aggregate bit rate of 1.472 Mbps, which is the sum of the 23 B channels (23×64 kbps = 1.472 Mbps) in a baseline PRI implementation in which channel 24 is devoted to the D channel. H_{10} applies in North America and Japan, and is based on T1 and J-1, respectively.
- **H_{11}**: An aggregate bit rate of 1.536 Mbps, the sum of all 24 B channels for the North American and Japanese versions, which is based on T1 and J-1, respectively. H_{11} relies on non-facility associated signaling (NFAS) to provide a D channel on an H_{10} facility for signaling and control.
- **H_{12}**: An aggregate bit rate of 1.920 Mbps, the sum of all 30 B channels for the European PRA, which is based on E-1.

See also *bandwidth, B channel, bonding, channel, D channel, DS-1, E-1, frame, J-1, NFAS, overhead, PRA, PRI,* and *T1.*

HCO (Hearing Carry Over) An offering of Telecommunications Relay Service (TRS) that allows a person with a speech disability to use his or her own hearing to listen to the called party, but to respond in text through the call administrator (CA), who acts as a facilitator. See also *TRS.*

HCV (High Capacity Voice) A voice compression technique that encodes analog voice signals based on a series of samples represented as a bit string, which is termed a vector. HCV expands the principles of vector coding used in vector quantizing code (VQC) to model the actual vocal process. See also *analog, compression, encode, VCQ,* and *vector.*

HDB3 (High Density Bipolar order 3) The line coding technique employed in E-1. HDB3 is a bipolar transmission method that reverses the polarity of alternate marks, or 1 bits, expressing the first as a positive voltage of +3V, the second as a negative voltage of -3V, the third as +3V, and so on. Zero bits are coded as 0V. HDB3 is based on a combination of alternate mark inversion (AMI) and Bipolar with Eight-Zeros Substitution (B8ZS) in T1 networking, but imposes a limit of three successive 0 bits. A fourth 0 bit triggers zeros suppression, substituting a known bit pattern with an intentional bipolar violation (BPV) known to the receiver, as illustrated in Figure H-1. See also *AMI*, *bipolar*, *B8ZS*, *BPV*, *E-1*, *mark*, *multiplexer*, *polarity*, and *T1*.

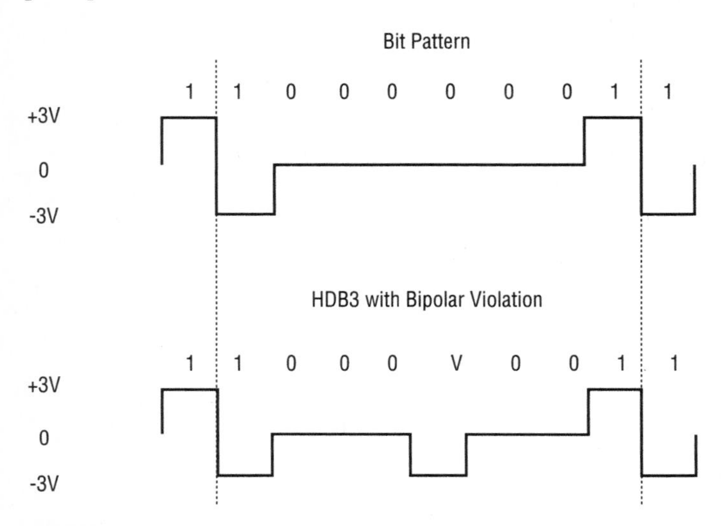

Figure H-1

HDLC (High-Level Data Link Control) A bit-oriented, synchronous data communications protocol developed by the International Organization for Standardization (ISO) as a superset of Synchronous Data Link Control (SDLC) and Advanced Data Communications Control Procedures (ADCCP). A version of HDLC is the Link Access Procedure-Balanced (LAPB), which is used in packet-switched networks conforming to the ITU-T X.25 Recommendation. HDLC also was imported into other standards, including ISDN as LAPD and frame relay as LAPF. See also *ADCCP*, *bit-oriented protocol*, *frame relay*, *ISDN*, *ISO*, *LAPB*, *LAPD*, *LAPF*, *SDLC*, *synchronous*, and *X.25*.

HDSL (High bit-rate Digital Subscriber Line) An access technology developed as a more cost-effective means of providing T1 local loop circuits over existing unshielded twisted pair (UTP). HSDL was developed by Bellcore (now Telcordia Technologies) at the request of the Regional Bell Operating Companies (RBOCs) in the United States, and was later standardized in 1999 by the American National Standards Institute (ANSI) as T1E-1.4. HDSL eliminates repeaters in the T1 local loop for distances up to 12,000 feet, which can be extended another 12,000 feet through the use of a line doubler, which essentially is an HDSL repeater.

In the North American implementation of HDSL, the upstream and downstream signals are split across both pairs, with each pair operating in full-duplex (FDX) mode at 784 kbps, which is half the T1 rate plus additional overhead. In the European implementation, each of two pairs operates at 1.168 Mbps, which is roughly half the E-1 rate, plus additional overhead.

The HDSL line coding scheme is 2B1Q, also known as *4 PAM* (4-level Pulse Amplitude Modulation). As 2B1Q impresses two bits on each symbol, each of which is in the form of one of four voltage levels. The symbol rate, therefore, is one-fourth the line rate, meaning that an HDSL T1 implementation at a line rate of 784 kbps across each of two pairs requires a carrier frequency of only 196 kHz, at least at the peak power level. At this relatively low frequency, issues of attenuation and crosstalk are mitigated.

HDSL2 is an HDSL variant that supports T1 and E-1 over a single twisted pair, with a maximum transmission span of 13,200 feet. A variant known as HDSL4 can run over two twisted pair in order to extend the maximum transmission span to as much as 16,500 feet. See also *attenuation, 2B1Q, crosstalk, E-1, FDX, HDSL2, HDSL4, line doubler, local loop, overhead, power, repeater, SDSL, symbol, T1, UTP,* and *voltage.*

HDSL2 An HDSL variant that supports T1 and E-1 spans over a single unshielded twisted pair (UTP) local loop of 24 AWG up to 13.2 kft., 768 kbps up to 17.7 kft., and 384 kbps up to 22.5 kft. An HDSL2 line doubler, i.e., repeater, can double the distance for each speed rating. HDSL2 employs a line coding technique known as *trellis-coded pulse amplitude modulation* (TC-PAM), also known as *trellis-coded modulation* (TCM). This technique places three bits on a baud, which is an improvement over the two bits per baud realized through the 2B1Q technique used in HDSL. Also used in HDSL4 and SDSL, TCM features an inherent forward error correction (FEC) mechanism to overcome issues of attenuation and interference. See also *2B1Q, attenuation, AWG, bit, baud, E-1, FEC, interference, line coding, line doubler, local loop, PAM, repeater, SDSL, T1, TCM,* and *UTP.*

HDSL4 An HDSL variant that extends the T1 and E-1 spans to a maximum of 16,500 feet over two unshielded twisted pair (UTP) local loops. HDSL4 employs a line coding technique known as *trellis-coded pulse amplitude modulation* (TC-PAM), also known as *trellis-coded modulation* (TCM). This technique places three bits on a baud, which is an improvement over the two bits per baud realized through the 2B1Q technique used in HDSL. Also used in HDSL2 and SDSL, TCM features an inherent forward error correction (FEC) mechanism to overcome issues of attenuation and interference. See also *attenuation, baud, bit, E-1, FEC, HDSL, HDSL2, interference, line coding, local loop, T1, TCM,* and *UTP.*

HDTV (High Definition TeleVision) A standard for digital television (DTV) that supports display formats that are larger and higher in resolution than either legacy analog TV or digital standard definition television (SDTV). Specifically, HDTV specifies two formats, as detailed in Table H-2.

Table H-2: SDTV Scanning Formats

Vertical Lines	*Horizontal Pixels*	*Aspect Ratio*	*Refresh Rate (fps)**
1080	1920	16:9	24p, 30p, 60i**
720	1280	16:9	24p, 30p, 60p**

*fps = frames per second
**i = interlaced, p= progressive

In comparison to analog TV, DTV offers improved reception, without the ghosting, snowy images, and generally poor audio quality. Issues of signal quality in DTV transmission manifest in artifacts such as blocking, or tiling, and stuttering. The ATSC standard specifies MPEG-2 compression, and the transport subsystem as ISO/IEC 13818. Packet transport involves a serial data stream of packets of 188 octets, one octet of which is a synchronization byte and 187 octets of which are payload. This packet approach is suitable for ATM switching, as each 188-octet MPEG-2 packet maps into the payload of four ATM cells, with only 4 octets of padding required. SDTV employs Reed-Solomon forward error correction (FEC) and 8-level vestigial sideband (8 VSB) RF modulation to support a bit rate of 19.28 Mbps over a 6 MHz terrestrial broadcast channel. Audio compression is based on the AC-3 specification from Dolby Digital and the ATSC. SDTV standards were developed by the Grand Alliance and reviewed, tested, and documented by the Advanced Television Systems Committee (*ATSC*) at the request of the United States Federal Communications Commission (FCC). See also *8-VSB, AC-3, analog, artifact, aspect ratio, ATM, ATSC, broadcast, byte, channel, compression, digital, DTV, FCC, FEC, fps, ghosting, Grand Alliance, interlaced scanning, modulation, MPEG-2, NTSC, octet, packet, padding, PAL, payload, pixel, progressive scanning, Reed-Solomon, refresh rate, resolution, RF, scanning, SDTV, SECAM,* and *synchronize.*

HDX (Half DupleX) **1.** A transmission path, circuit, or channel designed to support information transfer in both directions, but only one direction at a time. The physical circuit may be capable of supporting full duplex transmission, i.e., simultaneous transmission in both directions, but the protocol may be half duplex. **2.** A protocol, such as polling, that operates on a query/response basis, with a back-and-forth dialogue between the polling device and the polled devices. A full duplex (FDX) protocol differs in that it will support simultaneous operation in both directions. See also *FDX* and *simplex*.

headend The point of signal origin and centralized communications control in a CATV network or a traditional Dataphone Digital Service (DDS) network. The headend is the common point of control and connection in a point-to-multipoint network configured in a physical star. See also *CATV*, *DDS*, *point-to-multipoint*, and *star topology*.

header **1.** The portion of a data block, cell, frame, or packet that precedes the text field or payload and provides information such as the source address and destination address. The header often includes synchronization bits that serve to synchronize the operations of the transmit and receive devices across the link. Certain data protocols also use fields in the header to identify the length of the text field and the type of data, to indicate the level of tolerance for delay or loss during network transit, and any optional headers that might follow. For example, an Internet Protocol (IP) header might indicate that a Transmission Control Protocol (TCP) header follows. See also *block*, *cell*, *data format* (illustration), *frame*, *packet*, *payload*, *text field*, and *trailer*. **2.** The user header includes user-definable information such as system access (password), organization or department ID, operator ID, terminal ID, database or application ID, destination address, message sequence number, date/time ID, and message priority.

header compression See *payload header suppression*.

header error control (HEC) See *HEC*.

header suppression See *payload header suppression*.

headroom Attenuation-to-crosstalk ratio (ACR). See *ACR*.

Hearing Carry Over (HCO) See *HCO*.

heavy carrier A facilities-based carrier, i.e., one that owns the switching and transmission systems that compose the network it uses to provide services to its customers. See also *carrier* and *light carrier*.

heavy metal **1.** Also known as *big iron*. The colloquial term for early generations of mainframe computers, PBXs, and central offices (COs). Prior to the advent of electronic common control (ECC) and miniaturized semiconductor circuitry, such systems contained a lot of metal and were quite heavy, indeed. Fans of heavy metal are known as *old computer guys*. See also *forklift upgrade* and *mainframe computer*. **2.** A subgenre of really loud rock music dominated by really loud drums and really loud guitars, strong really loud rhythms, and a really loud bluesy style played by really loud bands like Iron Butterfly, Iron Maiden, Led Zeppelin, and Metallica. Fans of heavy metal are known as *head bangers* and *metalheads*.

HEC (Header Error Control) In asynchronous transfer mode (ATM), 8 bits in the cell header that provide error checking of the header, but not the payload. There is no provision for error correction. See also *ATM*, *cell*, *error control*, *header*, and *payload*.

Hellenologophobia The irrational fear of Greek (or Latin) words, or complex scientific terminology. Hellenologophobia is real. There are books written about Hellenologophobia, and there are therapists who specialize in its treatment. (I kid you not.) See also *logophobia*.

henry (H) The unit of inductance of a circuit in SI units. The inductance of a circuit is one henry if the electromotive force (emf) in volts (V) is numerically equal to the rate of change in amperes (A) per second. Named after American scientist Joseph Henry (1797–1878), who discovered the phenomenon of self-inductance, and whose work on the electromagnetic relay laid much of the foundation for the electric telegraph. See also *ampere*, *emf*, *inductance*, *SI*, and *volt*.

Hertz (Hz) See *Hz*.

Hertz, Heinrich Rudolf (1857–1894) The physicist who was the first to demonstrate the existence of electromagnetic radiation by constructing an apparatus that produced radio waves, Heinrich Rudolf Hertz proved that electromagnetic energy can transverse free space, can penetrate various materials, and is reflected by other materials. His experiments further explained reflection, refraction, the velocity of light, and other electromagnetic phenomena. In his honor, the International Electrotechnical Commission (IEC) established the Hertz (Hz) as the SI unit of measurement for frequency, which prior to 1930 was expressed as cycles per second (cps).

hex (hexadecimal notation) See *hexadecimal notation*.

hexadecimal notation (hex) From the Greek *hexadeca*, meaning *sixteen*. A base-16 numbering system. The first 10 numbers are indistinguishable from decimal notation, but the next six numbers use the letters A–F. The full complement of hexadecimal digits is 0, 1, 2, 3, 4, 5, 6, 7, 8, 9, A, B, C, D, E, and F. Programmers use hex, rather than decimal notation, because it easily translates into binary notation, which is the language of computers. It is also easier for a programmer to remember 1F1D than 00011111 00011101. As $2^4 = 16$, each hex number can represent four binary digits (bits). For example, 0010 (binary) = 2 (decimal) = 2 (hex). As a better example, 1101 (binary) = 13 decimal = D (hex). Table H-3 provides a brief comparison of hexadecimal, decimal, and binary notation. See also *binary notation* and *decimal notation*.

Table H-3: Notation Comparison

Hexadecimal	*Decimal*	*Binary*	*Hexadecimal*	*Decimal*	*Binary*
0	0	00000000	11	17	00010001
1	1	00000001	12	18	00010010
2	2	00000010	13	19	00010011
3	3	00000011	14	20	00010100
4	4	00000100	15	21	00010101
5	5	00000101	16	22	00010110
6	6	00000110	17	23	00010111
7	7	00000111	18	24	00011000
8	8	00001000	19	25	00011001
9	9	00001001	1A	26	00011010
A	10	00001010	1B	27	00011011
B	11	00001011	1C	28	00011100
C	12	00001100	1D	29	00011101
D	13	00001101	1E	30	00011110
E	14	00001110	1F	31	00011111
F	15	00001111	20	32	00100000
10	16	00010000	21	33	00100001

HF (High Frequency) HF radio is in the frequency range of 3 MHz–30 MHz and has a wavelength of 100 m–10 m. HF radio has applications in citizens band (CB) radio (also known as *shortwave radio*), mobile radio, and maritime radio systems. See also *electromagnetic spectrum*, *frequency*, *Hz*, and *wavelength*.

HFC (Hybrid Fiber/Coax) Referring to transmission facilities that comprise both optical fiber and coaxial cable. HFC generally refers to a CATV local loop of optical fiber from the provider's headend to a neighborhood node that acts as a media converter between the optical fiber and the embedded coax that runs to the customer premises. The corresponding terms in the telephone networks are fiber-to-the-neighborhood (FTTN) and fiber-to-the-node (FTTN), where the final link to the premises is unshielded twisted pair (UTP). See also *CATV*, *coax*, *local loop*, *optical fiber*, and *UTP*.

high bit-rate digital subscriber line (HDSL) See *HDSL*.

high capacity voice (HCV) See *HCV*.

high-cost area In the context of the United States federal Universal Service Fund (USF), an area where the cost of providing local telephone service is at least 115 percent of the national average. See also *USF*.

high definition television (HDTV) See *HDTV*.

high density bipolar order 3 (HDB3) See *HDB3*.

high frequency (HF) See *HF*.

High-Level Data Link Control (HDLC) See *HDLC*.

high-order mode A relatively highly transverse path taken by an optical signal through a waveguide. Some high-order modes can be so transverse as to be less than the critical angle and, therefore, penetrate the interface between the core and cladding and be permanently lost in the cladding. See mode for more detail. See also *cladding*, *core*, *critical angle*, *low-order mode*, and *waveguide*.

high-pass filter A device that passes all signals above a designated frequency (electrical) or wavelength (optical) band, but absorbs, attenuates, blocks, rejects, or removes all other signals. See also *absorption*, *attenuation*, *band*, *band-pass filter*, *electrical*, *frequency*, *low-pass filter*, *optical*, *signal*, and *wavelength*.

High Performance Parallel Interface (HIPPI) See *HIPPI*.

high speed The United States Federal Communications Commission (FCC) defines high speed services as supporting a data rate of at least 200 kbps in at least one direction and advanced telecommunications capability as at least 200 kbps in both directions. See also *broadband* and *FCC*.

high-speed channel (H channel) See *H channel*.

High-Speed Circuit Switched Data (HSCSD) See *HSCSD*.

High Speed Downlink Packet Access (HSDPA) See *HSDPA*.

High Speed Uplink Packet Access (HSUPA) See *HSUPA*.

high-tier In wireless telecommunications, referring to systems, such as cellular radio systems, that support high-speed vehicular traffic. See also *cellular radio* and *low-tier*.

high voltage (HV) See *HV*.

HiperACCESS (High performance radio ACCESS) A developing broadband wireless local loop (WLL) access technology specified in the Broadband Radio Access Networks (BRAN) project chartered by the European Telecommunications Standards Institute (ETSI). HiperACCESS is targeting frequencies in the 40.5–43.5 GHz band, and is intended to seamlessly interoperate with IEEE 802.16, also known as *WiMAX*. See also *802.16*, *BRAN*, *broadband*, *ETSI*, *IEEE*, *WiMAX*, and *WLL*.

HiperLAN/1 (High performance radio Local Area Network version 1) An ETSI standard (February 2000) for a wireless LAN (WLAN) operating in the 5.725–5.825 GHz range. HiperLAN1 operates at rates up to 20 Mbps, and HiperLAN2 at rates up to 54 Mbps. HiperLAN uses orthogonal frequency division multiplexing (OFDM) as the signal modulation technique. HiperLAN grew out of efforts to develop a wireless version of asynchronous transfer mode (ATM) and a European alternative to IEEE 802.11. HiperLAN, however, largely has been overwhelmed by 802.11a/b/g. See also *802.11, 802.11a, 802.11b, 802.11g, ATM, ETSI, IEEE, modulation, OFDM,* and *WLAN.*

HiperLAN2 (High performance radio Local Area Network version 2) A mobile short-range access network specified in the Broadband Radio Access Networks (BRAN) project chartered by the European Telecommunications Standards Institute (ETSI). HiperLAN/2, a competes directly with IEEE 802.11g/n, aka Wi-Fi. See also *802.11g, 802.11n, BRAN, ETSI, HiperLAN/1,* and *Wi-Fi.*

HiperMAN (High performance radio Metropolitan Area Network) A broadband wireless local loop (WLL) access technology specified in the Broadband Radio Access Networks (BRAN) project chartered by the European Telecommunications Standards Institute (ETSI). HiperMAN operates below 11 GHz, and mainly in the 3.5 GHz band, and is intended to seamlessly interoperate with IEEE 802.16, aka WiMAX. See also *802.16, BRAN, broadband, ETSI, IEEE, WiMAX,* and *WLL.*

HIPPI (HIgh Performance Parallel Interface) The ANSI (American National Standards Institute) specification (X3T9-3, 1991) for a high speed computer bus for the connection of storage devices. The original specification was for 50-pair twisted-pair cable (and huge connectors) supporting a data rate of 100 MBps (800 Mbps) in simplex mode over distances up to 25 meters. Subsequently, a specification was released for optical fiber supporting a data rate of 200 MBps over distances up to 10 km. The most recent specification was for HIPPI-6400, which runs at 6.4 Gbps over 50-pair twisted pair (50 meters) or optical fiber (1 km). HIPPI operates at the Physical Layer and a portion of the Data Link Layer of the OSI Reference Model. HIPPI has been overwhelmed by faster and more compact specifications such as Small Computer System Interface (SCSI) and Fibre Channel. See also *ANSI, bus, Data Link Layer, Fibre Channel, optical fiber, OSI Reference Model, Physical Layer, SCSI, simplex,* and *twisted pair.*

Hockham, George George Hockham and Charles Kao, while engineers at Standard Telecommunications Laboratories, an ITT subsidiary, developed the first conceptual breakthrough in the development of fiber optic transmission systems. In 1966, Kao and Hockham determined that optical fibers of fused silica could satisfy signal attenuation requirements by overcoming issues of absorption, diffusion, and bending loss. At the time, an attenuation of 20 dB per kilometer was considered satisfactory for a commercially viable system. See also *fiber optics.*

Holmes, E.T. The inventor of the first exchange for telephone service. Holmes's father had invented and installed the centralized burglar alarm system in 1858 in Boston, Massachusetts. In 1877, Holmes obtained telephone numbers 6 and 7, and attached them to a wire in his office. He then placed six box telephones on a new shelf in his office. During the daylight hours, the telephone exchange operator could switch any of these telephones into connection with the burglar alarm wires and any two of the six wires could be joined by a wire cord. At night, when the telephone operator was off duty, the telephone network reverted to a burglar alarm network. See also *central office.*

hold A voice telephone system (Centrex, KTS, or PBX) feature that enables a user to place an existing call in a suspended state simply by depressing the hold feature button, with a holding indication usually in the form of a blinking light next to the associated line. The user can reconnect the call at any time by depressing the button associated with the line on hold. In a KTS environment, any user can retrieve the held call from any telephone set where the line appears unless the primary user placed the call on exclusive hold, also known as *I-hold,* which often is initiated by depressing the hold button twice. See also *Centrex, KTS,* and *PBX.*

home page Also known as *page* and *Web page*. A document that serves as a starting point on a Web site. A home page typically contains hypertext and navigation buttons that allows the user to navigate the site by clicking them with a mouse and invoking hyperlinks to other pages and even other sites. See also *hyperlink*, *hypertext*, and *WWW*.

HomePlug A set of standards for in-house broadband over power line (In-house BPL). HomePlug allows a device to connect to a LAN directly through the in-building low voltage (LV) electric lines (110 volts at 50–60 Hz, or 220 volts at 50 Hz). Loosely based on Ethernet standards and using a variation of the CSMA/CA access control technique, HomePlug 1.0 supports up to 16 nodes sharing bandwidth up to a theoretical maximum of 14 Mbps. HomePlug-compatible devices include PCs, routers, bridges, switches, and any other devices that use RJ-45 or universal serial bus (USB) physical interfaces. The devices plug into a HomePlug adapter that is about the size of a typical LV transformer or power adapter, and that plugs into any electrical outlet on the premises. HomePlug uses of a version of orthogonal frequency division multiplexing (OFDM) specially tailored for powerline environments. HomePlug 1.0 specifies 84 equally spaced subcarriers, within each of which several differential modulation techniques are employed. Security is through 56-bit Data Encryption Standard (DES). Attenuation in HomePlug networks is influenced not only by signal propagation through the copper conductors, but also by splices and various components such as fuse boxes, surge suppressors, and circuit breakers. HomePlug currently offers a range of as much as 300 meters without repeaters and deals with issues of electromagnetic interference (EMI) through the mechanisms of forward error correction (FEC) and automatic repeat request (ARQ). See also *Access BPL*, *bandwidth*, *BPL*, *broadband*, *CSMA/CA*, *DES*, *EMI*, *Ethernet*, *FEC*, *In-house BPL*, *LAN*, *modulation*, *LV*, *OFDM*, *propagation*, *RJ45*, *subcarrier*, *transformer*, and *USB*.

home run An inside cable and wire star configuration in which each telephone or data jack connects directly to a common point, such as a demarcation point (demarc), wiring closet, or key service unit (KSU). The alternative is a shared loop that connects multiple jacks to one or two pairs that connect to the demarc or KSU. See also *daisy chain* and *loop*.

hook switch See *switch hook*.

hoot **1.** shout, or holler. **2.** Something or someone hilarious, as in "Billy Bob is a real hoot, don't y'all reckon?" See also *hoot 'n' holler*.

hoot and holler See *hoot 'n' holler*.

hoot 'n' holler An always-on two-way voice system that operates on a point-to-point or point-to-multipoint basis over physical four-wire dedicated circuits, with one transmit pair and one receive pair. A hoot 'n' holler system operates much like a full duplex paging system or intercom system. A typical application allows front desk or service counter personnel to press a button on a handset or a speakerphone and establish a point-to-multipoint connection to speakers, also known as *squawk boxes* or *holler horns*, positioned throughout the manufacturing or assembly floor, or the warehouse, where the personnel can respond by hooting (shouting) and hollering (shouting) into one of the microphones positioned around the area. Telephone companies use hoot 'n' holler systems to contact and converse with central office personnel who can just holler back rather than stopping what they are doing, perhaps climbing down a ladder, and running to answer a telephone, or perhaps just ignoring the call altogether. In such an application, multiple central office systems might be connected back to a central test and dispatch center by dedicated four-wire circuits. Financial and brokerage firms make extensive use of hoot 'n' holler systems for communicating market updates and instructions to brokers and traders on the trading floor. News agencies, publishers, power plants, refineries, and salvage yards all make extensive use of such systems.

hop **1.** A small, quick jump. **2.** In networks, the journey a signal makes across a transmission link between two devices such as bridges, hubs, switches, or routers. There often are multiple links in an end-to-end circuit between two devices. Therefore, a frame may make multiple hops as it transverses a physical path from one workstation to another. An IP packet typically makes a significant number of hops as it transverses the Internet. See also *bridge*, *circuit*, *frame*, *hub*, *Internet*, *LAN*, *link*, *packet*, *router*, and *switch*. **3.** In radio communications, one skip of a radio wave from an earth station to the ionosphere and back. See also *ionosphere*, *radio*, and *waveform*. **4.** In satellite communications, one roundtrip of a signal from an Earth station to a space station (i.e., satellite) and back to an Earth station. See also *satellite* and *signal*. **5.** In frequency hopping spread spectrum (FHSS) radio communications, a small, quick jump from one frequency channel to another in a carefully choreographed hop sequence. See also *channel*, *FHSS*, and *frequency*.

hop sequence In frequency hopping spread spectrum (FHSS) radio communications, the sequence of small, quick jumps between frequency channels. See also *channel*, *FHSS*, and *frequency*.

horizontal cable A type of inside cable designed for horizontal use in non-plenum areas. While horizontal cable must be fire retardant, the National Electrical Code (NEC) specifications are not as demanding as those governing the use of plenum cable or riser cable. See also *NEC*, *plenum*, *plenum cable*, and *riser cable*.

host **1.** The central computer in a mainframe or midrange computer environment to which the networks and terminals connect. See also *computer*, *mainframe computer*, *midrange computer*, *network*, and *terminal*. **2.** In telecommunications, local area networks (LANs), and networks, in general, a server that functions to provide programs or data files to client computers. See also *client*, *LAN*, *telecommunications*, *network*, and *server*. **3.** In the Internet, any computer that can serve as a source or destination for data transfers. An Internet host has a unique Internet Protocol (IP) address and unique domain name. See also *domain*, *Internet*, and *IP*.

hosted PBX Synonymous with IP Centrex. See *IP Centrex*.

hot spare Synonymous with hot standby. See *hot standby*.

hotspot A location where with a sufficiently strong signal from an accessible Wi-Fi wireless LAN (WLAN). Many thousands of public hotspots are available in the United States and many developed countries. Most hotspots are made available through for-profit companies that charge on a daily or monthly basis, although municipalities increasingly deploy free public Wi-Fi networks as a public service. Companies and individuals often unknowingly offer public hotspots by failing to activate Wi-Fi security options provided in 802.11i, more commonly known as Wi-Fi Protected Access (WPA). See also *802.11i*, *Wi-Fi*, *WLAN*, and *WPA*.

hot standby Referring to a redundant system or processor that is not only turned on and warmed up, but is active and prepared to immediately assume the responsibilities of the primary system in the event it suffers a catastrophic failure. Synonymous with *hot spare*.

HSCSD (High-Speed Circuit Switched Data) A 2G+ upgrade to GSM designed to improve data transmission rates. HSCSD improves channel throughput from 9.6 kbps to a maximum of 14.4 kbps in GSM host networks operating at 1800 MHz by lowering overhead through the use of improved mechanisms for forward error correction (FEC). HSCSD also supports the concatenation of multiple time slots per frame in support of higher speeds. As examples, two concatenated time slots yields a transmission rate of up to 28.8 kbps, three yields 43.2 kbps, and four yields 57.6 kbps. See also *2G*, *channel*, *concatenation*, *FEC*, *frame*, *GSM*, *overhead*, *throughput*, *time slot*, and *transmission rate*.

HSDPA (High Speed Downlink Packet Access) Sometimes characterized as a 3.5G cellular radio technology, HSDPA is an upgrade to Universal Mobile Telecommunications System (UMTS) that increases theoretical downlink data rates to 14.4 Mbps, although current implementations support speeds more typically in the range of 400–700 kbps, bursting up to 3.6 Mbps for short periods of time using an adaptive modulation technique to throttle bit rates up and down as the link permits. Work has begun in the 3rd Generation Partnership Project (3GPP) on standards for High Speed Uplink Packet Access (HSUPA). See also *3GPP*, *cellular radio*, *downlink*, *link*, *modulation*, and *UMTS*.

H Series The H series of ITU-T Recommendations addresses audiovisual and multimedia systems. See Table H-4 for selected H-series Recommendations. For a full listing of ITU-T Recommendations, see the contact information in Appendix A.

Table H-4: Selected ITU-T H-Series Recommendations

Recommendation	*Description*
H.100	Visual telephone systems
H.222.1	Multimedia multiplex and synchronization for audiovisual communication in ATM environments
H.223	Multiplexing protocol for low bit rate multimedia communication
H.231	Multipoint control units for audiovisual systems using digital channels up to 1920 kbit/s
H.242	System for establishing communication between audiovisual terminals using digital channels up to 2 Mbit/s
H.243	Procedures for establishing communication between three or more audiovisual terminals using digital channels up to 1920 kbit/s
H.244	Synchronized aggregation of multiple 64 or 56 kbit/s channels
H.245	Control protocol for multimedia communication
H.248	Gateway control protocol
H.261	Video codec for audiovisual services at px64 kbit/s
H.263	Video coding for low bit rate communication
H.310	Broadband audiovisual communication systems and terminals
H.320	Narrow-band visual telephone systems and terminal equipment
H.321	Adaptation of H.320 visual telephone terminals to B-ISDN environments
H.322	Visual telephone systems and terminal equipment for LANs that provide guaranteed QoS
H.323	Packet-based multimedia communications systems

HSUPA (High Speed Uplink Packet Access) A developing upgrade to Universal Mobile Telecommunications System (UMTS) cellular radio that is intended to increase theoretical uplink data rates. High Speed Downlink Packet Access (HSDPA) is already developed and in the process of implementation. HSUPA standards are the responsibility of the 3rd Generation Partnership Project (3GPP). See also *3GPP*, *cellular radio*, *HSDPA*, *UMTS*, and *uplink*.

HTCPCP (Hyper Text Coffee Pot Control Protocol) A protocol defined in IETF RFC 2324 by Larry Masinter of Xerox PARC on April 1, 1998, as an April Fools' Day joke. The specification describes a protocol for monitoring, controlling, and diagnosing networked coffee pots. See also *IETF* and *protocol*.

HTML (Hypertext Markup Language) A tag-based notation language used to format documents for the World Wide Web (WWW) in a manner that can be interpreted by a program known as a Web browser. An application of Standard Generalized Markup Language (SGML), HTML allows authors to insert hyperlinks that display another HTML document when a user clicks on them with a mouse. The Internet Engineering Task Force (IETF) RFC 1866 (1995) standardized HTML 2, the first version to be widely used on the WWW. HTML development currently is the responsibility of the World Wide Web Consortium (W3C). See also *browser*, *hyperlink*, *IETF*, *SGML*, *W3C*, and *WWW*.

hub A central point of interconnection for devices on a local area network (LAN) or LAN segment. Hubs act as passive LAN concentrators and repeaters, with a single internal collapsed backbone bus typically running at a signaling rate of 10/100 Mbps. LAN-attached devices such as workstations, peripherals, and servers typically connect to a hub via either unshielded twisted pair (UTP) or shielded twisted pair (STP). Hubs operate at Layer 1, the Physical Layer of the OSI Reference Model. Filtering hubs also operate at a portion of Layer 2, the Data Link Layer. See also *10Base-T, 100Base-T, concentrator, Data Link Layer, LAN, OSI Reference Model, Physical Layer, repeater, STP*, and *UTP*.

Huffman coding A relatively simple entropy coding technique that assigns codes to symbols, such as characters in an alphabet, numbers in a numbering scheme, and punctuation marks, with the length of the code corresponding to the probability of the occurrence of the symbol. The technique was developed by David A. Huffman when he was a student at the Massachusetts Institute of Technology (MIT). Huffman coding is the basis for Modified Huffman (MH), a run-length encoding compression technique. See also *MH* and *run-length encoding*.

Huh? An informal human-to-human error correction protocol used in voice over frame relay (VoFR), voice over Internet Protocol (VoIP), and voice over Wi-Fi (VoWiFi). As packet networks are designed for data communications applications rather than isochronous traffic, levels of latency, loss, and error are variable and unpredictable in nature. Toll quality, real-time voice communications is highly intolerant of latency, jitter, loss, and error. So voice over packet networks is a challenge. When quality is less than acceptable and the meaning is lost, the Huh? protocol — as in "Huh? What did you say?" — must be invoked. If that fails, the next level is the hang up and call back protocol. Both protocols have been used extensively in cellular networks for many years. See also *error control, isochronous, jitter, latency, protocol, real-time, toll quality, VoFR, VoIP*, and *VoWiFi*.

human A person. Humans are living, breathing entities capable of feeling and showing emotions, such as love, hate, compassion, and indifference. Some, but not all, humans are considerate and generous. Humans are fallible, although some deny that. Most, but not all, humans are forgiving of the faults of others. (I sincerely hope that you love this book passionately, forgive me for any errors you find in it, and generously give copies of it to all of your friends and acquaintances.) Humans are still necessary, although many of their functions have been automated by machines. See also *automated attendant* and *machine*.

hunt The process by which a switch or other device searches for a circuit within a group of lines or trunks in order to complete a connection. See also *hunt group*.

hunt group A group of lines or trunks through which a switch or other device is programmed to search in a predetermined order until it finds one available to complete a connection.

Hush-a-Phone Decision In the United States, the Federal Communications Commission (FCC) decision (1955) that supported AT&T's contention that, under the Communications Act of 1934, even acoustically coupled foreign (non-telco provided) devices cannot be connected to the network without special arrangement. The Hush-a-Phone Corporation marketed a cup-like mouthpiece that mounted on the telephone transmitter. The Hush-a-Phone acted like a megaphone, allowing the speaking party to speak more softly and, thereby reduce the likelihood of being overheard by other parties, while reducing the impact of ambient noise, The Hush-a-Phone came in two models — one for pedestal phones and another for hand-set phones. The decision stated that the device was "deleterious to the telephone system and injures the service rendered by it." The decision was later overturned on appeal. Note: In this era, all telephone equipment was owned by the monopoly telephone company and rented to the consumer. The definition of equipment included not only telephones and telephone systems, but also answering machines, cords, acoustic couplers, and even snap-on mouthpieces like the Hush-a-Phone, which had been sold since 1921. See also *Carterfone Decision, Communications Act of 1934, FCC*, and *monopoly*.

HV (High Voltage) A high amount of electromotive force (emf). The power utilities use HV, at 165,000–765,000 volts, in their main transmission lines. High Voltage is stepped down at substations to medium voltage (MV) of approximately 7,200 volts by transformers for transmission over distribution networks. Transformers in proximity to the subscribers' premises step that voltage down further to low voltage (LV) of 110 volts (or 205 or 220 volts) in North America, and 220 volts in Europe and most of the rest of the world. See also *Access BPL*, *emf*, *LV*, *MV*, *transformer*, *volt*, and *voltage*.

hybrid A circuit, device, or component that comprises multiple elements or performs multiple functions not normally associated with one another. A hybrid communications system might support both digital and analog signals or perhaps both circuit switching and packet switching.

Hybrid ARQ II (Hybrid Automatic Repeat reQuest II) Also known as *incremental redundancy* (IR). An enhanced ARQ technique employed in Enhanced General Packet Radio System (EGPRS), the packet-switched mode of Enhanced Data rates for GSM Evolution (EDGE) cellular radio networks. See also *ARQ*, *EDGE*, *EGPRS*, *IR*, and *packet*.

hybrid fiber/coax (HFC) See *HFC*.

hybrid KTS A key telephone system (KTS) that can function as either a KTS, with direct circuit selection, or as a PBX, with switched access to pooled facilities. Many hybrids can function simultaneously as both a KTS for one workgroup and as a PBX for another. See also *KTS* and *PBX*.

hybrid TDM/IP PBX A PBX that has both TDM and IP components co-existing, side-by-side. The TDM component comprises TDM line and trunk cards and ports and a TDM bus. The IP component comprises Ethernet ports, an Ethernet switch, a router, and IP trunk ports. A gateway interconnects the TDM and IP components, both of which are under the control of a telephony server running a commercial operating system (OS). See also *IPBX*.

hydroxyl (OH) **1.** A negative ion formed by the attachment of an oxygen (O) atom and a hydrogen (H) atom. **2.** In telecommunications, hydroxyl ions are a contaminant introduced into the single-mode optical fiber (SMF) during the manufacturing process. The hydroxyl ions cause water peak attenuation in several wavelength windows, rendering them unusable, which has considerable implications for high speed fiber optics transmission systems (FOTS) employing coarse wavelength division multiplexing (CWDM). See also *attenuation*, *CWDM*, *FOTS*, *ion*, *SMF*, *water peak*, *wavelength*, and *window*.

hyperlink A characteristic or property of an element (e.g., symbol, word, phrase, sentence, or image) in a document that points to and causes to display another document when the user clicks it with a mouse. In a hypertext system, such a linked element is underlined, **bolded**, or otherwise ***emphasized*** to indicate to the user that a link to another document is available. Hyperlinks are created through the use of programming languages such as SGML (Standard Generalized Markup Language) and HTML (Hypertext Markup Language). See also *HTML*, *hypertext*, and *SGML*.

hypermedia The extension of hypertext into a combination of media, including image, animation, video, audio, hyperlinks, and other elements that intertwine into a non-linear document presentation in the form typical of contemporary Web documents. The World Wide Web (WWW) is a global hypermedia system linked through the public Internet. The terms hypertext and hypermedia are attributed to Ted Nelson, who, along with Douglas Englebart, developed the Hypertext Editing System in 1968. See also *hyperlink*, *hypertext*, and *WWW*.

hypertext Text prepared and published in such a way that it is linked together in a non-sequential web of associations that allows the user to navigate through related topics, from one document to another. The author embeds hyperlinks in the text that the user can simply click on to view the related document associated with the link. The World Wide Web (WWW) is a global hypertext system of information residing on servers linked across the public Internet. If this dictionary were in electronic format with hypertext, you could simply click on just about any hyperlinked word (italicized in print) and instantly view the definition of that word, without having to flip pages to find it. Also, the last sentence in this paragraph would disappear, saving ink and paper in the process. The terms hypertext and hypermedia are attributed to Ted Nelson, who, along with Douglas Englebart, developed the Hypertext Editing System in 1968. See also *HTML, hyperlink, Internet, server, text,* and *WWW.*

Hyper Text Coffee Pot Control Protocol (HTCPCP) See *HTCPCP.*

Hypertext Editing System See *hypertext.*

Hypertext Markup Language (HTML) See *HTML.*

Hz **1.** (Hertz) The measurement of frequency, which previous to 1930 was expressed as cycles per second (cps). See *Hertz, Heinrich Rudolf.* **2.** In analog terms, bandwidth is measured in Hz, specifically as the difference between the highest and lowest frequencies over a circuit or within a channel. For example, a channel operating in the range between 4,000 Hz and 8,000 Hz has a bandwidth of 4,000 Hz, or 4 kHz (kiloHertz). Although some applications operate in very low capacity environments, measured in tens of Hz or hundreds of Hz, analog bandwidth more commonly is measured in kHz or kiloHertz (thousands of Hz), MHz or MegaHertz (millions of Hz), GHz or GigaHertz (billions of Hz), and even THz or TeraHertz (trillions of Hz).

I The symbol for current intensity, measured in amperes. See also *current*.

i.e. (id est) Translates literally from Latin as *that is* [to say], meaning *in other words*.

IA (Implementation Agreement) See *Implementation Agreement*.

IAB (Internet Architecture Board) Originally known as the Internet Activities Board. A technical advisory group of the Internet Society (ISOC) that provides oversight for the architecture for the protocols and procedures used by the Internet. The IAB supervises the activities the Internet Engineering Task Force (IETF) and the Internet Research Task Force (IRTF). In combination, those organizations set policy and direction. The IAB comprises 13 expert individuals who use the resources of their sponsoring companies to further the interests of the Internet. See also *architecture*, *IETF*, *Internet*, *IRTF*, *ISOC*, and *protocol*.

IAD (Integrated Access Device) A device installed at the customer premises that enables multiple services to share a single circuit. For example, an IAD might support simultaneous PSTN voice, packet voice or data, and video to share a single local loop. An IAD typically is installed by the telco or other service provider, and may run a combination of PSTN, Ethernet, IP, and frame relay or ATM protocols. See also *ATM*, *Ethernet*, *frame relay*, *IP*, *local loop*, *protocol*, and *PSTN*.

IANA (Internet Assigned Numbers Authority) Originally managed a group of functions performed by the Information Sciences Institute under contract with the United States Department of Defense Advanced Research Project Agency on the ARPANET. Those functions included the assignment of parameters for Internet protocols, management of the Internet Protocol (IP) address space, assignment of domain names, and management of root server functions. Internet protocol parameters managed by IANA include the assignment of Transmission Control Protocol (TCP) ports, which are logical points of connection. In fact, IANA initially was the responsibility of Jon Postel, who performed those functions until his death in 1998. At that time, Internet Corporation for Assigned Names and Numbers (ICANN), an independent not-for-profit organization, assumed the responsibility for managing IANA. See also *ARPANET*; *ICANN*; *Internet*; *IP*; *logical*; *port*; *Postel, Jon*; *protocol*; *root server*; and *TCP*.

ICANN (Internet Corporation for Assigned Names and Numbers) A not-for-profit organization formed in 1999 to assume the responsibilities for management of the Internet Assigned Numbers Authority (IANA). See also *domain name*, *IANA*, *Internet*, *IP*, *logical*, *port*, *protocol*, *root server*, and *TCP*.

ICASA (Independent Communications Authority of South Africa) The regulatory authority in South Africa responsible for broadcasting and telecommunications services.

ICE (In Case of Emergency) The directory name for an emergency contact number that users should enter into their cell phones. If you were to be clobbered by a train or otherwise hurt badly, others could quickly find and call an ICE number to alert friends or relatives.

ICEA (Insulated Cable Engineers Association) A professional organization dedicated to developing cable standards for the electric power, control, and telecommunications industries. ICEA has the objective of ensuring safe, economical, and efficient cable systems utilizing proven state-of-the-art materials and concepts. See Appendix A for contact information.

icky-pic (icky–plastic insulated cable) A type of outside telephone cable that comprises some number of twisted pair copper conductors protected from moisture by an unpleasantly sticky and gooey (i.e., icky) water-blocking gel, and surrounded by a plastic sheath. (Almost as icky is the citrus-based solvent that seems to be required to wash off the gel.) Icky-pic can be used in aerial and direct bury construction. See also *aerial cable* and *direct bury cable*.

ICMP (Internet Control Message Protocol) An extension to the original Internet Protocol (IP) that reports errors that may have occurred in the processing of datagrams. For example, a datagram may be undeliverable or an incorrect route may have been chosen. ICMP supports the testing of a path to a distant host computer through an echo function known as the ping utility. ICMP also supports the requesting of a subnet mask. ICMP is integral to IP and must be implemented in both hosts and routers. See also *datagram*, *echo*, *host*, *IP*, *ping*, and *utility*.

ICST (Information and Communication Science and Technology) An international term used to describe a blend of information services (IS) and telecommunications science and technology. See also *IS*.

ID **1.** Inside Diameter. **2.** Identification.

IDDD (International Direct Distance Dialing) A feature of the public switched telephone network (PSTN) that enables a subscriber to dial an international long-distance telephone number directly, i.e., without the intervention of an operator. Domestic DDD was introduced by the Bell System in 1951. DDD and IDDD are virtually ubiquitous today, although they are not available in some developing countries. See also *appointment call*, *long distance*, and *PSTN*.

IDF (Intermediate Distribution Frame) A distribution frame that serves as an intermediate point of inside cable and wire interconnection between the main distribution frame (MDF) and the terminal blocks or terminal outlets. See also *MDF*.

IDSL (ISDN Digital Subscriber Line) A DSL variant that employs ISDN BRI (Basic Rate Interface) technology to deliver symmetric transmission rates of 128 kbps or 144 kbps on unshielded twisted pair (UTP) local loops as long as 18,000 feet. IDSL bonds the two 64-kbps bearer (B) channels to provide a 128-kbps channel, and bonds the 16-kbps data (D) channel to bring the total to 144 kbps. Unlike ISDN, which is a circuit-switched network service for voice, data, fax, video, and multimedia, IDSL operates only at the local loop level to provide always-on Internet access. IDSL is virtually non-existent in the United States, where ISDN never enjoyed any significant success. As IDSL compares so unfavorably with asymmetric digital subscriber line (ADSL) and other DSL variants with respect to bandwidth, the very low IDSL penetration rates will only decrease. See also *ADSL*, *always on*, *bandwidth*, *bonding*, *BRI*, *circuit switch*, *Internet*, *ISDN*, *local loop*, *symmetric*, *transmission rate*, *UTP*, and *xDSL*.

IDU (Interface Data Unit) In asynchronous transfer mode (ATM), referring to a frame of data presented to an ATM interface for switching. ATM Adaptation Layer 3/4 (AAL3/4) supports message mode service, in which only one IDU is passed. AAL3/4 also supports streaming mode service, in which multiple interface data units (IDUs) are passed in a data stream. An IDU can be up to 65,535 octets in length, with a cyclic redundancy check (CRC) added as part of the trailer. See also *AAL3/4*, *ATM*, *CRC*, *frame*, *IDU*, *message mode service*, *octet*, *streaming mode service*, and *trailer*.

IE (Internet Explorer) See *Internet Explorer*.

IEC **1.** International Electrotechnical Commission. The IEC was formed in 1904 with the objective of standardizing the nomenclature and ratings of electrical apparatus and machinery. Among its accomplishments are the development of the International Electrotechnical Vocabulary and the Système international d'unités (SI). See Appendix A for contact information. See also *SI*. **2.** InterExchange Carrier. See *IXC*.

IEEE (Institute of Electrical and Electronics Engineers) A worldwide not-for-profit professional association for the advancement of technology. The IEEE establishes standards and otherwise serves as a leading technical authority in areas including aerospace systems, computers and telecommunications to biomedical engineering, electric power, and consumer electronics. In telecommunications, the IEEE is most notable for its 802 Working Group, which set local and metropolitan area network (LAN and MAN) standards. The

IEEE formed in 1963, with the merger of the American Institute of Electrical Engineers (AIEE), formed in 1884, and the Institute of Radio Engineers (IRE), formed in 1912. See also *LAN* and *MAN*.

IESG (Internet Engineering Steering Group) The group responsible for the day-to-day management of the activities of the Internet Engineering Task Force (IETF), including management of the standards process. The IESG provides the final technical review of standards submitted by the IETF. See also *IETF* and *standard*.

IETF (Internet Engineering Task Force) The group of the Internet Society (ISOC) that identifies, prioritizes, and addresses short-term Internet issues and problems, including protocols, architecture, and operations. The IETF publishes proposed Internet standards in the form of Requests for Comment (RFCs). Once the final draft of a standard is prepared, it is submitted to the Internet Engineering Steering Group (IESG) for approval. The IETF operates under the supervision of the Internet Architecture Board (IAB). See also *architecture*, *IESC*, *IAB*, *Internet*, *ISOC*, *protocol*, *RFC*, and *standard*.

I/F (Inter/Face) See *interface*.

IFP (Internet Fax Protocol) See *T.38*.

IGF (International Gateway Facility) A point of interconnection between an international carrier and a national carrier. An IGF commonly serves not only as a physical gate between the international and national networks, but also as a point-of-protocol conversion, perhaps interfacing an international E-carrier circuit to a domestic T-carrier circuit. The gateway also may serve as a point of media conversion, perhaps serving to interconnect a submarine fiber optic cable or satellite link to a microwave or copper wire circuit. See also *carrier*, *media converter*, *network*, and *protocol converter*.

IGP (Interior Gateway Protocol) An interior protocol for routing within an Autonomous System (AS), i.e., a group of routers within a given administrative domain. IGP was described in IETF RFC 1074 (1988). Common IGPs include the Routing Information Protocol (RIP) and the Open Shortest Path First (OSPF) protocol. See also *AS*, *domain*, *exterior protocol*, *IETF*, *interior protocol*, *OSPF*, *protocol*, *RIP*, *router*, and *routing*.

I-hold See *hold*.

III (International Information Infrastructure) Synonymous with Global Information Infrastructure (GII). The international version of the National Information Infrastructure (NII), or Information Superhighway. See also *Information Superhighway* and *NII*.

ILEC (Incumbent Local Exchange Carrier) A company providing local telephone service prior to the introduction of local competition, which introduced one or more competitive LECs (CLECs). In the United States, an ILEC is a company that as of February 8, 1996 provided telephone exchange services to the area in which it was authorized to provide service and was permitted to participate as a member of the National Exchange Carrier Association (NECA). See also *carrier*, *IXC*, and *NECA*.

ILF (InfraLow Frequency) ILF radio is in the frequency range of 300–3000 Hz (3 kHz) and has a wavelength of 1,000–100 km. ILF radio has no contemporary telecommunications applications. See also *electromagnetic spectrum*, *frequency*, *Hz*, and *wavelength*.

IM (Instant Messaging) A client/server messaging technology that is much like e-mail, but operates in near real time. Instant messaging originated in the 1970s on PLATO, a private online instructional system for schools and universities in the United States, and was popularized in 1996 by ICQ, an Israel-based company later acquired by AOL. (Note: Instant Message is a Service Mark (^{SM}) of AOL.) There are now a number of public Web-based IM services and enterprise systems, all of which are proprietary, i.e., non-standard. IM users create, by mutual consent, closed user groups (CUGs), commonly known as buddy lists, of correspondents. As IM occurs in near real time, it is necessary that both correspondents in a given message session be online at the same time. Therefore, IM systems include a presence mechanism to advertise all users of the status (e.g., available or unavailable) of all other users. Some IM systems now support one-way messaging if the recipient is not online. In this mode, the recipient can access the message at a later time, much like an e-mail communication. IM features typically include presence, privacy, contact lists (buddy lists), attachments, and message history. Some systems also include text, voice, and video and conferencing, and even whiteboarding. See also *client/server*, *CUG*, *e-mail*, *near-realtime*, *presence*, *proprietary*, *standard*, and *whiteboarding*.

IMA (Inverse Multiplexing over ATM) An inverse multiplexing technique that fans out an Asynchronous Transfer Mode (ATM) cell stream across multiple circuits between the user premises and the edge of the carrier network. In such a circumstance, multiple physical T1 circuits, for example, can be used as a single, logical ATM pipe. The IMA-compliant ATM concentrator at the user premises spreads the ATM cells across the T1 circuits in a round-robin fashion, and the ATM switch at the edge of the carrier network scans the T1 circuits in the same fashion in order to reconstitute the cell stream. See also *ATM*, *carrier*, *concentrator*, *multiplexing*, *network*, *switch*, and *T1*.

image A still photograph or other still visual representation of a person, place, or thing. Video comprises a series of still images presented in rapid succession.

image decimation Also known as *filtering*. See *filtering*.

IMAP (Internet Message Access Protocol) An IETF standard (RFC 2193) protocol for accessing e-mail. IMAP enables the client to manage mail much more effectively than POP3, an earlier standard that is widely deployed. IMAP enables the user to view the header of each mail message before deciding whether to download it, delete it, or take other action. IMAP also enables the user to create, manipulate, and delete individual mail folders and mailboxes on the server. However, IMAP requires that the connection be maintained between client and server continuously while working with mail, whereas POP3 enables the user to work with mail offline. Also, security is an issue with IMAP, as the remote client takes on the appearance of a remote virtual server. See also *client*, *e-mail*, *IETF*, *POP3*, *RFC*, *server*, and *virtual server*.

impedance (Z) The total passive opposition offered by a circuit to the flow of an alternating electric current (AC), impedance is a combination of resistance, inductive reactance, and capacitive reactance. See also *AC*, *capacitance*, *inductance*, and *resistance*.

Imperial Standard Wire Gauge Synonymous with British Standard Gauge (BSG). The measure of the diameter, or thickness, of a conductor in England. See also *gauge*.

i-Mode (Internet Mode). A microbrowser technology that supports text, graphics, audio, and video for Web access over the Japanese cellular network. In consideration of the inherently limited bandwidth of the cellular network, i-Mode employs Compact HTML (C-HTML), a simplified version of HTML similar to Wireless Markup Language (WML) used in WAP networks. Transmission between the handhelds and the i-Mode-enabled cell sites is via packet mode, using packets of 128 octets, at rates up to 9.6 kbps. i-Mode is a proprietary service developed by NTT DoCoMo, initially for the Japanese market. See also *browser*, *cellular radio*, *octet*, *packet*, *WAP*, and *WML*.

IMP (Instant Messaging and Presence) See *IMPP* and *SIMPLE*.

Implementation Agreement (IA) Consensus agreements, developed and promoted by the Frame Relay Forum (FRF), that address manufacturer interoperability issues. (Note: The Frame Relay Forum merged with the ATM Forum and MPLS Forum to form the MFA Forum.) Table I-1 shows relevant IAs. See also *frame relay* and *MFA Forum*.

Table I-1: Frame Relay Implementation Agreements

Implementation Agreement	*Description*
FRF.1.2	User-to-Network (UNI) (April 2000)
FRF.2.1	Network-to-Network Interface (NNI) (July 1995)
FRF.3.2	Multiprotocol Encapsulation (MEI) (April 2000)
FRF.4.1	SVC User-to-Network Interface (UNI) (January 2000)
FRF.5	Frame Relay/ATM PVC Network Interworking (December 1994)
FRF.6.1	Customer Network Management (MIB) (September 2002)
FRF.7	PVC Multicast Service and Protocol Description (October 1994)
FRF.8.2	Frame Relay/ATM PVC Service Interworking (February 2004)
FRF.9	Data Compression (January 1996)
FRF.10.1	Network-to-Network SVC (September 1996)
FRF.11.1	Voice over Frame Relay (May 1997, Annex J added March 1999)
FRF.12	Fragmentation (December 1997)
FRF.13	Service Level Definitions (August 1998)
FRF.14	Physical Layer Interface (December 1998)
FRF.15	End-to-End Multilink Frame Relay (August 1999)
FRF.16.1	Multilink Frame Relay UNI/NNI (May 2002)
FRF.17	Privacy (January 2000)
FRF.18	Network-to-Network FR/ATM SVC Service Interworking (April 2000)
FRF.19	Operations, Administration and Maintenance (March 2001)
FRF.20	IP Header Compression (June 2001)

Implicit Congestion Notification In frame relay, inference by user equipment that congestion has occurred in the network. The inference is triggered by realization of the receiving frame relay access device (FRAD) of transmission delays. Based on block, frame or packet sequence numbers, another protocol may recognize that one or more frames have been lost in transit. Control mechanisms at the upper protocol layers of the end devices then deal with frame loss by requesting retransmissions. See also *block*, *Explicit Congestion Notification (ECN)*, *FRAD*, *frame*, *packet*, and *protocol*.

IMPP (Instant Messaging and Presence Protocol) A group of specifications proposed by the Internet Engineering Task Force (IETF) and intended to define the protocols necessary to build an IM system that will scale to Internet size. The RFCs define presence requirements, and common semantics and data formats to facilitate the development of gateways between services. See also *gateway*, *IETF*, *IM*, *Internet*, *presence*, *protocol*, *RFC*, and *SIMPLE*.

Improved Mobile Telephone Service (IMTS) Also known as *specialized mobile radio* (SMR) and *trunk mobile radio* (TMR). See *SMR*.

impulse noise Noise on a circuit that can be caused by voltage spikes in equipment, voltage changes on adjacent pairs in a copper cable, tones generated for network signaling, maintenance and test procedures, lightening flashes during thunderstorms, and a wide variety of other phenomena. As impulse noise is short in duration ($\frac{1}{100}$ of a second, or so); it has little effect on voice communications, but can cause bit errors in a data transmission. See also *noise*.

IMS (IP Multimedia Subsystem) An architectural concept built around a packet core and providing an environment in which a user can access a wide range of multimedia services using any device and any type of network connection. IMS supports Internet Protocol (IP) sessions between devices over any type of connection and protocol, whether wireline or wireless in nature. IMS will support sessions between devices in the PSTN, Internet, WLAN, and cellular domains, recognizing the limitations of each and adjusting as required, even as a terminal device roams amongst them. IMS manages internetwork handoffs, bandwidth negotiation and quality of service (QoS), while it keeps peers engaged in the session advised via the Session Initiation Protocol (SIP) as to the level of multimedia presence. IMS originated in the 3rd Generation Partnership Project (3GPP), which was seeking a common means by which GSM cellular operators could deliver data services. IMS subsequently transcended the cellular domain, and is now being embraced by both wireless and wireline service providers. Industry groups such as the Multiservice Switching Forum (MSF), European Telecommunications Standards Institute (ETSI), and Alliance for Telecommunications Industry Solutions (ATIS) have adopted IMS as the foundation for their next-generation infrastructure strategies. See also *3GPP*, *architecture*, *ATIS*, *cellular radio*, *ETSI*, *Internet*, *IP*, *multimedia*, *packet*, *protocol*, *PSTN*, *QoS*, *session*, *SIP*, *wireless*, *wireline*, and *WLAN*.

IMT (InterMachine Trunk) A high-capacity, multichannel circuit that interconnects circuit switches in the core of a carrier network.

IMT-2000 (International Mobile Telecommunications-2000) An initiative of the ITU-R for a twenty-first century wireless network architecture that replaced Future Public Land Mobile Telecommunications System (FPLMTS) as that organizations vision for a single global standard for digital wireless networks. Specifications call for operation on the 2 GHz (2000 MHz) band and include high-mobility applications at 128/144 kbps, pedestrian speed applications at 384 kbps, and fixed wireless applications such as wireless local loop (WLL) and in-building applications such as wireless LANs (WLANs) at 2.048 Mbps. Technologies and standards falling under the IMT-2000 umbrella include 2.5G and 3G such as Enhanced Data rates for GSM Evolution (EDGE), General Packet Radio Service (GPRS), High-Speed Circuit Switched Data (HSCSD), and Universal Mobile Telecommunications System (UMTS). See also *2.5G*, *3G*, *cellular radio*, *digital*, *EDGE*, *GPRS*, *HSCSD*, *UMTS*, *wireless*, *WLAN*, and *WLL*.

IMTC (International Multimedia Telecommunications Consortium) A not-for-profit organization with the stated mission of promoting, encouraging, and facilitating the development and implementation of interoperable multimedia teleconferencing and telecommunications solutions through open standards. IMTC focus is on the T.120 and H.320 standards suites for data conferencing and video telephony, respectively. See Appendix A for contact information.

IMTS (Improved Mobile Telephone Service) Also known as *specialized mobile radio* (SMR) and *trunk mobile radio* (TMR). See *SMR*.

IN (Intelligent Network) A public switched telephone network (PSTN) that, at a minimum, provides for the switches to consult centralized, service- and customer-specific databases for routing instructions and authorization code verification. Intelligent Network Version 1 (IN/1) was conceived at Bell Labs and unveiled in 1976 with the introduction of INWATS (800) services and the first common channel signaling (CCS) system. IN/1 services included INWATS, calling card verification, and voice virtual private networks (VPNs). IN is dependent on the service creation element (SCE), a set of modular programming tools permitting services to be developed independently of the switch, thereby divorcing the service-specific programmed logic from the switch logic. This enables the service to be developed independently and be made available to all switches in the network. IN/1 was succeeded by the advanced intelligent network (AIN), developed by Bell Labs in the early 1980s. See also *AIN*, *Bell Labs*, *CCS*, *database*, *PSTN*, *routing*, *SCE*, *switch*, and *VPN*.

in-band signaling and control Signaling and control that takes places over the same physical path (i.e., through the same switches and across the same circuit) and either occupies the same frequencies or competes for the same time slots as the user payload. In-band signaling is intrusive, or disruptive, in nature. As examples, touch-tone signals can disrupt a voice conversation and call-waiting alerts can terminate a modem connection. T-carrier signaling clearly is in-band rather than out-of-band, as it involves bit robbing, which periodically replaces payload bits with signaling bits. E-carrier signaling and control occurs exclusively in time slots reserved for that purpose. See also *frequency*, *out-of-band signaling and control*, *payload*, *signaling and control*, and *time slot*.

incident angle See *angle of incidence*.

incremental redundancy (IR) See *IR*.

incumbent local exchange carrier (ILEC) See *ILEC*.

Independent Communications Authority of South Africa (ICASA) See *ICASA*.

independent telephone company Referring to United States telephone companies that are not part of the Bell System, i.e., are not owned by AT&T. The term was obsoleted when AT&T divested the operating telephone companies in 1984 as a result of the Modified Final Judgement (MFJ), also known as the Divestiture Decree. The term is now doubly obsolete as the AT&T of the twenty-first century bears no resemblance to that of the nineteenth and twentieth centuries. Nonetheless, more than 1,000 telephone companies continue to refer to themselves as independent. See also *MFJ*.

index-matching gel A gelatinous substance with an index of refraction (IOR) that closely matches that of the core of an optical fiber. Mechanical splices and connectors are filled with index-matching gel to reduce Fresnel reflections from a fiber end face. See also *connector*, *core*, *Fresnel reflection*, *IOR*, *mechanical splice*, and *optical fiber*.

index of refraction (IOR) See *IOR*.

inductance (L) The property of an electric circuit or device by virtue of which a varying current induces an electromotive force (emf), i.e., voltage (V), in that circuit or device, or in an adjacent circuit or device. See also *circuit*, *current*, *emf*, and *voltage*.

inductive reactance (X_L) The opposition to the flow of alternating electric current (AC) in an inductor. Inductive reactance is an inertial reaction to changes in the electromagnetic field created when an alternating voltage is applied. When AC passes through a component that contains reactance, energy is alternately stored in and released from a magnetic field or electric field. In the case of magnetic energy, the reactance is inductive. The greater the amount of inductance, the greater the inertial opposition. The faster the reversal of current, the greater the inertial opposition. Inductive reactance is measured in Ohms (Ω). See also *capacitive reactance*, *inductance*, and *reactance*.

inductor A device comprising one or more windings of a conductive material, around a core of air or a ferromagnetic material, for introducing inductance into an electric circuit. An inductor opposes changes in current, whereas a capacitor opposes changes in voltage. See also *capacitor*, *inductance*, and *inductive reactance*.

Industrial/Scientific/Medical (ISM) See *ISM*.

InfiniBand An architecture and specification for data flows between processors and high performance I/O devices such as servers in a storage area network (SAN), InfiniBand is a high-performance switched fabric interconnect standard. The baseline 1X InfiniBand specification supports a signaling rate of 2.5 Gbps using 8B10B coding, which yields a data rate of 2 Gbps. The more capable 4X specification quadruples the signaling rate to 10 Gbps and the data rate to 8 Gbps by spreading a datastream over four bonded links. The 12X specification supports a signaling rate of 30 Gbps and a data rate of 24 Gbps. Double-rate and quad-rate options effectively double and quadruple each of these theoretical speeds. See also *8B10B*, *architecture*, *bonding*, and *SAN*.

InFLEXion See *FLEX*.

.info (information) Pronounced dot info. The Internet generic Top Level Domain (gTLD) intended for, although not restricted to, informational sites. This is an unsponsored domain. See also *gTLD*, *Internet*, and *unsponsored domain*.

infobahn A play on the terms information superhighway and the German autobahn, an expressway known for its high speed limits. Actually, posted speed limits on the autobahn are more along the lines of suggestions. See *information superhighway*.

Information and Communication Science and Technology (ICST) See *ICST*.

information service As defined in the U.S. Telecommunications Act of 1996, the offering of a capability for generating, acquiring, storing, transforming, processing, retrieving, utilizing, or making available information via telecommunications, and includes electronic publishing, but does not include any use of any such capability for the management, control, or operation of a telecommunications system or the management of a telecommunications service. As interpreted by the Federal Communications Commission (FCC), broadband wireline services such as digital subscriber line (DSL) and cable modem service are information services, rather than telecommunications services. See also *broadband*, *cable modem*, *DSL*, *FCC*, *Telecommunications Act of 1996*, *telecommunications service*, and *wireline*.

Information Superhighway Also known as the infobahn. The Internet and its physical infrastructure, including access and transport circuits, switching and routing systems, public and private networks, and online services, with emphasis on high-speed and broadband capabilities. The term was popularized in the United States during the Clinton-Gore administration (1993–2001), and is generally associated with Vice President Al Gore. The Clinton-Gore administration is no longer in power and the term is now considered obsolete. The National Information Infrastructure (NII) is a similar concept promoted by the United States government. See also *access circuit, Internet, network, NII, router, service, switch, system, and transport circuit.*

information technology (IT) See *IT*.

infralow frequency (ILF) See *ILF*.

infrared (IR) See *IR*.

infrastructure mesh In wireless local area networks (WLANs), a node mesh, that is, a mesh by which the majority of access points (APs) interconnect on a peer-to-peer basis through wireless RF links, with only those at the logical edge of the mesh connecting back to the wired LAN domain. An infrastructure mesh eliminates the requirement for cabling from the APs or wireless routers to wired ports on switches, or for cabling between APs. See also *AP*, *mesh topology*, *pure mesh*, *RF*, and *WLAN*.

infrastructure mode In wireless local area networks (WLANs), a manner of operation that involves a centralized hub, or access point (AP), through which computers and peripherals interconnect and intercommunicate. See also *ad hoc mode* and *WLAN*.

In-house BPL (In-house Broadband over Power Line) A data communications transmission technology that allows a device to connect to a local area network (LAN) directly through the low voltage (LV) electric grid inside the premises. The LV grid runs at 110 volts at 50–60 Hz in North America, and 220 volts at 50 Hz in Europe and most of the rest of the world. HomePlug 1.0 standards support up to 16 nodes sharing theoretical bandwidth of up to 14 Mbps over a LAN based loosely on Ethernet standards.

HomePlug uses of a version of orthogonal frequency division multiplexing (OFDM) specially tailored for powerline environments. HomePlug 1.0 specifies 84 equally spaced subcarriers, within each of which several differential modulation techniques are employed. Security is through the 56-bit Data Encryption Standard (DES). See also *Access BPL*, *bandwidth*, *BPL*, *broadband*, *DES*, *HomePlug*, *LAN*, *node*, *OFDM*, *PLC*, and *voltage*.

in-house broadband over power line (In-house BPL) See *In-house BPL*.

injection fiber See *launch cable*.

initialism An unpronounceable abbreviation comprising the initial letters of a term and commonly used in place of that term. Each letter of an initialism is pronounced independently. Fox example, SDH is the initialism for Synchronous Digital Hierarchy. An acronym is a pronounceable word formed from the initials or other parts of several words that comprise a term. SONET, for example, is the acronym for Synchronous Optical NETwork, the North American standard for fiber optic transmission systems that later became internationalized as SDH. Acronyms and initialisms generally comprise all uppercase letters, although they sometimes are all lower case, or a mixture of upper and lower case. Examples include SONET, SDH, bit (binary digit), bps (bits per second), QoS (Quality of Service), and Sesame (Secure European Systems for Applications in a Multivendor Environment). See also *abbreviation*, *acronym*, and *contraction*.

input/output (I/O) See *I/O*.

INRIA (National de Recherche en Informatique et en Automatique) Translates from French as the National Institute for Research in Computer Science and Control. The French *National Institute for Research in Computer Science and Control* operates under the dual authority of the Ministry of Research and the Ministry of Industry, and dedicated to fundamental and applied research in information and communication science and technology (ICST).

insertion gain The gain, or increase in signal power, resulting from the insertion of a component, such as an amplifier or repeater, in a circuit. Insertion loss is measured as a comparison of signal power at the point the incident energy strikes the component and the signal power at the point it exits the component. Insertion gain typically is measured in decibels (dB), although it also may be expressed as a coefficient or a fraction. A negative gain is a loss. See also *attenuation*, *dB*, and *insertion loss*.

insertion loss **1.** The attenuation, or loss in signal power, resulting from the insertion of a component, such as a connector or splice, in a circuit. Insertion loss is measured as a comparison of signal power at the point the incident energy strikes the component and the signal power at the point it exits the component. Insertion loss typically is measured in decibels (dB), although it also may be expressed as a coefficient or a fraction. A negative loss is a gain. See also *attenuation*, *dB*, and *insertion gain*. **2.** In a fiber optic transmission system (FOTS), insertion loss is a measure of loss across a circuit due to all factors, including absorption, bending loss from both macrobends and microbends, diffusion, dispersion, Fresnel reflection, and leaky modes. See *absorption*, *bending loss*, *macrobend*, *microbend*, *diffusion*, *dispersion*, *Fresnel reflection*, *insertion gain*, and *leaky mode*.

inside plant (ISP) See *ISP*.

inside vapor deposition (IVD) See *IVD*.

inside wire and cable Referring to wire and cable systems inside a customer premises, often owned by the end user. See also *ISP* and *OSP*.

instant messaging (IM) See *IM*.

Instant Messaging and Presence (IMP) See *IMPP* and *SIMPLE*.

Instant Messaging and Presence Protocol (IMPP) See *IMPP*.

Institute of Electronic and Electrical Engineers (IEEE) See *IEEE*.

l'Institut National de Recherche en Informatique et en Automatique (INRIA) See *INRIA*.

Insulated Cable Engineers Association (ICEA) See *ICEA*.

insulator **1.** A material that does not conduct electricity, heat, light, etc. **2.** In the context of telecommunications, an insulator is a material that does not conduct electricity, i.e., a dielectric. See also *conductor* and *dielectric*.

insulation The dielectric material that surrounds a metal conductor and prevents it from touching another conductor or the ground and, thereby, shorting the circuit. Insulation typically is made of some sort of plastic material. Inside wire and cable standards vary according to the applications, but generally favor insulation that is characterized as low flame-spread, low-smoke, and low-toxicity.

insure **1.** To obtain an insurance policy on something. **2.** Used in error to mean to make certain of something of something by taking necessary precautions. I vote for #1, exclusively. It makes me crazy when people talk about how some process or some such thing insures some result. I think that the only thing that insures something is an insurance policy, and even that only insures that somebody gets paid when the barn burns down or some such thing and even that assumes that the insurance company will actually live up to the terms of the policy and pay up. Oh, well, maybe it's just me utilizing my editorial privilege and ranting on and on and on in unnecessarily complex sentences. See also *ensure* and *utilize*.

.int (international) Pronounced *dot i-n-t*. The Internet generic Top Level Domain (gTLD) reserved exclusively for organizations formed under international treaties between governments. This is an unsponsored domain. See also *gTLD*, *Internet*, and *unsponsored domain*.

integrated access device (IAD) See *IAD*.

integrated circuit Synonymous with microcircuit and semiconductor chip. In computer hardware, a miniaturized electronic circuit comprising many individual circuit elements, such as transistors, diodes, resistors, capacitors, and inductors, etched on a tiny wafer of semiconducting material such as silicon. See also *capacitor*, *circuit*, *diode*, *electronics*, *inductor*, *resistor*, and *transistor*.

integrated messaging Synonymous with multimedia messaging and unified messaging. See *unified messaging*.

integrated services digital network (ISDN) See *ISDN*.

Integrated Services LAN (ISLAN) Specified by the IEEE as 802.9 and also known variously as IsoEthernet and IsoEnet. ISLAN is considered obsolete. See also *802.9*, *IEEE*, and *IsoEthernet*.

intellectual property Property derived from the work of human intellect. Intellectual property laws cover a wide range of property created by artists, authors, inventors, and musician, and protect copyrights, patents, trademarks, and trade secrets. See also *copyright*, *patent*, and *trademark*.

intelligent network (IN) See *IN*.

intelligent peripheral (IP) See *IP*.

Interactive Video and Data Services (IVDS) See *218-219 MHz Service*.

interactive voice response (IVR) See *IVR*.

intercom An abbreviation of intercommunication, an intercom system is a closed (i.e., not networked) system that allows members of a group to converse with each other. An intercom system can be standalone, although Centrex systems, key telephone systems (KTSs), and PBXs commonly feature intercom capabilities. See also *Centrex*, *KTS*, and *PBX*.

interexchange carrier (IEC or IXC) See *IXC*.

interface **1.** The common physical point, boundary, surface, or plane where two things touch, meet, or come together. The interface between the core and cladding in a glass optical fiber (GOF) is an example of such a physical interface. See also *cladding*, *core*, and *GOF*. **2.** The device or component that serves to physically and logically interconnect two other devices or systems and that enables their interoperation. Such an interface may comprise a combination of hardware and firmware. A trunk interface that connects a trunk and a switch or router is an example of such an interface. See also *hardware* and *firmware*. **3.** Software that logically interconnects two computers or a computer and another device and allows them to interoperate. Such software generally is characterized as performing a gateway function. A gateway that accomplishes protocol conversion between X.25 and frame relay is an example. See also *gateway*, *protocol*, and *software*. **4.** Software that enables a user to work with a computer program perhaps in an intuitive way. Examples include a command-line interface and a graphical user interface (GUI). See also *program* and *software*. **5.** Software that enables a computer to work with another program, or with the computer hardware. Such software enables the computer to function as a whole. See also *API*, *hardware*, *program*, and *software*.

interface data unit (IDU) See *IDU*.

interference **1.** In general, spurious or extraneous energy that appears in the circuitry of a system or component and impedes the reception of desired signals. **2.** In radio communications, the negative impact of undesired energy by emission, radiation, or induction on the reception of desired signals.

interferometer See *etalon*.

interframe compression A step in video compression that considers and eliminates redundant information in successive video frames. The background of a movie scene, for example, might not change, even though the actors move around the set. See also *compression*, *frame*, and *video*.

Interim Standard 54 (IS-54) See *IS-54*.

Interim Standard 136 (IS-136) See *IS-136*.

Interim Standard 856 (IS-856) See *IS-856*.

Interim Standard 2000-A (IS-2000-A) See *IS-2000-A*.

Interior Gateway Protocol (IGP) See *IGP*.

interior protocol A protocol concerned with routing within a network. Interior Gateway Protocol (IGP) is an interior protocol. See also *exterior protocol*, *IGP*, *protocol*, and *routing*.

interlaced scanning The process of refreshing a video screen that is used with most analog TV systems. Interlaced scanning involves two fields. Odd lines (field 1) are refreshed in one scan, and even lines (field 2) in the next. Each set of odd and even lines refreshed constitutes a frame refreshed. See also *progressive scanning* and *scanning*.

interLATA Referring to a long distance call between Local Access and Transport Areas (LATAs). In the United States, LATA boundaries were defined in the Modified Final Judgement (MFJ) that broke up the AT&T Bell System on January 1, 1984. The Regional Bell Operating Companies (RBOCs) and their component Bell Operating Companies (BOCs) initially were prevented from offering interLATA toll services. See also *BOC*, *LATA*, *MFJ*, and *RBOC*.

interleaving See *byte interleaving*.

intermachine trunk (IMT) See *IMT*.

intermediate distribution frame (IDF) See *IDF*.

intermodulation noise Noise that is the result of modulation, demodulation, and any nonlinear characteristics of the transmission medium or transmission system components. See also *modulation* and *noise*.

internal reflection See *total internal reflection*.

International Direct Distance Dialing (IDDD) See *IDDD*.

International Electrotechnical Commission (IEC) See *IEC*.

international gateway facility (IGF) See *IGF*.

International Information Infrastructure (III) Synonymous with Global Information Infrastructure (GII). The international version of the National Information Infrastructure (NII), or Information Superhighway. See also *Information Superhighway* and *NII*.

International Mobile Telecommunications-2000 (IMT-2000) See *IMT-2000*.

international Morse code See *Morse code*.

International Multimedia Telecommunications Consortium (IMTC) See *IMTC*.

International Organization for Standardization (ISO) See *ISO*.

International Record Carrier (IRC) See *IRC*.

International Telecommunication Union (ITU) See *ITU*.

International Telecommunication Union-Telecommunication Standardization Sector (ITU-T or ITU-TSS) See *ITU-T*.

International Telecommunication Union-Radiocommunication Sector (ITU-R) See *ITU-R*.

International Telegraph Alphabet #2 (ITA #2) See *Baudot code*.

internet An interconnection of networks that is so seamless as to appear to the user as one network. The networks can include local area networks (LANs), metropolitan area networks (MANs), and wide area networks (WANs). See also *Internet*, *LAN*, *MAN*, *network*, *seamless*, and *WAN*.

Internet A massive, global network of packet data networks based on the Internet Protocol (IP) suite. The Internet is grounded in the U.S. Department of Defense ARPANET (Advanced Research Projects Agency NETwork), which began in 1969 as a means of linking personnel and systems involved in various computer science and military research projects. The Internet since has grown to comprise more than 400 million hosts connected to more than 60,000 academic, business, and governmental networks in more than 150 countries. The Internet also has evolved to support not only data, but also voice, image, video, facsimile, audio, and multimedia communications. Fundamental to the Internet is the Internal Protocol (IP) suite, which, in the context of the OSI Reference Model, includes the Internet Protocol (IP) at the Network Layer and Transmission Control Protocol (TCP) and User Datagram Protocol (UDP) at the Transport Layer. At the Application Layer are File Transfer Protocol (FTP), Simple Mail Transfer Protocol (SMTP), Simple Network Management Protocol (SNMP), and Telecommunications Network (TELNET). The physical infrastructure has evolved into one that is largely broadband in nature, comprising extremely high speed transmission systems and routers. The physical topology is organized in a hierarchical manner, as follows:

> **Level 1**: Network Access Points (NAPs) and MAEs serve as points of interconnection where national and regional carriers exchange traffic
>
> **Level 2**: National backbones comprising facilities-based, long haul carriers
>
> **Level 3**: Regional carriers comprising facilities-based, long haul carriers operating in a single state or province

Level 4: Internet Service Providers (ISPs)

Level 5: End users

See also *ARPANET, backbone, broadband, carrier, computer, data, end user, FTP, host, Internet2, ISP, IP, long haul circuit, MAE, NAP, network, Network Layer, OSI Reference Model, packet, router, SMTP, SNMP, system, TCP, TELNET, transmission system, Transport Layer,* and *UDP.*

Internet Activities Board The original name for the Internet Architecture Board (IAB) See *IAB.*

Internet Architecture Board (IAB) See *IAB.*

Internet Control Message Protocol (ICMP) See *ICMP.*

Internet Corporation for Assigned Names and Numbers (ICANN) See *ICANN.*

Internet Engineering Task Force (IETF) See *IETF.*

Internet Engineering Steering Group (IESG) See *IESG.*

Internet Exchange (IX) See *IX.*

Internet Explorer (IE) A highly capable Web browser developed by Microsoft (1995) and packaged with Microsofts Windows suite of software. See also *browser* and *WWW.*

Internet Fax Protocol (IFP) See *T.38.*

Internet Protocol (IP) See *IP.*

Internet Protocol next generation (IPng) The working name for what became Internet Protocol version 6 (IPv6). See *IPv6.*

Internet protocol suite See *TCP/IP protocol suite.*

Internet Protocol version 4 (IPv4) See *IPv4.*

Internet Protocol version 6 (IPv6) See *IPv6.*

Internet Protocol television (IPTV) See *IPTV.*

Internet Research Task Force (IRTF) See *IRTF.*

Internet Small Computer System Interface (iSCSI) See *iSCSI.*

Internet Society (ISOC) See *ISOC.*

Internetwork Packet Exchange (IPX) See *IPX.*

Internetwork Packet Exchange/Sequenced Packet Exchange (IPX/SPX) See *IPX/SPX.*

Internet2 A private Internet for the benefit of its member organizations, which include the U.S. National Science Foundation (NSF), the U.S. Department of Energy, more than 200 U.S. research universities, and over 60 private companies formed into a not-for-profit consortium. Internet2 is not a separate physical network, and it does connect to the present Internet, as required. Internet2 is a project of the University Corporation for Advanced Internet Development (UCAID), a not-for-profit entity created specifically to develop and manage the network. The Internet2 and its members are in the process of developing and testing technologies such as IPv6, multicasting, and quality of service (QoS) mechanisms in support of what they characterize as revolutionary Internet applications such as digital libraries, virtual laboratories, distance-independent learning, and tele-immersion. The Internet2 initiative is parallel and complementary to the Next Generation Internet (NGI) initiative funded by the United States federal government. The physical transmission infrastructure initially (1995) was in the form of the very-high-speed Backbone Network Service (vBNS), which was replaced by the Abilene Project. Géant2 is a similar European backbone project. See also *Abilene Project*, *Géant2*, *IPv6*, *multicast*, *NFS*, *NGI*, *QoS*, *tele-immersion*, *UCAID*, and *vBNS*.

intersymbol interference (ISI) See *ISI*.

in the clear Referring to radio frequency (RF) transmission with no encryption or scrambling mechanism for security purposes. See also *encryption*, *RF*, *scramble*, and *security*.

in the wild Referring to an application in the real world, rather than a laboratory experiment or demonstration or an unproven concept.

intraLATA Referring to a local long distance call within the boundaries of a Local Access and Transport Area (LATA). In the United States, LATA boundaries were defined in the Modified Final Judgement (MFJ) that broke up the AT&T Bell System on January 1, 1984. The Regional Bell Operating Companies (RBOCs) and their component Bell Operating Companies (BOCs) initially were prevented from offering interLATA toll services, but they and other local exchange carriers (LECs) had the exclusive right to offer local exchange service and intraLATA toll service. See also *BOC*, *LATA*, *LEC*, *MFJ*, and *RBOC*.

intranet A private network based on the TCP/IP protocol suite and designed to provide access to information resources within a company, university, or other organization. Designed to look much like a site on a private World Wide Web (WWW) and based on the same protocols, an intranet supports familiar client/server software such as browsers and e-mail. Intranets can be used for communications to and between employees to advise them of company policies, job postings, company events, product literature, press releases, and so on. On a password-protected basis, privileged users can access sensitive internal company information, including customer billing records and network usage data. Intranets support the transmission of images, video clips, and sound clips, as well as textual information. Hypertext links can be included to hot link to other sites and databases. An intranet can be confined to a campus environment or can extend across the wide area to link together multiple, geographically dispersed locations. Access from an intranet to the wider public Internet is possible through a security firewall. Extranets are intranets opened to select groups of users outside the company. See also *browser*, *database*, *e-mail*, *extranet*, *firewall*, *hypertext*, *Internet*, *protocol*, *protocol suite*, *TCP/IP*, and *WWW*.

intrinsic loss In fiber optics transmission, signal attenuation due to absorption and scattering resulting from internal forces or inherent characteristics of the optical fiber. See also *absorption*, *attenuation*, *extrinsic loss*, *optical fiber*, *scattering*, and *signal*.

inverse multiplexer A device that performs inverse multiplexing. See inverse *multiplexer*.

inverse multiplexing A technique that is the inverse, or opposite, of multiplexing. Traditional multiplexing folds together multiple low-speed channels onto a high-speed circuit. Inverse multiplexing spreads a high-speed channel across multiple low-speed circuits. Inverse multiplexing is used where an appropriately high-speed circuit is not available. A 6-Mbps data stream, for example, might be inverse multiplexed across four (4) T1 circuits, each running at 1.544 Mbps. Inverse multiplexing over ATM (IMA) fans out an ATM cell stream across multiple circuits between the user premises and the edge of the carrier network. In such a circumstance, multiple physical T1 circuits can be used as a single, logical ATM pipe. The IMA-compliant ATM concentrator at the user premises spreads the ATM cells across the T1 circuits in a round robin fashion, and the ATM switch at the edge of the carrier network scans the T1 circuits in the same fashion in order to reconstitute the cell stream. There is a similar implementation agreement (IA) for Frame Relay. Multilink point-to-point protocol (PPP) serves much the same purpose in the Internet domain.

inverse multiplexing over ATM (IMA) See *IMA*.

inverter **1.** A device that converts electrical energy from direct current (DC), typically 12 or 48 volts, to alternating current (AC), typically 120 volts. See also *AC* and *DC*. **2.** In digital systems, a logic gate that inverts the polarity of an electrical pulse from positive (+) to negative (–) or from negative (–) to positive (+). See also *digital*, *polarity*, and *pulse*.

INWATS (INward Wide Area Telecommunications Service) WATS service for incoming calls, only, INWATS was a discounted bulk long distance plan in the United States and Canada that reversed the charges to the called party, rather than the calling party. Although INWATS service, as such, has been obsoleted, toll-free services are common throughout the world. Such services are toll free only to the caller, of course. See also *toll free service* and *WATS*.

I/O (Input/Output) The computer component that receives data from or transmits data to a system or device, such as a buffer, peripheral, storage unit, or another computer across a network.

ion An atom or group of atoms that has acquired a positive (+) or negative (–) electric charge by gaining (+) or losing (–) one or more electrons through a chemical reaction or by the action of certain forms of radiant energy. The movement of electrons and ions constitutes electric current. See also *current*, *electron*, *hydroxyl*, and *ionosphere*.

ionosphere Four regions of the Earth's outer atmosphere, beginning at an altitude of approximately 55 kilometers (34 miles), ionized by solar radiation. The ionosphere contains high concentrations of ions and free electrons, and can serve to reflect certain radio signals back to Earth, depending on their frequency and angle of incidence. Long-range radio communications depend on this skywave propagation, or atmospheric skip. See also *angle of incidence*, *frequency*, and *skywave*.

IOR (Index Of Refraction) The ratio between the velocity of light in a vacuum to the velocity of light in a transmission medium, such as a given type of glass. Refraction refers to the phenomenon by which light changes velocity and changes direction (i.e., bends or refracts) as it exits one medium and enters another of different density. The IOR is mathematically expressed as

$$n = c \div v$$

where c is the velocity of light in a vacuum, and v is the speed of light in the given medium. So, IOR is the mathematical inverse of velocity of propagation (Vp). IOR is a convenient means for expressing the differences between the speed of light in different types of optical fiber, as well as between the core and cladding of a glass optical fiber (GOF). Table I-2 provides approximate IOR comparisons of various substances.

Table I-2: Index of Refraction (IOR)

Medium	*Signal Velocity (km/s)*	*Velocity of Propagation (Vp)*	*Index of Refraction (IOR)*
Vacuum	300,000	100.00	1.0000
Air	299,890	99.97	1.0003
Water	226,000	75.33	1.3275
Optical Fiber	203,910–209,910	67.57–68.97	1.4500–1.4800

Not all glass is created equal, by the way. The raw material for all glass is quartz sand, a very pure sand comprising nearly 100 percent crystalline quartz silica. During the manufacturing process, the glass is purified to reduce the slight amount of iron oxide that might be present, various dopants (i.e., impurities) are added to alter brittleness and other characteristics, and heat treatments can be applied to produce tempered glass. Glass optical fiber is typically doped with some amount of germanium oxide (GeO2) or other compounds, which increase the IOR and variously impact other operating characteristics of the fiber at certain wavelengths. Erbium-doped fiber amplifiers (EDFAs) are used extensively in long haul fiber optic transmission systems (FOTS) in place of more traditional optical repeaters.

It is extremely important to know the IOR of a given cable in order to calculate latency, i.e., the delay imposed on the signal, from end-to-end, by the medium. Latency has significant impact on the synchronization of transmitters, repeaters, multiplexers, and other active devices in a digital transmission system, and particularly in a high speed system such as a FOTS that runs at signaling speeds up to 40 Gbps.

The IOR also has considerable impact on the physical construction of a glass optical fiber, which consists of an inner core surrounded by one or more layers of cladding. The optical signal is intended to travel through the inner core, and the cladding serves in various ways to ensure that happens. Step-index fiber is characterized by a sharp decrease in the IOR between the core and cladding, i.e., the cladding is sharply lower in IOR than the core material. This sharp step of approximately one percent in IOR at the core/cladding interface causes any errant light rays to reflect back into the core in a phenomenon known as total internal reflection. Graded-index fiber is characterized by a gradual decrease in the refractive index of the cladding through a great many layers of glass. The approach causes the errant light rays to gradually gain in velocity and bend back towards the core. Step-index construction is used largely in single-mode fiber (SMF) and graded-index in multimode fiber (MMF). See also *core*, *cladding*, *graded-index fiber*, *latency*, *MMF*, *SMF*, *step-index fiber*, and *Vp*.

IP (Intelligent Peripheral) In the advanced intelligent network (AIN) architecture, a separate computing device that enhances the delivery of certain services by offloading processing demands from a service control point (SCP) and providing a basic set of services to the SCP. The role of the IP typically includes collection of digits, collection and playing of voice prompts, collection of voice responses and their conversion to digits, menu services, and database lookups. As examples, voice processing and voice recognition can be implemented on an IP in support of the processing of collect calls. Because the abbreviation IP is used so extensively to identify the Internet Protocol, the term special resource function (SRF) often is used to describe these peripherals and the functions they perform. See also *AIN* and *SCP*.

IP (Internet Protocol) In the context of the OSI Reference Model, a Network Layer (Layer 3) connectionless protocol for the routing of datagrams through gateways connecting networks and subnetworks, IPv4, the first version to be deployed, is defined in IETF RFC 791 and is the basic block of the Internet. IP can be characterized as datagram-oriented because each IP packet works its way through the network independently, with no thought of an individual packet belonging to a larger stream of packets. IP also can be characterized as a best effort protocol, as it offers no guarantees of delivery, no sequencing, and no error detection and correction mechanism. IP provides for packet segmentation and reassembly and provides specific addressing conventions in the form of dotted decimal notation. IP supports routing control, as well as status translation and communications. Although IP has no concept of the specific content of the packet or of its service requirements, it supports multiple service types, including low-delay, high-bandwidth, and high-reliability paths. Dial-up IP access protocols include Ethernet over Point-to-Point Protocol (Enet over PPP) Serial Line Internet Protocol (SLIP), and Point-to-Point Protocol (PPP). There are two versions of IP. The legacy version is IPv4, which is widely deployed. The most current version is IPv6, which is deployed sparingly, largely due to the fact that it is not compatible with older routers. Although IP can be used on a standalone basis, it more typically is used in conjunction with higher layer protocols in the TCP/IP protocol suite. See also *bandwidth*, *best effort*, *connectionless*, *datagram*, *dial-up access*, *dotted decimal notation*, *IETF*, *IPv4*, *IPv6*, *gateway*, *Internet*, *network*, *Network Layer*, *OSI Reference Model*, *packet*, *PPP*, *protocol*, *router*, *segmentation*, *SLIP*, and *TCP/IP protocol suite*.

IP address (Internet Protocol address) A binary number that uniquely identifies a host computer connected to the Internet. The IP packet header provides an originating address field so that a host can identify itself as the originator of a packet. The IP packet header also provides a destination address field so that an originating host can identity the target host for which a packet is intended. Based on that information, the Internet routers can act to deliver the packet to the target host, which can respond to the originating host, as appropriate. All IP addresses are written in dotted decimal notation. An IPv4 address, for example, comprises 4 fields separated by dots and expressed as xxx.xxx.xxx.xxx, with each field given a value in decimal notation of 0–255, the range expressed by a single octet in binary notation. See also *binary, binary notation, computer, dotted decimal notation, header, host, Internet, IPv4 address, IPv6 address, octet*, and *packet*.

IPBX (Internet Protocol Private Branch eXchange) Synonymous with IP PBX. A pure IPBX is based on a client/server architecture that generally is implemented on a switched Ethernet LAN platform running at 100 Mbps or more in support of VoIP. Client software residing on intelligent IP hardphones and softphones runs against one or more servers that can be distributed across an enterprise, perhaps in geographically diverse locations connected over the public Internet or, preferably, a private IP-based network. One or more telephony servers are responsible for all call control functions (e.g., call setup and teardown), telephony applications (e.g., feature content and class of service (CoS) assignments), and associated databases (e.g., IP and station numbering schemes and assignments, and directory information). Although the switched Ethernet LAN infrastructure is shared between voice and data, and perhaps video, each IP phone has a dedicated physical port on an Ethernet switch in order to minimize any issues of congestion at the port level. Further, VoIP generally runs in a logical virtual LAN (VLAN) partition to maintain an acceptable level of quality of service (QoS). A digital signal processor (DSP) embedded in each IP phone digitizes the analog voice signals in pulse code modulation (PCM) format and may compress the resulting voice data. Application software then forms VoIP packets, which are then inserted into Ethernet frames, and presented to the switch.

There are several hybrid IPBX approaches that allow a user organization variously to maximize the embedded investment in conventional TDM-based PBX equipment and ease into a pure IPBX mode. An IP-enabled PBX is a conventional TDM PBX circuit switch platform with a VoIP module. A hybrid TDM/IP PBX has both TDM and IP components co-existing, side by side. The TDM component comprises TDM line and trunk cards and ports and a TDM bus. The IP component comprises Ethernet ports, an Ethernet switch, a router, and IP trunk ports. A gateway interconnects the TDM and IP components, both of which are under the control of a telephony server running a commercial operating system (OS). See also *client/server, CoS, DSP, Ethernet, gateway, IP, IP Centrex, IP-enabled PBX, LAN, PCM, QoS, VLAN*, and *VoIP*.

IP Captioned Telephone Service A hybrid Telecommunications Relay Service (TRS) that involves a special telephone with a text display. Rather than using TTY or computer technology for text entry, the called partys speech is re-voiced by the communications assistant (CA), converted into text by a voice recognition system, and transmitted directly to the hearing-impaired callers display telephone over the Internet. The CA can simultaneously listen to one party and read the text of the other party. See also *Internet, TRS, TTY*, and *voice recognition*.

IP Centrex Synonymous with hosted PBX and virtual Centrex. Centrex service provided from a network-based carrier-class IPBX platform to which the user organization connects over the public Internet or a private IP network as though it were a remote Centrex central office (CO). See also *carrier-class, Centrex, Internet*, and *IPBX*.

IP-enabled Frame Relay Also known as *Private IP*. A frame relay variant that employs virtual routers to route frames at the Network Layer of the OSI Reference Model. IP-enabled Frame Relay typically makes use of MultiProtocol Label Switching (MPLS) to speed frame processing and support quality of service (QoS). See also *frame relay, IP, MPLS, Network Layer, OSI Reference Model, QoS*, and *router*.

IP-enabled PBX (Internet Protocol-enabled Private Branch eXchange) A PBX that couples VoIP (Voice over Internet Protocol) onto a conventional PBX based on time division multiplexing (TDM). The intelligent IP phones can take the form of either hardphones or softphones that connect either over a switched Ethernet LAN. The LAN-attached data terminals interconnect as usual. The IP phones connect to the PBX through an Ethernet port on a line card that includes an IP gateway that resolves the interface issues between the TDM bus and the switched Ethernet LAN that supports VoIP. Calls between the LAN-attached IP phones are conducted on a peer-to-peer basis using their LAN addresses, and are confined to the LAN. Calls between TDM phones also are on a peer-to-peer basis through the TDM switching matrix. Calls between a LAN-attached IP phone and a PBX-attached TDM phone go through the gateway, where protocol issues are resolved, including address translation between PBX extension numbers and Ethernet LAN addresses. See also *LAN*, *IPBX* and *PBX*.

IP Multimedia Subsystem (IMS) See *IMS*.

IPng (Internet Protocol next generation) The working name for what became Internet Protocol version 6 (IPv6). See *IPv6*.

IP PBX Synonymous with IPBX. See *IPBX*.

IP Relay A text-based Telecommunications Relay Service (TRS) that uses the Internet Protocol (IP) based Internet rather than a traditional public switched telephone service (PSTN) connection to communicate. A speech- or hearing-disabled person initiating an IP Relay TRS call uses a computer or other text input device to call the TRS relay center over an Internet connection and give the communications assistant (CA) the destination telephone number, which the CA uses to place a traditional telephone call. The CA then serves as a translator, relaying the text of the calling party in voice to the called party, and converting the voice of the called party into text for the benefit of the caller. See also *Internet*, *IP*, *PSTN*, and *TRS*.

IPsec (Internet Protocol security) A standards-based security suite that operates transparently and may eliminate the need for proprietary firewall mechanisms in some applications. IPsec also provides for encapsulation of the secured IPv6 packets inside IPv4 datagrams, in consideration of both the increasing need for security and the long-term transition process to IPv6. IPsec is described in IETF RFC 2401 and, in the context of the OSI Reference Model, runs at the Network Layer (Layer 4). See also *datagram*, *encapsulate,*, *firewall*, *IETF*, *IPv4*, *IPv6*, *Network Layer*, *OSI Reference Model*, and *packet*.

IP security (IPsec) See *IPsec*.

IPTV (Internet Protocol TeleVision) Referring to distribution of television programming over a network running the IP protocol. This approach delivers each TV channel only to the subscriber premises that select to view it, rather than broadcasting it to all stations on the network. The primary focus on IPTV is from telephone companies offering voice, Internet access, and TV programming over broadband fiber optic local loops conforming to passive optical network (PON) standards. See also *broadband*, *broadcast*, *channel*, *fiber optics*, *IP*, *local loop*, *PON*, and *TV*.

IPv4 (Internet Protocol version 4) The legacy version of IP, IPv4 is the widely deployed foundation protocol on which the Internet is built. Developed in the 1970s and documented in IETF RFC 791 (1981), IPv4 is a Network Layer (Layer 3) connectionless protocol for the routing of datagrams through gateways connecting networks and subnetworks. IPv4 also can be characterized as a best effort protocol, as it offers no guarantees of delivery, no sequencing, and no error detection and correction mechanism. IPv4 provides for packet segmentation and reassembly and provides specific addressing conventions in the form of dotted decimal notation. IPv4 supports routing control, as well as status translation and communications. While IPv4 has no concept of the specific content of the packet or of its service requirements, it also supports multiple service types, including low-delay, high-bandwidth, and high-reliability paths. In the 30+ years since IPv4 was introduced, the Internet (and the world, in general) has become a much more complicated, more intense, more populated, and less secure place. In a contemporary context, IPv4 is criticized for its inadequate address space and lack of inherent security mechanisms. IPv6, the current version resolves those issues, but is incompatible with older routers and other legacy Internet infrastructure. So, IPv4 remains the dominant protocol of the Internet. Although IPv4 can be used on a standalone basis, it more typically is used in conjunction with higher layer protocols in the TCP/IP protocol suite.

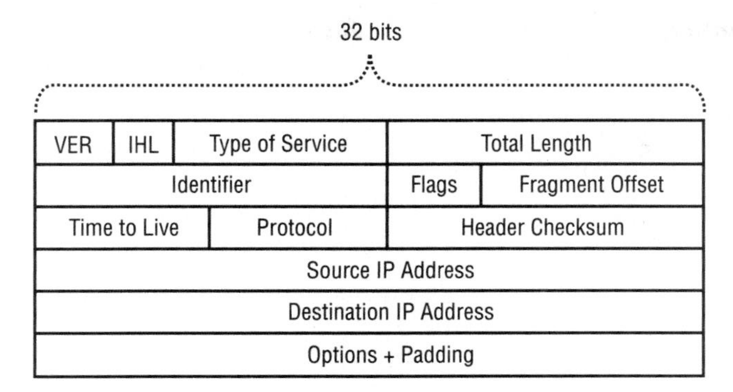

Figure I-1

The total size of the IPv4 datagram, including the header (shown in Figure I-1), can be up to 65,535 octets in length. At a minimum, all networks must support a packet of at least 576 octets. As illustrated in Figure I-1, the minimum size of the IP header is 20 octets. The IPv4 datagram contains the following fields:

- **VERS (VERSion):** 4 bits identifying the IP version number. The version number is 4.
- **IHL (Internet Header Length):** 4 bits indicating the length of the header, which has a minimum value of five 32-bit words, or 20 octets. The IHL also provides a measurement of where the TCP header, or other higher-layer, header begins.
- **Type of Service (ToS):** 8 bits indicating the quality of service (QoS) requested for the datagram. Although TCP/IP networks currently do not provide guaranteed QoS, the networks will attempt to honor QoS requests in terms of parameters that include packet precedence (i.e., priority), low delay, high throughput, and high reliability.
- **Total Length:** 16 bits describing the total length of the datagram, including the IP header. The maximum size is 65,535 octets (2^{16}–1, with 0 not considered as it has no value). All network hosts must be able to handle a datagram of at least 576 octets.
- **Identification:** 16 bits that are used in fragmentation control. In the event that the receiving network cannot accommodate a datagram of the specified total length, that datagram must be fragmented. Each fragment must contain a copy of the identification field and certain other fields in the IP header so they can be reassociated and the datagram can be reconstituted.
- **Flags:** 3 bits that define the manner in which the fragmentation occurs. The first bit always is set at 0. The second bit defines whether fragmentation is permitted. The third bit is used to identify the last fragment in a series of fragments.
- **Fragment Offset:** 13 bits that identify where the fragment fits in the complete set of fragments that comprise the original datagram. This field is used to sequence the fragments correctly, as they may arrive at the destination device out of sequence.
- **Time To Live (TTL):** 8 bits that specify the length of time in seconds that the datagram can live in the Internet system. The maximum length of time is 255 seconds (2^8–1, with 0 not considered, as it is the official time of death), or 4.25 minutes. From the instant the IP datagram enters the Internet, each gateway and host that acts on the datagram decrements the TTL by at least one second. When the TTL reaches 0, the datagram is declared dead and is discarded. The TTL mechanism prevents packets from wandering the Internet for eternity, at which point they would have no value, and would only contribute to overall network congestion. Over time, the TTL field has been redefined to indicate, as an option, the number of hops, that is the number of routers through which the packet travels. The default TTL is 64.

- **Protocol:** 8 bits identifying the higher-layer protocol that created the message contained in the Data field. Examples include Transmission Control Protocol (TCP) and User Datagram Protocol (UDP).
- **Header Checksum:** 16 bits used for error control in the header. The process is that of a cyclic redundancy check (CRC).
- **Source IP Address:** 32 bits containing the IP address of the source host.
- **Destination IP Address:** 32 bits containing the IP address of the destination host.
- **IP Options, If Any:** An optional, variable-length field used by gateways to control fragmentation and routing options.
- **Padding**: A variable-length field used only when necessary to ensure that the IP header extends to an exact multiple of 32 bits.
- **Data:** A variable-length field that contains the actual data content, or payload.

See also *bandwidth, best effort, connectionless, CRC, datagram, dial-up, dotted decimal notation, error control, IETF, IPv4, IPv4 address, IPv6, gateway, header, host, Internet, network, Network Layer, octet, OSI Reference Model, packet, payload, PPP, protocol, QoS, router, segmentation, SLIP, TCP, TCP/IP protocol suite*, and *UDP*.

IPv4 address (Internet Protocol version 4 address) A binary number that uniquely identifies a host connected to the Internet and running the IPv4 protocol suite. The IP packet header provides an originating address field so that a host can identify itself as the originator of a packet. The IP packet header also provides a destination address field so that an originating host can identity the target host for which a packet is intended. Based on that information, the Internet routers can act to deliver the packet to the target host, which can respond to the originating host, as appropriate. The IPv4 address field size is 32 bits, which yields the theoretical potential for 2^{32} or 4,294,967,296 unique addresses. All IP addresses are written in dotted decimal notation, which also is known as dotted quad format in the case of IPv4. Each IPv4 address field is divided into four fields separated by dots and expressed as

xxx.xxx.xxx.xxx

with each field given a value in decimal notation of 0–255, the range expressed by a single octet in binary notation ($2^8 = 256$, or 0–255). (*Note:* Leading zeros are suppressed.) The addressing architecture defines five address formats, each of which begins with one, two, three, or four bits that identify the class of the network (Class A, B, C, D, or E). The Network ID space identifies the specific network, and the Host ID space identifies the specific host computers on the network. Classes A, B, and C are each associated with a range of IP addresses that supports a limited number of networks per class and a limited number of hosts per network, as detailed in Table I-3. Class D addresses are reserved for multicast purposes and Class E addresses are reserved for future use.

Table I-3: IPv4 Address Classes

Class	*Address Range*	*Networks per Class*	*Hosts per Network*
A	1.0.0.1 to 126.255.255.254	126	16,777,214
B	128.0.0.1 to 191.255.255.254	16,384	65,534
C	192.0.0.1 to 223.255.255.254	2,097,152	254
D	Reserved for multicast purposes		
E	Reserved for experimental use		

The Internet Corporation for Assigned Names and Numbers (ICANN) assigns public IP addresses to organizations desiring to place computers on the Internet. Theoretically speaking, the size of the user organization determines the IP Class and, therefore, the number of available host addresses, although no more Class A addresses are being assigned and virtually all Class B addresses have already been assigned. The user organization assigns the host address numbers internally and can reassign them on the basis of either static or dynamic addressing. Static addresses are permanently or semi-permanently associated with a specific host. Dynamic addresses are assigned each time a connection is established. There are both public and private IP address spaces. Computers on private LANs running the TCP/IP protocol suite do not require public addresses, at least not for internal use within the LAN domain. Private IP addresses are set aside for such purposes and will never be used publicly. The Internet Protocol version 6 (IPv6) addressing scheme is quite different. See also *binary, binary notation, Class A, Class B, Class C, Class D, Class E, computer, decimal notation, dynamic address, header, host, ICANN, Internet, IPv6 address, multicast, packet,* and *static address.*

IPv6 (Internet Protocol version 6) The IPv6 specification grew out of the efforts of the IPng (IP next generation) Working Group to define a successor protocol to IPv4. Specified originally in IETF RFC 1883 (1995), since replaced by RFC 2460 (1998), IPv6 overcomes many of the limitations of IPv4, most notably with respect to address space, integration of application level protocols, quality of service (QoS), and security. The total size of the IPv6 datagram, including the header, is increased beyond the IPv4 limit of 65,535 octets to support jumbo payloads. At a minimum, all network links must support a maximum transmission unit (MTU) of at least 1,280 octets.

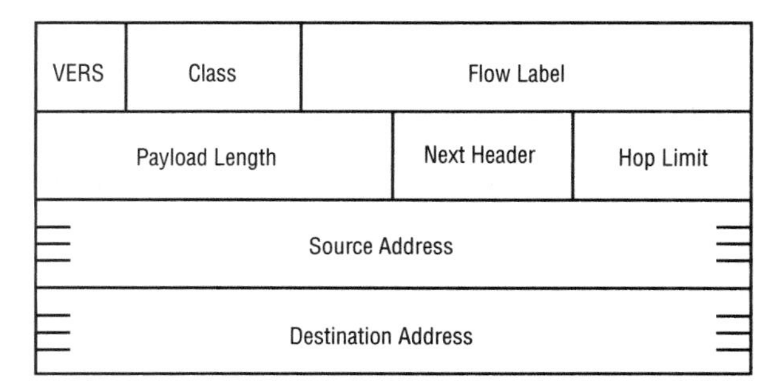

Figure I-2

As indicated in Figure I-2, the IPv6 header is 40 octets, compared to the IPv4 header of 20 octets, and can be extended as necessary through optional headers associated with higher layer protocols such as Transmission Control Protocol (TCP) and User Datagram Protocol (UDP). In a departure from IPv4, the IP header and any extension headers associated with the datagram are in addition to the payload length. The IPv6 datagram contains the following fields:

- **VERS (VERSion):** 4 bits identifying the IP version number. The version number is 6.
- **Class:** 8 bits used by originating nodes and/or forwarding routers to identify and distinguish between different packet classes or priorities, as set by upper-layer protocols. This field replaces the IPv4 Type of Service (ToS) field.
- **Flow Label:** 20 bits used by a source host indicating any special handling requested for a flow, or sequence, of datagrams. Real-time voice, audio, or video communications exemplify applications that require special flow treatment.
- **Payload Length:** 16 bits describing the total length of the datagram. This field is much like the Total Length field in IPv4, although the Payload Length field does not include the IP header. IPv6 supports jumbo payloads larger than the traditional 65,535 octets. At a minimum, all network links must support a maximum transmission unit (MTU) of at least 1280 octets. The IPv6 payload length includes any IPv6 header extensions, TCP or UDP headers, and any Application Layer headers that might be associated with the datagram.

- **Next Header:** 8 bits identifying the header immediately following the IPv6 header. Examples include TCP, UDP, Fragment, and Authentication.
- **Hop Limit:** 8 bits that specify the number of hops through which the packet can travel. Each router along the path decrements the field value by one until the value reaches zero, at which point the packet is discarded. This field is similar to the IPv4 TTL (Time To Live) field, with the exception that the seconds parameter has been eliminated and only the hops parameter is supported.
- **Source Address:** 128 bits (hexadecimal) containing the IP address of the source host.
- **Destination Address:** 128 bits (hexadecimal) containing the IP address of the destination host.

IPv6 supports multiple extension headers. RFC 1883 recommends that they be placed in the following order:

- **Hop-by-Hop Options:** Optional information that must be examined by every node along a path taken by a packet. This header carries information such as the type of extension header immediately following, and specific instructions as to what should be done if the processing node does not recognize the Option Type, and whether or not the option data may change en route. This header also identifies the length of the Hop-by-Hop header, and contains padding options.
- **Destinations Options:** Optional information that must be examined by the destination host. This header carries information such as the type of header immediately following, the length of the Destination header, and padding options.
- **Routing:** This IPv6 source node uses this header to list one or more intermediate modes to be visited along the path the packet takes to its destination.
- **Fragment:** The IPv6 source node uses this header to send a packet larger than the path MTU will accommodate. All fragmentation occurs at the source node.
- **Authentication:** As defined in RFC 2402, this header contains a mechanism for ensuring the connectionless integrity and data origin authentication of IP datagrams, as well as an anti-replay option. This header might also include a non-repudiation mechanism, which provides the origination node with confirmation of packet receipt. Included in this header may be the authentication algorithm and keys, the encryption algorithm and keys, and other security-related parameters.
- **Encapsulating Security Payload:** As defined in RFC 2406, this header provides confidentiality through encryption, and limited confidentiality of the traffic flow. It also may provide connectionless integrity and data origin authentication of IP datagrams, as well as an anti-replay option. IPsec (IP security) is a standards-based security suite that operates transparently and may eliminate the need for proprietary firewall mechanisms in some applications. IPsec also provides for encapsulation of the secured IPv6 packets inside IPv4 datagrams, in consideration of both the increasing need for security and the long-term transition process to IPv6.

IPv6 addressing is considerably enhanced, in comparison to IPv4. Not only is the address space increased to 128 bits, but IPv6 also supports multiple address types, including anycast, multicast, and unicast. IPv6 offers highly flexible address assignment through two approaches. Stateful autoconfiguration dynamically assigns unique addresses to devices as they require them, drawing from a pool of such addresses. Stateless autoconfiguration employs two IP addresses, and is particularly advantageous in mobile applications. One address is assigned permanently to the mobile device, and another address is used to route data to the network to which the mobile device is connected at the time. See also *algorithm*, *anycast*, *Application Layer*, *authentication*, *datagram*, *encryption*, *firewall*, *flow*, *hop*, *IETF*, *IPng*, *IPsec*, *IPv4*, *IPv6 address*, *jumbo payload*, *key*, *MTU*, *multicast*, *node*, *non-repudiation*, *payload*, *protocol*, *QoS*, *real-time*, *stateful autoconfiguration*, *stateless autoconfiguration*, *TCP*, *TTL*, *UDP*, and *unicast*.

IPv6 address (Internet Protocol version 6 address) A binary number that uniquely identifies a host computer connected to the Internet and running the IPv6 protocol suite. The IP packet header provides an originating address field so that a host can identify itself as the originator of a packet. The IP packet header also provides a destination address field so that an originating host can identity the target host for which a packet is intended. Based on that information, the Internet routers can act to deliver the packet to the target host, which can respond to the originating host, as appropriate. The IPv6 address field size is 128 bits, which yields the theoretical potential for 2^{128} or 340,282,366,920,938,463,463,374,607,431,768,211,456 unique addresses. At this level, the dotted decimal notation convention employed in IPv4 becomes cumbersome, so an IPv6 address comprises 8 sections separated by colons and expressed as

x:x:x:x:x:x:x:x

where each x represents 16 bits defined in hexadecimal notation. Following are example IPv6 addresses:

FEDC:BA98:7654:3210:FEDC:BA98:7654:3210

1080:0000:0000:0000:0008:0800:200C:417A

with each field given a value in decimal notation of 0–65,535, the range expressed by two octets in binary notation (2^{16} = 65,536, or 0–65,535). Leading zeros can be suppressed, which simplifies the latter above address to the following:

1080:0:0:0:8:800:200C:417A

If one or more groups (sections) of 16 zeros appear in an address, they can be compressed through the use of a :: (double colon) code. This compresses the above address to the following:

1080::8:800:200C:417A

Technically speaking, IPv6 addresses are assigned to interfaces, rather than nodes. The IPv6 addressing architecture provides for unicast, multicast, and anycast addresses.

- **Unicast:** A unicast address is associated with a single interface that is associated with a single node, and can, in effect, identify the node. A unicast address can be of several types and here are also special-purpose unicast subtypes, including IPv6 addresses with embedded IPv4 addresses. A global unicast address is a conventional, publicly routable address that can be used in the Internet or any public domain. A link-local address is similar to an IPv4 private IP address, as it is not meant to be routed, but confined to a single segment. A site–local address is used by an organization that has not yet connected to the Internet. A loopback address is used when a host needs to send a packet back to itself.
- **Multicast:** A multicast address identifies a group of nodes, each of which can belong to multiple groups. A packet sent to a multicast group is delivered to every interface in the group. Multicast addresses begin with the binary prefix 11111111. A multicast address essentially is a targeted version of an IPv4 broadcast address.
- **Anycast:** An anycast address is assigned to multiple interfaces, typically on multiple nodes. A packet with an anycast address is delivered to the closest interface with that address, as determined by the routing protocols measure of distance. Anycast addresses are allocated from the unicast address space.

IPv6 address types are defined in RFC 3513. The Internet Protocol version 4 (IPv4) addressing scheme is quite different. See also *anycast address*, *binary*, *binary notation*, *broadcast*, *compression*, *computer*, *dotted decimal notation*, *hexadecimal notation*, *global unicast address*, *header*, *hexadecimal notation*, *host*, *interface*, *Internet*, *IPv4 address*, *link-local address*, *loopback address*, *multicast address*, *node*, *packet*, *protocol suite*, *router*, *site–local address*, *suppression*, and *unicast address*.

IPX (Internetwork Packet eXchange) The Network Layer (Layer 3) protocol in Novell NetWare for exchanging packets between LANs. IPX is similar to the Internet Protocol (IP), but is not industry standard and has been overwhelmed by IP. Sequenced Packet Exchange (SPX) is the associated Transport Layer (Layer 4) protocol. See also *IP*, *IPX/SPX*, *LAN*, *Network Layer*, *protocol*, and *SPX*.

IPX/SPX (Internetwork Packet eXchange/Sequenced Packet eXchange) The Network Layer (Layer 3) and Transport Layer (Layer 4) protocols in Novell NetWare. IPX/SPX is similar to TCP/IP, but is not industry standard and has been overwhelmed by the TCP/IP protocol suite. Novell currently supports the TCP/IP protocol suite. See also *Network Layer*, *protocol*, *protocol suite*, and *TCP/IP*.

Ir (Iridium) A silver-white metallic element, iridium has 77 electrons around its nucleus and, therefore, is number 77 in the Periodic Table of Elements. (We used to refer to iridium as having an atomic weight of 77, but that characterization is only relative and not absolute; the term, therefore, is obsolete.) The name is from the Greek *iridis*, translating as *rainbow*, due to the fact that when iridium is dissolved in hydrochloric acid, its salts are of various colors of the rainbow. Iridium is a component of extremely hard, corrosion-resistant alloys used in pen nibs, jewelry, surgical instruments, watch and compass pivot bearings, electrodes, and chemical crucibles. At one time, the official measurement of the meter was recorded as the distance between two fine lines engraved on a platinum-iridium bar kept at the International Bureau of Weights and Measures in Paris. See also *Iridium*.

IR **1.** InfraRed. Infrared light is in the frequency range of 300 GHz–400 THz and has a wavelength of 1 mm–0.75 mm (750 nm). EHF radio has applications in fiber optics and free space optics (FSO) transmission systems. Infrared light is at the very lower limit of the radio spectrum, just above extremely high frequency (EHF) radio. See also *EHF*, *electromagnetic spectrum*, *fiber optics*, *frequency*, *FSO*, *Hz*, and *wavelength*. **2.** Incremental Redundancy. Also known as *Hybrid ARQ II*. An enhanced automatic repeat request (ARQ) technique employed in EGPRS (Enhanced General Packet Radio System), the packet-switched mode of Enhanced Data rates for GSM Evolution (EDGE) cellular radio networks. As transmission begins, IR initially transmits packets with little forward error correction (FEC) overhead in an attempt to maximize throughput. If the initial transmission cannot be successfully decoded by the receiver, IR ratchets up the FEC overhead until it finds a level at which the receiver can successfully decode the transmission. See also *ARQ*, *cellular radio*, *EDGE*, *EGPRS*, *FEC*, *overhead*, and *throughput*.

IRC (International Record Carrier) A carrier that offers record communications services, which are services that are designed or used primarily to transfer information that originates or terminates in written or graphic form. Examples of record communications services include telex and TWX. See also *record communications services*, *telex*, and *TWX*.

iridium (Ir) The silver-white metallic element with an atomic number of 77, Iridium is represented by the symbol Ir. See also *Ir*.

Iridium The first mobile satellite system (MSS) operating in low-Earth orbit (LEO). According to legend, the wife of a Motorola executive was vacationing in the Bahamas during 1987 and was irritated by her inability to place a cellular telephone call. As the story goes, she complained to her husband, and captured his imagination, so to speak. Motorola engineers subsequently determined that a constellation of 77 communications satellites in non-geosynchronous low-altitude orbits would be sufficient to provide cellular-like service to essentially all dry land on the Earth's surface. Motorola named its proposed 77-satellite constellation Iridium, after the element iridium (Ir), which has 77 electrons whizzing around its nucleus. Subsequently, the proposal was pared down to 66 operational satellites, although the name Iridium stuck. After all, Iridium seems to roll off the tongue better than dysprosium (Dy), the rare earth element with an atomic weight of 66. Dysprosium, by the way is from the Greek *dysprositos*, meaning hard to get at, which is hardly a good name for a communications technology. The Iridium constellation is now fully launched and is fully operational, with 11 operational satellites, and 1 spare, placed in each of the 6 orbital planes, at altitudes of 421.5 nautical miles. Connectivity between each satellite and the Earth is established via 48 highly focused spot beams, each of which has a footprint of approximately 30 miles (50 km) in diameter. Assuming line of sight (LOS) is available, the end user can connect directly from a satellite phone to a satellite using the L-band, at frequencies of 1.616–1.6265 GHz. Alternatively the user can connect from a landline through a local terrestrial gateway at Ka-band frequencies of 29.1–29.3 GHz on the uplink and 19.4–19.4 GHz on the downlink. When connected to the satellite, the inter-satellite links operate in the Ka-band, at frequencies of 23.18–23.38 GHz. Iridium is essentially a cellular telephone network in reverse. The satellites are much like cellular base station transceivers that whiz around the Earth whereas the user terminal device remains relatively stationary. Iridium applications go well beyond cellular telephony and currently include aviation and maritime applications such as aircraft and ship tracking and en-route voice, fax, and data communications between crews and terrestrial operations centers. See also *base station*, *cellular radio*, *downlink*, *footprint*, *GEO*, *Ka-band*, *L-band*, *LEO*, *LOS*, *MSS*, *spot beam*, *transceiver*, and *uplink*.

IRTF (Internet Research Task Force) The group of the Internet Society (ISOC) that deals with long-term issues. The work of the IRTF is accomplished in small, focused research groups that work on topics related to Internet protocols, applications, architecture, and technology. See also *application*, *architecture*, *Internet*, *ISOC*, *protocol*, and *technology*.

IS (Information Services) The formal term for a data processing department, also known as *Management Information Systems* or *Management Information Services* (*MIS*). An IS department typically is responsible for all computer, network, and storage systems and technologies, from design through implementation and support. The traditional term was *Information Technology* (*IT*).

IS-54 (Interim Standard 54) A 2G digital cellular radio standard from the Electronics Industry Alliance (EIA) and Telecommunications Industry Association (TIA) and later approved by the American National Standards Institute (ANSI) to officially become ANSI/TIA/EIA-567. IS-54 is an extension of the analog Advanced Mobile Phone System (AMPS) and is commonly known as Digital AMPS (D-AMPS) and United States Digital Cellular (USDC). IS-54 evolved into IS-136. The two standards jointly specified the time division multiple access (TDMA) method and are often referred to as TDMA. See also *2G*, *ANSI*, *cellular radio*, *D-AMPS*, *EIA*, *IS-136*, *TIA*, and *TDMA*.

IS-136 (Interim Standard 136) An improvement to the IS-54 standard that added a digital control channel (DCCH) for residential and in-building coverage, increased battery standby time, text messaging, over-the-air activation, and circuit switched data (CSD) capabilities. See also *CSD*, *DCCH*, *IS-54*, and *texting*.

IS-856 (Interim Standard 856) A 3G digital cellular radio standard from the Electronics Industry Alliance (EIA) and Telecommunications Industry Association (TIA) for what is more commonly known as *CDMA2000* (Code Division Multiple Access 2000). See also *3G*, *CDMA2000*, *cellular radio*, *EIA*, and *TIA*.

IS-2000-A (Interim Standard 2000-A) In cellular radio systems, also known as *3x*. An enhancement to CDMA2000 that uses three cdmaOne carriers for total bandwidth of 3.75 MHz. See also *bandwidth*, *carrier*, *cellular radio*, *CDMA2000*, and *cdmaOne*.

iSCSI (internet Small Computer System Interface) A network protocol specified by the Internet Engineering Task Force (IETF) in RFC 3347 (July 2002) for the use of the Small Computer System Interface (SCSI) protocol over TCP/IP networks. As data exits the computer, headed toward the storage device, a host bus adapter (HBA) converts the data to a SCSI format, enclosed in an Internet Protocol (IP) packet, and transmitted over an Ethernet network. An advantage of iSCSI is that it is transparent, as the server software sees what looks to be a SCSI controller and the network sees only IP traffic. The protocols employed at Layers 2, 3, and 4 are Ethernet, IP, and Transmission Control Protocol (TCP). Further, iSCSI is intended to run at speeds up to 10 Gbps over 10 Gigabit Ethernet (10GigE), over IP wide area networks (WANs), and at lower cost than Fibre Channel. See also *10 Gigabit Ethernet*, *IETF*, *IP*, *OSI Reference Model*, *protocol*, *SCSI*, *TCP*, *Transport Layer*, and *WAN*.

ISDN (Integrated Services Digital Network) A set of ITU-T Recommendations describing a set of interfaces for access to a digital public switched telephone network (PSTN) intended to provide ubiquitous access to a wide range of services, including voice, data, video, and multimedia. The I series ISDN Recommendations map into the OSI Reference Model at the bottom three layers:

- **Network Layer:** The Q.931 specifications include user-to-user and network-to-network call control messages for both circuit-switched and packet-switched networking.
- **Data Link Layer**: ISDN specifies the Link Access Procedure, D Channel (LAPD). Issues of packetization, error control, and flow control are addressed.
- **Physical Layer:** Mechanical and electrical issues including connectors, signaling rates, line coding, and synchronization.

ISDN Recommendations describe three types of channels:

- **Bearer (B):** Channels that bear the end user data, or payload.
- **Data (D):** Also known as *Delta channels*. Channels set aside for out-of-band signaling and control functions. D channels also can support low-speed end user packet data and telemetry applications.
- **High-speed (H):** Aggregations of B Channels (2 to 30) to accommodate bandwidth-intensive applications such as video.

The User Network Interface (UNI) is defined differently at two levels. Basic rate interface (BRI) for low-speed access at rates up to 128 kbps. Primary rate interface (PRI) is defined for high-speed access in North America at 1.544 Mbps. Primary rate access (PRA) is defined for high-speed access in European and other countries at 2.048 Mbps. Common Channel Signaling System #7 (SS7), which supports out-of-band signaling and control, is a fundamental requirement of ISDN. Broadband ISDN (B-ISDN) is set of specifications from the ITU-T for an ISDN requiring transmission channels capable of supporting rates greater than the primary rate. See also *bandwidth, B channel, B-ISDN, BRI, channel, circuit switch, Data Link Layer, D channel, H channel, I series, ITU-T, LAPD, Network Layer, OSI Reference Model, out-of-band signaling and control, packet switching, payload, Physical Layer, PRA, PRI, PSTN, SS7,* and *UNI.*

ISDN modem Synonymous with, and more correctly known as, *Terminal Adapter* (TA). See *TA*.

ISDN Digital Subscriber Line (IDSL) See *IDSL*.

I series The series of ITU-T Recommendations specifying protocols and interfaces relating to integrated service digital network (ISDN). See Table I-4 for selected I-series Recommendations. For a full listing of ITU-T Recommendations, see the contact information in Appendix A.

Table I-4: Selected ITU-T I-Series Recommendations

Recommendation	*Description*
I.113	B-ISDN Vocabulary
I.120	Integrated services digital networks (ISDNs)
I.121	Broadband aspects of ISDN
I.122	Framework for frame mode bearer services
I.150	B-ISDN ATM Functional Characteristics
I.211	B-ISDN service aspects
I.233.1	ISDN frame relaying bearer service

Table I-4: Selected ITU-T I-Series Recommendations *(continued)*

Recommendation	*Description*
I.233.2	ISDN frame switching bearer service
I.311	B-ISDN General Network Aspects
I.313	B-ISDN network requirements
I.320	ISDN protocol reference model
I.321	B-ISDN protocol reference model
I.324	ISDN network architecture
I.327	B-ISDN Functional Architecture Aspects
I.354	Network performance objectives for packet-mode communication
I.355	ISDN 64 kbit/s connection type availability performance
I.356	B-ISDN ATM layer cell transfer performance
I.361	B-ISDN ATM Layer Specification
I.362	B-ISDN ATM Adaptation Layer Functional Description
I.363	B-ISDN ATM Adaptation Layer (AAL) Specification
I.363.1	B-ISDN AAL Type 1
I.363.2	B-ISDN AAL Type 2
I.363.3	B-ISDN AAL Type 3/4
I.363.5	B-ISDN AAL Type 5
I.1413	B-ISDN User-Network Interface
I.420	Basic user-network interface (BRI)
I.421	Primary rate user-network interface (PRI)
I.430	Basic user-network interface: Layer 1
I.431	Primary rate user-network interface: Layer 1
I.432	B-ISDN User-Network Interface-Physical Layer
I.530	Network interworking between an ISDN and a PSTN
I.555	ATM and Frame Relay interworking
I.570	Public/private ISDN interworking
I.610	B-ISDN Operations and Maintenance Principles and Functions
I.761	Inverse multiplexing for ATM (IMA)
I.762	ATM over fractional physical links

ISI (InterSymbol Interference) Signal distortion caused by the overlap of symbols (i.e., the smallest units transmitted) in a digital transmission system. Intersymbol interference includes the overlap of bits in a purely digital transmission system, such as a fiber optic transmission system (FOTS), or sinusoidal waveforms in a digital system involving the modulation of a carrier waveform, such as a T-carrier or E-carrier system. See also *pulse dispersion*.

ISLAN (Integrated Services Local Area Network) Specified by the IEEE as 802.9 and also known as *IsoEthernet* and *IsoEnet*. ISLAN is now considered obsolete. See also *802.9*, *IEEE*, and *IsoEthernet*.

ISM (Industrial/Scientific/Medical) Radio frequency (RF) bands in the 902–928 MHz, 2.4–2.5 GHz, and 5.8–5.9 GHz ranges. In the United States and most countries, these bands require no licensing. As signal propagation characteristics are excellent at these relatively low frequencies (particularly in the lower of the two bands), they are employed extensively in a wide variety of applications, including Wireless Local Area Networks (WLANs). See also *propagation*, *RF*, and *WLAN*.

ISO (International Organization for Standardization) A network of the standards institutes of 157 countries. ISO is not an acronym. Rather, it is derived from the Greek *isos*, meaning equal, suggesting that all members of the organization have an equal voice. ISO is a non-governmental organization intended to serve as a bridge not only between countries, but between the governmental and private sectors. In the context of telecommunications, ISO is perhaps best known for the development of the OSI Reference Model. See Appendix A for contact information. See also *OSI Reference Model*.

ISOC (Internet Society) A voluntary organization that lends formal structure to the administration of the Internet. ISOC is the organizational home of the Internet Engineering Task Force (IETF), Internet Architecture Board (IAB), Internet Engineering Steering Group (IESG), and Internet Research Task Force (IRTF). The ISOC is active in such areas as censorship and freedom of expression, taxation, governance, and intellectual property. ISOC has granted the IESG formal authority to make decisions on standards. See also *IAB*, *IESG*, *IETF*, *Internet*, *IRTF*, and *standards*.

isochronous From the Greek *isos*, meaning *equal* or *uniform*, and *chronos*, meaning *time*. **1.** Equal in frequency, or periodicity. Uniform in time, having equal duration, or occurring at precise intervals. Isochronous communications, such as real-time audio and video, are stream-oriented, flowing at a constant and regular pace, with each audio and video element being of equal importance. Therefore, each element (e.g., instant of audio or pixel of color) must be delivered to the receiver in exactly the sequence in which it was presented to the transmitter and with no significant level of either latency (i.e., delay), or jitter (i.e., variation in delay). See also *jitter*, *latency*, *near-realtime*, and *store-and-forward*. **2.** A type of digital circuit in which the device on one end sets the bit rate for its own transmission and the receiving device copies that bit rate when responding, but there is no clock signal on the circuit interface.

IsoEnet (IsoEthernet) See *IsoEthernet*.

IsoEthernet (Isochronous Ethernet) Also known as *IsoEnet* and *Integrated Services LAN* (ISLAN). Specified by the IEEE as 802.9 (1995), IsoEthernet was intended to support isochronous traffic such as voice and videoconferencing over the same twisted pair as Ethernet. To traditional Ethernet at 10 Mbps, IsoEthernet added 96 ISDN B channels and a single D channel, all of which are 64 kbps channels and all of which share a cable using the 20 MHz signaling rate of 10BaseT Ethernet. The added B channels increased the theoretical Ethernet transmission rate by 6.144 Mbps, bringing the aggregate nominal rate to 16 Mbps. The individual B channels could be used for voice, or could be bonded for high speed videoconferencing. ISLAN is considered obsolete, having been overwhelmed by switched Ethernet at 100 Mbps, 1 Gbps, and even 10 Gbps. See also *10BaseT*, *802.9*, *B channel*, *bonding*, *D channel*, *Ethernet*, *IEEE*, *ISDN*, and *isochronous*.

ISO/IEC 11801 An international standard from the International Organization for Standardization (ISO) and the International Electrotechnical Commission (IEC), specifying general-purpose telecommunications structured cabling systems. The standards include both copper wire and optical fiber inside cabling systems. The standard specifies classes of twisted pair copper performance, as follows:

Class A: ≤ 100 kHz

Class B: ≤ 1 MHz

Class C: ≤ 16 MHz

Class D: ≤ 100 MHz

Class E: ≤ 250 MHz

Class F: ≤ 600 MHz

isotropic **1.** Having physical properties (e.g., conductivity, elasticity, and power density) that are the same in any direction of measurement. **2.** In telecommunications, an antenna, light source, or sound source that theoretically radiates a signal with equal power density in all directions. A purely isotropic source does not exist, except in theory. In telecommunications, as in all things natural and unnatural, perfection is purely theoretical.

ISP **1.** InSide Plant. All of the telecommunications apparatus, equipment, wiring, and systems housed in buildings. ISP includes the main distribution frame (MDF), intermediate distribution frames (IDFs), inside wire and cable systems, switches and routers, multiplexers, storage batteries, backup power generators, and related equipment. See also *outside plant.* **2.** Internet Service Provider. A company that provides Internet access services to end users. An ISP can be local, regional, national, or international in nature and can provide a wide range of access alternatives, including dial-up and ADSL. See also *ADSL* and *dial-up.*

IT (Information Technology) The traditional, formal term for a data processing department. The contemporary term is Information Services (IS).

ITA #2 (International Telegraph Alphabet #2) See *Baudot code.*

ITU (International Telecommunication Union) Chartered by the United Nations (UN), the ITU-T primarily is responsible for setting recommendations intended to ensure the interconnectivity of national telecommunications networks. Those recommendations are treated as standards in most countries. The original predecessor organization was the International Telegraph Union (ITU), which was formed in 1865 to ensure the interconnectivity of national telegraph networks. The ITU formed the International Telephone Consultative Committee (CCIF) in 1924, the International Telegraph Consultative Committee (CCIR) in 1925, and the International Radio Consultative Committee (CCIR) in 1927. The ITU changed its name in 1934 to the International Telecommunication Union and in 1956 the CCIT and CCIF merged to form the International Telephone and Telegraph Consultative Committee (CCITT). In 1992, the ITU formed into three sectors, the ITU-Telecommunication (ITU-T), Radiocommunication (ITU-R), and Telecommunication Development (ITU-D). See Appendix A for contact information. See also *ITU-D*, *ITU-R*, and *ITU-T.*

ITU-D (International Telecommunication Union-Development Sector) The sector of the ITU that works to further telecommunications development around the world, especially in developing countries. See Appendix A for contact information. See also *ITU.*

ITU-R (International Telecommunication Union-Radiocommunication) The sector of the ITU that studies technical questions relating to radio communications and manages international radio frequency (RF) spectrum and satellite orbits. The ITU-R originally (1925) was the Comité consultatif international des Radiocommunications (CCIR), which translates from French as the International Radio Consultative Committee. See Appendix A for contact information. See also *ITU* and *RF.*

ITU-T (International Telecommunication Union-Telecommunication Standardization Sector) The sector of the ITU that studies technical questions relating to telegraph and telephone communications and establishes international telecommunications standards to ensure the interconnectivity of national networks. The ITU-T previously (1956) was known as the Comité consultatif international télégraphique et téléphonique (CCITT), which translates from French as the International Telegraph and Telephone Consultative Committee. See Table I-5 for ITU-T Series Recommendations. See Appendix A for contact information. See also *ITU.*

Table I-5: ITU-T Series Recommendations

Series	*Description*
A	Organization of the work of ITU-T
B	Means of expression: definitions, symbols, classification
C	General telecommunication statistics
D	General tariff principles
E	Overall network operation, telephone service, service operation and human factors
F	Non-telephone telecommunication services
G	Transmission systems and media, digital systems and networks
H	Audiovisual and multimedia systems
I	Integrated services digital network (ISDN)
J	Cable networks and transmission of television, sound program and other multimedia signals
K	Protection against interference
L	Construction, installation and protection of cables and other elements of outside plant
M	Telecommunication management, including TMN and network maintenance
N	Maintenance: international sound programme and television transmission circuits
O	Specifications of measuring equipment
P	Telephone transmission quality, telephone installations, local line networks
Q	Switching and signaling
R	Telegraph transmission
S	Telegraph services terminal equipment
T	Terminals for telematic services
U	Telegraph switching
V	Data communication over the telephone network
X	Data networks, open system communications and security
Y	Global information infrastructure, Internet protocol aspects and next-generation networks
Z	Languages and general software aspects for telecommunication systems

IVD (Inside Vapor Deposition) A commonly used technique for the mass production of glass optical fiber, IVD begins with heating silica and germanium dopant to the point of vaporization. As the glass vapor cools, it is deposited as layers of soot on the inside of a rotating hollow glass cylinder, which typically remains as the outside cladding of the end product. The first layers deposited are cladding of relatively pure silica. The final layer is germanium-doped silica, which forms the core. If the end product is to be a step-index fiber, there is an abrupt change in the chemical composition between the core and cladding. If the end product is to be a graded-index fiber, there will be many graded layers of silica of slightly different chemical compositions deposited on the cylinder wall to yield slightly and successively less pure layers of cladding surrounding the fiber axis. The composition of the glass layers in a graded-index fiber is much like the arrangement of the annular rings of a tree. Once the deposition process is complete, the entire glass cylinder is sintered and collapsed into a preform cylinder, which is cooled and stored. The tip of the preform cylinder is reheated to a temperature of 2,500 degrees in a drawing tower. The resulting gob of molten glass is carefully drawn by gravity, in a process known as broomsticking, into a fiber as long as 20 kilometers. As the fibers cool, an acrylate coating is applied to protect the raw glass from physical damage. As is the case with all of these techniques, OVD takes place in a vacuum environment, as it is the exposure to oxygen that makes glass so brittle. Outside vapor deposition (IVD) is a similar process, with the soot deposited on the outside of a rotating ceramic bait rod that is slipped out of the glass cylinder prior to the formation of the preform cylinder. See also *cladding*, *core*, *dopant*, *graded-index fiber*, *IVD*, *sintering*, and *step-index fiber*.

IVDS (Interactive Video and Data Services) See *218–219 MHz Service*.

IVR (Interactive Voice Response) Interaction between a human and a computer in which the human caller inputs commands and requests to the computer, which responds in either pre-recorded or synthesized speech form. The human input can be the form of spoken words or as tones sent via the telephone keypad. If the input is speech, IVR is much like having a frustrating conversation with a dimwitted and highly inflexible human call center agent. In a database access application, a voice processing system with IVR capability is positioned as a front end to a general-purpose computer and multiple databases. Through speech recognition technology, and text-to-speech (TTS) capability, and perhaps voice print matching for security, a complete transaction can be accomplished on a voice basis without human involvement&except for the caller, of course. Reservations centers and financial institutions make heavy use of such capabilities in support of routine transactions, thereby reducing staffing levels and providing customer service on a 24 × 7 basis. Telephone companies increasingly use voice recognition technology to provide automated access to directory databases.

IX (Internet eXchange) An official Network Access Point (NAP) at which an Internet service provider (ISP) can access the Internet backbone and exchange traffic with other ISPs. Some NAPs are known as NAPs, some as Internet Exchanges (IXs), and some as MAEs. Tier 1 IXs are located in Amsterdam, The Netherlands (AMS-IX); London, England (LINX); Sophia-Antipolis, France (SFINX); Cape Town and Johannesburg, South Africa (CINX and JINX); Hong Kong, Peoples Republic of China (HKIX); and Tokyo, Japan (JPIX). See also *Internet*, *ISP*, *MAE*, and *NAP*.

IXC (IntereXchange Carrier, or IEC) A company providing long haul telephone service between local exchange carriers (LECs), that is, local telephony companies (telcos) or other providers of local telephone service. See also *carrier*, *LATA*, *LEC*, and *long haul circuit*.

J (Joule) See *joule.*

J-1 The first level of the Japanese J-carrier digital hierarchy. J-1 mimics the North American T1 system. A J-1 system comprises circuit terminating equipment in the form of a combination of a channel service unit (CSU) and a data service unit (DSU) that jointly serve to interface a device to a full duplex (FDX) four-wire digital circuit and to perform various signal formatting, signal timing, monitoring, and diagnostic functions. J-1 operates at a signaling rate of 1.544 Mbps, which supports a frame rate of 8,000 frames per second (fps), with each frame comprising a framing bit followed by 192 bits of user payload. The framing bits are used for synchronization and, in some cases, for monitoring, diagnostic, and other network management purposes. The 192 bits of user payload are organized into 24 time-division multiplexed (TDM) time slots, each of which is 8 bits wide. At a rate of 8,000 fps, each time slot is repeated 8,000 times per second, which translates into a DS-O channel at 64 kbps (8 bits × 8,000 per second = 64,000 bps). Taken together, the 24 8-bit TDM channels at 8,000 fps yields an aggregate payload transmission rate of 1.536 Mbps. Adding the 8,000 framing bits (one per frame) per second, yields the aggregate signaling rate of 1.544 Mbps. J-1 was initially designed to operate over a physical four-wire twisted-pair copper circuit, but is considered medium independent and will run over coaxial cable, optical fiber, microwave, satellite, and free space optics (FSO) just as well. J-1 generally is used in local loops and other short-haul applications. In long-haul applications, J-3 and other, higher speed, standards generally are employed. See also *carrier, CSU, DS-0, DSU, FDX, four wire, frame, framing bit, J-3, J-carrier, payload, signaling rate, synchronization, T1, T-carrier, TDM, time slot,* and *transmission rate.*

J-2 The second level in the Japanese J-carrier digital hierarchy. J-2 runs at two signaling rates. At 6.312 Mbps, J-2 mimics the North American T2, comprising four J-1s at 1.544 Mbps each, plus 132 kbps of overhead and justification, or bit stuffing, to adjust for variations in the clocking rates of the incoming J-1s. Multiplexing is performed by M12 (Multiplex J1-to-J2) terminals that multiplex four J-1 signals, which yields 96 DS-0 channels at 64 kbps per channel. At 7.786 Mbps, J-2 supports 120 DS-0 channels, which maps J-2 into E-2, the European version. Despite similarities in terms of signaling rates and channel capacities, J-carrier is not compatible with either E-carrier or T-carrier. See also *bit stuffing, digital signal hierarchy, DS-0, E-2, E-carrier, J-carrier, T2,* and *T-carrier.*

J-3 The third level in the Japanese J-carrier digital hierarchy. J-3 runs at a signaling rate of 32.064 Mbps in support of 480 DS-0 channels, which maps J-3 into E-3, the European version. Despite similarities in terms of signaling rates and channel capacities at various levels, J-carrier is not compatible with either E-carrier or T-carrier. See also *digital signal hierarchy, DS-0, E-3, E-carrier, J-carrier,* and *T-carrier.*

J-4 The fourth level in the Japanese J-carrier digital hierarchy. J-4 runs at a signaling rate of 97.728 Mbps in support of 1,440 DS-0 channels. See also *digital signal hierarchy, DS-0,* and *J-carrier.*

J-5 The fifth level in the Japanese J-carrier digital hierarchy. J-5 runs at a signaling rate of 565.148 Mbps in support of 7,680 DS-0 channels which maps J-5 into E-5, the European version. Despite similarities in terms of signaling rates and channel capacities at various levels, J-carrier is not compatible with either E-carrier or T-carrier. See also *digital signal hierarchy, DS-0, E-5, E-carrier, J-carrier, signaling rate,* and *T-carrier.*

jabber Meaningless data introduced into a local area network (LAN), usually as the result of a malfunctioning transceiver, transceiver cable, or network interface card (NIC). See also *LAN, NIC,* and *transceiver.*

Jabber An XML-based instant messaging (IM) system written as an open source application by Jeremie Miller in 1998. See also *IM, open source,* and *XML.*

jack A female connector or outlet with receptacles designed to receive the pins of a male plug in a plug-and-jack connection. See also *connector*, *plug*, and *RJ*.

jacket See *cable jacket*.

jailbreak See *voice mail jail*.

jamming signal **1.** A signal that intentionally introduces interference into a communication channel, either to intentionally prevent error-free reception or as a means of advising stations of some event. **2.** In local area networks (LANs), employing the carrier sense multiple access with collision detection (CSMA/CD) protocol, a station that detects a signal collision sends a jamming signal over a subcarrier frequency to advise all stations of that fact. See also *CSMA/CD*, *frequency*, *LAN*, *signal*, and *subcarrier*.

Japanese Digital Cellular (JDC) A Japanese 2G digital cellular radio standard now known as Personal Digital Cellular (PDC). See also *PDC*.

Japanese Total Access Communications System (JTACS) See *JTACS*.

Java Telephony Application Programming Interface (JTAPI) A cross-platform, multivendor computer telephony (CT) solution based on Java, JTAPI uses highly efficient applets, which are small sets of application program code, for network-based CT operation. JTAPI added Internet/intranet functionality to CT, thereby enabling the creation of Web-based applications that integrate browser applications with call center functionality. JTAPI was developed jointly by Sun Microsystems, Lucent Technologies, IBM, and Nortel.

J-band The range of radio frequencies from 10 GHz to 20 GHz in the super high frequency (SHF) range of the electromagnetic spectrum. The J-band is used for satellite communications and radar applications. See also *electromagnetic spectrum*, *radar*, *satellite*, and *SHF*.

JBOD (Just a Bunch Of Disks) A derogatory term for spanning, which is the use of a number of external physical hard drives organized into a single logical drive. JBOD is a simple storage technology that allows a computer to write to a large storage medium comprising multiple smaller drives. Unlike a redundant array of independent disks (RAID), JBOD does not provide any advantages in terms of redundancy or performance. See also *RAID*.

JDC (Japanese Digital Cellular) A Japanese 2G digital cellular radio standard now known as Personal Digital Cellular (PDC). See also *PDC*.

jitter **1.** Uncertain variation in the timing of a received signal as compared to the timing of the transmitted signal. All signals experience some amount of delay, or latency, as they propagate across a circuit as, even at the speed of light, it takes some amount of time to travel the distance from one point to another. The timing of the signal elements remains consistent, however, barring changes in the length or other physical characteristics of the circuit caused by variations in temperature or other external forces. As devices are added to a circuit, even relatively simple devices such as amplifiers and repeaters performing relatively simple processes, additional delay is introduced, and the potential for variability in delay is increased because of factors such as fluctuations in power sources and faulty internal components. Should variability exceed specified tolerances, the timing of the pulses can be unacceptably irregular, and the receiving device may be unable to interpret the received signal correctly. **2.** Variability in latency of a block, cell, frame, packet, or other message unit. Data message units can suffer jitter not only due to issues of signal jitter, but also because they may encounter different levels of congestion, which may cause them to spend different amounts of time in queues. These factors, and others, contribute to jitter. Some applications, such as e-mail, are tolerant of jitter, while other applications, such as real-time, uncompressed voice, are highly intolerant of jitter. **3.** Undesirable rapid or jumpy movement of images, such as those displayed on a television or computer monitor. Jitter can be caused by circuit instability or faulty system components. See also *latency*, *propagation delay*, and *velocity of propagation (Vp)*.

J-carrier (Japanese-carrier) A hierarchy of standards for digital transmission that essentially is the Japanese version of the United States T-carrier digital carrier system. At levels one and two, J-carrier mimics T-carrier with respect to the signaling rates, but diverges at level three and beyond. J-carrier also uses a different PCM companding technique (A-law rather than μ-law). As J-carrier is medium-independent, it can be provisioned over any of the transmission media, at least at the J-1 rate of 1.544 Mbps. (At the J-3 rate of 32.064 Mbps, twisted pair is unsuitable due to issues of signal attenuation.) The fundamental building block of J-carrier is a 64-kbps channel, referred to as DS-0 (Digital Signal level Zero). Through time division multiplexing (TDM), J-carrier interleaves DS-0 channels at various signaling rates to create the services that comprise the Japanese digital hierarchy, as detailed in Table J-1.

Table J-1: Japanese Digital Hierarchy: J-carrier

J-carrier Level	*Data Rate*	*64-kbps Channels (DS-0s)*	*Equivalent J1s*
0	64 kbps	1	Not applicable
J-1	1.544 Mbps	24	1
J-2	6.312 Mbps 7.786 Mbps	96 120	4 5
J-3	32.064 Mbps	480	20
J-4	97.728 Mbps	1,440	60
J-5	565.148 Mbps	7,680	320

See *digital signal hierarchy* for a side-by-side comparison of the North American, European, and Japanese digital hierarchies. See also *A-law, carrier, channel, companding, digital, DS-0, E-carrier, J-1, J-2, J-3, J-4, J-5, μ-law, PCM, signaling rate, T-carrier,* and *TDM.*

Joan/Eleanor The code name for the first walkie talkie system, which comprised a ground unit, Joan, and an airborne unit, Eleanor. The system allowed agents of the Office of Strategic Services (OSS), the predecessor to the Central Intelligence Agency (CIA), behind enemy lines to communicate with aircraft in a manner that virtually defied detection at the time. See also *walkie talkie.*

job security [This space intentionally left blank.] See also *offshoring* and *outsourcing.*

.jobs Pronounced *dot jobs.* The generic Top Level Domain (gTLD) reserved exclusively for job seekers and companies seeking employees. It is not now permissible for third parties (i.e., corporate recruiters) to use .jobs. This domain was created in 2005 under the sponsorship of an alliance between Employ Media, the Society for Human Resource Management (SHRM), Verisign, and ICANN. See also *gTLD, Internet,* and *sponsored domain.*

Joint Photographic Experts Group (JPEG) See *JPEG.*

joule (J) The work done when a force of one newton applied to a point moves that application point one meter in the direction of application. Joule is named for James Prescott Joule (1818–1889), a British physicist. See also *newton.*

JPEG (Joint Photographic Experts Group) A graphics file format for editing still images, as well as color facsimile, desktop publishing, graphic arts, and medical imaging. A symmetrical compression technique, JPEG is equally expensive, processor-intensive, and time consuming in terms of both compression and decompression. JPEG is a joint standard of the International Telecommunications Union (ITU-T T.81) and the International Organization for Standardization (ISO 10918-1). JPEG involves a lossy compression mechanism using discrete cosine transform (DCT). Compression rates of 100:1 can be achieved, although the loss is noticeable at that level. Compression rates of 10:1 or 20:1 yield little degradation in image quality. See also *compression, DCT, GIF, ISO, ITU-T, lossy compression, PNG,* and *symmetric.*

.jpg The file extension that identifies files in the Joint Photographic Experts Group (JPEG) format. See also *JPEG*.

JTACS (Japanese Total Access Communications System) A Japanese version of the TACS 1G analog cellular radio technology developed for use in the United Kingdom. See *1G*, *analog*, *cellular radio*, *narrowband*, and *TACS*.

JUGHEAD (Jonzys Universal Gopher Hierarchy Excavation And Display) An Internet browser similar in operation to Veronica, but limited to keyword searches in directory titles on a specific site in Gopherspace. Jughead indexes the keywords in directory titles, but does not index the files within the directories. Playing off Archie and Gopher, subsequent developers of search mechanisms tried to stay with the Archie comic book/rodent theme, proving once and for all that even acronyms can be fun. Unfortunately, there is not even a hint of humor reflected in the chosen names of more recent and more powerful browsers. See also *Archie*, *browser*, *Gopher*, *Gopherspace*, *Internet*, and *Veronica*.

jumbo payload In Internet Protocol version 6 (IPv6), a user data field that exceeds the IPv4 limit of 65,535 octets. A jumbo payload is indicated in the payload length field of the IPv6 header. See also *header*, *IPv4*, *IPv6*, *octet*, and *payload*.

jumper A short wire or optical fiber used to make semi-permanent connections between circuits on a distribution frame or patch panel.

Junk Fax Protection Act In the United States, a modification (2005) to the Telephone Consumer Protection Act (1991) and Federal Communications Commission (FCC) rules that defines an existing business relationship (EBR), codifies an existing EBR exemption to the prohibition to sending unsolicited fax advertisements, requires the sender to provide details as to how the recipient can opt out of future faxes, and specifies the circumstances under which a request to opt out complies with the Act. See also *fax*, *FCC*, and *Telephone Consumer Protection Act*.

junk mail See *spam*.

just a bunch of disks (JBOD) See *JBOD*.

justification See *bit stuffing*.

k **1.** kilo (k). From Greek *khilioi*, meaning thousand. **2.** In transmission systems, kHz (kiloHertz) is a thousand (10^3) Hertz, kbps (kilobit per second) is a thousand (10^3) bits per second, and km is a thousand (10^3) meters. In transmission systems, therefore, a thousand is exactly 1,000, since the measurement is based on a base 10, or decimal, number system. **3.** In computing and storage systems, a kB (kiloByte) is actually 1,024 (2^{10}) bytes, since the measurement is based on a base 2, or binary, number system. The term kB comes from the fact that 1,024 is nominally, or approximately, 1,000. So, 64 kB of memory is actually 65,536 (2^{16}) bytes. **4.** The only letter worth five points in the American version of Scrabble, the popular board game.

K The symbol for Kelvin, the SI unit of absolute temperature. Measured from absolute zero, zero (0) degrees Kelvin (K) = –273.16 degrees Celsius (C). The formula for converting Kelvin to Celsius is Kelvin (K) = Celsius (C) + 273.16. See also *SI.*

Ka band The portion of the microwave radio spectrum in the range of 20.0–30.0 GHz, as specified by the ITU-R. The K band is subdivided into Ku band and Ka band. Ka-band is so called as it is *above* the center of the K band. Applications include satellite microwave systems, with current emphasis on mobile voice and data. See also *electromagnetic spectrum*, *ITU-R*, *K band*, *Ku band*, *microwave*, and *satellite*.

Kao, Charles While engineers at Standard Telecommunications Laboratories, an ITT subsidiary, Charles Kao and George Hockham achieved the first conceptual breakthrough in the development of fiber optic transmission systems (FOTS). In 1966, Kao and Hockham determined that optical fibers of fused silica could satisfy signal attenuation requirements by overcoming issues of absorption, diffusion, and bending loss. At the time, attenuation of 20 dB per kilometer was considered satisfactory for a commercially viable system. Contemporary FOTS are designed around attenuation levels in the range of 1.5–3.0 db per kilometer, or so, depending on the specific type of fiber and the wavelength employed. See also *FOTS* and *optical fiber*.

kB (kilobyte) One thousand bytes. In computing and storage systems, a kB (kiloByte) is actually 1,024 (2^{10}) bytes, since the measurement is based on a base 2, or binary, number system. The term kB comes from the fact that 1,024 is nominally, or approximately, 1,000. See also *byte* and *k*.

K band The portion of the microwave radio spectrum in the range of 10.9–36.0 GHz, as specified by the ITU-R. The K band is subdivided into Ku band and Ka band. Ku band is so called as it is *under* the center of the K band, and Ka band is so called as it is *above* the center of the K band. Applications include satellite and terrestrial microwave systems. See also *electromagnetic spectrum*, *ITU-R*, *microwave*, and *satellite*.

kbps (kilobit per second) One thousand (10^3) bits per second. A measure of bandwidth in a digital transmission system. See also *bandwidth*, *bps*, and *k*.

keep alive bits In the event that a digital circuit is silent, i.e. there is no active data transmission, the channel service unit (CSU) that interfaces the circuit to the customer's data communications equipment (DCE) will regularly transmit a one (1) bit, collectively known as *keep alive bits*, to ensure that there are electrical pulses on the circuit with at least a minimal density. This process ensures that the various circuit terminating equipment and repeaters remain synchronized and, in some cases, powered. See also *CSU*, *DCE*, *ones density*, *repeater*, and *synchronize*.

Kelvin (K) See *K*.

Kerberos Authorization software that makes use of private-key authentication. Developed by the Massachusetts Institute of Technology (MIT), Kerberos is available for free, although commercial versions exist. Kerberos was named for the three-headed dog, also known as Cerberus, that guarded the gates of Hades in Greek mythology. Note: Although, according to Greek legend, Hercules defeated Kerberos, a hacker of Herculean proportions has yet to emerge victorious over this powerful security software. See also *Access Manager*, *authorization*, *security*, and *Sesame*.

Kermit **1.** Kermit the Frog. The green frog puppet star of *The Muppet Show*. **2.** An asynchronous file transfer protocol that organizes data into 128-byte blocks and employs a cyclic redundancy check (CRC) for excellent error control. Kermit operates over a wide variety of connections, including dial-up modem connections and TCP/IP connections. Although Kermit is not in the public domain, Columbia University generally allows its use at no charge, so most communications protocols support it. Kermit was developed by the Columbia University Computer center in 1981 and named for Kermit the Frog. See also *asynchronous*, *block*, *CRC*, *protocol*, *XMODEM*, and *ZMODEM*.

kernel The core of a computer operating system (OS). The kernel resides in memory and performs basic tasks such as managing internal memory, input and output operations, and peripheral devices. A kernel also is responsible for launching applications and allocating associated system resources such as processor time. See also *OS*.

KEVLAR® The registered name under which Dupont markets aramid fiber. See also *aramid*.

key **1.** A small mechanical device for opening, closing, or switching electrical circuits. A telegraph key, for example, is used to open and close an electrical circuit to send short and long pulses of electric current. See also keyboard, keypad, and KTS. **2.** A string of bits used for encrypting and decrypting information. A private key is known to only one person, typically the sender. A public key is known to more than one person, typically both the sender and the receiver, and is published and freely available from a public key infrastructure (PKI) or certificate authority (CA). See also *CA*, *encryption*, *PKI*, *private key encryption*, and *public key encryption*.

keyboard A simple computer input device comprising a set of alphabetic, numeric, punctuation, symbol, control, and function keys mounted on a board or other flat surface or control panel. When the user depresses a key or a combination of keys, a coded signal is sent to the computer. If the key is associated with a character, the computer echoes the signal by displaying the character on the monitor. See also *keypad*.

keypad A simple input device comprising a set of keys or buttons mounted on a keyboard or control panel associated with a computer, telephone, microwave oven, or other device or on a remote control device. See also *keyboard*.

key service unit (KSU) See *KSU*.

key telephone system (KTS) See *KTS*.

kHz (kiloHertz) One thousand (10^3) Hertz. A measure of bandwidth in an analog transmission system. See also *bandwidth*, *Hertz*, and *k*.

kilo- (k) See *k*.

kilobit per second (kbps) See *kbps*.

kiloByte (kB) See *kB*.

kiloHertz (kHz) See *kHz*.

kilometer (km) See *km*.

Kingsbury Commitment The commitment, expressed in a letter written by Nathan C. Kingsbury, an AT&T vice-president, to the United States Attorney General in December 1913, to resolve a number of antitrust issues. Kingsbury committed AT&T to dispose of its holdings in Western Union, to purchase no more independent (i.e., non-Bell) telephone companies without the approval of the Interstate Commerce Commission (ICC), and to make interconnection with the independent telephone companies. Eventually, national standards were established to govern the nature and rules of interconnection. The ITU-T governs interconnection at the international level. See also *AT&T*, *ITU-T*, and *Western Union*.

km (kilometer) One thousand (10^3, or 1,000) meters. See also *k* and *meter*.

knife-edge diffraction The phenomenon by which an electromagnetic waveform diffracts, or bends, as it strikes the sharp edge of an obstacle transverse to its direction of propagation. The portion of the signal that is not cut off by the knife edge continues to propagate, but the edge of the signal bends into the line-of-sight (LOS) shadow region as if to fill the void left by the portion of the signal cut off. Knife-edge diffraction can be used to advantage in radio communications when line-of-sight (LOS) cannot be achieved due to the presence of an obstacle, such as a mountaintop or building, that lies in the path of the transmit and receive antennas. See also *diffraction* and *LOS*.

KSU (Key Service Unit) The common control unit of a key telephone system (KTS). The KSU stores programmed features, which are served to individual users and stations based on their individual access privileges.

KTS (Key Telephone System) A voice-optimized telecommunications system designed for small business or small office applications (see Figure K-1), typically defined as involving no more than 50 stations. The term key refers to the buttons that mechanically opened and closed the line circuits on the early generation of KTSs. Although the buttons are no longer mechanical keys, KTSs remain relatively simple CPE systems that allow multiple station users to share a number of outside lines that the users select by depressing the button associated with the specific circuits they desire to access. This approach is unlike that of a PBX switch, which has the intelligence to accept a call request from a user station, determine the most appropriate circuit from a shared pool of circuits, and set up the connection through common switching equipment.

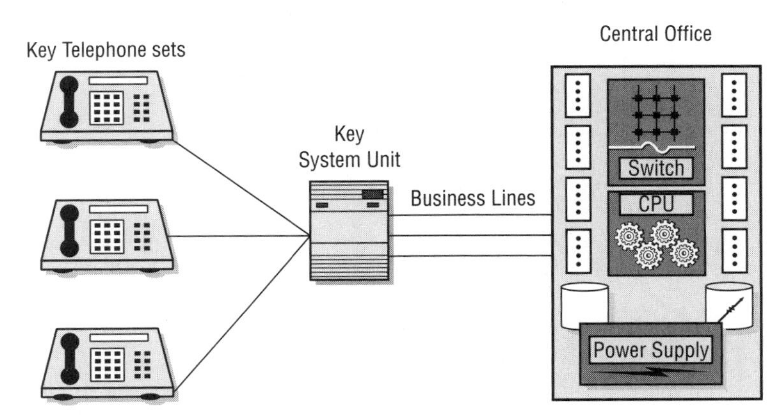

Figure K-1

Most small KTSs are squared, meaning that every key set is configured alike, with every outside line appearing on every set. Thereby, every station user can access every outside line for both incoming and outgoing calls, and all feature presentations are consistent. In larger systems, the physical size of the telephone sets required to maintain the squaring convention would be impractical, but departmental subgroups often are squared. As illustrated in Figure K-1, all but the simplest key systems have a software-based common control unit, known as a *Key Service Unit* (KSU), where programmed features are stored and provided to individual users and stations based on their individual access privileges. Introduced in 1938, early electromechanical KTS systems (1A, 1A1, and 1A2) were limited in feature content to hold, intercom, speakerphones, and auto dialers. Electronic KTS (EKTS) systems appeared in the 1970s, offering many of the same features as PBX systems. Most contemporary key systems are hybrids, meaning that they can operate either as a KTS or a PBX. Hybrids generally are limited to 200–250 or so ports, each of which often can serve to connect end user terminal equipment or an outside line to the KSU. See also *1A*, *1A1*, *1A2*, *CPE*, *line*, *PBX*, and *trunk*.

Ku band The portion of the microwave radio spectrum in the range of 12.0–18.0 GHz, as specified by the ITU-R. The K band is subdivided into Ku band and Ka band. Ku band is so called as it is *under* the center of the K band. Applications include satellite and terrestrial microwave systems. See also *electromagnetic spectrum*, *ITU-R*, *K band*, *Ka band*, *microwave*, and *satellite*.

L Symbol for Inductance. See *inductance*.

label **1.** A set of data attached to and providing identification or other information relative to a larger data unit, such as a packet or message. See also packet and message. **2.** In Multiprotocol Label Switching (MPLS), the set of data attached to a packet and used by a Label Switched Router (LSR) to select a link across which to forward that packet. The initial packet is inserted by a Label Edge Router (LER). Each LER along the path swaps the label associated with the incoming packet for a new label associated with the outgoing packet to be used by the adjacent downstream router in making the next link selection. See also *downstream*, *LER*, *link*, *LSR*, *MPLS*, *packet*, and *router*.

Label Distribution Protocol (LDP) See *LDP*.

Label Edge Router (LER) See *LER*.

Label Switching Router (LSR) See *LSR*.

Label Switched Path (LSP) See *LSP*.

LACNIC (Latin-American and Caribbean Network Information Center) The Regional Internet Registry (RIR) responsible for assigning Internet Protocol (IP) addresses variously to National Internet Registries (NIRs) or directly to Local Internet Registries (LIRs) in Latin America and the Caribbean. See also *IP*, *IP address*, *LIR*, *NIR*, and *RIR*.

Lamarr, Hedy (1913–2000) A famous actress and dancer of pre-war (WWII) fame, and co-inventor of spread spectrum (SS) radio. Born Hedwig Eva Maria Keisler in Vienna, Austria, her first husband (of six) was Friedrich Mandl, an arms manufacturer who socialized with Adolph Hitler and Benito Mussolini. She reportedly hated her overly possessive husband and his Nazi friends and escaped to London before WWII. Using knowledge she gained while married to Mandl, Lamarr developed spread spectrum radio, which she initially used to remotely synchronize multiple player pianos in radio-controlled piano concerts that reportedly were quite popular in those much simpler times. The United States Patent and Trademark Office issued patent #2,292,387 (1942) to Ms. Lamarr and George Antheil, a film-score composer to whom she had turned for help in perfecting the idea, for a Secret Communication System that was, in effect, a spread spectrum radio. In the Pacific Theater during World War II, the Allies used that patented technology extensively to prevent the Japanese from jamming radio-controlled torpedoes. This primitive system used a mechanical switching system much like a piano roll to shift frequencies faster than the Axis military could follow them. Subsequently, spread spectrum combined with digital technology for spy-proof and noise-resistant battlefield communications. During the 1962 Cuban nuclear missile crisis, Sylvania installed SS on U.S. warships sent to blockade Cuba, where the technology provided improved security as well as prevented signal jamming. Ms. Lamarr never requested nor received any royalties from the use of her invention. Ms. Lamarr was an exotic beauty and talented actress. She delighted and shocked prewar audiences by dancing nude in the movie *Ecstacy* (1933). Her greatest screen role was as Delilah, opposite Victor Mature as Sampson in Cecil B. DeMille's *Sampson and Delilah* (1949). Hedy Lamarr has a star on the Hollywood Walk of Fame at 6247 Hollywood Boulevard. In 1997, the Electronic Frontier Foundation (EFF) gave her an award for inventing SS radio, which is the basis for code division multiple access (CDMA), which is used in cellular and other radio networks. See *Antheil, George*. See also *CDMA*, *cellular radio*, *EFF*, *radio*, and *SS*.

lambda (λ) The Greek letter used by physicists to denote wavelength. See *wavelength*.

LAN (Local Area Network) A LAN is a packet network designed to interconnect host computers, peripherals, storage devices, and other computing resources within a local area, i.e., limited distance. LANs conform to the client/server architecture, a distributed computing architecture that runs applications on client microcomputers against one or more centralized servers, which are high-performance multiport computers with substantial processing power and large amounts of memory. A LAN might serve an office, a floor of a building, and entire building, or a campus area, but generally does not cross a public right-of-way such as a street. The distance limitation generally is in the range of a few kilometers, at most, although that is sensitive to the transmission media employed, which include coaxial cable, twisted pair, optical fiber, infrared (IR) light, and radio frequency (RF) systems. Raw bandwidth ranges up to 10 Gbps, although actual throughput generally is much less. LANs generally are private networks, although public wireless hotspots offering wireless Internet access currently are popular. Most LAN standards are set by the 802 Working Group of the Institute of Electrical and Electronic Engineers (IEEE), with examples being 802.3 (Ethernet) and 802.11a/b/g (Wi-Fi). A personal area network (PAN) such as Bluetooth, is much more limited in geographic scope than a LAN. LANs and LAN segments can be interconnected over a metropolitan area network (MAN) or wide area network (WAN). LANs operate at Layer 1, the Physical Layer, and Layer 2, the Data Link Layer, of the OSI Reference Model. See also *802.3*, *802.5*, *802.11*, *architecture*, *bandwidth*, *Bluetooth*, *client/server*, *coaxial cable*, *Data Link Layer*, *Ethernet*, *hotspot*, *IEEE*, *IR*, *MAN*, *optical fiber*, *OSI Reference Model*, *PAN*, *Physical Layer*, *RF*, *throughput*, *Token Ring*, *twisted pair*, *WAN*, and *Wi-Fi*.

landline **1.** A traditional telephone connected to the PSTN by a traditional wire (or fiber) local loop that terminates in a fixed location, rather than a cellular mobile telephone connected to a cellular network via radio technology. A cordless telephone is considered part of a landline as the local loop terminates in a fixed base station on the subscriber premises, even though the connection to the base station is wireless. A wireless local loop (WLL) is considered a landline, as it is terrestrial and connects two fixed points. See also *local loop* and *WLL*. **2.** A telecommunications system that uses traditional terrestrial cabled, or conducted, transmission media such as copper or fiber optics, and wireless systems such as microwave, rather than mobile wireless radio technologies such as cellular or, especially, non-terrestrial satellite.

LANE (LAN Emulation) A specification (January 1995) from the ATM Forum (since merged into the MFA Forum) for an ATM service in support of native Ethernet (802.3) and Token Ring (802.5) local area network (LAN) communications over an ATM network. Software in the end systems (e.g., ATM-based hosts or routers, known as *proxies*), of the ATM network emulates a native LAN environment. LANE acts as Layer 2 bridge in support of connectionless LAN traffic, with the connection-oriented ATM service being transparent to the user application. In LANE, a LAN emulation client (LEC) connects to the ATM network over a LANE user-to-network interface (LUNI). The network-based LAN emulation server (LES) registers the LAN medium access control (MAC) addresses and translates them into ATM addresses using the address resolution protocol (ARP). Each LEC is assigned to an emulated LAN (ELAN) by an optional network-based LAN emulation configuration server (LECS). Each LEC also is associated with a broadcast and unknown server (BUS) that handles broadcast and multicast traffic, as well as initial unicast frames before address resolution. LANE traffic generally is Class C variable bit rate (VBR) traffic in message mode, and is supported over ATM Adaptation Layer Type 5 (AAL5). See also *802.3*, *802.5*, *AAL5*, *ARP*, *ATM*, *ATM Forum*, *broadcast*, *BUS*, *Class C ATM traffic*, *connectionless*, *connection-oriented*, *ELAN*, *emulation*, *Ethernet*, *host*, *Layer 2*, *LEC*, *LECS*, *LES*, *LUNI*, *MAC*, *message mode service*, *MFA Forum*, *multicast*, *proxy*, *router*, *Token Ring*, *unicast*, and *VBR*.

LAN emulation (LANE) See *LANE*.

LAN emulation client (LEC) See *LEC*.

LAN emulation configuration server (LECS) See *LECS*.

LAN emulation server (LES) See *LES*.

LANE user-to-network interface (LUNI) See *LUNI*.

LAN switch An intelligent, active hub that establishes, maintains, and changes logical connections over physical circuits. Switches flexibly connect transmitters and receivers across networks of interconnected ports and links, thereby allowing network resources to be shared by large numbers of end users. LAN switches are packet switches that can support multiple simultaneous transmissions, reading the destination address of each frame and forwarding it directly to the port associated with the target device. LAN switches are classified as workgroup switches or backbone switches. A workgroup switch serves the needs of a group of users who work together and, therefore, share common resources and intercommunicate frequently. Through the use of workgroup switches, a great deal of traffic can be confined to relatively small LAN segments and overall LAN congestion can be reduced considerably. High capacity backbone switches interconnect workgroup switches and provide connections to routers, as required. LAN switch architectures are several. A shared bus switch has a single high-speed bus that is shared by all incoming and outgoing ports on a time-division multiplexing (TDM) basis. A matrix switch contains multiple interconnected high-speed internal buses in a multi-bus switching matrix that can provide full bandwidth to multiple, simultaneous transmissions. If there are congestion issues in a matrix switch, it may have the ability to subdivide its capacity, with the buses becoming shared buses. A cut-through switch quickly reads the address of the data frame and quickly flows it through the switching matrix, bit by bit. A store-and-forward switch temporarily buffers and examines each incoming frame for errors through a CRC check before forwarding it through the switching matrix to the output port. Fragment-free switching, is similar to cut-through except for the fact that the switch stores the first 64 octets of the frame before forwarding it. As most errors occur at the beginning of a frame, this approach eliminates the possibility that runt frames (i.e., truncated frames) will be transmitted.

LAPB (Link Access Procedure, Balanced) Pronounced *lap bee*. A bit-oriented protocol that is part of the X.25 protocol suite, LAPB runs at Layer 2, the Data Link Layer, of the OSI Reference Model. The LAPB frame comprises a header and trailer that encapsulate the packet formed by the X.25 packet layer protocol (PLP) at the Network Layer. The LAPB frame provides a mechanism for transporting packets across a link, ensuring that frames of data are ordered correctly and are free from error. LAPB is a balanced protocol that operates in asynchronous balanced mode (ABM), referring to the fact that the devices have a balanced, rather than a master/slave, relationship. LAPB is a derivative of the High-level Data Link Control (HDLC) protocol, which, in turn, is based on the IBM Synchronous Data Link Control (SDLC) frame. Figure L-1 illustrates the LAPB frame and its component fields.

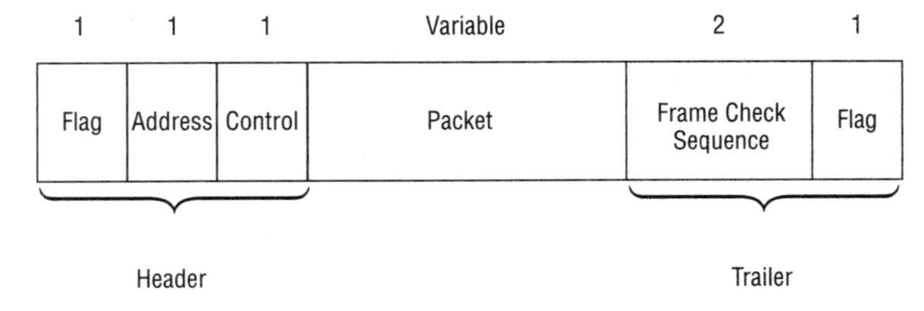

Figure L-1

The fields in the LAPB frame are as follows:

- **Flag:** A specific one-octet field that delimits (i.e., establishes the limits or boundaries of) the beginning and end of the frame. This octet is always 01111110 in binary notation (7E in hexadecimal notation), which is prevented from appearing in the payload by bit stuffing.
- **Address:** A one-octet field that indicates the sending station in command frames and the receiving station in response frames. User addressing is the responsibility of the logical channel identifier (LCI) field contained in the packet layer protocol (PLP), which is one layer above LAPB in the X.25 protocol stack.

- **Control:** A one-octet field that identifies the frame type. An information frame carries upper-layer information and some control data. A supervisory frame carries control information such as information-frame acknowledgement, request for retransmission, and flow control. An unnumbered frame carries control data such as disconnection request, acknowledgement, and frame rejection.
- **Data:** A variable-size field that contains an encapsulated packet.
- **Frame Check Sequence (FCS):** A two-octet cyclic redundancy check (CRC) field that provides error detection.

See also *ABM, asynchronous, bit-oriented protocol, CRC, Data Link Layer, flag, HDLC, header, master/slave, Network Layer, OSI Reference Model, PLP, SDLC, trailer,* and *X.25.*

LAPD (Link Access Procedure, D channel) Pronounced *lap dee.* A bit-oriented protocol that is part of the ISDN protocol suite, LAPD runs at Layer 2, the Data Link Layer, of the OSI Reference Model. LAPD defines the ISDN data (D) channels, which are designated for out-of-band signaling and control purposes in all ISDN implementations. LAPD evolved from the LAPB protocol used in X.25 networks. As the Signaling System 7 (SS7) signaling and control network specified for ISDN employs the X.25 packet format, consistency is maintained in the packet format, end-to-end. LAPD is a balanced protocol that operates in asynchronous balanced mode (ABM), referring to the fact that the devices have a balanced (i.e., peer-to-peer) relationship, rather than a master/slave relationship. Figure L-2 illustrates the LAPD frame and its component fields.

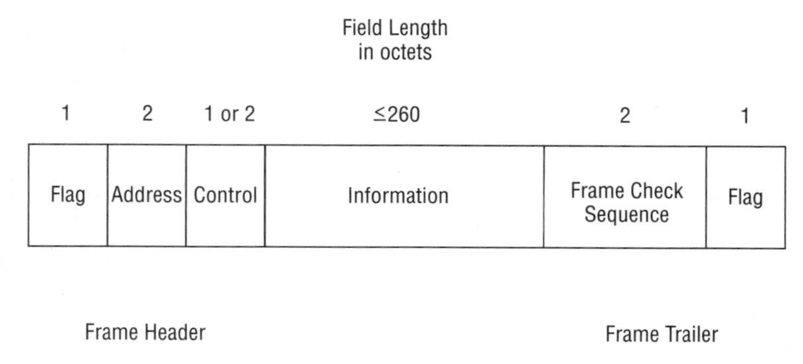

Figure L-2

The fields in the LAPD frame are as follows:

- **Flag:** A specific one-octet field that delimits (i.e., establishes the limits or boundaries of) the beginning and end of the frame. This octet is always 01111110 in binary notation (7E in hexadecimal notation), which is prevented from appearing in the payload by bit stuffing. Flags also fill idle time on the line.
- **Address:** A two-octet field known as the Data Link Connection Identifier (DLCI). The first octet is the Service Access Point Identifier (SAPI), which identifies the destination service access point, each of which can support multiple terminal devices. The second octet is the Terminal Endpoint Identifier (TEI), which is the address of the destination terminal device.
- **Control:** A one- or two-octet field that identifies the LAPD frame type. An information (I) frame carries upper-layer information and some control data. A supervisory (S) frame carries control information such as information-frame acknowledgement, request for retransmission, and flow control. An unnumbered (U) frame carries control data such as disconnection request, acknowledgement, and frame rejection.

- **Information:** A variable-size field with a maximum of 260 octets comprising upper-layer information. The size of the field is system-dependent. Only information frames include an information field.
- **Frame Check Sequence (FCS):** A two-octet cyclic redundancy check (CRC) field that provides error detection.

See also *ABM, asynchronous, bit-oriented protocol, CRC, Data Link Layer, flag, frame, D channel, ISDN, LAPB, master/slave, OSI Reference Model, out-of-band signaling and control, protocol suite, SS7,* and *X.25.*

LAPF (Link Access Procedure for Frame Mode Services) Pronounced *lap ef.* A bit-oriented protocol that is part of the frame relay protocol suite, LAPF is a subset High-level Data Link Control (HDLC) and runs at Layer 2, the Data Link Layer, of the OSI Reference Model. The LAPF frame comprises a header and trailer that frame the user payload. Figure L-3 illustrates the LAPF frame and its component fields.

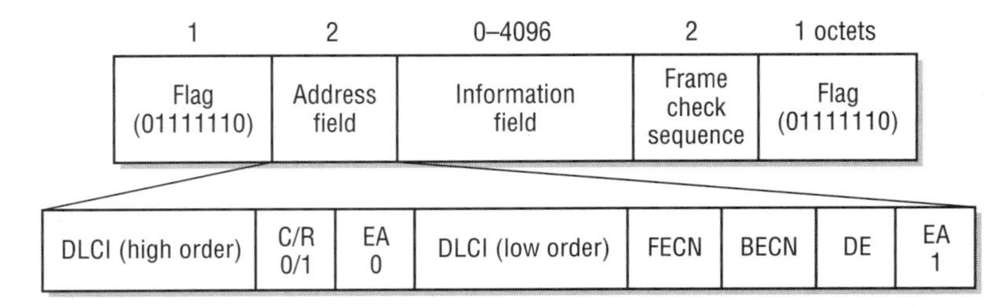

Figure L-3

The fields in the LAPF frame are as follows:

- **Flag:** A specific one-octet field that delimits (i.e., establishes the limits or boundaries of) the beginning and end of the frame. This octet is always 01111110 in binary notation (7E in hexadecimal notation), which is prevented from appearing in the payload by bit stuffing.
- **Address:** A two-octet (default) or four-octet (option) field that includes the necessary control information in the Data Link Connection Identifier. The address field also contains a Command/Response field, Address Field Extension, Forward and Backward Explicit Congestion Notification fields, and Discard Eligibility data.
- **Data Link Connection Identifier (DLCI, pronounced delsey):** 10 bits (or 20 in the 4-octet header) that identify the data link, the virtual circuit (VC), and its service parameters, which include frame size, Committed Information Rate (CIR), Committed Burst Size (B_c), Burst Excess Size (B_e), and Committed Rate Measurement Interval (T_c).
- **Command/Response (C/R):** 1 bit reserved for use of the frame relay access device (FRAD) in order to facilitate the transport of polled protocols (e.g., SNA), which require a command/response for signaling and control purposes.
- **Address Field Extension (EA):** 2 bits that signal the extension of the addressing structure beyond the 2-octet default. The use of EA must be negotiated with the carrier when the service is established.
- **Forward Explicit Congestion Notification (FECN):** Pronounced *feckon.* A 1-bit field available to the network to advise downstream devices that the frame has experienced congestion. Should the receiving FRAD determine that subsequent frames were discarded or corrupted in transmission, it is advised that recovery may be required in the form of requests for retransmission.

- **Backward Explicit Congestion Notification (BECN)** Pronounced *beckon*. A 1-bit field used by the network to advise devices of congestion in the direction opposite of the primary traffic flow. If the target FRAD responds to the originating FRAD in the backward direction, the BECN bit is set in a backward frame. If there is no data flowing in the backward direction, the frame relay network creates a frame in that direction, setting the BECN bit. If the originating FRAD is capable of reducing the frame rate of its transmissions, it is advised to do so on the indicated DLCI, as the network may discard frames once the notification is posted.

- **Discard Eligibility (DE):** A 1-bit field indicating the eligibility of the frame for discard under conditions of network congestion. Theoretically, the user equipment sets the DE in consideration of the acceptability of the application to loss. Should the user equipment not set the DE, the network switches may do so on a random basis or when traffic exceeds subscribed levels (CIR). SDLC and real-time voice and video traffic demand high priority and do not tolerate loss, and so must not be marked DE.

- **Information Field:** Contains user information, either in the form of payload data or internetwork control information. Although this field can be as much as 4,096 octets in length, it generally is restricted to a maximum size of 1,600 octets.

- **Frame Check Sequence (FCS):** A two-octet cyclic redundancy check (CRC) field that provides error detection.

See also *carrier*, *CRC*, *Data Link Layer*, *DLCI*, *downstream*, *FRAD*, *frame*, *frame relay*, *FCS*, *HDLC*, *header*, *octet*, *OSI Reference Model*, *payload*, *protocol suite*, *real-time*, *SDLC*, *signaling and control*, *SNA*, *trailer*, and *VC*.

LAPM (Link Access Procedure for Modems) Pronounced *lap em*. An error control mechanism defined in the ITU-T V.42 Recommendation, LAPM employs a cyclic redundancy check (CRC) for error correction and automatic repeat request (ARQ) for error correction. In the event the LAPM fails, a V.42 modem falls back to the similar, but slower, Microcom Networking Protocol version 4 (MNP4). See also *ARQ*, *CRC*, *error control*, *ITU-T*, and *V.42*.

large business enterprise (LBE) See *LBE*.

laser (light amplification by stimulated emission of radiation) A device that produces an intense, coherent, collimated, focused, and nearly monochromatic beam of radiated optical energy by stimulating electronic, ionic, or molecular transitions to lower energy levels. A laser comprises an active medium, or gain medium, and a resonant cavity. An external power source, or pump, in the form of electricity or another laser, energizes the gain medium, which absorbs the energy. Some of the particles in the gain medium are excited into quantum high-energy states. When a critical level of energy is achieved, a light signal passing through the medium produces more optical energy than is absorbed, and the signal is greatly amplified. The resulting radiated optical signal is highly coherent, i.e., consistent in phase and polarization, and virtually monochromatic. Through a resonating cavity and either mirrors or a diffraction grating, the signal is narrowly channeled and collimated, i.e., the rays are lined up so that they are virtually parallel. Fiber optic transmission systems (FOTS) in long haul applications employ semiconductor diode lasers, generally Fabry-Perot lasers or distributed feedback (DFB) lasers. Short haul transmission systems such as those associated with local area networks (LANs) more commonly employ light-emitting diodes (LEDs) or vertical cavity surface-emitting lasers (VCSELs) as light sources. (*Note:* The laser was patented by AT&T Bell Telephones in 1960 as the optical maser.) See also *coherence*, *collimation*, *DFB laser*, *Fabry-Perot laser*, *LED*, *maser*, *pump laser*, *radiation*, and *VCSEL*.

laser diode A type of light source that resembles a light-emitting diode (LED) in structure, although much more difficult and expensive to manufacture, much less durable, and much more capable. Laser diodes are associated with more expensive and complex supporting electronics, but generally have much faster cycle times and, therefore, offer much more bandwidth. Diode lasers offer significant mechanical and optical coupling efficiency. Therefore, they can mechanically couple to a singlemode fiber (SMF) and can tightly focus a high-speed optical signal for presentation to its core, which has a diameter of only 5–10 microns. Diode lasers also are capable of generating tightly defined optical signals in very small spectral ranges, or windows. Diode lasers also generate signals at wavelengths longer than 850 nm. In these higher transmission windows at 1310 nm or 1550 nm, the signals attenuate much less and, therefore, can travel much farther without being repeated or amplified. In long-haul, high speed, carrier-class fiber optic transmission systems (FOTS), these narrowly defined windows allow the multiplexing of a number of wavelengths through a process known as wavelength division multiplexing (WDM). Fabry-Perot lasers and distributed feedback (DFB) lasers are two specific types of diode laser commonly used in such networks. See also *attenuation, bandwidth, cycle time, DFB laser, Fabry-Perot laser, FOTS, laser, LED, SMF, VCSEL, wavelength, window*, and *WDM*.

last-in-first-out (LIFO) See *LIFO*.

last mile Generally referring to the telco local loop, which is the link between the central office (CO) at the edge of the telco network and the user premises. In a broader contemporary context, the term applies to the physical connection between the edge of any service provider's network and the end user's premises. In practice, the last mile is often much longer than a mile. In the United States, UTP local loops are generally 12,000 feet or less, but often are as long as 18,000 feet. Passive optical network (PON) standards allow for local loops as long as 12 miles (20 km). Whether it is the first mile or the last mile, which is a matter of perspective, it is seldom exactly a mile. Sometimes referred to as the first mile. See also *central office, local loop*, and *PON*.

LATA (Local Access and Transport Area) A geographical area defined by the United States Federal Communications Commission (FCC) as a result of the Modified Final Judgement (MFJ) that broke up the AT&T Bell System on January 1, 1984. The Regional Bell Operating Companies (RBOCs) and their component Bell Operating Companies (BOCs) were prevented from offering interLATA toll services, i.e., long distance calling services that crossed LATA boundaries. Initially, the BOCs had the exclusive rights to offer intraLATA toll service, also known as local long distance, within the confines of the 196 defined LATAs. LATAs now serve primarily as reference points for call rating and routing. See also *BOC, FCC, MFJ*, and *RBOC*.

latency Delay. The total time required for a signal to travel from one point to another, generally from a transmitter through a network to a receiver. Propagation delay, a fundamental factor in latency, is dependent on the nature of the electromagnetic signal, as not all signals travel through a medium at the same speed. Propagation delay also is influenced by the distance between the two points, the density of the medium, the presence of passive devices such as loading coils that might increase the impedance of the medium. Latency also is affected by any processing time associated with devices such as repeaters, transponders, concentrators, multiplexers, switches, and routers as they variously transmit and retransmit, amplify and reamplify, time and retime, shape and reshape, code and decode, compress and decompress, encrypt and decrypt, and otherwise process signals and manipulate data. Latency also is affected by any time that data packets spend in queues due to issues of network congestion and any time required to retransmit packets errored or lost in transit. See also *jitter*.

Latin-American and Caribbean Network Information Center (LACNIC) See *LACNIC*.

launch cable A jumper cable of known good quality that is attached to a light source on one end and a fiber under test on the other end, calibrated for output power, and used for testing optical loss in a fiber optic cable. The launch fiber provides an opportunity for the signal to achieve modal equilibrium, i.e., even power distribution across all modes, and eliminates any possibility that fiber anomalies near the light source can affect the test results. A mandrel wrap sometimes is used with a multi-mode fiber (MMF) launch cable to assist in the achievement of modal equilibrium. Synonymous with *injection fiber*. See also *jumper, mandrel wrapping, MMF, modal equilibrium, optical fiber, power*, and *source*.

law A statement of scientific fact, phenomena, or relationships that occur with unvarying uniformity under given conditions. See also *theory*.

lay length The physical length of something laid flat. The lay length of a twisted pair cable, for example, refers to the length of the individual conductors if the pairs in the cable were to be separated and untwisted, and the individual conductors were to be laid flat. In a Category 5 (Cat 5) data grade cable, for example, each of the four pairs has a slightly different twist ratio, i.e., each pair is twisted slightly more or less than any other pair in the cable. Twisted, the pairs are the same length; untwisted and laid flat, they are not. See also *Cat 5* and *twisted pair*.

layer Referring to the protocol or protocols operating at a particular level within a network architecture. Such an architecture commonly is detailed in a protocol stack, such as the OSI Reference Model, or protocol suite, such as the TCP/IP protocol suite. The OSI Reference Model is a full seven-layer stack, of which the top layer, the Application Layer, addresses applications and end user processes. The bottom layer, the Physical Layer, deals with physical and mechanical aspects of the interface between a device and a transmission medium. Although each layer addresses different functions and responsibilities, the layers work together, as a whole, to enable an application or end user process. See also *application*, *Application Layer*, *OSI Reference Model*, *Physical Layer*, *protocol*, and *TCP/IP*.

Layer 1 Referring to the Physical Layer of the OSI Reference Model. See also *OSI Reference Model* and *Physical Layer*.

Layer 2 Referring to the Data Link Layer of the OSI Reference Model. See also *Data Link Layer* and *OSI Reference Model*.

Layer 2 Tunneling Protocol (L2TP) See *L2TP*.

Layer 3 Referring to the Network Layer of the OSI Reference Model. See also *Network Layer* and *OSI Reference Model*.

Layer 4 Referring to the Transport Layer of the OSI Reference Model. See also *OSI Reference Model* and *Transport Layer*.

Layer 5 Referring to the Session Layer of the OSI Reference Model. See also *OSI Reference Model* and *Session Layer*.

Layer 6 Referring to the Presentation Layer of the OSI Reference Model. See also *OSI Reference Model* and *Presentation Layer*.

Layer 7 Referring to the Application Layer of the OSI Reference Model. See also *Application Layer* and *OSI Reference Model*.

L band **1.** In IEEE terminology, the portion of the microwave radio spectrum in the range of 1–2 GHz. See also *IEEE* and *microwave*. **2.** In ITU-R terminology, the portion of the microwave radio spectrum in the range of 1.610–1.6255 GHz. Applications include global positioning satellite (GPS), low-Earth orbiting (LEO) satellites, Search for Extraterrestrial Intelligence (SETI), and telemetry. See also *electromagnetic spectrum*, *GPS*, *ITU-R*, *LEO*, *SETI*, and *telemetry*.

L-Band (Long Wavelength Band) The ITU-T standard optical transmission window in the wavelength range of 1,565–1,625 nm. See also *wavelength* and *window*.

LBE (Large Business Enterprise) A large commercial (i.e., for-profit) organization. There is no formal definition of the size of an LBE, although it would be larger than a medium enterprise, which generally is defined as having less than 250 employees. See also *SME*.

LCP (Link Control Protocol) Referring to a protocol responsible for negotiating Link Layer (Layer 2) details. The Point-to-Point Protocol (PPP), for example, is based on a Link Control Protocol and a Network Control Protocol (NCP). The LCP is responsible for setting up a link between two computers over a circuit-switched telephone connection, and for resolving any issues of authentication. The NCP negotiates any parameters specific to the Network Layer. See also *authentication, Data Link Layer, IP, link, NCP, Network Layer, PPP,* and *protocol.*

LCR (Least Cost Routing) Synonymous with *automatic route selection* (ARS). See *ARS.*

LDAP (Lightweight Directory Access Protocol) A subset of X.500 that can run over TCP/IP networks, LDAP was developed to simplify the demands of the X.500 DAP. Described in the IETF RFCs 1777, LDAP can run as a standalone directory system or can be used as a means of accessing an X.500 directory or other directory. LDAPv3 (RFC 3377) supports non-ASCII and non-English characters for international directories, and can sort through multiple directories on the basis of a single request. See also *X.500.*

LD-CELP (Low Delay Code Excited Linear Prediction) A voice compression algorithm defined in ITU-T G.728, and used in voice over frame relay (VoFR) and voice over Internet Protocol (VoIP). LD-CELP is geared to a rate of 16 kbps, although bit rates as low as 12.8 kbps can be achieved. The low level of delay suggested by the designation is due to the fact LD-CELP accumulates only 5 PCM samples, representing 625μs of a voice stream, in a buffer. A key element of CELP and its variants is the construction and maintenance of a codebook, which comprises binary descriptions of sets of voice samples. CELP employs silence suppression to remove periods of silence and redundancy from the data set, and normalizes the volume level. LD-CELP compares the resulting data set to a set of candidate shapes in the codebook, selects the shape that most closely match the actual data. LD-CELP then transmits the index number of the selected code description and the average loudness level of the set of samples. Every 625μs, the code is sent across the network in a block of 10 bits, yielding a data rate of 16 kbps, which is a compression ratio of 4:1, compared with toll quality PCM voice over the public switched telephone network (PSTN) at 64 kbps. At the receiving end of the transmission, the transmitted code is compared to the codebook, the PCM signal is reconstructed, and, eventually, the analog signal is reconstructed. As compared to CELP, the more frequent transmission of the shorter data blocks yields lower levels of delay through faster processing by the digital signal processors (DSPs), and the compression technique yields more efficient use of bandwidth. LD-CELP yields quality that generally is considered to be on a par with Adaptive Differential Pulse Code Modulation (ADPCM). See also *ADPCM, algorithm, analog, binary, buffer, CELP, circuit switching, compression, CS-ACELP, DSP, PCM, PSTN, silence suppression,* and *toll quality.*

LDM (Limited Distance Modem) See *short haul modem.*

LDP (Label Distribution Protocol) In Multiprotocol Label Switching (MPLS), a protocol used by a Label Edge Router (LER) to distribute labels, or tags, to each Label Switching Router (LSR) in the network core, identifying the treatment that should be afforded all packets in the flow on that particular Label Switched Path (LSP). See also *core, flow, label, LER, LSP, LSR, packet,* and *protocol.*

lead balloon There is no such thing. There was, however, a great rock band in the 1960s and 1970s known as Led Zeppelin, but it's not the same thing. See also *paperless office.*

leaky mode A physical path traveled by a leaky wave in an optical waveguide, such as an optical fiber. A leaky mode has an electromagnetic field that decays monotonically (i.e., steadily) for a finite distance in the transverse direction (i.e., at a 90 degree angle to the plane or axis of the fiber), but becomes oscillatory beyond that finite distance. See also *leaky wave.*

leaky wave In an optical fiber, an electromagnetic wave that travels outside the core and in the cladding or even beyond, usually as a result of an incident wave injected into the core at a severe angle less than the critical angle. Leaky waves are detached from the main body of the signal, are not guided, and, therefore, are valueless. Leaky waves travel high-order modes; waves traveling low-order modes generally remain confined within the core. See also *cladding*, *core*, *critical angle*, *high-order mode*, and *low-order mode*.

least cost routing (LCR) Synonymous with *automatic route selection* (ARS). See *ARS*.

least significant bit (LSB) In a binary number or bit sequence comprising one or more bytes, the low-order (i.e., least) bit in the sequence. The LSB conveys the least amount of information, and usually is the right-most bit. The approach is known as big-endian. In some data representation schemes, the LSB is the left-most bit. This approach is known as little-endian. See also *big-endian*, *bit*, *bit robbing*, *byte*, *endianess*, and *little-endian*.

LEC **1.** Local Exchange Carrier. A company providing local telephone service. The incumbent LEC (ILEC) is the local telephony company (telco) in place prior to competition, which introduced one or more competitive LECs (CLECs). See also *carrier*. **2.** LAN Emulation Client. In LANE (LAN Emulation), an end system or endpoint. The LEC connects to the ATM network over a LANE user-to-network interface (LUNI). See also *ATM*, *LANE*, and *LUNI*.

LECS (LAN Emulation Configuration Server) In LANE (LAN Emulation), an optional network-based server that assigns a LAN emulation client (LEC) to an emulated LAN (ELAN). See also *LANE* and *LEC*.

LED (Light-Emitting Diode) A semiconductor light source in the form of a transistor with a positive and a negative layer of particular semiconducting materials and a *p-n* (positive-negative) junction between them. The negative layer is doped with impurities to create extra electrons, which are negatively charged. The positive layer is doped to create extra holes into which electrons can migrate when a charge is applied, which has the effect of adding extra positive particles. When current is applied and the electrons move across the junction, from the *n* semiconductor layer to the *p* semiconductor layer, and settle into the holes, they release energy in the form of photons, i.e., light. (*Note:* A diode conducts current in only one direction, like a one-way gate.) The composition of the semiconductor material determines the color of light, how much of it is absorbed, and how much of it is released. LEDs manufactured with aluminum gallium arsenide (AIGaAs) are used in infrared (IR) applications such as fiber optics. Various other compounds, most including gallium (Ga), to create other colors in the visible and ultraviolet (UV) spectrum. Visible light-emitting diodes (VLEDs) operate in the visible light spectrum, and are found in clocks, watches, calculators, gauges, meters, and a wide variety of other devices. The infrared LEDs used in fiber optic transmission are, of course, much more sophisticated. LEDs pulse on and off relatively slowly, as specified by the cycle time, i.e., rise and fall times of signal intensity. Therefore, LEDs are relatively bandwidth-limited. LEDs also generate signals of relatively broad spectral width in the 850 nm region (850 Band), which wavelength attenuates substantially over relatively short distances. Slower LEDs emit light from an area etched into the surface of a semiconductor chip, while the faster LEDs emit light from the edge of the chip. The physical design of LEDs is such that they couple efficiently only to the relatively broad inner core of multimode fiber (MMF), and they do not tightly focus a collimated beam of light as does a laser diode. LEDs are used in local area networks (LANs), where they support transmission rates of up to 1 Gbps over relatively short distances. LEDs also are used in certain other short-haul transmission systems, including some passive optical networks (PONs). LEDs are relatively inexpensive and durable. See also *850 Band*, *coupling efficiency*, *cycle time*, *diode*, *IR*, *LAN*, *laser diode*, *MMF*, *PON*, *spectral width*, *transistor*, *UV*, *VCSEL*, *visible light*, *VLED*, *wavelength*, and *window*.

leg A segment or portion of an end-to-end path associated with a call. For example, an international voice telephone call via the public switched telephone network (PSTN) has an originating leg supported by a local exchange carrier (LEC) in one nation, a terminating leg supported by a LEC in another nation, and an international leg supported by an interexchange carrier (IXC) that interconnects the two national LECs. See also *IXC*, *LEC*, and *PSTN*.

legacy Referring to a software or hardware component or element, a system, or a network that is technically outdated, although often compatible with current technology. In consideration of the evolutionary nature of technology, systems and standards development generally places considerable emphasis on backward compatibility, i.e., compatibility with legacy technology.

Lemple-Ziv (LZ) See *LZ.*

Lemple-Ziv-Welch (LZW) See *LZW.*

LEO (Low Earth Orbit) A satellite or satellite constellation (i.e., system) operating at an altitude of 644–2,415 kilometers. Although the term is not precisely defined, Little LEOs involve a relatively small number of satellites, and operate at frequencies below 1 GHz in support of low bit-rate data traffic, such as telemetry, vehicle messaging, and personal messaging. Big LEOs are bigger networks that operate at higher frequencies in support of voice and higher-speed data communications. Unlike a geosynchronous Earth orbiting (GEO) satellite, LEO and MEO (Middle Earth Orbit) satellites do not remain in a fixed position relative to the Earth's surface, so are referred to as mobile satellite systems (MSSs), as opposed to the fixed satellite systems (FSSs) in geostatic orbit. See also *FSS*, *GEO*, *MEO*, *MSS*, and *satellite.*

LER (Label Edge Router) In Multiprotocol Label Switching (MPLS), a router at the edge of the carrier's network. The ingress LER examines the packet header and attaches a label, or tag, that identifies the Label Switched Path (LSP) that the packet is to travel through the MPLS network. The label is distributed to Label Switching Routers (LSRs) in the network core to ensure that the packet travels that path, link by link. The egress LER strips the tag away as the packet exits the network. See also *header*, *label*, *link*, *LSP*, *LSR*, *MPLS*, *packet*, *path*, *router*, and *switch.*

LES (LAN Emulation Server) In LANE, a network-based server to which a LAN emulation client (LEC) connects to the ATM network over a LANE user-to-network interface (LUNI). The LES registers the LAN medium access control (MAC) addresses and translates them into ATM addresses using the address resolution protocol (ARP). Each LEC is assigned to an emulated LAN (ELAN) by an optional network-based LAN emulation configuration server (LECS). See also *ARP*, *ATM*, *ELAN*, *LANE*, *LEC*, *LECS*, *LUNI*, *MAC*, and *server.*

leverage A euphemism for *reuse.* A considerable number of definitions in this book are leveraged from other books I have written for Wiley. I spent so much time writing these beautifully worded definitions over the last 10 years that I figured there was no point in trying to reword them and twist them out of shape in the process. Some things just don't make sense. See also *euphemism.*

LF (Low Frequency) LF radio is in the frequency range of 30–300 kHz and has a wavelength of 10 km – 1 km. LF radio has applications in navigation, maritime communications, information and weather systems, and time systems. See also *electromagnetic spectrum*, *frequency*, *Hz*, and *wavelength.*

liberalize Referring to reforming the telecommunications environment by reducing or eliminating the monopoly of the national carrier and creating a competitive environment. Liberalization generally is associated with some level of privatization, which involves transferring all or some portion of the telephone utility from government to private ownership. See also *privatize* and *utility.*

lifeline service Referring to discounted basic telephone service provided to low-income subscribers that meet certain criteria. In the United States, the federal Universal Service Fund (USF) subsidizes the Link-Up Program (installation charges) and Lifeline Program (monthly service charges) for subscribers who have an income that is at or below 135 percent of the federal poverty guidelines or who participate in any of a number of federal assistance programs. See also *USF.*

lifetime control Referring to Network Layer protocol mechanisms that limit the life of a packet. The Internet Protocol (IP) and Connectionless Network Protocol (CLNP) packets both contain a time-to-live (TTL) field for this purpose. See also *CLNP*, *IP*, *Network Layer*, *packet*, *protocol*, and *TTL.*

LIFO (Last-In-First-Out) A buffering method in which the entity that first exits the buffer is the one that last entered. Thereby, the entity served (e.g., processed or switched) is the one that waited the shortest period of time. LIFO is commonly used in file systems and e-mail systems, as the last file or correspondence saved is the first retrieved, or at least appears first in the stack. See also *FIFO* and *queue*.

light Electromagnetic energy with a waveform having a frequency above the upper limit of the radio range of 300 GHz and equal or less than the lower limit of the X-ray range of 30 PHz. At the low end of the range is infrared (IR) light, which operates at 30–300 Hz, and at the upper end of the range is extremely high frequency (EHF) radio. See also *electromagnetic spectrum*, *frequency*, and *Hz*.

light carrier A company that leases, rather than owns, the network facilities it uses to provide telecommunications transport services. A heavy carrier is a facilities-based carrier, i.e., a company that owns the switching and transmission systems that compose the network it uses to provide transport services. See also *carrier*.

light detector A device used in an optical transmission system to detect an optical signal generated by a light source and propagating through a medium. A light detector essentially is an optical receiver that is paired with an optical transmitter, both of which are connected to electrically based devices or systems. So, the source converts electrons to photons and the detector converts photons to electrons. Detectors take several forms. A positive-intrinsic-negative (PIN) diode detector is paired with a light-emitting diode (LED) or vertical cavity surface-emitting laser (VCSEL) light source over a multimode fiber (MMF) in a fiber optic transmission system (FOTS) or a free space optics (FSO) system. An avalanche photodiode (APD) is paired with a laser diode in a single-mode fiber (SMF) system. See also *APD*, *FOTS*, *FSO*, *laser diode*, *LED*, *light source*, *PIN*, and *VCSEL*.

lightguide An optical fiber. See *optical fiber*.

light source The source of the optical signal in an optical transmission system, which can take the form of a fiber optic transmission system (FOTS) or free space optics (FSO) system. Light-emitting diodes (LEDs) are used in multimode fiber (MMF), which usually is glass optical fiber (GOF), but also can be plastic optical fiber (POF). Vertical cavity surface-emitting lasers (VCSELs) also can be used in MMF systems. Laser diodes are used in single-mode fiber (SMF) systems. Virtually all light sources emit signals in the infrared (IR) spectrum. The exception is LEDs used in POF systems, which operate in the red region of the visible light spectrum. A light source essentially is an optical transmitter that is paired with an optical receiver, both of which are connected to electrically based devices or systems. So, the source converts electrons to photons and the detector converts photons to electrons. See also *laser diode*, *LED*, *light detector*, *FOTS*, *FSO*, *GOF*, *MMF*, *POF*, *SMF*, and *VCSEL*.

light-emitting diode (LED) See *LED*.

Lightweight Directory Access Protocol (LDAP) See *LDAP*.

limited distance modem (LDM) See *short haul modem*.

linear predictive coding (LPC) See *LPC*.

line **1.** A station line refers to the circuit between a private branch exchange (PBX) switch and a station user's terminal equipment, which usually is in the form of telephone, although it could be a computer workstation, a printer, a facsimile machine, or some other device. **2.** In rate and tariff terminology, line refers to a local loop connection from the telephone company central office (CO) switch to the user premises in support of customer premises equipment (CPE) other than a switch. Such CPE can be in the form of a single-line residence or business set, a multiline set, or the common control unit of a key telephone system (KTS). Such a line is single-channel in nature, i.e., supports a single conversation and is voice grade, i.e., provides enough bandwidth to support a voice conversation, and has a single associated telephone number. A line may be thought of as a tributary of a trunk. See also *line side*, *trunk*, and *trunk side*.

line coding The manner in which data bits, or blocks of data bits, are represented on a line. Examples include 4B3T, 4B/5B, alternate mark inversion (AMI), and Manchester. See also *4B3T*, *4B/5B*, *AMI*, *Manchester coding*, and *quadbit*.

line discipline In the context of a protocol, the sequence of network operations between devices that actually transmits and receives the data, controls errors in transmission, deals with the sequencing of message sets, and provides for confirmation or validation of data received. See also *handshaking* and *protocol*.

line doubler Also known as *line extender*. A device that more or less doubles the maximum physical reach of a digital subscriber line (DSL) or ISDN service. A line doubler essentially is a repeater. See also *HDSL*, *repeater*, and *xDSL*.

line driver A type of interface converter used to extend the distance of a digital connection by converting the digital signal to a low-voltage, low-impedance signal that can transmit more effectively and over longer distances on dedicated, specially conditioned twisted-pair circuits. The RS-232 specification (more correctly known as EIA-232), for example, generally limits the distance between devices to 50 feet at transmission rates of 56 kbps. At lower speeds, line drivers can reshape the digital pulses to extend that distance considerably. At speeds of up to 9.6 kbps, for example, line drivers can extend that limitation to 500–5,000 feet over Category 3 (Cat 3) unshielded twisted pair (UTP). Line drivers are unidirectional and operate over simplex circuits. Line drivers can generically be classified as modems, as they change the format of the signal. See also *Cat 3*, *EIA-232*, *impedance*, *modem*, *signal*, *simplex*, and *voltage*.

line extender Also known as *line doubler*. A device that extends, and more or less doubles, the maximum physical reach of a digital subscriber line (DSL) service. A line extender essentially is a DSL repeater. See also *xDSL* and *repeater*.

line finder A component of an electromechanical step-by-step (SxS) or panel switch, or electromagnetic crossbar (Xbar) circuit switch, that identifies lines that go off-hook to request service. A line finder allows many lines (e.g., 100 or so) to share a bank (10 or so) of selectors, depending on the activity levels of the individual subscribers. See also *line*, *off-hook*, *panel switch*, *selector*, *SxS*, and *Xbar*.

line frequency Also known as *lines per inch* (lpi). See lpi.

line interface The total of hardware and firmware that serves to interconnect a line and a switch, router, or other device, and to facilitate their interoperation. Such interfaces are specific to the physical layer protocols, which address such factors as transmission medium, physical dimensions of the medium, signal frequency and wavelength, signal format, and signaling speed. See also *firmware*, *hardware*, *line*, *physical layer*, and *protocol*.

line number The trailing digits of a telephone number. As defined by the North American Numbering Plan (NANP), the line number can consist of any four digits (XXXX), and corresponds with a port on a switch that connects to a circuit or channel over a local loop that serves the physical premises of a subscriber. In the case of a cellular telephone number, there is no local loop, as such, and the number is associated with a handset, rather than a physical premises. See also *NANP*.

line of sight (LOS) See *LOS*.

Line Overhead (LOH) See *LOH*.

line powered In reference to equipment that is electrically powered by the telecommunications circuit to which it connects, thereby eliminating the need for local power. The analog single line local loop circuits in the public switched telephone network (PSTN) are powered from batteries in the central office (CO) at -48 volts DC (Direct Current), which is sufficient to power conversation over a simple analog telephone set. Ringing current is 110 volts AC (Alternating Current) at 20 Hz in North America. As the typical telephone company CO has an uninterruptible power supply (UPS) comprising multiple power utility circuits, substantial battery packs, and backup diesel generators, the typical analog residential telephone still works even if the lights go out. The telephone companies are said to be the largest power utilities in the world, although that is not the focus of their business. See also *UPS*.

line setup A very basic protocol issue addressing the manner in which a circuit is set up between devices. There are three alternatives: simplex, half-duplex, and full duplex. See also *simplex*, *half duplex*, and *full duplex*.

line sharing **1.** A technique by which two voice grade services and two different telephone numbers can coexist on the same local loop through frequency division multiplexing (FDM). The two numbers can be associated with two different voice numbers or one number can be a fax number and another a voice number, with distinctive ringing patterns distinguishing fax calls from voice calls. The fax machine can be programmed to recognize the fax ringing pattern and to answer those calls automatically. See also *distinctive ringing*, *FDM*, *local loop*, and *voice grade*. **2.** A technique defined by the Federal Communications Commission (FCC) by which an incumbent local exchange carrier (ILEC) provides voice grade telephone service over a local loop, and shares that loop with a competitive local exchange carrier (CLEC) that provides digital subscriber line (DSL) service. Line sharing is accomplished through frequency division multiplexing (FDM). See also *CLEC*, *FCC*, *FDM*, *ILEC*, *local loop*, and *voice grade*.

line side In telephone company (telco) terminology, line describes the user side or local loop side of the central office (CO). The customer-facing side of the public switched telephone network (PSTN). In other words, the line side is the side of the network to which users connect in order to access the network core, or backbone. The line side is synonymous with the local loop access circuits, with the demarcation point being the local loop interface at the CO. The trunk side involves the high-capacity trunks that serve to interconnect the various telco switching centers in the core of the carrier network. See also *CO*, *line*, *local loop*, *PSTN*, and *trunk side*.

lines per inch (lpi) See *lpi*.

line status lamp A lamp that indicates the status of a line, usually steadily lit to indicate a line in use and flashing to indicate a line on hold. Line status lamps are used in key telephone systems (KTSs), where they may appear on an individual telephone set or a busy lamp field. See also *busy lamp field* and *KTS*.

line-switched ring A SONET/SDH topology that employs either two or four fibers. The single-ring configuration comprises two fibers, one of which is active and the other of which is held in reserve. Traffic moves in one direction across the active fiber. In the event of a network failure, such as a node failure or a cable cut, the backup ring is activated to enable transmission in the reverse direction, as well. A four-fiber configuration duplicates this approach in a second ring. Line-switched rings up to 1,200 kilometers in route distance offer standard restoral intervals of 50 milliseconds or less. See also *optical fiber*, *path-switched ring*, *SDH*, *SONET*, and *topology*.

link **1.** A two-point segment of an end-to-end physical circuit. A circuit may consist of a single link, as would be the case between a host computer and a directly attached peripheral, such as a printer. A circuit commonly comprises multiple links. For example, a telephone set may connect across a link to a central office switch at the edge of the carrier network, that central office switch may connect to another central office switch across a link, and to yet another central office switch across a link, and finally to another telephone set across a link. In this scenario, two terminal devices connect via an end-to-end circuit that comprises four links interconnected by three central offices. Link sometimes is used interchangeably with line or circuit. **2.** A conceptual two-point segment of an end-to-end circuit that connects two end users and enables them to communicate, even when two separate physical paths are used. In a satellite radio link, for example, there is an uplink from the Earth station (i.e., antenna) to the satellite and a downlink from the satellite to the Earth station. In a cellular network, the uplink is the upstream radio link from the mobile station to the base station and the downlink is the downstream link from the base station to the mobile station. See also *antenna, circuit, downlink, downstream, uplink, physical,* and *upstream.* **3.** In hypertext, the hyperlink, or logical connection between discrete data elements. See also *hyperlink* , *hypertext,* and *link rot.* **4.** A logical connection, association, or relationship between two or more things.

link access procedure, balanced (LAPB) See *LAPB.*

link access procedure, D channel (LAPD) See *LAPD.*

link access procedure for frame mode services (LAPF) See *LAPF.*

link access procedure for modems (LAPM) See *LAPM.*

Link Control Protocol (LCP) See *LCP.*

link-local address In Internet Protocol version 6 (IPv6), a type of unicast address intended for local use, only. A link-local address is similar to an IPv4 private IP address, as it is not meant to be routed and not intended for use in a public domain such as the Internet, but confined to a single link, or LAN segment. See also *domain, Internet, IPv4, IPv6, LAN, private IP address, site–local address, unicast,* and *unicast address.*

link rot The decomposition of hyperlinks as the linked sites are renamed, moved, or withdrawn from the Web. As the Web is a dynamic application, hyperlinks must be continuously updated to keep them fresh and viable. See also *hyperlink.*

link-state protocol A routing protocol that calls for each router to build a database of the names of its neighboring routers and the cost to connect to each. Once all routers in a network have done so, each has a map of the entire network and can calculate the costs of each available route, from end to end. Open Shortest Path First (OSPF) is an example of a link-state protocol. See also *database, OSPF, protocol,* and *router.*

Linux (Linus Unix) A UNIX-like computer operating system (OS) developed by Linus Torvalds, and numerous collaborators worldwide, that was designed to run on PCs powered by Intel processors. Linux is free, open source software that anyone can modify, although at one's own risk. Many companies package the Linux kernel with a number of utilities and other programs into commercial versions that include documentation and support. See also *kernel, open source, OS, program, UNIX,* and *utility.*

LIR (Local Internet Registry) The local organization responsible for assigning Internet Protocol (IP) addresses to Internet Service Providers (ISPs). The LIR receives IP address assignments from a National Internet Registry (NIR) or Regional Internet Registry (RIR) that receives address assignments from the Internet Assigned Numbers Authority (IANA). See also *IANA, IP, IP address, ISP, NIR,* and *RIR.*

LIS (Logical IP Subnetwork) In Classical IP (CIP) over ATM, the term for a virtual LAN (VLAN). See also *CIP* and *VLAN.*

little-endian Referring to the orientation of a computer system, application, or network design with respect to the placement of most significant bit or digit in a coding scheme. Little-endian places the most significant bit, digit, or byte in the last, or rightmost, position. Big-endian places the most significant bit, digit, or byte in the first, or leftmost, position. Bi-endian systems can work either way. Motorola processors employ the big-endian approach, whereas Intel processors take the little-endian approach. Telephone numbers, for example, are big-endian, beginning with a country code, followed by an area code, a central office prefix, and a line number. Bit robbing in T1 systems involves the least significant bit (LSB), which is the eighth bit in a byte, which is a little-endian approach. The terms derive from Jonathan Swift's *Gulliver's Travels,* in which the Big-Endians were a faction of people on the islands of Lilliput and Blefuscu who defied the emperor's decree that soft-boiled eggs should be broken at the small end before being consumed. See also *bi-endian, big-endian, bit, byte, endianess,* and *LSB.*

liveware Slang for *people*, distinguishing them from software, firmware, and hardware. Synonymous with *grayware*. See also *firmware*, *grayware*, *hardware*, and *software*.

LLC (Logical Link Control) According to the IEEE model, the upper sublayer of the Data Link Layer, described in the OSI Reference Model. The LLC sublayer is concerned with issues of multiplexing, flow control, and detection and retransmission of dropped frames. The IEEE 802.2 Working Group sets standards and develops recommendations for LLC. See also *802.2*, *Data Link Layer*, *flow control*, *IEEE*, *multiplexer*, and *OSI Reference Model*.

LLU (Local Loop Unbundling) See *unbundled service*.

LMDS (Local Multipoint Distribution Service) A wireless local loop (WLL) technology developed by Bernard B. Broussard for wireless cable television (TV), referring to premium wireless subscription TV rather than traditional free broadcast TV or cable TV. Broussard, with Shant and Vahak Hovnanian, formed a firm that provided 49 TV channels in New York City, and later added high speed Internet access. The technical rights to LMDS technology later were spun off into a separate company, and the U.S. Federal Communications Commission (FCC) auctioned the first LMDS radio licenses in early 1998. The A Block has a width of 1.15 GHz in the frequency bands of 27.5–28.35 GHz, 29.1–29.25 GHz, and 31.0–31.15 GHz. The B Block has a width of 150 MHz in the spectrum between 31.15 GHz and 31.3 GHz. Outside of North America, LMDS operates in the 20 GHz and 45 GHz bands. LMDS requires line of sight (LOS), has a maximum cell diameter of 10–15 miles, and compensates for rain fade through the use of adaptive power controls. LMDS can carve a 360° cell into four quadrants of alternating antenna polarity, and supports both point-to-point and point-to-multipoint service configurations. LMDS is flexible enough to support local loops ranging from 1.544 Mbps (T1) to 622 Mbps, and in either symmetric or asymmetric configurations. Multiplexing access methods include frequency division multiple access (FDMA), time division multiple access (TDMA), and code division multiple access (CDMA). TDMA modulation options include phase modulation (BPSK, DQPSK, QPSK, and 8PSK) and amplitude modulation (QAM, 16-QAM, and 64-QAM). LMDS is considered obsolete in the United States, having been replaced by the IEEE 802.16 standards, commonly known as WiMAX. See also *8-PSK*, *802.16*, *asymmetric*, *antenna*, *BPSK*, *CDMA*, *DQPSK*, *FCC*, *FDMA*, *frequency band*, *LOS*, *phase modulation*, *point-to-multipoint circuit*, *point-to-point circuit*, *polarity*, *QPSK*, *rain fade*, *spectrum*, *symmetric*, *TDMA*, *WiMAX*, and *WLL*.

LMI (Local Management Interface) In frame relay, a protocol that provides operational support for the user network interface (UNI). The LMI is a polling protocol between the frame relay access device (FRAD) and the network, which periodically verifies the existence and availability of the permanent virtual circuit (PVC), as well as the integrity of the UNI link. See also *FRAD*, *frame relay*, *link*, *polling*, *protocol*, *PVC*, and *UNI*.

LNP (Local Number Portability) Referring to the ability to port, i.e., move, a local telephone number from one service or service provider, i.e., local exchange carrier (LEC), to another in a competitive environment. In the United States, the Telecommunications Act of 1996 mandated LNP, and established the Local Number Portability Administration (LNPA) to oversee the development and deployment of the necessary mechanisms. LNP is accomplished through the use of local routing numbers (LRNs) that point the originating exchange to the correct terminating carrier and exchange. LNP eventually is expected to support geographic portability. In 2003, the Federal Communications Commission (FCC) required wireless number portability (WNP), extending portability to cellular telephone numbers. LNP and LRNs are supervised by the Number Portability Administration Center (NPAC). See also *LEC*, *LRN*, *Telecommunications Act of 1996*, and *WNP*.

LNPA (Local Number Portability Administration) The organization formed to oversee the development and deployment of a mechanism for Local Number Portability (LNP). See also *LNP*.

load **1.** The amount of traffic placed on a circuit or system. **2.** In reference to a load coil, or loading coil, which acts as a lumped inductor to compensate for capacitance on a long twisted pair copper circuit. See also *loading coil*. **3.** The amount of force or weight on a cable. Dynamic load refers to the short-term load placed on a cable during installation. Static load refers to the long-term load placed on a cable, such as a riser cable, which hangs vertically. See also *bend diameter* and *tensile strength*.

load balancing The process of distributing a load across multiple resources. For example, traffic can be distributed across multiple circuits to multiple switches or routers and, therefore, across multiple physical paths in order to mitigate issues of network congestion and enhance overall network performance. Or, competition for computational resources can be mitigated if a computing load is distributed among multiple processors. Local area network (LAN) segmentation can reduce congestion and, therefore, improve overall network performance through the application of routing bridges, switches, and routers that variously restrict traffic to physical segments and distribute traffic across multiple physical paths. See also *network management* and *traffic*.

load coil See *loading coil*.

loading coil A toroidal (i.e., ring shaped or donut-shaped) device comprising a powdered iron core, or sometimes a soft iron wire core, around which copper wire is wound. A loading coil is spliced into an unshielded twisted pair (UTP) copper local loop, where it functions as a lumped inductor, which is to say that at a specific point in the circuit the process of inductance takes place, to compensate for the distributed capacitance between the two parallel wires. In effect, the loading coil tunes the copper circuit, optimizing it for mid-voiceband performance. The loading coil also functions as a low-pass filter, increasing loss above the voiceband cutoff frequency of 4 kHz, while reducing mid-voiceband attenuation by as much as 80 percent. Loading coils are passive, i.e., not electrically powered, devices commonly placed on local loops that exceed approximately 18,000 feet (5.5 km) in length. The first loading coil is placed approximately 3,000 feet (.9 km) from the central office (CO) and at intervals of 6,000 feet (1.8 km) or so, thereafter. The presence of loading coils renders local loops unusable for ADSL, ISDN, T-carrier and other loops operating at high data rates, as they filter out the high frequencies associated with those higher data rates. Where such services are to be deployed, the local loops must be properly conditioned, which entails removing the loading coils, bridged taps, and other impediments. The presence of a loading coil also has the effects of increasing the impedance of the circuit, which significantly reduces the velocity of propagation (Vp), i.e., speed of signal propagation, to 10,000–12,000 miles per second. This speed penalty is not of particular significance in short voice grade local loops. If the loops are long, however, loading coils can create unacceptable problems with echo, or signal reflection. See also *bridged taps*, *capacitance*, *conditioning*, *echo*, *impedance*, *inductance*, *propagation*, and *Vp*.

Local Access and Transport Area (LATA) See *LATA*.

local area network (LAN) See *LAN*.

local battery A battery that provides electrical power to stations, or terminal devices, as distinguished from the common battery that provides loop current, i.e., electrical energy to the line. See also *battery*, *common battery*, *electricity*, and *loop current*.

local call An imprecise term describing a telephone call that is local, rather than long distance, in nature. Such a call incurs no distance-sensitive long distance, or toll, charges, although there may be a charge per call or per minute, perhaps for calling activity that exceeds a threshold. A local call typically is confined to a metropolitan area or a central office exchange (COE) area.

local exchange Central office exchange (CO or COE). See *CO*.

local exchange carrier (LEC) See *LEC*.

Local Internet Registry (LIR) See *LIR*.

local long distance A rate and tariff term that describes an intraLATA long distance call. See also *intraLATA*, *LATA*, *long distance*, and *tariff*.

local loop An access circuit from the network edge to the customer premises, a local loop is a short haul circuit for access to a local exchange. The most common example of a local loop is an electrically based, two-wire, copper access circuit between a telephone company central office (CO) switching center and a residential or small business premises. Such a circuit is provisioned over a single unshielded twisted pair (UTP), within which two wires are required to complete the electrical circuit, with the current in one wire opposite to the current in the other, and with both wires carrying the information signal. The two conductors comprise an electrical loop, with one wire carrying the go signal and the other carrying the electrical return signal. In the broader contemporary sense, any access circuit between the customer premises and the edge of the telco network, or that of any other service provider, is termed a local loop, whether it is electrically based or employs optical or radio energy. See also *last mile*.

local loop unbundling (LLU) See *unbundled service*.

local management interface (LMI) See *LMI*.

Local Multipoint Distribution Service (LMDS) See *LMDS*.

local number portability (LNP) See *LNP*.

Local Number Portability Administration (LNPA) See *LNPA*.

local routing number (LRN) See *LRN*.

location-based services Services offered by cellular radio providers that are sensitive the physical location of the terminal device. Such services include descriptions of and directions to restaurants and other retail establishments in proximity. See *cellular radio*.

log (logarithm) In mathematics, the exponent expressing the power to which a fixed number (base) must be raised to equal a given number (antilogarithm). Logarithmic functions are the inverses of exponential functions. In an equation expressed as:

$$b^n = x$$

where b and x are known, a logarithm can be used to discover n, and is expressed as

$$n = \log_b (x)$$

For example, $\log_{10} (1000) = 3$ because $10^3 = 1{,}000$

logarithm (log) See *log*.

logical **1.** Based on facts, rational thought, and clear reasoning. **2.** Referring to something that does not exist, but has the appearance or effect of physical presence. In the context of telecommunications, a logical circuit, for example, does not have a physical presence in the sense that it is not tangible. Rather, it is defined as some amount of bandwidth provided over a physical, i.e., tangible, circuit that may support many logical circuits. Similarly, a channel may be in the form of regular time slots provided over a digital circuit. The time slots are brief moments in time during which electromagnetic signals convey bits of information associated with a given data transfer. Such a channel is purely logical, with no material presence and no physical presence except to the extent that the electromagnetic energy is the stuff of the science of physics. See also *channel*, *circuit*, *digital*, *physical*, *physics*, *signal*, and *time slot*.

logical IP subnetwork (LIS) See *LIS*.

logical reach The maximum length of a fiber optic link, without regard to the loss budget, or the allowable amount of signal attenuation that the system can withstand. See also *attenuation*, *fiber optic*, *link*, *loss budget*, *physical reach*, and *signal*.

logical topology In reference to the manner in which devices logically interconnect in a network. The logical topology of a network may differ considerably from the physical topology. LAN and WAN topologies variously include bus, mesh, partial mesh, ring, star, and tree. See also *LAN*, *logical*, *logical topology*, *mesh topology*, *partial mesh*, *physical*, *ring*, *star*, *tree topology*, and *WAN*.

Logical Unit (LU) See *LU*.

login Synonymous with *logon*. Referring to the process by which one identifies oneself to a computer, and typically comprising the entry of a user identification (user ID) name and password. See also *password*.

logodaedaly Pronounced *log-a-DEE-da-lee*. From the Greek *logos*, meaning *words*, and French *legerdemain*, meaning *sleight of hand*, hence, any artful deception or trick. The arbitrary or capricious coinage of words. For example, comatext is defined as text that pleases the person who wrote it, but puts others into a coma. Telecrastination is defined as letting the phone ring more than twice before you pick it up, even if it is only six inches away. I don't include such funsense in this dictionary.

logon Synonymous with *login*. See *login*.

logophobia From the Greek *logos*, meaning *words*, and *phobos*, meaning *fear*. Fear of words. Logophobia symptoms include breathlessness, excessive sweating, nausea, dry mouth, feeling sick, shaking, heart palpitations, inability to think clearly, a fear of dying, becoming mad or losing control, a sensation of detachment from reality, or a full blown anxiety attack. If you are experiencing any of these symptoms, put this book down immediately and seek medical attention. *Note:* Neither the author nor the publisher assumes any liability whatsoever for any ill effects, real or imagined, for the logophobic experience as a result of reading the words in this book. (Honestly, I don't make these things up. Well, OK, I made up the part about the liability, but I don't make all of these things up, at least not entirely.) See also *Hellenologophobia*.

LOH (Line OverHead) In a SONET or SDH frame, overhead of 18 octets that controls the reliable transport of payload data in the Synchronous Payload Envelope (SPE) between any two network elements (NEs). LOH and Section Overhead (SOH) between adjacent NEs compose Transport Overhead (TOH). See also *frame*, *link*, *octet*, *overhead*, *SDH*, *SOH*, *SONET*, and *TOH*.

long distance An imprecise rate-and-tariff term that describes a call or circuit that connects two relatively distant parties or systems. A long distance call terminates outside the central office exchange (COE) area, i.e., the serving area of a central office (CO), and outside the municipality. In the United States, an intraLATA long distance call is termed local long distance.

long haul circuit Within the core, or backbone, of a Wide Area Network (WAN), transport circuits, or long haul circuits, carry data over long distances. Long haul traditionally is defined as distances equal to or greater than 50 miles (80 kilometers). See also *access circuit*, *short haul circuit*, and *WAN*.

longitudinal redundancy checking (LRC) See *LRC*.

long wavelength (LW) See *LW*.

long wavelength band (L-Band) See L-*Band*.

loop **1.** An electrical loop, i.e. closed electrical circuit. The two conductors of an electrical loop compose one wire carrying the go signal and the other carrying the electrical return signal. The circuit is closed and the loop is completed when the conductors are connected. **2.** A local loop. In the broader contemporary sense, any access circuit between the customer premises and the edge of the telco network, or that of any other service provider, is termed a local loop, whether it is electrically-based or employs optical or radio energy. See also *local loop*. **3.** A physical configuration used in residential or small business inside wire installations. Such a configuration connects multiple voice telephone jacks to one or two pairs of wires in a continuous, shared electrical loop. See also *home run*.

loopback address In Internet Protocol (IP), a type of unicast address used when a host needs to send a packet back to itself for test purposes. See also *host*, *IPv6*, *packet*, *unicast*, and *unicast address*.

loopback test A test of a local loop initiated by a local exchange carrier (LEC) at the network edge and causing a contact closure at the network interface unit (NIU) on the customer premises. The loopback test thereby allows the testing of the local loop in the forward direction, through the NIU, and looping in the backward direction, all in isolation from the customer premises equipment (CPE) and wiring. See also *CPE*, *LEC*, *local loop*, and *NIU*.

loop current Referring to the electrical current provided by a common battery and powering the electrical loop in telecommunications applications. The term typically is used in the context of a local loop, and the current typically is 48 volts (V) direct current (DC). See also *battery*, *common battery*, *DC*, *local loop*, *power*, and *V*.

loop start A signaling technique used in single line, multiline telephones, and key telephone systems (KTSs) to start, or seize the line between the terminal and the central office switch. With loop start signaling, the telco central office switch provides battery. When the handset of the telephone is lifted, the electrical loop is completed and dc current flows across the circuit. The central office switch detects that fact, bridges the line, assigns a register, and returns dial tone for an outgoing call, or connects an incoming call, as appropriate. Loop start also is used in some PBXs, although it is unusual due to issues of glare, or collision between incoming and outgoing calls. See also *glare*, *ground start*, *loop*, and *register*.

loose-tube cable A fiber optic cable configuration involving a semi-rigid hollow plastic tube that houses and protects a number of optical fibers. The fibers can be either individually coated or organized into ribbons coated with protective acrylate and the tube is flooded with a water blocking gel. The tube or tubes are then helically stranded around a dielectric (e.g., fiberglass) or steel central strength member that supports the handing weight of the cable and prevents buckling. There also commonly is an aramid yarn strength member in the cable core. An outer polyethylene jacket surrounds the entire cable. If armoring is required for protection from rodents and other critters, a metallic tape or mesh is formed around the jacket and another jacket is formed to surround the armoring. Loose-tube cable is used in outside plant (OSP) applications, where extremes of temperature, rough handling, and mechanical disturbances make tight buffered cable unsuitable. See also *aramid*, *cable*, *dielectric*, *fiberglass*, *optical fiber*, *OSP*, *polyethelene*, *strength member*, *tight buffered cable*, and *water-blocking gel*.

LOS (Line-of-Sight) **1.** Optical LOS. Line of vision. A direct imaginary line between two points, as though it were from the center of the eye to the center of the object viewed. A direct non-guided path in the form of a straight line between a transmitter and receiver, uninterrupted by physical matter other than that suspended in the atmosphere. Opaque objects such as mountains, buildings, and trees interrupt optical LOS, as does the horizon over long distances due to the natural curvature of the Earth. Optical LOS is critical in free space optics (FSO) transmission systems and high-frequency radio system. Optical LOS is always preferable in radio systems, even those operating at low frequency. **2.** Radio LOS. A direct non-guided path between a transmitting antenna and a receiving antenna. The criticality of LOS is sensitive to the radio frequency (RF) employed. Very low frequency (VLF), and low frequency (LF) signals tend to be travel between the Earth and the ionosphere. LF and medium frequency (MF) signals propagate as ground waves, which tend to follow the curvature of the Earth. Signals at the high end of the MF range and in the high frequency (HF) range benefit from ionospheric refraction, a phenomenon in which the density gradient in the atmosphere acts like a lens and tends to bend radio beams back towards the Earth. At very high frequencies (VHF) and above (i.e., ? 30 MHz) true optical LOS is considered essential, absent special modulation techniques combined with space division multiplexing techniques such as multiple input/multiple output (MIMO). See also *electromagnetic spectrum*, *ground wave*, *HF*, *ionosphere*, *LF*, *MF*, *MIMO*, *near-LOS*, *NLOS*, *radio*, *refraction*, *RF*, *space division multiplexing*, *VHF*, and *VLF*.

loss **1.** The energy or power expended without having accomplished any useful work. In electrical systems, such loss is described in watts (W). See also *watt.* **2.** The attenuation of signal level in a communications medium. Such loss generally is expressed in decibels (dB). See also *attenuation, dB,* and *signal.* **3.** The disappearance of packets or other message units in network transit. Packet loss can be caused by errors introduced into address fields, which affects the ability of the various network switches, routers, and other devices to properly forward the message units. Loss also can be caused if device buffers overfill and message units are erased. See also *address, buffer, switch,* and *router.*

loss budget Referring to the calculated amount of signal attenuation that a transmission system will tolerate between two points. The calculation considers the power level of the transmitter and the gain, i.e., sensitivity level, of the receiver. In a fiber optic transmission system (FOTS), the loss budget establishes the total amount of tolerable loss attributable to the propagation of the signal through splices, connectors, and the optical fiber cable itself. See also *attenuation, connector, FOTS, gain, loss, optical fiber, power, propagation, signal,* and *splice.*

lossless compression Referring to compression techniques that enable faithful reproduction of the signal, with no data loss, although compression rates tend to be relatively low. See also *artifact, compression, lossy compression,* and *signal.*

lossy compression Referring to compression techniques that tend to produce artifacts, which are unintended and unwanted distortions or aberrations that result in a degraded signal, but supports very high compression rates. In video systems and communications, the artifacts often show up as jagged blockings or tiling effect known as aliasing, banding of colors, white spots, and even dropped frames. Although the picture is degraded as a result, the compression ratios can be as high as 200:1. The MPEG standards, for example, specify lossy compression in the form of discrete cosine transform (DCT). See also *artifact, compression, DCT, lossless compression, MPEG,* and *signal.*

low delay code excited linear prediction (LD-CELP) See *LD-CELP.*

low earth orbit (LEO) See *LEO.*

Low Income Consumers Program In the United States, a program that provides lifeline subsidies to reduce the installation and monthly costs of basic telephone service for low income consumers. The Low Income Consumers Program is one of four programs established by the Telecommunications Act of 1996, supported by the Universal Service Fund (USF), and administered by the Universal Service Administrative Company (USAC). See also *lifeline service, Telecommunications Act of 1996, USAC,* and *USF.*

low frequency (LF) See *LF.*

low-order mode A physical path taken by a signal or signal component that is either parallel to or relatively modestly transverse to the waveguide. Some signal components travel directly through the center of the waveguide, at least theoretically, and, therefore, travel the shortest possible distance between the point at which they enter the waveguide and the point at which they exit the waveguide. Other modes take more transverse paths, striking and reflecting off of the interface between the core and cladding as they propagate through an optical fiber, for example. Low-order modes take parallel or modestly transverse paths, while high-order modes take considerably more transverse paths. See mode for more detail. See also *cladding, core, critical angle, high-order mode,* and *waveguide.*

low-pass filter A device that passes all signals below a designated frequency (electrical) or wavelength (optical) band, but absorbs, attenuates, blocks, rejects, or removes all other signals. See also *absorption, attenuation, band, band-pass filter, electrical, frequency, high-pass filter, optical, signal,* and *wavelength.*

Low-Power Radio Service (LPRS) See *LPRS.*

low-tier In wireless telecommunications, referring to systems intended for pedestrian, in-building, on-campus, and wireless local loop (WLL) application. Examples include wireless LANs (WLANs) and wireless office telecommunications systems (WOTS). See also *high-tier*, *WLAN*, *WLL*, and *WOTS*.

low voltage (LV) See *LV*.

low-water-peak fiber (LWPF) See *LWPF*.

LPC (Linear Predictive Coding) A method of digitally encoding analog signals that predicts the value of a signal at a specific point in time to be a linear function of the past values of the quantized signal. Code excited linear prediction (CELP) and its derivatives are examples of LPC methods. See also *analog*, *CELP*, *digital*, *encode*.

lpi (lines per inch) Also known as *line frequency*. A measure of the vertical resolution of material printed in grayscale or halftone. See also *dpi*.

LPRS (Low-Power Radio Service) In the United States, a private one-way, short distance communication service designed to provide auditory assistance to persons with hearing disabilities, persons who require language translation services, and persons in educational, health care, law enforcement, and various other settings. Two-way communications are prohibited. LPRS is an unlicensed service operating in the 216.75–217.0 MHz band. The Federal Communications Commission (FCC) regulates LPRS, which is in the family of personal radio services. See also *FCC* and *personal radio services*.

LRC (Longitudinal Redundancy Check) A parity checking error control method that improves on the simple vertical redundancy check (VRC) by viewing data as a block, or data set. As shown in Table L-1, this approach is characterized in terms of the manner in which human beings add numbers, not only in columns, but also in rows across columns, as though the devices were viewing data set in a matrix format. This additional technique of checking the total bit values of the characters on a longitudinal (i.e., horizontal) basis employs the same parity (i.e., odd or even) as the vertical check technique in the example in Table L-1. LRC and VRC are easily and inexpensively implemented in devices employing asynchronous transmission. LRC adds a significant measure of reliability when used in conjunction with VRC, although compensating errors still can occur in non-adjacent characters. The LRC is sent as one or two extra characters, known as *block check characters* (BCCs), at the end of each data block. See also *error control*, *parity check*, and *VRC*.

Table L-1: Longitudinal Redundancy Check (LRC)

Bit/Value	*C*	*O*	*N*	*T*	*R*	*O*	*L*	*Odd Parity*
1	1	1	0	0	0	1	0	0
2	1	1	1	0	1	1	0	0
3	0	1	1	1	0	1	1	0
4	0	1	1	0	0	1	1	1
5	0	0	0	1	1	0	0	1
6	0	0	0	0	0	0	0	1
7	1	1	1	1	1	1	1	0
8 (Odd Parity)	0	0	1	0	0	0	0	

LRN (Location Routing Number) In the United States, a 10-digit number that identifies a switch port for a central office (CO), an LRN is used to provide local number portability (LNP). The LRN is maintained in a Services Management System (SMS) database in a Service Control Point (SCP), which is part of the Advanced Intelligent Network (AIN). When a caller dials a telephone number, the originating CO queries a routing SMS database in an SCP for the associated LRN. The network uses that number to direct the call to the proper physical exchange, which may be one of several operated by competing local exchange carriers (LECs). LRNs are supervised by the Number Portability Administration Center (NPAC). See also *AIN, CO, database, Directory Number, LEC, LNP, SCP,* and *SMS.*

LSB (Least Significant Bit) See *least significant bit.*

LSP (Label Switched Path) In Multiprotocol Label Switching (MPLS), the physical path selected by a Label Edge Router (LER) for a flow of packets between common endpoints and based on the Forwarding Equivalence Class (FEC) associated with the packet flow. Using a Label Distribution Protocol (LDP), the LER distributes to Label Switched Routers (LSRs) in the core the individual link selections that comprise the end-to-end path selection. The LER inserts labels into the headers of the individual packets in the flow. Each LSR in the core uses the label of the incoming packet to select the outgoing port and link, and swaps that label for a new label, which the downstream LSR uses to select the next port and link, and so on. See *downstream, FEC, flow, header, label, LDP, LER, link, LSR, MPLS, packet, path,* and *port.*

LSR (Label Switching Router) In Multiprotocol Label Switching (MPLS), a router in the network core that forwards packets along paths consistent with the labels inserted into packet headers by Label Edge Routers (LERs) as they entered the network. The LERs employ a Label Distribution Protocol (LDP) to distribute labels to the LSRs in the network core. As the LSRs receive the packets, they examine the short labels, compare them against a label database, switch the existing label for a new one, and quickly forward the packet across an appropriate link to the next LER. The process is repeated by each LSR until the packet reaches the egress LER, which strips the tag away as the packet exits the network. See also *core, header, link, LSP, LDP, LER, MPLS, packet, path, router,* and *switch.*

L2TP (Layer 2 Tunneling Protocol) A tunneling protocol used for secure node-to-node communications by Internet service providers (ISPs) and other virtual private network (VPN) service providers in support of multiple, simultaneous tunnels in the network core. End users gain access to the service provider on an unencrypted basis, with the service provider assuming the responsibility for encryption at the edge of the packet network. L2TP is an extension to the Point-to-Point Protocol (PPP) that evolved from a combination of Microsoft's PPTP and Cisco's Layer 2 Forwarding (L2F) protocol. L2TP is described in IETF RFC 2661. See also *encryption, ISP, PPP, PPTP, protocol, tunneling,* and *VPN.*

LU (Logical Unit) In the IBM Systems Network Architecture (SNA), a program that manages communications software for communications with end users. A logical unit session originally was defined as an end-to-end communication between an end-user terminal and the originating application residing in the host. Later versions, such as LU 6.2, support peer-to-peer communications between intelligent devices, without requiring the host to assume responsibility for communications support activities. LU 6.2 is also known as Advanced Program-to-Program Communications (APPC).

Luddite A pejorative term for someone who opposes technological or industrial innovation, especially if it automates manual functions or processes. The term originates in an unconfirmed folk legend about a feebleminded English (Nottinghamshire or Leicestershire) textile worker named Ted Ludd who smashed (c. 1779) two stocking knitting frames with a hammer to protest beatings by his master. The Luddites (1811–1816) in England were groups of textile workers who protested the use of automated wide-frame textile looms that could be operated by cheap, unskilled labor. The movement was so strong that it led to labor riots and battles with the British army. At least 17 Luddites were convicted of industrial sabotage and executed for that crime. Many more were convicted and shipped off to penal colonies in Australia. *Note:* I wrote this definition (and this entire book) on a computer that I seriously considered smashing with a hammer on multiple occasions.

luminance The measure of intensity, or brightness, as measured in candelas per square meter. Luminance can vary within an image. An analog video transmission varies the luminance by varying the power level, or amplitude, of the signal, with high power representing black and low power representing white. See also *candela* and *luminous flux*.

luminiferous ether See *ether*.

luminous flux Radiant flux evaluated with respect to its luminous (brightness) efficiency. See also *flux*, *luminance*, and *radiant flux*.

lumped amplification A type of amplification that occurs in a single, discrete location. Also known as *discrete amplification*. Erbium-doped fiber amplifiers (EDFAs), and most amplifiers, perform the process of amplification on a lumped, or discrete, basis. See also *amplifier*, *distributed amplification*, *EDFA*, and *Raman amplifier*.

LUNI (LANE User-to-Network Interface) Pronounced *loonee*. In the LANE (LAN Emulation) environment, the interface by which an end system, known as a LAN Emulation Client (LEC), connects to the ATM network. See also *ATM*, *LANE*, *LEC*, and *UNI*.

LV (Low Voltage) **1.** A relatively low amount of electromotive force (emf). The LV grid runs at 110 volts at 50–60 Hz in North America, and 220 volts at 50 Hz in Europe and most of the rest of the world. Access broadband over power line (Access BPL) technology can make use of those LV lines as a portion of a local loop for broadband data communications. See also *Access BPL*, *emf*, *HV*, *Hz*, *MV*, *volt*, and *voltage*. **2.** According to the International Electrotechnical Commission (IEC), alternating current (AC) voltage of 50–1,000V, or direct current (DC) voltage of 120–1,500V. Extra low voltage (ELV) is AC voltage less than 50V or DC voltage below 120V. Unlicensed personnel can safely install ELV wiring. See also *AC*, *DC*, and *IEC*.

LW (Long Wavelength) Referring to fiber optic systems operating in the 1300, 1310, and 1550 nm ranges, with the IEEE 802.3ae specification for 10GBase-SR, LW being one example. See also *10GBase-SR*, *LW*, and *SW*.

LWPF (Low-Water-Peak Fiber) Single-mode fiber (SMF) manufactured with low levels of hydroxyl (OH) ions in order reduce the attenuation peak in the 1400 nm window, which is in the E-band (1360–1460 nm). The traditional SMF manufacturing process introduces hydroxyl (OH) ions into the fiber core. Wavelengths in the region around 1400 nm attenuate about 2 dB/km as a result of their interaction with those ions. As traditional single-wavelength fiber optic transmission systems (FOTS) employing SMF operate in the 1310 nm or 1550 nm window, water peak attenuation does not affect them. However, 4 of the 18 channels in coarse wavelength division multiplexing (CWDM) systems fall within the E-band and, therefore, are rendered unusable by water peak attenuation. Zero-water-peak fiber (ZWPF) contains near zero hydroxyl contamination and, therefore, suffers near zero water peak attenuation. See also *attenuation*, *CWDM*, *dB*, *dB/km*, *E-band*, *FOTS*, *hydroxyl*, *SMF*, *water peak*, *wavelength*, and *ZWPF*.

LZ (Lemple-Ziv) A compression algorithm used in some modems, LZ achieves compression ratios of better than 5:1 for some forms of text and numerical data. LZ compression allows the data terminal equipment (DTE) to operate at speeds up to 128 kbps while the analog link between the modems remains at 38.4 kbps or less over a 4 kHz channel. LZ is named after its inventors, Abraham Lempel and Jacob Ziv. See also *analog*, *channel*, *compression*, *DTE*, *link*, *modem*, and *LZW*.

LZW (Lemple-Ziv-Welch) A lossless data compression algorithm used in the Graphics Interchange Format (GIF) and optionally used in Tagged Image File Format (TIFF) files. LZW is named after its inventors, Abraham Lempel, Jacob Ziv, and Terry Welch. See also *analog*, *channel*, *compression*, *DTE*, *GIF*, *link*, *lossless compression*, *modem*, *TIFF*, *LZ*, and *LZW*.

m meter. See *meter.*

M **1.** Mega. From Greek *megas*, meaning *great*, translates to *million*. **2.** In transmission systems, MHz (MegaHertz) is a million (10^6) Hertz, Mbps (Megabit per second) is a million (10^6) bits per second, and MBps (MegaByte per second) is a million (10^6) bytes per second. In transmission systems, therefore, a million is exactly 1,000,000, since the measurement is based on a base 10, or decimal, number system. **3.** In computing and storage systems, a MB (MegaByte) is actually 1,048,576 (2^{20}) bytes, because the measurement is based on a base 2, or binary, number system. The term MB comes from the fact that 1,048,576 is nominally, or approximately, 1,000,000.

M12 (Multiplex 1-to-2) A device used in a digital carrier system to multiplex level one bit streams into a level two bit stream. In a T-carrier system, for example, an M12 multiplexes four T1s into a T2 bit stream. See also *bit stream, carrier, multiplexer, T1, T2,* and *T-carrier.*

M13 (Multiplex 1-to-3) A device used in a digital carrier system to multiplex level one bit streams into a level three bit stream. In an E-carrier system, for example, an M13 multiplexes four E-1s into an E-2 bit stream, and then multiplexes four E-2s into an E-3 bit stream. The corresponding T1 version of an M13 combines 28 T1 bit streams into a T3 bit stream. See also *bit stream, carrier, E-1, E-2, E-carrier,* and *multiplexer.*

M24 Also known as *D4.* A T-carrier framing convention. See *D4* and *T-carrier.*

MAC **1.** Medium Access Control. The process employed to control the basis on which devices can access a shared medium. In a local area network (LAN), some method of control is required to ensure, or at least improve, the ability of all devices to access the network within a reasonable period of time. It also is important that some method exist to either detect or avoid and to recover from data collisions, caused by multiple transmissions placed on the shared medium simultaneously. Medium access control takes place at Layer 1, the Physical Layer, and Layer 2, the Data Link Layer, of the OSI Reference Model. MAC programmed logic is embedded in a device variously known as a *network interface unit* (NIU) or *network interface card* (NIC). Medium access control can be centralized or decentralized. Token Ring LANs centralize that function in a master control station. CATV networks centralize control in a headend. Ethernet LANs decentralize the function, distributing the responsibility among the attached devices. Medium access control also can be either deterministic (e.g., Token Ring) or non-deterministic (e.g., Ethernet) in nature. See also *CATV, Data Link Layer, deterministic, Ethernet, headend, LAN, NIC, NIU, non-deterministic, OSI Reference Model, Physical Layer,* and *Token Ring.* **2.** Move, Add and Change. Activity associated with relocating, activating, disconnecting, or changing the features associated with a station set or some other device or component associated with a voice or data telecommunications system, such as a PBX or LAN router. As vendors generally bill MAC activity typically at a much higher rate than activity associated with the initial installation of a system, plug 'n' play features such as automatic set relocation are highly desirable. See also *plug 'n' play* and *automatic set relocation.*

MAC bridge (Medium Access Control bridge) See *encapsulating bridge.*

machine **1.** A simple, unpowered instrument, such as a lever, pulley, or inclined plane, that is used for performing some kind of work by transmitting or changing the direction of energy. **2.** A powered mechanical device, such as an automobile or drill press, that consists of a structure and various moving and unmoving parts and is for doing some kind of work. **3.** An electronic device, such as a computer or facsimile machine, that can be thought of as operating on a mechanical basis. Such machines have automated many of the functions previously performed by humans. See also *human.*

macrobend A relatively large bend in an optical waveguide, such as a fiber optic cable. A technician may need to bend a cable around a corner, for example, or may need to coil some cable in a span to provide necessary slack in the event that the cable must be spliced in the future, perhaps to repair a break. A technician placing a macrobend in a cable must consider the minimum allowable bend diameter in order to prevent either damage to the cable or bending loss, i.e., loss of signal strength resulting from an excessive bend. See also *bend diameter*, *bending loss*, and *microbend*.

macrocell In radio systems, an imprecise term referring to a relatively large area of coverage, perhaps an entire metropolitan area of 50 miles in diameter. A macrocell is larger than a microcell and much larger than a picocell. See also *cell*, *microcell*, and *picocell*.

MAE (Merit Access Exchange or Metropolitan Area Exchange) The exact meaning of the acronym is lost in the mists of time; it means either Merit Access Exchange or Metropolitan Area Exchange. In either case, MAE now is just MAE, a registered trademark of MCI, now (September 2007) a Verizon company, and is an official Network Access Point (NAP) at which an Internet service provider (ISP) can access the Internet backbone and exchange traffic with other ISPs. Some NAPs are known as NAPs, some as Internet Exchanges (IXs), and some as MAEs. Tier 1 MAEs are located in San Jose, California (MAE West); Vienna, Virginia (MAE East); Miami, Florida; and Paris, France. Tier 2 MAE sites currently are located in Chicago, Illinois; Dallas, Texas; Los Angeles, California; and New York, New York. See also *Internet*, *ISP*, *IX*, and *NAP*.

mailbox A partition of computer memory designated for the temporary storage of messages intended for an individual, department, company, or other authorized user entity. Mailboxes can be associated with e-mail, facsimile mail, or voice mail systems.

mail transfer agent (MTA) See *MTA*.

mail user agent (MUA) See *MUA*.

main distribution frame (MDF) See *MDF*.

mainframe computer Also referred to colloquially as big iron and heavy metal. A large, expensive, and often highly redundant computer designed to support a large organization, handle intensive computational tasks, support a large number of users, and make use of large volumes of secondary storage. The largest mainframes are capable of supporting thousands of simultaneous users and use terabytes of secondary storage. Notably, mainframes are employed in a centralized computing architecture, which is opposite the distributed architecture of local area networks (LANs) and the Internet. The term originally described the main frame that contained the central processing unit (CPU) of computers in the days when all computers were heavy metal. See also *Internet*, *LAN*, *minicomputer*, and *personal computer*.

maintenance and administration terminal (MAT) See *MAT*.

make and break See *soft handoff*.

malware (malicious software) Software that is harmful or evil in intent. See also *spyware*, *Trojan horse*, *virus*, and *worm*.

MAN (Metropolitan Area Network) A public data network that serves an entire metropolitan area, or perhaps a portion of a metropolitan area such as a city or a suburb, commonly serving to interconnect Local Area Networks (LANs). A number of carriers offer Metropolitan Ethernet services, for example. MANs can be interconnected across a Wide Area Network (WAN). See also *LAN* and *WAN*.

managed service provider (MSP) See *MSP*.

management information systems (MIS) See *IS*.

management plane In the ATM reference model, the functions that involve the management of the ATM switch or hub. The management plane is divided into plane management and layer management. *Plane management* acts on the management of the switch as a whole, with no layered approach. *Layer management* acts on the management of the resources at each specific layer of the model, e.g., operation, administration, and maintenance (OA&M) information. See also *ATM reference model, control plane,* and *user plane.*

manager of managers (MOM) See *MOM.*

Manchester coding A technique for encoding both the clock and data pulses into a self-synchronizing bit stream. Manchester coding does not send data as a series of raw 1 bits and 0 bits. Rather, each data bit includes a midpoint voltage level transition from positive (+) to negative (–) or from negative (–) to positive (+), with the direction of the transition indicating whether the bit is a 1 bit or a 0 bit. The fact that each bit representation includes both a positive and a negative pulse ensures pulse density and, therefore, ensures proper synchronization. This characteristic also maintains DC balance on the line. Sometimes referred to as phase encoding (PE) and biphase encoding, Manchester coding is a special case of binary phase-shift keying (BPSK). Manchester coding is specified in early versions of IEEE 802.3, also known as Ethernet, and 802.4, also known as token bus. In contemporary high speed networking, more efficient coding schemes, such as 4B/5B and 8B/10B, largely have replaced Manchester coding, which requires two pulses (+/– or –/+) for each data bit. See also *4B/5B, 802.3, 802.4, 8B/10B, BPSK, Ethernet, line coding, ones density, phase, pulse, synchronous,* and *Token Bus.*

mandrel A rod or spindle around which material such as metal, wire, or glass is cast, molded, bent, shaped, or wrapped. See also *mandrel wrapping.*

mandrel wrapping A technique used in multimode fiber (MMF) optics to modify the modal distribution of an optical signal. The wrapping of the MMF around a mandrel results in intentional macrobends and forces modes into higher orders, i.e., away from direct paths through the core and towards the core/cladding interface. If the MMF is fully filled by the source, mandrel wrapping forces the higher-order modes into the cladding, where they are attenuated and lost. If the MMF is underfilled, mandrel wrapping forces some low-order modes into higher-order modes, which redistribution results in modal equilibrium, i.e., equal distribution of power across modes propagating in the core. Mandrel wrapping sometimes is used in jumper cables to intentionally attenuate high-power optical signals in order to prevent damage to optical receivers. Mandrel wrapping also is used in launch cables to achieve modal equilibrium for testing purposes. The diameter of the mandrel and the number of wraps or turns around it are sensitive to the fiber characteristics and the desired modal distribution. See also *attenuation, cladding, core, high-order mode, jumper, launch cable, low-order mode, mandrel, MMF, mode,* and *signal.*

Mann-Elkins Act In the United States, the act (1910) that granted the Interstate Commerce Commission (ICC) interstate regulatory authority.

Manufacturing Automation Protocol (MAP) See *MAP.*

Manufacturing Automation Protocol/Technical and Office Protocol (MAP/TOP) See *MAP.*

MAP (Manufacturing Automation Protocol) A local area network (LAN) protocol developed by General Motors (GM) in the early 1980s for the interconnection of computers and programmable machine tools in factory or assembly line operations, MAP is based on Token Bus (IEEE 802.4) running at 1, 5, 10, and 20 Mbps. MAP sometimes is referred to as Manufacturing Automation Protocol/Technical and Office Protocol (MAP/TOP). See also *802.4, LAN,* and *Token Bus.*

MAP/TOP (Manufacturing Automation Protocol/Technical and Office Protocol) See *MAP.*

mark Referring to a service mark or trademark. See also *service mark* and *trademark.*

Markov source A statistical model for predicting the occurrence frequencies of letter or word pairs and triplets. Markov source is used in some data compression mechanisms. See also *compression.*

maser (microwave amplification by stimulated emission of radiation) The maser is a device similar to the laser, but emitting microwave radio waves rather than light. In 1953, Charles H. Townes, James P. Gordon, and Herbert J. Zeiger invented the maser, which earned them the Nobel Prize in Physics in 1964. Research continued into the 1950s, leading to the optical maser, or laser, for which AT&T Bell Telephone Laboratories was awarded a patent in 1960. See also *laser.*

mashing A process of building new Web-based services from reusable components of other services, mashing is a technique defined in Web 2.0. See also *Web 2.0.*

master/slave Descriptive of a relationship in which one entity, the master, is in total control of another, the slave. In computer networking, master/slave is a network architecture and set of protocols in which one device or program, the master, exerts total control over one or more other devices, the slaves. The master determines the communications priorities of the slaves, for example. A master/slave architecture, such as IBM Token Ring or Synchronous Data Link Control (SDLC), is decidedly different from a peer-to-peer architecture, in which computers communicate as equals, sharing the same responsibilities and using the same programs to communicate. As the term master/slave can be offensive to some people, some computer manufacturers prefer the term primary/secondary. Bluetooth specifications provides for ad hoc piconets that can include as many as seven slaves under the control of a master, which assumes that responsibility when initiating the network. See also *ad hoc, Bluetooth, client/server, network architecture, PC, peer-to-peer, piconet, protocols, SDLC,* and *Token Ring.*

MAT (Maintenance and Administration Terminal) A PC or other data terminal connected to the maintenance port of a PBX or key telephone system (KTS) in order that authorized users can access the system software for purposes that might include Move, Add, and Change (MAC) activity; Class of Service (CoS) changes; automatic route selection (ARS) programming; requests for traffic and usage statistics; requests for status reports; and diagnostic testing and analysis. Remote maintenance generally can be accomplished over the PSTN via a modem connection.

material dispersion A type of dispersion that occurs in optical fiber due to the interaction of various wavelengths with the physical matter in the crystalline structure of the glass. The refractive index of the glass varies according to the wavelength of the optical signal, i.e., different wavelengths travel at different speeds in the medium. The longer the wavelength, the faster the signal travels. No pulse is perfectly defined, i.e. includes just one wavelength. Rather, an optical pulse emitted by a light source has a certain spectral width, i.e., includes a range of wavelengths of lesser power around the center wavelength. The effect of material dispersion is that the various wavelengths comprising the pulse travel at different velocities through the medium. So, the pulse can spread over a distance simply due to the interaction of various wavelengths with the matter in the crystalline core, which causes some portions of a pulse to arrive earlier than other portions. As the wavelength increases (and frequency decreases), material dispersion decreases. So, optical signals in the 1550 nm window suffer less from material dispersion than wavelengths in the 1310 nm window. Material dispersion, which is synonymous with intramodal dispersion and spectral dispersion, is one factor contributing to chromatic dispersion. Material dispersion and chromatic dispersion are issues in long haul fiber optic transmission systems (FOTS) employing single-mode fiber (SMF) of step-index construction. Multimode graded-index fibers suffer so much from modal dispersion over short distances that material dispersion and chromatic dispersion never become factors. See also *chromatic dispersion, dispersion, graded-index fiber, MMF, modal dispersion, refractive index, SMF, spectral width, step-index fiber,* and *window.*

Matthews, Gordon (1937–2002) The inventor of modern voice processing systems, Matthews filed the basic patents and first commercialized the systems. Matthews was on a business trip in the 1970s and was having trouble reaching his office to pick up his messages due to time zone differences. He mentioned the problem to his wife, Monika, who suggested that he invent a computer so that he and his employees could leave messages for each other. Matthews went to work on the project and founded VMX (Voice Message eXpress) in Dallas, Texas (United States) in 1979. Shortly thereafter, 3M bought the first commercial system, a standalone voice mail system with an interface to the PBX, from which the call was forwarded in the event of a busy or no-answer condition at the user station. Monika Matthews recorded the first greeting on this first commercial system. *Note:* In 1992, Matthews retired and sold VMX to Octel. Octel subsequently merged with the Enterprise Networks Group of Lucent Technologies, an AT&T spin-off. VMX and Octel systems are now a product line of Avaya, a Lucent spin-off, which is just as well, as Lucent was acquired by Alcatel, a French company, in 2006. See also *voice processor.*

matrix See *matrix switch* and *switch matrix*.

matrix switch A switch comprising an array of internal buses or circuits laid out in a grid of rows and columns so that connections can be set up anywhere the circuits intersect, and a path can be established between any input port and any output port. There are physical matrixes (with metallic contacts) such as voice crossbar switches, and logical matrixes used in packet switches. A matrix switch can provide full bandwidth to multiple, simultaneous transmissions on a port-to-port, point-to-point basis. If there are congestion issues in a matrix switch, it may have the ability to subdivide its capacity, with the buses becoming shared buses through a process of time division multiplexing (TDM). See also *bus, circuit, crossbar, cut-through switch, LAN switch, port, shared bus switch, switch*, and *TDM.*

MAU (Medium Access Unit) A device that provides a point of access for hosts, workstations, and peripherals into a local area network (LAN). An MAU typically is a multiport device that houses multiple network interface cards (NICs). See also *LAN* and *NIC.*

maxCTD (maximum Cell Transfer Delay) See *CTD.*

maximum burst size (MBS) See *MBS.*

maximum frame size (MFS) See *MFS.*

maximum transmission unit (MTU) See *MTU.*

MB (MegaByte) Million bytes. In computing and storage systems, a MB (MegaByte) is actually 1,048,576 (2^{20}) bytes, since the measurement is based on a base 2, or binary, number system. The term MB comes from the fact that 1,048,576 is nominally, or approximately, 1,000,000. See also *byte* and *M.*

Mbps (Megabit per second) Million (10^6) bits per second. A measure of bandwidth in a digital transmission system. See also *bandwidth, bps*, and *M.*

MBps MegaByte per second, or million (10^6) bytes per second. MBps is a measure of transmission system bandwidth in ESCON, Fibre Channel, FICON, and other select storage area network (SAN) solutions. See also *bandwidth, Bps, byte, ESCON, Fibre Channel, FICON, M*, and *SAN.*

MBS (Maximum Burst Size) In asynchronous transfer mode (ATM), a traffic parameter describing the maximum number of consecutive cells that can transmit within the peak cell rate (PCR), given the burst tolerance (BT), or burst cell tolerance (BCT), of the network. See also *ATM, BCT, BT, cell, PCR*, and *traffic parameter.*

MC (Multipoint Controller) See *MCU.*

McNutt, Emma The first female telephone operator, Emma McNutt was hired by New England Bell in Boston, Massachusetts, and began work on September 1, 1878. Previously, all operators had been young men who reportedly were given to lightening the tedium of their work by roughhousing, shouting at each other, and swearing at the customers. Young ladies of the time were brought up to be gracious, diffident, prim, proper, virtuous, and never profane. The Bell System also required them to be single. (*Note:* New

England Bell did not hire married women until 1942, with the thought being that, by definition, a married woman could not be virtuous, at least not from a male perspective. It was, after all, male chauvinists who did the hiring in those days.) Within a few years, the male operator was extinct, not to resurface until the 1960s. History lost track of Emma McNutt after she retired in 1911.

MCR (Minimum Cell Rate) In asynchronous transfer mode (ATM), a traffic parameter describing the minimum number of cells per second that the network agrees to support for a given originating endpoint across a user network interface (UNI). A service descriptor for the available bit rate (ABR) service class, MCR is that rate, expressed in cells per second, at which the originating endpoint can always transmit during the course of the connection. See also *ABR*, *ATM*, *cell*, *traffic parameter*, and *UNI*.

MCU (Multipoint Control Unit) In H.323-compliant multimedia networks, a device that supports conferencing among three or more participating terminals. The MCU comprises a Multipoint Controller (MC) and optional Multipoint Processors (MPs). The MC is responsible for call control negotiation to achieve common levels of communication. The MP may process either a single media stream or multiple media streams, depending on the nature of the conference. See also *H.323* and *multimedia*.

MDF (Main Distribution Frame) A distribution frame on one side of which main cable connect and on the other side of which lesser cable connect. An MDF may connect external lines and trunks from the public network on one side and internal cables on the other. In an end user environment, an MDF may connect lines and trunks from the public network on one side, or perhaps cables from a PBX, and on the other side connect internal cables connecting to intermediate distribution frames, terminal blocks, or directly to terminals within the customer premises. Short jumpers connect one side to the other. In a telecommunications environment, network trunks terminate on one side and local loop lines on the other. The MDF incorporates protectors for protection against power spikes and surges, and serves as a test point. See also *distribution frame*, *frame*, and *IDF*.

MDU (Multi-Dwelling Unit) An apartment building, condominium, residence hotel, or other building that is subdivided into multiple residences. An MDU has considerable implications for local loop deployments, particularly with respect to passive optical network (PON). See also *local loop* and *PON*.

mean opinion score (MOS) See *P.800*.

mean time between failures (MTBF) See *MTBF*.

mean time to repair (MTTR) See *MTTR*.

mechanical splice In optical fiber installation and repair, the semi-permanent joining of two fibers through the use of a plastic or metal crimp sleeve filled with index-matching gel. See also *fusion splice*, *index-matching gel*, and *splice*.

mechanical strength The ability to withstand the stress of physical forces. Cable and wire systems, for example, must be designed in consideration of the amount of twisting and bending (flex strength) they can tolerate and the amount of weight or longitudinal stress a cable or wire can support (tensile strength) without suffering deformation or breaking (break strength). See also *break strength*, *flex strength*, *strength member*, and *tensile strength*.

media converter A device that interfaces disparate transmission media, making the necessary signal conversions at the Physical Layer. For example, a media converter is required to interface a fiber optic transmission system such as Passive Optical Network (PON) to the copper unshielded twisted pair (UTP) used in inside wire applications on the customer premises. See also *Physical Layer*, *PON*, and *UTP*.

media gateway **1.** A protocol converter that interfaces a traditional public switched telephone network (PSTN), or device running PSTN protocols, with a device running the Internet protocol (IP) suite. **2.** A small IPBX system sized for small business enterprise (SBE) applications up to a dozen or so extensions, or perhaps small-to-medium enterprise (SME) applications up to 50 or so extensions. A media gateway also can be networked with a larger media server, media manager, or communication manager that serves a larger regional office. See also *IPBX*. **3.** In H.323-compliant multimedia networks, a gateway is an optional element used for various levels of protocol conversion. The gateway serves as a protocol converter between devices and networks that have native H.323 capability and those that do not. The gateway also may translate between audio, video, and data formats, and may perform signaling conversions between the H.225 packet protocol and external protocols such as SS7 and Q.931. Alternatively, signaling conversions may be performed by gatekeepers, call processors, or session border controllers. See also *H.225*, *H.323*, *multimedia*, *network*, *packet*, *protocol*, *protocol converter*, *Q.931*, *session*, and *SS7*.

Media Gateway Control (Megaco) See *Megaco*.

Media Gateway Control Protocol (MGCP) A predecessor to Media Gateway Control (Megaco). See *Megaco*.

Media Gateway Controller (MGC) In voice over Internet Protocol (VoIP) networks, a device responsible for the registration and management of resources at the media gateway (MG). Sometimes referred to as a call agent, call controller, or softswitch. See also *media gateway*, *softswitch*, and *VoIP*.

media manager See *media gateway*.

media server See *media gateway*.

mediation device Also known as a *set handler*. See *set handler*.

Medical Implant Communications Service (MICS) See *MICS*.

medium A substance that conveys something, i.e., through which something is carried or transmitted. In the context of telecommunications, a medium is something that passively supports or allows the conveyance or transmission of a signal, and is not necessarily a tangible thing that can be touched. See *transmission medium*.

medium access control (MAC) See *MAC*.

medium access control (MAC) bridge See *encapsulating bridge*.

Medium Access Unit (MAU) See *MAU*.

medium frequency (MF) See *MF*.

medium earth orbit (MEO) See *MEO*.

medium voltage (MV) See *MV*.

Mega- (M) See *M*.

Megabit per second (Mbps) See *Mbps*.

Megabyte per second (MBps) See *MBps*.

Megaco (Media gateway control) A joint standardization effort of the ITU-T (H.248) and the IETF (RFC 3525) that defines the call control protocols employed in a physically decomposed gateway with subcomponents distributed across multiple devices that may be in multiple physically distinct locations. Those subcomponents take the form of a Media Gateway (MG) and a Media Gateway Controller (MGC), also known as a *softswitch* or a *call agent*. A single MGC can control a large number of MGs, each of which is optimized for a particular gateway application function to convert the media format between a packet network and another form of network, such as a public switched telephone network (PSTN) or an asynchronous transfer mode (ATM) network. The call control and signaling logic are centralized in the MGC and can include features such as dial tone, collect dialed digits, call hold, call transfer, call forward, and call conference. The MGC signals the MGs, which then execute the feature commands and process the call, performing gateway functions as required to interface the incompatible networks or network elements (NEs). There is a master/slave relationship between the centralized MGC and decentralized MGs, much like that of a traditional PSTN, except that the MGs that execute the features and perform the switching are distributed across the network. See also *ATM, call transfer, conference call, dial tone, gateway, H.248, hold, IETF, ITU-T, master/slave, protocol, PSTN,* and *softswitch.*

MegaHertz (MHz) See *MHz.*

memory A device that stores computer data or programs for subsequent retrieval. In the general sense, the term refers to all forms of on-line storage, including hard disk drives and tape drives. In practice, the term generally refers to a computer's fast semiconductor-based main memory, or random access memory (RAM), as distinguished from its secondary storage, such as hard drives. Virtual memory is disk space pretending to be RAM. See also *flash memory, RAM,* and *ROM.*

MEMS (Micro-ElectroMechanical Systems) Tiny electromechanical systems on a silicon chip. MEMS comprise integrated mechanical elements, sensors, actuators, and electronics on a common silicon substrate through microfabrication technologies, including the micromachining of the mechanical elements. MEMS are employed in optical projectors as well as in purely optical switches used in fiber optics transmission systems (FOTS), and actually are more along the lines of optical cross-connects. As purely optical devices, they do not require that the incoming optical signal be converted to an electrical signal for processing and then be reconverted to an optical signal. In other words, MEMS are optical-optical-optical (OOO), rather than optical-electrical-optical (OEO). See also *cross-connect, FOTS,* and *switch.*

Merit Access Exchange (MAE) See *MAE.*

mesh topology A network topology characterized by the intertwining of nodes through links connecting them together directly, rather than through one or more intermediate points of interconnection. There are two types of mesh topologies: full mesh and partial mesh.

- **full mesh:** A topology that connects every node directly with every other node. A full mesh minimizes propagation delay, latency, and the potential for data errors and loss in transit, as the path is direct and involves no intermediate processing points. A full mesh also provides the greatest number of alternative paths between any two nodes, which has the advantage of extreme network redundancy and resiliency. In the event of a catastrophic failure or performance degradation anywhere in the network, sufficiently intelligent nodes can redirect traffic around the point of failure. A full mesh topology would be highly desirable if it were not for the large number of circuits and ports required. An eight-node full mesh network, as illustrated in Figure M-1, requires 28 circuits, each of which requires two ports, calculated as follows:

  $X = n(n - 1) / 2$

  $28 = 8(8 - 1) / 2$

  The cost and complexity of full mesh networks generally makes them impractical for more than three or four nodes, at least in wireline networks. Wireless networking can be quite another matter in some cases. Several Wireless LAN (WLAN) technologies operating in the unlicensed ISM bands support full mesh networking, as the cost of a establishing a link is essentially zero, as is the level of complexity.

- **partial mesh:** A topology that provides a path by which any node can connect with any node, but not necessarily a direct path, and which provides no common point of interconnection for all nodes. A partial mesh is characterized by a backbone that takes the form of a small mesh of centrally located nodes, through which the more remote end nodes interconnect. As illustrated in Figure M-1, a partial mesh provides some of the advantages of a full mesh, but without the extreme circuit and port requirements.

See also *infrastructure mesh*, *ISM*, *pure mesh*, *topology*, and *WLAN*.

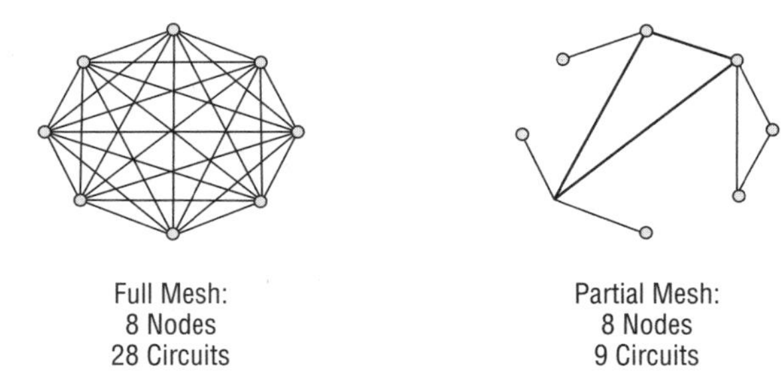

Figure M-1

message A complete thought or idea prepared for transmission. A message may consist of a single discrete set of data prepared for transmission as a whole, or it may be segmented, fragmented, or otherwise divided into multiple parts of the whole in the form of frames, blocks, packets, cells or other sets of data for enhanced effectiveness in transmission, switching or routing, format conversion, storage, etc. At the destination, the fragments or segments are reassembled into the complete message.

Message Handling Service (MHS) See *X.400.*

Message Handling Service Protocol (MHSP) See *X.400.*

message mode service In asynchronous transfer mode (ATM) a type of service for framed data in which only one interface data unit (IDU) is passed. In other words, it is a single-frame message. Message mode service is supported by ATM Adaptation Layer 3/4 (AAL 3/4). AAL 3/4 also supports streaming mode service, which is used for framed data in which multiple IDUs are passed in a data stream. See also *AAL3/4*, *ATM*, *IDU*, and *streaming mode service.*

message service Also known as a *messenger call.* A service offered in developing countries that involves the telephone company's sending a messenger to a remote village or other location without telephone service. The messenger carries a message advising a person to expect a telephone call at a particular pay station or telephone company office or agency in another village with telephone service. Messenger service is charged at a premium rate, with the cost of the messenger added to the cost of the call.

message switch A device that switches complete messages within a data network. A message switch is a store-and-forward device that receives, stores, and forwards messages.

message unit A collection of bits or characters that represent a fragment or segment (i.e., portion) of a message. A message unit can take the form of a block, cell, frame, or packet. See also *block*, *cell*, *frame*, *message*, and *packet.*

Message Telecommunications Service (MTS) Synonymous with Direct Distance Dialing (DDD) See *DDD.*

message waiting A condition in which a voice message has been deposited in a voice mail system and is available for the telephone system (key telephone system (KTS), PBX, or Centrex system) user to retrieve. The voice messaging system advises the user of the message waiting by one of several means such as lighting a message waiting lamp on the station set, causing the telephone system to provide stuttered dial tone when the user lifts the handset to place a call, or perhaps ringing the station periodically. See also *Centrex*, *KTS*, *PBX*, and *voice mail*.

messenger call See *message service*.

metadata From the Greek *meta*, meaning *beside* or *after*, and the Latin *datum*, meaning *what is given*. Data about data, that is, data that describes other data. For example, the title, subject, publisher, and author comprise metadata about its contents.

Metcalfe, Robert Melancton (1946–) Credited as the inventor of the local area network (LAN) concept and the enabling technology. Metcalfe and his associates at the Xerox Palo Alto Research Center (Xerox PARC) developed the first LAN, which originally was known as the Altos Aloha Network as it connected Altos computers through a network based on the AlohaNet packet radio system technology developed at the University of Hawaii. The network later (1973) became known as *Ethernet*, from luminiferous ether, the omnipresent passive medium once conjectured to pervade all space and to support the propagation of electromagnetic energy, even through a vacuum. See also *ether*, *Ethernet*, *LAN*, *Metcalfe's law*, and *Xerox PARC*.

Metcalfe's law The value of a telecommunications network is proportional to the square of the number of users (n^2) of the system. Robert Metcalfe developed the law to describe the value of Ethernet, which he invented in early 1973. Metcalfe's law is used to describe the value of network technologies such as telephones, fax machines, the Internet, and the World Wide Web (WWW). See also *Ethernet* and *Metcalfe, Robert Melancton*.

meter (m) The basic SI unit of length, a meter is equivalent to approximately 1.094 yard, or 39.37 inches. The meter was originally determined by Napoleonic scientists at the French Academy of Sciences as one ten millionth (10^{-7}) of the distance between the North Pole and the Earth's equator through Paris, France. The meter was then recorded as the distance between two fine lines engraved on a platinum-iridium bar kept at the International Bureau of Weights and Measures in Paris. The meter is now defined as the distance traveled by light in a vacuum in $1/299{,}792{,}458$ seconds. See also *SI*.

metre See *meter*.

metric gauge The measure of the diameter, or thickness, of a conductor, metric gauge is used outside of the United States and England, where American Wire Gauge (AWG) and British Standard Gauge (BSG) are used, respectively. In the metric gauge scale, the gauge is 10 times the diameter of the wire in millimeters (mm), so a 50 gauge metric wire would be 5 mm in diameter. Note that AWG is retrogressive, i.e., the larger the number, the thinner the conductor, but metric gauge is progressive. In order to avoid confusion, metric sized wire generally is specified in millimeters rather than metric gauge. See also *AWG*, *BSG*, and *gauge*.

metro Ethernet (metropolitan Ethernet) An Ethernet LAN-to-LAN metropolitan area network (MAN) service offering that involves centrally positioning one or more gigabit Ethernet (GbE) or 10-gigabit Ethernet (10 GbE) switches in a metro area. Metro Ethernet offers the advantage of carrying all traffic in native Ethernet format, with no requirement for introducing SDH/SONET, frame relay, ATM or other Physical Layer or Data Link Layer protocols that can increase both complexity and cost, while adding overhead. See also *10GbE*, *ATM*, *Data Link Layer*, *Ethernet*, *frame relay*, *GbE*, *LAN*, *overhead*, *Physical Layer*, *protocol*, *SDH*, and *SONET*.

Metropolitan Area Exchange (MAE) See *MAE*.

Metropolitan Area Network (MAN) See *MAN*.

metropolitan Ethernet (metro Ethernet) See *metro Ethernet*.

MEO (Medium Earth Orbit or Middle Earth Orbit) A satellite or satellite constellation (i.e., system) operating at an altitude of 10,062–20,940 kilometers. Unlike a geosynchronous Earth orbiting (GEO) satellite, MEO and LEO (Low Earth Orbit) satellites do not remain in a fixed position relative to the Earth's surface, so are referred to as mobile satellite systems (MSSs), as opposed to the fixed satellite systems (FSSs) in geostatic orbit. See also *FSS*, *GEO*, *LEO*, *MSS*, and *satellite*.

MF (Medium Frequency) MF radio is in the frequency range of 300 kHz – 3 MHz and has a wavelength of 1 km – 100 m. MF radio has applications in navigation, AM radio, and mobile radio. See also *electromagnetic spectrum*, *frequency*, *Hz*, and *wavelength*.

MFA Forum A special interest group formed of the MPLS Forum, Frame Relay Forum, and ATM Forum. The MFA Forum is a not-for-profit special interest group of manufacturers, vendors, carriers and others with interests in the development and promotion of multi-vendor, multi-service packet-based networks, associated applications, and interworking solutions. The emphasis, clearly, is on MPLS, frame relay, and ATM solutions. See also *ATM*, *frame relay*, and *MPLS*.

MFJ (Modified Final Judgement (MFJ) Also known as the *Divestiture Decree*. In the United States, a negotiated settlement (1982) between the Department of Justice (DOJ) and AT&T as a modification to the 1956 Consent Decree. The MFJ forced AT&T to divest its wholly owned Bell Operating Companies (BOCs), which it later spun off to form the seven Regional Bell Operating Companies (RBOCs). Bell Communications Research (Bellcore) was established for RBOC common R&D support. Local Access and Transport Areas (LATAs) were established, with AT&T and other interexchange carriers (IXCs) permitted to provide interLATA service. The BOCs and other local exchange carriers (LECs) were granted exclusive rights to provide local and intraLATA long distance services. The MFJ established requirements for equal access, which allows a telephone subscriber to access any IXC through the LEC network to place a long distance call with equal ease, i.e., by dialing 1+. The MFJ also removed restrictions on AT&T against computer and related businesses. AT&T retained Long Lines (long distance), Bell Telephone Laboratories (R&D), and Western Electric (manufacturing). AT&T retained the embedded base of customer premises equipment (CPE). The MFJ took full effect January 1, 1984. See also *Bellcore*, *BOC*, *Consent Decree*, *CPE*, *equal access*, *IXC*, *LATA*, *LEC*, *RBOC*, and *Telecommunications Act of 1996*.

MFR (Multilink Frame Relay) A service that bundles multiple T1 local loops to provide access to frame relay at speeds between T1 and T3. MFR essentially is a form of inverse multiplexing that also offers a significant measure of redundancy. See also *frame relay*, *inverse multiplexing*, *local loop*, *T1*, and *T3*.

MFS (Maximum Frame Size) In asynchronous transfer mode (ATM), a traffic parameter describing the maximum size of a protocol data unit (PDU), or frame, supported by the network. MFS relates specifically to the guaranteed frame rate (GFR) service category. See also *ATM*, *frame*, *GFR*, *PDU*, and *traffic parameter*.

MG (Media Gateway) See *Media Gateway*.

MGC (Media Gateway Controller) See *Media Gateway Controller*.

MGCP (Media Gateway Control Protocol) A predecessor to Media Gateway Control (Megaco). See *Megaco*.

MH (Modified Huffman) A relatively simple encoding and compression algorithm used in facsimile machines that eliminates signal redundancy using a one-dimensional run-length encoding process to digitize and compress a document prior to transmission. The transmitting machine scans a document from top-to-bottom and left-to-right, sensing dots of black and white color at an interval that depends on the resolution setting. Rather than transmitting a set of bits identifying the value (black or white, grayscale, or color) of each dot of each line, the scanning machine looks for redundancy, or runs, of dots of the same value. The machine then can transmit a set of bits identifying that value and the length of the run before the value changes from black to white, then transmit the length of the run of white, and so on. The receiving machine reverses the process, decompressing the data in order to reconstruct a facsimile of the original image. Modified Huffman is a lossless compression technique, as no data is lost during the compression and decompression processes. Group III fax machines operating at 9.6 kbps transmit documents at business letter quality at a rate of approximately 30 seconds per page, using the MH compression algorithm. MH is supported by all Group III devices as the lowest common denominator, yielding a compression ratio of about 20:1 in black and white. See also *compression*, *encode*, *facsimile*, *Group III*, *Huffman coding*, and *run-length encoding*.

mho The measure of conductance, which is the reciprocal of resistance (R), as measured in ohms (Ω). Conductance is measured in mhos, which is ohms spelled backwards. (Isn't that clever? Physicists call it whimsical.)

MHS (Message Handling Service) See *X.400*.

MHSP (Message Handling Service Protocol) See *X.400*.

MHz (MegaHertz) Million (10^6) Hertz. A measure of bandwidth in an analog transmission. See also *bandwidth*, *Hertz*, and *M*.

mic (microphone) See *microphone*.

microbend An abrupt discontinuity in the core/cladding interface of an optical fiber, a microbend can be in the form of a small bulge, dent, kink, or slight axial displacement. Microbends typically are caused by trauma to the fiber during the manufacturing process when the fibers are being drawn, coated, cabled, and reeled. Microbends also can be created by trauma during the installation process if, for example, metal clips are left on too long or tie wraps (i.e., zip ties) are cinched too tight and the fiber is compressed excessively. Although a microbend does not compromise the physical integrity of the fiber, it can cause a small amount of bending loss. See also *bending loss* and *macrobend*.

microbrowser A simplified version of a browser for use in Web-enabled cellular terminals. See also *browser*, *cellular*, and *Web*.

microbusiness An enterprise (i.e., for-profit commercial business) with fewer than 10 employees. Synonymous with SOHO. See also *SME* and *SOHO*.

microcell In radio systems, an imprecise term referring to a relatively small area of coverage, of perhaps 6–8 miles in diameter. A microcell is smaller than a macrocell, but larger than a picocell. See also *cell*, *femtocell*, *macrocell*, and *picocell*.

microchip Synonymous with integrated circuit and microcircuit. See *integrated circuit*.

microcircuit Synonymous with integrated circuit and microchip. See *integrated circuit*.

Microcom Networking Protocol (MNP) See *MNP*.

Microcom Networking Protocol version 4 (MNP4) See *MNP4*.

microcomputer Synonymous with *personal computer* (PC). A computer built around a single central processing unit (CPU), a microcomputer is less capable than a midrange computer, or minicomputer, and certainly much less capable than a mainframe computer. See also *CPU*, *mainframe computer*, and *minicomputer*.

micron (μ) One one-millionth ($\frac{1}{1,000,000}$) of a meter (m). Just to put it in perspective, a human hair is approximately 5–10μ in diameter. The core of a single-mode optical fiber is typically 5–10μ in diameter, and the core of a multimode fiber is typically either 50μ or 62.5μ in diameter. It is through the core of an optical fiber that the optical signal primarily is intended to travel. See also *meter* and *nm*.

microphone (mic) A device containing a transducer for converting sound waves into electrical signals that can then be amplified, transmitted, and output through a speaker. See also *speaker* and *transducer*.

microprocessor An integrated circuit contained on a single silicon chip, a microprocessor contains the arithmetic logic unit, control unit, internal memory registers, and other vital circuitry of a computer's central processing unit (CPU). Microprocessor commonly is used interchangeably with CPU and processor. See also *CPU.*

microwave A form of radio transmission that uses ultra-high frequencies, developed out of experiments with radar (radio detecting and ranging) during the period preceding World War II. Developed by Harald T. Friis and his associates at AT&T Bell Telephone Laboratories, microwave systems are point-to-point radio systems operating in the GigaHertz (GHz) frequency range. The wavelength is in the millimeter range, which is to say that each electromagnetic cycle or waveform is in the range of a millimeter, which gives rise to the term microwave. As such high-frequency signals are especially susceptible to attenuation due to interaction with the physical matter in the atmosphere, terrestrial microwave radio beams must be tightly focused and must be amplified or repeated frequently. Microwave is a Line-of-Sight (LOS) technology as such high-frequency radio waves will not pass through solid objects of any significance. See also *frequency, frequency spectrum, Friis, LOS, point-to-point, radar, radio,* and *wavelength.*

MICS (Medical Implant Communications Service) In the United States, an ultra-low power, mobile radio service for transmitting data in support of diagnostic or therapeutic functions associated with implanted medical devices, such as cardiac pacemakers and defibrillators. MICS requires no licensing, but MICS equipment must be operated by an authorized health care professional. MICS operates in the 402–405 MHz band. The Federal Communications Commission (FCC) regulates MICS, which is in the family of personal radio services. See also *FCC* and *personal radio services.*

middle earth orbit (MEO) See *MEO.*

middleware Programs that serve as intermediaries and translators between two different computing platforms, perhaps between client workstations requesting data or programs, and servers that provide them. Middleware is used in cross-platform situations where the clients and servers run on different operating systems (OSs) or where different database file structures are used. See also *client, database, OS, server,* and *software.*

midrange computer See *minicomputer.*

Military Network (MILNET) See *MILNET.*

millimeter (mm) See *mm.*

.mil (military) Pronounced *dot mill.* The Internet generic Top Level Domain (gTLD) reserved exclusively for the United States military. This is an unsponsored domain. See also *gTLD, Internet,* and *unsponsored domain.*

MILNET (MILitary NETwork) Also known as *Defense Data Network* (DDN). A packet network formed in 1983 of ARPANET users in European and Pacific Rim continents. See also *ARPANET.*

MIME (Multipurpose Internet Mail Extensions) Extensions to the Simple Mail Transport Protocol (SMTP) used in IP-based networks for e-mail. MIME was developed to overcome SMTP's ASCII limitation, which supports only plain text, i.e., unformatted text. MIME standards, as defined in the IETF RFC 2045, include a number of types and subtypes that support a range of data formats. Those types include the following:

- **text type** for textual messages. Subtypes include plain text for 7-bit ASCII, and rich text for enhanced text formatting.
- **image type** for image files. Subtypes include Graphics Interchange Format (GIF) and Joint Photographic Experts Group (JPEG).

- **video type** for time-varying picture images. The Moving Pictures Experts Group (MPEG) subtype is defined.
- **audio type** for basic audio data at 8 KHz.
- **application type** for executable code and any data that doesn't fit neatly into any of the other types. Subtypes include octet-stream for binary data and postscript for PostScript files.
- **message type** for encapsulated messages within e-mail. Partial subtype permits a long e-mail message to be fragmented at the transmitter and reassembled at the receiver.
- **multipart type** supports the combination of multiple types into a single e-mail message.

MIMO (Multiple-Input Multiple-Output) A technique for increasing wireless bandwidth by spatial antenna diversity, MIMO is incorporated into IEEE 802.11n specifications for wireless LAN (WLAN) and 802.16 specifications for broadband wireless access (BWA), more commonly known as WiMAX. As radio signals travel from transmitter to receiver in an enclosed space, they propagate along multiple paths. The signal elements traveling a direct path along a line of sight (LOS) arrive first and strongest. Those that travel the least direct paths, having reflected off walls, floors, ceiling, potted plants, people, and other obstructions, not only arrive last, but also suffer the greatest attenuation due to absorption, diffusion, and other contributing factors. MIMO technology employs multiple spatially diverse transmit antennas to actually encourage the signals to traverse multiple paths and multiple receive antennas to extract additional information from the signals that do so. MIMO algorithms in the receive device correlate and recombine the signals, realizing diversity gain, i.e., an increase in signal strength, in the process. MIMO technology doubles the spectral efficiency. The 802.11n MIMO technology, for example, is expected to yield a theoretical maximum signaling rate of 108 Mbps, compared to the 54 Mbps yielded by the earlier 802.11g technology. See also *802.11g*, *802.11n*, *802.16*, *absorption*, *attenuation*, *bandwidth*, *BWA*, *diffusion*, *gain*, *IEEE*, *LOS*, *spatial diversity*, *spectral efficiency*, *WiMAX*, and *WLAN*.

minicom In the United Kingdom, a printing telegraph service widely used by those with hearing or speech impairments. Also generically known as TTY (TeleTYpewriter), TDD (Telecommunications Device for the Deaf) in the United States, and textphone In Europe. See also *telegraph* and *teletype*.

minicomputer Synonymous with midrange computer. A medium-size computer that is less capable than a mainframe, but more so than a microcomputer. A minicomputer is designed to handle intensive and complex computational tasks, support a large number of users, and make use of large volumes of secondary storage. Minicomputers commonly are networked and share computational tasks in a distributed processing architecture. A minicomputer often is used as a front-end processor (FEP), managing the interface between a wide area network and a mainframe, and serving to manage user access privileges, deal with security issues, and otherwise relieve the mainframe of routine tasks. See also *Internet*, *LAN*, *mainframe computer*, and *personal computer*.

minimum cell rate (MCR) See *MCR*.

minimum point of entry (MPOE) See *MPOE*.

MIS (Management Information Systems) See *IS*.

MLT (Multi-Level Transition) A technique used in 100Base-TX Ethernet to reduce the signaling rate and thereby lower the carrier frequency in consideration of attenuation issues over twisted pair cable. The line coding technique used in 100Base-TX is 4B/5B, which increases the signaling rate to 125 Mbps in support of a datastream of 100 Mbps. MLT-3 is a ternary approach that cycles through three signal levels in the pattern +V, 0V, –V and adds a scrambling step before placing the signal on the line (see Figure M-2). In combination, these intermediate steps support a signaling rate of 125 Mbps while placing the main spectral energy at a frequency of only 31.25 MHz. That low frequency results in improved signal quality and a reduced potential for interference. See also *100Base-TX*, *4B/5B*, *attenuation*, *carrier*, *frequency*, *line coding*, *signal*, *signaling rate*, and *twisted pair.*

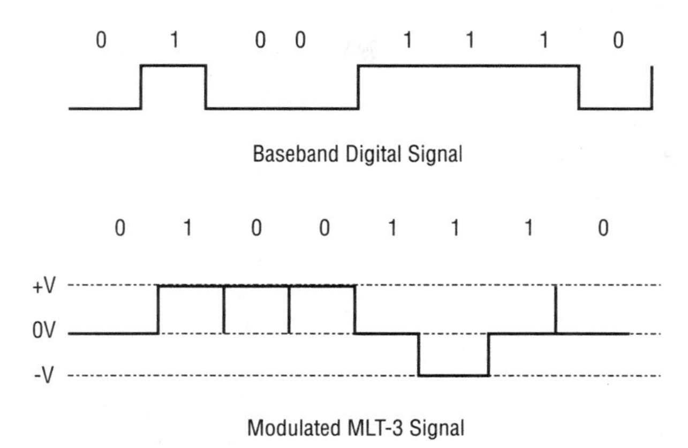

Figure M-2

mm (millimeter) One one-thousandth (10^{-3}, or $\frac{1}{1{,}000}$) of a meter. Your little fingernail is about 1 mm thick and it grows at a rate of about 1 nm (nanometer) per second. If a nanometer were scaled up to the width of your little fingernail, your fingernail would be about the size of Delaware. See also *meter.*

MMDS (Multichannel Multipoint Distribution Services) A point-to-multipoint microwave technology operating in five bands in the 2.15–2.68 GHz range in the United States and Canada and in the 3.5 GHz range elsewhere, MMDS was developed for television transmission. MMDS subsequently was tweaked for two-way Internet access applications but is bandwidth-limited at only 200 MHz, which does not compare favorably with most alternative access technologies. MMDS operates on the basis of a macrocell up to 31 miles (50 kilometers) miles in radius. The transmit antenna is placed on a hilltop, tall building, or other point with maximum lookdown in order to maximize line of sight (LOS). The antenna may be omnidirectional or may be sectorized in order to improve spectrum efficiency. MMDS has been rendered obsolete and incorporated into WiMAX. See also *bandwidth*, *Internet*, *LOS*, *macrocell*, *microwave*, *point-to-multipoint*, *spectral efficiency*, and *WiMAX.*

MMF (MultiMode Fiber) A type of optical fiber with a relatively thick inner core that allows light rays to propagate along multiple modes, or physical paths, through the fiber. The number of modes is sensitive to the core diameter, the numerical aperture (NA), and the wavelength. A core diameter that is large in relationship to the wavelength supports a large number of modes. MMF typically has a core diameter of 62.5 microns or 50 microns as illustrated in Figure M-3. The light source used in an MMF transmission system generally is either a light-emitting diode (LED) or a vertical cavity surface-emitting laser (VCSEL). LEDs operate in the 850 nm range, emit poorly focused light beams, and couple effectively to 62.5 micron MMF. VCSELs typically operate in the 1300 nm and 1310 nm ranges, emit well-focused light beams, and couple effectively to both 62.5 micron and 50.0 micron fiber. The numerical aperture mathematically describes the angle of acceptance, which defines the light-gathering ability of the fiber. These interrelated terms all establish the size of the angle at which the light source can effectively inject a signal into the fiber. The wider the angle from axial, the higher the mode, the more modes supported, the more light reaches the far end. So, a fiber optic transmission system (FOTS) employing an LED connected to a fiber with a 62.5 micron core and a relatively large numerical aperture supports a great many modes. A FOTS employing a VSCEL connected to a fiber with a 50.0 micron core and a relatively small numerical aperture supports many fewer modes of light. The latter example is preferable. There are two general categories of modes:

- **Low-order modes** are relatively straight paths that signals or signal elements travel as they propagate. As low-order modes are relatively direct, the signals encounter and interact with relatively little physical matter and attenuate relatively little. Such signals also reach a distant point in the fiber circuit relatively soon as they have traveled a short distance.
- **High-order modes** are relatively transverse paths signals or signal elements travel as they propagate through the fiber. As these paths are highly indirect, the signals encounter and interact with considerable physical matter and, therefore, attenuate to a greater extent than those propagating along low-order modes. Such signals also reach a distant point in the fiber circuit relatively late, as they have traveled a long distance.

Most MMF also is graded-index fiber, rather than step-index. Graded-index fiber compensates for high order modes by accepting errant light rays into the cladding, which comprises a great many layers of glass of slightly and progressively lower refractive index, or index of refraction (IOR). As a result, the errant signal components gradually increase in velocity and refract, or bend, back towards the core, where they rejoin other signal elements that propagated along other modes, i.e., took more or less direct paths. To some extent, graded-index fiber mitigates issues of pulse dispersion, a type of inter-symbol interference that occurs when various signal elements of various strengths overrun each other and that is especially problematic in high-bandwidth applications, as bit times (pulse lengths) are very short and pulses must be very close together. Issues of attenuation and pulse dispersion combine to limit MMF to relatively short-haul, low-speed applications such as local area networks (LANs) running at 1 Gbps or less. See also *cladding*, *core*, *graded-index fiber*, *IOR*, *LED*, *MMF*, *mode*, *numerical aperture*, *pulse dispersion*, *reflection*, *refraction*, *SMF*, *step-index fiber*, *VCSEL*, and *wavelength*.

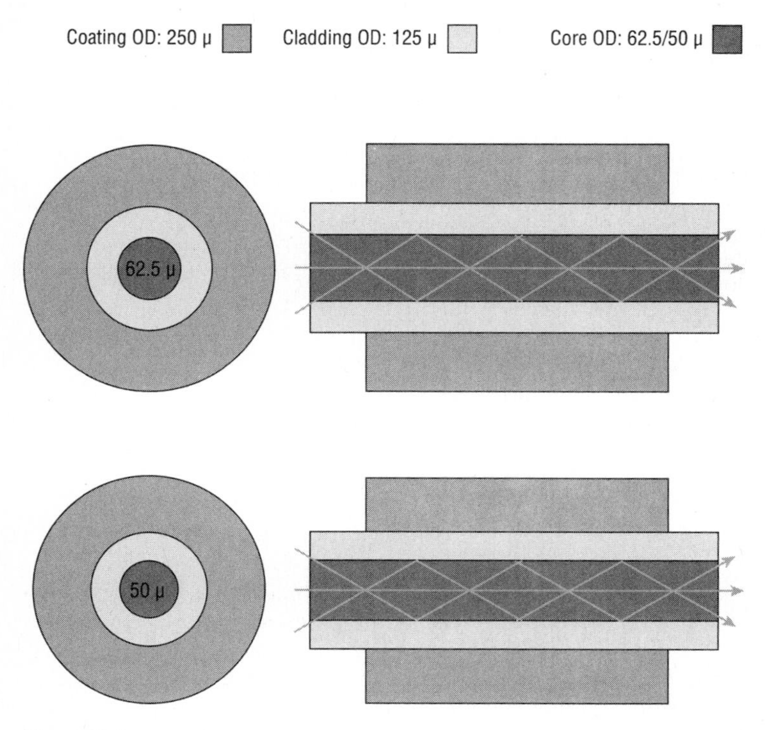

Figure M-3

MMR (Modified Modified Read) A compression algorithm used in Group III and Group IV facsimile machines operating at 14.4 kbps or better and specified in ITU-T Recommendation T.6, MMR uses a two-dimensional compression technique that permits the transmitting modem to view and consider multiple lines of data during the encoding process. At 28.8 kbps, Group III machines can transmit a page in approximately four (4) seconds using a V.34 modem employing Quadrature Amplitude Modulation (QAM). At 33.6 kbps, Group III machines can transmit a page in about three (3) seconds using a V.34bis (aka V.34+) modem employing Trellis Coding Modulation (TCM). See also *compression*, *encode*, *facsimile*, *Group III*, *Group IV*, *ITU-T*, *modem*, *QAM*, *T.6*, *TCM*, *V.34*, and *V.34bis*.

MMS (Multimedia Messaging Service) A messaging service offered on many digital cellular networks, MMS is an evolution of Short Message Service (SMS) that supports text messages with audio, image, and video attachments. See also *SMS*.

MNP (Microcom Networking Protocol) An early protocol for modem communications, MNP was among the first to include error correction. MNP subsequently was incorporated into the ITU-T V.42 Recommendation. See also *ITU-T*, *modem*, *protocol*, and *V Series*.

MNP 4 (Microcom Networking Protocol class 4) An enhancement to the MNP family of modem protocols, MNP 4 added variable packet assembly, which adjusts the packet size downward to compensate for line conditions that cause dropped packets. (*Note:* MNP supports packets as small as 64 bytes and as large as 256 bytes.) Reducing the packet size reduces the raw packet efficiency, as the ratio of payload to overhead drops, but increases throughput under such conditions, as smaller packets are more likely to transit the connection without error and, therefore, unlikely to require retransmission. See also *byte*, *MNP*, *modem*, *overhead*, *packet*, *payload*, *protocol*, and *throughput*.

.mobi Pronounced *dot mobi*. The generic Top Level Domain (gTLD) reserved for consumers and providers of mobile products and services. This domain was created in 2005 under the sponsorship of mTLD Top Level Domain, Ltd. See also *gTLD*, *Internet*, and *sponsored domain*.

mobile **1.** Easily movable. **2.** In telecommunications, able to maintain a connection while in motion. Some RF-based wireless technologies support mobile communications. Cordless telephony, for example, allows the user to establish and maintain a connection while in motion, as long as the telephone is within range of the base station (BS). Cellular telephony not only allows the user to establish and maintain a connection while in motion, as long as the mobile station (MS) is within range of a base station, but also can accomplish call hand-offs to seamlessly transfer the call between base stations as the user moves from one cell to another cell. See also *cellular*, *cordless telephony*, and *RF*. **3.** Referring to a mobile phone, or cellular telephone. In some parts of the world the slang term for such a phone is *mobile* (pronounced "MO-byle," at least in Great Britain). In other parts of the world, the slang term is simply *cell*. See also *cellular radio*.

mobile satellite system (MSS) See *MSS*.

mobile station (MS) See *MS*.

mobile telephone switching office (MTSO) See *MTSO*.

mobile traffic switching office (MTSO) See *MTSO*.

Mobitex (Mobile text) Low-speed, packet-switched wireless data networks deployed in Europe and the United States alongside cellular radio networks. Developed jointly by Ericsson and Swedish Telecom, Mobitex offers runs in Europe in the 400–450 MHz band and in the United States in the 800 and 900 MHz bands, although it will run in any band. Mobitex channel spacing of 12.5 kHz supports theoretical data transmission rates of 8 kbps through Gaussian minimum-shift keying (GMSK) modulation. Mobitex packets are up to 512 octets, including overhead, using a slotted Aloha protocol. Mobitex networks support short message service (SMS), e-mail, and Internet access. See also *cellular radio*, *channel*, *e-mail*, *GMSK*, *Internet*, *overhead*, *packet switch*, *protocol*, *slotted Aloha*, and *SMS*.

modal bandwidth The measure of the capacity of a multimode fiber (MMF) in Gbps applications, modal bandwidth is expressed in MHz/km. A MMF with a higher modal bandwidth will support data transmission at a given rate over a longer distance. Modal bandwidth is determined by the dispersion characteristics of the fiber, including both modal dispersion and chromatic dispersion. MMF core diameter (50µ vs. 62.5µ) is one of the fiber attributes that influence modal bandwidth. See also *chromatic dispersion*, *dispersion*, *MMF*, and *modal dispersion*.

modal dispersion The dispersion, or spreading, of an optical pulse in multimode fiber (MMF) due to the multiple modes, or paths, along which the signal components can propagate. Modal dispersion is a type of intersymbol interference (ISI). See pulse dispersion for more detail. See also *dispersion*, *ISI*, and *MMF*.

modal equilibrium More correctly known as *equilibrium mode distribution*. Referring to a state in which optical power is evenly distributed across all modes, i.e., physical paths in an optical fiber. See also *mode* and *optical fiber*.

modal noise **1.** Noise caused by interactions between the fiber and the connectors, modal noise results in power fluctuations at the receivers, increasing the signal-to-noise ratio (SNR) and limiting the length of the fiber link. Modal noise occurs only when high power lasers are used in conjunction with multimode fiber (MMF). See also *SNR*. **2.** Noise generated in a multimode fiber optic transmission system (FOTS) by the combination of mode-dependent attenuation and fluctuations in the distribution of optical energy, i.e., optical power density, among the guided modes or in their relative phases. Synonymous with *speckle noise*. See also *MMF* and *power density*.

mode The physical path a signal or signal component follows as it propagates through a waveguide. Some signal components travel directly through the center of the waveguide, at least theoretically, and, therefore, travel the shortest possible distance between the point at which they enter the waveguide and the point at which they exit the waveguide. Other modes take more transverse paths, striking and reflecting back and forth off of the interface between the core and cladding as they propagate through an optical fiber, for example. Low-order modes take modestly transverse paths, while high-order modes take considerably more transverse paths. Some modes at the transmitter can be so transverse as to strike the core-cladding interface at less than the critical angle and, therefore, penetrate the interface and be permanently lost in the cladding. Figure M-4 illustrates the differences between these paths. See also *cladding*, *core*, *critical angle*, and *waveguide*.

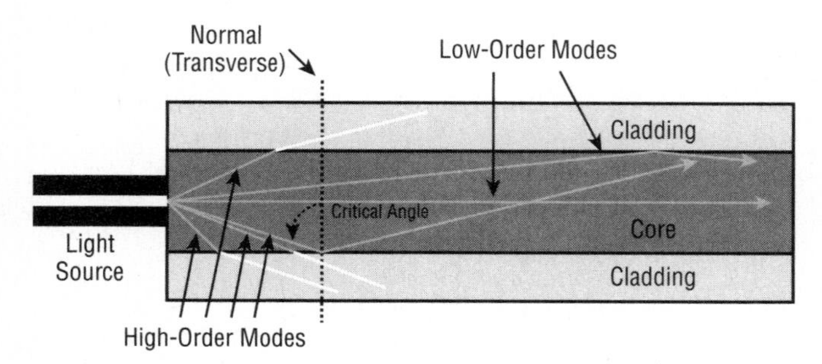

Figure M-4

mode field diameter A measure of the diameter of the area in which an optical signal propagates through a single-mode fiber (SMF). Although most of the signal is confined to the core, some of the signal travels in the cladding, so the mode field diameter is of greater significance than the core diameter. See also *cladding*, *core*, *optical fiber*, and *SMF*.

mode filter A device that attenuates high-order modes in an optical fiber. As high-order modes can be lost in the cladding and, therefore, serve no useful purpose, a mode filter essentially strips them out of the transmitted signal. See also *attenuation*, *high-order mode*, and *optical fiber*.

modem (modulator/demodulator) **1.** A device that comprises both a modulator that changes a signal in some way in the forward direction and a demodulator that changes the signal back to its original form in the backward direction, essentially reversing the modulation process. Modems operate in balanced and symmetrical pairs, with one at each end of the communications circuit and with both having the same capabilities, at least at a minimum level. There are many types of devices characterized as modems, including cable modems, conventional modems, ADSL modems, ISDN modems (terminal adapters, or TAs), line drivers, and short haul modems. See also *ADSL*, *cable modem*, *line driver*, *short haul modem*, and *TA*. **2.** A conventional modem is a signal conversion device that interfaces a digital device to an analog circuit or channel. On the transmit side of the connection, a modem accepts an incoming digital signal and modulates (i.e., changes or varies) the characteristics of an electromagnetic waveform in some way to represent that signal over an analog carrier. The modulation technique generally involves amplitude modulation (AM), frequency modulation (FM), phase modulation (PM), or some combination. On the receive side of the connection, a modem with matching capabilities accepts the modulated signal over the analog carrier and demodulates the signal to extract the information and recreate the original digital signal. Many modems are capable of operating in full duplex, simultaneously modulating signals as they transmit them and demodulating signals as they receive them. See also *AM*, *FM*, and *PM*.

Modem on Hold (MoH) A feature of V.92 modems that puts a data session on hold when it detects a voice call, either incoming or outgoing, through a call waiting indication, and gracefully resumes that session when the voice call is terminated. Thereby, V.92 allows a single analog line to be used for both voice and data. MoH requires that the line be equipped with call waiting. See also *call waiting*, *hold*, *modem*, and *V Series*.

Modified Final Judgement (MFJ) See *MFJ*.

Modified Huffman (MH) See *MH*.

Modified Modified Read (MMR) See *MMR*.

Modified Read (MR) See *MR*.

modulate **1.** Change or vary in some way. **2.** In acoustics, modulation involves varying the pitch, tone, or volume of an audio signal, such as the human voice. **3.** In physics, modulation involves varying the characteristics of an electromagnetic waveform, generally by varying the amplitude, frequency, or phase, or some combination. **4.** In telecommunications, modems perform signal modulation and demodulation processes to encode information from digital devices onto signals carried by analog circuits. Digital Service Units (DSUs) and various other devices modulate carrier waves to place digital signals on digital circuits. See also *AM*, *FM*, *modem*, *PM*, and *PSK*. **5.** In telecommunications, codecs code analog signals into digital format prior to transmission over a digital circuit, and decode them on the receiving end of the connection. Codecs use a variety of modulation techniques, including adaptive differential pulse code modulation (ADPCM), differential pulse code modulation (DPCM), and pulse code modulation (PCM). See also *ADPCM*, *analog*, *codec*, *digital*, *DPCM*, and *PCM*.

modulo **1.** An integer that leaves the same remainder when it is the divisor of two other integers. For example, 6 modulo 4 = 2 and 14 modulo 4 = 2. In other words, 6 divided by four results in a remainder of 2, and 14 divided by 4 leaves a remainder of 2. **2.** In telecommunications, referring to the window size, which identifies the maximum number of bytes, frames, or packets that a device can transmit without an acknowledgement from the receiver. As an example, the X.25 protocol has a windowing mechanism that specifies the number of packets that one router can transmit to another without receiving an acknowledgement. TCP has a similar mechanism that is expresses in bytes. Windowing serves as flow control and error control mechanisms. Modulo 8, for example, allows the transmission of 7 datasets without an acknowledgement, but requires an acknowledgement after the 8th dataset. See also *error control, flow control, TCP*, and *X.25.*

MoH (Modem on Hold) See *Modem on Hold.*

MOH (Music On Hold) See *music on hold.*

MOM (Manager Of Managers) A high-level network management system (NMS) that gathers and correlates alarms and alerts from multiple element management systems (EMSs). See also *EMS* and *NMS.*

monomode fiber Synonymous with single-mode fiber (SMF). See *SMF.*

monopoly A condition in which a single company is the exclusive manufacturer of a product or provider of a service or controls an entire industry, thereby allowing the company to fix prices. A monopolistic condition can arise when a company invents and patents a product that is so compelling that an entire industry builds up around it. If the company continues to develop and patent further versions of the product, it can extend that monopolistic condition, perhaps acquiring or overwhelming competitors. Natural monopoly, a concept developed by political economist John Stuart Mill (1806–1873), refers to utility services that are so capital-intensive and right-of-way intensive that it just doesn't make sense to have more than one provider. Natural monopolies include transportation infrastructure such as roads, bridges, ferries, subways, and railroads, as well as utility infrastructure such as water, natural gas, electricity, sewer, cable television, and wireline telephone. Well, cable television and wireline telephone infrastructure used to be considered natural monopolies, but it's not quite that simple any longer. See also *Graham-Willis Act.*

monotonic **1.** Unvarying in quality or characteristics, such as pitch, tone, style, or color. **2.** In mathematics, steadily increasing or decreasing in value. See also *oscillator.*

Moore, Gordon E. (1929–) A co-founder of Intel, Moore is best known for his 1965 prediction that the number of transistors on an integrated circuit would roughly double every year for the foreseeable future. Moore updated the prediction in 1975, stating that the number of transistors on an integrated circuit for minimum component cost doubles approximately every 2 years. The prediction, known as Moore's Law, has proved largely true for the last 40+ years. A chemist and physicist by training, Moore retired as Chief Executive Officer (CEO) of Intel in 1987. See also *Moore's Law.*

Moore's Law The observation of Gordon E. Moore (1929–), co-founder of Intel, that the complexity of semiconductor components had roughly doubled per year since the first prototype microchip and that the rate would remain nearly constant for at least 10 years. Moore stated the observation in an article entitled "Cramming more components onto integrated circuits" that appeared in the April 19, 1965 edition of *Electronics* magazine. In the article, Moore predicted that "With unit cost falling as the number of components per circuit rises, by 1975 economics may dictate squeezing as many as 65,000 components on a single chip." In 1975, Moore updated the prediction, stating that the number of transistors on an integrated circuit for minimum component cost doubles approximately every 2 years. As a point of interest, the first microprocessor had 2,200 transistors. In 2007, Intel placed more than 1,700,000,000 transistors on its Itanium chip.See also *Moore, Gordon E.*

Morse code The first widely accepted standard coding scheme for digital data communications. Morse code was invented by Samuel Morse sometime prior to 1844 for use with the electric telegraph. Friedrich Clemens Gerke invented the International Morse Code in 1848 out of necessity, as some of the spaces in letters created difficulty in radiotelegraphy. International Morse code was standardized by the International Telegraph Union (ITU) in 1865 and was widely used in radiotelegraphy through the early twentieth century. Morse code was the primary communication code for many years, until Emile Baudot invented the Baudot Distributor in the 1870s. International Morse code remains widely used by amateur radio operators, or hams, although proficiency is no longer required. Morse code uses series of short and long marks in the form of dots (short marks known as *dits* to radio operators) and dashes (long marks, or *dahs*), with spaces between them, to represent letters, numbers, punctuation marks, and procedural signals (prosigns). The length of the spaces varies, with a short space between dots and dashes within a character, a longer space between characters, an even longer space between words, and a yet longer space between sentences. In order to speed transmission, the fewest number of dots and dashes represent commonly used letters (e.g., E is •, T is —, A is • —). Commonly used words are abbreviated (e.g., Calling is abbreviated CG, or — • — • — — •, as are commonly used phrases (e.g., Love and Kisses is abbreviated 88, or — — — • • — — — • •. Table M-1 provides the International Morse Code for English letters, numbers, and select punctuation marks and prosigns.

Table M-1: International Morse Code

Letter	*Code*	*Letter*	*Code*	*Number*	*Code*	*Punctuation, Prosigns*	*Code*
A	• —	N	— •	1	• — — — —	Period (.)	• — • — • —
B	— • • •	O	— — —	2	• • — — —	Comma (,)	— — • • — —
C	— • — •	P	• — — •	3	• • • — —	Question (?)	• • — — • •
D	— • •	Q	— — • —	4	• • • • —	Colon (:)	— — — • • •
E	•	R	• — •	5	• • • • •	Semicolon (;)	— • — • — •
F	• • — •	S	• • •	6	— • • • •	Hyphen (-)	— • • • • —
G	— — •	T	—	7	— — • • •	Dollar ($)	• • • — • • —
H	• • • •	U	• • —	8	— — — • •	At Sign (@)	• — — • — •
I	• •	V	• • • —	9	— — — — •	Stop	• — • — •
J	• — — —	W	• — —	0	— — — — —	Wait	• — • • •
K	— • —	X	— • • —			Invitation to Transmit	— • —
L	• — • —	Y	— • — —			Received	• — •
M	— —	Z	— — • •				

Note: The @ sign, a combination of a and c, was added in 2004 in order that telegraphers could send e-mail addresses. There are extensions to International Morse Code to accommodate letters with diacritical marks used in non-English alphabets. See also *dash*, *dot*, *ham*, *ITU*, and *telegraph*.

Morse, Samuel Finley Breese (1791–1872) Generally recognized as the inventor of the electric telegraph and the Morse code cipher. Although not actually the inventor of the original electric telegraph, Morse partnered with Leonard Gale and Alfred Vail, who improved on Morse's early telegraph models, and in 1837 filed for a patent that was finally granted in 1840. Perhaps just as importantly, Morse is credited with developing the binary telegraphic cipher known as Morse code, although many claimed at the time that Vail actually invented the code. In 1844, Morse secured funding for the first intercity telegraph line in the United States, which ran from the railway depot in Baltimore, Maryland, to the Supreme Court Room in the United States Capitol in Washington, D.C. The first message sent over that line was the biblical phrase "What hath God wrought!," which was chosen by Annie Elllsworth, the daughter of a classmate at Yale University. See also *Morse code* and *telegraph*.

MOS (Mean Opinion Score) See *P.800*.

Mosaic An Internet browser that provides a consistent user interface available in versions to support Macintosh, Microsoft Windows, and Unix–X Windows. Mosaic can be used over dedicated or dial-up connections. In dial-up mode, the Internet service provider (ISP) must support either Serial Line Internet Protocol (SLIP) or Point-to-Point Protocol (PPP). Mosaic enables the easy browsing of WWW resources through menus that support hypertext. Selected files can include audio and graphics, both of which can be viewed without the requirement to download the subject file. Mosaic was developed by a team led by Marc Andreessen at the National Center for Supercomputing (NCSA) at the University of Illinois Urbana-Champaign campus. Mosaic technology is licensed by the NCSA for commercial application. Spyglass, Inc. licensed well over 12 million copies of Mosaic to IBM, DEC, and others who intended to resell the company's Enhanced Mosaic, which includes security mechanisms based on Secure Hypertext Transport Protocol (S-HTTP). Ultimately, Spyglass Mosaic was licensed to Microsoft, where it formed the basis for Internet Explorer (IE). Mosaic also is known as *NCSA Mosaic. Note:* Marc Andreessen later developed the Netscape Navigator browser, as well. See also *browser, hypertext, Internet, Internet Explorer, ISP, Mozilla, Netscape Navigator, PPP, S-HTTP, SLIP,* and *WWW.*

mote From Old English mot, and akin to Middle Dutch and Fris meaning sand or grit. **1.** A speck of dust or other tiny particle. **2.** In telecommunications, a wireless sensor so tiny as to compare to a speck of dust. About the size of a grain of sand, a mote comprises sensors, a processor, a bidirectional antenna, and a power supply. Also known as smart dust, future applications for motes are many. The military is developing motes that can be spread from the air to gather and transmit information about enemy troop movements. Civil engineers intend to embed motes in concrete to monitor and report on the condition of bridges and roadways. Spread around a warehouse housing weapons, financial records, or other sensitive materials, motes can be used to sense and report security violations. The ZigBee Alliance is involved in the development of standards for mote mesh networking. See also *ZigBee* and *ZigBee Alliance.*

Motion-JPEG (Motion Joint Photographic Experts Group) A version of JPEG designed specifically for use in editing of digital video. See also *JPEG.*

motion picture quality Referring to video quality at a frame rate of 24 frames per second (fps) or more. See also *broadcast quality* and *frame rate.*

mouse A palm-sized computer navigation device that enables a user to move it about on a flat surface in order to move a cursor on the monitor. The user can position the cursor over an area of text or a navigation button and click on objects through the use of one or more control buttons on the mouse in order select items or invoke commands.

move, add and change (MAC) See *MAC.*

Moving Picture Experts Group (MPEG) See *MPEG.*

Mozilla Originally a code name for Netscape Navigator, Mozilla is a contraction of Mosaic killer, referring to the hope that it would unseat Mosaic as the top Web browser, and Godzilla, referring to the fictional monster of Japanese science fiction movies. Mozilla now refers to an open source application suite based on the Netscape Navigator source code, which was released by Netscape in 1998 under an open source license. See also *browser, Firefox, Mosaic, Netscape Navigator,* and *Web.*

MP (Multipoint Processor) See *MCU.*

MPEG (Moving Pictures Expert Group) A group of standards for encoding and compressing audiovisual information such as movies, video, and music. MPEG is asymmetric in nature, as the compression process is time consuming and processor-intensive, whereas the decompression process is rapid and involves relatively inexpensive equipment. MPEG compression is as high as 200:1 for low-motion video of VHS-quality, and broadcast quality can be achieved at 6 Mbps. Audio is supported at rates from 32 kbps to 384 kbps for up to two stereo channels. MPEG specifies lossy compression in the form of discrete cosine transform (DCT). MPEG is a joint technical committee of the International Standards Organization (ISO) and the International Electrotechnical Commission (IEC). See also *asymmetric, broadcast quality, compression, DCT, encode, IEC, ISO,* and *lossy compression.*

MPEG-1 The ISO/IEC International Standard (IS 11172) for storage and retrieval of moving pictures and audio on storage media such as Compact Disc (CD). MPEG-1 provides VHS quality at 1.544 Mbps and is compatible with single-speed CD-ROM technology. MPEG-1 integrates synchronous and isochronous audio with video, and permits the random access required by interactive multimedia applications such as video games. Intended for limited-bandwidth transmission, it provides acceptable quality and output compatible with standard televisions. MPEG-1 supports video compression of about 100:1. MPEG-1 standards are the basis for the MP3 (MPEG-1 Audio Layer 3) audio data encoding system. See also *audio, bandwidth, compression, encode, IEC, ISO, isochronous, MP3, MPEG, synchronous,* and *video.*

MPEG-2 The ISO/IEC International Standard (IS 13818) for digital television (DTV). MPEG-2 was conceived as an encoding and compression standard for broadcast television based on interlaced scanning of images at 720 × 480 pixels at a rate of 30 frames per second (fps). Although MPEG-2 supports compression rates as high as 200:1, it is much more bandwidth-intensive (4–100 Mbps) than MPEG-1. MPEG-2, however, provides much better resolution and image quality. MPEG-2 is used in direct broadcast satellite (DBS) services and is the compression technique of choice in a convergence scenario. MPEG-3, designed for high definition television (HDTV) application, was folded into MPEG-2. See also *bandwidth, broadcast television, compression, convergence, DBS, DTV, fps, HDTV, IEC, interlaced scanning, ISO, MPEG, MPEG-1,* and *resolution.*

MPEG-3 The ISO/IEC international standard for high definition television (HDTV), MPEG-3 was folded into MPEG-2 in 1992. See also *HDTV, IEC, ISO,* and *MPEG-2.*

MPEG-4 The ISO/IEC International Standard (IS 14496) for multimedia applications. MPEG-4 is a low bit-rate standard for encoding and compression intended for application in broadcast television, videophones, and mobile phones and other small handheld devices. MPEG-4 is designed for IMT2000 wireless applications at rates of up to 384 kbps upstream and 2 Mbps downstream. MPEG-4 deals with the coded representation of audiovisual objects, both natural and synthetic, and their multiplexing and demultiplexing for transmission, playback, and storage. See also *compression, downstream, encode, IEC, ISO, MPEG, multimedia, multiplex,* and *upstream.*

MPEG-7 The ISO/IEC international standard officially known as the *Multimedia Content Description Interface*, MPEG-7 is intended to be the content representation standard for multimedia information search, filtering, management, processing, and retrieval. MPEG-7 essentially is a metadata standard based on XML (eXtensible Markup Language) for describing multimedia content features in order that one can easily search for multimedia content on the Web. See also *MPEG, multimedia, Web,* and *XML.*

MPEG-21 The ISO/IEC international standard officially is known as the *Multimedia Framework*, is an ongoing effort to determine how various multimedia components fit together and to identify new multimedia infrastructure standards that may be required. MPEC-21 also deals with issues of content identification, description, and security. In large part, the focus of MPEG-21 is on the protection of intellectual property through security mechanisms designed to prevent unauthorized access and modification of multimedia content. See also *intellectual property, MPEG,* and *multimedia.*

.mpg The file extension for video and audio data encoded in the Moving Picture Experts Group (MPEG) format. See also *MPEG.*

.mp2 The file extension for audio data encoded in the Moving Picture Experts Group (MPEG) Audio Layer 2 format. See also *MPEG-2.*

.mp3 The file extension for audio data encoded in the Moving Picture Experts Group (MPEG) Audio Layer 3 format. See also *MPEG-3.*

MPI (MultiPath Interference) See *multipath fading.*

MPI specter (MultiPath Interference specter) Referring to the ghosting effect, which occurs when the multiple paths of a television (TV) video signal are relatively long and when some reflected signals, therefore, arrive on a significantly delayed basis in a phenomenon known variously as multipath fading and delay spread. This ghosting effect is particularly evident in poorly installed coax-based CATV systems, and in traditional broadcast radio TV reception in mountainous areas. See also *broadcast, CATV, coaxial cable, delay spread, ghosting, multipath fading, path, radio, TV,* and *video.*

MPLS (MultiProtocol Label Switching) Defined by the IETF in RFC 2702 (1999) as a label-swapping framework with Network Layer (Layer 3) routing, MPLS integrates Data Link Layer (Layer 2) information about network links into Layer 3 routing logic. MPLS enables routers to make packet forwarding decisions very quickly on the basis of short labels, rather than making complex routing decisions after analyzing lengthy packet headers. MPLS is designed to work through routers at even higher speed than ATM switches, while realizing much of the flexibility of an Internet Protocol (IP) network. MPLS works on the basis of forwarding equivalence classes (FECs) and flows. A flow consists of packets between common endpoints identified by features such as network addresses and port numbers. An FEC is a class of packets, all of which are treated the same in terms of destination, priority level, and so on. At the ingress edge of the carrier network, a label edge router (LER) identifies the flow based on the IP header, the interface through which the packet arrives, the packet type (unicast, multicast, or anycast), or perhaps information in the IPv4 Type of Service (ToS) field. The LER attaches a 32-bit MPLS header that includes a 20-bit label, or tag, to that packet and to each subsequent packet of the flow.

The LER uses a Label Distribution Protocol (LDP) to distribute the labels or tags to each intervening Label Switching Router (LSR) in the network core, identifying the treatment that should be afforded all packets in the flows on that particular Label Switched Path (LSP). If the traffic engineering options are exercised, traffic is balanced between optimum and non-optimum paths, and congestion is minimized. Otherwise, the traffic takes the same paths that IP packets would take, as MPLS nodes use IP routing protocols, such as the Border Gateway Protocol (BGP), Open Shortest Path First (OSPF), and Routing Information Protocol (RIP) to distribute the labels. From edge-to-edge through the core of the network, each LSR makes note of the incoming port number and analyzes the label associated with each packet in order to select the appropriate Label Switched Path (LSP) over which the packet is to be forwarded on its way to the next LSR. The LSR then switches the existing label for a new one, and forwards it. Thereby, and through a series of links, the end-to-end path is set up and maintained for a given traffic flow. The more complex processes of complete header analysis and routing table lookup are performed only at the ingress edge of the network. In the core of the network, only the abbreviated MPLS label is analyzed in order to make a relatively simple and straightforward packet forwarding decision. All in all, MPLS simplifies the routing process and reduces latency. At the egress LER, the tag is stripped away, as it no longer has any purpose.

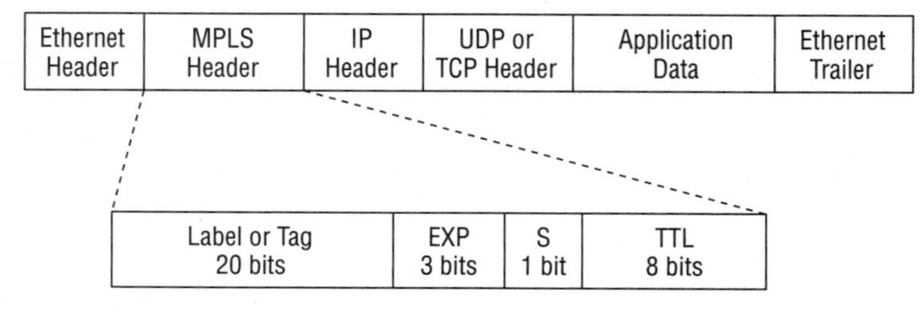

Figure M-5

The structure of the 32-bit MPLS header, as illustrated in Figure M-5, is as follows:

- **Label:** 20 bits matching the packet to the LSP.
- **Experimental (EXP):** 3 bits indicating the precedence, or packet queuing priority, for class of service (CoS) purposes.
- **Stack (S):** The stacking bit is set at 1 to indicate the last (i.e., innermost) MPLS header in a stack of headers. Outer tags carry a 0 bit in this position. MPLS virtual private networks (VPNs) involve hierarchical routing logic that requires multiple headers. As many as four MPLS headers can be contained within a stack.
- **Time To Live (TTL):** Copied from the IP TTL. The TTL is a hop count, with a default value of 64.

MPLS is based on several vendor-specific protocols, including Cisco's Tag Switching, Ipsilon's IP Switching, and IBM's ARIS technology. See also *anycast*, *ATM*, *BGP*, *congestion*, *CoS*, *Data Link Layer*, *endpoint*, *FEC*, *flow*, *header*, *hop*, *IETF*, *IP*, *IPv4*, *IPv6*, *label*, *latency*, *LDP*, *LER*, *link*, *LSP*, *LSR*, *multicast*, *Network Layer*, *OSPF*, *packet*, *port*, *queue*, *RIP*, *router*, *routing*, *ToS*, *traffic engineering*, *TTL*, *switch*, *unicast*, and *VPN*.

MPLS Forum A not-for-profit special interest group of manufacturers, vendors, carriers and others with interests in the development and promotion of Multiprotocol Label Switching (MPLS) technology. The MPLS Forum merged with the ATM Forum and Frame Relay Forum to form the MFA Forum. See also *MFA Forum* and *MPLS*.

MPOA (MultiProtocol Over ATM) A specification (July 1997) from the ATM Forum (now merged into the MFA Forum) designed to enhance LAN Emulation (LANE) by enabling interELAN communications without the intervention of a router, and the associated packet delay. MPOA provides high-performance, scalable routing functionality over an ATM platform, expanding on LANE, Classical IP over ATM (RFC 1577), and the IETF's Next Hop Resolution Protocol (NHRP) in order to create a standardized notion of a virtual router within an ATM network. See also *ATM*, *ATM Forum*, *Classical IP over ATM*, *ELAN*, *IETF*, *LANE*, *MFA Forum*, *NHRP*, *packet*, *router*, and *virtual*.

MPOE (Minimum Point of Entry) The closest logical and practical point within the customer domain for the placement of a demarcation point (demarc), which sets the boundary of responsibility between the PSTN carrier and the customer. In a high-rise office building, for example, the MPOE typically is defined as a point of the entrance cable 12 inches from the inside wall. See also *carrier*, *demarc*, and *PSTN*.

MPPP (Multilink Point-to-Point Protocol) Based on Internet Engineering Task Force (IETF) RFC 1990, MPPP supports the linking, or bonding, of multiple links between systems to increase the available bandwidth. MPPP applications include narrowband ISDN (N-ISDN) and analog modem connections. See also *AODI*, *bonding*, *IETF*, *modem*, and *N-ISDN*.

MR (Modified Read) An optional compression algorithm used in some Group III facsimile machines, MR scans and compresses the first line using Modified Huffman (MH) compression. Subsequent lines are scanned and compared to the first line, and only the differences (deltas) are encoded and transmitted. This process continues for some predetermined number of lines in a group, at which point the process is reset and the first line in another group is encoded using MH, and so on. MR is particularly effective if there are few differences between lines in a document. See also *compression*, *facsimile*, *Group III*, and *MH*.

MS (Mobile Station) A wireless portable terminal device used in a radio network designed to establish and maintain connections between mobile terminals or between mobile terminals and one or more fixed base stations (BSs) while in motion. Such networks include cellular radio and specialized mobile radio (SMR). See also *BS*, *cellular radio*, and *SMR*.

MS-DOS (MicroSoft Disk Operating System) The single-tasking, single-user operating system (OS) with a command-line user interface, released for use in IBM PCs in 1981. See also *OS* and *PC.*

MSF (Multiservice Switching Forum) A global association of manufacturers and service providers working to develop and promote interoperable multiservice switching systems based on an open architecture. The MS Forum, the European Telecommunications Standards Institute (ETSI), and the Alliance for Telecommunications Industry Solutions (ATIS) have adopted the IP Multimedia Subsystem (IMS) as the foundation for their next-generation infrastructure strategies. See Appendix A for contact information. See also *architecture*, *ATIS*, *ETSI*, and *IMS.*

MSO (Multiple System Operator) An operator of multiple CATV systems. The systems generally are standalone CATV islands rather than being interconnected. See also *CATV.*

MSP (Managed Service Provider) A company that delivers and manages network-based services, applications, and equipment for a fee. An MSP may load a company's application data on its servers, customizing it as necessary, and operating the service at a remote data center. An MSP might manage a company's entire wide area network (WAN), perhaps including dedicated leased lines, a frame relay network, and an IP-based virtual private network (VPN). See also *ASP*, *frame relay*, *IP*, *server*, and *VPN.*

MSS (Mobile Satellite System) A constellation (i.e., system) of satellites placed at altitudes and in orbits such that they move around the Earth at a different speed than the Earth rotates on its axis and, therefore, appear to be in motion. The MSS classification includes LEO (Low Earth Orbit) and MEO (Middle Earth Orbit) systems. An MSS is in contrast to an FSS (Fixed Satellite System), in which the satellites appear to be fixed in space due to their positioning in geosynchronous Earth orbit (GEO). See also *FSS*, *GEO*, *LEO*, *MEO*, and *satellite.*

MTBF (Mean Time Between Failures) An indicator of component or system reliability based on the known failure rate calculated as the average period of time, expressed in hours, between total breakdowns.

MTTR (Mean Time To Repair) An indicator of the performance of service personnel based on the known elapsed period of time required to perform corrective maintenance and restore a component or system to operational status.

MTS (Message Telecommunications Service) Synonymous with Direct Distance Dialing (DDD). See *DDD.*

MTS/WATS Decision (Message Telecommunications Service/ Wide Area Telecommunications Service Decision) In the United States, the Federal Communications Commission (FCC) decision (1978) that permitted MCI and others to offer switched MTS and WATS voice services in competition with AT&T. See also *FCC*, *MTS*, and *WATS.*

MTSO (Mobile Telephone Switching Office or Mobile Traffic Switching Office) In a cellular radio network, a switch that serves to interconnect fixed base station (BS) antennas and, thereby, to interconnect radio cells and the mobile stations (MSs) within them. See also *antenna*, *BS*, *cell*, *cellular radio*, *MS*, and *switch.*

MTU **1.** Multi-Tenant Unit. An office building or mixed-use (i.e., commercial/residential) building housing more than one entity. An MTU has considerable implications for local loop deployments, particularly with respect to passive optical network (PON). See also *local loop* and *PON.* **2.** Maximum Transmission Unit. In Internet Protocol version 6 (IPv6), all networks must support a packet of at least 1280 octets. The recommendation, however, is that all networks be configured to support an MTU of 1500 octets or greater in order to support encapsulation of Ethernet payloads without incurring fragmentation. See also *encapsulate*, *Ethernet*, *fragmentation*, *IPv6*, *octet*, *packet*, and *payload.*

mu (μ) The Greek letter μ, used to denote one-millionth, as in μm, a micrometer, or micron. See *micron* and *mu-law*.

MUA (Mail User Agent) e-mail client software responsible for effecting compatibility with the e-mail server, which provides message storage and transport through a Mail Transfer Agent (MTA). See also *client*, *e-mail*, *server*, and *software*.

mu-law (μ-law) A voice companding technique specified in the ITU-T G.711 Recommendation for pulse code modulation (PCM). This technique, which is used primarily in the North American and Japanese digital hierarchies (i.e., T-carrier and J-carrier, respectively), converts 14-bit linear PCM samples into 8-bit compressed samples. A-law is a similar, but not exactly compatible, companding technique used in the European hierarchy (i.e., E-carrier). In an international circuit, if one national network uses A-law and the other uses mu-law, a gateway is used to resolve the incompatibility. In the absence of a gateway, the voice signal is intelligible, if a bit odd-sounding. See also *A-law*, *companding*, *E-carrier*, *G.711*, *gateway*, *ITU-T*, *J-carrier*, *PCM*, and *T-carrier*.

multicast A transmission mode in which a signal or packet is sent to multiple receivers, but not all receivers on a network. An e-mail distribution list is an example of multicast transmission. See also *broadcast*, *multicast address*, *packet*, *signal*, and *unicast*.

multicast address In Internet Protocol version 6 (IPv6), an address that identifies a group of nodes, each of which can belong to multiple groups. A packet sent to a multicast group is delivered to every interface in the group. Multicast addresses begin with the prefix 11111111. In IPv4, a multicast address is any Class D address. See also *broadcast*, *interface*, *IPv4*, *IPv4 address*, *IPv6*, *IPv6 address*, *node*, and *packet*.

Multichannel Multipoint Distribution Services (MMDS) See *MMDS*.

multi-dwelling unit (MDU) See *MDU*.

multi-level transition (MLT) See *MLT*.

multilink frame relay (MFR) See *MFR*.

Multilink Point-to-Point Protocol (MPPP) See *MPPP*.

Multilink PPP (Multilink Point-to-Point Protocol) A variation on PPP that allows the combination of multiple links to increase bandwidth between two nodes, in a PPP version of inverse multiplexing. Multilink PPP is defined in Internet Engineering Task Force (IETF) RFC 1990. See also *bandwidth*, *IETF*, *IMoATM*, *inverse multiplexing*, *link*, *MLFR*, and *PPP*.

multilocation extension dialing In the advanced intelligent network (AIN) specifications, a virtual private network (VPN) service of the public switched telephone network (PSTN) that provides for network routing of calls based on abbreviated numbers. This VPN service resembles a coordinated dialing plan in a networked PBX environment. See also *AIN*, *coordinated dialing plan*, *PBX*, *PSTN*, and *VPN*.

multimedia A method of communications incorporating a combination of media, such as audio, video, text, graphics, and whiteboarding, perhaps on an interactive basis. See also *Smell-O-Vision*.

multimedia conference A call involving two or more parties and incorporating not only voice and video, but also other media such as text, graphics, and whiteboarding, perhaps on an interactive basis. See also *collaborative computing* and *whiteboarding*.

multimedia messaging **1.** Synonymous with integrated messaging and unified messaging. See *unified messaging*. **2.** Multimedia Messaging Service (MMS). See *MMS*.

Multimedia Messaging Service (MMS) See *MMS*.

multimode fiber (MMF) See *MMF*.

multipath fading Also known as *multipath interference* (MPI). Signal attenuation and distortion due to multipath propagation. Wireless radio or optical signals bounce off of physical obstructions they encounter between a transmitter and a receiver. Those signal elements that travel the most direct routes not only arrive soonest, but also suffer less absorption and diffusion attenuate the least and, therefore, are the strongest. Those that travel the least direct routes arrive last and are weakest. In broadcast television and poorly installed cable television, ghosting is the result of multipath fading. In broadcast television and poorly installed cable television, MPI specter, or ghosting, is the result of multipath fading. Signals that travel different paths but arrive at approximately the same time can cancel each other. All of these factors contribute to multipath fading. See also *attenuation*, *diffusion*, *distortion*, *ghosting*, *MPI specter*, *multipath absorption*, and *propagation*.

multipath interference (MPI) See *multipath fading*.

multipath propagation Referring to the fact that wireless radio or optical signals take multiple physical paths between transmitter and receiver as they bounce off of various physical obstructions. See also *propagation*.

multiple-input multiple-output (MIMO) See *MIMO*.

multiple system operator (MSO) See *MSO*.

multiplexer (mux or MUX) See *MUX*.

Multipoint Controller (MC) See *MCU*.

Multipoint Control Unit (MCU) See *MCU*.

Multipoint Processor (MP) See *MCU*.

Multiprotocol Label Switching (MPLS) See *MPLS*.

Multiprotocol Lambda Switching The term once applied to what is now known as *Generalized MPLS* (GMPLS). See also *GMPLS*.

MultiProtocol over ATM (MPOA) See *MPOA*.

Multipurpose Internet Mail Extensions (MIME) See *MIME*.

multirate ISDN A feature of ISDN primary rate access (PRA) and primary rate interface (PRI) that aggregates or bonds multiple bearer (B) channels to yield a transmission rate in excess of the single-channel rate of 64 kbps. The resulting channels are known has *high-speed (H) channels*, which are specified at a number of levels. Multirate ISDN also is known as *bonding*, *channel aggregation*, and *Nx64*. See *bonding*, *H channel*, *ISDN*, *PRA*, and *PRI*.

Multiservice Switching Forum (MSF) See *MSF*.

multislot service class In General Packet Radio (GPRS) cellular radio networks, the multislot service class determines the number of time slots in each direction, with each time slot supporting a theoretical transmission rate of 21.4 kbps. Of the 12 classes, the simplest is service class 1, which supports one timeslot in each direction. The most capable is service class 12, which supports four time slots in each direction. See also *cellular radio*, *GPRS*, *time slot*, and *transmission rate*.

multi-step fiber A type of step-index optical fiber comprising multiple layers of cladding with sharp steps in the index of refraction (IOR) at the boundaries. Multi-step fiber compounds the effect of step-index fiber. See also *cladding*, *IOR*, and *step-index fiber*.

multi-tenant service An optional PBX software feature package that allows a single PBX to serve multiple tenants, or user organizations. Logical software partitions allow each entity to have its own attendant console, trunk groups, and blocks of telephone numbers. Thereby, each organization has a unique identity through a logical PBX that is part of a single physical PBX.

multi-tenant unit (MTU) See *MTU*.

Multi-Use Radio Service (MURS) See *MURS*.

munitions Military supplies such as weapons and ammunition. The government of the United States and many other countries classified cryptology as munitions for many years. Those policies were relaxed in the 1990s, due to the rise of the Internet, which made the policies impossible to enforce. See also *encryption* and *Internet*.

MURS (Multi-Use Radio Service) In the United States, a Citizens Band (CB) Radio Service established in 2002 for private, two-way, short distance voice or data communications service for personal or business activities of the general public. MURS operates in the 151.820, 151.880, 151.940, 154.570, and 154.600 MHz bands. See also *CB Radio Service*.

.museum Pronounced *dot museum*. The generic Top Level Domain (gTLD) reserved exclusively for the museum community. This domain was created in 2002 under the sponsorship of the Museum Domain Management Association. See also *gTLD*, *Internet*, and *sponsored domain*.

music on hold (MOH) A voice telecommunications system feature that interfaces the system to an external audio source that provides background music, promotional messages, or other audio content while callers are either in queue or on hold.

must-carry rules In the United States, FCC rules that require CATV providers to provide channels for all commercial and public local television broadcasters within a 50-mile radius of the cable companies' service area. The CATV providers are not required to support multiple stations with redundant programming. The broadcast stations are allowed a choice of being carried under the must-carry rules or under more recent regulations that require cable companies to obtain retransmission consent prior to carrying a broadcast signal. The latter rules afford the broadcasters more freedom to negotiate favorable terms of carriage. See also *broadcast TV*, *CATV*, and *FCC*.

mute The voluntary silencing of a microphone or speaker.

mux or MUX (Multiplexer) The term *multiplex* has its roots in the Latin words *multi*, which translates as *many*, and *plex* which translates as *fold*. So, a multiplexer folds, or interleaves, many transmissions onto a single circuit or channel. A typical multiplexer acts as both a concentrator and a contention device that enables multiple, relatively low speed terminal devices to share a single, high-capacity circuit between two points in a network. Multiplexers allow carriers and end users to take advantage of the economies of scale. Although the cost of the multiplexers and the high-bandwidth circuit that interconnects them involves additional cost, they can support a number of channels at a relatively low cost per channel. This approach is analogous to building a multi-lane superhighway. The cost of the high-speed highway is considerable, but the cost per lane and per vehicle mile is much more reasonable than the cost of building multiple single-lane, low-speed roads. There are a number of multiplexing techniques that can be employed, including Add/Drop Multiplexing (ADM), Frequency Division Multiplexing (FDM), Time Division Multiplexing (TDM), Statistical Time Division Multiplexing (STDM), and Wavelength Division Multiplexing (WDM).

Inverse multiplexing performs the inverse process, spreading a high bandwidth signal across multiple lower-bandwidth circuits. See also *ADM*, *FDM*, *inverse multiplexer*, *STDM*, *TDM*, and *WDM*.

MV (Medium Voltage) An amount of electromotive force (emf) between low voltage (LV) and high voltage (HV). The power utilities use MV, generally at 7,200 volts, in their distribution networks. Access broadband over power line (Access BPL) technology can make use of those MV lines as local loops for broadband data communications. See also *Access BPL*, *emf*, *HV*, *LV*, *volt*, and *voltage*.

n (nano-) One billionth (10^{-9}, or $\frac{1}{1{,}000{,}000{,}000}$). A nanometer, or one-billionth of a meter, is the measurement of an optical wavelength.

N **1.** In mathematics, an indefinite whole number. In telecommunications, for example, N × 64 refers to some number of 64-kbps channels. **2.** The symbol for newton. See *newton*.

N × 64 (Number times 64 kbps channels) Bandwidth of N DS-0 channels of 64 kbps each. Synonymous with bonding, channel aggregation, dynamic bandwidth allocation. See *bonding* and *DS-0*.

NA (Numerical Aperture) See *numerical aperture*.

nailed-up circuit Dedicated circuits sometimes are referred to as nailed-up circuits because, in bygone days, the twisted-pair copper physical circuits were literally hung from nails driven in the walls of the carrier's wire centers. See also *dedicated circuit* and *wire center*.

NAK (Negative AcKnowledgement) **1.** A transmission control character sent by a station indicating that it is not ready to receive a transmission. In ASCII, NAK is represented by the bit pattern 1010100. See also *ASCII*. **2.** A negative acknowledgement that a message, block, or frame has been received in errored condition across a communications circuit, and that it must be retransmitted. See also *ACK*, *block*, *circuit*, *data set*, *frame*, *message*, and *station*.

.name Pronounced dot name. The Internet generic Top Level Domain (gTLD) reserved exclusively for individuals. This is an unsponsored domain. See also *Internet*, *gTLD*, and *unsponsored domain*.

N-AMPS (Narrowband Advanced Mobile Phone Service) An improvement on the AMPS 1G analog cellular radio technology that tripled system capacity, by splitting each 30 kHz channel into three 10 kHz voice channels. Very few cellular providers in the United States deployed N-AMPS. See also *1G*, *analog*, *cellular radio*, *channel*, and *AMPS*.

NANC (North American Numbering Council) An impartial body chartered in 1995 with oversight responsibility for the North American Numbering Plan (NANP). See also *NANP*.

nano- (n) A metric prefix meaning one billionth (10^{-9}). See *n*.

nanometer (nm) See *nm*.

NANP (North American Numbering Plan) The scheme of telephone numbers for 19 countries in the area officially known as World Zone 1, which excludes Mexico, and includes the Continental United States, Hawaii, Canada, Puerto Rico, the Virgin Islands, and parts of the Caribbean. AT&T and Bell Telephone Laboratories established the NANP in 1947 as a means of integrating the area codes and central office (CO) exchange codes in the area loosely known as North America. The North American Numbering Council (NANC), chartered in 1995, is an impartial body with oversight responsibility for the NANP. The Canadian Radio-television and Telecommunications Commission (CRTC) is responsible for the numbering plan in that country. In the Caribbean, some of the various governments delegate administration, while others retain that responsibility. In the United States, the Federal Communications Commission (FCC) has appointed NeuStar, a commercial enterprise, as the NANP administrator. The U.S. number format comprises fixed-length, ten-digit national numbers, which fit into the international NPA dialing scheme (+CC.NPA.NXX.xxxx). See also *area code*, *AT&T*, *Bell Labs*, *central office exchange code*, *CRTC*, *FCC*, *NANC*, *NeuStar*, and *NPA*.

NAP (Network Access Point) An official Tier 1 site at which an Internet service provider (ISP) can access the Internet backbone and exchange traffic with other ISPs. Some NAPs are known as Internet Exchanges (IXs) and some as MAEs. NAP locations include San Francisco, California; Chicago, Illinois; Miami, Florida; and New Jersey. See also *Internet*, *ISP*, *IX*, and *MAE*.

narrowband **1.** Voice grade bandwidth. In analog transmission systems, a narrowband channel has nominal bandwidth of 4 kHz, which is the standard for analog voice. In digital systems, a narrowband channel is 64 kbps, which is the fundamental standard for PCM digitized voice. **2.** A channel or circuit of less than voice grade bandwidth. **3.** Narrowband also is used to describe some number of 64 kbps channels (N × 64 kbps), but less than a T1 or E-1. T1 is a North American standard for a transmission system comprising 24 64-kbps channels and with a total signaling rate of 1.544 Mbps. E-1 is a European standard for a transmission system comprising 30 64-kbps information-bearing channels and with a total signaling rate of 2.048 Mbps. Narrowband ISDN (N-ISDN), for example, comprises two information bearing channels of 64 kbps, plus a signaling and control channel of 16 kbps, for a total of 144 kbps. **4.** A circuit or channel offering relatively little bandwidth. See also *bandwidth*, *broadband*, and *wideband*.

Narrowband AMPS (N-AMPS) See *N-AMPS*.

Narrowband ISDN (N-ISDN) See *N-ISDN*.

Narrowband PCS (Narrowband Personal Communications Services or N-PCS) See *PCS*.

Narrowband TACS (NTACS) See *NTACS*.

Narrowband Total Access Communications System (NTACS) See *NTACS*.

narrowcast Referring to transmission from one device to a limited number of other devices on a network. Cable television (CATV), direct broadcast satellite (DBS), and satellite radio use narrowcast transmission, as only subscribers to various channels, especially premium channels, are able to receive those transmissions. Internet content providers that operate on a subscription basis use a narrowcast model. The term is applied to Internet content providers who use push technology to send content to a subset of users on a network who have subscribed to or registered for a service. The terms narrowcast and multicast sometimes are used interchangeably. However, narrowcast refers to the business model, whereas multicast refers to the underlying transmission mode. See also *broadcast*, *CATV*, *channel*, *DBS*, *Internet*, *multicast*, *push*, and *satellite*.

NAS **1.** Network Access Server. A host computer on a local area network (LAN) dedicated to serve the needs of end users seeking access to internal computer resources or perhaps to the Internet through an Internet service provider (ISP). An NAS generally is associated with a Remote Authentication Dial-In User Service (RADIUS) server that performs authentication and accounting functions to ensure network security. See also *authentication*, *ISP*, *LAN*, *RADIUS*, *RAS*, and *security*. **2.** Network-Attached Storage. A simple storage technology in which one or more disk arrays or other storage devices are associated with a server that exists as a node on a LAN. The storage server assumes the responsibility for all data storage and for making the data available to all users with access privileges. A storage area network (SAN) is a much more sophisticated approach. See also *LAN*, *node*, *SAN*, and *server*.

NAT (Network Address Translation) A protocol that translates a private Internet Protocol (IP) address used in private domain, such as a LAN, into a public IP address that can be used in a public domain, such as the Internet. The translation process takes place in a router that interfaces to both domains, and operates on a symmetric basis, with translations taking place in both directions for the duration of a public session. NAT software allows the LAN-attached host to protect the privacy of its local identity as it accesses the Internet. NAT is specified in RFC 3022 from the Internet Engineering Task Force (IETF). See also *domain*, *IETF*, *Internet*, *LAN*, *private IP address*, *protocol*, *public IP address*, *router*, and *symmetric*.

NA-TDMA (North American Time Division Multiple Access) A digital cellular radio standard better known as *Digital Advanced Mobile Phone Service* (D-AMPS). See *D-AMPS.*

National Bureau of Standards (NBS) Now the National Institute of Standards and Technology (NIST). See *NIST.*

National Do-Not-Call Registry See *do-not-call registry.*

National Electrical Code (NEC) See *NEC.*

National Exchange Carrier Association (NECA) See *NECA.*

National Information Infrastructure (NII) See *NII.*

National Institute of Standards and Technology (NIST) See *NIST.*

National Internet Registry (NIR) See *NIR.*

nationalize To transfer ownership of a commercial enterprise from private interests to the government. Also known as *legalized theft of private property*. See also *privatize.*

National Research and Education Network (NREN) See *NREN.*

National Science Foundation (NSF) See *NSF.*

National Television Standards Committee (NTSC) See *NTSC.*

native **1.** Pertaining to something in its original, natural form. As examples, voice and video are analog in their native forms, whereas computer-to-computer communications are digital. **2.** Referring to program code or an application written for a specific operating environment, such as an operating system (e.g., DOS or UNIX) or processor.

natural monopoly A concept developed by political economist John Stuart Mill (1806–1873) and referring to utility services that are so capital-intensive and right-of-way intensive that it just doesn't make sense to have more than one provider. Natural monopolies include transportation infrastructure such as roads, bridges, ferries, subways, and railroads, as well as utility infrastructure such as water, natural gas, electricity, sewer, cable television, and wireline telephone. In the United States, the Graham-Willis Act (1921) established telephone companies as natural monopolies. Cable television (CATV) and wireline telephone infrastructure used to be considered natural monopolies, but it's not quite that simple any longer. See *CATV*, *monopoly*, *right of way*, *Telecommunications Act of 1996*, *utility*, and *wireline.*

NBS (National Bureau of Standards) Now the National Institute of Standards and Technology (NIST). See *NIST.*

NCC (Network Control Center) Also known as *network operations center* (NOC). See *NOC.*

NCP **1.** Network Control Program. In a network switch or node, software designed to enable the transfer of frames between nodes. **2.** Network Control Protocol. Referring to a protocol responsible for negotiating Network Layer (Layer 3) details. The Point-to-Point Protocol (PPP), for example, is based on a Link Control Protocol (LCP) and a Network Control Protocol. The LCP is responsible for setting up a link between two computers over a circuit-switched telephone connection, and for resolving any issues of authentication. The NCP negotiates any parameters specific to the Network Layer. The Internet Protocol Control Protocol (IPCP), for example, is used when the Internet Protocol (IP) is employed at the Network Layer. See also *authentication*, *IP*, *link*, *Network Layer*, *PPP*, and *protocol.*

NCSA Mosaic See *Mosaic.*

NDSF (Non Dispersion-Shifted Fiber) The earliest type of single-mode fiber (SMF). NDSF improved considerably on multimode fiber (MMF) with respect to distance limitations. However, chromatic dispersion and material dispersion were discovered to be issues. NDSF runs in the O-Band, specifically in the 1,300–1320 nm range, where the effects of material dispersion are lowest in a standard single-mode fiber. Dispersion-shifted fiber (DSF) shifts the optimal dispersion point. See also *dispersion, chromatic dispersion, DSF, material dispersion, MMF, O-Band, SMF,* and *window.*

NE (Network Element) **1.** An elemental, or fundamental, unit of a network, such as a transmitter, amplifier, repeater, multiplexer, switch, router, copper or fiber optic transmission link, microwave antenna, or receiver. See also network. **2.** In network management, a manageable device. See also *network management.*

near-end cross talk (NEXT) See *NEXT.*

near line-of-sight (near LOS) See *near-LOS.*

near-LOS (near Line-Of-Sight) Not LOS, but nearly so. Microwave and other high frequency terrestrial radio communications generally require LOS. WiMAX and some other microwave radio technologies can tolerate near-LOS.

near-realtime (nrt) Referring to communications that does not occur in realtime (i.e., not at the precise moment as the event, itself) but nearly so. See *nrt.*

NEC (National Electrical Code) A set of standards published by the National Fire Protection Association (NFPA) for the safe installation of electrical wiring and optical fiber and equipment on the premises. The NEC is approved by the American National Standards Institute (ANSI) as ANSI NFPA 70, and its use is commonly mandated by state and local law. The NEC is updated and published every three years. See also *ANSI.*

NECA (National Exchange Carrier Association) A not-for-profit organization of U.S. incumbent local exchange carriers (ILECs), NECA administers the FCC's access charge plan that determines the fees that interexchange carriers (IXCs) pay ILECs to complete calls. NECA also administers the Telecommunications Relay Service (TRS) fund that supports telecommunications services for individuals who are speech and/or hearing impaired. The subsidiary Universal Service Administrative Company (USAC) administers programs that provide support for ILEC companies in high-cost areas, assistance for low income subscribers, discounts for telecommunications service to schools and libraries, and discounts to rural health care providers. See also *FCC, ILEC, IXC, TRS,* and *USAC.*

negative **1.** Something with the same charge (–) as an electron, opposite of the positive (+) charge of a proton. **2.** The part of a circuit towards which electrons flow from the positive point. (The flow is opposite inside the current source.) See also *null, positive,* and *potential.* **3.** No.

negative acknowledgement (NAK) See *NAK.*

nerd Someone, usually a techie gone over the edge, who is so overly intellectual and abstruse as to be regarded as socially inept, unsophisticated, awkward, unattractive, or otherwise exceedingly obnoxious, odd, or unpleasant. The word first appeared in Dr. Seuss's book *If I Ran the Zoo* (1950) as the name of an imaginary animal. In the book, the narrator Gerald McGrew plans to collect a number of imaginary animals for his imaginary zoo. McGrew states, "And then, just to show them, I'll sail to Ka-Troo And Bring Back an It-Kutch a Preep and a Proo A Nerkle a Nerd and a Seersucker, too!". A good techie can be invaluable. A nerd can also be invaluable, even though obnoxious. See also *geek* and *techie.*

nest egg **1.** A real or artificial egg placed in a hen's nest or nesting box to encourage it to continue laying eggs in the same place. **2.** Money put aside as a reserve for future expenses or leaner times.

net (network) See *network.*

.net (network) Pronounced *dot net*. Originally, the Internet generic Top Level Domain (gTLD) reserved exclusively for network access providers, such as Internet Service Providers (ISPs). The domain is now unrestricted. This is an unsponsored domain. See also *Internet*, *gTLD*, *ISP*, and *unsponsored domain*.

Net (InterNet) See *Internet*.

Net neutrality (InterNet neutrality) The principle that the cost of Internet access should not be sensitive to the nature of the content. As Internet content has become increasingly stream-oriented and bandwidth-intensive, the demands on the network for bandwidth and congestion management have become more pronounced. Service providers have responded with increased bandwidth and various quality of service (QoS) mechanisms that prioritize various traffic types, such as audio, video, and voice. Net neutralists oppose price differentials associated with prioritized traffic types. Those opposed suggest that neutrality will discourage innovation and investment. See also *bandwidth*, *congestion*, *Internet*, *QoS*, and *stream-oriented*.

Netscape Communicator A software suite that includes the Netscape Navigator Web browser, an editing program, voice conferencing, an e-mail client, and a calendaring system. See also *browser*, *client*, *conference call*, *e-mail*, *Netscape Navigator*, and *Web*.

Netscape Navigator A Web browser featuring simultaneous image loading and continuous document-streaming speed performance. The software was developed at Netscape Communications Corp. by a team led by Marc Andreessen, the creator of the original Mosaic browser, and was first commercially available in late 1994. Netscape Navigator subsequently was packaged in the Netscape Communicator application software suite and lost its individual identity. In 1998, Netscape began work on an open source version known as Mozilla. In 1999, Netscape was acquired by America Online (AOL), which packaged Netscape software as part of AOL service. See also *browser*, *Mosaic*, *Mozilla*, *software*, and *Web*.

network (net) A fabric of elements that work together much as the fabric or mesh of a net to support the transfer of information. A network includes all links, amplifiers and repeaters, multiplexers, switches, routers, and other intermediate devices involved in establishing, maintaining, and terminating a session between a transmitter and a receiver. A network may take the form of a local area network (LAN), metropolitan area network (MAN), personal area network (PAN), or wide area network (WAN). See also *LAN*, *MAN*, *PAN*, and *WAN*.

Network Access Point (NAP) See *NAP*.

network access server (NAS) See *NAS*.

Network Address Translation (NAT) See *NAT*.

network architecture The design and framework of a network, including the characteristics of individual hardware, software, and transmission system components and how they interact in order to ensure the reliable transfer of information. Prior to the development of such architectures, interoperability between the various systems of a single manufacturer was unusual, and it certainly did not exist between the products of multiple manufacturers. IBM's Systems Network Architecture (SNA) and the Digital Equipment Corporation's (DEC's) Digital Network Architecture (DNA), aka DECnet, corrected these shortcomings within the IBM and DEC domains, but they still did not interoperate. Truly open systems architectures still remain in the distant future, although great strides have been made in this regard through the Open Systems Interconnection (OSI) model fostered by the International Organization for Standardization (ISO). Network architectures tend to be layered, which serves to enhance their development and management. While they primarily address issues of data communications, they also include some data processing activities at the upper layers. These upper layers address application software processes, presentation format, and the establishment of user sessions. Each independent layer, or level, of a network architecture addresses different functions and responsibilities. The layers work together, as a whole, to maximize the performance of the process. See also *ISO*, *OSI Reference Model*, and *SNA*.

network-attached storage (NAS) See *NAS.*

network control center (NCC) Also known as *network operations center* (NOC). See *NOC.*

Network Control Program (NCP) See *NCP.*

Network Control Protocol (NCP) See *NCP.*

network element (NE) See *NE.*

network interface card (NIC) See *NIC.*

network interface device (NID) Synonymous with network interface unit (NIU). See *NID.*

network interface unit (NIU) Synonymous with network interface device (NID). See *NID.*

Network Layer Layer 3 of the seven-layer Open Systems Interconnection (OSI) Reference Model. Software at the Network Layer is responsible for addressing and sequencing the message units and transporting them to their ultimate destinations, setting up the appropriate paths between the nodes. At this layer, message routing, error detection, and control of internodal traffic are managed. The Internetwork Protocol (IP) operates at this layer. See also *IP, layer, network architecture, OSI Reference Model, protocol, routing,* and *software.*

network management Referring to a broad range of functions associated with the management and control of a network, particularly one large and complex in nature. Network management functions include fault management, configuration management, accounting management, performance management, and security management. Some of those functions are real-time, whereas others are purely administrative. Real-time network management functions include the aspects of fault management, configuration management, performance management, and security management that require virtually instantaneous identification, analysis, and action to ensure the performance and integrity of the network. Accounting management is a purely administrative function that includes inventory management, data collection, cost allocation, and billing. Some aspects of configuration management and performance management are non-realtime and administrative in nature, as well, including data collection, traffic analysis, and network optimization. Telemanagement is a subset of network management concerned with the administrative management of voice networks. Real-time network management systems (NMSs) include element management systems (EMSs) that deal with alarms and alerts from devices of the same type and manufacturer, and managers of managers (MOMs) that deal with alarms and alerts from multiple types of network elements (NEs). See also *accounting management, bandwidth, configuration management, EMS, fault management, load balancing, MOM, NE, network optimization, performance management, security management, TMN,* and *traffic analysis.*

network management system (NMS) See *NMS.*

network neutrality Referring to Internet neutrality. See *Net neutrality.*

network operating system (NOS) See *NOS.*

network operations center (NOC) Also known as network control center (NCC). See *NOC.*

network optimization The process of striking the best possible balance between network performance and network costs, in consideration of grade of service (GoS) requirements. See also *GoS.*

Network 10 address More commonly known as a *private Internet Protocol* (IP) address, a Network 10 address is so called in reference to the first field in the first address range, 10.0.0.0 to 10.255.255.255. See also *IP, IP address,* and *private IP address.*

Network Termination (NT) See *NT.*

Network Termination 1 (NT1) See *NT1.*

Network Termination 2 (NT2) See *NT2.*

network-to-network interface (NNI) See *NNI.*

network topology See *topology.*

NeuStar A commercial enterprise that provides certain administrative and clearinghouse services. NeuStar was appointed by the Federal Communications Commission (FCC) in 1997 as the administrator of the North American Numbering Plan (NANP). NeuStar also currently administers a number of Internet domains, including .biz (bizness), .tw (Taiwan), and .us (United States). See also *FCC* and *NANP.*

newb (*newb*ie) In reference to someone new at something such as computer usage, programming, or, most especially, the Internet.

newsgroup A forum on the Internet for textual discussions of specific subjects. A newsgroup essentially is a discussion group that generally involves an article that serves as a stimulus for message postings that form conversational threads.

newton (N) The unit of force that will impart an acceleration of one meter per second to a mass of one kilogram. The newton is named for Sir Isaac Newton (1642–1727), the English scientist who invented calculus, discovered gravitation, formulated the laws of motion, and discovered that white light is a combination of all colors in the visible light spectrum. (Not bad for an alchemist.)

NEXT (Near-End CROSS Talk) The unwanted coupling of energy between two circuits or channels occurring at the near end of the link, that is, at the end closest to the point of signal origin. Since at that point the outgoing downstream signal is at maximum strength and the incoming upstream signal is at minimum strength, the signals can couple quite easily if the attenuation-to-crosstalk ratio (ACR) is not maintained at acceptable levels. NEXT is a particularly significant issue at the network side of the connection for V.90 modems and ADSL services, as the pair count of the cables is quite high at the central office (CO) and, therefore, there are many opportunities for co-channel interference. However, ACR can be managed much more effectively at the edge of the telco network than at the customer premises. Far-end crosstalk (FEXT) occurs at the far end of the link. It is in consideration of the phenomenon of crosstalk and the differences between NEXT and FEXT, that V.90 modems and most DSL services are asymmetric, with the higher frequencies on the downstream side (i.e., from the edge of the telco network to the customer premises) in support of greater bandwidth in that direction. See also *ADSL, asymmetric, bandwidth, channel, circuit, CO, crosstalk, downstream, frequency, interference, link, FEXT, signal, upstream,* and *V.90.*

Next Generation Internet (NGI) See *NGI.*

Next Hop Resolution Protocol (NHRP) See *NHRP.*

NFAS (Non-Facility Associated Signaling) A signaling and control technique by which the signaling and control information associated with a communications channel is not carried within the communications channel, itself, or in a separate channel permanently associated with that communications channel. Rather, that signaling and control information is carried in a separate channel, perhaps over a separate subnetwork. NFAS typically refers to an ISDN technique in which signaling and control functions for multiple primary rate interface (PRI) trunks is accomplished using a common Signaling System 7 (SS7) D channel associated with one of them, thereby freeing the D channels associated with the other trunks and allowing them to be used as B channels for user payload transmission. SS7 is a Common Channel Signaling (CCS) system. Contrast with CAS (Channel-Associated Signaling). See also *B channel, CCS, channel, D channel, ISDN, NFAS, PRI, signaling and control, SS7,* and *trunk.*

NFPA 70 National Fire Protection Association (NFPA) publication 70. Synonymous with National Electrical Code (NEC). See also *NEC.*

NGI (Next Generation Internet) A initiative of the U.S. federal government intended to develop the technologies and applications that will form the future Internet. The NGI initiative is in parallel to the university-led Internet2. See also *Internet* and *Internet2.*

NHRP (Next Hop Resolution Protocol) A method by which a host computer or router connected to a non-broadcast multi-access (NBMA) network can discover the internetworking layer addresses and subnetwork addresses of the NBMA next hop towards a destination station. The source station sends a query packet, to which the destination station responds if on the same network. If the destination station is not on the same network or subnetwork, the subnetwork egress node nearest to the destination station responds. Thereby, the source host or router dynamically discovers the most direct path to the host or egress router and to avoid unnecessary intermediate hops. NHRP is described in IETF RFC 2332. See also *IETF* and *subnet.*

nibble Sometimes spelled nybble. **1.** Half a byte. **2.** A four-bit byte. (I swear that I don't make this stuff up.) See also *byte*, *quadbit*, and *quartet.*

NIC (Network Interface Card) Also known as a network interface unit (NIU). Traditionally, a device in the form of firmware on a printed circuit board that fits into an expansion slot of a device (e.g., workstation, printer, or server) and provides the necessary programmed logic for that device to interface to a local area network (LAN). Alternatively, multiple cards traditionally were contained within a multiport device that supported multiple devices on a pooled basis. Contemporary computers commonly include NIC functions on the main circuit board, or motherboard. NICs are responsible for medium access control (MAC). Each NIC has a theoretically unique logical MAC address, assigned by the manufacturer, for purposes of identification. See also *LAN* and *MAC.*

NID (Network Interface Device) Also known as a *network interface unit* (NIU). A device that serves as the demarcation point (demarc) and interface between the public carrier network and the customer premises. An NID includes some form of protector that insulates the premises from potentially disastrous high voltage current caused by lightening strikes on outside copper cable plant. An NID also generally contains a chipset that supports remote testing of the local loop from a centralized Network Operations Center (NOC).

Nigerian Connection See *419 Advance Fee Fraud.*

Nigerian Scam See *419 Advance Fee Fraud.*

NII (National Information Infrastructure) A program of the United States government to encourage the development of an advanced and seamless high-speed web of public and private communications networks, interactive services, interoperable hardware and software, databases, and other elements in support of voice, data, video, fax, and multimedia transmission, storage, processing, and display. The NII encompasses a wide range of interactive databases, functions, and services interconnected in a technology-neutral manner. See also *Information Superhighway.*

NIR (National Internet Registry (NIR) The national organization responsible for assigning Internet Protocol (IP) addresses to Local Internet Registries (LIRs) that, in turn, assign them to Internet Service Providers (ISPs). The NIR receives address assignments from a Regional Internet Registry (RIR). In some nations, there is no NIR, so an RIR assigns IP addresses directly to the LIRs. See also *IANA*, *IP*, *IP address*, *ISP*, *LIR*, and *RIR.*

N-ISDN (Narrowband ISDN) In the terminology of the integrated services digital network (ISDN), narrowband refers to bandwidth at the level of DS-1 or less. According to North American standards, a

DS-1 translates to an ISDN primary rate interface (PRI), which runs over a T1 facility, which has a signaling rate of 1.544 Mbps. According to European standards, a DS-1 translates to ISDN primary rate access (PRA), which runs over an E-1 facility, which has a signaling rate of 2.048 Mbps. N-ISDN is in contrast to broadband ISDN (B-ISDN), which operates at a rate greater than the primary rate. See also *B-ISDN, broadband, DS-1, E-1, ISDN, narrowband, PRA, PRI, signaling rate*, and *T1*.

NIST (National Institute of Standards and Technology) nee National Bureau of Standards (NBS). An agency of the United States Department of Commerce, NIST has the stated mission of promoting U.S. innovation and industrial competitiveness by advancing measurement science, standards, and technology in ways that enhance economic security and promote the quality of life. As an example, NIST developed the Advanced Encryption Standard (AES), a highly secure encryption algorithm used the United States government for sensitive unclassified documents. AES also is used in a number of commercial applications. See Appendix A for contact information. See also *AES*.

NIU (Network Interface Unit) Synonymous with network interface device (NID). See *NID*.

NLOS (Non-Line-Of-Sight) A radio transmission system that can support effective communications without optical LOS. The higher the frequency of the signal, the more important is LOS. Cellular telephones and pagers, for example, operate in MHz (e.g., 800 and 900 MHz ranges) and the lower end of the GHz ranges (e.g., 1.8 and 1.9 GHz), and operate quite effectively in NLOS mode. Higher frequency signals, such as microwaves, generally require LOS, yet WiMAX, for example, is a wireless local loop technology operating in the 2-11 GHz range in both LOS and non-LOS (NLOS) modes. See also *LOS* and *near-LOS*.

nm (nanometer) One billionth (10^{-9}, or $\frac{1}{1,000,000,000}$) of a meter. The wavelength of an optical signal is measured in nanometers. The wavelength of a typical optical signal in a long haul fiber optic transmission system (FOTS) is in the 1550 nm window, for example. Just to put it in perspective, an average molecule is about a nanometer in diameter, a human DNA molecule is about 2.5 nm wide, and a human hair is in the range of 5,000–10,000 nm in diameter. A human fingernail grows about 1 nm per second. If you watch your fingernails grow for about 25 minutes and 50 seconds, you'll have a sense for how long a wavelength is at 1550 nm. *Note:* It is much more exciting to watch grass grow, or even paint dry.

NMS (Network Management System) A system comprising software, firmware, and hardware used to manage a large and complex network. An NMS, also known as a manager of managers (MOM), may receive alarms and alerts from multiple element management systems (EMSs), each of which manages one or more specific intelligent network elements (NEs), i.e., devices of the same type (e.g., modems or multiplexers) and generally of the same manufacturer. See also *EMS, firmware, hardware, MOM, NE, network management*, and *software*.

NMT (Nordic Mobile Telephone) A 1G analog cellular radio technology developed and placed into service in the early 1980s in Scandinavian countries, including Denmark, Finland, Norway, and Sweden. NMT 450 operates in the 450 MHz band, which yields excellent signal propagation and, therefore, is especially appropriate for sparsely populated areas supported by few cell sites. NMT 900 operates in the 900 MHz range. NMT largely has been replaced by GSM. See also *1G, analog, cellular radio, propagation, GSM, NMT 450*, and *NMT 900*.

NMT 450 (Nordic Mobile Telephone 450 MHz) A 1G analog cellular radio technology developed and placed into service in the early 1980s in Scandinavian countries. NMT 450 operates in the 450 MHz band, employing frequency division multiple access (FDMA) to derive 200 channels with a width of 25 kHz. NMT 450 employs frequency division duplex (FDD) to achieve bidirectional communications, with the downlink in the 463–468 MHz band and the uplink in the 453–458 MHz band. Within each channel, frequency modulation (FM) is employed. See also *1G, analog, cellular radio, channel, downlink, FDD, FDMA, FM*, and *uplink*.

NMT 900 (Nordic Mobile Telephone 900 MHz) A 1G analog cellular radio technology developed and placed into service in the early 1980s in Scandinavian countries. NMT 900 operates in the 900 MHz band, employing frequency division multiple access (FDMA) to derive 1999 channels with a width of 12.5 kHz. NMT 900 employs frequency division duplex (FDD) to achieve bidirectional communications, with the downlink in the 935-960 MHz band and the uplink in the 890-915 MHz band. Within each channel, frequency modulation (FM) is employed. See also *1G*, *analog*, *cellular radio*, *channel*, *downlink*, *FDD*, *FDMA*, *FM*, and *uplink*.

NN (Network Neutrality) Referring to Internet neutrality. See *Net neutrality*.

NNI (Network-to-Network Interface) The boundary or point of interaction between network service providers. The NNI is both a physical and logical point of demarcation. The NNI serves the technical boundary where protocol issues are resolved and as the point of division between the responsibilities of the individual service providers. NNIs are defined for asynchronous transfer mode (ATM) and frame relay, as examples. See also *ATM*, *logical*, *frame relay*, *physical*, and *protocol*.

NOC (Network Operations Center) (pronounced *knock*) Also known as *network control center* (NCC). A centralized location from which a large, complex network and its component subnetworks and network elements can be monitored, and faults or performance failures can be identified, diagnosed, isolated, and often corrected. See also *network*, *network element*, and *subnet*.

node **1.** A junction point at which two or more circuits interconnect in a data network. A bridge, for example, interconnects two or more segments of a local area network (LAN). See also *bridge*, *circuit*, *LAN*, and *network*. **2.** In a switched network, a switching point that comprises a point of interconnection for circuits, a data switch, and control facilities. In the public switched telephone network (PSTN), for example, a great many circuits terminate in a central office (CO) and a tandem office, each of which comprises one or more switches and Signaling System 7 (SS7) network control logic, multiplexers, a wide variety of other devices. See also *access node*, *CO*, *PSTN*, *service node*, *SS7*, *switch*, and *tandem switch*. **3.** A device such as a station, bridge, computer, repeater, server, switch, or other device that connects to a network. **4.** In the IBM Systems Network Architecture (SNA), a physical device such as a computer, communications processor (e.g., FEP), terminal controllers, or terminal. See also *SNA*.

noise Unwanted disturbances superimposed on a signal and interfering with its integrity. Noise can be introduced by equipment or can be the result of natural phenomena. Noise can take a number of forms, including amplitude noise, cross-talk, echo, intermodulation noise, harmonic distortion, impulse noise, random noise, and white noise.

nominal In name only, but not in reality. A T1 circuit sometimes is described as having a nominal transmission rate of 1.5 Mbps, although in reality the signaling rate is 1.544 Mbps and its payload is 1.536 Mbps. It is easier to say one point five Megabits per second or one and a half Megabits per second than one point five four four Megabits per second. It is not exact, but it is close enough for all but the most obnoxious purists. Similarly, an E-1 is generally described as having a transmission rate of 2 Mbps, although in reality the signaling rate is 2.048 Mbps and the payload is 1.92 Mbps. See also *payload*, *signaling rate*, and *transmission rate*.

non-blocking A switch that provides a guaranteed talk path for every terminal; in other words, there exists a 1:1 (one-to-one) relationship between ports and time slots. Such a configuration is expensive, generally considered excessive and, therefore, unusual in all but the most intense applications scenarios, such as call center ACDs and backbone data switches. See Figure N-1. See also *ACD*, *backbone*, *blocking*, and *call center*.

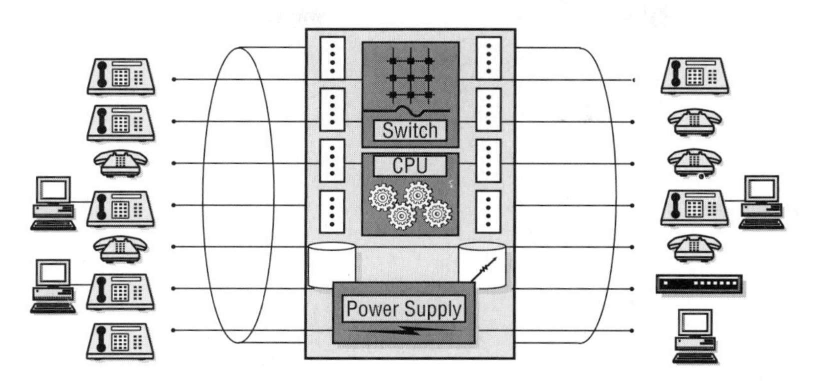

Figure N-1

non-deterministic **1.** Non-predictive. Referring to the inability to objectively predict an outcome or result of a process due to lack of knowledge of a cause and effect relationship or the inability to know initial conditions. **2.** In telecommunications switching and routing, the unpredictability of a path between nodes. See also *node*, *path*, *route*, and *switch*. **3.** In local area networks (LANs) a contentious medium access control (MAC) technique that does not allow a node to predict when it will be allowed to gain access to the network. Carrier sense multiple access (CSMA), which is used in 802.3 Ethernets, is non-deterministic. See also *802.3*, *CSMA*, *deterministic*, *LAN*, *MAC*, and *node*.

non-dialable toll points A location that a caller cannot dial directly, but must call with operator assistance. Such locations are very remote and may be beyond the reach of cable systems and may be too low and sheltered to be reached via satellite, but can be reached via special radio circuits. In the United States, there remain a large number of non-dialable toll points, generally in deep valleys and canyons.

non dispersion-shifted fiber (NDSF) See *NDSF*.

non-facility associated signaling (NFAS) See *NFAS*.

non-line-of-sight (NLOS) See *NLOS*.

nonpersistent carrier sense multiple access (Nonpersistent CSMA) See *CSMA*.

Nonpersistent CSMA (Nonpersistent Carrier Sense Multiple Access) See *CSMA*.

non-real-time (nrt) See *nrt*.

non real-time variable bit rate (nrt-VBR) See *nrt-VBR*.

non-repudiation Referring to a mechanism that proves that the originating node sent a message and that the receiving node received it. Therefore, the sender cannot deny sending the message and the receiver cannot deny having received it. Non-repudiation can be effected through digital signatures, confirmation receipts, and timestamps. See also *digital signature*.

non-return-to-zero (NRZ) See *NRZ*.

non-wireline carrier Also known as an *A Carrier*. A provider of cellular radio service that is not a traditional landline telecommunications services. The distinction between wireline and non-wireline carriers was made primarily for purposes of segregating bidders for radio spectrum assignment during the FCC cellular spectrum auctions. The initial approach toward spectrum assignment was designed to ensure that there was one wireline (i.e., telephone company) and one non-wireline carrier per market. See also *carrier*, *cellular*, *FCC*, *landline*, *radio*, *spectrum*, and *wireline carrier*.

Nordic Mobile Telephone (NMT) See *NMT.*

normal **1.** Conforming to an accepted, usual, or typical form, model, or pattern. **2.** In geometry and mathematics, the transverse or perpendicular, i.e., a right angle, which is 90 degrees from a plane or surface. Frankly, this seems decidedly abnormal to me, as I have known a lot of people who seemed to be at right angles compared to normalcy. See also *counterintuitive*, *critical angle*, *out-of-phase*, and *total internal reflection.*

North American Numbering Council (NANC) See *NANC.*

North American Numbering Plan (NANP) See *NANP.*

North American Time Division Multiple Access (NA-TDMA) A digital cellular radio standard better known as *Digital Advanced Mobile Phone Service* (D-AMPS). See *D-AMPS.*

NOS (Network Operating System) Software that provides a local area network (LAN) with multi-user, multitasking capabilities, facilitates communications and resource sharing, and thereby provides the basic framework for LAN operation. The NOS comprises modules distributed throughout the LAN, with some residing in the servers and others in the clients. See also *client*, *client/server*, *LAN*, and *server.*

notation The use of signs or symbols to represent numbers, words, phrases, or even complete concepts in fields such as language, mathematics, chemistry, and music. See also *binary notation* and *hexadecimal notation.*

NPA (Numbering Plan Administration) The administration of the scheme of logical addresses, i.e., telephone numbers, used in the global switched telephone network (GSTN) and the national public switched telephone networks (PSTNs) that compose it. The ITU-T is responsible for international numbering plan administration, and individual nations or regions have similar responsibilities within their domains. The North American Numbering Plan (NANP), for example, defines the telephone numbering scheme in the area loosely described as North America. The ITU-T E.164 recommendation (The International Public Telecommunication Numbering Plan) specifies the current international NPA convention at a maximum of 15 digits, although the number of digits required for calling within a nation varies. In many cases, numbering schemes vary within the same country; for example, six- and seven-digit telephone numbers coexist in Namibia and many other countries. In the countries within the NANP, the dialing scheme is +CC.NPA.NXX.xxxx, with the fields defined as follows:

- \+ The plus sign indicates that there may be leading digits for international dialing. In the United States, the caller dials 011 as an international access code.
- **CC (Country Code)** The country code is one, two, or three digits, established by the ITU-T. As examples, the United States is 1, South Africa is 27, and Luxembourg is 352.
- **NPA (Numbering Plan Area)** The NPA, or area code, is a three-digit number that corresponds to a geographic area. The NPA follows the pattern NXX, with N indicating that only numbers 2–9 are allowed, as 0 or 1 would confuse the network, and X indicating that any number is allowed. The area code is used only when a call crosses an area code boundary. In such a case, the dialing sequence is 1.NNX.NNX.xxxx.

N-PCS (Narrowband Personal Communications Services) See *PCS.*

NPAC (Number Portability Administration Center) In the United States, the Federal Communications Commission (FCC) established the NPAC to supervise and perform clearinghouse functions in support of local number portability (LNP) and local routing numbers (LRNs). See also *LNP* and *LRN.*

NREN (National Research and Education Network) The first (1990) asynchronous transfer mode (ATM) network in the United States, NREN was a test-bed gigabit network sponsored by the Advanced

Research Project Agency (ARPA) and the National Science Foundation (NSF). Previously (1987), a consortium of European carriers, end users, and universities sponsored a similar project known as the Research for Advanced Communications in Europe (RACE) project 1022 (1987). See also *RACE.*

nrt **1.** near-realtime. Referring to a quality of service (QoS) level designed for applications that do not require transmission to take place in real time, but nearly so. That is to say the transmission must take place within a reasonably short time from the exact moment as the event itself takes place in the real world. Near-realtime QoS is essential in many applications directly involving humans and their perception of time. Transaction processing, for example, must take place in near-realtime to avoid customer dissatisfaction. See also *QoS* and *realtime.* **2.** non-realtime. Referring to a quality of service (QoS) level designed for applications that do not require transmission to take place in real time, that is to say that the transmission need not take place at the exact moment and in the exact sequence as the event itself takes place in the real world. Internet access, for example, need not be available instantly on demand. e-mail need not be sent across a network to be received instantly, but can tolerate a considerable level of error, latency, loss of sequence, loss of data, and retransmission. See also *latency, QoS,* and *realtime.*

nrt-VBR (non real-time Variable Bit Rate) In asynchronous transfer mode (ATM), a class of traffic that is bursty, with periods of intense activity and periods of low or no activity, and of a non real-time nature that is not dependent on loss or delay because there is time to recover through retransmission. Traffic parameters include peak cell rate (PCR), cell delay variation tolerance (CDVT), sustainable cell rate (SCR), maximum burst size (MBS), and burst tolerance (BT). The quality of service (QoS) parameter is cell loss ratio (CLR). ATM also defines available bit rate (ABR), constant bit rate (CBR), real-time Variable Bit Rate (rt-VBR), unspecified bit rate (UBR), and variable bit rate (VBR) traffic classes. Examples of nrt-VBR traffic include data traffic such as X.25, frame relay, transaction processing, LAN-to-LAN, and non real-time buffered voice and video traffic. See also *ABR, ATM, BT, CBR, CDVT, CLR, compression, frame relay, LAN, MBS, PCR, QoS, realtime, rt-VBR, SCR, time slot, UBR,* and *VBR,* and *X.25.*

NRZ (Non-Return-to-Zero) A binary line coding technique in which 1 bits are represented by a high value significant condition (e.g., +V) and 0s are represented by a low value significant condition (e.g., –V), that is, opposite polarity, with no neutral or rest condition (e.g., 0V). See also *line coding, Manchester coding,* and *polarity.*

NSF (National Science Foundation) An independent agency of the United States government formed in 1950 "to promote the progress of science; to advance the national health, prosperity, and welfare; to secure the national defense, according to the NSF..." The mission of the NSF includes support for all fields of fundamental science and engineering. In telecommunications, the NSF has taken the initiative projects such as the development of the very-high-speed Backbone Network Service (vBNS), which formed the initial backbone infrastructure for Internet2. See also *Internet2* and *vBNS.*

NT (Network Termination) In ISDN networks, a set of functions accomplished through the use of programmed logic variously embedded in the carrier network and the customer premises equipment (CPE). NT devices operate to interface the four-wire customer wiring to the physical two-wire UTP local loop, performing functions similar to those provided by digital service units (DSUs) and channel service units (CSUs) in non-ISDN digital networks. See also *CPE, CSU, DSU, four-wire circuit, ISDN, NT1, NT2,* and *two-wire circuit.*

NT1 (Network Termination 1) In ISDN, the logical interface to the carrier side of the connection, performing such functions as signal conversion, synchronization, multiplexing, frame alignment, echo cancellation, line maintenance, and performance monitoring of the local loop. Such functions correspond to Layer 1 (Physical Layer) of the OSI Reference Model. See also *Echo canceller, ISDN, local loop, multiplexing, NT, NT2, OSI Reference Model,* and *synchronization.*

NT2 (Network Termination 2) An interface to an intelligent ISDN-compatible device (e.g., PBX or router) responsible for the user side of the connection to the network, performing such functions as multiplexing and switching. Such functions correspond to Layer 2 (Data Link Layer) of the OSI Reference Model. An NT2 commonly is actually an NT1/2 device, performing the combined functions, and operating at Layers 1, 2, and 3 of the OSI Reference Model. See also *ISDN, NT, NT1,* and *OSI Reference Model.*

NTACS (Narrowband Total Access Communications System) A narrowband version of the TACS 1G analog cellular radio technology developed for use in the United Kingdom. NTACS operates in the 900 MHz band, employs frequency modulation (FM), and supports 400 channels of 12.5 kHz. As an analog system, TACS derives channels using frequency division multiple access (FDMA) and bidirectional communications is achieved through frequency division duplex (FDD) with the downlink in the 860–870 MHz band and the uplink in the 915–925 MHz band. See also *1G, analog, cellular radio, downlink, ETACS, FDD, FDMA, FM, narrowband, TACS,* and *uplink.*

NTSC (National Television Standards Committee) The initial standard (1953) for broadcast television, NTSC was named for the committee that established it in the United States. NTSC is characterized as analog in nature, with 525 interlaced scan lines. There are 640 pixels per line, 485 of which are dedicated to the active picture. The frame rate is 30 fps, 60 fields interlaced, and the aspect ratio is 4:3. As an early analog standard that is viewed by some as overly complex and ineffective in a contemporary digital context, NTSC sometimes is referred to by its detractors in the pejorative as Never The Same Color. NTSC is defined in ITU-R Recommendation 1125 and served as the baseline for subsequent standards, Phase Alternate Line (PAL) and SECAM (**SÉ**quential **C**ouleur **A**vec **M**émoire). See also *analog, aspect ratio, broadcast television, frame, frame rate, interlaced scanning, ITU-R, PAL, pixel,* and *SECAM.*

null **1.** Valueless; amounting to nothing; zero. **2.** In some computer programs, a field into which nothing is entered, not even a zero (0). In such programs, even a 0 can affect calculations. See also *negative* and *positive.*

null modem Referring to a metallic wire cable used to connect two computers directly without the use of modems. A null modem cable simply crosses the transmit and receive wires so that the wire used by one machine for signal transmission is used by the other machine for signal reception. A null modem cable can be created by manually crossing the wires. Alternatively, a null modem adapter can be used to accomplish the necessary crosslinks. (Note: A null modem is much less expensive and much prettier.) See also *cable, modem,* and *wire.*

numbering plan administration (NPA) See *NPA.*

Numbering Plan Area (NPA) See *NPA.*

number portability Referring to the ability to port a number across carriers, i.e., move a telephone number from one carrier to another in a competitive environment. The United States first required number portability, initially with respect to toll-free numbers. The Telecommunications Act of 1996 required local number portability (LNP). In 2003, the Federal Communications Commission required wireless number portability (WNP), extending portability to cellular telephone numbers. Number portability is possible in many countries with respect to toll-free numbers and landlines, in general. Portability is possible in fewer countries with respect to cellular service. There are restrictions, however. For example, it generally is not possible to port a number across landline and cellular domains. Neither can numbers be ported across countries. See also *LNP.*

Number Portability Administration Center (NPAC) See *NPAC.*

numerical aperture (NA) The light-gathering ability of an optical fiber, as determined by the square root of the difference of the squares of the refractive indexes of the core (n_1) and the cladding (n_2), and as expressed in the equation:

$$NA = (n_1^2 - n_2^2)^{1/2}$$

Fiber optic transmission systems (FOTS) are based on the principle of total internal reflection, meaning that all light injected into the fiber is retained in the fiber. The objective is to retain all components of the optical signal in the core. However, a light source naturally injects some light rays into the core at angles less than the critical angle, which is perpendicular to the plane of the core/cladding interface. At such severe angles, the incident light rays penetrate the interface and enter the cladding, where they may be lost. The numerical aperture essentially is an indication of how well an optical fiber accepts and propagates light. As illustrated in Figure N-2, optical fiber with a small NA (top) requires more directional, i.e., collimated, light, whereas fiber with a large NA (bottom) does not. The higher NA allows the fiber to accept more light and propagate more modes. The NA is mathematically equal to the sine of the angle of acceptance. *Note:* The NA is important in multimode fiber (MMF). It is not, however, a critical measurement in single-mode fiber (SMF), as the small core supports only a single mode of propagation and, therefore, the light is neither reflected nor refracted. The light-accepting ability can also be defined in terms of the cone of acceptance, which is the maximum angle at which the fiber will accept incident light, represented in three dimensional view. See also angle of *acceptance*, *collimation*, *cone of acceptance*, *critical angle*, *MMF*, *SMF*, and *total internal reflection*.

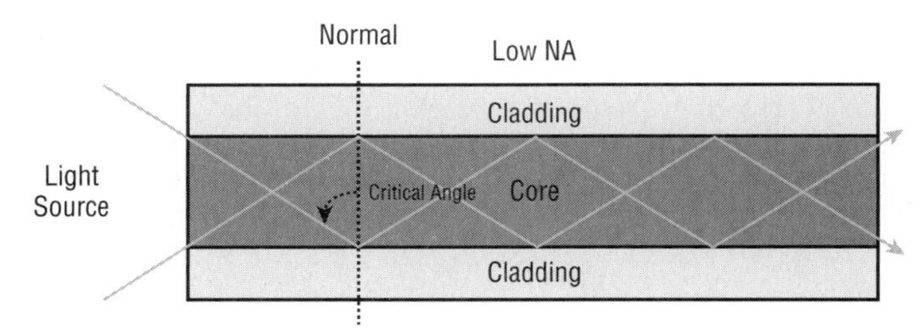

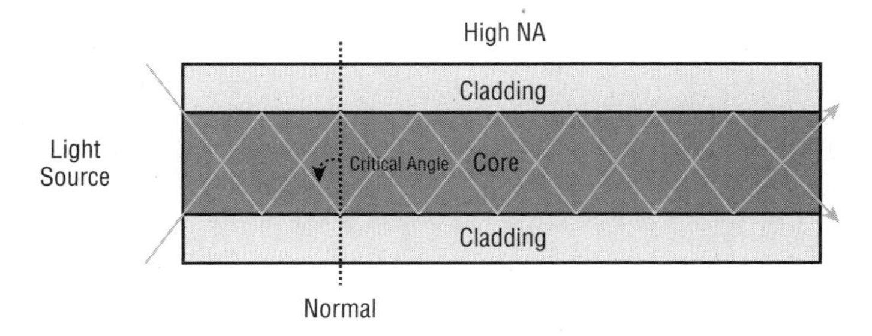

Figure N-2

NXX The central office (CO) prefix, also known as the central office exchange (COE) code is a three-digit number that identifies the central office and the associated geographic carrier serving area (CSA). The CO prefix specifies the dialing pattern NXX, with N indicating that only numbers 2–9 are allowed, as 0 or 1 would confuse the network, and X indicating that any number is allowed. A call within an area code requires that the caller enter only the CO prefix and the line number, i.e., NXX.xxxx, unless there is an overlay area code.

The line number can consist of any four digits (xxxx), and corresponds with a port on a switch that connects to a circuit or channel over a local loop that serves the physical premises of a subscriber. In the case of a cellular telephone number, there is no local loop, as such, and the number is associated with a handset, rather than a physical premises.

See also *CO*, *CSA*, *GSTN*, *NANP*, *overlay area code*, and *PSTN*.

Nyquist, Harry (1889–1976) A physicist and engineer at AT&T Bell Telephone Laboratories, where he did important work in areas such as thermal noise, telegraphy, facsimile systems, and television. Nyquist also did significant work that laid the foundations for information theory, as was subsequently developed by Claude Shannon. In 1928, Nyquist published a paper entitled Certain Topics in Telegraph Transmission Theory, in which he expressed what is now known as the Nyquist theorem. In telecommunications, the Nyquist theorem forms the basis for pulse code modulation (PCM), the fundamental method for converting analog voice to digital format. See also *Nyquist theorem* and *PCM*.

Nyquist theorem The theorem developed by Harry Nyquist and published in his 1928 paper entitled "Certain Topics in Telegraph Transmission Theory." The Nyquist theorem states that an analog signal waveform can be converted to digital format and be reconstructed without error from samples taken at equal time intervals if the sampling rate is equal to, or greater than, twice the highest frequency component in the analog signal. The Nyquist theorem forms the basis for pulse code modulation (PCM), the fundamental method for converting analog voice to digital format. See also *Nyquist, Harry* and *PCM*.

NZDF (Non Zero Dispersion-Shifted Fiber) A type of dispersion-shifted single-mode fiber (SMF) that shifts the optimal dispersion point by adjusting the refractive index profile of the core and the cladding. There are two types of dispersion-shifted fiber (DSF). Zero Dispersion-Shifted Fiber (ZDSF) shifts the point of zero dispersion by increasing material dispersion to the point that it cancels out chromatic dispersion at 1550 nm, rather than 1310 nm. Dense Wavelength Division Multiplexing (DWDM) and Erbium-Doped Fiber Amplifiers (EDFAs) both work in this higher window, which can create yet another noise problem in the form of four-wave mixing (FWM), a phenomenon by which wavelengths interact to create additional wavelengths. The EDFAs amplify those signals, and superimpose them on the DWDM channels. Non Zero Dispersion-Shifted Fiber (NZDF) addresses this issue by shifting the optimal dispersion point slightly above the range in which EDFAs operate. See also *chromatic dispersion*, *cladding*, *core*, *dispersion*, *DSF*, *DWDM*, *EDFA*, *FWM*, *material dispersion*, *noise*, *refractive index*, *SMF*, *wavelength*, *window*, and *ZDSF*.

O **1.** Ohm. The ASCII symbol for ohm. The preferred symbol is Ω, but it doesn't print easily with some software programs. **2.** O+ (O positive). The most common blood type. Give blood, please. Assuming that the technician is skilled and the equipment is properly sterilized, it doesn't take long, it doesn't hurt, and it doesn't make you ill, but it may very well save a life.

O&M (Operations and Maintenance) Referring to all functions associated with the ongoing operations and maintenance of a system or network.

OADM (Optical Add/Drop Multiplexer) In wavelength division multiplexing (WDM) a device that is able to add or drop individual wavelengths without separating all wavelengths and certainly without converting them to electrical format. An OADM is a purely optical version of an ADM. See also *ADM*, *wavelength*, and *WDM*.

OAM&P (Operations, Administration, Management, and Provisioning) Referring to all functions associated with the ongoing operation, administration, and management of a system or network, as well as its initial installation.

O-Band (Original Band) The ITU-T standard optical transmission window in the wavelength range of 1,260–1,360 nm. See also *wavelength* and *window*.

oblique Sloping, slanting, or indirect, i.e., neither perpendicular nor parallel. See also *angle of incidence* and *obtuse*.

OBRA 93 See *Omnibus Budget Reconciliation Act of 1993*.

obsessive compulsive disorder (OCD) See *OCD*.

obtuse **1.** In mathematics, an angle greater than 90 degrees, but less than 180 degrees. **2.** A person who is slow to comprehend or understand, or who is dull or insensitive. We all know such people. See also *abstruse*.

OC (Optical Carrier) The optical signal as defined in Synchronous Optical Network (SONET) for North America and Synchronous Digital Hierarchy (SDH) standards for international applications. All SONET levels are expressed in multiples of OC-1 (51.84 Mbps), which is a T3 signal (44.736 Mbps), or the equivalent, plus SONET signaling and control overhead, converted from electrical to optical format. The SONET digital hierarchy begins at OC-1 (51.84 Mbps) and the SDH hierarchy at OC-3 (155.52 Mbps). From OC-3, the progression of SONET and SDH levels are the same, culminating in OC-768 (39.813 Gbps). OC-1536 (79.626 Gbps) and OC-3072 (159.252 Gbps) have yet to be fully defined. See SONET for a table of OC-N levels. See also *overhead*, *SDH*, *signaling and control*, *SONET*, and *T3*.

OCC (Other Common Carrier) The term applied in the United States prior to the breakup of the Bell System (January 1, 1984) to describe the long-distance carriers that competed with AT&T Long Lines. As a result of the Modified Final Judgement (MFJ) that caused the reorganization of the Bell System, AT&T and the OCCs became known as interexchange carriers (IXCs, or IECs) and initially were not allowed to provide intraLATA long distance service. See also *Bell System*, *IXC*, *LATA*, *long distance*, and *MFJ*.

OCD (Obsessive Compulsive Disorder) A type of psychiatric anxiety disorder characterized by a person's recurrent, unwanted thoughts (obsessions) and related ritualistic and repetitive behaviors (compulsions) such as counting, checking, cleaning, or handwashing in an attempt to neutralize the obsessions. The rituals, however, provide only temporary relief and, therefore, must be repeated, again and again. By the way, there are 54 words and 406 characters in this definition, including spaces and not including this sentence, or the following sentences in this definition. There are 84 words and 503 characters in the previous definition. Obsessive compulsive disorder comprises 3 words and 30 characters, with spaces. OCD is one word, although it really is not a word, and 3 letters, with no spaces. Personally, I prefer phrases that contain an even number of words and words that contain an even number of letters. Thank you. (By the way, thank you is 2 words of 5 and 3 letters, which is OK because 5 + 3 = 8, which is an even number.)

OCR (Optical Character Recognition) Application software that allows a computer to recognize printed or written characters, e.g., letters, numbers, punctuation marks, and pictograms using an optical scanner for input. Optical mark recognition (OMR) also employs an optical scanner, but does not employ a character recognition engine. See also *OMR*.

octet A unit of data eight bits in length. An octet may comprise multiple fields of data or perhaps a small portion of a field and, therefore, does not necessarily bear any relationship to a byte. A byte generally comprises eight (8) bits that represent a letter in an alphabet (e.g., a, A, z, or Z), a diacritical mark (e.g., ~ or `), a single digit number (e.g., 0, 1, 2, or 3), a punctuation mark (e.g., , ., or !), or a control character (e.g., paragraph break, page break, or carriage return). A byte also may be four (4) bits, five (5) bits, 16 bits, or even 32 bits in length. See also *byte*.

octothorpe See #.

OD (Outside Diameter) The measure of the width of the outer surface of a circular or cylindrical object such as a hollow pipe or tube.

ODU (Optical Data Unit) An entity for processing in an Optical Transport Network (OTN). An ODU comprises a frame or series of frames in their native protocol, encapsulated in a digital wrapper for network management purposes. See also *digital wrapper*, *encapsulate*, *frame*, *network management*, *OTN*, and *protocol*.

OEM (Original Equipment Manufacturer) A company that builds components or products that are incorporated into products or systems sold by another company commonly referred to as a value-added reseller (VAR). An Ethernet network interface card (NIC), for example, might be built by an OEM to the specifications of a manufacturer of laptop or tablet computers.

OEO (Optical-Electrical-Optical) Optical repeaters are characterized as being OEO in nature. Such a repeater receives an attenuated optical signal, converts it to an amplified electrical signal, reshapes and retimes it, converts it back to optical signal, and retransmits it. See also *OOO* and *repeater*.

OFC (Optical Fiber, Conductive) The designation by the National Fire Protection Association (NFPA) for fiber optic cables that contain at least one electrically conductive, non–current-carrying component, such as a metallic strength member or vapor barrier, and are not certified for use in interior plenum or riser applications. See also *cable*, *conductor*, *current*, *optical fiber*, *plenum cable*, and *riser cable*.

OFCP (Optical Fiber, Conductive, Plenum) The designation by the National Fire Protection Association (NFPA) for fiber optic cables that contain at least one electrically conductive, non–current-carrying component, such as a metallic strength member or vapor barrier, and are certified for use in interior plenum applications. See also *cable*, *conductor*, *current*, *optical fiber*, and *plenum cable*.

OFCR (Optical Fiber, Conductive, Riser) The designation by the National Fire Protection Association (NFPA) for fiber optic cables that contain at least one electrically conductive, non–current-carrying component, such as a metallic strength member or vapor barrier, and are certified for use in interior riser applications. See also *cable*, *conductor*, *current*, *optical fiber*, and *riser cable*.

OFDM (Orthogonal Frequency Division Multiplexing) Synonymous with discrete multitone (DMT). See also *COFDM* and *DMT*.

offered load In frame relay, the data rate, as measured in bits per second (bps) offered to the network for delivery. The aggregate offered load can be less than the access rate supported by the access link and/or the port speed of the frame relay network device (FRND), but can never exceed whichever is less. See also *access rate*, *data rate*, *frame relay*, *FRND*, *link*, and *port*.

off-hook A condition that exists when a telephone receiver or handset is lifted or removed from its cradle, thereby completing a circuit and placing the telephone in use. The term refers to the fact that early telephone handsets hung from a switch hook, a hook that activated a switch. See also *on-hook*.

off-hook voice announce A key telephone system (KTS) feature that enables the system attendant to announce another incoming call even if the station user is off-hook and engaged in another call. In some implementations, the announcement is through the telephone speaker, with the station microphone muted so that the other party does not hear the announcement. In other implementations, the announcement is through the handset receiver so that the other party is not muted, as the muting is noticeable. See also *KTS*.

off-premises extension (OPX) Synonymous with off-premises station (OPS). See *OPX*.

off-premises station (OPS) Synonymous with off-premises extension (OPX). See *OPX*.

offset quadrature phase shift keying (OQPSK) See *OQPSK*.

offshoring The exporting of certain business functions to foreign countries, where either company employees or independent contractors perform the work. Offshoring commonly involves the exporting of work from developed countries to undeveloped foreign countries for reasons that include cost reduction, tax avoidance, legal liability avoidance, and strategic market expansion. If work is shifted to independent contractors, the process becomes one of offshore outsourcing. See also *job security* and *outsourcing*.

OFN (Optical Fiber, Nonconductive) The designation by the National Fire Protection Association (NFPA) for fiber optic cables that contain no electrically conductive components and are not certified for use in interior plenum or riser applications. See also *cable*, *conductor*, *optical fiber*, *plenum cable*, and *riser cable*.

OFNP (Optical Fiber, Nonconductive, Plenum) The designation by the National Fire Protection Association (NFPA) for fiber optic cables that contain no electrically conductive components and are certified for use in interior plenum applications. See also *cable*, *conductor*, *optical fiber*, and *plenum cable*.

OFNR (Optical Fiber, Nonconductive, Riser) The designation by the National Fire Protection Association (NFPA) for fiber optic cables that contain no electrically conductive components and are certified for use in interior riser applications. See also *cable*, *conductor*, *optical fiber*, and *riser cable*.

OH The chemical symbol for hydroxyl. See also *hydroxyl*.

ohm (Ω) The unit of both resistance (R) and impedance (Z). In the International System of Units (SI), one (1) ohm is the resistance such that a difference in potential of one (1) volt (V) between the positive (+) end and the negative (–) end of a conductor produces a constant current of one (1) ampere (A). See also *ampere*, *current*, *impedance*, *mho*, *Ohm's Law*, *resistance*, *SI*, and *volt*.

Ohm, Georg Simon (1787–1854) The German physicist whose research on electric currents led to the formulation of Ohm's Law. See also *Ohm's Law*.

Ohm's Law $V = I \times R$ or $I = V/R$ The law of physics that defines the relationships between power, voltage, current, and resistance in linear constant-current circuits. Ohm's Law states that the current (I) in an electric circuit is directly proportional to the electromotive force (emf), or voltage (E), applied to a circuit, and inversely proportional to the resistance (R) of the circuit. Another way of thinking of Ohm's Law is that the current flowing through a wire is directly proportional to its cross-sectional area and inversely proportional to its length. So, for a circuit of a given metal at a given constant temperature, the thicker and shorter the wire, the less the resistance. Ohm's Law is named for Georg Simon Ohm, its inventor. See also *current*, *emf*, *resistance*, and *voltage*.

OLT (Optical Line Terminal) In a passive optical network (PON), the device that terminates the optical local loop at the edge of the network. In a telco PON, the OLT is housed in the central office (CO). In a CATV PON, the OLT is housed in the headend. The OLT can either generate downstream optical signals on its own, or can pass optical signals from the optical backbone through a collocated optical crossconnect or multiplexer. The OLT also receives upstream signals from the optical network terminals (ONTs) at the customer premises and optical network units (ONUs) in remote nodes. See also *backbone*, *CATV*, *CO*, *cross-connect*, *downstream*, *headend*, *local loop*, *multiplexer*, *OLT*, *ONU*, *PON*, *telco*, and *upstream*.

Omnibus Budget Reconciliation Act of 1993 (OBRA 93) Also known as the *Deficit Reduction Act of 1993* and the *Revenue Reconciliation Act of 1993*. In the United States, an act of Congress that included a provision authorizing the Federal Communications Commission (FCC) to auction domestic radio spectrum. See also *FCC* and *spectrum*.

OMR (Optical Mark Recognition) The process of gathering data with an optical scanner by measuring the reflectivity of light at predetermined positions on a surface. OMR differs from optical character recognition (OCR), which requires a recognition engine in order to make sense of written characters. See also *OCR*.

ones density The density of one (1) bits in a digital bit stream. Depending on specific nature of the T-carrier network, 15–80 zero (0) bits can be transmitted in a row as long as the density of ones is at least 12.5 percent (1 in 8) over a specified interval of time. Ones density ensures that there are electrical pulses on a circuit with at least a minimal density in order to keep the various circuit terminating equipment and repeaters synchronized. Ones density is maintained by a channel service unit (CSU), a type of data communications equipment (DCE) that provides the customer interface to a digital circuit. In the event that the circuit is silent, i.e., there is no active data transmission, the CSU will regularly transmit one (1) bits, known as *keep alive bits*, for the same purpose. See also *CSU*, *DCE*, *pulse*, *repeater*, *synchronize*, and *T-carrier*.

on-hook A condition that exists when a telephone receiver or handset is in its cradle and available to receive an incoming call. The term refers to the fact that early telephone handsets hung from a switch hook, a hook that activated a switch. See also *off-hook*.

ONT (Optical Network Terminal) In a passive optical network (PON), the device that terminates the optical local loop at the customer premises in a fiber-to-the-premises (FTTP) scenario. The ONT serves as a media converter, interfacing the optical fiber to the copper-based inside wire. The ONT is an addressable device that recognizes and accepts downstream data addressed to it specifically, ignoring all other data. The ONT is synchronized with the optical line terminal (OLT) at the network edge and is assigned time slots for user-generated upstream data, which it buffers as necessary. See also *buffer*, *downstream*, *FTTP*, *local loop*, *media converter*, *OLT*, *optical fiber*, *PON*, *synchronize*, and *upstream*.

ONU (Optical Network Unit) In a passive optical network (PON), the device that terminates the optical circuit in a remote network node in a fiber-to-the-curb (FTTC) or fiber-to-the-neighborhood (FTTN) scenario. The ONU serves as a media converter, interfacing the fiber circuit to embedded copper in the form of unshielded twisted pair (UTP) in a telco network and coaxial cable in a CATV network. The ONU physically is positioned between the optical line terminal (OLT) at the network edge and the optical network terminal (ONT) at the customer premises. See also *CATV*, *circuit*, *coaxial cable*, *FTTC*, *FTTN*, *media converter*, *node*, *OLT*, *ONT*, *PON*, *telco*, and *UTP*.

OOO (Optical-Optical-Optical) A device that is purely optical in nature, rather than having electrical or electronic as well as optical components. The repeaters used in fiber optic transmission systems (FOTS), for example, are Optical-Electrical-Optical (OEO). Optical switches used in wavelength division multiplexed (WDM) transmission systems sometimes are purely optical. See also *FOTS*, *OEO*, *optical*, and *WDM*.

opaque OXC (opaque Optical CROSS-Connect) A digital cross-connect (DXC) employed in an optical system. This approach requires that the optical signal be converted into electrical format before being switched by the electronic DXC, then converted back into optical format before being placed back on the fiber optic transmission system (FOTS). This optical-electrical-optical (OEO) signal conversion process adds some amount of delay to the signal, but offers the advantage of signal regeneration to deal with issues of signal dispersion and attenuation. A transparent OXC is an optical-optical-optical (OOO) device that includes an optical switching module. A translucent OXC is a hybrid that includes both electronic and optical switching modules. See also *attenuation*, *dispersion*, *DXC*, *FOTS*, *OEO*, *OOO*, *OXC*, *translucent OXC*, and *transparent OXC*.

open circuit **1.** A signal path, line, or channel that is available for use. **2.** In electrical engineering, an electrical loop or path that contains infinite impedance. An open may indicate a fault, as in a severed cable, or may be intentional, as in a switch. See also *impedance*.

The Open Group An inter-industry forum dedicated to the development and promotion of electronic messaging, including e-mail, fax, electronic data interchange (EDI), based on open standards and global interoperability. Previously known as the Electronic Messaging Association (EMA), The Open Group is technology-neutral and vendor-neutral. See Appendix A for contact information. See also *EDI*, *e-mail*, and *fax*.

open-loop algorithm In frame relay, a congestion control mechanism (or lack thereof) that permits the frame relay network device (FRND) to accept incoming frames with no prior knowledge of the likelihood of the network's ability to deliver them successfully. See also *closed-loop algorithm*, *congestion*, *frame relay*, and *FRND*.

Open Shortest Path First (OSPF) See *OSPF*.

open source Also known as *open source software* (OSS). Software distributed under a license that makes the source code (i.e., program instructions) freely available to the end user. Such a license often encourages the user to modify the source code as long as the modifications are made freely available to other users, as well. Open source software generally is available at no charge, i.e., free. Linux is a classic example of successful open source software. See also *software*, *Linux*.

open source software (OSS) See *open source*.

open standard A set of specifications that are standardized by a formal body and are then published and made freely available to the technical community. See also *standard*.

Open Systems Interconnection Model See *OSI Reference Model*.

open wire Referring to a metallic wire pair circuit comprising uninsulated electrical conductors. The conductors are physically separated, attached to insulators of glass, ceramic, or plastic mounted on crossarms that are attached to poles or towers. Open wire circuits are unusual in contemporary telecommunications, where they historically were used to traverse open country in rural outside plant (OSP) applications. See also *circuit*, *conductor*, *insulation*, and *OSP*.

operating system (OS) See *OS*.

operations, administration, management, and provisioning (OAM&P) See *OAM&P*.

operations and maintenance (O&M) See *O&M*.

operations support system (OSS) See *OSS*.

operator **1.** Someone who operates something, such as a machine. A telephone operator, for example, originally operated a switchboard, which at one time was thought to be a highly complex machine. A contemporary operator mans a PBX console or perhaps a complex traffic service position system (TSPS) console. **2.** A mathematical symbol or term that performs or describes an operation or process. Examples include multiplication, addition, and subtraction signs.

OPS (Off-Premises Station) Synonymous with off-premises extension (OPX). See *OPX.*

optical In physics, relating to the study of light in the infrared (IR), visible, and ultraviolet (UV) regions of the electromagnetic spectrum. See also *electromagnetic spectrum*, *IR*, *light*, *UV*, and *visible light.*

optical add/drop multiplexer (OADM) See *OADM.*

Optical Carrier (OC) See *OC.*

optical character recognition (OCR) See *OCR.*

optical cross-connect (OXC) See *OXC.*

optical data unit (ODU) See *ODU.*

optical-electrical-optical (OEO) See *OEO.*

optical fiber A slender strand of transparent glass or plastic specially constructed to serve as a dielectric conductor, or waveguide, of infrared (IR) light in a fiber optic transmission system (FOTS). The fiber generally is one of many types of glass optical fiber (GOF) although plastic optical fiber (POF) is sometimes used. GOF offers the advantage of very low signal attenuation over long distances, in support of signaling rates that currently are as high as 40 Gbps per lambda, or wavelength. POF sometimes is used over short distances where its flexibility, general durability, and low cost are advantageous. See also *GOF* and *POF.*

optical fiber, conductive (OFC) See *OFC.*

optical fiber, conductive, plenum (OFCP) See *OFCP.*

optical fiber, conductive, riser (OFCR) See *OFCR.*

optical fiber, nonconductive (OFN) See *OFN.*

optical fiber, nonconductive, plenum (OFNP) See *OFNP.*

optical fiber, nonconductive, riser (OFNR) See *OFNR.*

optical isolator **1.** A device used in a high-speed, high-power fiber optic transmission system (FOTS) to isolate erbium-doped fiber amplifiers (EDFAs), laser diode light sources and other devices. The optical isolators act as diodes, preventing signals from propagating in the upstream direction and confusing the downstream signal or device generating it. See also *diode*, *EFDA*, *FOTS*, and *laser diode.* **2.** A device containing a short length of optical fiber and inserted into an electrified communications link to electrically isolate the attached devices from ground loops, power spikes, and surges. Such an optical isolator can act as a repeater, as well as a protector. See also *link*, *optical fiber*, *protector*, and *repeater.*

optical line-of-sight (optical LOS) See *LOS.*

optical line terminal (OLT) See *OLT.*

optical mark recognition (OMR) See *OMR.*

optical network terminal (ONT) See *ONT.*

optical network unit (ONU) See *ONU.*

optical-optical-optical (OOO) See *OOO.*

optical transport network (OTN) See *OTN.*

optical transport unit (OTU) See *OTU.*

optics The branch of physics dealing with the nature and properties of electromagnetic energy in the light spectrum and the phenomena of vision. In the broadest sense, optics deals with infrared light, visible light, and ultraviolet light. See also *electromagnetic spectrum, infrared light, physics, ultraviolet light*, and *visible light.*

OPX (Off-Premises eXtension) Synonymous with off-premises station (OPS). A key telephone system (KTS) or PBX station or extension that terminates on a telephone physically located off-premises, typically on a non-contiguous property. In addition to a special line card, an OPX requires a dedicated private line, typically leased from the local exchange carrier (LEC). The OPX enjoys all of the features of an extension located on-premises, but at considerable cost. OPXs are, and always were, uncommon, but sometimes are used to connect a distant guard shack or security building to the PBX or perhaps to provide a technically challenged key executive with a PBX extension at home. See also *KTS, LEC*, and *PBX.*

OQPSK (Offset Quadrature Phase Shift Keying) A variant of QPSK that uses a half-symbol timing offset to prevent large amplitude fluctuations in the modulated signal. See also *amplitude, modulate, QPSK, signal*, and *symbol.*

orange hose A term used to describe the thick coaxial cable specified by the IEEE as 10Base5, for use in early Ethernet networks. The term was in reference to the orange cable sheath used by some manufacturers. The cable also was about as thick as a garden hose. See also *10Base5, coaxial cable, Ethernet*, and *IEEE.*

orderwire A circuit or channel used by technical personnel for coordination and control functions relating to activation, deactivation, reconfiguration, reporting, and maintenance of communications systems, networks, and services. See also *channel* and *circuit.*

.org (organization) Pronounced *dot org.* Originally, the Internet generic Top Level Domain (gTLD) reserved exclusively for noncommercial (not-for-profit) organizations. The domain is now unrestricted. This is an unsponsored domain. See also *Internet, gTLD*, and *unsponsored domain.*

original band (O-Band) See *O-Band.*

original equipment manufacturer (OEM) See *OEM.*

orthogonal **1.** In mathematics, at right angles to or perpendicular to. **2.** In telecommunications, describing radio frequency (RF) signals that are independent and mutually exclusive and, therefore, avoid intersymbol interference. Thereby, a receiver can recognize a legitimate signal and reject an unwanted signal or signal element. The concept applies to orthogonal frequency division multiplexing (OFDM), also known as discrete multitone (DMT). See also *DMT, interference, RF, signal*, and *symbol.*

orthogonal frequency division multiplexing (OFDM) Synonymous with discrete multitone (DMT). See *DMT.*

OS (Operating System) The master software program that controls the allocation and usage of all internal resources (e.g., memory, queuing, input/output, central processing unit [CPU], time, disk space, and transmission and reception processes) and, thereby, enables and controls the operation of the entire computer system. The OS forms the foundation for application software that performs end-user tasks such as word processing and mathematical calculation. An OS is largely software, although there are firmware components. See also *firmware, program*, and *software.*

oscillate Vary predictably, or rhythmically, between two extremes, usually within a set period of time. An alternating current (AC) waveform, for example, oscillates between maximum and minimum electrical values, which are positive voltage (+V) and negative voltage (–V), respectively. See also *AC, voltage*, and *waveform.*

oscillator An electronic circuit designed to produce an ideally stable alternating current (AC) or voltage. See also *AC, circuit, crystal oscillator, current, electronic, oscillate*, and *voltage*.

OSI Reference Model (Open Systems Interconnection Model) A layered architecture consisting of a set of international networking standards developed by the International Organization for Standardization (ISO) in 1983, and now known collectively as ITU-T Recommendation X.200. The ISO promoted the model as a full standard, and the United States federal government and many computer manufacturers invested heavily (billions in US$) in compliance. The initiative failed within a few years, at least in the United States, as SNA, TCP/IP, and a few other standards seemed to have satisfied most people's appetite for standards and OSI seemed too complex and redundant. The European community, however, embraced OSI, at least in part because of the confusion caused by the multinational nature of the region and, therefore, the plethora of national standards. The European Union (EU) actually legally imposed the model for some applications. Eventually, TCP/IP pushed OSI aside as a standard, but the reference model remains valuable, and most manufacturers relate their products to the model in order to put them in context. So, the OSI Reference Model continues to have great value. The model is a layered architecture that defines a set of common rules that computers of disparate origin can use to exchange information. The layers serve to segment functions, so that each layer can be considered independently, yet all are interrelated, with supporting software embedded in each node providing the interface between layers.

In a typical scenario, a transmitting device uses the top layer, at which point the data is placed into a packet, prepended by a header. The data and header, known collectively as a *Protocol Data Unit* (PDU), are handled by each successively lower layer as the data works its way across the network to the receiving node, typically with each layer adding a header. At the receiving node, the data works its way up the layered model; successively higher layers strip off the header information. While in transit, the data may work its way up and down the model as it transits different networks and subnetworks running different protocols. The seven layers of the OSI Reference Model and their functions are organized in Table O-1.

Table O-1: OSI Reference Model Layers

Layer	*Functional Focus*
7 Application	Semantics: Applications and end user processes such as e-mail, file transfer, and authentication
6 Presentation	Syntax: Data format (coding) and display, code conversion, encryption, compression
5 Session	Dialog coordination: Establishing, maintaining, coordinating, and terminating dialogues and data exchanges
4 Transport	Reliable data transfer: End-to-end error detection and correction, and flow control, ensuring the integrity of the complete datastream
3 Network	Routing and relaying: Message routing, error detection, and control of internodal traffic
2 Data Link	Technology-specific transfer over a link or channel: Framing, error control, flow control, data sequencing, time-out levels, and data formatting (encoding and decoding)
1 Physical	Physical connections: Electrical and mechanical aspects of the interface of a device to a transmission medium

See also *application layer, data link layer, ISO, ITU-T, network, network architecture, Network Layer, Physical Layer, Presentation Layer, protocol, Session Layer, SNA, software, standard, subnetwork, TCP/IP, Transport Layer, X.200*, and *x series*.

OSP (OutSide Plant) All of the telecommunications apparatus and cable systems outside (i.e., not housed in buildings) such as central offices or customer premises. OSP includes all the components of cable systems such as the aerial, buried, and underground cables, amplifiers and repeaters, cross-connect boxes, and remote neighborhood nodes, some of which may be located in vaults or sheds. See also *inside plant.*

OSPF (Open Shortest Path First) A link-state protocol that routers use to exchange IP packets over the shortest available end-to-end path based on link-state advertisements from other routers. Router advertisements identify the direction, availability, and cost of links to other routers. With that information, the originating router is able to determine its exit port that leads to the shortest available path to the destination router. OSPF is an Interior Gateway Protocol (IGP) used to exchange path information between routers in the same network domain. See also *domain, IGP, IP, link-state protocol, packet, path, port,* and *router.*

OSS (Operations Support System) **1.** Referring to the system or systems that perform management, inventory, engineering, planning, and repair functions for communications service providers. Specific OSS functions can include inventory management, cost allocation, billing, service order management, trouble ticket management, traffic analysis, capacity planning, and network optimization. **2.** Open source software. See *open source software.*

Ostrofsky, Marc The world record holder, to the best of my knowledge, with respect to the sale price of a domain name. In the mid-1990s, Ostrofsky put a number of his employees to work searching for domain names that had e-commerce potential. They searched for just about any word in the English language, especially words preceded with e- or e. As a result of that effort, he identified and registered `www.eflowers.com`. In 1999, he rejected an offer of $1,000,000 from Flowers Direct, preferring to sell it to them for $25,000, plus $0.50 for every transaction generated over the Web site, plus free flowers for his wife (now ex-wife), Sarah, for the rest of her life. Given the projections of 500,000 transactions per year, Ostrofsky realized an excellent return on an investment of $70.00. Ostrofsky later sold the rights to `www.business.com` for $7,000,000, a domain name he had acquired a few years earlier for $250,000 for use in connection with a business he later sold for many millions of dollars. See also *domain name.*

OTA (over the air) See *over the air.*

Other Common Carrier (OCC) See *OCC.*

OTN (Optical Transport Network) Described in the ITU-T Recommendation G.709 (2003), OTN adds operations, administration, maintenance, and provisioning (OAM&P) functionality to optical carriers, specifically in a multi-wavelength system such as dense wavelength division multiplexing (DWDM). OTN specifies a digital wrapper, which is a method for encapsulating an existing frame of data, regardless of the native protocol, to create an optical data unit (ODU), similar to that used in SDH/SONET. OTN provides the network management functionality of SDH and SONET, but on a wavelength basis. A digital wrapper, however, is flexible in terms of frame size and allows multiple existing frames of data to be wrapped together into a single entity that can be more efficiently managed through a lesser amount of overhead in a multi-wavelength system. The OTN specification includes framing conventions, nonintrusive performance monitoring, error control, rate adaption, multiplexing mechanisms, ring protection, and network restoration mechanisms operating on a wavelength basis. A key element of a digital wrapper is a Reed-Solomon forward error correction (FEC) mechanism that improves error performance on noisy links. Digital wrappers have been defined for 2.5-, 10-, and 40-Gbps SDH/SONET systems. SDH/SONET operation over an OTN involves additional overhead due to encapsulation in digital wrappers. The resulting line rates are defined as optical transport units (OTUs). See also *digital, DWDM, error control, FEC, ITU-T, link, multiplexer, ODU, OTN, OTU, rate adaption, Reed-Solomon,* and *wavelength.*

OTU (Optical Transport Unit) In an optical transport network (OTN), an optical channel for transporting optical data units (ODUs), which comprise native frames encapsulated in digital wrappers. ITU-T Recommendation G.709 specifies three line rates, which are listed in Table O-2, alongside their corresponding SDH/SONET line rates.

Table O-2: G.709 Line Rates and Corresponding SDH/SONET Line Rates

G.709		*SDH/SONET*	
Interface	*Line Rate*	*OC/STM Level*	*Line Rate*
OTU-1	2.666 Gbps	OC-48/STM-16	2.488 Gbps
OTU-2	10.709 Gbps	OC-192/STM-64	9.953 Gbps
OTU-3	43.018 Gbps	OC-768/STM-256	39.813 Gbps

G.709 also specifies an interface for 10 GbE clients, utilizing the same digital wrapper, which results in a line rate of 11.095 Gbps. See also *10GbE, channel, digital wrapper, encapsulate, frame, ITU-T, line rate, ODU, OTN, SDH,* and *SONET.*

out-of-band signaling and control Signaling and control that takes place over frequencies (e.g., in guard bands) or in time slots separate from those that carry user payload. In an analog context, out-of-band signaling can take place over the guard bands that separate channels. T-carrier signaling clearly is in-band rather than out-of-band, as it involves bit robbing, which periodically replaces payload bits with signaling bits. E-carrier signaling and control occurs exclusively in time slots reserved for that purpose. Packet technologies such as frame relay, Internet Protocol (IP), asynchronous transfer mode (ATM), and X.25 variously include headers and trailers that support out-of-band signaling, and use separate signaling packets for various network management purposes. Truly out-of-band signaling takes place over an entirely separate path and even a separate network, as is the case with Signaling System 7 (SS7). In any case, out-of-band signaling and control does not directly compete with payload for bandwidth. See also *analog, ATM, digital, E-carrier, frame relay, guard band, header, in-band signaling and control, IP, network management, packet, payload, SS7, T-carrier, trailer,* and *X.25.*

out-of-phase **1.** A signal that has suffered phase distortion so that the sinusoidal waveform has been unintentionally altered in phase, or periodic angle. See also *phase* and *sine wave.* **2.** Someone who is out-of-sync or out-of-step with the normal world, whatever that is. Such people may be crazy as loons, but they seem to get by, and some even thrive, in spite of it. See also *normal.*

outside plant (OSP) See *OSP.*

outside vapor deposition (OVD) See *OVD.*

outsourcing The transferring of certain business functions from internal staff to outside contractors. Outsourcing commonly is applied to non-core functions, such as accounting, information technology, human resources, facilities management, fleet management, parts manufacturing, payroll, press relations, and real estate management for reasons that include lowering costs, avoiding liabilities, and allowing management to focus on the core business. Outsourcing also is an excellent way for management to shift or even avoid responsibility. See also *job security* and *offshoring.*

OVD (Outside Vapor Deposition) A commonly used technique for the mass production of glass optical fiber, OVD begins with heating silica and germanium to the point of vaporization. As the glass vapor cools, it is deposited as layers of soot on the outside of a rotating hollow ceramic bait rod to create a glass cylinder. The first layer is the core material of germanium-doped silica. On top of the core material, many layers of slightly purer silica soot, i.e., silica with lower levels of dopants, are deposited to form the cladding. If the end product is to be a step-index fiber, there is an abrupt change in the chemical composition between the core and cladding. If the end product is to be a graded-index fiber, there will be many graded layers of silica of slightly different chemical compositions deposited on the core to yield slightly and successively purer layers of cladding surrounding the fiber axis. The composition of the glass layers in a graded-index fiber is much like the arrangement of the annular rings of a tree. When the deposition process is complete, the bait rod is slipped out of the glass cylinder, which is then sintered and collapsed into a preform cylinder, which is cooled and stored. The tip of the preform cylinder is reheated to a temperature of 2,500 degrees in a drawing tower. The resulting gob of molten glass is carefully drawn by gravity, in a process known as *broomsticking*, into a fiber as long as 20 kilometers. As the fibers cool, an acrylate coating is applied to protect the raw glass from physical damage. As is the case with all of these techniques, OVD takes place in a vacuum environment, as it is the exposure to oxygen that makes glass so brittle. Inside vapor deposition (IVD) is a similar process, with the soot deposited on the inside of a rotating glass tube that becomes the outside cladding. See also *cladding*, *core*, *graded-index fiber*, *IVD*, *sinter*, and *step-index fiber*.

overhead Data that is not part of the user data, but that is stored or transmitted with it. Overhead can be used for a wide variety of purposes, such as circuit monitoring, channel separation, addressing, error control, priority indication, and congestion management. Although overhead is essential to the integrity of data storage and transmission, it reduces the amount of user data that can be stored or transmitted. In a circuit or channel used to maximum capacity, bandwidth less overhead equals throughput. See also *bandwidth* and *throughput*.

overlay In telecommunications, a deployment in which the new infrastructure parallels that of existing infrastructure. In a passive optical network (PON) deployment, for example, this approach enables the service provider to construct the new system and provide enhanced broadband service to PON subscribers as required, while continuing to serve others subscribing to more basic services from the old cable plant. See also *brownfield*, *greenfield*, *infrastructure*, and *PON*.

overlay area code An area code that covers the same geographical area as an existing area code. As a result, callers within the area must dial a full 10-digit number (NNX-NNX-xxxx) to reach any other number within the area. The concept of the overlay area code was developed as a means of avoiding forcing subscribers to change telephone numbers after an area code split. Such telephone number changes cause great disruption, result in lost business and otherwise cost businesses great amounts of money for advertising, re-printing of business cards and letterhead stationary, and so on. See also *area code*.

overlay carrier A carrier that builds a network that overlays, or approximately follows the same physical layout as, the traditional PSTN. Competitive carriers sometimes deploy microwave or other wireless systems, for example, that duplicate the physical topology of the incumbent carrier.

over-over Synonymous with push-to-talk (PTT). See *PTT*.

oversubscribe To place potentially greater demands on a device or circuit than it is capable of handling at one time. A T1 circuit, for example, supports a data rate of 1.536 Mbps, which commonly is subdivided into 24 voice grade channels of 64 kbps. So, the circuit can support 24 voice or data calls of 64 kbps each. If more than 24 calls are offered to the T1, it is said to be oversubscribed. If the calls are all uncompressed voice calls truly requiring continuous bandwidth of 64 kbps, the 25th call must be denied. If the calls are data calls supporting bursty applications, such as e-mail, that do not require continuous bandwidth of 64 kbps, the TDM multiplexer may be able to manage the contention by buffering some of the data and sharing bandwidth among multiple such calls. Thereby, the circuit may be able to support considerably more than the 24 calls supported by a more rigid approach. Oversubscription is the economic foundation of carrier services. For example, a neighborhood of 96 phone lines may be served well by two oversubscribed T1 lines because the probability of all users placing calls at one time is very small (less than 1 percent). See also *bandwidth*, *carrier*, *channel*, *circuit*, and *voice grade*.

over the air (OTA) **1.** Referring to broadcast airwave radio and television transmission. See also *airwave*, *broadcast radio*, *broadcast television*, and *broadcast transmission*. **2.** Referring to methods of distributing software updates and application settings to cellular telephones.

OXC (Optical CROSS-Connect) A network device used to switch high-speed optical signals (e.g., SDH/SONET OC-3, OC-12, and OC-48). An OXC can switch the signal in its entirety, or can demultiplex it and switch its component signals. For example, an OXC can receive an OC-48 signal and demultiplex it into four constituent OC-12 signals, each of which it forwards through a separate OC-12 port. An opaque OXC is essentially a digital cross-connect (DXC) employed in an optical system. This approach requires that the optical signal be converted into electrical format before being switched by the electronic DXC, then converted back into optical format before being placed back on the fiber optic transmission system (FOTS). This optical-electrical-optical (OEO) signal conversion process adds some amount of delay to the signal, but offers the advantage of signal regeneration to deal with issues of signal dispersion and attenuation. A transparent OXC, also known as a *photonic cross-connect* (PXC), is characterized as optical-optical-optical (OOO), as it performs the switching function without converting the signal to electronic format. This approach does not impose the same level of signal processing delay as the OEO process employed in an opaque OXC, but neither does it provide signal visibility. In other words, it is not possible to monitor the signal quality or determine the nature of the higher layer protocols employed. A translucent OXC is a hybrid that includes both optical and electronic switching modules and operates in both opaque and transparent mode. If signal visibility or regeneration is desirable, the electronic module is employed. See also *attenuation*, *cross-connect*, *dispersion*, *DXC*, *OC-3*, *OC-12*, *OC-48*, *OEO*, *OOO*, *optical*, *protocol*, *regenerator*, *SDH*, *signal*, and *SONET*.

P **1.** Peta (P), from the Greek *penta*, meaning *five*, translates to *quadrillion*, referring to the fact that, in terms of order of magnitude in base 1,000, Peta is $1{,}000^5$. In order, that puts it right behind kilo (thousand), Mega (million), Giga (billion), and Tera (trillion). **2.** In terms of the electromagnetic spectrum, PHz (PetaHertz) is a quadrillion (10^{15}) Hertz, which is in the range of visible light, ultraviolet (UV) light, and x-rays, none of which currently have any application in telecommunications. A Pbps would be a quadrillion (10^{15}) bits per second. In transmission systems, therefore, a quadrillion would be exactly 1,000,000,000,000,000, since the measurement is based on a base 10, or decimal, number system. That definitely would be broadband, if it were possible, but it is difficult to imagine a contemporary application for that level of bandwidth. **3.** In computing and storage systems, a PB (PetaByte) is actually 1,125,899,906,842,624 bytes ($1{,}024^5$, or 2^{50}) bytes, as the measurement of internal computer memory is based on a base 2, or binary, number system. The term PB comes from the fact that 1,125,899,906,842,624 is nominally, or approximately, 1,000,000,000,000,000. Until recently, a petabyte was rarely even mentioned. A very few supercomputers and supercomputer centers have access to a PB of networked storage. Google, the Web search company, in 2004 reportedly had 100,000 or so servers that shared a distributed, fault-tolerant file system on the order of a PB. To put a PB in further perspective, some sources suggest that the total volume of information contained in 20,000,000 four-drawer file cabinets full of 250,000,000,000 pages of text would equal approximately 1 PB of storage. See also *byte*, *electromagnetic spectrum*, and *Hertz*. **4.** Power. See also *power*.

P1024B A Unisys protocol used in airline reservations systems such as American Airlines' SABRE System and United Airlines' APOLLO. Both P1024B and Airline Link Control (ALC), the IBM version, employ a non-standard six-bit coding scheme. See also *coding scheme* and *protocol.*

P.563 The ITU-T Recommendation for Single-Ended Method for Objective Speech Quality Assessment in Narrow-Band Telephony Applications, an automated method for evaluating the quality of voice transmissions. P.563 is characterized as non-intrusive, as it does not require inserting a reference signal into the device being tested and does not require the generation of test traffic. See also *P.800*, *P.861*, *P.862*, and *narrowband.*

P.800 The ITU-T standard, Measurements for Subjective Determination of Transmission Quality. Recommendation P.800 provides a standard subjective transmission quality evaluation process in the Conversation Opinion Test. The mean opinion score (MOS) is a mean average of the subjective evaluations of a group of trained volunteer listeners who rate voice transmission in terms of listening quality and listening effort on a scale of 1.0 (bad) to 5.0 (excellent). An MOS score of 4.0 (good) or better is considered toll quality. P.800 is an effective means of quantifying such a highly subjective perception, but it is expensive and time-consuming. In the mid-1990s, the ITU-T began the process of automating the objective measurement and testing of end-to-end voice quality across both circuit-switched and packet-switched networks. Those techniques include the following:

- Perceptual Analysis/Measurement System (PAMS)
- Perceptual Speech Quality Measurement (PSQM), standardized as P.861
- Perceptual Evaluation of Speech Quality (PESQ), standardized as P.862
- Single Ended Method for Objective Speech Quality Assessment in Narrow-Band Telephony Applications, standardized as P.563

See also *ITU-T*, *P.862*, *PAMS*, *PESQ*, and *toll quality*.

P.861 The ITU-T standard defining the Perceptual Speech Quality Measurement (PSQM) method of objectively evaluating voice transmission quality. PSQM is an automated method developed to measure the perceived quality of real-time voice as impacted by compression codecs. An enhanced version known as PSQM+ was later developed to measure the effects of packet loss and other network impairments. PSQM compares a distorted voice sample to an original, clear voice sample and uses a complex analytical process to evaluate the difference in terms of factors that influence the perceptions of human listeners. The ultimate distortion score corresponds closely with the mean opinion score (MOS) yielded by the panel of human listeners employed using the P.800 method. P.861 was withdrawn and replaced by P.862, defining the Perceptual Evaluation of Speech Quality (PESQ). See also *codec, compression, MOS, P.800, P.862,* and *PSQM.*

P.862 The ITU-T standard defining the Perceptual Evaluation of Speech Quality (PESQ) method of objectively evaluating transmission quality. A companion to the more subjective P.800, PESQ is an automated method that addresses the effects of filters, jitter, and coding distortions. PESQ replaced the Perceptual Speech Quality Measurement (PSQM) method, which was viewed as limited in certain applications. PESQ is considered to be an intrusive test method as it inserts a reference signal into the device under test. See also *ITU-T, P.800, PSQM,* and *toll quality.*

PABX (Private Automatic Branch eXchange) Generally synonymous in contemporary usage with private branch exchange (PBX), PABX refers to an automatic PBX, as compared to a manual PBX, or cordboard. The term *PBX* is preferred in North America, and PABX in much of the balance of the world. See *PBX* for more detail.

packet **1.** In the generic sense, referring to the manner in which data are organized into discrete units for transmission and switching through a data network. The data unit can be known as a block, frame, cell, or packet, depending on the protocol specifics. The packet comprises a header, payload, and sometimes a trailer, again depending on protocol specifics. The packet can be a user packet containing user data, or a signaling and control packet for various network monitoring, alerting and alarming, maintenance, and other administrative purposes. The payload can be a complete message, a fragment or segment of a message, or an aggregation of bits or bytes that form a short portion of a long data stream associated with a voice or video call. See also *bit, block, byte, cell, data stream, fragment, frame, header, message, payload, protocol, segment,* and *trailer.* **2.** In a technology-specific sense, a packet is a data unit in an internetwork, such as the Internet or other packet-switched network in which routers interconnect networks and subnetworks to exchange traffic between nodes. In terms of the OSI Reference Model, a packet is defined in Layer 3, the Network Layer. Blocks, cells, and frames are defined in Layer 2, the Data Link Layer, and have local significance, only. See also *block, cell, datagram, Data Link Layer, frame, Internet, Network Layer, OSI Reference Model, packet switch,* and *router.*

packet assembler/disassembler (PAD) See *PAD.*

packet-filtering firewall A security firewall that examines all data packets, forwarding or dropping individual packets based on predefined rules that specify where a packet is permitted to go, in consideration of both the authenticated identification of the user and the originating address of the request. See also *authentication, firewall, proxy firewall, security,* and *stateful inspection firewall.*

packet Internet groper (ping) See *ping.*

packet layer protocol (PLP) See *PLP.*

Packet over SONET (POS) See *POS.*

packet switch A device that switches data organized into packets, discrete sets of data that may take the specific form of packets, frames, or cells depending on the network technology specifics. For example, packet switches switch packets in networks based on the Internet Protocol (IP), frames in networks based on the Frame Relay or Ethernet protocol, and cells in those based on the Asynchronous Transfer Mode

(ATM) protocol. Packet switches were initially developed for interactive networking of host computers and, therefore, in support of computer-to-computer data transfer. Packet switches can support other forms of data, as well, although with varying degrees of success. See also *packet* and *switch*.

PACS (Personal Access Communications Services) A digital cordless telephony total system standard that addresses both the air interface and network infrastructure. A modification of the WACS (Wireless Access Communication System) specification, PACS was developed in the U.S. for licensed Personal Communications Services (PCS) applications, although several unlicensed versions exist as well. The licensed version of PACS employs frequency division duplex (FDD) to support downstream transmission in the 1930–1990 MHz band and upstream transmission in the 1850–1910 band, with channel spacing at 300 kHz. Each channel will support a data rate of 384 kbps with µ/4 quaternary phase shift keying (µ/4 QPSK) as the modulation technique. Each RF channel supports seven users through time division multiple access (TDMA). Voice encoding is adaptive differential pulse code modulation (ADPCM) at 32 kbps. See also *ADPCM, air interface, channel, cordless telephone, digital, downstream, encode, FDD, modulation, PCS, µ/4 QPSK, TDMA*, and *upstream*.

PAD (Packet Assembler/Disassembler) A functional unit that organizes user data into packets according to the X.25 packet layer protocol (PLP) and encapsulates each in an LAPB frame before presenting it to the network. A PAD also may be responsible for password protection and performance reporting. In contemporary X.25 networks, a PAD generally is in the form of software installed on a terminal or a communications server. ITU-T Recommendation X.3 is the specification for a PAD. See also *frame, LAPB, packet, PAD, password, PDN, PLP, server, software, terminal, X.3*, and *X.25*.

padding **1.** In storage, irrelevant material, usually zero (0) bits, added to a data block in order to fill it to a minimum size, to force certain fields of control data or user data into certain positions or sizes, or to prevent the user data from duplicating a bit pattern that has a specific control meaning. See also *block*. **2.** In transmission, irrelevant material added to a data block, packet, or frame in order to fill it to a minimum size, to force certain fields of control data or user data into certain positions or sizes, or to prevent the user data from duplicating a bit pattern that has a specific control meaning. See also *block, frame*, and *packet*. **3.** In transmission, redundant, irrelevant bits, usually one (1) bits, appended to a bit stream in order to increase the bit rate or to maintain a session or connection during periods of inactivity. See also *keep alive bits*. **4.** In some synchronous protocols, such as binary synchronous communications (Bisync, or BSC), one or more optional characters that alert the receiving device of the transmission of a block of data and ensure that the receiving device is in sync with the data bits. See also *BSC, protocol*, and *synchronous*.

page **1.** Of uncertain origin, but likely from the Greek *paidion*, a diminutive of *pais*, and meaning *little boy* or *slave boy*. To find, notify, or summon someone by using a loudspeaker system or radio system comprising a base station and small terminals known as pagers or beepers. Such systems have largely replaced human pages, who were young, uniformed attendants who performed tasks such as running errands and carrying messages in a hotel or legislature. Historically, pages were apprenticed to knights as an initial phase of their training for the knighthood. See also *pager* and *paging system*. **2.** From the Latin *pagina*, meaning *fastened together* and referring to strips of papyrus fastened together, as in a book. See *home page*.

pager Also known as a *beeper*, after the beeping sound some use as an alert. A small terminal device that receives radio signals from a base station. Pagers generally are receive-only devices for alerts or text messaging, although some can receive recorded voice messages, and some have two-way text messaging capabilities. Tone-only pagers cause the device to emit an audible tone or perhaps to vibrate or blink in non-disruptive silent mode. Numeric pagers are capable of receiving and displaying numeric characters. Alphanumeric pagers are capable of receiving and displaying both alphabetic and numeric characters. Relatively few pagers currently can support the storage of voice messages. See also *paging system*.

pager identification number (PID) See *PID*.

paging system **1.** A public address, or loudspeaker, system used to make announcements and notify or summon people. In large buildings, paging systems commonly are divided into a number of zones, or coverage areas. Key telephone systems (KTSs) commonly feature voice-over paging, which allows an authorized user to page through the intercom system, which works through the speakers built into the telephone sets. See also *intercom* and *KTS*. **2.** A radio system designed for alerting or sending messages to individuals. The radio paging system was invented by Al Gross, who also invented the walkie talkie, CB radio, and cordless telephone. The first system, which Gross sold to New York's Jewish Hospital in 1950, employed a centralized antenna that could broadcast alerts to small, inexpensive pagers, or beepers. A page simply transmitted a unique pager identification number (PID), which was recognized only by the pager being addressed. If that pager were in range, it beeped, hence the term *beeper*. The response to the page was in the form of a telephone call to the paging company to retrieve a message. The Federal Communications Commission (FCC) approved pagers for consumer use in the United States in 1958. The first consumer pager was the Motorola Pageboy I, which was based on proprietary standards. During the 1970s, an international team developed a standard set of code and signaling formats that evolved into the Post Office Code Standardization Advisory Group (POCSAG) code. Contemporary digital paging systems in the European Union (EU) are based on the European Radio Message System (ERMES) standard. In the United States, the FLEX set of proprietary solutions from Motorola largely has replaced POCSAG and has become the de facto set of standards throughout most of the world, with the exception of Western Europe. Paging systems generally operate over 25 kHz channels in the 900 MHz band. Radio common carriers (RFCs) are regulated providers of public services and are restricted to designated frequencies. Private paging operators (PPOs) are unregulated, but must share unlicensed spectrum with other users in the VHF and UHF bands.

A typical page begins with a text message transmitted via e-mail to a centralized network operations center (NOC). The NOC forwards the page to a satellite, which forwards it to a terrestrial network of centralized antennas that forward it to the target pager. Various types of two-way paging (TWP) systems support duplex transmission. See *Gross, Al*. See also *duplex*, *FCC*, *page*, *pager*, *POCSAG*, *PPO*, *radio*, *RCC*, *TWP*, *UHF*, and *VHF*.

pair-gain Referring to a local loop transmission system that uses concentrators or multiplexers to serve one or more subscribers with fewer twisted pairs than would otherwise be required. Digital loop carrier (DLC) and asymmetric digital subscriber line (ADSL) are examples of pair-gain systems. See also *ADSL*, *concentrator*, *DLC*, *local loop*, *multiplexer*, and *twisted pair*.

PAL (Phase Alternate Line) A television standard established in Western Germany, The Netherlands, and the United Kingdom in 1967. PAL addresses problems of uneven color reproduction that affect the NTSC standard due to phase errors associated with electromagnetic signal propagation. PAL inverts the color signal by 180° on alternate lines, hence the term Phase Alternate Line. PAL currently is used in much of Western Europe, Australia, and Africa. PAL is characterized as analog, with 625 interlaced scan lines. There are 640 pixels per line, with 576 dedicated to the active picture. The frame rate is 25 fps, and the aspect ratio is 4:3. Competing standards are NTSC (National Television Standards Committee) and SECAM (**S**Équential **C**ouleur **A**vec **M**émoire). See also *analog*, *aspect ratio*, *frame*, *frame rate*, *interlaced scanning*, *NTSC*, *phase*, *pixel*, *propagation*, *SECAM*, *signal*, and *television*.

Palo Alto Research Center (PARC) See *Xerox PARC*.

PAM (Pulse Amplitude Modulation) A form of signal modulation in which the amplitude of the digital pulse carrier is modulated according to the amplitude level of the original signal. PAM samples an incoming analog signal, for example, measures its amplitude, and outputs a digital pulse of a representative amplitude. The outgoing pulse closely matches the amplitude of the incoming signal, but when digitized in an A-to-D converter, the digital output is the nearest of a number of standard amplitude values. PAM was used in this manner in early channel banks to interface analog PBXs and central offices (COs) to DS-1 digital circuits. The PAM-encoded signal subsequently was further encoded using pulse code modulation (PCM) before the signal was placed on the circuit. PAM is considered obsolete in this application, having been replaced by direct

PCM-encoding of the analog signal, with the sampling performed by the same chips as contained in the A-to-D converter. However, variations on PAM remain widely used. Quadrature amplitude modulation (QAM) is such a variation. See also *amplitude, channel bank, DS-1, modulation, PCM,* and *QAM.*

PAMS (Perceptual Analysis Measurement System) An automated testing mechanism designed to objectively measure to quality of speech transmission. PAMS uses a sensory model to compare the degraded signal received to the original, unprocessed analog signal prior to encoding and transmission. PAMS is analogous to the mean opinion score (MOS) defined in P.800, which is a much more subjective approach involving a panel of human listeners. See also *P.800.*

PAN (Personal Area Network) A wireless personal area network (WPAN). See *WPAN.*

panel switch A type of electromechanical circuit switch developed by the Bell System for use in large metropolitan areas, where it was felt that the more conventional step-by-step (SxS) would not scale properly. As the first common control switch, the panel switch used a store-and-forward technique in which the digits dialed by the end user were stored in a register, or temporary buffer. Once the complete telephone number had been dialed, the register sent the dialed digits either to a translator for translation into routing instructions, or directly to the sender, which then sent the appropriate dialed digits across a trunk or to local switching equipment. A panel switch was so called because the line selector system operated on a system of ladders. As a selector received each dialed number, it would rise up a ladder in a vertical panel mounted on a frame. Panel switches were first placed into service in the 1920s, through the 1930s, and remained in service until the late 1970s, at which point they were replaced by electronic common control (ECC) switches. See also *ECC* and *SxS.*

PANS (Pretty Advanced New Services, Pretty Advanced Network Services, Peculiar And Novel Services) A term that appeared in the 1970s to distinguish new PSTN services from POTS (Plain Old Telephone Service). PANS includes custom calling services such as caller ID and name ID, call waiting, conference calling, and three-way calling. The term really never caught on. See also *POTS* and *PSTN.*

PAP (Password Authentication Protocol) A commonly used mechanism for password protection in support of remote users attempting to log on to a Point-to-Point Protocol (PPP) server. While PAP is easy to use, passwords typically are sent to the remote access server (RAS) in plain text (i.e., in the clear, or unencrypted). PAP demands the user's login name and password, and continues to do so until they both are supplied correctly. A hacker who knows the login name can use password-guessing programs to keep trying, in hopes that he can hit on the right combination and gain access. See also *CHAP, hacker, in the clear, PPP,* and *RAS.*

paper A flexible material consisting of thin, flat, felted sheets made of pulped wood, rags, or other fibrous materials laid down on a fine screen from a water suspension. Applications for paper include packaging, structural material, fabric substitute, and wall coverings. In days of yore, before the paperless office, people actually wrote and printed words and drew images on paper. See also *paperless office.*

paperless office In 1975, *Business Week* magazine predicted that the office of the future would be entirely paperless due to the impact of computers, in general, and particularly, the personal computer. The thought was that information would be communicated electronically, and that paper would be redundant. During that time, many companies, such as Southwestern Bell Telephone Company, converted paper business records to microfilm or microfiche. Microfilm soon became obsolete, as computer systems, computer networks, and electronic storage technologies began to truly impact business operations in the office of the early 1980s. Paper mail gave way to facsimile and e-mail in the 1980s and 1990s. Many print magazines largely converted to electronic format. In truth, contemporary business and commerce would not exist as we know it (not to mention the forestlands that would have been turned into deserts due to the untold billions of trees that would have been ground into pulp), if we had not advanced beyond the paper office of the 1960s. However, we still print hard copies of electronic documents, and seem to consume more paper with each passing year. So much for the paperless office. See also *lead balloon.*

parabolic antenna An antenna comprising a parabolic reflector with a transmitting and receiving element positioned at or near its focal point. The parabolic reflector is made of a reflective material, usually metal, and is shaped like a parabola, which looks much like a cross-section of a bowl. A parabolic reflector shapes a transmitted radio signal, focusing it into a collimated beam, with increased power density, or signal strength. Similarly, the reflector gathers an incoming signal and focuses it on the receiving element with greater intensity, i.e., gain. A fully circular parabola, or paraboloid, is a three-dimensional parabola that looks much like a shallow bowl. A paraboloid antenna shapes a radio signal much as the mirror in a flashlight shapes an optical signal. Paraboloid antennas are used on satellites to create spot beams. See also *collimation*, *gain*, *satellite*, and *spot beam*.

parallel communications The exchange of digital data between devices through the simultaneous transmission of multiple signal elements, e.g., bits, over separate channels or circuits. In microcomputers, parallel transmission refers to the simultaneous transfer of the eight (8) bits of a byte, each over a separate circuit. Parallel communications is much faster than serial communications, which transfers information one bit at a time over the same channel or circuit. The cost of provisioning multiple channels or circuits, however, limits parallel communications to very short distance applications, such as within a computer or between a computer and a local printer. Parallel communications is used in some wireless local area network (WLAN) technologies to improve signal quality, rather than increase raw transmission rate. A parallel-input, serial-output (PISO) shift register is used to interface a serial circuit to a parallel circuit. See also *channel*, *circuit*, *serial*, *shift register*, and *transmission rate*.

parallel-in, serial-out (PISO) See *PISO*.

PARC (Palo Alto Research Center, Inc.) See *Xerox PARC*.

Pareto principle Also known as the *80/20 rule*, the Pareto principle states that 80 percent of the results come from 20 percent of the actions. The principle is named for Vilfredo Pareto, who observed, sometime between 1890 and 1906, that 80 percent of the income in Italy accrued to 20 percent of the population, or that 80 percent of the land was owned by 20 percent of the population, or that 20 percent of the peapods in his garden yielded 80 percent of the peas, or something of the sort, depending on which source you believe. The Pareto principle essentially is the law of the vital few versus the trivial many.

parity **1.** The state or quality of equality, i.e., being the same, or identical. **2.** In the context of mathematics, the quality of being equally odd or equally even. If two numbers are both odd in value, or both are even in value, parity exists. If one number is odd and the other is even, no parity exists. **3.** In error control, the quality of being equally odd or equally even. If the dataset, as transmitted, has an odd (or even) number of 1 bits and the dataset, as received, has an odd (or even) number of 1 bits, parity exists. If the dataset, as transmitted, has an odd (or even) number of 1 bits and the dataset, as received, has an even (or odd) number of 1 bits, parity does not exist. See also *error control*, *parity bit*, and *parity check*.

parity bit In asynchronous communications and primary storage, an extra bit in the form of a check bit appended to an array of information bits for error control purposes. The parity bit is added to ensure that the total value of the bits, including the parity bit, is always either odd or even, depending on whether the communicating stations are set for odd parity or even parity. One or more parity bits are generally associated with a character, word, or block for purposes of checking the integrity of data after transmission or after storage and retrieval. When using ACSII code, for example, an eighth bit is always appended to each character as a parity check bit. In ASCII, the upper case letter A is represented by the binary code 1000001, which comprises an even number (2) of 1 bits. If the machine is set for even parity, a 0 bit is appended in the eighth bit position to retain the even value. If the machine is set for odd parity, a 1 bit is appended in the eighth bit position to create an odd value. The default is odd parity. See also *ASCII*, *asynchronous*, *binary*, *block*, *code*, *error control*, *parity*, and *parity check*.

parity check A common method for error control in asynchronous communications and primary storage. Prior to transmitting or storing an array of bits, such as an ASCII character, a device, such as a PC, adds the marks, or 1 bits, that compose the seven-bit character to determine if their total value is an odd or even number. If the number is even and the machine is set for the default odd parity, it will append a 1 bit in the eighth bit position. The device that either retrieves or receives the character performs the same operation on the data. If the parity is odd, the machine assumes that the character is unerrored. There are two variations on the theme of parity checking. Vertical redundancy checking (VRC) is the simplest and least reliable. Longitudinal redundancy checking (LRC) is more complex and much more reliable. Neither, however, provides any inherent error correction mechanism. See also *error control*, *LRC*, *recognition and flagging*, and *VRC*.

parse To analyze the grammatical structure of a sentence, a character string, or a line of code and separate it into its parts. In order for the routing logic of a system to determine how to route a telephone call originating in the United States, for example, it must parse the dialed digits to determine if the telephone number series begins with 011, indicating that the call is international, or 1, indicating that the call is domestic long distance crossing an area code boundary. See also *area code*.

Part 68 The section of Title 47 of the Code of Federal Regulations (1975) of the Federal Communications Commission (FCC) that provides the technical and procedural standards under which customer premises equipment (CPE) can be connected directly to the United States public switched telephone network (PSTN) without causing harm and without a requirement for protective circuit arrangements. Part 68 allows manufacturers of foreign (i.e., non-telco) equipment to certify that it will cause no harm to the network, thereby eliminating any need for special interface devices incorporating electrical protection. Previously, foreign CPE required individually tested and registered interface devices in order to connect to the PSTN. See also *CPE*, *FCC*, and *PSTN*.

party line A local loop shared by perhaps two, four, eight, or as many as 16 parties. Distinctive ringing comprising various combinations of short and long rings distinguishes a call intended for each individual residence on the shared line. There is no privacy on a party line, as any party can pick up the phone and answer the call or listen in on it. Placing outgoing calls is a free-for-all, as the caller must pick up the phone to determine if the line is available before placing the call. If someone else is using the line, the process must be repeated again and perhaps again and again in hopes that the line eventually will be available. Party lines are now rare in the United States, but not uncommon in rural areas in developing countries.

passband Referring to the frequencies in a band supported by a band-pass filter. Such a filter allows signals in the passband to pass through, but absorbs, attenuates, blocks, rejects, or removes signals above or below the designated cutoff frequencies. See also *absorption*, *attenuation*, *band*, *band-pass filter*, and *frequency*.

passive Not active or energized, i.e., not electrically powered. A reflective satellite, for example, serves only to reflect a signal, rather than act on it to respond to requests for transponder access, amplify the received signal, detect and perhaps correct for errors in the data, resynchronize the signal, shift between uplink and downlink frequencies, or retransmit the signal. A passive optical network (PON) involves passive splitters that only split an incoming signal into two or more outgoing signals at proportionately lower signal strength.

passive optical network (PON) See *PON*.

password A security mechanism in the form of an authentication tool used to identify a user of a device, program, or network and to identify the user's level of privilege on a site, host, application, screen, and field level. A password usually comprises a string of characters that the user enters through a keypad, ideally a reasonably lengthy combination of letters, numbers, punctuation marks, and other characters that an intruder cannot easily guess. Passwords should be changed frequently. A password usually is associated with a user ID, or username, and both must be entered during the logon process. See also *authentication* and *security*.

Password Authentication Protocol (PAP) See *PAP*.

patch A quick, unscheduled fix for a bug, security hole, or other deficiency or defect in a program. Patches are incorporated into subsequent program upgrades. See also *bug*, *bug fix*, *update*, and *upgrade*.

patch cord A short cable assembly used for establishing semi-permanent interconnections between the cable terminations on a patch panel. See also *patch panel*.

patch panel A vertical panel with multiple points of termination for cables. Short patch cords are used to establish semi-permanent interconnections between the cable terminations. Multiple types of cables (e.g., coaxial and unshielded twisted pair) can terminate on a single patch panel. See also *distribution frame*.

patent Intellectual property protection for inventors of products that are deemed novel, useful, and not obvious to one reasonably skilled in that particular art. A United States patent prevents others from making, using, offering for sale, or selling an invention throughout the United States or from importing the invention into the United States. In order to receive United States patent protection, which extends for 20 years from application, the invention must be submitted to the United States Patent and Trademark Office for examination. United States patent law protects not only physical devices, but also software, mathematical algorithms, business processes, and other inventions that involve the use of a computer. Note: The first patent issued in the United States was U.S. Patent No. 1X, *Method of producing pot ash and pearl ash*, issued to Samuel Hopkins on July 31, 1790, and signed by G. Washington. See also *intellectual property*.

path The physical route of a circuit. See also *circuit* and *virtual path (VP)*.

Path Overhead (POH) See *POH*.

path-switched ring A SONET/SDH topology that employs two active optical fibers in a dual counter-rotating ring configuration. All traffic moves in both directions. Should the primary ring fail, the secondary ring is already active in the reverse direction, thereby providing protection from network failure. A path-switched approach also improves error performance because the receiving stations examine both data streams and select the better signal. See also *line-switched ring*, *optical fiber*, *SDH*, *SONET*, and *topology*.

path terminating equipment (PTE) See *PTE*.

payload **1.** Also known as the *text field*. The user data within a block, cell, frame, or packet. See also *block*, *cell*, *data format*, *frame*, *packet*, and *text field*. **2.** The user data that traverses a circuit or network, in contrast to signaling and control information and overhead. Signaling and control information is used to monitor, supervise, maintain, and otherwise manage and control a network and its components. Overhead data is used for purposes such as addressing, congestion management, error control, priority indication, and routing calls, blocks, frames, packets, and cells containing payload. See also *block*, *cell*, *frame*, *overhead*, *packet*, and *signaling and control*.

payload header suppression Also known as *header compression* and *header suppression*. Referring to various techniques used to eliminate or reduce the number of headers associated with voice over Internet Protocol (VoIP) packets. In applications where VoIP packets associated with multiple conversations are grouped into an Ethernet or frame relay frame, much of the header information is redundant and, therefore, eligible for suppression. Payload header suppression improves the efficiency of the link and increases throughput. For example, the FRF.20 Implementation Agreement describes an algorithm that examines the 44 octets of VoIP header, looking for redundancy and other opportunities to send only a reference to the header, rather than the entire header. This technique reduces the header to as few as 2–4 octets, with the process reversed at the receiving end of the circuit. Thereby, as many as five VoIP calls can be supported over the same 64-kbps permanent virtual circuit (PVC). See also *compression*, *Ethernet*, *frame*, *frame relay*, *header*, *Implementation Agreement*, *link*, *packet*, *PVC*, *throughput*, and *VoIP*.

payload type indicator (PTI) See *PTI*.

pay-per-view (PPV) A system by which cable television (CATV) subscribers can purchase the right to view sporting events, feature-length films, and other special programming. Pay-per-view generally is available only over CATV systems, with the greatest selections over broadband CATV systems. See also *broadband* and *CATV*.

payphone See *pay telephone*.

payphone service provider (PSP) See *PSP*.

pay station See *pay telephone*.

pay telephone A telephone available for public use and requiring that the customer pay for that use by inserting money, usually in the form of coins or a debit card or credit card of some sort. Pay telephones traditionally have been provided by the local telephone company, although deregulation allows other parties to provide them in the United States and some other countries. A customer-owned coin-operated telephone (COCOT) is owned by the end user that owns or occupies the premises in which it is located. A payphone service provider (PSP) is a third party that may own a great many payphones. The pay telephone was invented by William Gray, who installed it in a bank in Hartford, Connecticut, in the United States. Gray previously had invented the chest protector for use in the game of baseball. A pay telephone is no longer a common sight in North America and many developed regions of the world, as the low cost and wide availability of cellular telephones have made pay phones unprofitable. Synonymous with *payphone* and *pay station*.

PBX (Private Branch eXchange) A voice-optimized switching system physically located on the customer premises, serving the internal station-to-station communications requirements of one or more user organizations and with trunk circuits connecting to the public switched telephone network (PSTN) via one or more central office (CO) switches, and perhaps one or more other PBXs composing a private network. The term refers to the fact that the PBX originated as a switching system located on the subscriber's private premises and serving the subscriber's private communications requirements, while functioning as a branch (i.e., partition) of the public exchange. The first PBX was placed into service in the Old Soldiers' Home in Dayton, Ohio, in 1879. The first systems were non-standard modifications of CO switches. AT&T offered the first standard PBX, the No. 1 PBX, in 1902. The evolution of PBX technology can be organized along generational lines, as shown in Table P-1.

Table P-1: Evolution of PBX technology

Generation	*Designation*	*Nature of Technology*
0	Cordboard	Manual Switchboard
1	PBX (Private Branch eXchange)	Electro-Mechanical Step-by-Step (SxS)
2	PABX (Private Automatic Branch eXchange)	Electro-Magnetic Crossbar (XBar) or Crossreed
3	EPABX (Electronic Private Automatic Branch eXchange)	Electronic Common Control (ECC) Analog or Digital Stored Program Control (SPC)
4	IP PBX or IPBX (Internet Protocol Private Branch eXchange or Intranet Private Branch eXchange)	Digital Stored Program Control (SPC) LAN-based Voice over Internet Protocol (VoIP)

PBX generations 0–3 were all circuit switches, optimized for voice. It wasn't until the 1980s that the third generation of PBXs supported data communications, and even then on a circuit-switched basis through special, and very expensive, data modules. The primary physical and logical components of a 3rd-generation PBX include power supply, common control, memory, switching matrix, trunk interfaces, line interfaces, and terminal equipment, as illustrated in Figure P-1.

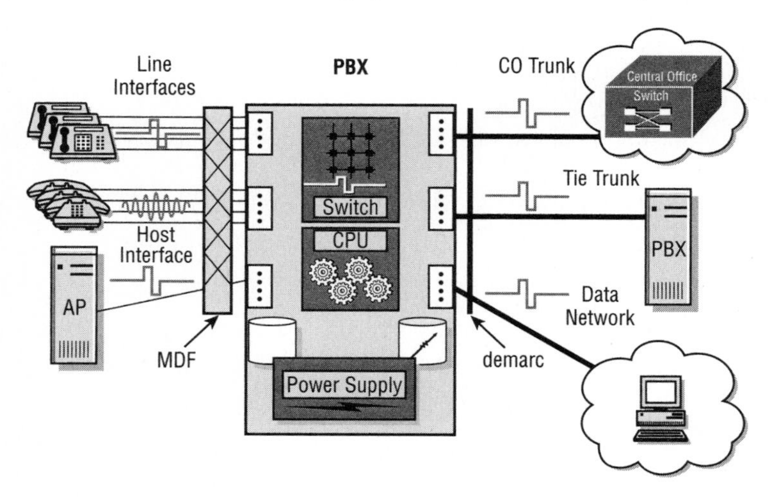

Figure P-1

While some PBX systems serve the small enterprise market, that market is largely addressed by key telephone systems (KTSs). PBXs are largely intended for applications over 50–100 lines, and the very large can serve 10,000 or more lines. Such large PBXs essentially are COs with software loads providing PBX features. The 4th generation of PBXs, the IPBXs, are switched Ethernet local area networks (LANs) that support packet-switched voice, as well as data communications applications. *Note:* Outside North America, the term *PABX* (Private Automatic Branch Exchange) is often used in a generic sense to refer to a PBX. See also *circuit switch*, *CO*, *common control*, *ECC*, *IPBX*, *KTS*, *line interface*, *power supply*, *SPC*, *switch matrix*, *SxS*, *trunk interface*, and *Xbar*.

PC **1.** Personal computer. A microcomputer, including its own operating system (OS), application software, and other components necessary for its operation in service of a single user. See also *mainframe* and *minicomputer*. **2.** Politically correct. In reference to language, humor, or behavior crafted to be minimally offensive to anyone and everyone, but most especially to the humorless, developmentally deprived, and overly sensitive poor souls who take offense (or feign offense) and often threaten legal action at any reference to gender, race, religion, disability, or other human attribute they choose to target at the moment. Thanks to these poor souls, linemen are now lineworkers, manholes are now personholes, manhole covers are now maintenance covers, master/slave is now primary/secondary, dikes are now wire clippers, waiters and waitresses are now servers, stewards and stewardesses are now flight attendants, and fishermen are now fishers. Further, *he or she* is now *they*, which may be PC, but certainly is GI (Grammatically Incorrect), and so I refuse to use it in this book or any other. (My editors may override my decision, of course.) Believe it or not, *disabled* is moving towards *differently-abled*. Holy Moley! (Or should I say Good Gravy?) See also *Bless his heart* and *euphemism*.

PCB (Printed Circuit Board) See *printed circuit board*.

PC card More correctly, *PCMCIA card*. An add-in integrated circuit card that conforms to specifications developed and promoted by the Personal Computer Memory Card International Association (PCMCIA). PC cards are approximately the length (86.5 mm) and width (54 mm) of a credit card, but much thicker, and fit into a slot built into a laptop or tablet personal computer (PC) or peripheral. Type I cards are 3.3 mm thick and are used for add-in random access memory (RAM). Type II cards are 5.0 mm thick and are commonly used for add-on modems, fax modems, and Ethernet network interface cards (NICs). Type III cards are 10.5 mm thick and are used for supplemental rotating hard disk drives. See also *fax*, *modem*, *NIC*, *PC*, *PCMCIA*, and *RAM*.

PCM (Pulse Code Modulation) A modulation technique in which an analog signal is encoded, i.e., converted from analog to digital format. The term generally is applied to the conversion of voice from a continuous analog waveform to digital pulses as specified in ITU-T Recommendation G.711, which is based on the Nyquist theorem. That theorem states that an analog signal waveform can be converted to digital format and be reconstructed without error from samples taken at equal time intervals, if the sampling rate is equal to, or greater than, twice the highest frequency component in the analog signal. As the voice band is 0–4,000 Hz, the highest frequency component is 4,000 cycles per second (cps), and the sampling rate is 8,000 per second. At that rate, a codec measures the amplitude of the audio sine wave and encodes (i.e., quantizes) that value as a 14-bit number, which it compresses by assigning each sample to an eight-bit binary (byte) approximate value, based on a table of 256 (2^8) standard values of amplitude according to the non-linear PCM scale. The standard values are closer together at low volume, or loudness, levels and spread further apart for high loudness. The process yields a digital voice signal at a rate of 64 kbps (8,000 cps × 8 bits = 64,000 bps). The individual byte values then are transmitted over a digital circuit, such as a channelized E-1 or T1. If a T1, a time division multiplexer (TDM mux) byte interleaves the eight-bit (byte) samples with those of 23 other conversations into a frame, and repeats that process at the very precise pace of one each 125μs (8,000 times a second). The process is reversed on the receiving end of the connection as the encoded amplitude samples are expanded (i.e., decoded, or decompressed), to reconstitute an approximation of the original analog voice signal. A sampling rate that is too low, i.e., below the Nyquist rate, results in a phenomenon known as aliasing, in which the reconstructed signal is inaccurate, or even unintelligible. The twin processes of compressing and expanding the signal are known as companding, and are illustrated in Figure P-2. Recommendation G.711 specifies two companding techniques that define the assignment of 14-bit values to 8-bit values: μ-law (mu-law), which is used in the North American T-carrier systems, and A-law, which is used in the European E-carrier systems and the Japanese J-carrier systems. PCM was invented and patented in 1925 by P. M. Rainey of Western Electric, the manufacturing arm of the Bell System. See also *ADPCM*, *A-law*, *analog*, *byte*, *byte interleaving*, *codec*, *digital*, *DPCM*, *E-carrier*, *frequency*, *ITU-T*, *J-carrier*, *μ-law*, *modulation*, *Nyquist theorem*, *quantizing noise*, *sine wave*, *T-carrier*, *TDM*, *voice band*, and *waveform*.

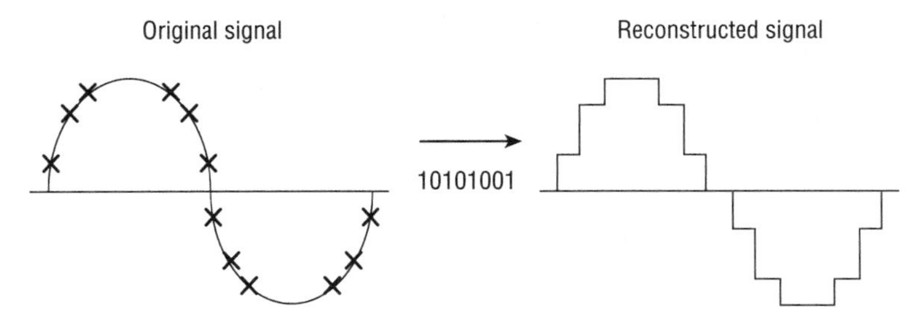

Figure P-2

PCMCIA (Personal Computer Memory Card International Association) An international standards body and trade association of manufacturers and vendors that sets standards for interoperable add-in integrated circuit cards known as PCMCIA cards or, more commonly, PC cards. See also *PC card*.

PCMCIA card See *PC card*.

PCN (Personal Communications Network) **1.** Also known as *DCS 1800* (Digital Cellular System 1800). An upbanded version of GSM, the pan-European digital cellular radio system. PCN operates in the 1800 MHz (1.8 GHz) range, rather than the 900 MHz range of GSM. See also *GSM.* **2.** A concept originally developed in the United Kingdom, and known generically as PCS (Personal Communications Services). See *PCS.*

PCR (Peak Cell Rate) In asynchronous transfer mode (ATM), a traffic parameter defined as the maximum number of cells per second, in a burst, that the network agrees to accept and transfer for a given user network interface (UNI). Excess cells may be discarded by the ingress switch, or marked as eligible for discard. For constant bit rate (CBR) service, the PCR is the guaranteed CBR for the virtual circuit. Enforcement of the PCR enables the network to allocate sufficient resources to ensure that the quality of service (QoS) parameters are met. See also *ATM, CBR, cell, QoS, traffic parameter, UNI,* and *virtual circuit.*

PCS **1.** Personal Communications Services. Originating in the United Kingdom, where it is known as PCN (Personal Communications Network), PCS is a service concept for a full range of wireless services accessible through a single device and utilizing one telephone number that will work anywhere and anytime, for life. While the service is technology-dependent and the spectrum allocation varies by country or region, the designated U.S. band designations and frequency allocations are as follows:

- Narrowband PCS spectrum is allocated in the 900–901 MHz, 930–931 MHz, and 940–941 MHz bands. This spectrum is intended to extend the capabilities of pagers and cell phones to include acknowledgment paging, two-way messaging, and digital voice.
- Broadband PCS spectrum is allocated 120 MHz in the 1.85–1.91 GHz and 1.93–1.99 GHz bands. This spectrum is intended for the delivery of high-tier wireless communications, including wireless local loop (WLL), voice and data services, and cellular-like services for pedestrian traffic in high-density areas.
- Unlicensed PCS spectrum is allocated in two bands, with 20 MHz in the 1.91–1.93 GHz range, and 10 MHz in the 2.39–2.40 GHz range. This spectrum serves low-tier applications such as wireless LANs (WLANs) and wireless office telecommunications systems (WOTS).

As the spectrum assignments will vary by country, the terminal devices must be multimode, i.e., multiband, and even cognitive in nature. See also *band, broadband, cognitive radio, frequency, high-tier, low-tier, narrowband, spectrum, WLAN, WLL,* and *WOTS.* **2.** Personal Communications Services. A cordless telephony service in the United States, based on Wireless Access Communication System (WACS), which subsequently was modified to an industry standard known as Personal Access Communications Services (PACS). PCS derives 32 carriers with spacing of 300 kHz in the 1850–1910 MHz and 1930–1990 bands, and employs frequency division duplex (FDD) to pair 16 downstream and 16 upstream carriers. Within each frequency channel, PCS derives eight time division multiple access (TDMA) channels employing $\pi/4$ Quadrature Phase Shift Keying ($\pi/4$ QPSK). Voice encoding is adaptive differential pulse code modulation (ADPCM) at 32 kbps, and the theoretical data rate is 384 kbps per channel. PCS has never gained any traction due largely to the high availability and low cost of cellular telephony. See also *ADPCM, carrier, cellular, cordless telephone, downstream, encode, FDD, $\pi/4$ QPSK, TDMA,* and *upstream.* **3.** Personal Communications System. A U.S. term for digital cellular radio systems based on EIA/TIA IS-95a, and also known as cdmaOne and CDMA Digital Cellular. PCS is a 2G upgrade to the analog Advanced Mobile Phone System (AMPS), and the two systems can coexist. The implementation of PCS requires that one or more AMPS frequency bands of 1.25 MHz be converted and subdivided into 20 carriers, each of which supports as many as 798 channels and aggregate bandwidth of up to 1.288 Mbps. The specified signal modulation methods are binary phase shift keying (BPSK) and offset quadrature phase shift keying (OQPSK). Bidirectional communication is supported via frequency division duplex (FDD), with the downlink operating in the 869–894 band and the uplink in the 824–849 band. The variable-rate speech encoding algorithms run at maximum rates of 8 kbps using code excited linear prediction (CELP) or 13 kbps using enhanced variable rate vocoder (EVRC), and varies the rate downward to as low as one-eighth rate if the level of speech activity permits. The basic user channel rate is 9.6 kbps, although various channel rates can be achieved, depending on the carrier implementation. As compared to AMPS, CDMA offers the advantages of improved bandwidth utilization (as much as 10:1 or even 20:1) and time division multiple access (TDMA) (as much as 6:1), soft handoff, variable-rate speech coding, and support for both voice and data. The IS-95-B specification supports symmetric data rates of 4.8 kbps and 14.4 kbps per channel; as many as eight channels can be aggregated to support a data rate up to 115.2 kbps. CDMA offers additional advantages in terms of maximum cell size due to improved antenna sensitivity, and battery time due to precise power control mechanisms. The first commercial systems were installed in South Korea and Hong Kong in 1995, and in the United States in 1996. See also *2G, AMPS, analog, band, bandwidth, BPSK, carrier, CDMA, cellular radio, CELP, channel, digital, downlink, encode, EIA, EVRC, FDD, frequency, handoff, modulation, OQPSK, signal, symmetric, TDMA, TIA*, and *uplink*.

PCS 1900 An upbanded and modified version of the pan-European GSM cellular radio standard specified by the American National Standards Institute (ANSI) as ANSI J-STD-007, 1995 for PCS at 1900 MHz (1.9 GHz). See also *cellular radio, GSM*, and *PCS*.

PD (Powered Device) In power over Ethernet (PoE), a device receiving power over twisted-pair cable plant from power sourcing equipment (PSE). See also *PoE* and *PSE*.

PDC (Personal Digital Cellular) Previously known as *Japanese Digital Cellular* (JDC). A Japanese standard for a 2G digital cellular radio system operating in the 800 MHz, 900 MHz, and 1400 MHz frequency bands. PDC employs frequency division duplex (FDD) to support bidirectional communications with the downlink in the 810–826 and 1429–1453 bands and the uplink in the 940–956 and 1477–1501 bands. PDC derives 1,600 carriers of 25 kHz, and employs time division multiple access (TDMA) to derive three channels per carrier. The channel bit rate is 42 kbps, with the modulation method being $\pi/4$ differential quaternary phase shift keying ($\pi/4$ DQPSK). Full-rate voice encoding is at 9.6 kbps using vector sum excited linear predictive coding (VSELP) and half-rate at 5.6 kbps using code excited linear prediction (CELP). PDC has not found acceptance outside Japan, with the exception of a few Asian countries under Japanese economic influence. PDC is being phased out in Japan, in favor of cdma2000. See also *2G, band, carrier, cdma2000, cellular radio, CELP, channel, digital, downlink, FDD, frequency, modulation, $\pi/4$ DQPSK, TDMA, uplink*, and *VSELP*.

PDH (Plesiochronous Digital Hierarchy) Referring to the digital network hierarchy of T-carrier and E-carrier systems. Originally, the PDH involved highly only link layer synchronization, where the device on one end sent at its own not very precise rate, and the device at the other end sent return data at the same rate. Multiplexing, as when combining 28 T1 circuits into a T3, provided stuff bits to compensate for the variations in clock rates among T1 lines. When digital cross connect systems (DACS) were introduced, they required all attached devices to clock off them. Quickly there was a need for a precise master clock, off which all digital network elements slaved in a hierarchical fashion to sync up, i.e., all operate at the same precise bit rate. In contemporary T/E/J-carrier networks, the switches and muxes sync up with a Global Positioning System (GPS) master clock, which is distributed by radio rather than land line. See also *E-carrier, GPS, J-carrier, stuff bits, synchronization*, and *T-carrier*.

PDN (Public Data Network) A switched data network available for general public use. The core of such a network generally is digital in nature, although the local loop access may be analog. PDN examples include Switched 56, the Internet, and X.25. See also *Internet* and *Switched 56*.

PDU (Protocol Data Unit) **1.** Information that is formatted as a distinct element to be exchanged between network peers. A PDU has a specific function or set of functions, such as user data and control functions, including addressing, sequencing, and error detection. **2.** In the context of a layered protocol stack or suite, a unit of data specified in a given layer, including payload and overhead. At the Data Link Layer, a PDU is a frame, at the Network Layer, a packet, and at the Application Layer, a message. See also *bit stream, frame, message, overhead, packet, payload, protocol stack*, and *protocol suite*.

PE (Phase Encoding) More commonly known as *phase-shift keying* (PSK). See *PSK*.

peak cell rate (PCR) See *PCR*.

pedant A person who puts unnecessary emphasis on minor or trivial rules or points of learning, thereby displaying a scholarship lacking in proportion or judgment. Such a display often is purely ostentatious, if occasionally fun among friends. However, one should always eschew obfuscation.

peer-to-peer Descriptive of a relationship between equals. A peer-to-peer architecture is a network design based on computers that share the same responsibilities and use the same programs to communicate. Machines operating in such a network communicate with each other as equals, with each acting as a server for the others. Such a network differs markedly from a client/server network, where some machines are designated as servers to serve the needs of client machines. A peer-to-peer network is also markedly different from a network characterized by master/slave relationships between machines. See also *client/server* and *master/slave*.

people A group of persons with common historic, linguistic, national, racial, religious, or traditional ties. There are three basic types of people: those who can do math and those who can't.

Perceptual Analysis Measurement System (PAMS) See *PAMS*.

Perceptual Evaluation of Speech Quality (PESQ) See *PESQ*.

Perceptual Speech Quality Measurement (PSQM) See *PSQM*.

Perceptual Speech Quality Measurement plus (PSQM+) See *PSQM+*.

performance management An element of network management, performance management comprises the collection and analysis of data regarding the performance of network elements (NEs), systems, and networks. A network management system (NMS) responsible for an IP network, for example, might gather and analyze performance data from multiple routers with respect to errored packets, lost packets, queue lengths, and packet jitter. Based on that analysis, the NMS might reroute packet traffic between two edge routers. See also *jitter*, *NE*, *network management*, *NMS*, and *queue*.

per-hop behavior (PHB) See *PHB*.

peripheral In computing, a hardware device, such as a printer or external disk drive, that is separate from but connected to and controlled by the computer system's central processing unit (CPU). The computer system generally makes use of a program known as a driver to deal with device-specific peripheral operations.

permanent virtual circuits (PVC) See *PVC*.

persistence of vision The theory that electrochemical processing delays in the human eye or human brain create the illusion of continuity of motion if a series of images is presented in rapid succession. The theory is based on the discoveries of Paul Nipkow, a German engineer who developed the first true television mechanism in 1884. Nipkow used a scanning disk, lenses, mirrors, a selenium cell, and electrical conductors to transmit images in rapid succession to a lamp that changed in brightness according to the strength of the currents received. Using this mechanical scanning technique, Nipkow demonstrated that portions of a full image viewed in rapid succession created the illusion of viewing the full image. It later was discovered that viewing 15 or more images per second created the illusion of full motion. The theory of persistence of vision is controversial. See also *phi phenomenon*.

Personal Access Communications Services (PACS) See *PACS*.

personal area network (PAN) A wireless personal area network (WPAN). See *WPAN*.

personal communications network (PCN) See *PCN.*

personal communications services (PCS) See *PCS (1).*

Personal Communications Services (PCS) See *PCS (2).*

Personal Communications System (PCS) See *PCS* (3).

Personal Computer Memory Card International Association (PCMCIA) See *PCMCIA.*

Personal Digital Cellular (PDC) See *PDC.*

Personal Handyphone System (PHS) See *PHS.*

Personal Locator Beacon (PLB) See *PLB.*

personal operating space (POS) See *POS.*

personal radio services In the United States, a class of short-range, low-power radio services for personal and business communications and signaling applications. Some personal radio services must be licensed by the Federal Communications Commission (FCC), while others only require the use of authorized equipment. Personal radio services include 218-219 MHz Service, Citizens Band (CB) Radio Service, Family Radio Service (FRS), General Mobile Radio Service (GMRS), Low Power Radio Service (LPRS), Medical Implant Communications Service (MICS), Multi-Use Radio Service (MURS), Personal Locator Beacons (PLB), Radio Control Radio Service (R/C), and Wireless Medical Telemetry Service (WMTS). See also *218-219 MHz Service, CB Radio Service, FCC, FRS, GMRS, LPRS, MICS, MURS, PLB, radio, R/C,* and *WMTS.*

Personal Wireless Telecommunications (PWT) See *PWT.*

PESQ (Perceptual Evaluation of Speech Quality) An objective measurement tool defined in the ITU-T Recommendation P.862 for the evaluation of transmission quality. PESQ is an automated measurement tool that addresses the effects of filters, jitter, and coding distortions. PESQ is analogous to the mean opinion score (MOS) defined in P.800, which is a much more subjective approach involving a panel of human listeners. See also *ITU-T, P.800, P.862,* and *toll quality.*

PGP (Pretty Good Privacy) A public key encryption program based on the RSA algorithm. Developed by Phil Zimmerman as freeware, PGP is also available in commercial versions. PGP was under a cloud for some time because there was concern that it was so powerful as to violate U.S. technology export laws. Note: Encryption technology technically is classified under U.S. law as a form of munitions. See also *encryption, freeware, public key encryption,* and *RSA.*

phantom circuit A third circuit derived from two pairs of electrified metallic circuits known as side circuits, arranged in parallel on the line side of the network. Each side circuit is a circuit unto itself, and at the same time it acts as one conductor of the phantom circuit, which is derived by means of center-tapped transformers known as repeating coils. See also *circuit* and *transformer.*

pharming An Internet scam that involves misdirecting a user to a fraudulent Web site or proxy server by exploiting weaknesses in DNS (Domain Name Server) server software and hijacking transactions, or by changing certain files in the client software on a victim's computer. The term pharming is a play on farming and phishing. (Hackers commonly replace f with ph, phor reasons that are entirely unphathomable to the rest of us.) See also *e-mail, hyperlink, Internet, phishing, pretexting, scam, social engineering,* and *spam.*

phase In a periodic function such as the cyclic variation of an electromagnetic waveform, a relative measurement that describes the relationship between the positions of a signal at two instants in time. In other words, phase is a measurement of the relative position of a waveform at a significant instant of the signal relative to a time scale. Phase is measured in degrees (°), with a full oscillation expressed as 360°, as illustrated in Figure P-3. The phase of a signal can shift unintentionally, which causes signal distortion. Phase modulation (PM), more commonly known as phase-shift keying (PSK) is a signal modulation technique that modems often use to impress digital bits on an analog carrier. See also *analog, carrier, digital, distortion, modulation, signal, PSK,* and *waveform.*

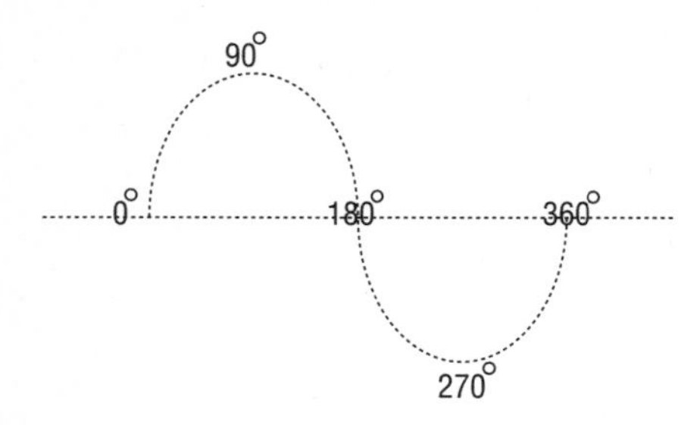

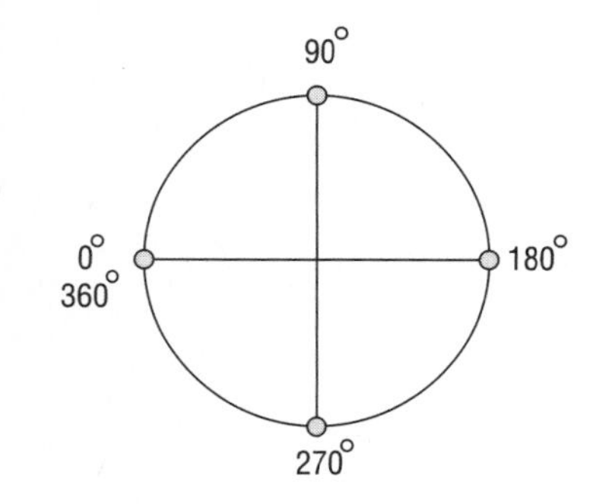

Figure P-3

phase alternate line (PAL) See *PAL.*

phase encoding (PE) More commonly known as *phase-shift keying* (PSK). See *PSK.*

phase modulation (PM) More commonly known as *phase-shift keying* (PSK). See *PSK.*

phase-shift keying (PSK) See *PSK.*

PHB (Per-Hop Behavior) In the Differentiated Services (DiffServ) protocol, a service level, or forwarding classes. Expedited Forwarding (EF), for example, provides minimal delay, jitter, and loss. Assured Forwarding (AF) comprises four classes, each of which contains three drop precedences and allocates certain amounts of buffer space and bandwidth. DiffServ operates on a packet-by-packet and hop-by-hop basis. See also *AF, bandwidth, buffer, delay, DiffServ, EF, hop, jitter, packet,* and *protocol.*

phi phenomenon A perceptual illusory phenomenon that explains the fact that people hallucinate, or believe they see, a continuous action where a series of still images is presented in rapid succession. Psychologist and film theorist Hugo Münsterberg explained the process in 1916. Modern television and video systems create the illusion of motion by refreshing screens in rapid succession, a dot at a time and a line at a time. Film projectors refresh the entire screen at once. See also *persistence of vision.*

phishing Also known as *brand spoofing* and *carding.* A popular Internet e-mail scam that involves unsolicited e-mail (i.e., spam) contact in which the scam artist attempts to gain valuable information from the target by gaining that person's confidence through various social engineering techniques and technical subterfuge. The term phishing was coined in the 1996 timeframe by crackers (malicious computer hackers) to describe the process of fishing for suckers by using some sort of lure or bait. (Hackers commonly replace f with ph, phor reasons that are entirely unphathomable to the rest of us.) Phishing commonly involves phony e-mails from banks, credit card companies, e-tailers, insurance companies, mortgage brokers, or other financial institutions warning that your account has been subjected to fraud or perhaps that your credit card is due to expire, and that you must confirm certain information such as an account number and password, or perhaps your social security number. The mail includes a hyperlink to a phony website that quite closely matches the legitimate website. If the scam is successful, the unsuspecting target clicks on the link and divulges information necessary for the scam artist to perhaps wipe out a bank account, max out a credit card, or even steal a person's identity, incur extraordinary debts in his name, and generally ruin his credit. See also *e-mail*, *hyperlink*, *Internet*, *pharming*, *pretexting*, *scam*, *social engineering*, and *spam*.

phone book **1.** A book, printed annually on paper, that contains listings of telephone subscribers who wish to be listed. The white pages list subscribers in alphabetical order, with larger books typically separating residence and business listings. The yellow pages list business subscribers alphabetically by business type and feature advertisements that contain additional information about those businesses. Local phone books are free to subscribers in reasonable quantities. Additional phone books and out-of-area phone books used to be free, but that was many years ago. Subscribers who connected service after the publishing deadline are not listed in the phone book, but are listed in supplements available to information operators. Calls to information used to be free, but that was many years ago. Use the phone book when you can — it's much less expensive. Oh dear, I almost forgot: Be nice to your mother, floss daily, and recycle. **2.** An electronic version, or soft copy, of a printed phone book. Electronic white pages and yellow pages are available online on the World Wide Web. See also *soft copy* and *World Wide Web*. **3.** A user-programmable electronic directory of frequently called numbers stored in system memory and accessible for autodialing applications.

Phone Power An AT&T program designed in the 1970s to promote long-distance telemarketing, specifically to promote WATS service. See also *telemarketing* and *WATS*.

phonetic alphabet The military and the aviation industry use a phonetic English alphabet that has many other uses, such as in the identification of time zones. The full alphabet is listed in Table P-2. See also *Zulu Time*.

Table P-2: Military and Aviation Phonetic Alphabet

A: Alpha	B: Bravo	C: Charlie
D: Delta	E: Echo	F: Foxtrot
G: Golf	H: Hotel	I: India
J: Juliet	K: Kilo	L: Lima
M: Mike	N: November	O: Oscar
P: Papa	Q: Quebec	R: Romeo
S: Sierra	T: Tango	U: Uniform
V: Victor	W: Whiskey	X: X-Ray
Y: Yankee	Z: Zulu	

photon **1.** A quantum of electromagnetic energy, a photon is a subatomic particle that has energy and momentum, but no mass or electric charge. A photon demonstrates wave and particle properties when in motion, but demonstrates no physical properties when at rest. See also *electromagnetic spectrum* and *wave*. **2.** A unit of light intensity at the retina equal to the illumination of one candle per square meter received through a pupil area of one square millimeter. See also *candle* and *light*.

photonic cross-connect (PXC) See *PXC*.

photonics The science of the creation, control, and detection of radiant energy, the fundamental element of which is the photon. In other words, the study of the use of light. See also *energy* and *photon*.

photophone Alexander Graham Bell, in early 1880, invented the photophone, a system utilizing mirrors to focus modulated sunlight onto a selenium cell. He was successful in transmitting voice over a distance of 700 feet on sunny days, and was granted four patents for the invention. Although it was clearly impractical, Bell nonetheless felt the photophone to be his greatest achievement. During World War II, the Nazi military experimented with similar but more advanced systems in tank warfare applications, but the technology remained impractical. Contemporary free space optics (FSO) transmission systems do not use mirrors and selenium cells, but undoubtedly have conceptual roots in the photophone. Bell later renamed the photophone the radiophone. See also *FSO*.

PHS (Personal Handyphone System) A digital cordless telephony system developed in Japan and used in China, Taiwan, and throughout Asia. PHS carves out 75 carriers with spacing of 300 kHz in the 1895–1918.1 MHz band, the upper half of which is used for public systems and the lower half for home office applications. PHS divides the band into 75 carriers spaced at 300 kHz, each of which supports four channels through time division multiple access (TDMA) and time division duplex (TDD). PHS employs $\pi/4$ differential quaternary phase shift keying ($\pi/4$ DQPSK) to yield a theoretical data rate of 384 kbps per carrier. Voice is encoded at 32 kbps through adaptive differential pulse code modulation (ADPCM). See also *ADPCM*, *carrier*, *channel*, *cordless telephone*, *$\pi/4$ DQPSK*, *TDD*, and *TDMA*.

PHY (PHYsical Layer) See *Physical Layer*.

physical **1.** Something that exists in the reality of the material world, rather than in logic, as an idea or notion. Something physical is tangible and can be seen and touched. A physical circuit, for example, might consist of metal, glass, or plastic. **2.** Associated with the sciences such as chemistry and physics that deal with nonliving things such as energy and matter, rather than the sciences such as biology and physiology that deal with living organisms. So, a radio circuit is physical, even though it is not tangible.

Physical Layer (PHY) Layer 1, the bottom layer, of the seven-layer Open Systems Interconnection (OSI) Reference Model. The Physical Layer defines the electrical and mechanical aspects of the interface between a device and a physical transmission medium, such as twisted pair, coaxial cable, optical fiber, or radio link. Communications hardware and software drivers are found at this layer, as are electrical and optical specifications. E/J/T-Carrier runs at the Physical Layer. Synchronous Optical Network (SONET) and Synchronous Digital Hierarchy (SDH) run at the Physical Layer and Data Link Layer. See also *coaxial cable*, *Data Link Layer*, *driver*, *E-carrier*, *electrical*, *hardware*, *J-carrier*, *layer*, *link*, *mechanical*, *network architecture*, *optical fiber*, *OSI Reference Model*, *physical*, *protocol*, *radio*, *SDH*, *software*, *SONET*, *T-carrier*, and *twisted pair*.

Physical Medium (PM) See *PM*.

physical reach The maximum length of a fiber optic link, in consideration of the loss budget, or the allowable amount of signal attenuation that the system can withstand. See also *attenuation*, *fiber optic*, *link*, *logical reach*, *loss budget*, and *signal*.

physical topology In reference to the physical layout of a network, specifically the physical positioning of the nodes and the circuits that interconnect them. LAN and WAN topologies variously include bus, mesh, partial mesh, ring, star, and tree. The logical topology of a network may differ considerably from the physical topology. See also *LAN*, *logical topology*, *mesh topology*, *partial mesh*, *physical*, *ring*, *star*, *topology*, *tree*, and *WAN*.

Physical Unit (PU) See *PU.*

physics The science of matter and energy, and their properties and interactions in fields including mechanics, acoustics, optics, heat, electricity, magnetism, radiation, and atomic and nuclear science. Physics is the science of how things work.

pi (π) The sixteenth letter of the Greek alphabet, π is used to mathematically represent the ratio of the circumference of a circle to its diameter, which is equal to 3.14159265358979323846. . . . Pi is an infinite decimal, at least as far as we know. Pi has been calculated to many millions of decimal places, but to no end. Further, no repeating pattern has ever been discovered, and many mathematicians maintain that a repeating pattern is impossible. Pi has many important mathematical uses. For example, the area of a circle is pi times the square of the length of the radius, or $A = \pi r^2$, or pi r squared. Everyone, from middle school, knows that to be impossible, of course — pies are round, some cakes are squared.

π/4 Differential Quaternary Phase Shift Keying (π/4 DQPSK) See *π/4 DQPSK.*

π/4 DQPSK (π/4 Differential Quaternary Phase Shift Keying) A variation on the QPSK modulation technique. π/4 DQPSK can be viewed as the superposition of two QPSK constellations offset by 45 degrees relative to each other. π/4 DQPSK is specified in cordless telephony standards, including Personal Wireless Telecommunications (PWT and PWT/E), Personal Handyphone Service (PHS), and Personal Communications Service (PCS). π/4 DQPSK also is specified in cellular radio standards, including Digital Advanced Mobile Phone Service (D-AMPS) and Personal Digital Cellular (PDC). See also *cellular radio, cordless telephone, D-AMPS, modulation, PCS, PDC, PHS, PWT, PWT/E,* and *QPSK.*

picker In power over Ethernet (PoE), a tap that acts as a splitter, picking off the 48V DC voltage, making it available to the non-PoE-compliant device at 5V, 6V, or 12V DC, for example. See also *DC, PoE, splitter,* and *voltage.*

picocell In radio systems, an imprecise, relative term referring to a very small area of coverage, perhaps comprising only a few city blocks, a tunnel, walkway, or single floor of a parking garage. A picocell is smaller than a microcell and much smaller than a macrocell. See also *cell, femtocell, macrocell,* and *microcell.*

piconet An imprecise, relative term used to describe a very small wireless network. Cellular radio networks, for example, are described, in order from largest to smallest, as macronets, micronets, and piconets. Bluetooth devices can establish ad hoc piconets with a maximum radius of 100 meters at maximum power level, although they are generally much smaller. The Bluetooth device that initiates the piconet acts as the master station and can include as many as seven slaved devices in the network. A device can participate in multiple piconets and act as a bridge between them, forwarding packets from one piconet to another in a scatternet configuration. See also *ad hoc mode, Bluetooth, cellular, macrocell, master/slave, microcell,* and *scatternet.*

PID (Pager Identification Number) A unique code assigned to each pager in a paging system. See also *pager* and *paging system.*

piezoelectric effect From the Greek *piezo-*, meaning *press*, and the Greek *elektor*, meaning *shining* or *the sun.* In electronics, the phenomenon by which certain natural and synthetic crystals become mechanically deformed under the influence of an electric field or, conversely, generate a voltage when subjected to pressure. See also *crystal oscillator, electric,* and *voltage.*

PIN (Positive Intrinsic Negative) A type of diode used as a light detector in optical transmission systems. A PIN comprises three layers of semiconducting material in the forward current-carrying direction. The first layer is chemically doped, i.e., infused with impurities, to create a positive (*p*) electromagnetic region. The second layer is either undoped or lightly doped to retain its intrinsic (*i*) properties and, therefore, is neither strongly positive nor strongly negative. The third layer is doped to create a negative (*n*) electromagnetic region. So, the diode structure is Positive, Intrinsic, and Negative, or PIN. A PIN generates a single electron from each photon received and, therefore, does not provide a significant gain, or increase in signal strength. However, PINs are fairly rugged and inexpensive. PINs generally are matched with light-emitting diode (LED) and vertical cavity surface-emitting laser (VCSEL) light sources and multimode (MMF) glass optical fiber (GOF) or plastic optical fiber (POF) in relatively low speed, short haul fiber optic transmission systems (FOTS). See also *diode, dopant, FOTS, GOF, laser diode, LED, light detector, MMF, POF,* and *VCSEL.*

ping **1.** A utility used to test a path from one host computer to another across an IP-based network in what is essentially a command to echo the packet from the remote host back to the originating host. Ping is an application of the Internet Control Message Protocol (ICMP). Ping was invented by Mike Muuss of the Army Research Laboratory in 1983 to diagnose an IP network problem. Muuss had done a considerable amount of work in college on sonar and radar modeling and was inspired by the principle of echo location. He named the utility after the sound that sonar makes when a signal returns. A lot of people think that ping is an acronym for Packet INternet Groper. According to Muuss, "packet internet groper" is a *backronym* reverse-engineered by Dr. David L. Mills. Note: Mills warns people that the clock on his wall runs backwards. I guess that explains it. See also *acronym, backronym, host, ICMP, IP, radar, sonar,* and *utility.* **2.** Slang for bouncing something off someone, i.e., asking someone a question, as in "Hold on just a minute while I ping Margaret and see if she'd like to join us."

pipe **1.** A circuit, in the vernacular, as in a broadband pipe. See also *circuit.* **2.** A coaxial cable configuration. See also *coaxial cable.* **3.** A cable conduit or ductwork. See also *conduit.* **4.** A waveguide for microwave transmission. See also *waveguide.*

PISN (Private Integrated Services Network) A private version of Integrated Services Digital Network (ISDN). A PISN makes use of the QSIG (Q Signaling) standard that defines services and signaling protocols for interconnecting TDM-based PBXs based on the ITU-T ISDN standard Q.931. See also *ISDN, PBX,* and *QSIG.*

PISO (parallel-in, serial-out) A shift register with parallel input ports and serial output ports. See also *parallel, port, serial,* and *shift register.*

pixel (*pics el*ement) From *pics,* which is slang for *pictures.* The basic unit that makes up an image displayed on a video screen. Pixels are dots of picture elements, similar to the dots in half-tone printing. The greater the number and areal density of the pixels, the better the resolution. See also *resolution.*

pizza box **1.** A directional, rate-adaptive, passive array antenna, so called because it is about the size and shape of a pizza box. Such an antenna possesses beamforming properties that permit it to adjust its logical focus to maximize the strength of the incoming signal. These adjustments in focus are accomplished passively, with no physical, i.e., mechanical, reorientation required. Pizza box antennas are used in wireless local loop (WLL) technologies such as WiMAX, and are placed at the customer premises. See also *antenna, rate adaption, signal, WiMAX,* and *WLL.* **2.** A rack-mounted enclosure 1 rack unit (RU) high (1.75″) that holds some network device or server.

PKI (Public Key Infrastructure) A formal structure that enables the user of an inherently insecure public network, such as the Internet, to electronically transfer information, funds, and other sensitive materials through the use of encryption key pairs obtained from and shared through a trusted entity. A certificate authority (CA) issues and verifies digital certificates that contain an encryption key and attest to the authenticity of the transaction party. A registration authority (RA) verifies the CA prior to the issuance of a digital certificate to the requesting party. See also *authentication, CA, digital certificate, encryption, Internet, key, private key, public key,* and *RA.*

PKM (Privacy Key Management) A technique developed for managing encryption keys in cable modem networks. PKM also is the foundation for the security protocol developed for 802.16, more commonly known as WiMAX. See also *802.16, cable modem, encryption, key, protocol, security,* and *WiMAX.*

plain old telephone service (POTS) See *POTS.*

plain text Data comprising standard characters (e.g., letters, numbers, and punctuation marks), with no formatting codes. Such unformatted text usually is coded according to the ASCII standard, which adequately represents the characters and symbols used to form the words used in most spoken languages and which, therefore, lends itself to text-to-speech (TTS) conversion. See also *ASCII, data, rich text,* and *TTS.*

plastic optical fiber (POF) See *POF.*

platform independent Capable of performing in a computing environment that connects *computers* made by different manufacturers and running different operating systems (OSs). See also *computer* and *OS.*

PLB (Personal Locator Beacon) A small transmitter device designed to provide a distress and alerting capacity for use by the general public in life-threatening situations in remote environments, after all other means (e.g., telephone and radio) of notifying search and rescue responders have been exhausted. Although it is not necessary to license a PLB, it is mandatory to register it with the National Oceanographic and Atmospheric Administration (NOAA). Thereby, search and rescue personnel will know who you are, where you are, and whether there are any pre-existing medical conditions that may require treatment when they reach you. PLBs transmit in the 406.0–406.1 MHz band, which is an internationally recognized distress frequency to the COSPAS-SARSAT satellite system, to which 36 countries belong. The satellites fix on the location using a Doppler Shift location method or GPS data from the terminal. The coordinates are then relayed to a local tracking station, of which there are a great many throughout the world. Each PLB transmits a unique 15-digit alphanumeric code that identifies the user and allows search and rescue personnel to access personal information maintained in an international database. In the United States, the Federal Communications Commission (FCC) regulates LPRS, which is in the family of personal radio services. See also *FCC, GPS,* and *personal radio services.*

PLC (PowerLine Carrier) A technology that uses electric power transmission lines and inside wire for telecommunications transmission. The technology was invented in 1928 by AT&T Bell Telephone Laboratories, and has been used since that time by the electric power utilities in select internal low speed data communications applications. Some customer premises equipment (CPE), such as small key telephone systems and intercom systems, based on PLC technology has made use of embedded inside electrical wire to avoid the cost of special wiring. In Europe and most of the rest of the world, PLC standards allow for communications over the 240-volt power grid at frequencies from 30 kHz to 150 kHz. In the United States, the standards for the 120-volt grid allow the use of frequencies above 150 kHz, as well. Power utilities use the frequencies below 490 kHz for internal applications such as telemetry and monitoring and control of equipment at remote substations. In the 1990s, development began on broadband over power line (BPL), which since has been standardized. BPL includes Access BPL and In-House BPL. See also *Access BPL, BPL,* and *In-house BPL.*

plenum An air-handling space such as those used for air conditioning and heating systems. Plenums commonly are between walls, under raised floor structures, and above drop (suspended) ceilings. While plenums are convenient places to run cables, they also are conducive to the spreading of fires within buildings, as they are primarily intended to support air flow. See also *plenum cable.*

plenum cable A type of inside cable intended for use in plenums. While plenums are convenient places to run cables, they also are conducive to the spreading of fires within buildings, as they are primarily intended to support air flow. Therefore, the National Electrical Code (NEC) specifies that the insulation on plenum cables must be fire retardant, low-smoke and low-toxicity. See also *horizontal cable, plenum, riser cable,* and *NEC.*

plesiochronous From the Greek *plesio,* meaning *near,* and *chronos,* meaning *time,* plesiochronous communications involves devices running at nominally the same rate. Plesiochronous networks comprise subnetworks and devices that are free-running, although at approximately the same rate and within defined parameters of tolerance for variation. Much as clocks and watches run at approximately the same rate, devices in a digital network are free-running, with some running at a slightly faster or slower pace.

Plesiochronous Digital Hierarchy (PDH) See *PDH*.

PLP (Packet Layer Protocol) The X.25 protocol that forms packets and manages packet exchanges between physical data terminal equipment (DTE) across a network of virtual circuits. PLP also can run on LANs and ISDN interfaces running link access procedure on the D channel (LAPD). PLP is responsible for call setup, synchronization, data transfer, and call clearing. In data transfer mode, PLP transfers data between DTE across both permanent virtual circuits (PVCs) and switched virtual circuits (SVCs), segmenting data blocks on the transmit side and reassembling them on the receive side of the communication. PLP also handles bit padding, flow control, and error control. PLP maps into Layer 3, the Network Layer, of the OSI Reference Model. Figure P-4 illustrates the PLP packet and its constituent fields, which are as follows:

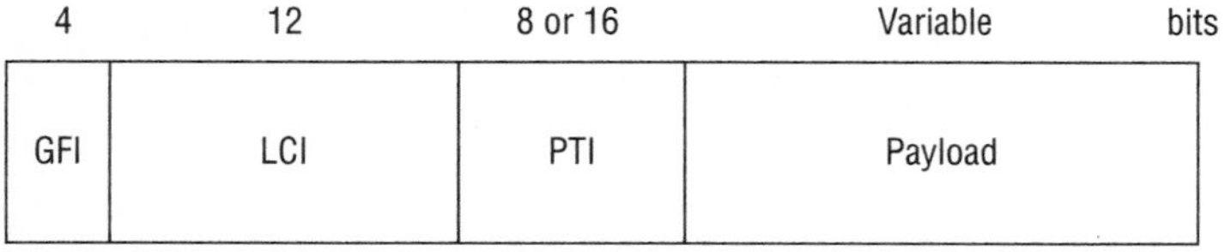

4	12	8 or 16	Variable bits
GFI	LCI	PTI	Payload

Figure P-4

- **General Format Identifier (GFI):** A 4-bit field that identifies packet parameters, including payload type (e.g., user data or control data), windowing information, and whether or not delivery confirmation is required.
- **Logical Channel Identifier (LCI):** A 12-bit field that identifies the logical channel group and channel number of the virtual circuit that connects to the destination data terminal equipment (DTE).
- **Packet Type Identifier (PTI):** An 8-bit or 16-bit field that identifies the PLP packet type, of which there are 17. Packet types include various call setup and call clearing, data and interrupt, flow control and reset, restart, and diagnostic packets.
- **User Data or Control Data:** The payload of a data packet comprises encapsulated higher-layer application information. The payload of a control packet comprises various information relating to call setup and clearing, flow control and reset, and so on. The default maximum payload size is 128 octets, which every network must support. Public X.25 networks variously accommodate packets with maximum payloads of 16, 32, 64, 128, 256, 512, and 1,024 octets. Custom networks can accommodate packet sizes of up to 4,096 octets.

See also *block, channel, DTE, error control, flow control, ISDN, LAN, LAPB, LAPD, Network Layer, OSI Reference Model, packet, padding, PVC, SVC, virtual circuit, window,* and *X.25*.

plug A male connector with pins designed to be inserted into the sockets of a female jack in a plug-and-jack connection. See also *connector* and *jack*.

Plug and Play (PnP) A set of specifications developed jointly by Intel and Microsoft that allows a PC to automatically recognize the presence of a new peripheral hardware device as it is connected to the PC, locate the necessary support software (e.g., driver), and configure the device interface. PnP requires a BIOS that supports PnP, and a PnP expansion card. PnP has been succeeded by the Internet-capable Universal Plug and Play (UPnP). See also *BIOS, driver,* and *UPnP*.

plug 'n' play (plug and play) A reference to the ability of a computer system to automatically recognize the presence of a new peripheral hardware device, locate the necessary support software (e.g., driver), and configure the device interface. See also *Plug and Play*.

PM **1.** Phase Modulation. More commonly known as *phase-shift keying* (PSK). See *PSK.* **2.** Physical Medium. In the ATM reference model, and other OSI-like models, a Physical Layer sublayer that specifies the physical and electro-optical interfaces with the transmission media. The PM also provides timing functions. See also *ATM reference model* and *Physical Layer.*

PMD (Polarization Mode Dispersion) A type of optical signal distortion caused by anomalies in the cross-section of an optical fiber. A light signal travels through a single-mode fiber along two planes that are orthogonal, i.e., perpendicular. If the fiber is perfectly round, light will travel along both planes at exactly the same speed, and both planes of light will arrive at exactly the same time, barring other dispersion phenomena. PMD is caused by the fact that fibers are never perfectly round, but are always inherently somewhat asymmetric; in other words, they are slightly elliptical in cross-section. Also, some additional asymmetry is caused as the fibers become somewhat misshapen during installation, as they are bent around corners, twisted, coiled, and so on. Further, transient asymmetry can occur due to vibration and temperature changes at various places along the link, or even from aerial fibers swaying in the wind. As the timing differences are so slight as to be measured in picoseconds (10^{-12}), PMD is not an issue at speeds of 2.5 Gbps or less over short distances. At contemporary speeds of 10 Gbps and 40 Gbps, however, bit times are so short that PMD results in unacceptable bit error rates. PMD is especially an issue over long-haul cable routes, as cascading asymmetry issues compound the slight effects and they reach noticeable levels. Closer spacing of regenerators will overcome the effects of PMD, although that solution tends to be expensive. PMD compensators have been developed to control the effects of PMD at speeds up to 40 Gbps by physically squeezing the fiber to counter-stress it.

PMI (Property Management Interface) An optional PBX interface for a property management system (PMS), which is an application software system intended for the hospitality industry, which includes hotels and motels, dormitories, cruise ships, hospitals, and extended care facilities such as nursing homes and retirement homes. PMI allows the coordination of front office and back office management features with PBX communications functions. For example, a housekeeper can use the telephone in the guest room to indicate room status, such as "clean and ready for occupancy." Front desk personnel can disable a guest telephone when a room is vacant, or if the guest's credit card has been rejected. Charges for calls through the PBX can be associated with the guest account and a single, all-inclusive bill can be rendered.

PN (Pseudorandom Noise or PseudoNoise) Noise that satisfies one or more standard tests for statistical randomness. Although pseudorandom noise seems to lack any pattern, it contains a sequence of pulses that repeat after a long period of time or after a long sequence of pulses. Spread spectrum (SS) radio systems modulate the carrier so that it appears as pseudorandom noise to receivers that are either not locked onto the transmitter frequencies or are incapable of generating a pseudorandom code sequence that correlates with the received signal. See also *carrier, code, frequency, noise, signal,* and *SS.*

PNG (Portable Network Graphics) A bitmapped graphic format, similar to Graphics Interchange Format (GIF). PNG also uses a lossy compression mechanism, but is open source. See also *bit map, compression, GIF,* and *lossy compression.*

PNNI **1.** Private Network-to-Network Interface. A specification for the interface between two private asynchronous transfer mode (ATM) networks. See also *ATM.* **2.** Public Network-to-Network Interface. A specification for the interface between two public asynchronous transfer mode (ATM) networks. See also *ATM.*

PnP (Plug and Play) A set of specifications developed jointly by Intel and Microsoft that allows a PC to automatically recognize the presence of a new peripheral hardware device, locate the necessary support software (e.g., driver), and configure the device interface. PnP requires a BIOS that supports PnP, and a PnP expansion card. PnP has been succeeded by the Internet-capable Universal Plug and Play (UPnP). See also *BIOS, driver,* and *UPnP.*

PoC (Push-to-talk over Cellular) Referring to the incorporation of push-to-talk (PTT) technology in a cellular radio handset. Although the PTT CB radio-like technology is relatively primitive by cellular standards, it offers considerable advantages in group communications applications, as the transmission is broadcast to all devices on the channel. PoC and PTT, in general, are especially appropriate for short, immediate voice interactions and can be characterized as voice instant messaging (IM). See also *cellular radio*, *IM*, and *PTT*.

POCSAG (Post Office Code Standardization Advisory Group) A standard set of code and signaling formats for radio paging systems developed in the 1970s by an international team of engineers. The name was derived from the fact that the British Post Office (BPO), which was the Post, Telegraph, and Telephone (PTT) agency for the United Kingdom at the time, chaired the effort. The POCSAG standard, which is in the public domain, provides for transmission speed of up to 2,400 bps, using channels of 25 kHz in the 150–170 MHz band. The CCIR (now ITU-R) standardized that code internationally in 1981, and most nations quickly adopted it. POCSAG can support as many as 2 million individual pager addresses. POCSAG supports simplex transmission, downstream only, and tone-only, numeric, and alphanumeric pagers. See also *CCIR*, *downstream*, *ITU-R*, *pager*, *paging system*, *PTT*, *public domain*, *radio*, and *simplex*.

PoE (Power over Ethernet) The IEEE 802.3af (June 2003) standard that defines how power is delivered over twisted pair to devices also using 10Base-T, 100Base-T, and 1000Base-T technologies. PoE provides electrical over two Cat 5e or Cat 6 wire pairs separate from those used for data transmission, except in the case of 1000Base-T. PoE not only provides an alternative to expensive electrical cabling in hard-to-reach places, but also alleviates concerns about power outages if an uninterruptible power supply (UPS) is available. PoE specifies that the power sourcing equipment (PSE), or power injectors, provide output of 48V DC power over the cable plant to terminal units that provide 12V DC output to PoE-compliant devices known as powered devices (PDs). The 802.3af standard also specifies four different power draw levels of up to 3.84W, 6.49W, 12.95W, and 15.4W for attached devices. The PSE automatically senses the power requirements of the PDs, including IP phones typically drawing 3-5W, wireless access points (APs) typically drawing 6-10W, and security cameras typically drawing 9-12W. PoE typically involves power supplied directly from an Ethernet switch to a client device, although there are provisions for midspan power injectors. See also *10Base-T*, *100Base-T*, *1000Base-T*, *AP*, *Cat 5e*, *Cat 6*, *DC*, *Ethernet*, *LAN switch*, *UPS*, and *volt*.

POF (Plastic Optical Fiber) A type of multi-step step-index, multimode optical fiber (MMF) manufactured of plastic rather than glass, POF has a large diameter core and a thin layer of cladding comprising multiple steps in index of refraction (IOR). The outside diameter of the fiber typically is approximately one millimeter (1.0 mm). The core is very large in diameter (960 microns), and the cladding only 40 microns or so. POF is designed to couple to a light-emitting diode (LED) emitting red light in the 650 nm window, which is in the visible light spectrum. In combination, the large mode field created by the large diameter core, the short wavelength of the signal, the use of LEDs as light sources, and the high IOR of the plastic relegate POF to short-haul, low-bandwidth applications. POF offers advantages in certain applications due to its flex strength, ability to withstand extremes of temperature (–40 to +85 degrees Celsius), and ability to withstand a bend radius of down to 25 mm with no break or damage. Consumer POF, which is used in consumer electronics, automobile wiring, and cold lighting, comprises a core of polymethyl methacrylate (PMMA), a type of acrylic, and cladding of fluorinated polymers. Perfluorinated polymer fibers are commonly used for much higher-speed applications such as data center wiring and local area network (LAN) wiring. Running at a wavelength of 650 nm, POF can support data rates of up to 400 Mbps over distances up to 100 meters, which compares very favorably to standardized Cat 5 and Cat 5e UTP, as well as Cat 6 and the developing Cat 7 standards. While its speed rating does not compare favorably with glass optical fiber (GOF), the MFA Forum has approved POF as a viable medium for use in ATM applications at speeds up to 155 Mbps for horizontal links up to 50 meters in length. The IEEE included POF in the 1394b FireWire standard, also for links up to 50 meters. See also *bend radius*, *Cat 5*, *Cat 5e*, *Cat 6*, *Cat 7*, *cladding*, *core*, *FireWire*, *flex strength*, *GOF*, *IOR*, *LED*, *MMF*, *short haul circuit*, *step-index fiber*, and *window*.

POH (Path OverHead) In a SONET or SDH frame, nine octets of overhead contained within the Synchronous Payload Envelope (SPE). The POH field is dedicated to the relay of operations, administration, management, and provisioning (OAM&P) information in support of end-to-end network management. See also *frame*, *network management*, *overhead*, *SDH*, *SONET*, and *SPE*.

pointer An identifier that indicates the position of an item of data within a frame. In a SONET frame, for example, a pointer indicates the position of a Virtual Tributary (VT) within a Virtual Tributary Group (VTG). The pointer also serves a synchronization function. See also *data*, *frame*, *SDH*, *SONET*, *synchronize*, *VT*, and *VTG*.

point of presence (POP) See *POP*.

point of termination (POT) See *POT*.

point-to-multipoint circuit A dedicated circuit that connects a single device (i.e., point) to multiple devices. A simple bridge connects the link from the primary device to the links that connect to each secondary device. Such a circuit is also known as a multi-drop circuit and a fantail circuit, as the circuit fans out like a tail from the headend. Point-to-multipoint circuits generally are phrased in the context of a wide area network (WAN), and generally are provided as a carrier service. As an example, Digital Data Service (DDS) circuits are configured as either point-to-point or point-to-multipoint. See Figure P-5. See also *bridge*, *circuit*, *DDS*, *link*, and *point-to-point*.

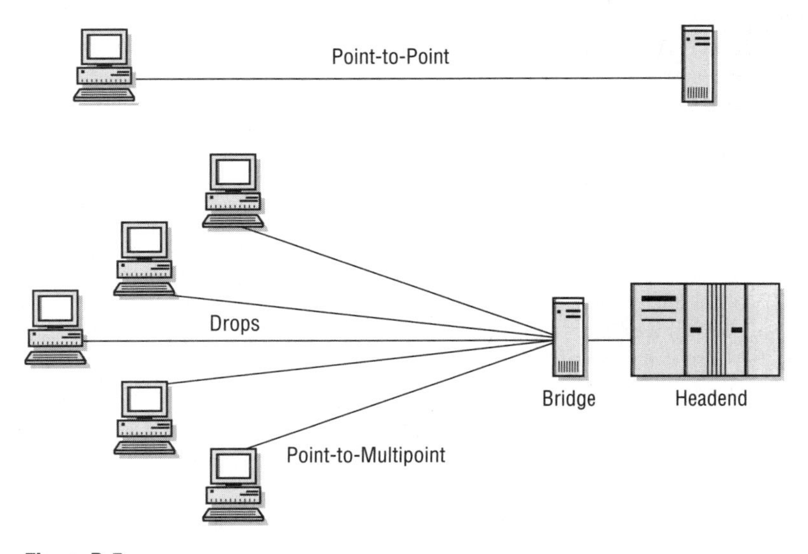

Figure P-5

point-to-point circuit A dedicated circuit that directly connects exactly two devices. Point-to-multipoint circuits generally are phrased in the context of a wide area network (WAN), and generally are provided as a carrier service. As an example, Digital Data Service (DDS) circuits are configured as either point-to-point or point-to-multipoint. See Figure P-5. See also *point-to-multipoint circuit* and *unicast*.

Point-to-Point Protocol (PPP) See *PPP*.

Point-to-Point Tunneling Protocol (PPTP) See *PPTP*.

Poisson distribution A probability theory that expresses the probability of events occurring in a fixed period of time if each event is independent of the previous event. In traffic engineering, that theory means that each call is completely independent of any previous call. There are a number of other techniques based on other assumptions underlying other formulas that yield different results. See also *Erlang*; *Erlang B*; *Erlang C*; *Equivalent Queue Extended Erlang B*; *Extended Erlang B*; *Poisson, Siméon-Denis*; and *traffic engineering*.

Poisson, Siméon-Denis (1781–1840) The French mathematician and physicist who discovered the *Poisson distribution*, which he published together with his probability theory in 1838, in his work *Recherches sur la probabilité des jugements en matières criminelles et matière civile*, translated as *Research on the Probability of Judgments in Criminal and Civil Matters*. See also *Poisson distribution*.

polarity The positive (+) or negative (–) property of an electrical charge. Opposite electrical conditions determine the direction in which discharged electrical current tends to flow, from positive to negative. Direct current (DC) flows in one direction along a circuit, as the positive and negative poles are fixed at either end of the circuit. Alternating current (AC) flows first in one direction and then in the other, as the polarity reverses. See also *AC* and *DC*.

polarization The direction of the electric field vector of an electromagnetic wave. All electromagnetic waves have electric and magnetic fields that are perpendicular to the direction of propagation. The electric field and the magnetic fields are orthogonal, i.e., perpendicular to each other. The simplest graphical representation is of one field along the x-axis and one along the y-axis, with the direction of propagation plotted along the z-axis, as illustrated in Figure P-6. The electric and magnetic fields are at the same frequency, although their amplitude may not be the same and they may be out of phase. Polarization can be planar (linear), circular, or elliptical. If the emitter, i.e., transmitter, causes the wave to take place in one plane, the wave is said to be a plane-polarized wave. If the amplitude of the electric and magnetic fields is constant and the fields are in phase, the polarization is said to be linear and is graphically traced in a straight line. If the two component fields are of the same amplitude and exactly 90 degrees out of phase, polarization is said to be circular, rotating in either a clockwise or counterclockwise direction, depending on which field is ahead of the other. All other conditions create what is known as elliptical polarization. The design of an electromagnetic transmitter determines the polarization of the emitted signal, the propagation characteristics of the signal, and the design of the receiver. The signal from a transmitter begins as planar, or linear, although reflections and other interactions with physical matter can change the polarization. Vertical polarization is used in AM and FM radio, and horizontal polarization in television. Satellites and terrestrial microwave systems use alternating horizontal and vertical polarization in adjacent frequency bands, which results in orthogonal signals that minimize the potential for mutual interference. Automobile headlights are horizontally polarized to provide a better view of the road. Sunglasses are vertically polarized to block glare (i.e., reflection) from water and snow, which reflect sunlight on a horizontal plane. See also *amplitude*, *frequency*, *phase*, *propagation*, and *sine wave*.

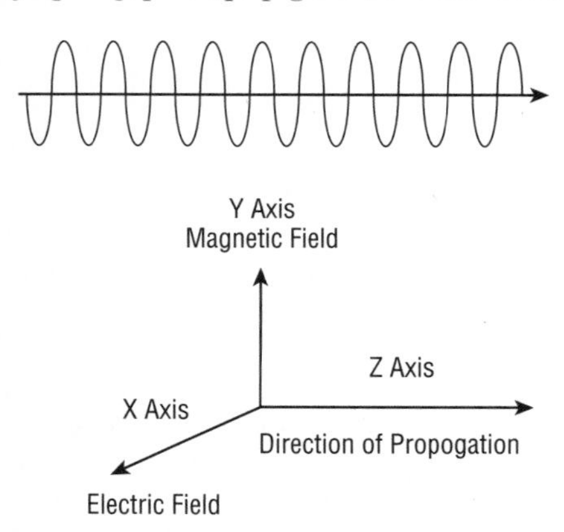

Figure P-6

polarization mode dispersion (PMD) See *PMD*.

pole **1.** A long, slender, tapering shaft of wood, metal, or concrete with a round cross-section that is planted in the ground and stands vertically, and to which cables, antennas, transformers, and other devices can be secured for support of applications including telecommunications and television transmission, and electrical power distribution and transmission. **2.** Either of two points at which opposing or differentiated forces are concentrated, such as the ends of a magnet or the terminals of a battery, motor, or dynamo.

pole attachment Referring to the physical connection of a communications cable or antenna used to provide communications services to a pole or in an underground conduit (even though there is no pole involved). In consideration of the substantial capital and maintenance costs associated with poles and their impact on limited network access corridors (easements, and rights of way), sharing of poles is considered essential. Prior to the emergence of the cable television (CATV) industry in the United States in the 1960s and 1970s, telephone companies and electric companies entered into reciprocal arrangements for shared pole usage. As the number of providers needing the pole resource has expanded, governments have established regulations regarding sharing of this critical infrastructure that may include local governments, which, in turn, may need to attach street lights, traffic lights, traffic counters, and other devices to utility poles. CATV providers also are allowed to attach cables to poles owned by telcos and power utilities, as are providers of wireless services such as Wi-Fi and WiMAX. The costs for such pole attachments are a subject of much controversy. In the United States, for example, there are not only different formulas for calculating costs for telephone companies and CATV providers to attach to power poles, but also different sets of laws that may apply to these assets, depending on the legal definition of the pole-owning entity (e.g., private, government-owned, or cooperatively organized) and the state in which these assets are located, among other things. See also *conduit*, *easement*, *pole*, *Pole Attachment Act*, *right of way*, *Wi-Fi*, and *WiMAX*.

Pole Attachment Act In the United States, an Act of Congress (1978) that required utilities that had made their poles available for use by communications usage to provide cable television (CATV) providers with just, reasonable, and nondiscriminatory access to any pole, duct, conduit, or right of way owned or controlled by it. The Telecommunications Act of 1996 expanded the jurisdiction of this act to include certain (non-incumbent local exchange) carriers, and subsequent court decisions confirmed the applications of this act to CATV providers (no matter what kind of service was carried over their wires), wireless telecommunications service providers, and high-speed Internet access services, as well as television and voice communications. See also *CATV*, *ILEC*, *right of way*, *Telecommunications Act of 1996*, and *utility*.

policing Also known as *traffic shaping*. See also *traffic shaper*.

polling **1.** A procedure by which a machine accesses a network, calls another machine, and downloads documents, files, or other data stored in the remote machine, commonly with security provided on the basis of a programmable poll code. **2.** A procedure by which a central computer controls network channel access by querying other computers or terminal devices to invite them to transmit data. Polling typically is in a predetermined query sequence, and the invitation is for transmission for a limited time, perhaps measured in milliseconds.

polyethene See *polyethylene*.

polyethylene Also known as *polyethene*. A dielectric plastic polymer of ethylene used in manufacturing packaging and electrical insulation. See also *dielectric*.

PON (Passive Optical Network) A fiber optic local loop network without active electronics, such as repeaters or multiplexers. A PON uses inexpensive passive optical splitters and couplers that require no electrical power and that perform no processes other than to split downstream signals and combine upstream signals. PON standards originated as a set of Physical Layer (Layer 1) specifications developed by an international consortium of vendors. The resulting Full Service Access Network (FSAN) initiative for an ATM-based Passive Optical Network (APON) scheme, was ratified by the ITU-T within the G.983.1 standard. Elements of a PON comprise an Optical Line Terminal (OLT), an Optical Network Terminal (ONT), an Optical Network Unit (ONU), a passive splitter, and optical fiber.

- **Optical Line Terminal (OLT)** The OLT terminates the optical local loop at the edge of the network. In a telco PON, the OLT is housed in the central office (CO). In a CATV PON, the OLT is housed in the headend. The OLT can either generate downstream optical signals on its own, or can pass optical signals from the optical backbone through a collocated optical crossconnect or multiplexer. The OLT also receives upstream signals from the ONTs and ONUs.
- **Optical Network Terminal (ONT):** The ONT terminates the optical circuit at the customer premises in a fiber-to-the-premises (FTTP) scenario. The ONT serves as a media converter, interfacing the optical fiber to the copper-based inside wire. The ONT is an addressable device that recognizes and accepts downstream data addressed to it specifically, ignoring all other data. The ONT is synchronized with the OLT, and assigned time slots for user-generated upstream data, which it buffers as necessary.
- **Optical Network Unit (ONU):** The ONU terminates the optical circuit in a remote network node in a fiber-to-the-curb (FTTC) or fiber-to-the-neighborhood (FTTN) scenario. The ONU serves as a media converter, interfacing the fiber circuit to embedded copper in the form of unshielded twisted pair (UTP) in a telco network and coaxial cable in a CATV network.
- **Splitter:** The passive optical splitter is positioned between the OLT and the ONUs or ONTs, depending on the scenario. The splitter divides the single downstream signal into multiple, identical signals at reduced power levels. As a passive network, PON has no provision for amplifying the signal. The splitter serves as a passive signal concentrator with respect to upstream transmissions.
- **Optical Fiber:** The ITU-T specifies two types of optical fiber to be employed in a PON. G.652 describes a standard type of single-mode fiber (SMF). G.652c/d describes low-water-peak fiber (LWPF) or zero-water-peak fiber (ZWPF). The fiber link can be organized in either simplex or duplex configuration. Simplex refers to a single-fiber configuration supporting transmission in both directions. Duplex refers to a configuration of two fibers, one supporting downstream transmissions, and another supporting upstream transmissions.

See also *APON*, *backbone*, *buffer*, *CATV*, *CO*, *coaxial cable*, *coupler*, *cross-connect*, *downstream*, *duplex*, *FSAN*, *FTTC*, *FTTN*, *FTTP*, *headend*, *ITU-T*, *local loop*, *LWPF*, *multiplexer*, *passive*, *Physical Layer*, *repeater*, *simplex*, *SMF*, *splitter*, *telco*, *time slot*, *upstream*, *UTP*, and *ZWPF*.

Pony Express The Pony Express began service on April 3, 1860, with the promise that it could deliver a letter from St. Louis, Missouri, to Sacramento, California (a distance of 1,966 miles), in 10 days or less, and that is exactly what it did. Tough young riders, each of whom weighed no more than 125 pounds, rode 75 to 100 miles at an average speed of about 10 miles per hour, changing mounts every 10 to 15 miles. Riders earned about $100 per month for risking their lives on a daily basis. The cost for sending a half-ounce letter by Pony Express initially was $5.00, and was later reduced to $1.00, which is approximately $21.00 in present-day dollars, considering inflation. Eighteen months later, the completion of the Western Union transcontinental telegraph (October 24, 1861) network put the Pony Express out of business. See also *telegraph* and *Western Union*.

POP **1.** Point Of Presence. The point at which an interexchange carrier (IXC) establishes a physical presence in a geographic area, and at which the local exchange carriers (LECs) terminate access services. The term GigaPOP is sometimes used in reference to a POP with bandwidth in the Gbps range. See also *access circuit*, *GigaPOP*, *LEC*, and *IXC*. **2.** Post Office Protocol. An IETF standard for a store-and-forward service used by a client to retrieve e-mail from a remote server over an IP network. The current version is POP3 (Post Office Protocol version 3). See also *client*, *e-mail*, *IETF*, *IP*, *POP3*, and *server*.

POP3 (Post Office Protocol version 3) An IETF standard (RFC 2449) for a store-and-forward service used by a client to retrieve e-mail from a remote server over an IP network. Prior to accessing a mailbox at the remote server and downloading all mail to the client, the user can elect either to delete downloaded mail from the server or to leave a copy of each on the server. After downloading mail, the user can disconnect from the remote server and work with the mail offline. POP3 is used only when downloading mail from the mailbox. When uploading mail, client access is to a server running Simple Mail Transfer Protocol (SMTP), which simply forwards mail after looking up the proper IP addresses on a Domain Name Server (DNS) server. While POP3 is widely implemented, the more recent Internet Message Access Protocol (IMAP) offers certain advantages. See also *client, DNS, e-mail, IETF, IMAP, IP, RFC, server,* and *SMTP.*

port **1.** A point of physical access or physical interface between a circuit and a device or system at which signals are injected or extracted. A given computer system may be equipped with a number of ports for various, specific purposes and with various attributes appropriate to the application. Ports can be defined as analog or digital and as optical or electrical, for example. The port speed is always defined, as well. A PBX, for example, may include a number of printed circuit boards (PCBs), each of which contains a number of specific interfaces. A PBX digital line card commonly contains 4, 8, 16, or 32 digital line ports, each of which provides the electrical interface between a digital telephone set and the PBX common equipment, and operates either at 64 kbps or, if ISDN, at 144 kbps. The line ports also are designed to support the appropriate signaling and control protocols, which commonly are proprietary protocols defined by the PBX manufacturer. A PBX digital trunk card may include one or more trunk ports specifically designed to interface with T1 (1.544 Mbps) or E-1 (2.048 Mbps) electrical circuits. A router may have a number of ports, some of which may be optical interfaces to fiber optic transmission systems (FOTS) running at speeds of 2.5 Gbps or perhaps 10 Gbps, and others of which may be electrical interfaces for connection to copper circuits running at T1 or E-1 speeds. **2.** A logical connection, identified by a protocol address in a packet header, associated with a Transmission Control Protocol (TCP) or User Datagram Protocol (UDP) service. Ports provide a mechanism by which computers running the TCP/IP protocol suite can multiplex a number of concurrent connections at a single Internet Protocol (IP) address. In combination, the IP address and the port number identify a socket, with a source socket and destination socket defining a TCP connection. Port numbers are 16-bit values ranging from 0 to 65,535. Well-known ports are numbered 0 through 1,023 for the use of system (root) processes or of programs executed by privileged users. Examples of well-known ports include 25 for SMTP (Simple Mail Transfer Protocol), 80 for HTTP (HyperText Transport Protocol), and 107 for Remote TELNET Service. Registered ports, which are registered by the Internet Assigned Numbers Authority (IANA) as a convenience to the community, can be used by ordinary user processes or programs on most systems and can be executed by ordinary users. These same port assignments, numbered in the range 1,024 through 49,151, are used with the User Datagram Protocol (UDP) to the extent possible. Dynamic ports and/or private ports are those from 49,152 through 65,535. **3.** To modify or translate a software program so that it will run on a different computer operating system (OS). **4.** To move files from one computer to another.

portability **1.** The quality of being easily movable from one physical place to another. In the context of transmission systems, wireless technologies offer the advantage of portability. Radio systems and free space optics (FSO) systems are inherently somewhat portable, and many are lightweight, collapsible, and otherwise specifically designed to be highly portable. RF (Radio Frequency) systems operating in licensed frequency bands must be licensed in each geographic area in which they are intended to operate. See also *mobile* and *wireless.* **2.** Referring to the ability to move a telephone number from one carrier or service provider to another in a competitive environment. See *number portability.*

Portable Network Graphics (PNG) See *PNG.*

portal A site on the World Wide Web (WWW) that serves as a gateway or port of entry to the Internet. A portal typically includes hyperlinks to news, weather reports, stock market quotes, entertainment, chat rooms, and so on. See also *chat room, gateway, hyperlink,* and *WWW.*

portmanteau **1.** An old type of leather suitcase that usually opened into two compartments. The word originates from the French *portemanteau*, from *porter* (to carry) and *manteau* (cloak), and inspired Lewis Carroll's use of the word with a new meaning. **2.** A combination word, or blend, that combines the sounds and meanings of two words. *Modem*, for example, is a portmanteau that combines the words modulate and demodulate, describing a device that performs both functions. Similarly, a *codec* both codes and decodes data, and a *transceiver* is both a transmitter and receiver. Lewis Carroll coined this usage of the word in his famous book *Through the Looking-Glass and What Alice Found There* (1871). In the book, Humpty Dumpty explains to Alice the construct and meaning of words from the poem "Jabberwocky," telling her that "Well, '*slithy*' means 'lithe and slimy.' . . . You see it's like a portmanteau — there are two meanings packed up into one word."

POS **1.** Packet Over SONET. A MAN/WAN technology that supports packet data (e.g., IP) through either a direct optical interface to a router, or through a SONET demarc in the form of a terminating multiplexer (TM). POS uses SONET as the Physical Layer protocol, encapsulating packet traffic in High-level Data Link Control (HDLC) frames and using Point-to-Point Protocol (PPP) at the Data Link Layer, with the IP packet traffic running at the Network Layer. The result is that the combined overhead factor (SONET + PPP + IP) is only approximately 5 percent for a 1,500-byte IP datagram. This level of efficiency compares very favorably with asynchronous transfer mode (ATM), which involves an overhead factor of approximately 11 percent for the same IP datagram. This level of performance can be achieved only if packet data is to be transmitted, and only if the service is provided over the equivalent of a SONET-based, point-to-point private line, provisioned in the form of a Virtual Tributary (VT). The use of a VT enables POS traffic to bypass any ATM switches that might be in place in the carrier network. See also *ATM, datagram, Data Link Layer, demarc, encapsulate, HDLC, IP, MAN, Network Layer, overhead, packet, Physical Layer, point-to-point, PPP, protocol, router, TM, VT*, and *WAN*. **2.** Personal Operating Space. The area in the near vicinity of a device or individual. POS is fundamental to the concept of a wireless personal area network (WPAN). See also *WPAN*.

positive **1.** Something with the same charge (+) as a proton and opposite the negative (-) charge of an electron. **2.** Something with higher electric potential than the ground or a defined null or neutral point. Electrons flow from the positive point in a circuit. See also *negative, null*, and *potential*. **3.** Yes.

positive intrinsic negative (PIN) See *PIN*.

Postel, Jon (1943–1998) A significant contributor in the area of standards during the development of the Internet, Postel was the Editor of the *Request for Comments* (*RFC*) document series. Postel also served as the Internet Assigned Numbers Authority (IANA) until his death, at which time IANA management transferred to the Internet Corporation for Assigned Names and Numbers (ICANN). Postel is particularly well known for the Robustness Principle, also known as Postel's Law, which he stated in RFC 793 (1981) and again in RFC 1122 (1989), *Requirements for Internet Hosts – Communications Layers*. The Robustness Principle states, "Be liberal in what you accept, and conservative in what you send," which essentially advises the Internet community to design host software in such a way as to (1) be prepared for malevolent incoming packets and (2) be prepared for deficiencies on other hosts that can make it unwise to exploit legal but obscure protocol features that can cause disruption if the other hosts misbehave. See also *host, IANA, ICANN, Internet, packet, protocol, RFC, software*, and *standard*.

Postel's Law See *Postel, Jon*.

Post Office Code Standardization Advisory Group (POCSAG) See *POCSAG*.

Post Office Protocol (POP) See *POP3*.

Post, Telegraph, and Telephone (PTT) See *PTT*.

POT (Point Of Termination) Synonymous with *POP* (Point Of Presence). The point at which an interexchange carrier (IXC) establishes a physical presence in a geographic area, and at which the local exchange carriers (LECs) terminate access services. See also *LEC, IXC*, and *POP*.

potential **1.** The work required to bring a unit of positive electric charge from a reference point (as at infinity) to a specified point in an electric field. **2.** Something that could happen but has not happened yet. Someone who could do something but has yet to do it, as in "He has a lot of potential." Some people never advance beyond the potential. **3.** Maybe. See also *negative*, *null*, and *positive*.

POTS (Plain Old Telephone Service) Basic analog telephone service supporting full duplex (FDX) conversational voice communications and operator services. See also *basic service* and *PANS*.

pound sign See #.

power (P) **1.** The amount of current (*I*) times the voltage (*E*) at a given point in a circuit. Power is measured in watts (*W*), equivalent to joules per second. **2.** Strength or intensity. 3. The rate at which work is performed.

power back-off A feature of asymmetric digital subscriber line (ADSL) modems that dynamically reduces the modem power level at the customer premises in order to eliminate the potential for interference with the analog voice channel resulting from near-end crosstalk (NEXT). Power back-off occurs automatically and reduces the upstream data transmission rate for the duration of the phone call. Fast retrain enables rapid recovery of the upstream transmission rate when the phone call is terminated. See also *ADSL*, *channel*, *fast retrain*, *interference*, *modem*, *NEXT*, and *power*.

power density The power flow rate of an electromagnetic wave at a specific point in a medium, e.g., watts per square meter. The power density of a signal degrades over distance from the source due to absorption, diffraction, diffusion, dispersion, and scattering.

powered device (PD) See *PD*.

power failure cut-through See *power failure transfer*.

power failure transfer A voice telephone system feature that provides for an emergency mode of operation in the event of a commercial power failure. Also known as power failure cut-through, this feature allows an analog central office (CO) trunk to cut through the PBX switch to a predetermined analog emergency phone in a location such as the receptionist's desk, the security office, or the data center. There may be many such lines if a large PBX serves many users. Because analog CO lines and trunks are line-powered, i.e., powered from the CO, and the telco generally has several levels of emergency power backup in place, they are not highly susceptible to commercial power failures. This feature is particularly important if the user organization has no power backup such as an Uninterruptible Power Supply (UPS). Large organizations that consider telecommunications to be mission-critical generally have UPS systems in place that may include not only battery backup but also backup diesel power generators that serve to recharge the batteries during long outages. Many PBXs offer optional internal battery backup that is sufficient for one to eight hours of operation. See also *UPS*.

powerline carrier (PLC) See *PLC*.

power over Ethernet (PoE) See *PoE*.

power sourcing equipment (PSE) See *PSE*.

p-persistent carrier sense multiple access (P-Persistent CSMA) See *CSMA*.

P-Persistent CSMA (P-Persistent Carrier Sense Multiple Access) See *CSMA*.

P-phone (Proprietary phone) A proprietary voice terminal used with Centrex service, P-phones are proprietary to the central office (CO) switch manufacturer. P-phones are required for maximum ease of feature access, although most systems support generic sets, as well.

PPO (Private Paging Operator) An operator of a radio paging system for private use. PPOs are unregulated, but must share unlicensed spectrum with other users in the VHF and UHF bands. Enterprises operating in large campus environments may operate private paging systems. Restaurants often operate private paging systems to notify servers when orders are ready or to notify diners when a table is available. See also *paging system*, *radio*, *spectrum*, *UHF*, and *VHF*.

PPP (Point-to-Point Protocol) A Link Layer protocol based on a Link Control Protocol (LCP) and a Network Control Protocol (NCP). The LCP is responsible for setting up a link between two computers over a circuit-switched telephone connection, and for resolving any issues of authentication. The NCP negotiates any parameters specific to the Network Layer. The Internet Protocol Control Protocol (IPCP), for example, is used when the Internet Protocol (IP) is employed at the Network Layer. PPPdefines a sequence of characters that frame Internet Protocol (IP) datagrams on a serial line. In that respect, it is much like the Serial Line Internet Protocol (SLIP). However, PPP also supports a variety of authentication and encryption methods for enhanced security and a variety of compression options that can eliminate unused or redundant data in the headers of long sequences of packets in a transmission stream. PPP incorporates an error correction mechanism in the form of a cyclic redundancy check (CRC). PPP supports multiple native machine and network protocols, and supports subnet routing. PPP is defined in RFC 1661, and numerous other RFCs define various PPP implementations. Internet service providers (ISPs) commonly use PPP to support dial-up customers. Multilink PPP allows the combination of multiple links to increase bandwidth between two nodes. PPP over Ethernet (PPPoE) encapsulates PPP in Ethernet frames and is commonly used by ISPs to support Asymmetric Digital Subscriber Line (ADSL) customers. PPP over ATM (PPPoATM) is sometimes used by ISPs to support customers with ATM-based ADSL routers. See also *ADSL*, *ATM*, *authentication*, *CRC*, *datagram*, *dial-up*, *encryption*, *Ethernet*, *IP*, *ISP*, *LCP*, *Link Layer*, *Multilink PPP*, *NCP*, *packet*, *PPPoA*, *PPPoE*, *protocol*, *router*, *routing*, *serial*, and *subnet*.

PPPoA (PPP over ATM) A protocol that encapsulates PPP frames in ATM Adaptation Layer 5 (AAL5). PPPoA is used primarily in cable modems, wireless devices, and ADSL broadband local loops for Internet access and is specified in RFC 2364. In an ADSL implementation, PPPoA is employed in the local loop between the customer premises equipment (CPE) and the DSL access multiplexer (DSLAM). See also *AAL5*, *ADSL*, *ATM*, *broadband*, *cable modem*, *CPE*, *DSLAM*, *encapsulate*, *Internet*, *local loop*, and *PPP*.

PPPoE (PPP over Ethernet) A protocol that encapsulates PPP frames in Ethernet frames to allow multiple hosts on a shared Ethernet to open PPP sessions to multiple destinations via one or more bridging modems. PPPoE is used primarily in cable modems, wireless devices, and ADSL broadband local loops for Internet access and is specified in RFC 2516. See also *ADSL*, *broadband*, *cable modem*, *encapsulate*, *host*, *Internet*, *local loop*, and *PPP*.

PPP over ATM (PPPoA) See *PPPoA*.

PPP over Ethernet (PPPoE) See *PPPoE*.

PPTP (Point-to-Point Tunneling Protocol) A tunneling protocol that operates at the Data Link Layer (Layer 2) of the OSI Reference Model. PPTP is a proprietary technique that encapsulates Point-to-Point Protocol (PPP) frames in Internet Protocol (IP) packets using the Generic Routing Encapsulation (GRE) protocol. Packet filters provide access control, end-to-end and server-to-server. PPTP was initially conceived by Ascend, developed by Microsoft, and described in IETF RFC 2637. See also *Data Link Layer*, *frame*, *GRE*, *IETF*, *IP*, *IPsec*, *L2TP*, *OSI Reference Model*, *packet*, *PPP*, *protocol*, *server*, *SOCKS*, and *tunneling*.

PPV (Pay-Per-View) See *pay-per-view*.

PRA (Primary Rate Access) The user interface to an integrated services digital network (ISDN) intended for large business applications in European countries and elsewhere outside of North America and Japan. Also known as 30B+D, PRA offers 30 bearer (B) channels for user payload, plus one data (D) channel for signaling and control, and is backward-compatible with E-1 transmission. Primary rate interface (PRI), the North American version, offers 23 bearer (B) channels, plus one data (D) channel, and is backward-compatible with T1 and J-1 transmission systems. PRI and PRA both provide a full-duplex (FDX) point-to-point connection through an NT2-type intelligent CPE switching device, such as a PBX or router, for interfacing with the central office (CO) switch. The B and D channels are clear channels of 64 kbps, as signaling and control are out-of-band. The B channels can be used individually to connect on demand to any other ISDN device, or multiple B channels can be bonded and treated as a single fast connection for bandwidth-intensive applications such as data file transfers, videoconferencing, and any multimedia combination. Although the D channel is reserved exclusively for signaling and control, those functions are fairly light. ISDN standards, therefore, provide for non-facility associated signaling (NFAS), which allows a D channel to support up to five PRA/PRI circuits. See also *B channel, bond, BRI, CO, CPE, D channel, E-1, FDX, ISDN, J-1, out-of-band, PBX, PRI, router, signaling and control,* and *T1.*

predictive dialer A feature of standalone automatic call distributors (ACDs) that supports outgoing calling. A predictive dialer monitors the status of incoming calling activity and the level of availability of the agent pool. When the level of incoming calling activity drops to a level such that agents are idle for long periods, the system introduces outgoing calls. The predictive dialer statistically predicts the availability of an agent, searches a database of customers to be called, selects one and dials the associated telephone number, detects when the call is answered, and connects the call to an available agent. If no agent is available, the system serves the call to the first agent to become available. If the predictive dialer encounters a busy signal, does not get an answer after a predetermined number of rings, or senses an answering machine or voice processor at the target telephone number, it can simply hang up and dial the next number in the database. If the pool of agents is large enough, the predictive dialer can predict the availability of an agent quite accurately, and serve that agent a connected outgoing call within seconds. See also *ACD.*

preform cylinder A cylinder of very dense, very pure glass that is formed as a preliminary step in the manufacture of optical fiber. The cylinder is formed of vaporized glass and dopants that have cooled and settled on a bait rod in a process known as outside vapor deposition (OVD). The bait rod is then separated, and the cylinder is heated and collapsed to form the preform cylinder, which then is heated in a drawing tower. A gob of molten glass then forms and is drawn by gravity into an optical fiber. See also *OVD.*

premise An assumption, proposition, or presupposition that serves as the basis for an argument. The word is often confused with *premises.* For example, CPE is the initialism for Customer Premises Equipment, which is equipment physically located on the customer premises — at least that is the premise. See also *premises.*

premises A building, or part of a building, including its grounds. See also *premise.*

presence See *presence technology.*

presence technology An application that enables a user to post or advertise his availability status. Presence technology is a feature of instant messaging (IM) systems and some IPBX systems. Through presence technology a user can indicate his availability as online and available for e-mail or IM but unavailable for telephone calls, unavailable for IM or telephone calls, out to lunch, out of the office on business or vacation but returning on a certain date, and so on. See also *IM* and *IPBX.*

Presentation Layer Layer 6 of the seven-layer Open Systems Interconnection (OSI) Reference Model. Software at the Presentation Layer performs functions related to data format and display (e.g., fonts, tabs and line and page breaks), code (e.g., Baudot, ASCII, EBCDIC, Unicode, and HTML) and code conversion (e.g., HTML to ASCII), text compression and decompression (e.g., WinZip), and encryption (e.g., AES and DES). See also *code, compression, encryption, layer, network architecture, OSI Reference Model,* and *software.*

press-to-talk (PTT) Synonymous with *push-to-talk* (PTT). See *PTT.*

presubscription A procedure by which an end user designates an interexchange carrier (IXC) as the primary carrier for intraLATA or interLATA toll calls on an equal access (i.e., 1+ dialing) basis. See also *equal access*, *interLATA*, *intraLATA*, and *IXC*.

pretexting A means of social engineering in which a person gains unauthorized access to information under false pretenses, perhaps by pretending to be a person authorized to have such access.

Pretty Good Privacy (PGP) See *PGP*.

PRI (Primary Rate Interface) The user interface to an integrated services digital network (ISDN) intended for large business applications, primarily in North America. PRI offers 23 bearer (B) channels for user payload, plus one data (D) channel for signaling and control, and is backward-compatible with T1 and J-1 transmission systems. The European version is known as primary rate access (PRA) or 30B+D. PRA offers 30 B channels, plus one D channel, and is backward-compatible with E-1 transmission. PRI and PRA both provide a full-duplex (FDX) point-to-point connection through an NT2-type intelligent CPE switching device, such as a PBX or router, for interfacing with the central office (CO) switch. The B and D channels are clear channels of 64 kbps, as signaling and control are out-of-band. The B channels can be used individually to connect on demand to any other ISDN device, and multiple B channels can be bonded and treated as a single fast connection for bandwidth-intensive applications such as data file transfers, videoconferencing, and any multimedia combination. Although the D channel is reserved exclusively for signaling and control, those functions are fairly light. ISDN standards, therefore, provide for non-facility associated signaling (NFAS), which allows a D channel to support up to five PRI connections. PRI requires the extended superframe (ESF) format. Line coding is alternate mark inversion (AMI) with bipolar with eight-zeros substitution (B8ZS). This process is the same combination of techniques used in contemporary T1 circuits. See also *AMI*, *B8ZS*, *B channel*, *bonding*, *CO*, *CPE*, *D channel*, *E-1*, *ESF*, *FDX*, *ISDN*, *J-1*, *out-of-band*, *PBX*, *PRA*, *router*, *signaling and control*, and *T1*.

primary rate access (PRA) See *PRA*.

primary rate interface (PRI) See *PRI*.

primary/secondary The politically correct (PC) term for *master/slave*, a network architecture in which one device is designated as a *master* that is in control of other devices designated as *slaves*. See also *master/slave* and *PC*.

printed circuit An *electrical circuit, generally of copper foil, printed on a substrate dielectric* material. A sheet of copper foil is adhered to the substrate, and the unwanted portions of the foil are then removed by a process of silk screen printing, photoengraving, or mechanical milling. The remaining traces of copper foil form the circuit that connects components such as silicon chips mounted on the board. See also *printed circuit board*.

printed circuit board (PCB) A substrate of dielectric material on which a printed circuit, silicon chips, and other components are mounted, and to which connectors are affixed so that the board can plug into a system chassis. A PCB serves a defined purpose. In telecommunications systems, for example, a PCB may provide the physical and logical interface to one or more digital or analog lines or trunks, may provide voice mail service, or may provide call processing logic. See also *printed circuit*.

priority ringing See *distinctive ringing*.

privacy **1.** A feature of a key telephone system (KTS) or PBX that prevents a station user from accessing an outside line if another user has already engaged that line, unless the primary user chooses to override the restriction and to allow a conference call. See also *conference call*. **2.** A feature of a voice mail system that allows a sender to mark a message as private or confidential, thereby denying the recipient the ability to forward it to another user.

privacy key management (PKM) See *PKM*.

private automatic branch exchange (PABX) Synonymous with *private branch exchange* (PBX). See *PBX.*

private branch exchange (PBX) Synonymous with *private automatic branch exchange* (PABX). See *PBX.*

Private Integrated Services Network (PISN) See *PISN.*

Private IP See *IP-enabled Frame Relay.*

private IP address An Internet Protocol (IP) address set aside for use within a LAN, intranet, or other private network and not for use in a public network such as the Internet. Also known as Network 10 addresses, in reference to the first field in the first address range, private IP addresses fall into the following ranges:

- 10.0.0.0 to 10.255.255.255
- 172.16.0.0 to 172.31.255.255
- 192.168.0.0 to 192.168.255.255

Routers are not supposed to forward any packets to the public Internet from IP addresses in those private ranges. When a LAN-attached computer requires access to the public Internet, it is necessary that the private IP address be translated into a public IP address. The process takes place in a router that interfaces to both domains and runs the Network Address Translation (NAT) protocol, as defined in RFC 3022. See also *Internet, IP, IP address, LAN, NAT, network, packet, protocol,* and *router.*

private key encryption Also known as *single-key* or *secret-key encryption.* A symmetric encryption method that uses the same secret key to encrypt and decrypt data. See also *encryption, key, public key encryption,* and *security.*

private paging operator (PPO) See *PPO.*

private port Synonymous with *dynamic port.* A port that can be used by any computer application program to communicate with any other application program running Transmission Control Protocol (TCP) or User Datagram Protocol (UDP), with no registration requirements. Dynamic ports are numbered from 49,152 through 65,535. See also *port, registered port,* and *well-known port.*

Private Signaling System No. 1 (PSS1) See *PSS1.*

Private UNI (Private User Network Interface) In asynchronous transfer mode (ATM), the specifications for the interface for an ATM endpoint and an ATM switch in a private network, such as an ATM-based LAN. Data Exchange Interface (DXI) is a Private UNI for end-user access to an ATM network from equipment such as a bridge, router, or digital service unit (DSU). See also *ATM, ATM reference model, DXI, Public UNI,* and *UNI.*

privatize To transfer ownership of a public utility from the government to private interests. See also *liberalize, nationalize,* and *utility.*

.pro (professional) Pronounced *dot pro.* The Internet generic Top Level Domain (gTLD) reserved exclusively for certified professionals, professional companies, and professional associations. Individual professional domains include .med.pro for doctors, .law.pro for lawyers, and .cpa.pro for accountants. This is an unsponsored domain. See also *gTLD, Internet,* and *unsponsored domain.*

procedure Synonymous with *routine.* See also *routine.*

program A sequence of instructions written to be executed by a computer. See also *routine.*

program file An electronic file containing commands and instructions for execution by a computer. The contents of a program file are program code, generally written in a high-level language in which a human can compose quickly, and compiled into machine-readable code that a computer can execute quickly. See also *data file.*

progressive scanning The process of refreshing a video screen that involves displaying all horizontal scan lines sequentially in one frame, rather than refreshing all odd lines then all even lines. Progressive scanning avoids the problem of interline flicker associated with interlaced scanning. See also *interlaced scanning* and *scanning.*

propagation The movement of waves, such as electromagnetic waves, through a medium or through free space.

propagation delay The time required for a signal to travel from one point to another, generally from a transmitter through a medium to a receiver. Propagation delay is dependent on the nature of the electromagnetic signal, as not all signals travel at the same speed through a medium. Propagation delay also is influenced by the distance between the two points, the density of the medium, and the presence of passive devices such as loading coils that might increase the impedance of the medium. See also *impedance, loading coil, medium,* and *velocity of propagation (Vp).*

property management interface (PMI) See *PMI.*

proprietary Privately owned or controlled. Generally referring to a technology that has been developed by an individual, corporation, or other legal entity that considers the specifications to be a trade secret of value and that retains the exclusive right to exploit it, either for its own use or that of legitimate licensees that pay for the privilege. Proprietary technologies generally are patented, which prevents others from copying or otherwise duplicating them. See also *patent* and *public domain.*

protector A device that protects equipment and facilities from abnormally high voltage or current surges or spikes such as those caused by lightening strikes. The original carbon-based protectors have been obsoleted by gas-tube protectors and semiconductor resistor-based protectors. Good electrical protection requires good electrical grounding, and is essential in high-lightening areas. Synonymous with *surge protector.*

protocol From the Greek *protokollon,* for a leaf of paper glued to a manuscript volume and describing its contents. The rules and conventions for exchanging information between computers or across computer networks. Protocols comprise conventions that, at a basic level, commonly include the dimensions of line setup, transmission mode, code set, and non-data exchanges of information such as error control. Protocols have two major functions: handshaking and line discipline.

- **Handshaking:** The sequence that occurs between the devices over the circuit, establishing the fact that the circuit is available and operational. The handshaking process also establishes the level of device compatibility, and determines the speed of transmission by mutual agreement.
- **Line discipline:** The sequence of network operations that actually transmits and receives the data, controls errors in transmission, deals with the sequencing of message sets (e.g., packets, blocks, frames, and cells), and provides for confirmation or validation of data received.

As there exists a wide range of protocols that address various aspects of the communications between devices, protocols often are organized into protocol suites, such as the TCP/IP protocol suite, or layered architectures, such as the OSI Reference Model. In such a tiered approach, the lowest layer addresses physical issues of connectivity and signal format, and the highest layer in the most complete suite addresses applications issues dealing with the user interface. See also *handshaking, Hellenologophobia, line discipline, OSI Reference Model,* and *TCP/IP protocol suite.*

protocol analyzer A tool for identifying, analyzing, and diagnosing communications network problems. Analyzers enable technicians, engineers, and managers to test the performance of the network to ensure that the systems and the network function according to specifications. LAN managers, for example, use protocol analyzers to identify protocols in use, to identify protocol mismatches and other performance faults, to perform network maintenance and troubleshooting, and to plan network upgrades and expansions. See also *LAN*, *network*, and *protocol*.

protocol converter Synonymous with *protocol translator*. A collection of software and firmware that converts the protocols used in one network, sub-network, or network element (NE) into those used by another. Protocol conversion is the responsibility of a gateway. See also *gateway* and *protocol*.

protocol data unit (PDU) See *PDU*.

protocol stack A conceptual organization of protocol groupings into a vertical stack. In such a stack, the lowest level protocols are placed at the bottom, forming the foundation upon which higher levels, or layers, are placed. The OSI Reference Model, for example, comprises seven layers, each of which is interrelated, providing services to those above and below. The lowest is Layer 1, the Physical Layer, which contains specifications for the electrical, optical, and mechanical aspects of the interface of the device to a physical transmission medium, such as twisted pair, coax, or fiber. The highest is Layer 7, the Application Layer, which contains specifications for services that support user and application tasks such as file transfer, interpretation of graphic formats and documents, and document processing. See also *OSI Reference Model* and *protocol suite*.

protocol suite A set of related protocols that work together in the context of a larger protocol stack. For example, the TCP/IP protocol suite, which is the basis of the Internet, is a collection of related protocols that are used in various combinations to serve specific applications. See also *protocol*, *protocol stack*, and *TCP/IP*.

protocol translator See *protocol converter*.

provider of last resort In telecommunications, the local carrier designated by the regulatory authority to serve the least desirable residential end user in the most remote location and requiring only the most basic service, but without the ability to pay for service at market rates. In developed countries with a competitive telecommunications environment, some entity must be so designated in order to preserve the concept of universal service, and that designee is generally the incumbent local exchange carrier (ILEC). The provider of last resort generally is reimbursed for allowable associated costs from a universal service fund (USF). See also *carrier*, *ILEC*, *universal service*, and *USF*.

proxy In telecommunications, a device or program empowered to act for another. See also *proxy firewall* and *proxy server*.

proxy firewall A firewall in the form of security software installed on a proxy server to act as a barrier between internal and external networks and, thereby, to both prevent unauthorized entities from gaining access to internal company resources and block internal users from gaining access to unauthorized external resources. A proxy firewall presents a single network address to the Internet, rather than exposing the true addresses of internal users. Network address translation (NAT) software is required to make the address translations in order that authorized communications can take place in a conversational manner. See also *firewall*, *NAT*, *proxy server*, and *security*.

proxy server A hardware device that acts on behalf of other devices for purposes such as data storage and security. A proxy server can locally cache frequently accessed documents in order to reduce the level of Internet traffic to a remote server. A proxy server also may support a proxy firewall, thereby serving as both a logical and physical barrier. See also *proxy*, *proxy firewall*, and *security*.

PSAP (Public Safety Access Point) In the United States, a local agency, usually at the municipal or county level, responsible for answering 911 calls for emergency assistance from police, fire, and ambulance services. See also *911* and *E911*.

PSC (Public Service Commission) Synonymous with *PUC* (Public Utility Commission). See *PUC*.

PSE (Power Sourcing Equipment) In power over Ethernet (PoE), equipment that serves as power injectors to provide output of 48V DC power over the twisted-pair cable plant to terminal units with PoE-compliant devices known as powered devices (PDs). For devices not PoE-compliant, splitters inserted into the Ethernet cabling provide 12V or 6V DC output. See also *DC*, *PoE*, and *splitter*.

pseudonoise (PN) Also known as *pseudorandom noise*. See *PN*.

pseudoquaternary coding The line coding format used in ISDN basic rate interface (BRI) at the S/T interface, which generally resides in a terminal adapter (TA). Pseudoquaternary coding is similar to alternate mark inversion (AMI) used in T1, with respect to the polarity of the framing bit and the requirement for bipolar violations (BPV) to compensate for long strings of zero (0) bits. The only difference is that pseudoquaternary coding has two levels of positive and negative voltage. See also *AMI*, *BPV*, *BRI*, *framing bit*, *ISDN*, *Reference Point S*, *Reference Point T*, and *T1*.

pseudorandom noise (PN) Also known as *pseudonoise*. See *PN*.

pseudowire A point-to-point virtual circuit, so called because it is not a genuine wire (i.e., a point-to-point physical circuit), but it behaves much like one. See also *circuit*, *point-to-point*, *pseudowire emulation*, *virtual circuit*, and *wire*.

pseudowire emulation (PWE) See *PWE*.

pseudowire emulation edge to edge (PWE3) See *PWE3*.

PSK (Phase-Shift Keying) Also known as *PM* (Phase Modulation). A signal modulation technique that involves the carefully synchronized shifting of the phase (i.e., position) of the sine wave in order to represent digital bits across an analog carrier. The transmitting computer outputs a baseband electrical signal, defining 1 bits and 0 bits as discrete voltage levels. Binary phase-shift keying (BPSK) is a unibit technique in which a continuous sine wave pattern is interrupted and restarted at the baseline with a 180° phase shift to indicate a change in value from a 1 bit to a 0 bit or from a 0 bit to a 1 bit. See Figure P-7. Differential phase-shift keying (DPSK) is a unibit PSK scheme in which each 1 bit triggers a 180° phase shift, but 0 bits have no effect. Quadrature phase-shift keying (QPSK), also known as quaternary phase-shift keying and quadriphase keying, is a dibit technique achieved by defining four phase shifts separated by 90 degrees (0°, 90°, 180°, and 270°) and thereby impressing two bits on every phase shift. PSK can be used in conjunction with amplitude modulation (AM) and frequency modulation (FM). See also *AM*, *analog*, *baseband*, *bit*, *carrier*, *FM*, *modulation*, *phase*, *signal*, *sine wave*, and *synchronize*.

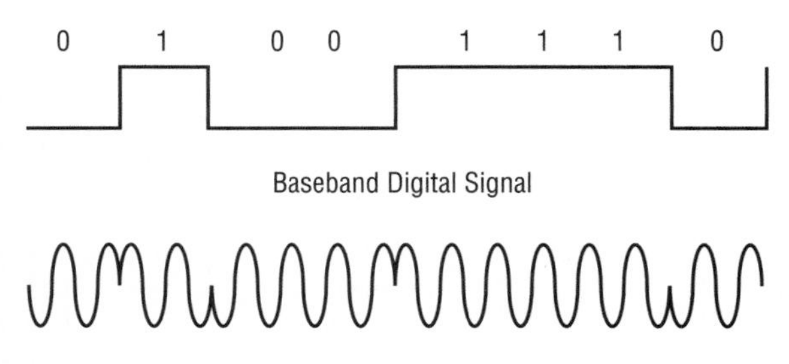

Figure P-7

PSP (Payphone Service Provider) A third party that owns and operates payphones connected to the public switched telephone network (PSTN) and located on another party's premises. See also *pay telephone* and *PSTN*.

PSQM (Perceptual Speech Quality Measurement) A method, defined in ITU-T P.861, for objectively evaluating voice transmission quality. PSQM is an automated method developed to measure the perceived quality of real-time voice as impacted by compression codecs. PSQM compares a distorted voice sample to an original, clear voice sample and uses a complex analytical process to evaluate the difference in terms of factors that influence the perceptions of human listeners. The ultimate distortion score corresponds closely with the mean opinion score (MOS) yielded by the panel of human listeners employed using the P.800 method. See also *codec*, *compression*, *MOS*, *P.800*, and *PSQM+*.

PSQM+ (Perceptual Speech Quality Measurement plus) An enhanced version of PSQM developed to measure the effects of packet loss and other network impairments on compressed real-time voice transmission. See also *PSQM*.

PSS1 (Private Signaling System No. 1) The ISO (International Organization for Standardization) mirror standard for the ITU-T Q Signaling (QSIG) standard. Formally known as Q.931, QSIG is a standard that defines services and signaling protocols for interconnecting TDM-based PBXs. QSIG is a Common Channel Signaling (CCS) protocol that runs over the ISDN D-channel for signaling between nodes in a Private Integrated Services Network (PISN). See also *ISO*, *PBX*, *PISN*, and *QSIG*.

PSTN (Public Switched Telephone Network) Synonymous with *PSTS* (Public Switched Telephone System). The generic term for the domestic public telephone network, which is traditionally is a public utility providing a circuit-switched network optimized for voice communications. The PSTN comprises all of the telecom infrastructure, including inside plant (ISP) and outside plant (OSP) owned by the carrier or carriers. Customer premises equipment (CPE) and other equipment and facilities owned by end users are considered private and, therefore, not part of the PSTN. As voice increasingly shifts to IP-based networks employing packet switching, and to alternative competitive carriers, such as cable television (CATV) providers, the term PSTN becomes somewhat vague. Global switched telephone network (GSTN) is the generic term for the international public telephone network that comprises the national PSTNs and the facilities that interconnect them. See also *CATV*, *circuit switch*, *GSTN*, *IP*, *ISP*, *OSP*, *packet switch*, and *utility*.

PSTS (Public Switched Telephone System) Synonymous with *PSTN* (Public Switched Telephone Network). See *PSTN*.

PTE (Path Terminating Equipment) In SDH or SONET, equipment that provides user access to the network. PTE operates in a manner similar to an E-3 or T3 time division multiplexer (TDM). See also *E-3*, *SDH*, *SONET*, *T3*, and *TDM*.

PTI (Payload Type Indicator) In asynchronous transfer mode (ATM) 3 bits in the cell header that distinguish between cells carrying user information and cells carrying service information. See also *ATM*, *cell*, *header*, and *payload*.

PTT **1.** Post, Telegraph, and Telephone. Referring to the government agencies in many countries that traditionally operated the public postal, telegraph, and telephone services. PTTs were the rule outside the United States, which never owned or operated the telegraph or telephone networks. Most nations later divided the responsibilities for postal, telegraph, and telephone services, and assigned telegraph and telephone services to what generically was known as a *telecommunications organization* (*TO*). More recently, most developed nations have fully or partially privatized telecommunications services. **2.** Push-To-Talk or Press-To-Talk. Also known as over-over. In reference to the process used to key a microphone or transmitter to send a message over a half-duplex radio system, such as Citizens Band (CB). The full cycle is push-to-talk and release-to-listen. As the half-duplex radio system will support transmission in only one direction at a time, the parties must negotiate a dynamic protocol that considers the demands of the conversation and the various conversationalists. See also *CB Radio Services*, *key*, *protocol*, and *radio*.

PU (Physical Unit) In the IBM Systems Network Architecture (SNA), a device that manages the communications hardware and software, participating in the controlling and routing of network communications. All physical devices are assigned a PU Type (1, 2, 3, or 5) that identifies the level of the device (terminal, controller, communications processor, or host node) and its origin (IBM/SNA or non-IBM/SNA). See also *hardware*, *SNA*, and *software*.

public data network (PDN) See *PDN*.

public domain Software, or any work of intellectual property, that was never protected by patent or copyright, on which such protection has expired, or that has been declared by its creator to be released unconditionally for public use, including distribution and modification, in any manner and under any circumstances. Such work is entirely free of any copyright restrictions or any other property protections. See also *copyright*, *freeware*, *intellectual property*, and *software*.

public IP address An Internet Protocol (IP) address that is designated for use in a public domain, such as the Internet. A public IP address is in contrast to a private IP address, which is in an address range designated for use only in a private domain, such as a local area network (LAN). See also *domain*, *Internet*, *IP*, *IP address*, *LAN*, and *private IP address*.

public key encryption An asymmetric encryption method with a freely available RSA encryption (encoding) key specific to each receiver that can be used by any sender and a decryption (decoding) key that is kept secret by the receiver. Public key encryption is available freely on the Internet via a program known as Pretty Good Privacy (PGP). See also *encryption*, *key*, *PGP*, *private key encryption*, *RSA*, and *security*.

public key infrastructure (PKI) See *PKI*.

Public Safety Access Point (PSAP) See *PSAP*.

Public Service Commission (PSC) Synonymous with *Public Utility Commission* (PUC). See *PUC*.

public switched telephone network (PSTN) See *PSTN*.

public switched telephone system (PSTS) Synonymous with *public switched telephone network* (PSTN). See *PSTN*.

Public UNI (Public User Network Interface) Also known as the Broadband UNI (B-UNI). In asynchronous transfer mode (ATM), the specifications for the interface between an ATM endpoint or private ATM switch and an ATM switch in a public network. See also *ATM*, *ATM reference model*, *Private UNI*, and *UNI*.

Public Utility Commission (PUC) Synonymous with *Public Service Commission* (PSC). See *PUC*.

PUC (Public Utility Commission) Synonymous with *Public Service Commission* (PSC). In the United States, each of the 50 states has formed a PUC or PSC to regulated public utilities. In the context of telecommunications, the PUCs regulate incumbent local exchange carriers (ILECs), certify competitive local exchange carriers (CLECs), register interexchange carriers (IXCs), adopt and enforce rules relative to competition, oversee emergency services, administer the universal service fund (USF), and monitor service quality. See also *CLEC*, *ILEC*, *IXC*, *LEC*, *USF*, and *utility*.

pull In the World Wide Web, a technology that requires that the user initiate access to a Web site to download content. See also *push* and *World Wide Web*.

pulse A brief, temporary change in a quantity or value from its normal or initial level for a period of time, and then a decay of that value back to the original level. Purely digital systems that do not rely on the modulation of an underlying carrier fit this definition. Telegraphy, for example, relies on the making and breaking of an electrical current so that the normal or initial level is a *current off* (no current) condition and the pulse is either a short or long *current on* (yes current) condition (dot or dash, respectively), separated by a short *current off* (no current) condition (space). Similarly, digital fiber optic transmission systems (FOTS) operate on the basis of *light on* (1 bit) and *light off* (0 bit).

pulse amplitude modulation (PAM) See *PAM*.

pulse code modulation (PCM) See *PCM*.

pulse dispersion A type of intersymbol interference (ISI) created when optical pulses overrun each other in a fiber optic transmission system (FOTS). Pulse dispersion is the result of modal dispersion, which is an issue in systems employing multimode fiber (MMF). Such systems permit optical signals and signal components to propagate along multiple modes, or physical paths, within both the core and the cladding of a fiber. Some light rays take a relatively direct path through the center of the fiber core. Other rays veer towards the edge of the core, where they reflect off the interface between the core and cladding on one side of the fiber, and then the other side of the fiber, and so on as they propagate across the link. Because some paths are more direct than others, and because the time of arrival is directly related to the distance traveled, some portions of the signal arrive before others. As the distance of the circuit increases, the differences in distances traveled by the various portions of each light pulse become greater as the effects of modal dispersion become more pronounced. As the speed of transmission increases, the bit time decreases, and the separation between bits is lost. The overall impact is that the pulses of light tend to smear together as they lose their shape and overrun each other, as illustrated in Figure P-8. See also *bit time*, *chromatic dispersion*, *dispersion*, *MMF*, *material dispersion*, *modal dispersion*, *mode*, *polarization mode dispersion*, and *waveguide dispersion*.

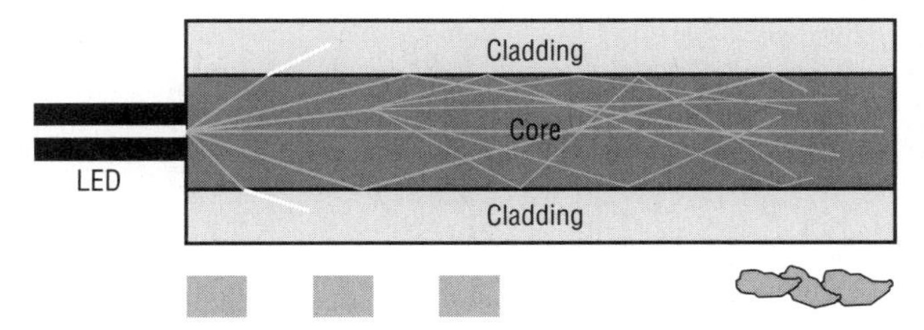

Figure P-8

pulse shape In reference to the physical appearance of an electromagnetic pulse as it is displayed on a monitor (oscilloscope). The pulse characteristics displayed include amplitude (i.e., signal strength) on the vertical axis, frequency (i.e., the number of cycles per unit time) on the horizontal axis, and phase (i.e., relative measure of position). Power requirements, co-interference, signal attenuation, spectral efficiency, and other considerations often require that pulses be shaped within certain minimum and maximum amplitude and frequency boundaries. As examples, T1 and ISDN pulses must fit a shape template to comply with standards. See also *amplitude*, *attenuation*, *frequency*, *interference*, *ISDN*, *phase*, *power*, *pulse*, *spectral efficiency*, and *T1*.

pulse shaping The process of forming an electromagnetic pulse so that it conforms to upper and lower system limits with respect to amplitude or frequency. For example, Bluetooth and DECT radio systems employ Gaussian frequency-shift keying (GFSK) to smooth out frequency deviations by limiting their spectral width. See also *amplitude*, *Bluetooth*, *DECT*, *GFSK*, *frequency*, *pulse*, *pulse shape*, and *spectral width*.

pump laser A laser used as an energy pump in an optical amplifier or other laser by transferring or injecting energy from one wavelength to another in a medium, often a doped fiber. Raman amplification uses a high-energy pump wavelength sent in the reverse direction from that of the operating laser from the output end of the optical fiber span, where the incoming signal is weakest. The pump wavelength, which generally is in the 1450 nm (E-band) window, interacts with atoms in the crystalline lattice of the fiber core. The atoms absorb the photons and, when stimulated by the counter-flowing signal, quickly release photons with energy equal to the signal photon plus/minus atomic vibration. In other words, a frequency/wavelength shift occurs as the pump wavelength propagates along the fiber in the reverse direction. The energy lost in the pump wavelength excites atoms that shift energy to longer-wavelength signals, typically in the 1550 nm window (C-band), in the forward direction, thereby serving to amplify them. See also *amplifier, C-band, E-band, frequency, laser, optical fiber, photon, propagate, Raman amplifier, signal, span*, and *wavelength.*

pure mesh In wireless local area networks (WLANs), a mesh network in which any and all devices can interconnect with any and all other devices on a purely wireless basis. Also known as a client mesh, this approach is not highly scalable. A pure mesh is much like a Bluetooth piconet. See also *Bluetooth, mesh, piconet*, and *WLAN.*

purist One who is preoccupied with and insists on the strict and excessive adherence to a tradition or set of formal, often pedantic, rules, especially with respect to maintaining the purity of language from foreign or altered forms. Photographers who persist in capturing images on film and printing them with a wet chemical process on silver emulsion paper also qualify. See also *bit/s, bps*, and *pedant.*

push In the World Wide Web, a technology that initiates content transmissions to users who have registered or subscribed to a service, relieving them of the requirement to initiate access to a Web site to retrieve that content, or "pull" it down. See also *pull* and *World Wide Web.*

push-to-talk (PTT) See *PTT.*

push-to-talk over cellular (PoC) See *PoC.*

PVC **1.** Permanent Virtual Circuit. A shared path established between two hosts through a packet network on a permanent basis. PVCs are preprogrammed in the routing tables of the transmission nodes throughout the network and are invoked based on various channel, channel group, or address information contained in the header of frames or packets. PVCs are defined on a permanent basis, until such time as they are permanently redefined, perhaps when the service provider rebalances the network to improve overall performance in consideration of changing usage patterns. Because the paths are predetermined and preprogrammed, network switches and routers can identify and exercise them quickly. PVCs are employed in frame relay and X.25 networks, as examples. A switched virtual circuit (SVC) is not preprogrammed, but is set up as the call is placed. See also *channel, circuit, frame, frame relay, packet, path, SVC, virtual circuit*, and *X.25.* **2.** PolyVinyl Chloride. Actually polychloroethene, a thermoplastic polymer. PVC's durability, flexibility, and dielectric properties, along with its low cost, make it useful as an insulating material in copper cables. See also *dielectric.*

PWE (PseudoWire Emulation) In Multiprotocol Label Switching (MPLS), referring to a Data Link Layer (Layer 2) Virtual Private Network (VPN). Such a network emulates a point-to-point virtual circuit connection, or pseudowire, between two routers or switches. Also commonly referred to as a Draft-Martini VPN. See also *connection, Data Link Layer, Draft-Martini VPN, MPLS, point-to-point, pseudowire, virtual circuit*, and *VPN.*

PWE3 (PseudoWire Emulation Edge to Edge) A PWE that operates from the ingress edge to the egress edge of a shared Internet Protocol (IP) or MultiProtocol Label Switching (MPLS) network. PWE3 makes use of existing mechanisms specified by the IETF, exerting no control over the network other than using existing quality-of-service (QoS) or path control mechanisms. See also *PWE.*

PWT (Personal Wireless Telecommunications) A U.S. standard for digital cordless telephony, based on the pan-European Digital Enhanced Cordless Telecommunications (DECT) standard. Through frequency division multiplexing (FDM), PWT provides 10 carriers in the unlicensed 1910–1920 MHz band, with channel spacing at 1.25 MHz. Each channel will support 1.152 Mbps using $\pi/4$ differential quaternary phase shift keying ($\pi/4$ DQPSK) modulation. Each channel supports 12 users through time division multiple access (TDMA) and time division duplex (TDD), for a total system load of 120 users. Voice encoding is adaptive differential pulse code modulation (ADPCM) at 32 kbps. PWT supports call handoff, so users can roam from cell to cell at pedestrian speeds as long as they remain within range of the system. PWT antennas can be equipped with optional spatial diversity to deal with multipath fading. Security is provided through authentication and encryption mechanisms. PWT/E is an extension into the licensed bands of 1850–1910 MHz and 1930–1990 MHz. The PWT specification was developed by the Telecommunications Industry Association (TIA). See also *ADPCM, antenna, authentication, carrier, channel, cordless telephone, DECT, digital, encode, encryption, ETSI, FDM, handoff, modulation, multipath fading, π/4 DQPSK, PWT/E, spatial diversity, TDD, TDMA,* and *TIA.*

PWT-E (Personal Wireless Telecommunications-Enhanced) A U.S. standard for digital cordless telephony, based on the pan-European Digital Enhanced Cordless Telecommunications (DECT) standard and essentially an enhancement of PWT. PTW-E extends PWT into the licensed bands of 1850–1910 MHz and 1930–1990 MHz, and tightens the channel separation to 1 MHz. See also *channel, cordless telephone, DECT, digital,* and *PWT.*

P × 64 (P times 64) An ITU-T videoconferencing specification more correctly known as *H.320* and sometimes referred to as H.261, which actually specifies the video encoding technique. Designed for videoconferencing applications, P × 64 supports *p* channels of 64 kbps, up to a maximum of 30 channels, which is equivalent to E-1. P × 64 video formats include Common Intermediate Format (CIF), which is optional, and Quarter-CIF (QCIF), which is mandatory in compliant codecs. H.261 CIF supports 352 × 288 = 101,376 pixels per frame and 30 frames per second (fps), although lower frame rates also are supported. QCIF supports 176 × 144 = 25,344 pixels per frame, exactly ¼ the resolution of CIF. See also *CIF, codec, encode, frame, frame rate, fps, H.261, H.320, ITU-T, pixel, video, videoconference,* and *QCIF.*

PXC (photonic cross-connect) Also known as *transparent optical cross-connect* (transparent OXC). See *transparent OXC.*

Q **1.** One of the two letters, along with Z, that traditionally did not appear on a telephone dial or keypad. The thought was that Q could be confused with O, and that Z could be confused with 2. Q now appears with P, R, and S on number 7. Alphanumeric dialing was, and remains, a North American practice. Telephones in most other countries do not sport letters. **2.** Q interface or Reference Point Q in ISDN. See *Reference Point Q*.

Q.931 The ITU-T Recommendation for the user network interface (UNI) for integrated services digital network (ISDN) basic call control, such as call setup and teardown. The specifications include user-to-user and network-to-network call control messages for both circuit-switched and packet-switched networking. ISDN signaling and control takes place over the D channel. Q.931 also is included in the ITU-T H.323 protocol suite for multimedia communications over packet networks. See also *D channel*, *H.323*, *ISDN*, and *ITU-T*.

QAM (Quadrature Amplitude Modulation) A signal modulation technique that splits the carrier into two waveforms that are 90° out of phase, and specifies two possible amplitude values for each of four phase shifts separated by 90° (0°, 90°, 180°, and 270°). This yields eight distinct signal states, as illustrated in the signal constellation graph in Figure Q-1. Thereby, each signal impulse, or symbol, carries one of eight possible signal combinations and represents three bits ($2^3 = 8$). As a result, the transmission rate is thrice the signaling rate, or baud rate. At a signaling rate of 2400 baud, for example, this tribit modulation scheme yields a transmission rate of 7200 bps. The ITU-T V.29 recommendation is for 16-QAM. More complex schemes include 64-QAM, 128-QAM, 256-QAM, and 512-QAM. Trellis-coded modulation (TCM) uses the same modulation scheme as QAM, but adds forward error correction (FEC) to overcome the increased susceptibility to signal impairments which make it harder for the receiver to judge correctly which state is signaled with each baud. QAM applications include asymmetric digital subscriber line (ADSL). See also *16-QAM*, *64-QAM*, *128-QAM*, *256-QAM*, *512-QAM*, *carrier*, *FEC*, *modulation*, *phase*, *signal*, *signaling rate*, *symbol*, *TCM*, *transmission rate*, and *tribit*.

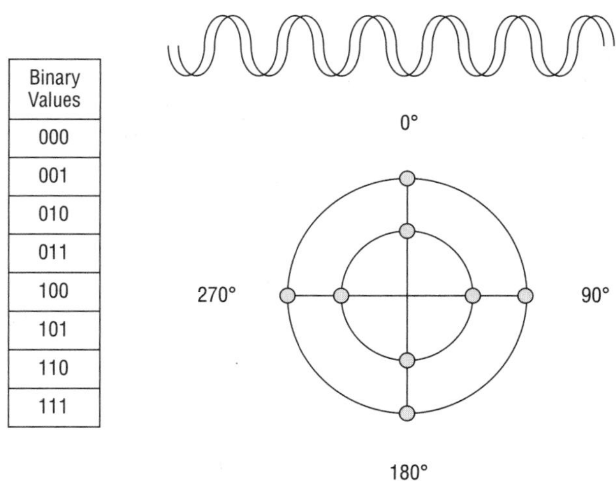

Figure Q-1

Q band The portion of the radio spectrum in the range of 36–46 GHz, as specified by the ITU-R. Applications are military in nature. See also *electromagnetic spectrum* and *ITU-R*.

QCIF (Quarter Common Intermediate Format) In the ITU-T H.320 umbrella standard for videoconferencing and multimedia communications over narrowband ISDN (N-ISDN), a mandatory video format that supports resolution of 176 × 144 pixels, which is exactly ¼ the resolution of CIF at 352 × 288 pixels. See also *CIF, H.320, ITU-T, multimedia, N-ISDN, pixel, resolution, video,* and *videoconferencing.*

QoS (Quality of Service) A measure of parameters that affect the level of performance a network offers a specific type of traffic. QoS parameters include access priority, bandwidth availability, latency, jitter, and packet loss. Toll quality, real-time compressed and uncompressed voice and video require fairly immediate network access, guaranteed availability of bandwidth throughout the call, low latency, zero jitter, and zero loss. E-mail is at the opposite end of the QoS spectrum, as it is highly tolerant of a low level of priority; high levels of latency, jitter, and loss; and does not require any bandwidth availability guarantees during the course of a mail transfer. QoS also must ensure that granting a QoS level to one traffic type or call does not violate the data flow requirements of another traffic type or call. The circuit-switched public switched telephone network (PSTN) offers all applications the highest level of QoS. Asynchronous transfer mode (ATM) offers multiple QoS levels appropriate for different traffic types. Frame relay, Internet Protocol (IP), and Ethernet networks can employ various priority mechanisms to offer differential grade of service (GoS) levels, but cannot guarantee QoS. See also *ATM, compression, Ethernet, frame relay, GoS, IP, jitter, latency, loss,* and *PSTN.*

QPSK (Quadrature, Quadriphase, or Quaternary Phase-Shift Keying) A modulation technique that achieves dibit transmission by defining four phase shifts separated by 90 degrees (0°, 90°, 180°, and 270°). IEEE 802.11a (Wi-Fi5) wireless LAN (WLAN) standards call for binary phase-shift keying (BPSK) at 6 Mbps and QPSK at 12 Mbps. IEEE 802.11b (Wi-Fi) calls for BPSK at 1 Mbps and QPSK at 2 Mbps. Bluetooth specifies BPSK for use in the 868 and 915 MHz bands, and QPSK for use in the 2.4 GHz band. μ/4 differential quaternary phase shift keying (μ/4 DQPSK) can be viewed as the superposition of two QPSK constellations offset by 45 degrees relative to each other. See also *802.11a, 802.11b, Bluetooth, BPSK, modulation, phase, PSK, μ/4 DQPSK,* and *WLAN.*

QPSX (Queued Packet Synchronous eXchange) A technology that formed the basis for the Distributed Queue Dual Bus (DQDB) defined in the IEEE 802.6 standard for metropolitan area networks (MANs). QPSX was developed at the University of Western Australia. See also *802.6, DQDB, IEEE,* and *MAN.*

quadbit **1.** A set of four bits. Some line coding techniques encode blocks of 4 bits of data at a time, rather than 1 or 2 bits. The 4B/5B technique, for example, encodes a 4-bit block of data into a 5-bit block of signal in order to provide sufficient clocking pulses and signal transitions to synchronize the network and to provide some level of error detection. 4B/5B is used in 100Base-TX, 100Base-FX, and Fiber Distributed Data Interface (FDDI) LANs. Compare to *nibble.* See also *4B/5B, line coding,* and *synchronization.* **2.** Referring to a modulation technique that impresses 4 bits on a baud, so that the bit rate is quadruple the baud rate. Such a technique employs 16 signal states. 16-QAM is a quadbit technique achieved by defining two amplitude values for each of eight phase shifts. See also *16-QAM, amplitude, amplitude modulation, baud, baud rate, bit, bit rate, dibit, modulation, phase, QAM, quartet, signal, tribit,* and *unibit.*

quadraplex A circuit or device that supports simultaneous transmission or reception of four independent signals. Quadraplex communications technology is a simple form of multiplexing that improved on diplex (two independent signals) and was considered quite revolutionary in the early days of telegraphy, when the four signals could be transmitted in one direction, only. The term is generally considered obsolete. See also *diplex* and *multiplex.*

Quadrature Amplitude Modulation (QAM) See *QAM.*

quadrature phase-shift keying (QPSK) See *QPSK.*

quadriphase phase-shift keying (QPSK) See *QPSK.*

quadruple play A marketing term used by broadband service providers to describe the triple play combination of voice, high speed data, and television services over a single local loop, plus wireless services. While there is no standard approach, quadruple play wireless services generally are designed to operate as cordless telephony when within range of a base station on the subscriber premises and as cellular telephony when out of range. As contemporary high-end cellular terminal devices are capable of supporting voice, data, image, and video, and as broadband cellular and other wireless networks are capable of supporting transmission rates in the range of hundreds of kbps and even Mbps, wireless capability is a significant addition to an integrated suite of service. The term triple play is a baseball analogy, referring to the very rare act in which the defense makes three outs on the same play. See also *broadband*, *cellular radio*, *cordless telephone*, *local loop*, and *triple play*.

quality of service (QoS) See *QoS*.

quantization and compaction encoding Referring to a step in the video compression process that reduces the number of bits required to represent a color pixel. Compaction techniques include run-length encoding, Huffman coding, and arithmetic coding.

quantize To express in multiples of a quantum number, in other words an integer or basic unit. In telecommunications, the term refers to the conversion of the amplitude of an analog sine wave into a digital signal, which necessarily requires expressing the amplitude value in binary terms. See also *quantizing noise*.

quantizing noise A type of distortion that occurs when an analog waveform is encoded into a digital signal and then decoded back into an analog signal. The digital-to-analog conversion process occurs at sampling intervals and always involves some amount of approximation as the amplitude of the waveform is quantized, which involves converting each sample amplitude value to the nearest of 256 (2^8) standard approximate binary values. When the approximate digital values are reconverted and the analog waveform is reconstructed, the effect of the approximation manifests as quantizing noise. If the sampling rate is too low (i.e., infrequent) and/or the approximation is too extreme, the result is a phenomenon known as aliasing in which the reconstructed signal is inaccurate, or even unintelligible, and the resulting voice quality unacceptable. See also *aliasing*, *amplitude*, *analog*, *binary*, *digital*, *distortion*, *noise*, *quantize*, and *waveform*.

quantum The elementary quantity of radiant energy, a photon. See also *photon*.

quantum leap **1.** In physics, an abrupt, or step, change in the energy state of an elemental unit, such as a molecule, atom, or subatomic particle. Such particles do not smoothly transition from one energy state to another; rather, they jump or leap from a state of rest to an excited state, for example, accompanied by the absorption or emission of a particle carrying the equivalent energy. Such changes are dramatic and instantaneous, but small in magnitude. **2.** In the vernacular, a change that is abrupt and large in magnitude.

Quarter Common Intermediate Format (QCIF) See *QCIF*.

quaternary phase-shift keying (QPSK) See *QPSK*.

quatraplex A synonym for quadraplex. See *quadraplex*.

queue A list, string, or stack of things constructed so that items are added to one end and relieved from one end or the other. Generally speaking, items are added to one end, known as the tail, and relieved from the other end, known as the head. In the absence of some priority mechanism for purposes of establishing and maintaining quality-of-service (QoS) differentiation, items are relieved from the head of the queue in the order they entered the tail. This approach is known as *first-in-first-out* (FIFO). Incoming call centers employ automatic call distributors (ACDs) that queue incoming calls, serving them to agents as they become available. Fax servers can queue documents for transmission during non-prime time hours, when international calling costs are lowest. PBX systems commonly have the capability to queue outgoing calls for expensive long distance circuits. Switches and routers queue packets in buffers until internal resources are available to process them or until bandwidth is available to forward them. Systems may support multiple queues for different types of calls or packets. Priority mechanisms can cause a call or packet to move up in the queue or even advance to the head of the queue in order that it can be served more quickly. See also *ACD*, *call*, *facsimile*, *packet*, *PBX*, *QoS*, *router*, and *switch*.

Queued Packet Synchronous Exchange (QPSX) See *QPSX*.

Q series The series of ITU-T Recommendations specifying protocols relating to switching and signaling. See Table Q-1 of selected Q-series Recommendations. For a full listing of ITU-T Recommendations, see the contact information in Appendix A.

Table Q-1: Selected ITU-T Q-Series Recommendations

Recommendation	*Description*
Q.20	Comparative advantages of in-band and out-of-band signaling systems
Q.21	Systems recommended for out-band signaling
Q.22	Frequencies to be used for in-band signaling
Q.23	Technical features of push-button telephone sets
Q.71	ISDN circuit mode switched bearer services
Q.700	Introduction to CCITT Signaling System No. 7 (SS7)
Q.716	SS7 - Signaling connection control part (SCCP) performance
Q.721	Functional description of SS7Telephone User Part (TUP)
Q.824.1	ISDN basic and primary rate access
Q.824.2	ISDN supplementary services
Q.824.3	ISDN optional user facilities
Q.824.4	ISDN teleservices
Q.921	ISDN user-network interface - Data link layer specification
Q.931	ISDN user-network interface layer 3 specification for basic call control
Q.2763	SS7 B-ISDN User Part (B-ISUP) - Formats and codes
Q.2764	SS7 B-ISDN User Part (B-ISUP) - Basic call procedures

QSIG (Q Signaling) A standard that defines services and signaling protocols for interconnecting TDM-based PBXs based on the ITU-T ISDN standard Q.931. QSIG is a Common Channel Signaling (CCS) protocol that runs over the ISDN D-channel for signaling between nodes in a Private Integrated Services Network (PISN). QSIG supports call setup, call teardown, and transparency of features such as message waiting, camp-on, and callback. The ISO (International Organization for Standardization) has adopted QSIG as Private Signaling System No. 1 (PSS1). See also *CCS*, *ISDN*, *ISO*, *Q.931*, and *Q series*.

Q Signaling (QSIG) See *QSIG*.

quality of service (QoS) See *QoS*.

quartet A four-bit byte. Also known as a *nibble*. Compare with *quadbit*. See also *bit*, *byte*, and *nibble*.

QuickConnect A feature of V.92 modems that reduces the time required for handshaking by approximately 50 percent, to about 10–15 seconds. QuickConnect trains the modem on the first call and remembers the characteristics of the circuit. Assuming that the circuit is the same on the next call, the circuit characteristics do not have to be relearned, which results in faster connect times, yielding obvious advantages to the end user and Internet service provider (ISP), alike. See also *handshaking*, *ISP*, *modem*, and *V Series*.

quintet A five-bit byte. See also *bit* and *byte*.

QWERTY The standard layout for English-language computer keyboards, so named for the top left six alphabetical characters. The QWERTY layout was patented by Christopher Sholes in 1868 for use with the first mechanical typewriter, which he also invented. Sholes sold the patent rights to Remington in 1873. The original layout was in alphabetical order, which caused the typebars to become entangled frequently once the typist gained proficiency and speed. Although the original justification is lost in time and there have developed a number of theories about it, the QWERTY layout certainly split up commonly used pairs of letters (e.g., s and t) and mitigated the issue of typebar entanglement.

R **1.** The symbol for Resistance. See *resistance*. **2.** R interface or Reference Point R in ISDN. See *Reference Point R*.

R1022 ATM Technology Testbed (RATT) The result of the 1987 RACE sponsored project 1022. See also *RACE*.

RA (Registration Authority) In a public key infrastructure (PKI), an entity that verifies the certificate authority (CA) prior to the issuance of a digital certificate to the requesting party. See also *CA*, *digital certificate*, and *PKI*.

RACE (Research for Advanced Communications in Europe) A consortium of European carriers, end users, and universities. In 1987, RACE sponsored project 1022 to demonstrate the feasibility of asynchronous transfer mode (ATM). The result of the RACE initiative was the R1022 ATM Technology Testbed (RATT). RACE project 2061, also known as EXPLOIT, is a more recent RACE project intended to prove the viability of integrated broadband communications (IBC) in the European Union (EU). The National Research and Education Network (NREN) was the first (1990) test-bed ATM network in the United States. Advanced Communications Technologies and Services (ACTS) was developed as the successor program to RACE, and continues that work on ATM networking and some 200 other projects. See also *ATM*.

radar (radio detecting and ranging) A microwave radio technology that uses reflected energy to detect and determine the direction of and distance to remote objects. Multiple return signals can be correlated over time to determine the velocity and direction of moving objects. Applications include navigation, targeting, and tracking for civilian and military purposes. See also *microwave*, *radio*, and *sonar*.

radian From the Latin radius. A unit of plane angular measurement equivalent to the angle between two radii that enclose a section of a circle's circumference (arc) equal in length to the length of a radius. There are 2π radians in a circle. See also *frequency* and *radius*.

radiant flux The time rate of energy flow of radiant energy as measured in watts or joules per second. See also *flux*, *joule*, *radiation*, and *watt*.

radiation **1.** The act or process of the spreading out of energy in rays. **2.** The emission, or outward flow, of energy in the form of electromagnetic waves, including radio waves and photons. See also *electromagnetic*, *photon*, and *waveform*.

radio Electromagnetic energy with a waveform having a frequency above the upper limit of the audio range of 3 kHz and equal or less than the lower limit of the infrared light range of 300 GHz. At the low end of the range is extremely low frequency (ELF) radio, which operates at 30–300 Hz, and at the upper end of the range is extremely high frequency (EHF) radio, which operates at 30–300 GHz. See also *electromagnetic spectrum*, *frequency*, and *Hz*.

Radio Act of 1927 In the United States, the act that established the Federal Radio Commission to regulate all radio spectrum, except bands owned by federal government. The Communications Act of 1934 replaced the Federal Radio Commission with the Federal Communications Commission (FCC). See also *band*, *Communications Act of 1934*, *FCC*, and *spectrum*.

radio access network (RAN) See *RAN*.

radio area network (RAN) Synonymous with *wireless radio area network* (WRAN). See *WRAN*.

radio common carrier (RCC) See *RCC*.

Radio Control (R/C) Radio Service See *R/C Radio Service.*

radio frequency (RF) See *RF.*

radio guide (RG) See *RG.*

radio line-of-sight (radio LOS) See *LOS.*

radiophone See *photophone.*

radius A straight line extending from the center of a circle to its edge, or from the center of a sphere to its surface. See also *bend radius* and *radian.*

RADIUS (Remote Authentication Dial-In User Service) An Internet protocol used for authentication, authorization, and accounting of end users seeking to gain access to internal computer resources, generally through a network access server (NAS) or, for remote users, by dialing into a remote access server (RAS). Originally developed by Livingston Enterprises, RADIUS was later described by the IETF in RFCs 2058 and 2059 and is currently described in RFCs 2865 and 2866. See also *authentication, authorization, IETF, NAS, RAS,* and *RFC.*

Radio Frequency Interference (RFI) See *RFI.*

RAID (Redundant Array of Independent Disks or Redundant Array of Inexpensive Disks) A storage technology that distributes data across a group of physically separate hard drives configured as a single logical memory unit. As RAID stores all data on redundant drives, it provides a considerable level of fault tolerance. RAID may involve drives on multiple servers in a cluster connected via a storage area network (SAN). A simpler and less expensive approach is known as just a bunch of disks (JBOD), which essentially is a bunch of disk drivers not configured as a RAID. See also *JBOD, SAN,* and *server.*

rain attenuation See *rain fade.*

rain-barrel effect The echo effect caused by signal reflection. In a real-time voice application, the effect is much like talking into a rain barrel. If you would like to experience the effect, but do not have a rain barrel handy, any barrel will do. See also *echo.*

rain fade Radio signal attenuation caused by rain. Rain fade is a factor at frequencies above 8 GHz and can be especially serious at frequencies above 11 GHz. Rain fade is sensitive to the rate of rainfall, the size of the raindrops, and the length of exposure as related to the length of the transmission path. See also *attenuation.*

rake receiver An antenna system that comprises a set of four receivers, or fingers, that work in a coordinated way to gather signal elements much like the tines of a garden rake work together to gather leaves. Each finger gathers a faded, or attenuated, signal element at a separate moment in time. The receiver employs spatial diversity and time diversity, combining and correlating the results of all four fingers to optimize the signal, thereby countering the effects of multipath fading and delay spread. Code-division multiple access (CDMA) systems employ rake receivers to deal with issues of multipath interference (MPI). See also *antenna, attenuation, CDMA, delay spread, MPI, multipath fading, spatial diversity,* and *time diversity.*

RAM (Random Access Memory) Semiconductor-based computer memory that stores program code and data in locations that can be accessed in any order. As the primary working memory of a computer, RAM stores program code and data that can be accessed, read, and written to by the central processing unit (CPU) and other hardware devices. RAM is characterized as read/write memory to distinguish it from ROM (Read Only Memory), which is the primary storage memory. RAM is volatile, meaning that any data stored in RAM is lost and unrecoverable if power is lost. Many programs set aside some amount of RAM as a temporary workspace for data until it can be printed, transmitted, or stored on a hard drive, floppy disk, or other permanent or semi-permanent medium. See also *ROM.*

Raman amplifier A type of amplifier used in long haul, single-mode (SMF) fiber optic transmission systems (FOTS). Raman amplification usually occurs throughout the length of the transmission fiber itself in a process known as distributed amplification, rather than in a discrete amplification, or lumped amplification configuration such as that employed by an erbium-doped fiber amplifier (EDFA). Raman amplification occurs as a high-energy pump wavelength is sent in the reverse direction from the output end of the fiber span, where the incoming signal is weakest. The pump wavelength, which generally is in the 1450 nm range (E-Band), interacts with and excites atoms in the crystalline lattice of the fiber core. The atoms absorb the photons, and quickly release photons with energy equal to the original photon, plus or minus atomic vibration. In other words, a frequency/wavelength shift occurs as the pump wavelength propagates along the fiber in the reverse direction. The energy lost in the pump wavelength shifts to longer-wavelength (within about 100 nm) signals, generally in the 1550 nm window (C-Band), in the forward direction, thereby serving to amplify them. Raman amplifiers offer the advantage of amplifying signals in the broad range extending from 1300 nm to 1700 nm. Further, they perform better than EDFAs in terms of signal-to-noise ratio (SNR). Raman amplifiers often are used as preamplifiers to enhance the performance of EDFAs in dense wavelength division multiplexing (DWDM) systems. See also *core*, *discrete amplification*, *DWDM*, *C-Band*, *E-Band*, *EDFA*, *FOTS*, *photon*, *propagation*, *SMF*, *SNR*, *wavelength*, and *window*.

RAN **1.** Radio Access Network. Referring to the wireless RF-based portion of a network providing access from a mobile terminal device (transmitter/receiver) to the core, or backbone, network of the radio service provider and ultimately to the public switched telephone network (PSTN) or the Internet or other IP-based network. A RAN comprises a base station, a controller, and the radio links between them. A RAN may be in the form of a 2G TDM-based cellular service (e.g., D-AMPS or GSM), a 3G cellular service (e.g., EDGE, GPRS, and UMTS), or other licensed and unlicensed services (e.g., WiMAX). See also *2G*, *3G*, *cellular radio*, *Internet*, *PSTN*, and *WiMAX*. **2.** Radio Area Network. Synonymous with wireless radio area network (WRAN). See *WRAN*.

random access memory (RAM) See *RAM*.

random noise Noise comprising large numbers of frequent, transient impulses, or disturbances, occurring at statistically random time intervals. Thermal noise is a form of random noise.

RAS **1.** Registration/Admission/Status. In H.323-compliant multimedia networks, the protocol that supports communications between terminals (i.e., endpoint devices) and the gatekeeper. See also *H.323*, *gatekeeper*, *multimedia*, and *terminal*. **2.** Remote Access Server. A host computer on a local area network (LAN) and equipped with modems to serve the needs of end users for dial-up access to internal computer resources through the public switched telephone network (PSTN). An RAS also can take the form of purpose-built hardware with either integral modems or ISDN interfaces. An RAS generally is associated with a Remote Authentication Dial-In User Service (RADIUS) server that performs authentication, authorization, and accounting functions to ensure network security. See also *authentication*, *authorization*, *ISDN*, *LAN*, *modem*, *NAS*, *PSTN*, *RADIUS*, and *security*.

raster The pattern of uniformly spaced horizontal scan lines that cover the display space of a device, such as a computer monitor or television monitor. Within each line are pixels (picture elements) that can be illuminated individually.

rasterize To scan a document to convert an image into a form suitable for display on a computer monitor or printout. In telecommunications, hard copy documents are rasterized by a facsimile machine prior to transmission. In computing, documents are rasterized prior to electronic processing or storage.

rate adaption See *dynamic rate adaption*.

rates and tariffs See *tariff*.

RATT (R1022 ATM Technology Testbed) See *RACE*.

ray A thin beam of radiant energy, especially light.

Rayleigh scattering The deflection of a light ray as it encounters matter while propagating in a physical medium. Named for Lord Rayleigh, a British physicist, the phenomenon is due to the interaction of light and matter at the atomic or molecular level. The closer the size of the particles to the wavelength of the light, the more scattering takes place. As scattering varies as the reciprocal of the fourth power of the wavelength (Scattering = λ^{-4}) the phenomenon decreases rapidly as the wavelength increases. As the light scatters it also variously is absorbed and attenuated by interaction with density changes and compositional variations in the crystalline structure of an optical fiber and the impurities that are always present to some extent. So, the longer wavelengths (e.g., 1550 nm) suffer less attenuation over a distance than the shorter wavelengths (e.g., 850 nm). Rayleigh scattering is the reason that the sky is blue in the day and red at sunset. The shorter blue wavelengths are scattered by matter in the atmosphere more than the green and red wavelengths, so we see blue, rather than the black of space, when the sun is overhead. During the sunset, however, the sun is at such a low angle and the sunlight passes through so much atmosphere that the shorter blue wavelengths are scattered and absorbed so much that we see little of them. The longer red wavelengths suffer less attenuation and, therefore, reach our eyes. See also *atmosphere*, *light*, *medium*, *physical*, *propagation*, *ray*, and *wavelength*.

RBOC (Regional Bell Operating Company) Also known as *Regional Holding Company* (RHC). In the United States, each of the seven regional companies formed by the Modified Final Judgement (MFJ), which broke up the AT&T Bell System effective January 1, 1984. Each RBOC comprised one or more of the 22 Bell Operating Companies (BOCs) that previously were wholly owned by AT&T. Over time, each RBOC fully absorbed its component BOCs, creating a single legal entity with a centralized management structure. As Cincinnati Bell and Southern New England Telephone (SNET) were not wholly owned by AT&T, they were divested as standalone operating telephone companies. Table R-1 maps the BOCs into the RBOC organizations.

Table R-1: Before and After: Bell Operating Companies (BOCs) and Regional Bell Operating Companies (RBOCs)

Bell Operating Companies (Primary States of Operation), Pre-divestiture	*Regional Bell Operating Companies (Headquarters), Post-divestiture*
Illinois Bell (Illinois)	Ameritech (Illinois). Acquired by SBC Communications (October, 1999); now AT&T.
Indiana Bell (Indiana)	
Michigan Bell (Michigan)	
Ohio Bell (Ohio)	
Wisconsin Telephone (Wisconsin)	
Bell of Pennsylvania (Pennsylvania)	Bell Atlantic (Pennsylvania); now Verizon Communications.
Diamond State Telephone (Delaware)	
The Chesapeake and Potomac Companies (District of Columbia, Maryland, Virginia, and West Virginia)	
New Jersey Bell (New Jersey)	
South Central Bell (Alabama, Kentucky, Louisiana, Mississippi, and Tennessee)	BellSouth (Georgia). Acquired by AT&T (January 2007).
Southern Bell (Florida, Georgia, North Carolina, and South Carolina)	

Table R-1: Before and After: Bell Operating Companies (BOCs) and Regional Bell Operating Companies (RBOCs) *(continued)*

Bell Operating Companies (Primary States of Operation), Pre-divestiture	*Regional Bell Operating Companies (Headquarters), Post-divestiture*
New England Telephone (Massachusetts, Maine, New Hampshire, Rhode Island, and Vermont) New York Telephone (New York)	NYNEX (New York). Acquired by Bell Atlantic (August 1997); now Verizon Communications.
Pacific Bell (California) Nevada Bell (Nevada)	Pacific Telesis (California). Acquired by SBC (April 1997); now AT&T.
Southwestern Bell (Arkansas, Kansas, Missouri, Oklahoma, and Texas)	Southwestern Bell Corporation (Texas), then SBC Communications; now AT&T (November 2005).
Mountain Bell (Arizona, Colorado, Idaho, Montana, New Mexico, Utah, and Wyoming) Northwestern Bell (Iowa, Minnesota, North Dakota, Nebraska, and South Dakota) Pacific Northwest Bell (Oregon and Washington)	US West (Colorado). Acquired by Qwest (June 2000).

RCC (Radio Common Carrier) A regulated provider of radio services to the general public at published rates available to all. RCCs are restricted to designated frequencies. See also *common carrier, frequency*, and *radio*.

R/C (Radio Control) Radio Service. In the United States, a one-way, short distance, radio service for on/off operation of remote devices. An R/C unit is not authorized to communicate voice or data. R/C operates in the 72.0–73.0 MHz and 75.4–76.0 MHz bands. The Federal Communications Commission (FCC) regulates R/C, which is in the family of personal radio services. See also *FCC* and *personal radio services*.

RDF (Resource Description Framework) A W3C specification that integrates a variety of applications, using XML as an interexchange syntax. See also *RDF Site Summary, W3C*, and *XML*.

RDF Site Summary (RSS) See *RDF* and *RSS*.

reactance (X) A form of opposition to the flow of alternating electric current (AC) because of capacitance or inductance, reactance is an inertial reaction to the flow of AC, and is measured in Ohms (Ω). See also *capacitive reactance* and *inductive reactance*.

read-only memory (ROM) See *ROM*.

Really Simple Syndication (RSS) See *RSS*.

real-time (rt) See *rt*.

Real Time Control Protocol (RTCP) See *RTCP*.

Real-Time Streaming Protocol (RTSP) See *RTSP*.

Real Time Transport Protocol (RTP) See *RTP*.

real-time variable bit rate (rt-VBR) See *rt-VBR*.

receiver Also known as a *sink*, a receiver is a target device, or destination device, that receives an information transfer originated by a transmitter. Receivers include telephones, facsimile machines, data terminals, host computers, and video monitors. See also *transceiver* and *transmitter*.

recognition and flagging An error control mode in which detected errors are flagged by the receiving device, but there is no mechanism for error correction. Rather, error correction requires a human-to-machine request for retransmission. Recognition and flagging is primarily used in networks involving dumb terminals with no means of buffering or retaining information transmitted and which, therefore, are unable to retransmit errored data. Parity checking is an example of recognition and flagging. See also *buffer*, *dumb terminal*, *error control*, *FEC*, *parity check*, and *recognition and retransmission*.

recognition and retransmission An error control mode that provides for retransmission of errored data packets. The error detection logic can be implemented not only in the receiving device but also in intermediate routers, switches, and other intelligent nodes. The device detecting an error issues a retransmission request to the device immediately upstream, or perhaps to the original transmitter, which holds some amount of data in a buffer until it has received an indication that the data either was received correctly or that the data was received in an errored state. If the upstream device receives no indication either way within a specified time interval, it assumes that the data was lost in transit and automatically initiates a retransmission through a protocol known as *automatic repeat request* (ARQ). As examples, recognition and retransmission is used in X.25 networks, and by applications running the Transmission Control Protocol (TCP) over Internet Protocol (IP) networks. Block parity is an example of recognition and retransmission. See also *ARQ*, *block parity*, *buffer*, *checksum*, *error control*, *FEC*, *IP*, *recognition and flagging*, *TCP*, *upstream*, and *X.25*.

recommended standard (RS) See *RS*.

record communications service A service designed or used primarily to transfer information that originates or terminates in written or graphic form. Examples of record communications services include telex and TWX.

rectifier A type of diode that converts alternating current (AC) to direct current (DC). Rectifiers originally were in the form of electron tubes but now are semiconductors or semiconductor arrays. See also *diode*.

redirected PVC In frame relay, an inactive permanent virtual circuit (PVC) that can be activated very quickly to direct traffic around a point of failure in the network or in the access loop. A redirected PVC also can be used to redirect traffic to a backup data center should the primary data center suffer a failure. See also *frame relay*, *local loop*, and *PVC*.

red sunset See *Rayleigh scattering*.

Redundant Array of Independent Disks (RAID) See *RAID*.

Redundant Array of Inexpensive Disks (RAID) See *RAID*.

Reed-Solomon (RS) A block coding algorithm for forward error correction (FEC). Reed-Solomon works by viewing the data as a polynomial, which it analyzes and organizes into symbols, which it groups into blocks, to each of which it adds parity bits to form codewords. For example, RS (255,223) is byte-oriented, working with 8-bit symbols. Each codeword contains 255 code word bytes comprising 223 data symbols and 32 parity symbols. See also *algorithm*, *block*, *block code*, *byte*, *code*, *FEC*, *parity bit*, and *symbol*.

Reference Model The ITU-T specifications for integrated services digital network (ISDN) use an alphabetical reference model to describe the various Reference Points (i.e., connection points or interfaces) on the customer side of the network. Those points are R, S, T, U, and Q, as illustrated in Figure R-1. See also *Reference Point R*, *Reference Point S*, *Reference Point T*, *Reference Point U*, and *Reference Point Q*.

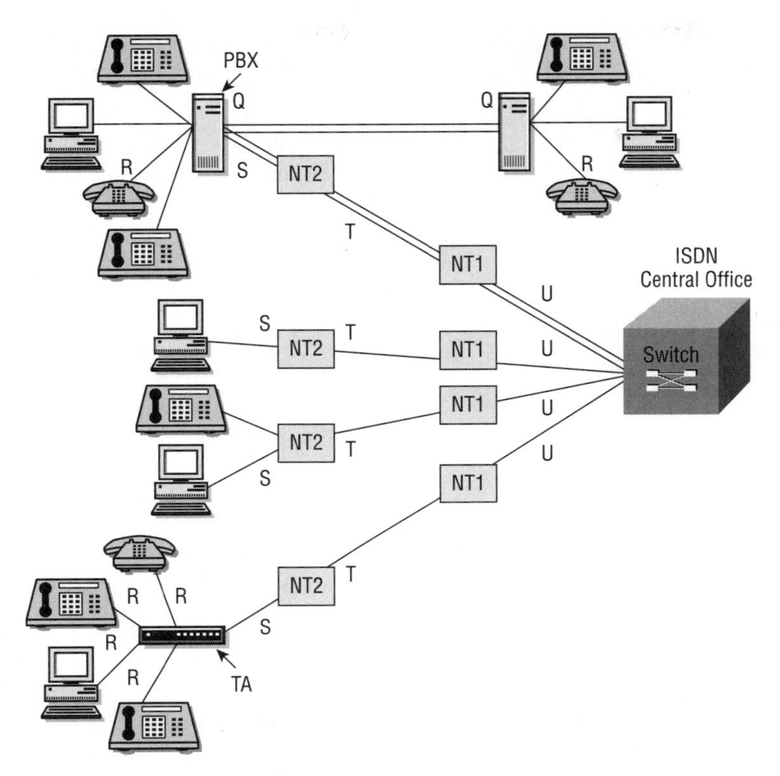

Figure R-1

Reference Point Q Also known as the *Q interface*. The ISDN point of reference for the QSIG interface between two ISDN PBXs in a private network. See also *ISDN*, *PBX*, and *QSIG*.

Reference Point R Also known as the R interface. The ISDN point of reference for the interface between non-ISDN compatible terminal equipment (TE2) and a terminal adapter (TA) that connects on the network side to an S interface of an NT2. See also *ISDN*, *NT2*, *TA*, and *TE2*.

Reference Point S Also known as the *S interface*. The ISDN point of reference for the interface between an ISDN terminal and the NT2, which is the user side of the local loop. The S interface serves to distinguish between terminal equipment and network-related functions. The S interface is defined as a passive bus on Cat 5 cable for up to 8 NT2 devices, which are intelligent and ISDN-compatible. See also *Cat 5*, *ISDN*, and *NT2*.

Reference Point T Also known as the *T interface*. The ISDN point of reference for the four-wire connection between NT1 and NT2. See also *four-wire*, *ISDN*, *NT1*, and *NT2*.

Reference Point U Also known as the *U interface*. The ISDN point of reference for the demarcation point (demarc) between the public network local loop and the customer premises equipment (CPE) or NT1. See also *demarc*, *CPE*, *ISDN*, *local loop*, and *NT1*.

reflection The act or process of redirecting electromagnetic energy along a new path, using a conductive surface or impedance discontinuity. At a surface, the angle of the reflection equals the angle of incidence, i.e., the angle at which the incident signal strikes the plane surface of the obstacle. At the extreme, the signal reflects back towards its point of origin. See also *angle of incidence* and *refraction*.

reflection grating A type of diffraction grating comprising grooves ruled into a surface that can be either plane or concave. In a distributed feedback laser (DFB laser), a diffraction grating with a concave surface serves to focus light without affecting the spectra. This approach is much more effective than the mirror technique employed with a Fabry-Perot laser and vertical cavity surface-emitting laser (VCSEL). See also *DFB laser*, *diffraction*, *diffraction grating*, *Fabry-Perot laser*, and *VCSEL*.

reflector A passive device that simply redirects radiant energy, rather than amplifying it or otherwise acting on the signal.

Reference Point V Also known as the *V interface*. The ISDN point of reference for the point of interface at the network side of the connection between the line termination or loop termination and the exchange termination. In other words, it is the point between the circuit terminating equipment and the ISDN central office (CO). As the V interface exists only if the CO does not have embedded circuit terminating equipment, it is unusual in contemporary ISDN-compatible COs. See also *CO* and *ISDN*.

ReFLEX See *FLEX*.

refraction The bending of electromagnetic waves caused by a change in the velocity of propagation (Vp) as they pass from a medium of a given density into a medium of another density at an oblique angle. The extent to which this phenomenon occurs is termed the *index of refraction* (IOR). See also *IOR*, *reflection*, and *Vp*.

refractive index Synonymous with index of refraction (IOR). See *IOR*.

refresh rate The rate at which a video monitor is completely renewed or updated.

regenerative repeater See *repeater*.

Regional Bell Operating Company (RBOC) See *RBOC*.

Regional Internet Registry (RIR) See *RIR*.

register A high-speed buffer, or region of memory, that is used to store digits or characters for a specific purpose. In telecommunications, a Central Office (CO) switch register stores dialed digits until the caller completes dialing the outgoing telephone number. At that point, the switch analyzes the number and determines how to process the call. If the call is within the CO switch's domain and the target telephone is available, the switch rings that phone and connects the call. If the call is a local call intended for a destination telephone outside the CO switch's domain, it will hand the call off to another CO switch within the local calling area, or perhaps to an intermediate tandem switch. A tandem switch is always involved if the call is long distance or international in nature. See also *buffer* and *shift register*.

registered jack (RJ) See *RJ*.

registered port A port that can be used by ordinary user processes or programs on most systems and can be executed by ordinary users. Registered port assignments, numbered in the range 1,024 through 49,151, are used with the User Datagram Protocol (UDP) to the extent possible. Registered ports are registered by the Internet Corporation for Assigned Names and Numbers (ICANN) as a convenience to the Internet community. See also *dynamic port*, *ICANN*, *port*, *UDP*, and *well-known port*.

Registration/Admission/Status (RAS) See *RAS*.

registration authority (RA) See *RA*.

regular pulse excitation linear predictive coding (RPELPC) See *RPELPC*.

regulation From the Latin regula meaning rule. Rule or order established by governmental bodies and having the force of law.

remote access server (RAS) See *RAS*.

Remote Authentication Dial-In User Service (RADIUS) See *RADIUS.*

reorder tone Synonymous with fast busy. See *fast busy signal.*

repeat dial Synonymous with continuous redial. See also *continuous redial.*

repeater A device that amplifies, reshapes, and retimes an input digital signal for retransmission. In an electrically based system, the repeater essentially guesses the binary value (1 or 0) of the attenuated incoming signal, including any accumulated noise, based on its relative voltage level and the relative time, and regenerates a stronger signal of the same value without the noise. The repeater also reshapes and retimes the signal, essentially redefining the distinct bit values and restoring the bit pace. In combination, these processes considerably enhance the signal quality, as compared to the simple amplification process performed by an amplifier. In a fiber optic transmission system (FOTS), the repeater comprises a light detector and a light source, positioned back-to-back. The detector receives an attenuated optical signal, converts it to an amplified electrical signal, reshapes and retimes it, converts it back to optical signal, and retransmits it. Such optical repeaters are characterized as being optical-electrical-optical (OEO) in nature. A repeater also may perform other signal processing functions. A satellite repeater, for example, also performs frequency translation, or frequency shifting, in order to differentiate in frequency the uplink and downlink signals, thereby to avoid their mutual interference. The spacing of repeaters is sensitive to a variety of factors, including the specifics of the transmission medium and the frequency of the carrier signal. See also *amplifier, attenuation,* and *signal.*

Request for Comment (RFC) See *RFC.*

request to send (RTS) See *RTS.*

Réseaux IP Européens Network Coordination Center (RIPE NCC) See *RIPE NCC.*

Resilient Packet Ring (RPR) See *RPR.*

resistance (R) A measure of the opposition by a circuit, component, material, or free space to the flow of an electric current. Resistance is the value of R in the Ohm's Law equation $I = V/R$, where I is the electric current, and V is voltage, or the difference in electric potential between two points in a circuit. Resistance is the real part of impedance. The SI unit of measurement of resistance is the Ohm (Ω). The reciprocal of resistance is conductance, the official measurement of which is the mho, which is Ohm spelled backwards. In an electrical circuit, resistance results in attenuation, or loss of signal strength. See also *conductance, current, Ohm,* and *voltage.*

resistor An element within a circuit that is specifically designed to restrict the flow of electric current when a potential difference occurs across it. See also *circuit* and *current.*

resolution The definition, sharpness, or level of detail of the reproduction of an image. Resolution is directly related to the number and density of the dots of color (black, white, and perhaps other colors). Group III facsimile specifications, for example, provide a number of options, expressed as horizontal lines per inch (lpi) in terms of scanning (input), and linear dots per inch (dpi) in terms of sensing and printing (output). The actual (and nominal) fax industry standards are as follows:

- **Standard**: 98 × 203 (100 × 200)
- **Fine**: 196 × 203 (200 × 200)
- **Superfine**: 392 × 203 (400 × 200)

In video images, resolution is determined by the number and areal density of the pixels, or pels (picture elements), which essentially are dots of picture similar to the dots in half-tone printing.

resource management cell (RM-Cell) See *RM-Cell.*

Resource Reservation Protocol (RSVP) See *RSVP*.

RF (Radio Frequency) The frequencies of the electromagnetic spectrum associated with radio waves, rather than electricity, light, x-rays, gamma rays, or cosmic rays. See also *electromagnetic spectrum* and *frequency*.

RFC (Request For Comment) The official document by which the Internet Activities Board (IAB) and the Internet Engineering Task Force (IETF) publish standards, protocols, best practices, or other information relative to the operation of the Internet. The format dates to the early days of the ARPANET in which authors circulated hard copies of their proposals to their colleagues and requested their comments. In the more formal context of contemporary Internet administration, requests for comments actually are made in an Internet Draft document. See also *ARPANET*, *IAB*, *IETF*, and *Internet*.

RFI (Radio Frequency Interference) Electromagnetic interference (EMI) that is within the radio frequency range of the electromagnetic spectrum. See also *electromagnetic spectrum* and *EMI*.

RG (Radio Guide) The RG numbering system of coaxial cable (coax) refers to the fact that the RF (Radio Frequency) signal is guided down the center conductor of the cable system. The RG numbering system dates to WWII United States military specifications and has no real contemporary significance other than type designators. Each RG number does, however, specify impedance, core conductor gauge (AWG) and type, outside diameter (OD), and other physical attributes of the cable. Table R-2 compares example coaxial cables. See also *coaxial cable*.

Table R-2: Coaxial Cable Types

RG Number	*Center Wire Gauge*	*Impedance (Ω)*	*Outside Diameter (OD)*	*Core Type*	*Example Applications*
RG-6/U	18 AWG	75 ohms	.332 in.	Solid	CATV, cable modems, DBS TV
RG-8/U	10 AWG	50 ohms	.405 in.	Solid	10Base5 Ethernet, Ham radio
RG-58/U	20 AWG	53.5 ohms	.116 in.	Solid	10Base2 Ethernet, Ham radio
RG-58C/U	20 AWG	50 ohms	116 in.	Solid	RG-58A/U Military Spec
RG-58A/U	20 AWG	50 ohms	116 in.	Stranded	10Base2 Ethernet, CATV
RG-59/U	20 AWG	75 ohms	146 in.	Solid	Ham Radio, CCTV
RG-62/U	22 AWG	93 ohms	.146 in.	Solid	ARCnet, IBM cabling system

RHC (Rural Health Care Corporation) See *RBOC* and *Rural Health Care Program*.

ribbon cable **1.** A type of horizontal cable comprising many metallic wires lying side by side, in parallel, forming a flat, ribbon-like structure. Ribbon cables are used indoors under carpeting, for reasons of safety and aesthetics, as they lie flat. Ribbon cables can be used only in straight cable runs, as they do not flex sideways. See also *horizontal cable*. **2.** A type of outside plant (OSP) fiber optic cable comprising unbuffered (i.e., uninsulated) acrylate-coated glass optical fibers (GOF) lying side by side, in parallel, and encased in a plastic material to form a flat, ribbon-like structure. A distribution cable typically contains a single ribbon of 6 or 12 fibers. A long haul cable may contain many such ribbons, stacked on top of each other. Along a high traffic physical cable route, it is not unusual to find a cable containing 12 ribbons of 12 fibers, each, for a total of 144 fibers. The advantage of ribbon fiber is in its ease of handling and splicing. As the fibers are not individually buffered and sheathed, they are less bulky, therefore, more manageable.

Also, a technician can splice the entire ribbon at once, rather than having to splice each individual fiber. See also *optical fiber.*

Rich Site Summary (RSS) See *RSS.*

rich text Textual data (letters, numbers, and punctuation marks), including formatting, such as *italics*, **bold**, and color other than the black and grayscale in which this book is printed. Rich text often conforms to the Rich Text Formatting (RTF) standard developed by Microsoft Corporation. RTF allows a word processing program to create a rich text file encoded with all necessary formatting instructions, and without any hidden codes. An RTF-encoded file also can be transmitted between applications on a computer and across a network without loss of formatting because it consists only of standard text characters. See also *plain text.*

Rich Text Formatting (RTF) See *rich text.*

right of way The right, established by common or statutory law, of passage over an area of land. A public right of way grants passage to all and essentially is a public easement that allows the construction of roads over it and public utilities (e.g., electrical, gas, telephone, sewer, and water) over and through a narrow strip of land. See also *utility.*

ring The electrically negative (–) wire of a cable pair. The central office (CO) feeds talking battery to the customer premises equipment (CPE) over the ring side. See also *ring topology*, *talk battery*, and *tip and ring.*

ring again A CLASS service feature of the public switched telephone network (PSTN). The feature allows a calling party who encounters a busy signal to request network notification when the called line becomes available. When both lines become available, the network calls the calling party back with a distinctive ring tone. When the caller lifts the handset, the network automatically redials the call. See also *CLASS* and *PSTN.*

ringback tone An intermittent audible indication to the calling party that a dialed telephone number is ringing. A ringback tone is a status indicator that the dialed number is available (i.e., not busy), that all connections through the appropriate network or networks between the originating and destination devices either have been made or are available to be made, and that the call can be connected if someone or something answers the call. The call can be answered by a human being if the destination telephone number is not engaged, or if the destination telephone number is engaged, has call waiting service, and a call waiting indicator tone prompted a human being to either abandon the first call or place it on hold in order to answer the second. The call can be answered by a premises-based answering machine or a local or network-based voice processor in the event that the called number does not answer, is busy, or does not answer within a programmable number of rings. The nature of the ringback tone varies by region and country. In the United States, Canada, and other countries in the North American Numbering Plan (NANP), the standard PSTN ringback tone is generated by summing a 440-Hz tone with a 480-Hz tone and applying these to the telephone line in a two-second on and four-second off cadence. The tone combination produces a warbling "ring ... ring ... ring" sound, caused by the 40-Hz beat, or interference due to the difference in frequency, between the two tones. The ringback tone may be generated by the switch serving either the called party or the calling party, but it is not generated by the called telephone instrument or PBX. The ringback tone generally starts and stops at the same rate as the ringing tone of the called telephone, but generally is out of phase, i.e., staggered in time.

Personalized ringback tones recently have become popular, especially with respect to cellular telephones. Dozens of music genres and hundreds of selections are commonly available for both the ringback tone and the ringing tone. The service generally is on a subscription basis and carries an additional charge per tone selected. It is possible to select up to 100 or so active ringback tones at any given time, to play different tones for different callers, and to vary the tone by time of day, perhaps to have one tone for business hours and another for all other times. See also *ring tone.*

ringlet In Resilient Packet Ring (RPR) networks, a small local ring where the larger ring drops traffic off to nodes. See also *node*, *ring topology*, and *RPR*.

ringing signal See *ring tone*.

ring tone Also known as *ringing signal* and *ringing tone*. An audible indication to the called party of an incoming call. The nature of the ring tone varies. In the United States, Canada, and other countries in the North American Numbering Plan (NANP), the traditional ring tone is generated by two metal bells mounted inside the telephone set. The ringer is activated by a current of approximately 90–110 volts at 20 Hz generated by a central office-based ringing machine and sent across the copper local loop at a cadence of approximately two seconds on and four seconds off. Contemporary telephones generally dispense with metal bells in favor of various microprocessor-generated ringing signals, and there may be several from which to choose. Personalized ring tones recently have become popular, especially with respect to cellular telephones. Dozens of music genres and hundreds of selections are commonly available for both the ring tone of the cell phone and the ringback tone. See also *ringback tone*.

ring topology A network structure in which the nodes are laid out in a physical ring, or closed loop, configuration, as illustrated in Figure R-2. Information travels around the ring in only one direction, with each attached station or node serving as a repeater. Rings generally employ coaxial cable or optical fiber as transmission media. In the local area network domain, rings are characterized as being deterministic in nature, employing token passing as the method of medium access control (MAC) to ensure that all nodes can access the network within a predetermined time interval. Priority access is recognized. A master control station controls access to the transmission medium by generating tokens, without which a station cannot access the network. Generally, any station can assume backup control responsibility in the event of a master failure. The IEEE 802.5 standard is a specification for LANs based on an electrical ring topology. Fiber Distributed Data Interface (FDDI) is a specification from the American National Standards Institute (ANSI) for a LAN based on a fiber optic dual counter-rotating ring. In the metropolitan area network (MAN) and wide area network (WAN) domains, Resilient Packet Ring (RPR), Synchronous Digital Hierarchy (SDH), and Synchronous Optical Network (SONET) each specifies several fiber optic dual ring configurations. See also *802.5*, *ANSI*, *deterministic*, *FDDI*, *IEEE*, *MAC*, *node*, *ringlet*, *RPR*, *SDH*, *SONET*, *token passing*, *token-passing ring*, *Token Ring*, and *topology*.

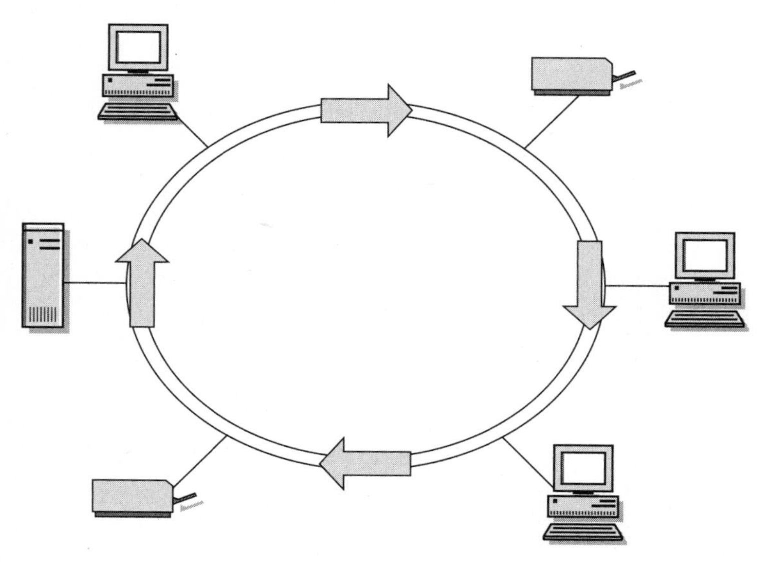

Figure R-2

R interface See *Reference Point R.*

RIP (Routing Information Protocol) A distance-vector routing protocol that employs the hop count metric for selecting the shortest path between an originating and a destination router. Each router in a network builds a database of the other routers to which it connects, and advertises that database to its neighboring routers every 30 seconds (RIP is very chatty), or when topology changes occur. Based on that information, the originating router selects the path with the lowest hop count. RIP is an Interior Gateway Protocol (IGP) used to exchange path information between routers in the same network domain. The initial RIPv1 was specified by the Internet Engineering Task Force (IETF) in RFC 1058, RIPv2 in RFC 2453, and RFCng for IPv6 in RFC 2080. RIP maps into Layer 3, the Network Layer of the OSI Reference Model, and can operate over heterogeneous networks. See also *domain, hop, IGP, Network Layer,* and *OSI Reference Model.*

RIPE NCC (Réseaux IP Européens Network Coordination Center) The Regional Internet Registry (RIR) responsible for assigning Internet Protocol (IP) addresses variously to National Internet Registries (NIRs) or directly to Local Internet Registries (LIRs) in Europe. See also *IP, IP address, LIR, NIR,* and *RIR.*

RIR (Regional Internet Registry) The regional organization responsible for assigning Internet Protocol (IP) addresses to either the National Internet Registry (NIR) or Local Internet Registries (LIRs) that, in turn, assign them to Internet Service Providers (ISPs). The RIR receives IP address assignments from the Internet Assigned Numbers Authority (IANA). RIRs include the following:

- African Network Information Center (AfriNIC)
- Asia Pacific Network Information Center (APNIC)
- American Registry for Internet Numbers (ARIN)
- Latin-American and Caribbean Network Information Center (LACNIC)
- Réseaux IP Européens Network Coordination Center (RIPE NCC)

In addition to general number administration, each RIR is responsible for maintaining one or more of the root servers, maintaining a Whois database for IP ownership lookups, deployment of a routing database, co-ordination of ENUM delegations, and network measurement and statistical reporting. See also *ENUM, IANA, IP, IP address, ISP, LIR, NIR,* and *Whois.*

rise and fall time See *cycle time.*

riser cable A type of inside cable intended for use in non-plenum vertical applications such as between floors of a building. In consideration of the longitudinal stress placed on a riser cable, load-bearing strength members are incorporated to increase the tensile strength, as well as to relieve the conductors of the load. Although riser cables must be fire retardant, the National Electrical Code (NEC) specifications for riser cable are not as demanding as those for plenum cable. See also *horizontal cable, NEC, plenum, plenum cable,* and *strength member.*

Rivest-Shamir-Adleman (RSA) See *RSA.*

RJ (Registered Jack) Referring to a series of jacks, described in the U.S. Code of Federal Regulations, Title 47, part 68, providing a standardized physical and electrical interface to the public switched telephone network (PSTN). See also *jack, Part 68,* and *PSTN.*

RJ11 A six-pin, two-conductor physical interface for connecting single-line telephone sets to the public switched telephone network (PSTN). See also *PSTN* and *RJ.*

RJ45 An eight-pin, eight-conductor physical interface used primarily in Ethernet local area networks (LANs), ISDN, and T1. See also *Ethernet*, *ISDN*, *LAN*, *RJ*, and *T1*.

RM-Cell (Resource Management Cell) In asynchronous transfer mode (ATM), a flow control feedback mechanism that communicates to the originating end-user device to change the transfer characteristics of the connection during periods of congestion. The network buffers cells and advises the sender to throttle back on the rate of transmission. The available bit rate (ABR) class of service makes use of RM-Cells. ABR is a best-effort category in which the network attempts to pass the maximum number of cells, but with no absolute guarantees. See also *ABR*, *best effort*, *buffer*, *cell*, *congestion*, and *flow control*.

road warrior Someone who travels extensively on business, conducting business warfare on the road. The term originates in the Mel Gibson movie *Mad Max 2: The Road Warrior* (1981).

Robustness Principle Also known as *Postel's Law*. A principle stated by Jon Postel in RFC 793 (1981) and again in RFC 1122 (1989), Requirements for Internet Hosts – Communications Layers. The Robustness Principle states, "Be liberal in what you accept, and conservative in what you send," which essentially advises the Internet community to design host software in such a way as to 1) be prepared for malevolent incoming packets and 2) be prepared for deficiencies on other hosts that can make it unwise to exploit legal but obscure protocol features that can cause disruption if the other host misbehaves. See also *host*; *Internet*; *packet*; *Postel, Jon*; *protocol*; *software*; and *standard*.

ROM (Read-Only Memory) Semiconductor-based computer memory that stores program code that the central processing unit (CPU) can read, but not write to, i.e., change or modify in any way. Programs are stored in ROM on semiconductor chips, also known as *firmware* or *hardware*, during the manufacturing process. Such programs are said to be hard-coded to distinguish them from software. The term read-only memory distinguishes it from random access memory (RAM), which also is stored on semiconductors, but is read/write memory. Unlike RAM, ROM is not volatile, i.e., the programs are not lost when the electric power is lost or turned off. Therefore, programs required during system start-up commonly are stored in ROM. *Note:* Many ROM chips can be reprogrammed after erasing the previous content either electrically or with an external source of ultraviolet (UV) light. See also *CPU*, *firmware*, *hardware*, *RAM*, *semiconductor*, and *software*.

root **1.** In a hierarchically organized structure of entities, the main level from which all other levels branch out. Such a structure can take the form of a root with multiple branches, each of which may have multiple leaves. **2.** In a hierarchical network tree topology, the central bus from which all other busses branch out. See also *bus topology* and *tree topology*. **3.** In a hierarchically structured database, a record at the first level, from which all other records branch out. Such a structure is known as a *tree*.

root server A server that serves as the central point in a hierarchical structure of database hosts. In the Internet, for example, root domain name servers (DNSs) are positioned as the authoritative sources of all domain names. The Regional Internet Registries (RIRs) maintain the root servers, to which they post domain names directly. Individual Internet service providers (ISPs) download updates from the root servers. See also *database*, *DNS*, *Internet*, *RIR*, and *server*.

router An intelligent switch capable of deciding where to forward packets based on a view of the network as a whole. A router is a programmable device that works with other routers, via a routing protocol, to establish the best path on which to forward a packet with a given address. A router can consider the network as a whole in determining the route for a given call. A router can be programmed to consider a number of factors including the addresses of the originating and destination devices, the least-cost route, the least congested route, the route with the fewest number of hops, and the geographically shortest route. Routers operate at least at Layer 3, the Network Layer, of the OSI Reference Model. Simple switches operate at Layer 2, the Data Link Layer, seeing only an individual link, and having no sense of the larger network. Depending on the applications, routers can operate at higher layers, as well, including Layer 7, the Application Layer. Routers can be capable of performing the gateway functions associated with protocol conversions such as code conversions or those necessary to connect dissimilar networks, such as circuit-switched and packet-switched networks. See also *Data Link Layer, gateway, hop, Network Layer, OSI Reference Model, packet, protocol,* and *switch.*

routine Synonymous with *procedure.* A program module, or section of code, that executes a specific task.

routing Referring to the process of deciding where to forward packets based on a view of the network as a whole. See also *router.*

routing by rumor See *distance-vector routing protocol.*

Routing Information Protocol (RIP) See *RIP.*

RPELPC (Regular Pulse Excitation Linear Predictive Coding) A speech encoding technique that uses regular pulses in an excitation frame and a long-term predictor, based on long-term correlation of voice samples, to model the speech pitch. RPELPC at 13 kbps is specified for use in cellular radio networks conforming to the pan-European GSM standard. See also *cellular radio, encode, frame,* and *GSM.*

RPR (Resilient Packet Ring) The IEEE 802.17 specifications for a medium access control (MAC) layer protocol that uses Ethernet switching and a dual counter-rotating ring topology to optimize the transport of Ethernet/IP packet data traffic over optical fiber rings. RPR is designed to maintain the resiliency of SONET/SDH, but at a much reduced level of overhead. As RPR is independent of the Physical Layer, it can be implemented over existing SONET/SDH physical rings or can run on a standalone basis. RPR calls for dual counter-rotating local ringlets that interconnect nodes where data traffic is intended to drop. RPR also uses statistical multiplexing, which allows bandwidth to be oversubscribed, while establishing Committed Information Rate (CIR) and peak-rate thresholds per application. The nodes negotiate bandwidth requirements among themselves based on fairness algorithms and in consideration of a classification scheme that recognizes and provides higher priority access to traffic sensitive to latency and jitter while ensuring that best effort traffic, such as Internet traffic, is afforded equal access and a fair share of the remaining bandwidth. RPR supports the following class of service (CoS) levels:

- Class A traffic is intolerant of latency and jitter. RPR addresses Class A traffic through a high CIR that ensures the availability of an average level of bandwidth appropriate for high priority traffic such as real-time voice and video.
- Class B is more tolerant of latency and jitter. RPR addresses Class B traffic through either a lower CIR that ensures the availability of an average amount of bandwidth appropriate for medium priority applications that have less stringent QoS requirements, or through an Excess Information Rate (EIR) option. In the event of network congestion, Class B traffic is subject to fairness-based flow control. Class B is intended for business-class data traffic such as transaction processing.
- Class C traffic is best effort traffic with no latency or jitter requirements and, therefore, is strictly EIR traffic. In the event of network congestion, Class B traffic is subject to fairness-based flow control. Class B traffic includes low priority applications such as consumer-level Internet access.

In the event of a node or link failure, the RPR protection scheme can restore the network in 50 ms or less, which is the SONET/SDH benchmark. There are two restoral mechanisms: wrapping and steering. The wrap option calls for data to travel around the ring until it reaches the node nearest the break. That node turns the traffic around and sends it in the reverse direction over the counter-rotating ring. The steer option calls for the originating station to exercise sufficient intelligence to avoid the failed ring and place the traffic on the ring that retains continuity. Traffic continuously travels over both fibers of the dual counter-rotating ringlets. See also *802.17, bandwidth, best effort, CIR, CoS, EIR, Ethernet, flow control, IEEE, IP, jitter, latency, MAC, node, optical fiber, overhead, QoS, real-time, ring topology, SDH, SONET,* and *STDM.*

RS 1. Recommended Standard **1.** A designation for specifications used by various organizations, including the Electronic Industries Alliance (EIA). RS-328, for example is a set of standards for facsimile machines published by the EIA. See also *EIA*. **2.** Reed-Solomon. A block coding algorithm for forward error correction (FEC). See *Reed-Solomon*.

RS-232 See *EIA-232*.

RS-328 (Recommended Standard 328) A set of standards for facsimile (fax) machines published by the Electronic Industries Alliance (EIA) in 1966, and subsequently accepted by the ITU-T as T.2. Machines conforming to the specifications later became known as *Group I*. See also *EIA*, *facsimile*, *Group*, and *I ITU-T*.

RSA (Rivest-Shamir-Adleman) A public key encryption algorithm developed by Ronald Rivest, Adi Shamir, and Leonard Adleman in 1978 that became a de facto standard. RSA formed the basis for a number of encryption programs, including Pretty Good Privacy (PGP). Current versions of RSA employ a 128-bit encryption algorithm, which is computationally infeasible to decode without the key. The 40-bit export version is not considered highly secure. See also *encryption*, *PGP*, *public key encryption*, and *standard*.

RSS (Really Simple Syndication or Rich Site Summary) A metadata push technology, i.e., a technology that can identify changes in data and initiate a content push to the end user, without the user having to search it out and pull it from the site. The term RSS is an umbrella term variously used to describe a number of versions of several data Web feed formats specified in Extensible Markup Language (XML) and used for syndication of Web content. Those standards include Really Simple Syndication, Rich Site Summary, and RDF Site Summary. More recently, the IETF adopted the Atom Publishing Protocol (APP), which builds on the previous RSS work. The Atom Syndication Format is described in IETF RFC 4287 (2005). See also *metadata*, *push*, *Web*, and *XML*.

RSVP (Resource Reservation Protocol) In the TCP/IP protocol suite, a Transport Layer (Layer 4) control protocol that operates on a hop-by-hop basis in order to signal quality of service (QoS) requirements for unicast and multicast data flows to each node and, thereby, reserve the necessary per-session resources from end-to-end across an Internet Protocol (IP) network. RSVP can operate in conjunction with other QoS protocols, including DiffServ and MPLS, to effect service discrimination. See also *DiffServ*, *hop*, *IP*, *MPLS*, *multicast*, *node*, *protocol*, *protocol suite*, *QoS*, *session*, *TCP/IP*, *Transport Layer*, and *unicast*.

rt (real-time) Referring to a quality of service (QoS) level designed for applications that require transmission to take place in real time, that is to say that the transmission must take place at the exact moment and in the exact sequence as the event itself takes place in the real world. Real-time QoS is essential in many applications directly involving humans and their perception of time, and particularly those involving human-to-human interaction. Voice conversations and videoconferences take place in real time and demand real-time QoS. See also *non-real-time*, *nrt*, and *QoS*.

RTCP (Real Time Control Protocol) In the TCP/IP protocol suite, a companion protocol to the Real Time Transport Protocol (RTTP), RTCP allows monitoring of the data delivery in a manner scalable to large multicast networks, and provides minimal control and identification functionality. RTCP and RTP both are defined in IETF RFC 1889 (1996). See also *IETF*, *multicast*, *protocol*, *protocol suite*, *RTP*, and *TCP/IP*.

RTF (Rich Text Formatting) A standard developed by Microsoft Corporation for formatting text files in such a way that the formatting survives transfer not only between applications on a computer, but also between computers on a network. See also *plain text* and *rich text*.

RTM (Ready To Market) A term indicating that a product has passed prescribed tests and is ready to go to market as a general release. Pre-release tests usually include an alpha test and a beta test.

RTP (Real Time Transport Protocol) In the TCP/IP protocol suite, a mechanism for providing end-to-end network transport functions suitable for applications transmitting real-time data, such as audio, video, or simulation data, over multicast or unicast network services. Defined in IETF RFC 1889 (1996), RTP provides end-to-end delivery services including payload type identification, sequence numbering, and timestamping. In combination, the sequence numbering and timestamping provide the receiving node with sufficient information to resequence them as necessary. RTP does not address resource reservation and does not guarantee quality of service (QoS) for real-time services. RTP does not either guarantee delivery through the network or prevent out-of-order delivery, and it does not assume that the underlying network is reliable and delivers datagrams in sequence to the receiving machine. RTP does, however, prevent out-of-order delivery to the application. Applications such as voice over Internet Protocol (VoIP) generally run RTP on top of the User Datagram Protocol (UDP), which provides multiplexing and checksum services. In the context of the OSI Reference Model, RTP falls into both the Session Layer (Layer 5) and the Presentation Layer (Layer 6). RTP Control Protocol (RTCP) is an upper-layer companion protocol that allows monitoring of the data delivery. See also *application, checksum, datagram, IETF, multicast, multiplex, payload, Presentation Layer, protocol, protocol suite, QoS, real-time, RTCP, Session Layer, TCP/IP, transport, UDP, unicast,* and *VoIP.*

RTS (Request To Send) **1.** A message sent from a device seeking access to a wireless network. Demand assigned multiple access (DAMA), for example, is a protocol that assigns available channel capacity to an Earth station from a pool of bandwidth, on demand and as available. The Earth station transmits request to send (RTS) messages to the satellite until it responds with a clear to send (CTS), at which time the message transmission ensues. See also *DAMA.* **2.** An electrical line on a serial interface (e.g., a COM port) that a terminal asserts to indicate it has information to send. The other device may give permission by asserting clear to send (CTS). See also *serial communications.*

RTSP (Real-Time Streaming Protocol) In the TCP/IP protocol suite, an Application Layer (Layer 7) protocol for control over the delivery of data that has real-time properties, such as audio and video and including both live data feeds and stored clips. Defined in IETF RFC 2326 (1998), RTSP is intended to control multiple data sessions, provide a means for choosing delivery channels such as the User Datagram Protocol (UDP), multicast UDP, and the Transmission Control Protocol (TCP). RTSP also provides a means for choosing delivery mechanisms based upon the Real Time Transport Protocol (RTP). There is no notion of an RTSP connection. Rather, an RTSP server maintains a session labeled by an identifier. RTSP establishes and controls one or more time-synchronized streams of continuous media such as audio and video. It does not typically deliver the continuous streams itself, although it is possible to interleave the media stream with the RTCP control stream. RTSP can be thought of as a network remote control for multimedia servers. See also *Application Layer, multicast, multimedia, protocol, protocol suite, real-time, RTP, server, session, TCP, TCP/IP,* and *UDP.*

rt-VBR (real-time Variable Bit Rate) In asynchronous transfer mode (ATM), a class of traffic that takes place in real time and that requires access to time slots at a rate that can vary significantly from time to time. Traffic parameters include peak cell rate (PCR), cell delay variation tolerance (CDVT), sustainable cell rate (SCR), maximum burst size (MBS), and burst tolerance (BT). The quality of service (QoS) parameter is cell loss ratio (CLR). Real-time compressed audio, voice, and video encoded at variable rates are examples of rt-VBR traffic. ATM also defines available bit rate (ABR), constant bit rate (CBR), non real-time Variable Bit Rate (nrt-VBR), unspecified bit rate (UBR), and variable bit rate (VBR) traffic classes. See also *ABR, ATM, BT, CBR, CDVT, compression, MBS, nrt-VBR, PCR, QoS, real-time, SCR, time slot, UBR,* and *VBR.*

rule of thumb A rule, or guideline, based on experience and sound judgement rather than scientific knowledge. The origin of the term is vague, but some suggest that it originated in the practice of carpenters, seamstresses, artists, and other craftsmen of using their thumb to approximate measurements. The fact that the measurement of an inch is based on the distance between the thumbnail and the first joint certainly lends credence to this theory.

rules of engagement Mutually agreed-upon rules that define how military forces should behave during times of war, including the treatment of prisoners and civilians. Rules of engagement differ from a warrior's code, which is a less formal set of defined limits on what warriors can and cannot do if they want to continue to be regarded as warriors, rather than murderers or cowards. For the warrior who adheres to such an informal code or to the formal rules of engagement, certain actions are unthinkable, even in the most dire or extreme circumstances. Rules of engagement date at least to the Middle Ages in Europe, when highly trained and well-paid gentleman knights spent years in apprentice and training to prepare them to wage war. They wielded heavy weapons and they, and their steeds, wore heavy armor for protection against those same weapons wielded by their noble opponents in battles fought on the field of honor. The invention of the crossbow upset the balance, however, as one small bolt from a crossbow fired by even the least skilled, most common peasant farmer could topple even the mightiest and most gentlemanly knight wearing the heaviest armor. Once toppled to the ground, the knight became immobile and, therefore, an easy kill for a common peasant with a stiletto. This innovation was considered so disgraceful that Pope Innocent II in 1139 declared the crossbow "hateful to God and unfit for Christians." The second Lateran Council of churches stated that, "We prohibit under anathema that murderous art of crossbowmen and archers, which is hateful to God, to be employed against Christians and Catholics from now on." This decree did not prohibit the use of the crossbow against infidels, who apparently weren't considered to be gentlemen, much less worthy of the protection of the Church. The Church of England also attempted to outlaw the crossbow. See also *Geneva Convention* and *warrior's code.*

run-length encoding A string coding compression technique used in facsimile machines to digitize and compress a document prior to transmission by identifying color redundancy in the original document. The transmitting machine scans a document from top-to-bottom and left-to-right, sensing dots of black and white, grayscale, or color at an interval that depends on the resolution setting. Rather than transmitting a set of bits identifying the value of each dot of each line, the scanning machine looks for redundancy, i.e., strings or runs, of dots of the same value. The machine then can transmit a set of bits identifying that value and the length of the run before the value changes. Modified Huffman (MH) is a run-length encoding compression technique used in Group III facsimile machines. See also *compression, facsimile, Group III, MH, resolution*, and *string coding.*

runt An unintentionally truncated frame or packet. A runt is either malformed or errored in transit and must be ignored if received.

Rural Health Care Corporation (RHC) See also *RBOC* and *Rural Health Care Program.*

Rural Health Care Program In the United States, a program administered by the Rural Health Care Corporation (RHC) to subsidize telecommunications and Internet service for eligible rural health care providers in high-cost areas. The Rural Health Care Program is one of four programs established by the Telecommunications Act of 1996, supported by the Universal Service Fund (USF), and administered by the Universal Service Administrative Company (USAC). See also *RBOC, Telecommunications Act of 1996, USAC*, and *USF.*

Rural and High Cost Program In the United States, a program that subsidizes basic telephone service in high-cost areas of the United States and its territories. The Rural and High Cost Program is one of four programs established by the Telecommunications Act of 1996, supported by the Universal Service Fund (USF), and administered by the Universal Service Administrative Company (USAC). See also *Telecommunications Act of 1996, USAC*, and *USF.*

S **1.** The symbol for entropy. See *entropy*. **2.** S interface or Reference Point S in ISDN. See *Reference Point S.*

SAN (Storage Area Network) A high-speed, special purpose, dedicated network that supports communications between computers and storage servers in support of data-intensive applications such as inventory management, credit and billing management, receivables management, customer relationship management, and supply chain management. A SAN is much more complex than simple network-attached storage (NAS) and provides application users with much faster access to databases, SANs also provide for centralized management of critical data, including accessibility, security, and backup. SAN protocols include 100Base-T, Gigabit Ethernet (GigE and 10GigE), asynchronous transfer mode (ATM), IBM's Enterprise Systems Connectivity (ESCON) and Fibre Connection (FICON), several versions of Fibre Channel (FC), Serial Storage Architecture (SSA), Small Computer Systems Interface (SCSI) and Internet Small Computer Systems Interface (iSCSI). The storage technologies include Just a Bunch Of Disks (JBOD), Redundant Array of Inexpensive Disks (RAID), a cluster of servers on a network, or a more complex and expensive host storage server such as mainframe computer storage. SAN applications include disk mirroring, data backup and restoration, data archival and retrieval, data transfer between storage devices, and data sharing between servers. See also *10 GigE, 100Base-T, ATM, ESCON, Fibre Channel, FICON, GigE, iSCSI, JBOD, LAN, NAS, RAID, SCSI, server,* and *SSA.*

SAPI (Service Access Point Identifier) In ISDN, a one-octet address field that identifies the destination service access point. The SAPI and terminal endpoint identifier (TEI) jointly compose the data link connection identifier (DLCI), which is the two-octet address field in an LAPD frame. See also *DLCI, ISDN,* and *LAPD.*

SAR (Segmentation And Reassembly) In the ATM reference model, a sublayer of the ATM Adaptation Layer (AAL) that functions to segment the user data into payloads for insertion into cells, prior to transmission. At the receiving end of the communication, the SAR extracts the payload from the cells and reassembles the data into the information stream as originally transmitted. See also *AAL, ATM reference model, cell,* and *payload.*

satellite From the Latin *satelles*, meaning *attendant.* **1.** A attendant celestial body that revolves or is intended to revolve around a larger celestial body. For example, a planet orbits a sun and a moon orbits a planet. The Earth's moon orbits the Earth. Actually, that's not quite true. The moon and the Earth orbit each other as they both revolve around their common center of gravity. As the Earth is the larger of the two and, therefore, exerts more gravitational force, the moon does most of the revolving. **2.** A manmade object that orbits the Earth or some other celestial body. Satellites in Geosynchronous Earth Orbit (GEO) are placed in orbital slots such that they maintain their positions relative to the Earth's surface. Medium Earth Orbiting (MEO) and Low Earth Orbiting (LEO) satellites do not. See also *GEO, LEO,* and *MEO.*

S band The portion of the radio spectrum in the range of 2.310–2.360 GHz, as specified by the ITU-R. Applications include civil defense radio, direct broadcast satellite (DBS) television, and satellite broadcast radio. See also *DBS, electromagnetic spectrum, ITU-R,* and *satellite.*

S-Band (Short [Wavelength] Band) The ITU-T standard optical transmission window in the wavelength range of 1,460–1,530 nm. See also *wavelength* and *window.*

SBE (Small Business Enterprise) A small commercial organization. See also *SOHO.*

SCAI (Switch-to-Computer Applications Interface) An application program interface (API) developed in the early phases of computer telephony and subsequently incorporated by the European

Computer Manufacturers Association (ECMA) into the Computer Supported Telephony Applications (CSTA) standard. See also *computer telephony*, *CSTA*, and *ECMA*.

scaling A step in the video compression process that deals with the creation of the digital image according to the presentation resolution scale associated with the display device. Rather than digitizing the video signal in large scale, the codec is tuned to the scale of presentation in terms of horizontal and vertical pixels, thereby reducing the amount of data that must be digitized. In consideration of this factor, the aspect ratio must be standardized. See also *aspect ratio*, *codec*, *compression*, *digital*, *pixel*, *resolution*, *signal*, and *video*.

scam A scheme for making money by deceptive, dishonest, or fraudulent means, i.e., a swindle or fraud. See also *pharming*, *phishing*, *pretexting*, and *social engineering*.

scanning The process of refreshing a video screen. Interlaced scanning, which is used with most analog TV systems, involves two fields. Odd lines (field 1) are refreshed in one scan, and even lines (field 2) in the next. Each set of odd and even lines refreshed constitutes a frame refreshed. The scanning rate is a function of the power source of the receiver. For example, the American NTSC standard provides for 30 fps, involving 60 scans, which relates directly to the 60 Hz of the United States power source. The European PAL standard provides for 25 fps, involving 50 scans, which relates directly to the 50 Hz of the European power source. Progressive scanning involves displaying all horizontal scan lines in one frame at the same time, which avoids the problem of interline flicker. See also *frame rate*, *NTSC*, and *PAL*.

scanning rate The rate of frame refreshment. See *scanning*.

scattering The deflection of a beam of radiant energy as it encounters physical matter in a medium. See also *Rayleigh scattering*.

scatternet In a Bluetooth personal area network (PAN), an ad hoc network formed of multiple piconets, or very small ad hoc networks. A device participating in multiple piconets acts as a bridge between them, forwarding packets from one piconet to another. See also *ad hoc mode*, *Bluetooth*, *bridge*, *packet*, *PAN*, and *piconet*.

SCC (Specialized Common Carrier) A common carrier offering a limited type or class of service or serving a unique or limited market. See also *common carrier* and *SCC Decision*.

SCC Decision (Specialized Common Carrier Decision) In the United States, the Federal Communications Commission (FCC) decision (1971) that cleared the way for MCI and other specialized common carriers (SCCs) to construct and operate networks.

SCE (Service Creation Element) In the intelligent network (IN), a set of modular programming tools permitting services to be developed independently of the switch, thereby divorcing the service-specific programmed logic from the switch logic. This enables the service to be developed independently and be made available to all switches in the network. The concept of a sparse network is one of dumb switches supported by centralized intelligence with connectivity between distributed switches and centralized logic provided over high-speed digital circuits. See also *IN*.

S-CDMA (Synchronous Code Division Multiple Access) A modulation technique that transmits 128 orthogonal codes simultaneously. S-CDMA is used in combination with 128-point quadrature amplitude modulation trellis-coded modulation (128-QAM TCM) in cable modems specified in the Data Over Cable Service Interface Specification (DOCSIS). See also *128-QAM*, *cable modem*, *DOCSIS*, *modulation*, *orthogonal*, *QAM*, and *TCM*.

Schools and Libraries Corporation (SLC) Administers the Schools and Libraries Program. See also *Schools and Libraries Program*.

Schools and Libraries Program Also known as the *E-rate Program*. In the United States, a program administered by the Schools and Libraries Corporation (SLC) to make telecommunications service, Internet access, and internal connections affordable for eligible schools and libraries. The Schools and Libraries

Program is one of four programs established by the Telecommunications Act of 1996, supported by the Universal Service Fund (USF), and administered by the Universal Service Administrative Company (USAC). See also *Telecommunications Act of 1996*, *USAC*, and *USF*.

SCN (Switched Circuit Network) More commonly known as circuit-switched network. A network based on circuit switching, rather than packet switching. The traditional public switched telephone network (PSTN) is an SCN, although it rapidly is transitioning to packet switching based on the Internet Protocol (IP), the fundamental protocol of the Internet. See also *circuit switch*, *Internet*, *IP*, *packet switch*, and *PSTN*.

SCO (Synchronous Connection-Oriented) A Bluetooth link option intended for real-time packet voice, which certainly benefits from a pre-defined path over a synchronous transmission facility. Bluetooth specifications also include an asynchronous connectionless link (ACL) for packet data. See also *ACL*, *Bluetooth*, *connection-oriented*, *link*, *packet*, and *synchronous*.

scope creep Referring to the gradual broadening of the scope of a project, and often due to some combination of poor project design, documentation, and management. Scope creep also can be attributed to opportunistic vendors attempting to increase the scope of a project and their associated revenues or opportunistic clients seeking to get a vendor to provide additional products or services at no charge.

SCP (Service Control Point) In the advanced intelligent network (AIN) architecture, an intelligent node that contains customer information in a database residing on a centralized network server. An SCP provides routing and other instructions to a service switching point (SSP). See also *AIN*, *database*, *node*, *server*, and *SSP*.

SCR (Sustainable Cell Rate) In asynchronous transfer mode (ATM), a traffic parameter defined as the maximum average rate at which the network agrees to accept cells and support their transfer from end to end for each user network interface (UNI). In other words, SCR is the average throughput. Enforcement of the SCR enables the network to allocate sufficient resources to ensure that the quality of service (QoS) parameters such as cell loss ratio (CLR) and cell transfer delay (CTD) are met over a period of time. SCR applies to variable bit rate (VBR) services. *ATM*, *cell*, *CLR*, *CTD*, *QoS*, *throughput*, *traffic parameter*, *UNI*, and *VBR*.

scramble To transpose, invert, displace, or otherwise modify a signal so as to render it unintelligible without the special receiving equipment to unscramble the signal. See also *encrypt* and *signal*.

screen pop A feature of customer contact systems that brings up a customer record or profile on the computer screen of a call center agent as an incoming call from the customer is connected. As the customer call connects to the call center, the automatic call distributor (ACD) receives the calling telephone number through Calling Line Identification (CLID) or Automatic Number Identification (ANI). In the absence of that information, the system can request that the customer enter an account number or some other Personal Identification Number (PIN). The ACD can request a database search to locate the customer's profile and any associated records. If the search is successful, the ACD then coordinates the presentation of the customer records in a screen pop as it connects the telephone call to the agent. See also *ACD*, *ANI*, *call center*, *call vectoring*, and *CLID*.

screened twisted pair (ScTP) See *ScTP*.

scripting Referring to the use of a simple scripting language to instruct a computer to perform a specific task, such as mimicking the log-on procedures of an e-mail program.

SCSI (Small Computer System Interface) Pronounced *scuzzy*. A high speed parallel interface defined by the X3T9.2 committee of the American National Standards Institute (ANSI) for connecting minicomputers to peripherals, to other computers, and to local area networks (LANs). The several SCSI versions have bus widths of 8 or 16 bits and support data transfer rates of 5–640 Mbps. A single SCSI port can support as many as 7 devices, such as hard disk drives, optical drives, scanners, laser printers, and digital cameras, in a daisy chain. See also *ANSI*, *daisy chain*, *iSCSI*, *LAN*, *microcomputer*, and *peripheral*.

ScTP (Screened Twisted Pair) The simplest form of shielded twisted pair (STP) copper cable, ScTP comprises multiple insulated pairs enclosed in a common metallic shield that is encased in a thermoplastic cable jacket. The shield typically consists of helically or longitudinally applied plastic and aluminum laminated solid tape, although it may consist of a woven mesh, and steel or copper also may be used. An uninsulated steel or tinned copper conductor in contact with the shield serves as a drain wire, ensuring that the continuity of the shield remains intact in the event that the tape is broken or cracked. The shield absorbs ambient energy and conducts it to ground through the drain wire, thereby protecting the signal transmitted through the center conductors. The shield also serves to confine the electromagnetic field associated with the transmitted signal within the core conductors, thereby reducing signal loss and maintaining signal strength over a longer distance. See also *Cat 6*, *STP*, and *UTP*.

SDDN (Software-Defined Data Network) Synonymous with SDN and virtual private network (VPN). See *VPN*.

SDH (Synchronous Digital Hierarchy) In 1988, the CCITT, predecessor to the ITU-T, accepted the SONET standards, with modifications that were mostly at the lower multiplexing levels in order to accommodate the complexities of internetworking the disparate national and regional networks. The ITU-T Recommendations referenced are G.707, G.708, and G.709. See also *SONET*.

SDLC (Synchronous Data Link Control) A bit-oriented, synchronous data communications protocol. SDLC supports high-speed transmission (56 kbps or better) over dedicated circuits in point-to-point or point-to-multipoint network configurations and operates in either half-duplex (HDX) or full duplex (FDX) mode. SDLC uses a cyclic redundancy check (CRC) error detection mechanism, specifically known as *Frame Check Sequence* (FCS) in this case. SDLC allows as many as 128 frames to be sent in a string, with each frame containing up to 7 blocks, each up to 512 characters. Each block within each frame is checked individually for errors. Errored blocks must be identified as such to the transmitting device within a given time limit, or they are assumed to have been received error-free. The SDLC frame consists of synchronizing bits, data bits, and control characters sent in a continuous data stream, frame-by-frame. The SDLC protocol was developed by IBM in the mid-1970s as a key element of its System Network Architecture (SNA). SDLC corresponds the Layer 2, the Data Link Layer of the OSI Reference Model. Figure S-1 illustrates the SDLC frame and its component fields. The same frame format applies to High-level Data Link Control (HDLC) frames and X.25 packets.

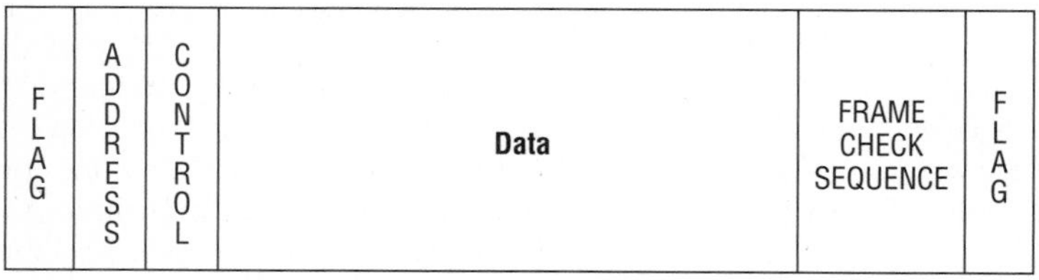

FLAG	ADDRESS	CONTROL	Data	FRAME CHECK SEQUENCE	FLAG

Figure S-1

The fields in the SDLC frame are as follows:

- **Flag:** A specific eight-bit pattern that alerts the receiving device to the transmission of the frame. The most commonly used flag character is 01111110, in binary notation (7E in hexadecimal). Flags also fill idle time on the line with a flag between each frame. A flag is also included at the end to alert the receiving device to the end of the transmission.
- **Address:** An eight-bit field that identifies the address of the target station, a group address for multiple target stations, or a broadcast address to all stations. This field also can be used to distinguish commands from responses.

- **Control:** An eight-bit control field that identifies the type of frame being transmitted. An information frame is used for the transfer of messages, frame numbering of contiguous frames in a message, and so on. A supervisory frame is used for purposes such as to indicate a detected error in transmission, acknowledge that frames have been received without error, request the retransmission of specified frames, and order the transmitting device to cease transmission.
- **Data:** A variable-length data field (aka payload or text field) that contains the data or request being transmitted. This field also can include a format identifier, logical channel group number, packet-type identifier, and packet sequence numbers. In total, the Information field can contain as many as seven blocks, each of which can contain as many as 512 octets of data, for a total of 4,096 octets.
- **Frame Check Sequence (FCS):** This 16- or 32-bit field contains the CRC character sequence used to check the integrity of the transmitted address and control information, as well as the data.
- **Flag:** A specific eight-bit pattern that alerts the receiving device to the end of transmission of the frame. The most commonly used flag character is 01111110 (7E in hexadecimal). Only one flag is needed between frames.

See also *binary notation, bit-oriented protocol, block, CRC, Data Link Layer, FDX, frame, HDLC, HDX, hexadecimal notation, OSI Reference Model, payload, point-to-multipoint circuit, point-to-point circuit, protocol, SNA, synchronous*, and *X.25*.

SDN (Software-Defined Network) Synonymous with SDDN and Virtual Private Network (VPN). See *VPN*.

SDP (Session Description Protocol) A protocol intended for the description of multimedia sessions for purposes of session announcement, session invitation, and other forms of session initiation. Described in IETF RFC 2327, SDP is used to advertise multimedia conferences and communicate the conference addresses and tool-specific information required for participation. SDP defines the characteristics and parameters of a multimedia session, but does not incorporate a transport protocol. Rather, SDP relies on transport protocols such as Hypertext Transport Protocol (HTTP), Session Initiation Protocol (SIP), and Real-Time Streaming Protocol (RTSP). See also *HTTP, IETF, multimedia, protocol, RTSP, session*, and *SIP*.

SDR (Software-Defined Radio) A developing type of radio equipment that can be reprogrammed quickly to transmit and receive on any frequency within a wide range of frequencies, and using virtually any transmission format and any set of standards. Theoretically, a device such as a wireless LAN (WLAN) network interface card (NIC) or cellular telephone with an SDR chipset could seek out frequency bands and native protocols supported by available networks, lock in on the signals, and negotiate access to the desired network, downloading any necessary supplemental software required to effect network compatibility. In the process, SDR-equipped devices would resolve any conflicts between networks sharing a given band (e.g., 802.11b/g and Bluetooth overlap in the 2.4 GHz ISM band). SDR originated in the early 1990s in the U.S. military SpeakEasy program. The U.S. Federal Communications Commission (FCC) began hearings on SDR in March 2000, with the intent that the development of SDR could promote more efficient use of spectrum, expand access to broadband wireless communications, and increase competition among service providers. Military equipment manufacturers have products they tout as being SDR, although they do not conform to open standards. Manufacturers of radio equipment for emergency response agencies are working to develop SDR products that can bridge multiple emergency radio protocols so that fire and police department system, for example, can intercommunicate during an emergency. See also *802.11b, 802.11g, Bluetooth, broadband, cellular, FCC, NIC, radio*, and *WLAN*.

SDSL (Symmetric Digital Subscriber Line) Also known as *Single-line DSL*. An umbrella term for nonstandard variations of high-bit-rate digital subscriber line (HDSL) operating over a single unshielded twisted pair (UTP). SDSL uses the same 2B1Q line coding as HDSL, and runs at rates from 128 kbps to 2.32 Mbps. Typical signaling rates are 400 kbps for local loops of up to 18,000 feet in length, and 784 kbps for loops up to 12,000 feet. At those signaling rates, SDSL yields a payload transmission rates of 384 kbps and 768 kbps, respectively. See also *2B1Q, HDSL, HDSL2, HDSL4, payload, signaling rate, transmission rate*, and *UTP.*

SDTV (Standard Definition TeleVision) A standard for digital television (DTV) that supports display formats that are relatively consistent with legacy analog TV formats with respect to resolution and other specifics. SDTV also is distinguished from high definition television (HDTV). Specifically, SDTV specifies two formats, as detailed in Table S-1.

Table S-1: SDTV Scanning Formats

Vertical Lines	*Horizontal Pixels*	*Aspect Ratio*	*Refresh Rate (fps)**
480	704	4:3, 16:9	24p, 30p, 60p, 60i**
480	640	4:3	24p, 30p, 60p, 60i**

* fps = frames per second
** i = interlaced, p= progressive

DTV in 4:3 aspect ratio has the same appearance as analog TV based on National Television Standards Committee (NTSC), Phase Alternate Line (PAL), and **S**équential **C**ouleur **A**vec **M**émoire (SECAM) standards, but without the ghosting, snowy images, and generally poor audio quality. Issues of signal quality in DTV transmission manifest in artifacts such as blocking, or tiling, and stuttering. The ATSC standard specifies MPEG-2 compression, and the transport subsystem as ISO/IEC 13818. Packet transport involves a serial data stream of packets of 188 octets, 1 octet of which is a synchronization byte and 187 octets of which are payload. This packet approach is suitable for ATM switching, as each 188-octet MPEG-2 packet maps into the payload of 4 ATM cells, with only 4 octets of padding required. SDTV employs Reed-Solomon forward error correction (FEC) and 8-level vestigial sideband (8 VSB) RF modulation to support a bit rate of 19.28 Mbps over a 6 MHz terrestrial broadcast channel. Audio compression is based on the AC-3 specification from Dolby Digital and the ATSC. SDTV standards were developed by the Grand Alliance and reviewed, tested, and documented by the Advanced Television Systems Committee (ATSC) at the request of the United States Federal Communications Commission (FCC). See also *8-VSB, AC-3, analog, artifact, aspect ratio, ATM, ATSC, broadcast, byte, channel, compression, digital, DTV, FCC, FEC, fps, ghosting, Grand Alliance, HDTV, interlaced scanning, modulation, MPEG-2, NTSC, octet, packet, padding, PAL, payload, pixel, progressive scanning, Reed-Solomon, refresh rate, resolution, RF, scanning, SECAM*, and *synchronize.*

seamless Referring to the integration of hardware or software components in such a carefully designed and executed manner that they blend smoothly, without apparent discontinuities or disparities.

search engine **1.** A program that searches a database of documents and files for information. The search looks for keywords or exact phrases found in titles or text. See also *database.* **2.** Application software that develops and maintains resource directories of documents and files found on the Internet that it searches using key words or phrases. A search engine can be restricted to a single website, or can search across many sites on the public Internet. See also *Internet* and *software.*

SECAM (Séquential Couleur Avec Mémoire) Translates from French as *sequential color with memory.* A variation of the Phase Alternate Line (PAL) television standard, SECAM was developed in France. In addition to its use in France, it also is the standard in regions once under French influence, including areas of Asia and the Middle East. See also *NTSC* and *PAL.*

SECBR (Severely Errored Cell Block Ratio) In asynchronous transfer mode (ATM), a dependability parameter expressed as the number of severely errored cell blocks (some number of cells transferred in a sequence) compared with the total number of cell blocks sent over a period of time. A severely errored cell block outcome is realized when more than some number of cells in a block is errored, lost, or misinserted. See also *ATM, block*, and *cell.*

second **1.** One sixtieth (1/60) of a minute. **2.** In the International Systems of Units (SI), the duration of 9,192,631,770 periods of the radiation corresponding to the transition between the two hyperfine levels of the ground state of the cesium-133 atom. This is the base unit of time for atomic clocks, which are the basis for the UTC time scale and the calibration of other clocks, frequency counters, and frequency standards. This refers to a cesium atom at rest, at sea level, at a temperature of 0 K (0 Kelvin, or absolute zero). This measurement is highly stable, unlike the natural frequency standards such as the rotation of the Earth on its axis, the rotation of the Earth around the sun, and electrical and mechanical vibrations such as pendulums, tuning forks, and vibrations of quartz crystals. See also *SI* and *UTC*.

secondary domain In the Internet Domain Name System (DNS), the rightmost portion of the address is the Top Level Domain (TLD), which identifies the type of entity owning or sponsoring the address, or the country in which the address is located. The several types of Top Level Domains (TLDs) include generic Top Level Domains (gTLDs) and country codes. The gTLD is always used. If a country code is also used in the address, it becomes the TLD and the gTLD becomes the secondary domain. See also *country code*, *DNS*, *domain*, *gTLD*, *Internet*, and *TLD*.

Second Computer Inquiry Also known as *Computer Inquiry II* (CI II). In the United States, the Federal Communications Commission (FCC) inquiry (1982) that re-examined the rules it established in the First Computer Inquiry (1971). CI II allowed AT&T to sell customer premises equipment (CPE), but required that it do so through a separate subsidiary. AT&T subsequently formed American Bell to satisfy that order. (As a point of interest, American Telephone and Telegraph (AT&T) was established in 1885 as a subsidiary of the American Bell Telephony Company to operate the long distance business.) The FCC also allowed the resale of public switched telephone network (PSTN) services like Message Telecommunications Service (MTS) and Wide Area Telecommunications Service (WATS). See also *CPE*, *FCC*, *First Computer Inquiry*, *MTS*, *PSTN*, and *WATS*.

secret-key encryption Also known as *private key* or *single-key encryption*. See *encryption* and *private key encryption*.

Section Overhead (SOH) See *SOH*.

section terminating equipment (STE) See *STE*.

Secure European System for Applications in a Multivendor Environment (Sesame) See *Sesame*.

Secure Hypertext Transport Protocol (S-HTTP) See *S-HTTP*.

Secure Multipurpose Internet Mail Extension (S/MIME) See *S/MIME*.

Secure Sockets Layer (SSL) See *SSL*.

security In telecommunications, protection of system and network resources from external attack or internal subversion, thereby ensuring that those resources are available only to those who have the legitimate right to use them and, further, to ensure that they are used only for legitimate purposes. Security encompasses the following dimensions.

- **Physical security** involves control over the personnel who have access to the facilities in which the hardware systems (e.g., terminals, servers, and PBXs) reside. Access control includes security guards, locks and keys, electronic combination locks, and electronic card key systems. Physical security also entails low-tech tools such as document shredders and burn bags, which jointly serve to make paper documents and electronic media unusable after they have served their purposes. Physical security also encompasses all measures required to protect wireline and wireless transmissions from physical wiretaps and other means of signal interception.
- **Authentication** includes all security measures designed to verify or validate the identity of a user or station prior to granting access to resources.

- **Authorization** is the process of granting approval or permission to an authenticated person or device seeking access to a resource. Authorization involves complex software that resides on every secured computer on the network.
- **Encryption** involves scrambling and perhaps compressing the data prior to transmission; the receiving device is provided with the necessary logic to decrypt and decompress the transmitted information.

See also *authentication*, *authorization*, *encryption*, *job security*, and *wiretap*.

security management An element of network management, security management comprises the functions that control and protect access to system and network resources. Security management functions include authentication, authorization, and encryption key management. Security management also entails the detection and reporting of network and system attacks and intrusions, and the initiation of protective measures such as port disabling. See also *authentication*, *authorization*, *encryption key*, and *network management*.

seed rod Synonymous with bait rod. See *bait rod*.

segment The Transmission Control Protocol (TCP) unit of data transfer between host computers. Segments are used to establish connections, transfer actual data, acknowledge packet receipt and request retransmissions, and terminate connections. See also *host*, *packet*, and *TCP*.

segmentation **1.** The act of dividing something into smaller units. **2.** In local area networks (LANs), the act of dividing a LAN into smaller physical units on a geographical basis through filtering bridges, hubs, switches, or routers. Segmentation relieves congestion by limiting the scope of broadcast messages and by confining unicast and multicast traffic to the segment or segments in the physical path between the originating device and each destination device. See also *broadcast*, *filtering bridge*, *hub*, *LAN*, *multicast*, *switch*, *router*, *unicast*. **3.** In asynchronous transfer mode (ATM), the process of dividing an incoming data unit — such as a block, frame, or packet — into cells. As the cells exit the ATM network, they are reassembled in order to reconstitute the original block, frame, or packet. The entire process is known as *segmentation and reassembly* (SAR). See also *ATM*, *block*, *cell*, *frame*, *packet*, and *SAR*. **4.** In packet networks, the breaking of large packets into multiple smaller packets (or IP frames) to meet limits on packet length. Transmission Control Protocol (TCP), for example supports the segmentation of files prior to transmission, and their reassembly upon receipt. See also *frame*, *IP*, *packet*, *segment*, and *TCP*.

segmentation and reassembly (SAR) See *SAR*.

seizure A signal over a line or trunk between connected equipment indicating a service or access request. The first step in establishing a stable call, a seizure or series of seizures creates a signaling path between the components, allowing supervision signaling. A seizure involves a physical state change in the circuit as a device goes off-hook and the circuit transitions from an idle state to a busy state and current flows, for example. See also *line*, *signal*, *signaling and control*, and *trunk*.

selective call acceptance A CLASS service feature of the public switched telephone network (PSTN). Selective call acceptance screens incoming calls against a user-defined list of acceptable telephone numbers. Calls from those numbers are accepted. Calls from all other numbers are diverted to a recorded message. See also *CLASS* and *PSTN*.

selective call block See *selective call rejection*.

selective call forwarding A CLASS service feature of the public switched telephone network (PSTN). Selective call forwarding screens incoming calls against a user-defined list of telephone numbers. Calls from only those numbers are forwarded to another telephone number. Calls from all other numbers ring through, as usual. See also *CLASS* and *PSTN*.

selective call rejection Also known as *call block* and *selective call block*. A CLASS service feature of the public switched telephone network (PSTN). Selective call rejection screens incoming calls against a user-defined list of acceptable telephone numbers. Calls from those numbers are diverted to a recorded message. Calls from all other numbers ring through, as usual. See also *call block*, *CLASS*, and *PSTN*.

selective ringing See *distinctive ringing*.

selector Also known as a *wiper*, a selector is a component of a step-by-step (SxS) electromechanical circuit switch. In the most basic implementation, the user dials digits on a rotary telephone, which opens and closes a contact, sending electrical pulses across an electrified copper local loop circuit. The central office (CO) or PBX switch detects the pulses, which cause the selector to step (i.e., move one step per pulse) across contacts. As the user dials each digit in the telephone number, the selector steps up to the next level, step-by-step. So, a typical seven-digit local telephone number corresponds to seven levels in a line selector. Originally, each subscriber line was associated with a line selector. The subsequent invention of the line finder allowed many lines (e.g., 100 or so) to share a bank (10 or so) of selectors, depending on the activity levels of the individual subscribers. See also *SxS*.

selenium A gray, non-metallic element, selenium is 34 on the periodic table of elements and has an atomic weight of 78.96. Selenium usually is obtained as a by-product of lead, copper, and nickel refining, and is used in photoelectric cells, TV cameras, light meters, copy machines, and anti-dandruff shampoos, as well as to color glass red and to give black and white silver prints greater image stability. Although toxic to humans in excess, selenium is considered to be an essential mineral in small amounts. See also *photophone*.

self-healing Referring to a network or subnetwork that has the ability to sense, diagnose, isolate, and at least temporarily correct a fault or performance condition without human intervention. Networks with redundant, intelligent network elements (NEs) may be self-healing in nature. SONET and SDH fiber optic networks are characterized as self-healing. See also *NE*, *network*, *SDH*, *SONET*, and *subnet*.

self-learning bridge A bridge that is able to build its own address table, rather than having to be programmed. Such a bridge typically broadcasts a query to all attached devices. When the devices respond, the bridge associates the originating address of each response frame with the port over which it entered. In this fashion, the bridge builds medium access control (MAC) address tables on a port-by-port and, by implication, segment-by-segment basis. Subsequently, the bridges view the destination MAC addresses of transmitted frames, consult the address table, and forward the frames only over the link connected to the proper port. Most bridges are self-learning, filtering bridges. Self-learning bridges are standardized in IEEE 802.1D, which describes the spanning tree protocol (STP). See also *bridge*, *broadcast*, *filtering bridge*, *frame*, *MAC*, *port*, and *STP*.

semiconductor A material that is neither a good conductor of electricity nor a good insulator, but has properties of electrical conductivity somewhere between the two. Silicon and germanium are good semiconductor materials. Dopants such as arsenic and antimony sometimes are introduced during the manufacturing process to alter the performance characteristics of the semiconductor. See also *conductor*, *dopant*, and *insulator*.

send and pray Referring to the uncertainty of vertical redundancy checking (VRC) as an error control technique for data transmission. See also *error control* and *VRC*.

sender **1.** Also known as a *source* or a *transmitter*, a sender is a device that originates, or generates, an information transfer to one or more receivers. See also *receiver* and *transmitter*. **2.** A component of an electromagnetic crossbar (Xbar) or panel switch, a sender receives dialed digits from a register or routing information from a translator, and sends the appropriate dialed digits across a trunk or to local station equipment. See also *panel switch*, *register*, *translator*, and *Xbar*.

Sequenced Packet Exchange (SPX) See *SPX*.

Séquential Couleur Avec Mémoire (SECAM) See *SECAM*.

serial communications The exchange of digital data between devices one character at a time and one character element, i.e., bit, at a time over a single circuit and channel. Serial communications can be in either asynchronous or synchronous mode. Serial, rather than parallel, communications are used exclusively in wireline transmission, except over extremely short distances, primarily due to the lower cost of provisioning a single circuit. A serial-input-parallel-output (SIPO) shift register is used to interface a serial circuit to a parallel circuit. See also *asynchronous*, *channel*, *circuit*, *parallel communications*, *shift register*, and *synchronous*.

serial-in, parallel-out (SIPO) See *SIPO*.

Serial Line Internet Protocol (SLIP) See *SLIP*.

serial line protocol A protocol that provides for communications over serial or dial-up links. Examples include High-level Data Link Control (HDLC), as specified in ISO 3309, Serial Line Internet Protocol (SLIP), as specified in IETF RFC 1055, and Point-to-Point Protocol (PPP), as specified in RFCs 1548, 1661 and 1662. The term also applies to serial interfaces such as RS-232 and V.35. See also *dial-up circuit*, *HDLC*, *link*, *PPP*, *protocol*, *RS-232*, *serial*, *SLIP*, and *V.35*.

Serial Storage Architecture (SSA) See *SSA*.

server In a client/server network architecture, a machine designated as to serve the needs of client machines. A server can be a mainframe, minicomputer, or personal computer that operates in a time-sharing mode to provide for the needs of many clients for application and file storage, network administration, security, and other critical functions. See also *architecture*, *client*, *client/server*, and *time-sharing*.

service From the Latin servitium, meaning slavery, referring to a person or device with the function of giving good by providing usefulness to others. A service is something done by a person or device for the benefit of another. A service may be provided for a fee or for free. Think of this book as providing a fee-based service. Please do not make copies of this book, as that would destroy the profit motive, which largely is why I wrote this monster. It also would violate my copyright, which would anger me greatly. See also *copyright*.

service access point identifier (SAPI) See *SAPI*.

service creation element (SCE) See *SCE*.

Session Initiation Protocol (SIP) See *SIP*.

Service Level Agreement (SLA) See *SLA*.

service management system (SMS) See *SMS*.

service mark (SM) Intellectual property comprising a word, name, logo or other graphic symbol, or other device used by a company to distinguish its services. A service mark is much like a trademark, but is used to distinguish services, rather than products. In order to receive federal legal protection, a service mark must be distinctive, affixed to a service in the marketplace, and must be registered with the United States Patent and Trademark Office. Service marks are perpetual, as long as they are in continued use. The terms trademark and mark are often used to refer to both trademarks and service marks. See also *intellectual property* and *trademark*.

service node Synonymous with access node and edge switch. In packet networks, a switching point that comprises a point of end user access to the network and network services. See also *edge switch* and *node*.

service-oriented architecture (SOA) See *SOA*.

service profile identification (SPID) See *SPID*.

service set identifier (SSID) See *SSID*.

service switching point (SSP) See *SSP.*

Sesame (Secure European System for Applications in a Multivendor Environment) Authorization software developed by the European Computer Manufacturers Association (ECMA) for large, complex, heterogeneous network computing environments. Sesame is a flexible and open standard. See also *Access Manager*, *authorization*, *ECMA*, *Kerberos*, and *security.*

session Referring to a period of time during which there is a logical association, i.e., communication, between two or more stations, terminals, or systems. A session may be established through a log-on procedure and terminated by a log-off procedure. A network administrator might limit the duration of a session, at the expiration of which time the session will be timed-out, or involuntarily terminated, in order to ensure that network and system resources are shared fairly among the universe of active users. A session also may time out if no activity is detected for a predetermined period of time.

Session Description Protocol (SDP) See *SDP.*

Session Initiation Protocol (SIP) for Instant Messaging and Presence Leveraging Extensions (SIMPLE) See *SIMPLE.*

Session Layer Layer 5 of the seven-layer Open Systems Interconnection (OSI) Reference Model. Software at the Session Layer performs functions related to establishing, maintaining, coordinating, and terminating dialogues and data exchanges from end-to-end. See also *layer*, *network architecture*, *OSI Reference Model*, *session*, and *software.*

set handler Also known as a *mediation device.* A device that acts as a protocol converter to interface a high-performance third-party electronic station set to a Centrex system. Centrex systems are designed to support either generic, featureless station equipment or feature-rich proprietary station sets known as *P-phones.* Third-party electronic station sets can offer the same (or better) performance as P-phones but require a set handler to resolve proprietary signaling and control protocols. See also *Centrex*, *protocol*, and *signaling and control.*

set-top box A small computing device that interfaces a television (TV) set or computer to a cable TV (CATV) network, cable modem network, or satellite TV dish and, perhaps, telephone network. A set-top box is responsible for functions such as decoding digital TV signals for display on an analog TV set, compression and decompression, buffering, security management, and various signaling and control communications. See also *analog*, *buffer*, *cable modem*, *CATV*, *compression*, *digital*, *encoding*, *security*, and *signaling and control.*

severely errored cell block ratio (SECBR) See *SECBR.*

SFTP (Shielded Foil Twisted Pair) Synonymous with shielded twisted pair (STP). See *STP.*

SGCP (Simple Gateway Control Protocol) A predecessor to Media Gateway Control (Megaco). See *Megaco.*

SGML (Standard Generalized Markup Language) A metalanguage used to describe markup languages, such as Hypertext Markup Language (HTML). SGML can be used to describe the structure and elements of a document and the tags that will be used to display those elements in a document. SGML supports the creation of documents that are platform- and application-independent and that, therefore, retain their information formatting, indexing, and linkages. SGML is standardized by the International Organization for Standardization (ISO) as ISO 8879:1986. The Extensible Markup Language (XML) is essentially a condensed version of SGML. See also *XML.*

Shannon, Claude Elwood (1916–2001) An American mathematician and electrical engineer generally known as the father of information theory. While a student at the Massachusetts Institute of Technology (MIT), Shannon discovered the relationship between Boolean algebra and binary arithmetic and applied that knowledge to the design of the electromechanical relays used in telephone switching at the time. He then demonstrated that such relays could be used to solve Boolean algebra problems. While at Bell Labs, Shannon continued that work and published "A Mathematical Theory of Communication," which became the foundation for digital circuit design and provided the basis for determining the capacity of a channel in bits per second (bps). Shannon later made significant contributions in the areas of cryptology and sampling theory. See also *Boolean logic* and *Shannon's Law*.

Shannon's Law A statement defining the theoretical maximum rate at which error-free digits can be transmitted over a finitely bandwidth-limited channel in the presence of Gaussian noise. Shannon's Law is mathematically expressed as $C = W \log_2 (1 + S/N)$, where C is the channel capacity in bits per second (bps), W is the bandwidth in Hertz, and S/N is the signal-to-noise ratio (SNR). Shannon's Law also is known as the *Shannon-Hartley theorem*, as Shannon developed the theorem in collaboration with R.V.L. Hartley, a colleague at Bell Labs. See also *bandwidth*, *bps*, *channel*, *Gaussian noise*, *Hertz*, *law*, *SNR*, and *theory*.

shared bus switch A switch with a single high-speed bus that is shared by all incoming and outgoing ports on a time-division multiplexing (TDM) basis. This is a relatively low cost approach commonly used in smaller, local area network (LAN) workgroup switches where issues of congestion typically are relatively modest. See also *bus*, *LAN*, *LAN switch*, *port*, *switch*, *TDM*, and *workgroup switch*.

Shared Registry System (SRS) See *SRS*.

SHDSL (Symmetric High-bit-rate Digital Subscriber Line) Specified by the ITU-T as G.991.2 (February 2001), SHDSL is a business-class DSL technology that supports symmetric rate-adaptive transmission rates ranging range from 192 kbps at 20,000 feet (6 kilometers) to 2.312 Mbps at 10,000 feet (3 kilometers) in increments of 8 kbps over a single unshielded twisted pair (UTP) local loop. An optional two-pair mode supports transmission rates ranging from 384 kbps to 4.624 Mbps in increments of 16 kbps. In either mode, line doublers can approximately double the distance. SHDSL uses the same trellis-coded pulse amplitude modulation (TC-PAM) line coding technique as HDSL2 and HDSL4. SHDSL also operates in a fixed-rate mode at 784 kbps and 1.544 Mbps. See also *DSL*, *G Series*, *ITU-T*, *line coding*, *line doubler*, *local loop*, *modulation*, *rate adaption*, *symmetric*, *TC-PAM*, *transmission rates*, *UTP*, and *xDSL*.

sheath See *cable sheath*.

SHF (Super High Frequency) SHF radio is in the frequency range of 3–30 GHz and has a wavelength of 10–1 cm. SHF radio has applications in microwave and satellite radio, and wireless LANs. See also *electromagnetic spectrum*, *frequency*, *Hz*, and *wavelength*.

shield A covering, sheath, or screen designed to protect a signal traveling over a conductor from electromagnetic interference (EMI) due to the unwanted coupling of extraneous signals. The shield is in the form of a conductive metal foil, screen, or braid that surrounds a conductor or group of conductors. A dielectric material insulates the conductors from each other and from the shield, which essentially is an outer conductor that intercepts extraneous signals and conducts them to ground. Therefore, the shield must be continuous and must be grounded via a ground wire, or drain wire. To ensure electrical continuity, the shield must be bonded at any splice points and the shield must be bonded to the ground wire, strap, or rod. Coaxial cable, screened twisted pair (ScTP), and shielded twisted pair (STP) are examples of shielded transmission media in telecommunications applications. See also *coaxial cable*, *dielectric*, *EMI*, *ground wire*, *ScTP*, and *STP*.

shielded foil twisted pair (SFTP) Synonymous with shielded twisted pair (STP). See *STP*.

shielded twisted pair (STP) See *STP*.

shift register An area of high speed temporary memory used to interface parallel and serial circuits. A PISO (parallel-in, serial-out) shift register has parallel input ports and serial output ports. A SIPO (serial-in, parallel-out) shift register has serial input ports and parallel output ports. See also *parallel communications*, *register*, *serial communications*, and *shift register*.

Shockley, William Bradford (1910–1989) The U.S. physicist who, in collaboration with John Bardeen and Walter Britain, invented the transistor in 1947 at AT&T Bell Telephone Laboratories. Shockley shared the Nobel Prize in physics in 1956. See also *transistor.*

short haul circuit Within the core, or backbone, of a Wide Area Network (WAN), transport circuits, or long haul circuits, carry data over long distances. Long haul traditionally is defined as a distance equal to or greater than 50 miles (80 kilometers). A short haul circuit, therefore, is less than 50 miles. Access circuits, aka local loops, are short haul circuits. A local area network (LAN) definitely involves short haul circuits. See also *access circuit*, *LAN*, *local loop*, and *WAN.*

short haul modem Synonymous with limited-distance modem. A type of modem used in applications where line drivers fail in terms of either bandwidth or distance. Short-haul modems can work at distances between 5,000–10,000 feet, with distance sensitive to signaling speed, i.e., the higher the signaling speed, the shorter the allowable distance. Short haul modems are generally used over private lines and hardwired links but can also operate over non-loaded local loop facilities. See also *bandwidth*, *line driver*, *loading coil*, *modem*, and *signaling speed.*

Short Message Service (SMS) See *SMS.*

Short Message Service Center (SMSC) See *SMSC.*

shortwave Referring to the short wavelength of high frequency (HF) radio, which is in the frequency range of 3–30 MHz and has a wavelength of 100–10 m, respectively. The term *shortwave* was popularized in the early days of radio, when the HF radio was unusual, but AM broadcast radio, which is in the medium frequency (MF) range and has a much longer wavelength, was popular. HF radio has applications in citizens band (CB) radio, mobile radio, and maritime radio systems. See also *CB Radio Service*, *electromagnetic*, *frequency*, *HF*, *Hz*, *MF*, *radio*, *skywave*, *spectrum*, and *wavelength.*

short wavelength (SW) See *SW.*

short wavelength band (S-Band) See *S-Band.*

S-HTTP (Secure Hypertext Transport Protocol) An extension to HTTP that supports various encryption and authentication mechanisms to secure end-to-end commercial transactions over the Internet. S-HTTP is an Application Layer protocol standardized by the IETF in RFC 2660 (1999). S-HTTP was developed by Enterprise Integration Technologies. As an Application Layer protocol, S-HTTP cannot support the secure, encrypted exchange of other types of data, including file transfer (FTP) and e-mail (SMTP). See also *Application Layer*, *authentication*, *encryption*, *FTP*, *HTTP*, *IETF*, *Internet*, *protocol*, *SMTP*, and *SSL.*

Si Symbol for silicon. See *silicon.*

SI Système international d'unités, which translates from French as *International System of Units*, is the contemporary form of the metric system, based on international agreements reached by the General Conference on Weights and Measures. The SI metric system is a system of measurement (with fundamental units) of length (meter, or m), mass (kilogram, or kg), time (second, or s), electrical current (ampere, or A), thermodynamic temperature (Kelvin, or K), amount of substance (mole, or mol), and luminous intensity candela, or cd). All SI multiples are integer powers of ten. From these base units, other units, such as frequency (Hertz, or Hz) have been derived. In writing SI units, the number value is always followed by a space; units are never capitalized when spelled out (e.g., hertz, Kelvin, and seconds), but units that are names are capitalized when abbreviated (e.g., Hz, K, s).

sideband The band of frequencies, above or below the carrier, created by the process of modulation. The sidebands have considerable value in certain transmission techniques, such as single sideband (SSB), double sideband (DSB), and vestigial sideband (VSB). See also *carrier*, *DSB*, *frequency*, *modulate*, *SSB*, and *VSB.*

side circuits Two pairs of electrified metallic circuits arranged in parallel on the line side of the network and from which a phantom circuit is derived by means of center-tapped transformers known as repeating coils. Each side circuit is a circuit unto itself at the same time it acts as one conductor of the phantom circuit. See also *circuit, line side, phantom circuit*, and *transformer.*

signal An electromagnetic impulse or wave transmitted to convey information in telecommunications, telegraphy, television, radio, radar, etc.

signal control point (SCP) See *SCP.*

signal transfer point (STP) See *STP.*

signaling and control Referring to the internal communications functions that must take place within a network in order to ensure that it operates properly, signaling and control communications are in contrast to the communications of user data, or payload. The various elements of a network must have the capability to signal (i.e., alert and inform) each other, indicating their status and condition. Typical status indications include available (dial tone), unavailable (busy), and alerting (ringing signal). The endpoints (i.e., terminal devices) or end offices also must pass identification information and instructions, such as the originating address, the target address, and the pre-selected carrier. Switches and routers within the carrier network must pass information such as route preference and route availability. Signaling and control systems and networks also handle billing matters, perhaps querying centralized databases in the process. Network management information often is passed over signaling and control links. Such information is used for remote monitoring, diagnostics, fault isolation, and network control. There are two basic types of signaling and control: in-band and out-of-band.

- In-band signaling and control takes places over the same physical path (i.e., through the same switches and across the same circuit), occupies the same frequencies, and competes for the same time slots as the user payload. In-band signaling is intrusive, or disruptive, in nature.
- Out-of-band signaling and control can take place over the same physical path as the user payload, or can involve an entirely separate path and even a separate network. In either case, out-of-band signaling and control takes place over frequencies (e.g., in guard bands) or in time slots separate from those that carry user payload. In either case, out-of-band signaling does not directly compete with payload for bandwidth.

See also *end office, endpoint, database, guard band, in-band signaling and control, network management, out-of-band signaling and control*, and *payload.*

signaling rate The total rate at which data, including overhead, i.e., signaling and control information, can be sent across a circuit. That compares with the transmission rate, which is the theoretical rate at which user data can be transmitted across a circuit. Therefore, the transmission rate is always less than the signaling rate. For example, a typical T1 circuit has a signaling rate of 1.544 Mbps, but a transmission rate of only 1.536 Mbps, as 0.008 Mbps is required for overhead. See also *circuit, data, overhead, signaling and control*, and *transmission rate.*

signaling speed See *signaling rate.*

Signaling System 7 (SS7) See *SS7.*

signal-to-noise ratio (SNR) See *SNR.*

silence suppression A technique used to improve bandwidth utilization of voice circuits. Through silence suppression, a mechanism senses periods of inactivity in a voice conversation and simply ceases sending data associated with that conversation, rather inserts the conversation of another speaker into that channel. Time-assignment speech interpolation (TASI) used this technique on long haul frequency-division multiplexed (FDM) analog voice circuits. In contemporary digital networks, digital speech interpolation (DSI) uses silence suppression. See also *DSI, FDM, suppression*, and *TASI.*

silica See *silicon dioxide.*

silicon (Si) A brittle nonmetallic element (No. 14 in the Periodic Table of Elements) found in abundance in nature. When combined with oxygen, the only element that is more abundant on Earth, silicon forms silicon dioxide (SiO_2), which is used in the manufacture of semiconductors used in transistors, rectifiers, solar cells, glass optical fiber (GOF), etc. Silicon is used in the manufacture of, but is not to be confused with, silicone.

silicon dioxide (SiO_2) Also known as *silica*, silicon dioxide is the oxide of silicon. Silicon dioxide is found in a number of natural forms, including quartz. Quartz sand is used in food additives, silica gel (a desiccant), and ceramics, and is the primary raw material used in the manufacture of glass optical fiber (GOF).

silicone A silicon-based synthetic substance, usually with various organics added. Silicone is characterized by relatively high resistance to heat and water, and is used in oils, greases, polishes, plastics, resins, adhesives, sealants, paints, etc. See also *silicon.*

SIM (Subscriber Identification Module) A smart card that serves as a security mechanism in GSM cellular radio networks. The SIM, which plugs into a card slot in the GSM handset, stores user-profile data, a description of access privileges and features, and identification of the cellular carrier that hosts the home registry. The SIM also can store telephone numbers entered by the user into a personal directory. The SIM can be used with any GSM set, thereby providing complete mobility across nations and carriers supporting GSM, assuming that cross-billing relationships are in place. See also *carrier*, *cellular radio*, and *GSM.*

SIMPLE (Session Initiation Protocol [SIP] for Instant Messaging and Presence Leveraging Extensions) The predominant IM standard. SIMPLE is an extension of SIP (RFC 3261) to address the suite of services the IETF refers to as Instant Messaging and Presence (IMP). SIMPLE is compliant with IETF specifications for Instant Messaging/Presence Protocol Requirements (RFC 2779) and the Common Profile for Instant Messaging (CPIM) specification (RFC 3860). See also *CPIM*, *IETF*, *presence*, *protocol*, and *SIP.*

Simple Gateway Control Protocol (SGCP) A predecessor to Media Gateway Control (Megaco). See *Megaco.*

Simple Mail Transfer Protocol (SMTP) See *SMTP.*

simple mode A mode defined by the ITU-T T.37 standard for Fax over Internet Protocol (FoIP). Simple mode uses the Tagged Image File Format-Fax (TIFF-F) S-profile and restricts fax transmission to the most popular fax machine formats (e.g., standard or fine resolution and standard page size). Simple-mode provides no confirmation of delivery. See also *facsimile*, *FoIP*, *full mode*, *resolution*, *T.37*, and *TIFF-F.*

Simple Network Management Protocol (SNMP) See *SNMP.*

simplex **1.** A transmission path or circuit designed to support information transfer in one direction only. Physically, the circuit may be able to support duplex, or two-way, information transfer, but the equipment on either end will not allow it. In other words, the protocol is simplex. Burglar alarm, intrusion alarm, and fire alarm circuits are simplex in nature, as they are designed to support information transfer from a sensor to a central alarm answering point, but will not support information transfer in the reverse direction. It would be unwise to reset such an alarm remotely without a physical inspection. Rather, the cause of the alarm must be analyzed by a human being who is physically present at the alarm point. The human being can then reset the alarm locally. Two simplex circuits, one in each direction, can be used to provision a full duplex circuit, i.e., a circuit that supports simultaneous transmission in both directions. See also *duplex*, *full duplex*, *half duplex*, and *protocol.* **2.** A fiber optic transmission system (FOTS) using a single single-mode fiber (SMF) for transmission in both directions and perhaps supporting multiple wavelengths. Passive optical network (PON) standards, for example, include a simplex option. In this configuration, the downstream transmissions are supported in the 1480–1580 nm window, with voice and data over a wavelength of

1490 nm, and video at 1550 nm. The upstream transmissions are supported in the 1260–1360 nm window, with voice and data supported at 1310 nm. See also *downstream*, *duplex*, *FOTS*, *nm*, *PON*, *upstream*, *wavelength*, and *window*. **3.** In telegraphy, a single channel circuit that supports transmission in one direction, only. See also *channel*, *circuit*, *diplex*, *quatraplex*, *SMF*, and *telegraph*.

simplified message desk interface (SMDI) A PBX feature that provides a simple interface for communications between a phone system and a voice messaging system. The PBX sends an SMDI message in advance of each call, advising the voice messaging system of the line that it is using, the type of call that it is forwarding (e.g., call forward, call forward busy or call forward no answer), and the source (calling number) and destination (called number) of the call. SMDI enables the voice messaging system to process the forwarded call more intelligently by accessing the correct personalized greeting associated with the correct mailbox. If the phone is so equipped, the voice mail system can turn on a message waiting light. Alternatively, the voice mail system can send stuttered dial tone in place of normal dial tone when the station user picks up the receiver to place a call, can ring the phone periodically, or use some other method. See also *dial tone* and *PBX*.

sine wave A sinusoidal waveform, a sine wave is perhaps best explained and illustrated using electrical current. Current is the flow of electrons through a metallic circuit, with the direction of flow being from positive (+) pole to negative (–) pole. Direct current (dc) travels in one direction, only, while alternating current (AC) travels in both directions across the circuit. A continuous flow of AC current travels first in one direction and then reverses polarity and flows in the opposite direction. The number of times the polarity reverses direction per second is termed the frequency (f or ν), and is measured in cycles per second (cps) or, in more contemporary terms, Hertz (Hz). Voltage (V) is the measure of the electromotive force (emf) and potential difference between the positive and negative ends of the circuit (or a battery). Voltage, i.e., signal strength, or amplitude, is expressed as +V and –V, depending on the direction of the current, with examples being +3V and –3V, or +1.5V and –1.5V. The frequency of the continuous AC waveform is the number of times per second that the waveform makes a complete cycle from zero voltage (0V) to its maximum positive voltage (+V), back to zero (0V), to its maximum negative voltage (–V), and back to zero (0V), as illustrated in Figure S-2.

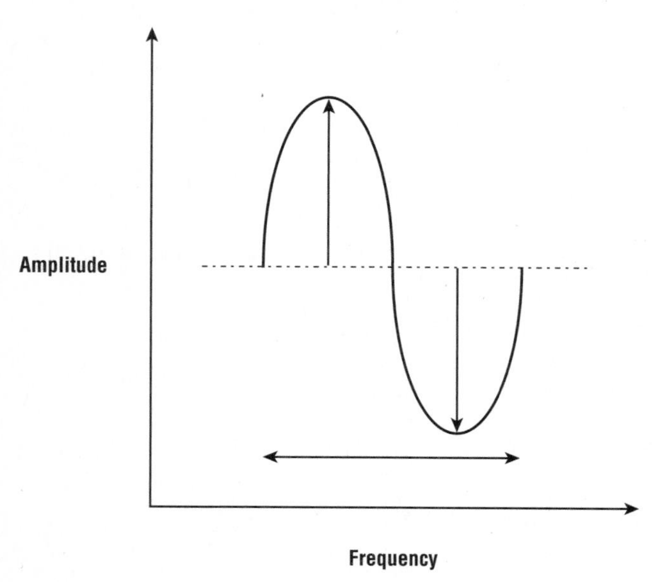

Figure S-2

The AC waveform is a sinusoidal waveform, or sine wave, which is a wave that can be expressed as the sine of a linear function of time, or space, or both. The sine wave makes perfectly smooth transitions over time from zero to its maximum positive voltage, through zero, to its maximum negative voltage, and back again. An unmodulated sine wave continuously makes the same smooth transitions at the same frequency, and the same angle (phase) to the same maximum positive and maximum negative voltage levels. A sine wave can be modulated, or changed, through amplitude modulation (AM), frequency modulation (FM), or phase modulation (PM). Figure S-3 illustrates sine waves at various frequencies, but constant amplitude.

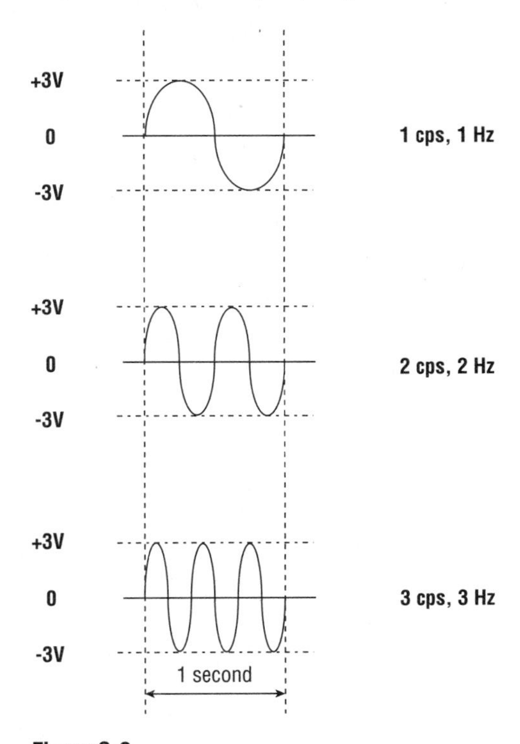

Figure S-3

See also *AC, AM, DC, FM, frequency, Hz, PM, polarity, voltage,* and *waveform.*

Single-Ended Method for Objective Speech Quality Assessment in Narrow-Band Telephony Applications See *P.563.*

single-key encryption Also known as *private key* or *secret-key encryption.* See *encryption* and *private key encryption.*

single-line digital subscriber line (SDSL) See *SDSL.*

single-line DSL (SDSL) See *SDSL.*

single-mode fiber (SMF) See *SMF.*

single number dialing Also known as *store locator service,* a service of the advanced intelligent network (AIN) that provides the ability to advertise a single number. The network routes calls to the closest store location in terms of either geography or time zone, based on the originating address (i.e., telephone number) of the caller. Single number dialing might be used by a chain of pizza parlors so that a caller is automatically routed to the closest parlor and, therefore, the pizza can be delivered in the shortest time, at the lowest cost, and in the hottest condition. (I like anchovies on mine, but my wife hates them. She thinks anchovies are at the root of all evil, at least in the pizza domain.) See also *AIN* and *pizza box*.

Single Point of Termination (SPOT) See *SPOT*.

single sideband (SS or SSB) See *SSB*.

sink Also known as a *receiver*, a sink receives an information transfer originated by a transmitter. See also *receiver*, *transceiver*, and *transmitter*.

S interface See *Reference Point S*.

sintering The process of using heat below the melting point and pressure to bond and partially fuse particles together. The process is used in the manufacture of glass optical fibers to fuse the core and cladding. See also *IVD*, *optical fiber*, and *OVD*.

sinusoidal waveform See *sine wave*.

SiO_2 Symbol for silicon dioxide. See *silicon dioxide*.

SIP (Session Initiation Protocol) An Application Layer signaling protocol for establishing, modifying, and terminating multimedia sessions or calls over an IP network. The IETF defined SIP in RFC 2543 (1999), which was replaced by RFC 3261 (2002). SIP is a modular component of IP telephony, although it can function over any network. SIP offers considerable advantages over H.323, which is criticized for being overly complex and highly centralized. SIP was built specifically for an IP environment in which intelligence is highly decentralized in a large number of client agent servers. SIP identifies clients through a hierarchical URL similar to an e-mail address. A calling client can initiate a call in several ways. If the SIP address of the destination SIP client is known, the calling client simply sends the destination client an invite message, in care of a local proxy server. The proxy server sends the invite message to the distant proxy server, inviting the destination endpoint to join the session, and providing it with enough information to do so. If the SIP address of the destination SIP client is unknown, the calling proxy server sends the invite message to a redirect server, which consults the location server for address information. The redirect server passes that information to the calling proxy server, which then issues an invite message to the distant proxy server, including the information required to join the call. If the call is to a call center, such information might include a request to employ H.261 video, G.728 audio, and Spanish as the preferred language. The proxy server on the receiving end might consult an optional SIP location server on the receive end to determine the exact location of the called client and connect the call. This approach is much simpler and faster than the back-and-forth process involved in H.323, although layers of complexity are being added as the standards process works to enhance SIP to match H.323 and PSTN functionality. When the called client receives the invitation to join the session, it can either accept the call, or forward it to a messaging system or a user, perhaps a Spanish-speaking call center agent. Assuming that the call is a multimedia call comprising both video and voice, the called client (or messaging system) can elect either to accept the composite call or to accept only one of the datastreams, perhaps rejecting the video call but accepting the voice call. SIP also supports call forking, or splitting, so that several client extensions can be rung at once. See also *Application Layer*, *call*, *client*, *endpoint*, *G.728*, *H.261*, *IETF*, *IP*, *multimedia*, *protocol*, *proxy server*, *PSTN*, *server*, *session*, and *URL*.

SIPO (serial-in, parallel-out) A shift register with serial input ports and parallel output ports. See also *shift register*.

site–local address In Internet Protocol version 6 (IPv6), a type of unicast address intended for local use, within a single end-user site, and not for use in a public domain such as the Internet. See also *domain*, *Internet*, *IPv6*, *IPv6 address*, *link-local address*, *unicast*, and *unicast address*.

skills-based routing In call centers, routing an incoming call to an agent in consideration of the customer's identity and requirements and the special skills and skill levels of the available agents in the agent pool in order to maximize the benefits of the customer contact. Such skills can include language fluency and technical expertise.

skin effect The phenomenon by which electrical signals tend to distribute themselves within a conductor so that the current density is greater near the surface, or skin, of the conductor than at the core. The higher the frequency, the more pronounced the phenomenon. For example, through a 24-gauge conductor with a diameter of 0.0201 inches, a signal at 20 kHz travels at a skin depth of 0.0181 inches, approximately the diameter of the wire. At 20 kHz, therefore, the skin effect is negligible. A signal at 25 MHz travels at a depth of only 0.00052 inches. In fact, much of the signal at this frequency is in the form of an electromagnetic field surrounding the conductor and traveling through the dielectric insulation.

skip Skywave propagation of radio signals. See also *skywave*.

Skipjack An encryption algorithm developed for the Clipper Chip. See also *Clipper Chip* and *encryption*.

skywave An electromagnetic wave that propagates upward in space, rather than either close to or through the ground. A skywave may continue upward through the ionosphere and into space, or may be returned to Earth by either ionospheric reflection or refraction. Radio signals in the medium frequency (MF) range and above are considered skywaves. See also *ground wave* and *surface wave*.

SLA (Service Level Agreement) A portion of a service contract that addresses service parameters such as availability (uptime and downtime), mean time to respond (MTTR), mean time to repair (MTTR), and overall network throughput, all of which can be crucial in a mission-critical and time-sensitive application environment. See also *downtime*, *throughput*, and *uptime*.

SLC **1.** Schools and Libraries Corporation (SLC). See *Schools and Libraries Program*. **2.** Subscriber Line Charge or Service Line Charge. Synonymous with Customer Access Line Charge (CALC) and now known as End User Common Line Charge (EUCL). In the United States, an access charge approved by the Federal Communications Commission (FCC) and billed by the incumbent local exchange carrier (ILEC), the SLC is intended to compensate the ILEC for the costs of connecting a call to competitive local exchange carrier (CLEC) or an interexchange carrier (IXC) through local exchange facilities (the local loop, central office, and associated equipment), maintaining the equal access database, and other related costs. The SLC applies to all ILEC local loops, but varies by type of facility (e.g., residence line, business line, and CO trunk). If the access circuit is digital in nature, a Digital Port Line Charge (DPLC) also applies. See also *CLEC*, *DPLC*, *equal access*, *EUCL*, *FCC*, *ILEC*, and *IXC*.

SLC-96 (Subscriber Line Carrier-96) A digital loop carrier (DLC) system introduced by Western Electric (now Alcatel-Lucent), in 1979. SLC-96 comprises 1–4 T1s, or 1 T2, from the central office (CO) to a remote neighborhood node, which serves as a remote CO line shelf and time division multiplexer (TDM mux). On the other side of the node terminate 96 voice-grade analog circuits from individual customer premises. The SLC-96 DLC thereby supports 96 voice grade, two-wire local loops over 1–4 four-wire (i.e., 4 two-pair) circuits, which results in a savings of at least 88 pairs (96 – 8 = 88) from the CO to the node. *Note:* For fewer than 4 T1s, not all subscriber lines can be active at once. See also *carrier*, *DLC*, *four-wire circuit*, *local loop*, *node*, *multiplexer*, *TDM*, *two-wire circuit*, and *voice grade*.

SLIP (Serial Line Internet Protocol) A Link Layer packet framing protocol that defines a sequence of characters that frame IP packets on a serial line. SLIP provides no addressing, packet type identification, error control, or compression mechanisms. SLIP is defined in RFC 1055 as a nonstandard protocol for transmission of IP datagrams, in the formal sense, although it has become a de facto standard. RFC 1144 defines Compressed SLIP (CSLIP), a method for improving TCP/IP performance over low-speed (300 bps to 19.2 kbps) serial lines by compressing the TCP/IP headers. Point-to-Point Protocol (PPP) performs the same basic functions as SLIP, plus compression and other functions. See also *compression*, *datagram*, *error control*, *header*, *IP*, *Link Layer*, *packet*, *PPP*, *protocol*, *serial communications*, *standard*, and *TCP/IP*.

slotted Aloha From the Hawaiian *aloha*, meaning *hello* and *goodbye*. A protocol developed at the University of Hawaii in the early 1970s as an improvement on the pure Aloha protocol. Slotted Aloha improves contention management through the use of beaconing, in which a receiver transmits signals at precise intervals, indicating to each source when the channel is clear to send a frame of data. Slotted Aloha essentially advertises the availability of a time slot in a channel. See also *Aloha*, *AlohaNet*, *channel*, *clear to send*, and *time slot*.

small and medium enterprise (SME) See *SME*.

small business enterprise (SBE) See *SBE*.

Small Computer System Interface (SCSI) See *SCSI*.

small office/home office (SOHO) See *SOHO*.

SMDR (Station Message Detail Record) A detailed record of incoming and outgoing calling activity generated by a Centrex system, key telephone system (KTS), or PBX. These records provide a call accounting system with the data necessary to generate reports for purposes such as client billback, cost allocation, cost control, and fraud detection. SMDR data typically includes the originating or target telephone number and station number, the time of day the call originated and terminated or the total elapsed time of the call, the access code used, the line or trunk employed, and the account code, if any. Synonymous with call detail record (CDR).

SMDS (Switched Multimegabit Data Service) An offshoot of the Distributed Queue Dual Bus (DQDB) technology defined by the IEEE 802.6 standard for metropolitan area networks (MANs), as a means of extending the reach of a local area network (LAN) across a metropolitan area. SMDS was developed by Bellcore (now Telcordia Technologies), at the request of the Regional Bell Operating Companies (RBOCs), which at the time were confined to operating within the boundaries of relatively small Local Access and Transport Areas (LATAs). SMDS was originally described by Bellcore as a high-speed, connectionless, public, packet-switching service that extends LAN-like performance beyond the subscriber's premises, across a metropolitan or wide area. SMDS is a MAN network service based on cell-switching technology. Generally delivered over an SDH/SONET ring, SMDS has a maximum effective serving radius of approximately 30 miles (50 Km). SMDS enjoyed limited, short-lived success in the United States through deployment by most of the RBOCs and GTE. Under the name Connectionless Broadband Data Service (CBDS), SMDS also enjoyed moderate success in Western Europe, where the nations tend to be small in geographic terms and where the population density of large businesses is high in the major metropolitan areas. In the mid-1990s, SMDS was overwhelmed by frame relay and various metropolitan Ethernet offerings, and is now considered obsolete. IEEE 802.6 has been withdrawn. See also *802.6*, *Bellcore*, *cell switch*, *connectionless*, *DQDB*, *Ethernet*, *frame relay*, *IEEE*, *LATA*, *MAN*, *packet switch*, *RBOC*, *SDH*, and *SONET*.

SME **1.** Small and Medium Enterprise. The European Union (EU) defines a small enterprise (i.e., commercial organization, or business) as one with fewer than 50 employees and a medium enterprise as one with fewer than 250 employees. There are no standard United States definitions for small or medium enterprises, although a small business generally is described as one with fewer than 100 employees. An enterprise with fewer than 10 employees is generally classified as a small office/home office (SOHO) or a microbusiness. A large business enterprise (LBE) is larger than a medium enterprise. See also *LBE* and *SOHO*. **2.** (Pronounced *smee*) Subject Matter Expert. An expert on a given subject, generally of a technical or scientific nature.

Smell-O-Vision A system that released odors during the projection of a motion picture, thereby allowing the audience to not only see the film but also to smell it, so to speak. Created by Hans Laube, Smell-O-Vision was used in only one film, *Scent of Mystery* (1960), produced by Mike Todd, Jr. On cue, the system released odors (e.g., garlic, pipe smoke, and shoeshine wax) from small tubes hidden under the audience seats. The result apparently was awful, as the odors did not reach all members of the audience at the same time and at the same strength. The odors also lingered and mixed with other odors to create unwanted and unpleasant combinations. The concept of a *smellie* was reintroduced briefly by filmmaker John Waters in the film *Polyester* (1980). Waters used scratch and sniff cards in his Odorama process. Smell-O-Vision was revived by some Japanese theaters for the movie *The New World* (2005), with the scent generators controlled and synchronized through the Internet by NTT Communications Corp., the Japanese telecommunications carrier. Various systems have been proposed to work in conjunction with *television* but have encountered technical difficulties thus far. While Smell-O-Vision can be considered a form of multimedia, odor as a communications medium thankfully seems to be an evolutionary dead end. See also *multimedia*.

SMF (Single-Mode Fiber) A type of glass optical fiber (GOF) with a very thin inner core that forces light rays to propagate along a single mode, or physical path, through the fiber. SMF has a core diameter of 5–10 microns, as illustrated in Figure S-4. As light sources, SMF transmission systems employ highly sophisticated laser diodes, which can couple efficiently to such a thin inner core and can fire collimated optical signals in the 1310 nm and 1550 nm windows. As SMF supports only a single mode, graded-index fiber is unnecessary, and only step-index fiber is used. The single mode obviates the issues of modal dispersion and pulse dispersion that plague multimode fiber (MMF). Optical signals do, however, suffer over long distances from chromatic dispersion and material dispersion. SMF is used in long haul, high speed fiber optic transmission systems (FOTS) such as those used in carrier backbone networks. See also *chromatic dispersion*, *collimation*, *coupling efficiency*, *FOTS*, *GOF*, *graded-index fiber*, *laser diode*, *long haul circuit*, *material dispersion*, *MMF*, *pulse dispersion*, and *step-index fiber*.

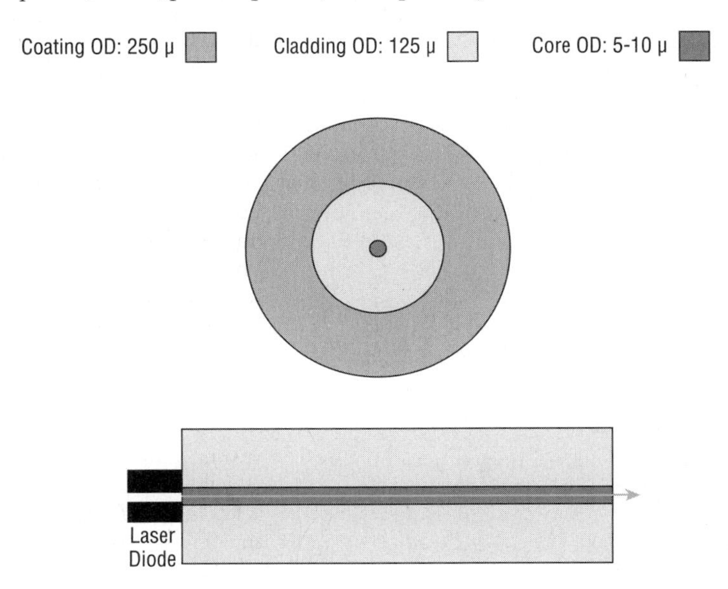

Figure S-4

S/MIME (Secure Multipurpose Internet Mail Extension) A version of MIME that supports encryption. See *encryption* and *MIME*.

SMR (Specialized Mobile Radio) Also known as *Trunk Mobile Radio* (TMR), SMR was introduced in the 1960s. Mobile radio has its roots early in the century, however. The United States Army Signal Corps mounted early spark transmitters in vehicles in 1904 and experimented with air-to-ground communications in 1908. The Detroit, Michigan (United States), Police Department placed the first experimental one-way mobile radio dispatch system into service in 1921, operating with the call letters KOP. The first two-way mobile system was installed by the Bayonne, New Jersey (United States), Police Department in the early 1930s. In 1946, AT&T was granted the first commercial license for two-way mobile service. That first system in St. Louis, Missouri (United States), was based on frequency modulation (FM) and employed a centralized antenna with a range of 50 miles. The system served not only to interconnect mobile phones, but also provided connection to the public switched telephone network (PSTN). The early systems were relatively inexpensive and proved very convenient, and were soon oversubscribed. In 1976, for example, service in the New York metropolitan area consisted of 20 channels supporting 543 subscribers out of a total population of approximately 20 million. There was a waiting list of approximately 3,700. In the 1960s, SMR was made commercially available, marketed as Improved Mobile Telephone Service (IMTS). SMR made better use of FM bandwidth through narrowband communications involving smaller frequency channels and enabled users to manually search multiple frequency channels. Shortly thereafter, intelligent mobile sets were developed that searched channels automatically.

SMR involves omni-directional transmit/receive antennas placed on a radio tower positioned on the highest possible point in a geographic area for maximum line of sight (LOS) and transmitting at the maximum allowable power level for maximum geographic coverage in what is termed a macrocell configuration. Although some SMR systems support full duplex (FDX) communications, many are only half-duplex (HDX), which supports transmission in only one direction at a time and, therefore, requires that the parties take turns talking. The talker must depress a key or button on the microphone to talk and must release it to listen. This procedure is commonly known as the *push-to-talk* (PTT) protocol. SMR largely has been supplanted by cellular service offerings, although it remains widely used in dispatch and fleet applications such as police, fire, and emergency vehicles, as well as taxi and utility fleets.

Enhanced Switched Mobile Radio (ESMR) is a technology developed by Nextel Communications and Geotek Communications for the development of a voice and data, cellular-like network using legacy SMR networks operating in the 800 MHz and 900 MHz range. Nextel acquired and linked a large number of SMR networks throughout the United States, in February 1999 acquired the 191 900-MHz licenses of the bankrupt Geotek, and later added the 1.5 GHz band to the mix. Time division multiple access (TDMA) divides each frequency channel into multiple time slots to support multiple conversations. ESMR also supports call handoff so mobile users can maintain connectivity as they travel from cell to cell. The network offers data throughput of 7.2 kbps, with coverage, including most major metropolitan areas in the United States. Nextel terminal equipment supports integrated voice, data, paging, and Internet access. See also *antenna, band, channel, FDX, FM, frequency, HDX, Internet, LOS, macrocell, paging system, PSTN, PTT, TDMA*, and *time slot*.

SMS **1.** Service Management System. In the advanced intelligent network (AIN) architecture, a network control interface that enables the service provider to vary the parameters of the AIN services. Under certain circumstances, the end user organization may be provided access to a partition of the SMS. See also *AIN*. **2.** Short Message Service. A text messaging service initially defined in the standards for Global System for Mobile Communications (GSM) and now available on most digital cellular telephone networks and some paging systems. SMS was originally designed to support one-way information transfer for applications such as weather reports, sports scores, traffic reports, and stock quotes, as well as short e-mail-like messages, which may be entered through the service provider's website. Most service providers also allow cellular users to receive and respond to e-mail messages, using the cell phone keypad for message input. SMS is a store-and-forward messaging technology that generally includes a chat option that operates in near-realtime mode, much like instant messaging (IM), and many IM systems support mobile communications via interfaces to SMS systems. SMS messages use the same Simple Mail Transfer Protocol (SMTP) specified in the TCP/IP protocol suite for Internet e-mail. SMS messages between the mobile phone, or other device, and the centralized message center, known as a Short Message Service Center (SMSC), are in packet format. In packet-based Code Division Multiple Access (CDMA) cellular networks, SMS packet data transport is relatively straightforward. In Time Division Multiple Access (TDMA) and GSM networks, SMS packet transport is over the signaling channel via Signaling System 7 (SS7). *Note:* The SS7 protocol limits the packet size to 140 octets, which translates to 160 ASCII 7-bit characters ($140 \times 8 = 1120 / 7 = 160$). Where other coding schemes are employed, the number of characters per packet varies. For example, a double-byte coding scheme such as those (e.g., Unicode-16) used in support of complex alphabets (e.g., Chinese and Japanese), limits the packet size to 70 characters. SMS has evolved into Multimedia Messaging Service (MMS), which has been widely available in Japan and select other countries since 2001. MMS allows the user to create multimedia mail comprising text messages with image, audio, and video attachments. See also *ASCII, CDMA, GSM, IM, MMS, nrt, packet, paging system, SMTP, SS7, store-and-forward, TCP/IP,* and *Unicode-16.*

SMSC (Short Message Service Center) The centralized message center in an SMS system. See also *SMS.*

SMTP (Simple Mail Transfer Protocol) An Application Layer protocol of the TCP/IP protocol suite, SMTP was defined by the IETF in RFC 821 (1982) for passing e-mail to and through the Internet between clients and servers. SMTP operates over the connectionless User Datagram Protocol (UDP) and supports text-oriented e-mail between any two devices that support Message Handling Service (MHS). SMTP was developed to support the 7-bit ASCII code, which accommodates plain text, including letters, numbers, punctuation marks, and a reasonable set of control characters. ASCII, however, does not support the rich text formatting (e.g., italics, bold, and color, other than the default black in which this document is printed) supported by most contemporary word processing programs. This 7-bit format also prevents the transmission of 8-bit binary data found in executable files. The Multipurpose Internet Mail Extensions (MIME) to SMTP are used to support binary files, and Secure MIME (S/MIME) supports encryption. See also *Application Layer, ASCII, binary, client/server, connectionless, e-mail, IETF, Internet, MHS, MIME, plain text, POP3, protocol suite, RFC, rich text, S/MIME, TCP/IP,* and *UDP.*

SNA (Systems Network Architecture) A five-level design architecture developed by IBM in 1974, SNA comprises software and hardware interfaces that permit various IBM systems and software to communicate. SNA that has grown into a seven-layer model that more closely corresponds to the OSI Reference Model, an internationally recognized open model, although the two remain incompatible. SNA includes network nodes, physical units, and logical units , defined as follows:

- **Nodes:** Physical devices including computers, communications processors (e.g., FEPs), terminal controllers, and terminals.
- **Physical units (PUs):** Physical devices that manage the communications hardware and software, participating in the controlling and routing of network communications.
- **Logical units (LUs):** Programs that manage communications software for communications with end users. A logical unit session is an end-to-end communication between an end-user terminal and the originating application residing in the host.

See also *hardware, network architecture, OSI Reference Model,* and *software.*

SNAFU (Situation Normal All Fouled Up) Referring to a normal situation gone awry, the term is generally attributed to U.S. military slang circa WWII. In contemporary usage, the spelled-out phrase is a bit more colorful and much less polite. See also *FUBAR.*

SNMP (Simple Network Management Protocol) As Application Layer protocol of the TCP/IP protocol suite, SNMP supports the exchange of network management information between host computers, typically including one or more centralized network management consoles that manage larger numbers of network elements (NEs) in real time. Defined in RFC 1157 (1990), SNMP operates over the User Datagram Protocol (UDP), thereby avoiding the overhead associated with the Transmission Control Protocol (TCP). There are three versions: SNMPv1, v2, and v3. See also *Application Layer, host, NE, network management, overhead, protocol, protocol suite, real time, TCP, TCP/IP,* and *UDP.*

SNR (Signal-to-Noise Ratio) The ratio of the power of the wanted signal power to the power of the unwanted noise at a given point in a given system at a given time. SNR is expressed in decibels (dB). Impulse noise generally is measured in peak-signal to peak-noise ratio, and random noise in root-mean-square (RMS) signal to root-mean-square noise. See also *dB, noise, power,* and *signal.*

SOA (Service-Oriented Architecture) An imprecise term referring to a means for integrating a mesh of collaborating services on a platform-independent basis through middleware that treats each disparate application as a service.

social engineering The use of deceptive or fraudulent means that rely on expected human behavior, often used to gain access to information such as a computer access number, password, or user ID. See also *pharming, phishing,* and *pretexting.*

socket **1.** An operating system (OS) abstraction that permits application programs to access communications protocols automatically. Bolt Beranek and Newman (now BBN Technologies) developed the concept in conjunction with the company's early work on TCP/IP. See also *application, OS, program, protocol,* and *TCP/IP.* **2.** Comprising a node address and a port number, a socket is an identifier for a service on a node. In combination a socket on an originating node and a socket on a destination node establish an application session. An Internet Protocol (IP) addressable client and server, for example, might engage in an e-mail transfer using SMTP (Simple Mail Transfer Protocol), which is well-known port 25. See also *application, client, IP, node, port, server, session, SMTP, SOCKS,* and *well-known port.*

SOCKS (SOCKetS) An internal development name that remained after general release. An IETF standard (RFC 1928, 1929, and 1961) networking proxy protocol that enables clients on one side of a firewall server to gain full access to clients on the other side without a direct Internet Protocol (IP) connection. The SOCKS server authenticates and authorizes requests from one side of the server, establishes a proxy connection to the other side, and relays data between the two through the use of secure sockets negotiated between client and server over a virtual circuit on a session-by-session basis. This process enables clients behind the firewall to gain full access to the Internet, but prevents unauthorized access from the Internet to the internal hosts. SOCKSv5, the current version, supports a variety of authentication methods and supports User Datagram Protocol (UDP) proxy. SOCKS runs at the Session Layer (Layer 5) of the OSI Reference Model. SOCKSv5 is a cross-platform technique, working across multiple operating systems (OSs) and browsers. SOCKSv5 also interoperates on top of IPv4, IPsec, PPTP, L2TP, and other lower-level protocols. See also *authentication, authorization, browser, client, firewall, host, IETF, Internet, IP, IPsec, IPv4, IPv6, L2TP, OS, OSI Reference Model, PPTP, protocol, proxy, server, session, Session Layer, socket, tunneling, UDP,* and *virtual circuit.*

soft copy Data in computer memory, on a computer disk, either internal or external, a magnetic tape, or some other temporary or semi-permanent medium, and perhaps viewed on a computer monitor. Information in soft copy form can easily be changed. See also *hard copy.*

soft handoff In cellular radio networks, a handoff process in which the connection is established by the second base station (BS) before being broken by the first as a mobile station (MS) moves out of the range of the first and into the range of the second. This technique is also known as *make and break.* See also *BS, cellular radio, handoff, hard handoff, MS,* and *radio.*

softkey A programmable button on the keypad of a telephone set or other device. A softkey commonly is display-based and context sensitive. See also *keypad.*

softphone A software-based telephone comprising a desktop, laptop, or tablet computer equipped with a microphone, a speaker, and software that enables it to emulate a hardphone, i.e., conventional telephone set. See also *computer* and *software.*

softswitch A switch that offloads call processing functions (e.g., signaling and call control) to industry-standard server hardware, essentially decomposing the call control logic from the switching platform. This function allows the placement of the call control logic at some geographically centralized location from which it can control multiple switching platforms, as well as the separate devices that provide for interconnection of circuit and packet networks. In addition to controlling a protocol conversion function, softswitches can support multiple QoS (Quality of Service) and GoS (Grade of Service) mechanisms and levels. Softswitches tend to be more flexible, less expensive, and more compact than traditional hard-coded switches. See also *GoS*, *hardware*, *Media Gateway Controller (MGC)*, *QoS*, *protocol converter*, *signaling and control*, *software*, and *switch.*

software The programs and routines for a computer system. System software includes programs and routines required to run the computer. Application software includes programs that enable users to perform tasks that use computer resources. Synonymous with software program. See also *firmware*, *grayware*, *hardware*, and *freeware.*

software-defined data network (SDDN) Synonymous with SDN and virtual private network (VPN). See *VPN.*

software-defined network (SDN) Synonymous with SDDN and virtual private network (VPN). See *VPN.*

software-defined radio (SDR) See *SDR.*

software program See *software.*

SOH (Section OverHead) In a SONET or SDH frame, overhead of nine octets dedicated to framing, span performance, error monitoring, and the transport of status, messages, and alarm indications for the maintenance of links connecting section terminating equipment (STE), which can be repeaters, add/drop multiplexers, or anything else that attaches to either end of a fiber link. The repeaters can be standalone or built into switches, such as digital cross connect systems (DCCSs). SOH and Line Overhead (LOH) comprise Transport Overhead (TOH). See also *ADM*, *frame*, *link*, *LOH*, *octet*, *SDH*, *SONET*, and *TOH.*

SOHO (Small Office/Home Office) An enterprise (i.e., for-profit commercial business) with fewer than 10 employees. Synonymous with microbusiness. See also *SME.*

sonar (sound navigation and ranging) A sonic device used for the detection and location of underwater objects. Active sonar emits sound waves and measures the characteristics of the waves reflected from objects. Passive sonar simply measures the characteristics of sound waves emitted by objects. Sonar determines the range and position of such objects through the use of the Doppler effect, radial component of velocity measurement, and triangulation. See also *ping* and *radar.*

SONET (Synchronous Optical NETwork) A set of North American standards for broadband digital transmission over single-mode fiber (SMF) optic transmission systems (FOTS), SONET grew out of the SYNTRAN (SYNchronous TRANsmission) standard developed at Bellcore. SONET was initially developed in 1984 and finally standardized by the American National Standards Institute (ANSI) in 1988. Also in 1988, SONET was internationalized by the ITU-T as Synchronous Digital Hierarchy (SDH), which differs primarily with respect to low-level line rates and some terminology. SONET/SDH essentially is a broadband optical version of T-carrier, defined in multiples of T3 bandwidth plus overhead for additional signaling and control functions. Primarily intended for carrier backbone implementation, SONET/SDH specifies a network-to-network interface (NNI), also known as *network node interface*, that supports the interconnection of national and regional networks into a cohesive global network. The user network interface (UNI) provides a standard basis for connection from the user premises to a SONET/SDH local loop.

Although SONET is a broadband optical specification, the signal originates in electrical format as a T3 signal to which additional signaling and control overhead is added to for the Synchronous Transport Signal level-N (STS-N). The term Optical Carrier-N (OC-N) applies once the electrical signal is converted to optical format. The SONET hierarchy begins with is OC-1 at a signaling rate of 51.84 Mbps, which is the equivalent of a T3 (44.736 Mbps) plus overhead, and currently tops out at the OC-768 nominal rate of 40 Gbps, which is the equivalent of 768 T3s plus overhead. Table S-2 details the SONET/SDH hierarchy.

Table S-2: SONET/SDH Signal Hierarchy

Optical Carrier (OC) Level	*SONET STS Level*	*SDH STM Level*	*Signaling Rate*	*Equivalent DS-3 Channels*	*Equivalent DS-0 Channels*
OC-1	STS-1		51.84 Mbps	1	672
OC-2[1]	STS-2		103.68 Mbps	2	1,344
OC-3[2]	STS-3	STM-1	155.52 Mbps	3	2,016
OC-4	STS-4	STM-3	207.36 Mbps	4	2,688
OC-9[1]	STS-9	STM-3	466.56 Mbps	9	6,048
OC-12	STS-12	STM-4	622.08 Mbps	12	8,064
OC-18[1]	STS-18	STM-6	933.12 Mbps	18	12,096
OC-24	STS-24	STM-8	1.24416 Gbps	24	16,128
OC-36[1]	STS-36	STM-12	1.86624 Gbps	36	24,192
OC-48	STS-48	STM-16	2.48832 Gbps	48	32,256
OC-96[1]	STS-96	STM-32	4.976 Gbps	96	64,512
OC-192	STS-192	STM-64	9.953 Gbps	192	129,024
OC-768	STS-768	STM-256	39.813 Gbps	768	516,096
OC-1536[3]	STS-1536	STM-512	79.626 Gbps	1536	1,032,192
OC-3072[3]	STS-3072	STM-1024	159.252 Gbps	3072	2,064,384

[1] OC-2, OC-9, OC-18, OC-36, and OC-96 are considered to be orphaned rates.
[2] OC-3 was defined by the CCITT as the basic transport rate for B-ISDN.
[3] This level is not fully defined.

The STS-1 frames are organized into 9 rows of 90 octets transmitted every 125µs, which is the requirement for voice encoded using the standard pulse code modulation (PCM) technique employed in the public switched telephone network (PSTN). The STS-1 frame, as illustrated in Figure S-5, comprises a Synchronous Payload Envelope (SPE) of 783 octets, plus 27 octets of overhead comprising Section Overhead (SOH), Line Overhead (LOH), and Path Overhead (POH). The payload in each frame comprises one or more Virtual Tributaries (VTs) organized into Virtual Tributary Groups (VTGs). A process known as *concatenation* allows multiple STS-1 frames to be multiplexed, switched, and transported over the network as a single entity.

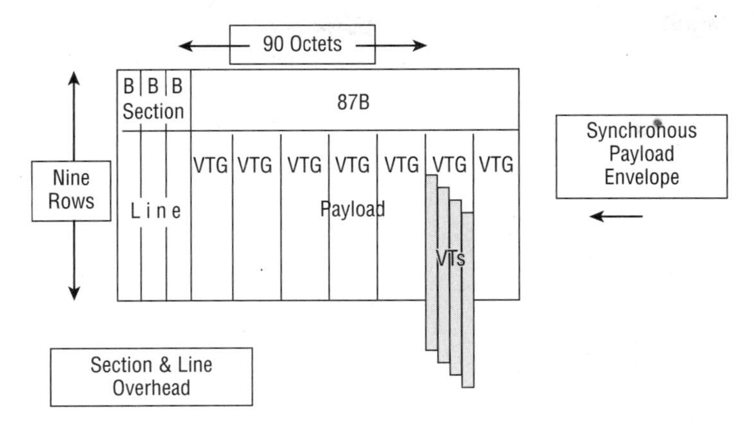

Figure S-5

SDH and SONET standards specify two physical configuration options. A path-switched ring employs a dual counter-rotating ring, with both fibers active. One ring transmits in the clockwise direction and the other in the counter-clockwise direction. This approach offers zero downtime in the event of the failure of a fiber. A line-switched ring features one active and one inactive fiber. In the event of the catastrophic failure of a node or fiber, a line-switched ring offers sub-50 millisecond restoral times for rings up to 1,200 kilometers in route distance. Figure S-6 illustrates a hierarchy of local, metropolitan, and backbone rings operating at example OC-N rates.

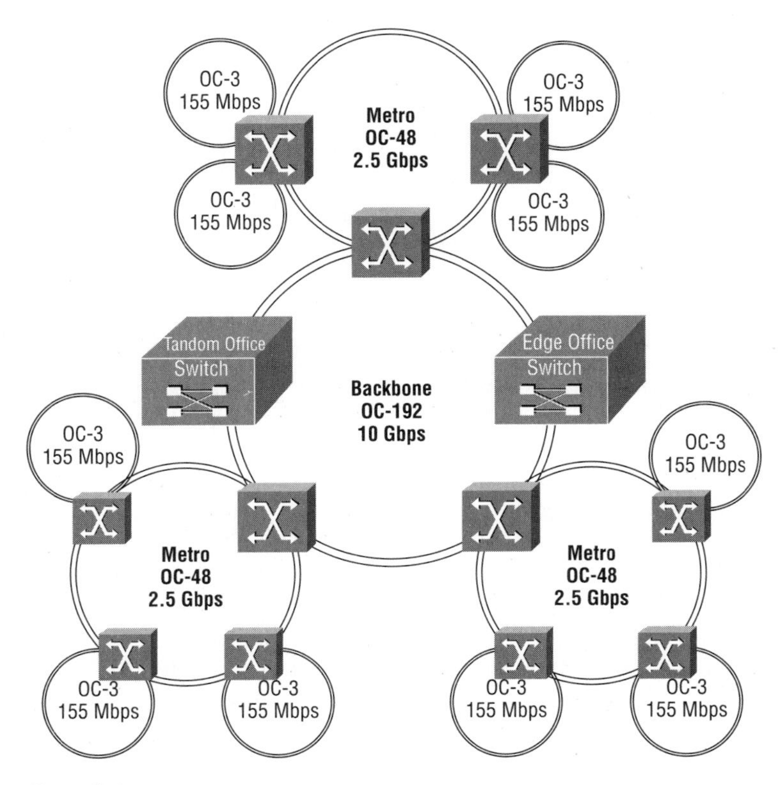

Figure S-6

See also *ANSI, bandwidth, Bellcore, B-ISDN, broadband, concatenation, FOTS, ITU-T, line-switched ring, local loop, LOH, overhead, path-switched ring, PCM, POH, PSTN, signaling and control, signaling rate, SMF, SOH, SPE, T3, T-carrier, VT,* and *VTG.*

source Also known as a *sender* or a *transmitter,* a source is a device that originates, or generates, an information transfer to one or more receivers. See also *transceiver, transmitter,* and *receiver.*

source routing protocol (SRP) See *SRP.*

source routing transparent (SRT) See *SRT.*

space division multiplexing A method by which metallic, radio, or optical transmission media are physically separated by insulation, waveguides, or space in order to maintain channel separations. Within each physically distinct channel, multiple channels can be derived through frequency, time, or wavelength division multiplexing. Some Passive Optical Network (PON) implementations employ space division multiplexing, with the downstream transmissions occurring over one fiber of a duplex fiber optic cable and upstream transmission occurring over the other fiber. See also *FDM, TDM,* and *WDM.*

space division switch A type of switching system in which all of the links, contacts, matrix crosspoints, and switches are physically separated. Early switchboards, cordboards, and panel switches are examples.

spam Unsolicited e-mail, or junk mail. The term refers to Hormel's ever-popular canned meat product made of processed pork and ham. The analogy supposedly is that junk e-mail is broadcast all over, just as the Hormel meat product spatters when hurled against a solid object with sufficient force. No one likes spam, the junk mail. Lots of people, on the other hand, love SPAM®, the lunch meat. SPAM was introduced in 1937 as the Miracle Meat. The name, an initialism of SPiced hAM, was submitted in a contest, with the winner receiving the sum of US$100.00. SPAM became a staple during WWII (World War II), both in the military and at home, at least partially because it required no refrigeration and was not rationed. There is even a SPAM Web site, fan club, gift catalog, museum, and three (count 'em, 3) promotional SPAMMOBILEs™ that tour the United States, making stops at grocery stores, festivals, sporting events, and other venues. Fried SPAM is an essential dish in any local food menu in the Hawaiian Islands, Guam, and the Commonwealth of the Northern Mariana Islands (CNMI), where it gained popularity as a result of the U.S. military presence. The junk mail type of spam is outlawed in the United States under the federal CAN-SPAM Act. There are no federal, state, or local laws against SPAM, the canned lunch meat. See also *CAN-SPAM Act, e-mail,* and *SPIM.*

spam over instant messaging (SPIM) See *SPIM.*

span See *T-span.*

spanning A storage technology that involves the use of a number of external physical hard drives organized into a single logical drive. See *JBOD* and *RAID.*

spanning tree protocol (STP) See *STP.*

sparse network A network comprising dumb switches supported by centralized intelligence with connectivity between distributed switches and centralized logic provided over high-speed digital circuits. The intelligent network (IN) and advanced intelligent network (AIN) are sparse networks. See also *AIN, IN,* and *network.*

spatial diversity The use of multiple radio antennas to improve signal integrity. Microwave transmission systems sometimes employ multiple spatial diverse receive antennas, vertically separated on a tower, for example. As the likelihood is that signal will not suffer the same level of attenuation as it disperses slightly and propagates along slightly different paths, the receiver with the strongest signal assumes control of the transmission. Wireless LAN (WLAN) systems based on the 802.11n specification employ multiple transmit and receive antennas in a technique known as multiple-input multiple-output (MIMO). Code division multiple access (CDMA) systems use a similar approach involving rake receivers with multiple fingers. See also *802.11n*, *antenna*, *attenuation*, *CDMA*, *dispersion*, *MIMO*, *propagation*, *rake receiver*, and *time diversity*.

speaker A device containing a transducer that converts electrical signals (electric current) into sound waves (acoustic energy) for the production of sound. See also *acoustics*, *current*, *microphone*, and *transducer*.

SPC (Stored Program Control) In reference to a common control unit consisting of one or more microprocessors operating under a stored program. See also *common control*.

SPE (Synchronous Payload Envelope) The portion of a SONET or SDH frame that carries the user payload data. The SPE consists of 783 octets in 87 columns and 9 rows. See also *frame*, *payload*, *SDH*, and *SONET*.

specialized common carrier (SCC) See *SCC*.

Specialized Mobile Radio (SMR) See *SMR*.

special resource function (SRF) See *SRF*.

specification A description of something that is sufficiently detailed to provide someone with all the information necessary to manufacture it. See also *standard*.

speckle noise Synonymous with modal noise. The cause of speckling in fiber optic transmission systems (FOTS). See *modal noise* and *speckling*.

speckling Graininess that appears in a degraded optical signal caused when perfectly coherent light waves travel slightly different physical paths, or modes, and interfere with each other. See also *coherence*.

spectral efficiency In reference to the extent to which a transmission system uses the available or assigned frequency spectrum. A more efficient system uses bandwidth more efficiently by optimizing coding, power, etc. A theoretically completely efficient system uses all bandwidth, wasting none. See also *bandwidth*, *frequency*, and *spectrum*.

spectral width The range of frequencies or wavelengths emitted by a transmitter and surrounding the center frequency or wavelength at a power level equal to half the maximum power level. The ITU-T has defined a number of wavelength bands, or windows, for use in standards-based fiber optics transmission systems (FOTS). As fairly crude light sources, light-emitting diodes (LEDs) emit signals of the greatest spectral width. More sophisticated diode lasers emit very narrowly defined signals that may be only 1 nm wide, or less. The more narrowly defined the spectral width, the tighter the windows can be packed and the greater the spectral efficiency. Some radio systems are similarly concerned with spectral efficiency, as they must pack multiple channels into a relatively narrow radio frequency (RF) band. Bluetooth and DECT both employ Gaussian frequency-shift keying (GFSK) as a means of both signal modulation and pulse shaping. See also *Bluetooth*, *channel*, *DECT*, *DFB laser*, *Fabry-Perot laser*, *FOTS*, *GFSK*, *laser diode*, *LED*, *modulation*, *pulse shaping*, *RF*, and *window*.

spectrum Generally referring to frequency spectrum. See *electromagnetic spectrum*.

spectrum allocation See *spectrum management*.

spectrum management The designation of certain frequency bands in the electromagnetic spectrum in support of certain applications in order to avoid interference between various applications using the same, or overlapping, frequency ranges. See also *electromagnetic spectrum* and *frequency band*.

speech The communication of thoughts and feelings by spoken words.

speech recognition The capability of a computer to recognize and understand a human speech signal in order that it might act on spoken commands. Speech recognition software relies on highly complex mathematical algorithms to accomplish the process of converting human speech into a form that a computer can understand. Speech recognition allows a caller to interact with a computer system in natural conversational voice mode, i.e., by speaking to the machine just as you would speak to another person. Applications include data entry, menu navigation, document creation, and telephone dialing. (*Note:* Computers are programmed not to respond well to cursing, which pretty much makes speech recognition useless, in my opinion.) See also *IVR* and *speech-to-text*.

speech-to-text The process of converting speech input into digital text, based on speech recognition. See also *speech*, *speech recognition*, *text*, and *text-to-speech*.

Speech-to-Speech (STS) See *STS*.

speed of light See *Vp*.

SPID (Service Profile IDentification) In integrated services digital network (ISDN), a Network Layer identification for each terminal device associated an NT1, NT2, or B channel. The carrier assigns up to eight SPIDs per telephone number, which may be the 10-digit directory number (i.e., telephone number), perhaps with a prefix or suffix. See also *B channel*, *ISDN*, *NT1*, and *NT2*.

spike A voltage fluctuation of very short duration, but very high voltage. See also *surge* and *voltage*.

spim or SPIM (SPam over Instant Messaging) Unwanted instant messages, SPIM is much like e-mail spam, and can be a vehicle for viruses and other malware. Enterprise-level systems typically support strong security measures, including authentication and encryption. SPIM is covered under the same federal CAN-SPAM Act that outlaws e-mail spam in the United States. See also *authentication*, *CAN-SPAM Act*, *e-mail*, *encryption*, *malware*, *spam*, and *virus*.

splice A permanent or semi-permanent connection between two conductors or cables. A splice in a twisted pair cable is accomplished by mechanical means involving various types of crimps and crimping tools, and is permanent. A splice in a coaxial cable employs barrel connectors and is semi-permanent. A splice in a fiber optic cable can be either a semi-permanent mechanical splice or a permanent fusion splice. A mechanical splice involves a gel-filled plastic or metal crimp sleeve and special crimping tool. A fusion splice entails fusing the glass fibers together at high temperature. See also *fusion splice* and *mechanical splice*.

splice case A heavy-duty weatherproof case designed to protect a splice in a cable. A splice case commonly includes a splice tray for organizing the splices between the metallic or glass conductors. See also *splice* and *splice tray*.

splice tray A plastic or metal tray used to organize splices between metallic or glass conductors at a splice point where two cables are spliced, or joined together. In an outside plant (OSP) application, the splice tray typically is housed in a protective splice case. See also *splice* and *splice case*.

splitter **1.** A digital subscriber line (DSL) modem, also referred to as a DSL filter. A DSL splitter is a multiplexer that combines upstream voice, data, and sometimes video signals at the customer premises prior to transmission over a single local loop. Asymmetric digital subscriber line (ADSL), the most common consumer-oriented DSL variant, supports voice over an analog channel and high speed Internet access over a digital channel, so the DSL modem is a simple frequency division multiplexer (FDM mux). The DSL modem also typically acts as a gateway, converting Internet Protocol (IP) data packets in Ethernet frames into asynchronous transfer mode (ATM) format prior to upstream transmission. The splitter acts as a combiner for upstream purposes. The splitter lives up to its name with respect to downstream transmissions, serving to split (i.e., demultiplex) the voice and data channels and to convert the data channels from ATM to IP packets inside Ethernet frames. See also *ADSL*, *analog*, *ATM*, *channel*, *digital*, *downstream*, *DSL*, *Ethernet*, *FDM*, *frame*, *gateway*, *IP*, *local loop*, *multiplexer*, *packet*, and *upstream*. **2.** An electrical device that separates

DC power from data signals on the end of a PoE (Power over Ethernet) cable. See also *DC* and *PoE*. **3.** A passive optical device that splits an incident optical signal, or beam, into two or more beams with total power equal to or less than the incident beam. Splitters commonly are signal paths etched into a dielectric material. Splitters also can be made of glass optical fibers twisted together and fused. Passive optical network (PON) local loops use cascading optical splitters, each of which evenly splits the incoming signal into two outgoing signals, each of which is at approximately 50 percent of the power level of the incident signal. As a passive network, PON does not amplify the signal. The splitter illustrated in Figure S-7 comprises three 1:2 splits for a total split ratio of 1:8, with the outgoing signal over each distribution fiber having a power level equivalent to approximately 12.5 percent (⅛) that of the incoming signal. As each split approximately halves the signal power, there are limits to the number of signal splits that can be tolerated over a circuit of a given length. A PON splitter might have a split ratio of 1:2 (one split), 1:4 (two splits), 1:8 (three splits), 1:16 (four splits), 1:32 (five splits), 1:64 (six splits), or 1:128 (seven splits). See also *amplifier, dielectric, local loop, passive, PON, power,* and *signal.*

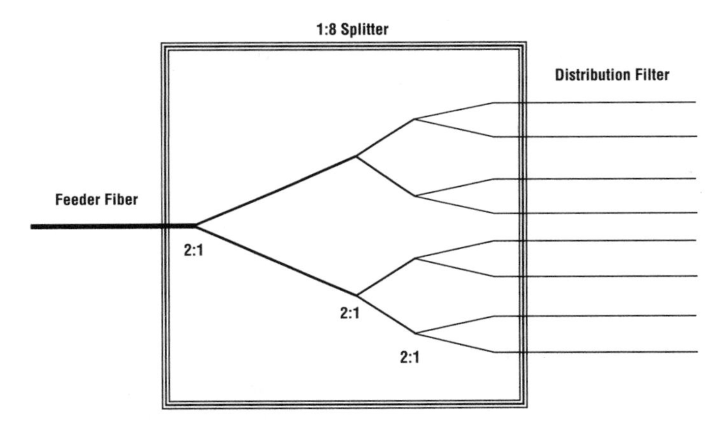

Figure S-7

splitterless ADSL (splitterless Asymmetric Digital Subscriber Line Lite) See *G.lite.*

spoofing **1.** The process of deception by which an individual or system alters its identity or creates additional identities, thereby causing another person or system to act incorrectly. **2.** The process of deception by which an unauthorized person causes a transmission or message to appear to come from an authorized user in order to gain privileged access to computer or network resources. IP spoofing, an integral element of many types of network attacks, involves creating TCP/IP packets that use false addresses, perhaps stolen from others. **3.** The process of deception by which a system alters its identity or creates additional identities, to impersonate another device in a communications session or transaction. Spoofing may be desirable, as when a router spoofs a fax protocol to ensure delivery over a connection with high latency. See also *fax spoofing*.

sponsored domain In the Internet Domain Naming System (DNS), a Top Level Domain (TLD) that has a sponsor representing the narrower community that is most affected by the TLD. Examples include .aero, sponsored by Société Internationale de Télécommunications Aéronautiques (SITA), and .jobs, sponsored through an alliance between Employ Media, the Society for Human Resource Management (SHRM), Verisign, and ICANN. See also *.aero, DNS, Internet, .jobs, TLD,* and *unsponsored domain.*

SPOT (Single Point Of Termination) In the context of digital subscriber line (DSL), a frame used where a competitive local exchange carrier (CLEC) or Internet service provider (ISP) is leasing dry copper pairs from the incumbent LEC (ILEC) for purposes of provisioning xDSL data services. The SPOT frame, and other hardware, is collocated in the ILEC central office (CO), generally in separately secured leased space. The ILEC cross-connects the individual leased circuits at the main distribution frame (MDF), and terminates them in the SPOT frame, where the CLEC or ISP connects them to the DSL access multiplexer (DSLAM). See also *CLEC, CO, cross-connect, dry copper pair, DSL, DSLAM, frame, ILEC, ISP, MDF,* and *xDSL.*

spot beam A tightly focused satellite radio beam designed to cover an area more limited than that of the entire satellite footprint. The tight focus improves signal power through the use of a parabolic antenna. See also *footprint* and *satellite.*

spread spectrum (SS) See *SS.*

spyware Software that collects personal information about users and their activities without their knowledge or consent. Spyware uses a number of techniques such as logging keystrokes, recording Internet Web browsing activities, and searching hard drives. Adware is a type of spyware that records such information and forwards it to an advertising agency or market research firm that later uses it to tailor pop-up ads for delivery to users without their knowledge or consent. See also *malware.*

SPX (Sequenced Packet eXchange) The Transport Layer (Layer 4) protocol in Novell NetWare for exchanging packets between LANs. Internetwork Packet Exchange (IPX) is the associated Network Layer (Layer 3) protocol in the IPX/SPX protocol suite. SPX is similar to the Transmission Control Protocol (TCP), but is not industry standard and has been overwhelmed by the TCP/IP protocol suite. See also *IPX/SPX, LAN, Network Layer, protocol, protocol suite, TCP,* and *Transport Layer.*

squared Referring to the configuration of a key telephone system (KTS), meaning that every key set is configured alike, with every outside line appearing on every set. Thereby, every station user can access every outside line for both incoming and outgoing calls and all feature presentations are consistent. In larger systems, the physical size of the telephone sets required to maintain the squaring convention would be impractical, but departmental subgroups often are squared. See also *call, KTS,* and *line.*

square wave A periodic wave that assumes one of two fixed, discrete values for equal lengths of time, with each value or group of values representing a digital bit. For example, a given signaling protocol might represent a one (1) bit as +3V (volts) and a zero (0) bit as –3V, with each bit measured at a precise moment in time for a specific duration known as a bit time. This is in stark contrast to an analog signal, which flows smoothly and continuously from zero voltage (0V) to its maximum positive voltage (+3V), back to zero (0V), to its maximum negative voltage (–3V), and back to zero (0V). Figure S-8 shows an analog waveform that varies in frequency and the digital representation of that waveform, with each set of low frequency signals depicted as a 0 bit and each set of high frequency signals as a 1 bit. See also *digital* and *sine wave.*

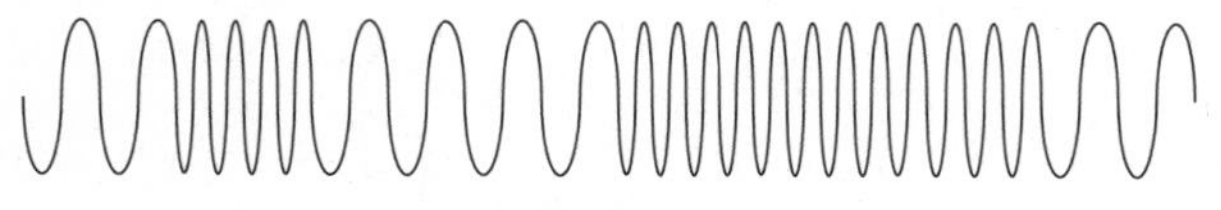

Analog Signal: Sine Wave

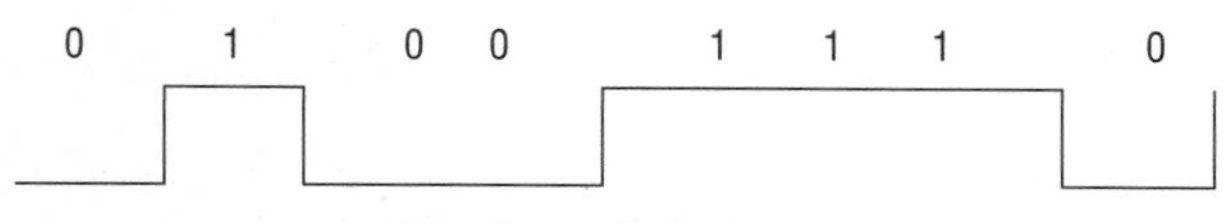

Digital Signal: Square Wave

Figure S-8

squatter **1.** Someone who illegally occupies another person's vacant house or building, or settles on another person's vacant land. **2.** A person who registers a domain name on the World Wide Web (WWW) that mimics the copyrighted, trademarked, or service-marked name of a company specifically for the purpose of selling it to the rightful owner. See also *domain name* and *WWW*.

SRDL (SubRate Digital Loop) A synonym for Dataphone Digital Service (DDS) operating at transmission rates less than 56 kbps. See also *DDS*.

SRF (Special Resource Function) In the advanced intelligent network (AIN) architecture, referring to intelligent peripherals (IPs) and the functions they perform. See also *AIN* and *IP*.

SRP (Source Routing Protocol) A bridge protocol developed for Token Ring local area networks (LANs) and used in other LANs, as well. In SRP, packets are programmed with specific routes, comprising lists of bridges based on considerations such as the physical location of the nodes and the capacity of the links involved. The maximum number of bridges hopped is 13. See also *bridge*, *hop*, *LAN*, *link*, *node*, *protocol*, *route*, and *Token Ring*.

SRS (Shared Registry System) A system of databases shared by Internet domain registrars for coordination of domain name assignment to ensure that duplicate names are not assigned. All domain names are maintained in mirrored databases on root Domain Name Servers (DNSs) distributed around the world. See also *DNS* and *root*.

SRT (Source Routing Transparent) As defined in the IEEE 802.1 standard, a bridge protocol that is essentially a combination of source routing protocol (SRP) and spanning tree protocol (STP). The SRT bridge can connect LANs or LAN segments by either method, using source routing with hosts that support it and transparent spanning tree bridging (STP) otherwise. See also *bridge*, *LAN*, *protocol*, *SRP*, and *STP*.

SS **1.** Spread Spectrum. A wideband radio transmission technology that spreads the bandwidth of the transmitted signal over a spectrum of radio frequencies that is much wider than that required to support the native narrowband transmission. Thereby, multiple transmissions can simultaneously use the entire system wideband, rather than just individual time slots or frequency channels. Direct sequence spread spectrum (DSSS) is a radio technique in which the narrowband signal is spread across a wider carrier frequency band. Frequency hopping spread spectrum (FHSS) transmits short bursts of data over a range of frequency channels within the wideband carrier, with the transmitter and receiver hopping from one frequency to another in a carefully choreographed hop sequence under the control of the centralized base station. Hedy Lamarr, the famous actress and dancer of pre-war (WWII) fame, and George Antheil, a composer of music, invented spread spectrum and were granted a U.S. patent in 1942. As the story goes, Lamarr developed spread spectrum radio in order to remotely synchronize multiple player pianos in radio-controlled piano concerts. Spread spectrum is the basis of code division multiple access (CDMA), which is employed in cellular radio and other networks. See also *Antheil, George*; *bandwidth*; *carrier*; *CDMA*; *channel*; *DSSS*; *FHSS*; *frequency*; *Lamarr, Hedy*; *narrowband*; *radio*; *signal*; *spectrum*; *time slot*; and *wideband*. **2.** Single Sideband. See *SSB*.

SS7 (Signaling System 7) Also known as Common Channel Signaling System 7 (CCS7), SS7 protocols were developed by AT&T in 1975 for use in the public switched telephone network (PSTN). The protocols were adopted and enhanced by the ITU-T in 1981, and are defined in the Q.7XX series of Recommendations. SS7 in a common channel signaling (CCS) system that uses channels separate from the communications channels for signaling and control purposes. In the United States, at least, SS7 operates not only over separate channels, but generally over a physically distinct network or subnetwork. SS7 significantly speeds call setup and call completion processes, allows access to databases that enable toll-free calling and number portability, and supports the CLASS (Customer Local Access Signaling Services) services such as caller ID, name ID, selective ringing (or priority ringing), selective call forwarding, call block (or call screen), repeat dial, call trace, and automatic call-back (call return). SS7 is fully deployed in all major TDM-based interexchange carrier (IXC) networks. While SS7 is largely deployed in the major incumbent local exchange carrier (ILEC) networks in developed countries, older central office (CO) switches do not support it. See also all of the features listed above. See also *CCS*, *CLASS*, *CO*, *ILEC*, *IXC*, and *signaling and control.*

SSA (Serial Storage Architecture) An IBM interface specification for a serial transport protocol based on a ring topology and operating in full duplex (FDX) at a maximum of 20 MBps per channel, with as many as two channels per cable. SSA maps into the pre-existing Small Computer System Interface (SCSI) and SSA devices are SCSI devices. The Transport Layer protocol is non-return-to-zero (NRZ) and utilizes 8B/10B encoding. See also *8B/10B*, *FDX*, *NRZ*, *protocol*, *ring topology*, *SCSI*, *serial*, and *Transport Layer.*

SSB (Single SideBand) The process of amplitude modulation (AM) results in the creation of two sidebands. An upper sideband is above the carrier frequency and a lower sideband is below the carrier frequency. SSB transmission suppresses one of the sidebands. See also *AM*, *amplitude*, *carrier*, *DSB*, *frequency*, *modulation*, *sideband*, and *VSB.*

SSID (Service Set IDentifier) In IEEE 802.11b wireless LAN (WLAN) specifications, a security mechanism in the form of an authorization code established by the system administrator. A device seeking to gain access must be in possession of the SSID. See also *802.11b*, *authorization*, and *WLAN.*

SSL (Secure Sockets Layer) A security protocol developed by Netscape Communications Corporation, SSL includes authentication and negotiates point-to-point security between client and server, including type of encryption scheme and exchange of encryption keys. SSL sends messages over a socket, which is a secure channel at the connection layer and existing in virtually every TCP/IP application. Although SSL can accommodate a number of encryption algorithms, Netscape has licensed RSA end-to-end public key encryption, as well as key creation and certification. Unlike S-HTTP, SSL is application independent and works with all Internet tools, not just the World Wide Web (WWW). SSL has emerged as a de facto standard. See also *authentication*, *client*, *de facto*, *encryption*, *Internet*, *protocol*, *public key encryption*, *RSA*, *server*, *S-HTTP*, *socket*, *standard*, *TCP/IP*, and *WWW.*

SSP (Service Switching Point) In the advanced intelligent network (AIN) architecture, a public switched telephone network (PSTN) switch that acts on the instructions dictated by centralized AIN databases. An SSP can be an end office or tandem switch. See also *AIN*, *database*, *end office*, *PSTN*, and *tandem switch.*

Standard Wire Gauge (SWG) Synonymous with British Standard Gauge (BSG). See also *BSG* and *gauge.*

standard A rule, principle, or measure established as a model or example by authority, custom, or general consent. Standards generally are in the form of baseline specifications according to which manufacturers can develop products with the assurance that they will interconnect and interoperate with those of other manufacturers, at least at a fundamental level. Standards typically allow for options that manufacturers can exercise in various fashions peculiar to their own product development philosophies, strategies, and so on, thereby distinguishing those products from others. Although standards have been criticized as common denominator or consensus solutions that stifle creativity, they in fact provide a common framework of technical specifications within which manufacturers can exercise a considerable level of creativity. Standards serve to create the technical basis for a competitive market that offers buyers a choice of products, while ensuring interconnectivity and interoperability at a fundamental level. Standards take several forms.

- **De jure:** From Latin, literally meaning *from the law.* Formal specifications that do not have the force of law, but often have considerable weight as they are set by formal standards bodies that generally are established by governmental or regulatory bodies, or at least by industry consensus. Such formal bodies include the American National Standards Institute (ANSI), the European Telecommunications Standards Institute (ETSI), and the International Telecommunications Union (ITU). Governments sometimes give these standards the force of law, as in requiring new buildings to comply with the National Electrical Code (NEC).

- **De facto:** From Latin, literally meaning *from what is done*, that is, *in fact*. Standards not established by such formally constituted bodies, that may even be established by a dominant vendor in its own self-interest and often for its own internal use in the context of an ad hoc solution. De facto standards take on the effect of formal standards simply because they become so widely accepted. Hayes, IBM, and Microsoft, for example, have developed numerous specifications that have become de facto industry standards.
- **Du jour:** From French, meaning *of the day*. The popular standard of the day. One day 10 years ago, ATM was really hot and a lot of people made a lot of money talking about ATM and selling products based on ATM. It seemed like only the next day that IP was really cool. (I made this one up.)

standard definition television (SDTV) See *SDTV*.

Standard Generalized Markup Language (SGML) A language used by Web developers and designers for creating declarative markup languages like Hypertext Markup Language (HTML). Extensible Markup Language (XML) is a condensed form of SGML that is published and maintained by the World Wide Web Consortium (W3C). See *HTML*, *W3C*, *WWW*, and *XML*.

start bit In asynchronous transmission, a bit that alerts the receiving computer of the arrival of a character. A stop bit, or sometime two stop bits, signals the end of the character. See also *asynchronous transmission* and *bit*.

star topology A network structure comprising a central node to which all other devices attached directly and through which all other devices intercommunicate. As illustrated in Figure S-9, the central node is in the form of a hub, switch, or router with multiple ports to which devices connect, usually through unshielded twisted pair (UTP) or shielded twisted pair (STP). In the public switched telephone network (PSTN), each carrier serving area (CSA) is a star, with local loops radiating from the central office (CO). Star configurations include 100Base-T and 1000Base-T local area networks (LANs). See also *100Base-T*, *1000Base-T*, *CO*, *CSA*, *hub*, *node*, *PSTN*, *router*, *STP*, *switch*, *Token Ring*, *topology*, and *UTP*.

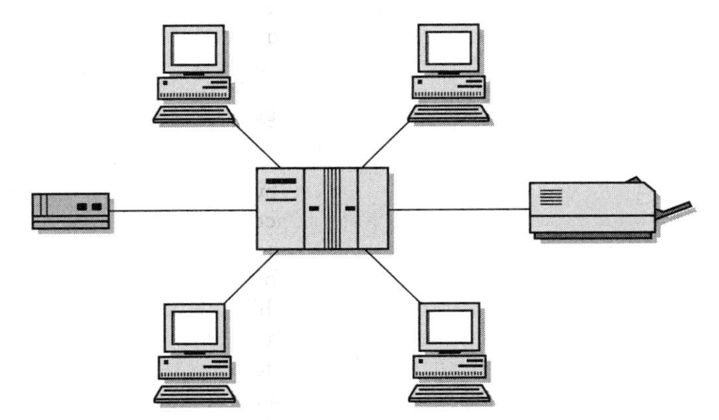

Figure S-9

start-stop transmission See *asynchronous transmission*.

stateful Referring to a system or process that is aware of the status or condition, perhaps in detail, of an activity in which it participates.

stateful autoconfiguration An IPv6 address assignment approach in which the configuration servers dynamically assign unique addresses to devices as they require them, drawing from a pool of such addresses.

This approach resembles Dynamic Host Configuration Protocol (DHCP) used in IPv4. See also *DHCP*, *IPv4*, *IPv6 address*, and *stateless autoconfiguration*.

stateful inspection firewall A security firewall that examines packets, notes the port numbers that they use for each connection, and shuts down those ports once the connection is terminated. See also *authentication*, *authorization*, *firewall*, and *security*.

stateless Referring to a system or process that is not aware of the status or condition of an activity in which it participates.

stateless autoconfiguration An IPv6 address assignment approach that employs two IP addresses, one that is assigned permanently to the mobile device and the other that is used to route data to the network to which the mobile device is connected at the time. This stateless approach is much like sending a datagram to a device in care of a network and is useful in the context of mobile devices that move among pager, cellular, packet radio, wireless LAN (WLAN), and other wireless networks. See also *cellular radio*, *datagram*, *IPv6*, *IPv6 address*, *pager*, *stateless autoconfiguration*, and *WLAN*.

static address Referring to an Internet Protocol (IP) address permanently or semi-permanently assigned to a specific host. See also *dynamic address*, *host*, *IP*, and *IP address*.

static bend The long term bend in a cable at rest, i.e., after installation. See also *bend diameter*.

static load The long term load, i.e., force or weight, placed on a cable, such as a riser cable, which hangs vertically. See also *load*.

station A terminal or endpoint on a network, such as a telephone set or data terminal.

station message detail record (SMDR) See *SMDR*.

statistical time division multiplexer (STDM mux or stat mux) A device that performs statistical time division multiplexing (STDM), an STDM MUX is commonly known as a *stat mux*. See also *mux* and *STDM*.

statistical time division multiplexing (STDM) See *STDM*.

stat mux (statistical time division multiplexer) A device that performs statistical time division multiplexing (STDM). See also *mux* and *STDM*.

STDM (Statistical Time Division Multiplexing) An improved TDM method that makes use of intelligent muxes, or stat muxes, that can dynamically adapt to the changing nature and associated requirements of the load placed on it in consideration of the available capacity of the circuit. STDM muxes can allocate bandwidth in consideration of the device and application priorities. An STDM can oversubscribe a trunk, supporting aggregate port speeds that can be multiples of the trunk speed, exercising flow control by buffering data during periods of high activity, restraining low-priority transmissions in favor of those of higher priority. STDM muxes may perform data compression, error detection and correction, and reporting of traffic statistics.

As shown in Figure S-10, STDMs typically divide a high-speed, four-wire digital circuit into multiple time slots to carry multiple voice conversations or data transmissions. Channelized T1 (North America), for example, commonly provides 24 time slots of 64 kbps. Channelized E-1 (European) commonly provides 30 time slots. Additionally, the individual channels can be grouped to yield higher transmission rates (superrate) for an individual, bandwidth-intensive communication such as a videoconference. The individual channels also can be subdivided into lower-speed (subrate) channels to accommodate many more, less bandwidth-intensive communications, such as low speed data. Also, many muxes allocate bandwidth on a priority basis, providing delay-sensitive traffic, such as real-time voice or video, with top priority. See also *buffer*, *channel*, *FDM*, *flow control*, *oversubscribe*, and *TDM*.

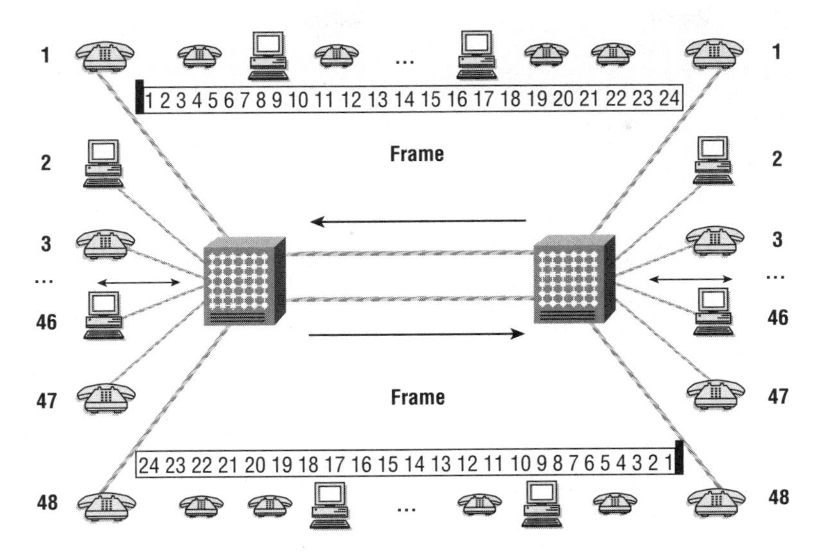

Figure S-10

STE (Section Terminating Equipment) In an SDH or SONET network, a repeater, add/drop multiplexer (ADM), or anything else that attaches to either end of an optical fiber link. See also *ADM*, *link*, *repeater*, *SDH*, and *SONET*.

steer A restoral mechanism employed in Resilient Packet Ring (RPR). In the event of a node or link failure, the steer option calls for the originating station to exercise sufficient intelligence to avoid the failed ring and place the traffic on the ring that retains continuity. See also *dual counter-rotating ring*, *fiber optics*, *link*, *node*, *RPR*, and *wrap*.

steganography From Greek and translating as *covered writing* or *hidden writing*, and dating to 440 B.C., steganography is the art or science, or system, of hiding the existence of a message. In *The Histories of Herodotus*, the Greek historian Herodotus mentions several examples. Into the wood backing of a wax tablet, Demeratus carved a message warning his countrymen of an impending attack. He then applied the wax, which hid the message from view until it was removed by the intended recipient. Another method involved shaving the head of a slave and tattooing a message on his scalp. After the hair grew back enough to cover the message, the slave could be sent through enemy lines, and his head could be shaved again to read the message. More recently, microfilm dots have been hidden under postage stamps, or disguised as punctuation marks in typewritten letters. Contemporary stenography takes more technologically sophisticated forms, such as a message hidden in a data file, for example, in an HTML file, a JPEG file, or an MP3 file. Such a hidden file also is typically encrypted for additional security. See also *encryption* and *watermark*.

step-by-step (SxS) See *SxS*.

step-index fiber A type of glass optical fiber (GOF) characterized by a sharp difference, or step, in the index of refraction (IOR) at the interface between the core and the cladding. The layer of cladding has a uniform IOR that is sharply lower (typically one percent or more), which causes errant light rays striking the interface to reflect back into the core, which is the primary light-conducting medium. Light rays striking the interface at extreme angles less than the critical angle can be lost in the cladding, as illustrated in Figure S-11. Multi-step fibers comprise multiple layers of cladding with sharp steps in IOR to compound the effect. See also *critical angle*, *GOF*, *graded-index fiber*, *IOR*, *reflection*, and *total internal reflection*.

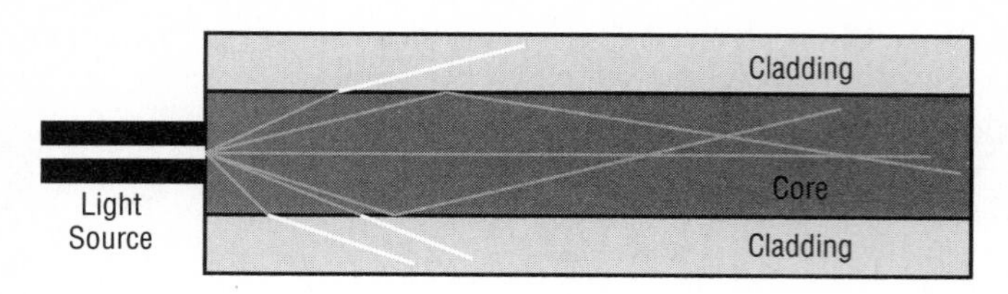

Figure S-11

STM (Synchronous Transport Module) The electrical equivalent of the Synchronous Digital Hierarchy (SDH) optical signal, according to ITU-T international standards. The STM is known as Synchronous Transport Signal (STS) in Synchronous Optical Network (SONET) terminology, according to North American standards. The signal begins in electrical format as three T3 signals plus SDH signaling and control overhead and converts to optical format for transmission over the SDH optical fiber facilities. Each STM-1 frame is transmitted in 125μs, yielding raw bandwidth of 155.52 Mbps. The STS frame includes five elements: Synchronous Payload Envelope (SPE), Section Overhead (SOH), Line Overhead (LOH), Path Overhead (POH), and Payload. See also *bandwidth*, *frame*, *ITU-T*, *LOH*, *overhead*, *payload*, *POH*, *SDH*, *signaling and control*, *SOH*, *SONET*, *SPE*, and *T3*.

stop bit In asynchronous transmission, a bit, or sometimes two bits, that signals the end of a character. A start bit alerts the receiving computer of the arrival of a character. See also *asynchronous transmission*, *bit*, and *start bit*.

Storage Area Network (SAN) See *SAN*.

store-and-forward A transmission method by which a device receives a complete message or protocol data unit (PDU) and temporarily stores it in a buffer before forwarding it toward the destination. Having the whole message allows the device to check for errors and discard an errored frame or packet before forwarding it and wasting bandwidth on the next hop. A switch or router, for example, may have buffers to store incoming frames or packets of data until internal computational resources are available to process them and buffers to store outgoing frames or packets until bandwidth is available on a circuit in the forward direction. That way the device can mitigate issues of switch and circuit congestion. Messaging systems add significant value by storing voice, e-mail, and image (e.g., fax) messages when the intended recipient is unavailable and forwarding them on demand when the recipient is available. Facsimile systems also may store international fax messages until off-peak hours, when calling rates are lowest.

stored program control (SPC) See *SPC*.

store locator service Also known as *single number dialing*, a service of the advanced intelligent network (AIN) that provides the ability to advertise a single number. The network routes calls to the closest store location in terms of either geography or time zone, based on the originating address (i.e., telephone number) of the caller. See also *AIN*.

STP **1.** Shielded Twisted Pair. Synonymous with Shielded Foil Twisted Pair (SFTP), STP is a copper cable configuration comprising a metallic foil shield that surrounds each insulated pair, of which there may be several. An uninsulated steel or tinned copper conductor in contact with each inner shield serves as a drain wire, ensuring that the continuity of the shield remains intact in the event that the foil is broken or cracked. The core of shielded pairs is then surrounded by an overall metallic shield of metallic tape or braid, or both, which is encased in a thermoplastic cable jacket, as illustrated in Figure S-12. The outer shield typically consists of helically or longitudinally applied plastic and aluminum laminated solid tape, although it may comprise a woven mesh, and steel or copper may also be used. Each shield absorbs ambient energy and conducts it to ground through the drain wire, thereby protecting the signal transmitted through the center conductors. The shield also serves to confine the electromagnetic field associated with the transmitted signal within the core conductors, thereby reducing signal loss and maintaining signal strength over a longer distance. Screened twisted pair (ScTP) is a simpler version with only an outer shield. See also *Cat 6*, *Cat 7*, *ScTP*, and *UTP*. **2.** Signal Transfer Point. In the advanced intelligent network architecture (AIN) architecture, a packet switch that routes signaling and control messages between a service switching point (SSP) and a service control point (SCP), and between STPs. See also *AIN*, *packet*, *SCP*, *signaling and control*, *SSP*, and *switch*. **3.** Spanning Tree Protocol. A bridge protocol for learning bridges, as defined in IEEE 802.1D standards. Spanning tree bridges are self-learning, filtering bridges for use in connecting LANs or LAN segments on a point-to-point basis. The bridge can be programmed or can teach itself the addresses of all devices on the network; subsequently, the network tree of the bridge provides only one span connection. Some spanning tree bridges also have the capability to provide security by denying access to certain resources based on user and terminal ID. Bridges that support the spanning tree algorithm have the ability to automatically reconfigure themselves for alternate physical paths if a network segment fails, thereby improving overall reliability. Radia Perlman invented STP while working for Digital Equipment Corporation (DEC). See also *algorithm*, *bridge*, *filtering bridge*, *LAN*, *path*, *protocol*, *segmentation*, and *self-learning bridge*.

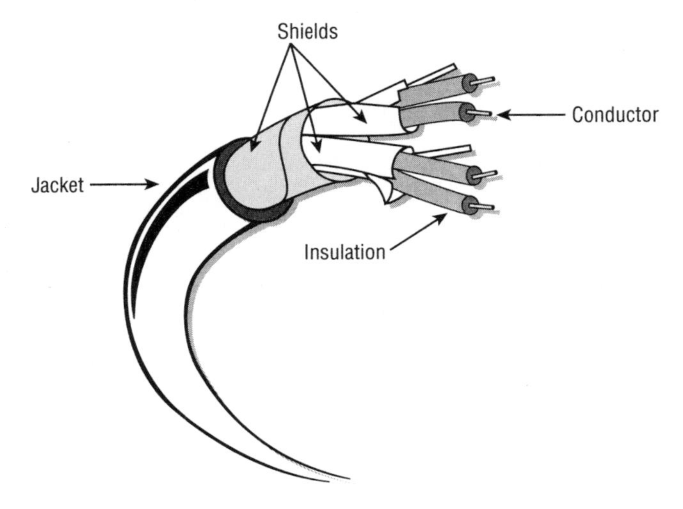

Figure S-12

streaming mode service In asynchronous transfer mode (ATM), a type of service used for framed data in which multiple interface data units (IDUs) are passed in a data stream. Streaming mode service is supported by ATM Adaptation Layer 3/4 (AAL3/4). AAL3/4 also supports message mode service, a type of service for framed data in which only one IDU is passed. See also *AAL3/4*, *ATM*, *IDU*, and *message mode service*.

stream-oriented Referring to an application that generates a continuous flow of data, rather than bursts of intense data activity interspersed with periods of inactivity. Real-time voice and video are stream-oriented. See also *application*, *bursty transmission*, *data*, *flow*, *real-time*, *video*, and *voice*.

strength Toughness or durability. The ability to withstand force, pressure, strain, or stress. See *bend radius*, *break strength*, *flex strength*, *mechanical strength*, *strength member*, and *tensile strength*.

strength member A load-bearing component of cables, particularly aerial and riser cables, designed to increase the overall tensile strength of the cable and to relieve the wires and fibers from bearing the load directly. As optical fiber, in particular, stretches very little before breaking, the strength members also must have very low elongation at the expected tensile loads. The strength members in optical fibers commonly are of aramid fiber yarn, invented and marketed by Dupont as Kevlar®, while those in copper cables commonly are rods of fiberglass epoxy (FGE). As both aramid and fiberglass are dielectrics, they provide strength without contributing to issues of electromagnetic interference (EMI). Steel strength members were used historically, but are electrically conductive, which presents problems not only in terms of EMI, but also with respect to power surges such as those caused by lightening strikes. As steel and fiberglass, however, are more thermally stable than aramid fiber, they are preferred when extreme cold temperature performance is required. See also *aramid*, *dielectric*, *EMI*, *FGE*, and *fiberglass*.

string A linear series of things, such as bits or characters.

string coding A compression technique that replaces long strings of redundant data with code words of much shorter fixed length. See also *compression* and *run-length encoding*.

Strowger, Almon B. A Kansas City undertaker who invented the step-by-step (SxS) switch, also known as the *Strowger switch*, which was the first automatic circuit switching system. See also *SxS*.

Strowger switch Step-by-step (SxS) switch. See *SxS*.

structured wiring plan A comprehensive, documented plan for inside wire and cable systems in new building construction, incorporating voice, data, audio, video, security, and any other applications, from switch to jack. Such a plan considers placement of repeaters, hubs, switches, routers, and other network devices, as well as terminal equipment. A structured wiring plan must also consider power requirements and should address intermediate-term and even relatively long-term requirements.

STS **1.** Speech-To-Speech. A Telecommunications Relay Service (TRS) offering that enables a person with a speech disability to use his or her own voice to speak to the called party through a call administrator (CA) specially trained to understand speech affected by a variety of disorders. The CA acts as a facilitator, repeating the spoken words in a clear and understandable manner. See also *TRS*. **2.** Synchronous Transport Signal. The electrical equivalent of the Synchronous Optical Network (SONET) optical signal, according to North American standards. The STS is known as Synchronous Transport Module (STM) in Synchronous Digital Hierarchy (SDH) terminology, according to ITU-T international standards. The signal begins in electrical format as a T3 signal plus SONET/SDH signaling and control overhead, and converts to optical format for transmission over the SONET optical fiber facilities. Each STS-1 frame is transmitted in 125μs, yielding raw bandwidth of 51.84 Mbps. The STS frame includes five elements: Synchronous Payload Envelope (SPE), Section Overhead (SOH), Line Overhead (LOH), Path Overhead (POH), and Payload. See also *bandwidth*, *frame*, *ITU-T*, *LOH*, *overhead*, *payload*, *POH*, *SDH*, *signaling and control*, *SOH*, *SONET*, *SPE*, and *T3*.

stuff bit A bit added into a bit stream during a process known as *bit stuffing*, in order to 1) ensure synchronization technique used in time division multiplexing (TDM) by avoiding long streams of 0 bits, 2) adjust for slight timing discrepancies between incoming bit streams when being multiplexed into faster links (e.g., multiplexing T1s into a T3 using an M13 multiplexer), and 3) prevent the appearance of the 0x7e flag character within an HDLC frame (4-zero suppression). See also *bit*, *bit stream*, *bit stuffing*, *flag*, *HDLC*, *M13*, *multiplexer*, *synchronize*, *T1*, *T3*, and *TDM*.

stuttered dial tone See *dial tone*.

subcarrier A frequency channel that occupies only a portion of RF bandwidth allocated to the carrier and, therefore, has a smaller information capacity. A subcarrier sometimes is used for signaling between stations on a network. See also *bandwidth*, *carrier*, *channel*, *frequency*, *RF*, and *signaling*.

submarine cable Cable designed to be placed underwater. Such cables must be specially protected against moisture. At shallow depths on continental shelves, submarine cables commonly are plowed in and armored to protect them against ship anchors, trawler nets, and sharks, which are attracted to the electromagnetic fields and like to gnaw on the cables and repeaters. See also *cable*.

subnet (subnetwork) A network, either physical or logical, that operates as part of a larger network. In a local area network (LAN), for example, there may be many virtual LANs (VLANs), each of which may comprise many users on separate physical segments. The users are grouped in VLAN domains by physical port number, Transmission Control Protocol (TCP) port address, medium access control (MAC) address, or Internet Protocol (IP) address. Each VLAN operates as a subnet. See also *LAN* and *VLAN*.

subnetting A technique that enables a network administrator to divide a single private Internet Protocol (IP) network into multiple smaller logical subnetworks by subdividing the host address into a subnetwork address and host address. Routers establish borders between subnets. See also *IP*, *router*, and *subnet mask*.

subnet mask In Internet Protocol version 4 (IPv4), an address mask, i.e., address filter, that selectively includes or excludes certain values to distinguish between the subnetwork address and the host address in order to enable a router to forward packets correctly in a network that has been subnetted. See also *IPv4*, *router*, *subnet*, and *subnetting*.

subnetwork (subnet) See *subnet*.

subrate A rate lower than the normal rate. A channel bank typically derives multiple 64-kbps voice-grade channels from a circuit. If multiple low-speed data applications require less bandwidth, a sufficiently sophisticated time division multiplexer (TDM mux) can subdivide a channel into multiple subrate data channels. See also *bandwidth*, *channel*, *channel bank*, *superrate*, *TDM*, and *voice grade*.

Subrate Digital Loop (SRDL) See *SRDL*.

subscriber In telecommunications, an entity (individual, company, or other organization) that leases a circuit or contracts to use a public telecommunications service.

subscriber identification Module (SIM) See *SIM*.

Subscriber Line Carrier-96 (SLC-96) See *SLC-96*.

subscriber line charge (SLC) See *SLC*.

suit A mildly derisive term for an anonymous business executive or bureaucrat, referring to the fact that such people typically wear suits of clothes and may lack individuality. A suit, especially an empty suit, is in sharp contrast to a techie. See also *empty suit* and *techie*.

superframe In the T-carrier D2, D3, and D4 framing conventions, a 12-frame sequence. Extended superframe (ESF) defines a 24-frame sequence. See also *D2*, *D3*, *D4*, *ESF*, *frame*, and *T-carrier*.

supernetting The aggregation of multiple Internet Protocol version 4 (IPv4) address blocks. See *CIDR*, *IPv4*, and *IPv4 address*.

surface wave An electromagnetic wave that propagates close to the surface of the Earth. See also *ground wave* and *skywave*.

surge A strong, sudden, and transient spike in voltage or current. See also *current* and *voltage*.

surge protector See *protector*.

Super High Frequency (SHF) See *SHF*.

superrate A rate higher than the normal rate. A time division multiplexer typically derives multiple 64-kbps voice grade channels from a circuit. If a data application requires more bandwidth, a sufficiently sophisticated mux can group multiple channels into a superrate channel. See also *subrate* and *TDM*.

supervision A basic signaling function that indicates the status of a component, such as trunk idle or busy, telephone on-hook or off-hook. In early switchboard operation, a human operator put a receiver across a line to monitor the status of a call, to determine if the call was in progress or had been terminated. See also *call*, *signaling and control*, *switchboard*, and *trunk*.

suppression Forceful constraint, prevention, or subduing. In electronics, the elimination or intentional attenuation of an unwanted oscillation, such as a sideband, a carrier, or an echo. In some voice encoding mechanisms, silence suppression senses periods of inactivity in a voice conversation and simply ceases sending data associated with that conversation. See also *attenuation*, *carrier*, *echo*, *encode*, *oscillate*, *sideband*, and *silence suppression*.

surge An elevated voltage level lasting longer that a spike. See also *ground loop*, *spike*, and *voltage*.

sustainable cell rate (SCR) See *SCR*.

SVC (Switched Virtual Circuit) A shared path established between two hosts through a packet network on command, i.e., via signaling as the call is placed. Once the path is selected, all packets in a given session travel the same path, which is selected in consideration of both the condition of the network and the load on it at the instant the connection is required. Thereby, an SVC bypasses failed and congested switches and circuits and improves overall performance through automatic load balancing. Although SVCs are defined in frame relay specifications, they are unusual in public networks due to the carrier's fear that frame relay SVCs would cannibalize more expensive services like ISDN and long distance voice. Rather, frame relay networks employ permanent virtual circuits (PVC), which are predetermined, preprogrammed paths. Globally, X.25 networks largely are based on SVCs. See also *channel*, *circuit*, *frame*, *frame relay*, *load balancing*, *packet*, *path*, *PVC*, *virtual circuit*, and *X.25*.

SW (Short Wavelength) Referring to fiber optic systems operating in the 850 nm range, with the IEEE 802.3ae specification for 10GBase-SR, SW being one example. See also *10GBase-SR*, *SW*, and *LW*.

SWG (Standard Wire Gauge) Synonymous with British Standard Gauge (BSG). See also *BSG* and *gauge*.

switch **1.** A mechanical, electromechanical, or electronic device that opens, closes, or changes the connections in an electrical circuit. **2.** A device that establishes, maintains, and changes logical connections over physical circuits. Switches flexibly connect transmitters and receivers across networks of interconnected links, thereby allowing network resources to be shared by large numbers of end users. Without switches, each transmitter/receiver pair would require a dedicated circuit in order to transfer data. There are a number of types of switches. In terms of switching technology, there are circuit switches and packet switches. **a.** Circuit switches establish connections between circuits, on demand and as available. Those connections are temporary, continuous, and exclusive in nature. Circuit switches were developed for voice communications, but will support any type of information transfer. Common examples of circuit switches include private branch exchanges (PBXs) and central office exchanges (COs or COEs). **b.** Packet switches switch data organized into packets, discrete sets of data that may take the specific form of packets, frames, or cells depending on the network technology specifics. For example, packet switches switch packets in networks based on the Internet Protocol (IP), frames in networks based on the frame relay or Ethernet protocols, and cells in those based on the asynchronous transfer mode (ATM) protocol. Packet switches were initially developed for data networking, but can support other forms of data, as well, although with varying degrees of success.

With respect to physical placement, there are edge switches and core switches. **c.** Edge switches are positioned at the physical edge of a public network. The user organization gains access to an edge switch via an access link, or local loop. A central office (CO) is an example of an edge switch in the context of the circuit-switched public switched telephone network (PSTN). In a Local Area Network (LAN), a workgroup switch is the equivalent of an edge switch in a public network. **d.** Core switches, also known as *tandem switches* and *backbone switches*, are high-capacity switches positioned in the physical core, or backbone, of a network and serving to interconnect edge switches.

Although switches can be very intelligent in many respects, they operate only at the Layer 2, the Data Link Layer of the OSI Reference Model. That is to say that they operate link-by-link, or hop-by-hop, generally under the control of a centralized set of logic that can coordinate their activities in order to establish end-to-end connectivity across a multi-link circuit. A switch has no concept of the network as a whole, from end-to-end. See also *ATM, backbone switch, cell, CO, core switch, Data Link Layer, edge switch, Ethernet, frame, frame relay, IP, LAN switch, OSI Reference Model, PSTN, router,* and *tandem switch.*

switchboard The first switching device, the switchboard literally was a series of small, mechanical switches mounted on a board. The operator manually switched the wires from one contact to another to establish a unique physical and electrical path or circuit to connect two parties in order that they might engage in a voice conversation. As all of the links, contacts, and switches are physically separated, a switchboard is a type of space division switch. Although the switchboard was superseded by the cordboard, the term remains widely used to refer to an operator console. See also *cordboard* and *switch.*

Switched 56 (Switched 56 kbps Service) More formally known as *Digital Switched Access* (DSA). A switched digital data service that operates much like the public switched telephone network (PSTN) operates for voice calls. Switched 56 service operates over a public data network (PDN) that actually is a physical and logical partition of the PSTN. Where the PDN supports out-of-band signaling and control, the service sometimes is known as *Switched 64*, as the full 64 kbps bandwidth of a DS-0 channel is available to support end user data transmission. Switched-1536 service supports a full ISDN PRI of 24 channels, each of which provides the full 64 kbps of DS-0 bandwidth, with all signaling and control taking place out-of-band on another PRI circuit through a technique known as *non-facility associated signaling* (NFAS). See also *DS-0, ISDN, NFAS, out-of-band signaling and control, PDN, PRI,* and *PSTN.*

Switched 64 See *Switched 56.*

switched circuit A circuit established through one or more intermediate switching devices, such as circuit switches or packet switches. A typical switched circuit can comprise a dedicated circuit from an originating device to an ingress switch port, or point of interface, a switch matrix through which a path is established to an egress port, and a dedicated circuit to a destination device. There may be many intermediate switches in a more complex scenario. Switched networks are highly shared, as a number of users contend for access to limited network resources through switches, which serve as points of contention, with connectivity between transmitters and receivers provided through the network on demand and as available. This sharing of limited network resources clearly allows the network providers to realize significant operational efficiencies, which are reflected in lower overall network costs. The end users realize the additional advantages of flexibility and resiliency, as the network generally can provide connectivity between any two physical locations through multiple alternate transmission paths. A switched circuit is in marked contrast to a dedicated circuit, which is dedicated to connecting two or more physical locations. Such a dedicated circuit is highly available, offers reliable levels of performance, and provides guaranteed bandwidth, but is inflexible and susceptible to catastrophic failure. See also *circuit, circuit switch, dedicated circuit, packet switch,* and *switch.*

switched circuit network (SCN) See *SCN.*

Switched Multimegabit Data Service (SMDS) See *SMDS.*

switched virtual circuit (SVC) See *SVC.*

switch hook An early telephone handset hung from a hook that activated a switch. When the telephone was not in use, the handset hung on the hook, or was on-hook. When the telephone was in use, the handset was off of the hook, or off-hook. When the user lifted the handset off the hook, a spring lifted the hook, which closed a switch and closed a circuit, drawing current from the central office (CO). The term now refers to the mechanical buttons or plungers that are mounted in the cradle of a telephone set, but the process remains essentially the same. Synonymous with hook switch. See also *off-hook* and *on-hook.*

switch hook flash A method of signaling a central office, key telephone system, or PBX by quickly depressing and releasing the switch hook on a telephone set, perhaps to answer another incoming call. Some Centrex and PBX telephone sets have special buttons that implement this function. See also *switch hook*.

switch matrix A set of buses interconnected in such a way that traffic from an input port can find a path to an output port. See also *matrix switch*.

Switch-to-Computer Applications Interface (SCAI) See *SCAI*.

SxS (Step-by-Step) SxS refers to the electromechanical circuit switches that improved on earlier manual cordboards. The SxS switch was invented and patented in 1891 by Almon B. Strowger, a Kansas City undertaker frustrated with the behavior of the local telephone company operator. According to legend, the operator was directing Mr. Strowger's incoming calls to a competing undertaker, who also happened to be her husband. Strowger responded by inventing an automated system that served 99 subscribers. The telephones that worked with that first automatic switch had two buttons. In order to reach subscriber 99, for example, the caller slowly and deliberately pressed the first button nine times and then the second button nine times. As the caller pressed a button, it completed an electrical circuit and as the caller released the button, it broke the circuit. Making and breaking the circuit caused a mechanical wiper, or selector, to rotate from one switch contact to another. As the user dialed each digit in the telephone number, the wiper would step up to the next level, step-by-step. Strowger's SxS patent served as the cornerstone for the company he founded, Automatic Electric Company, which later became the manufacturing subsidiary of General Telephone and Electric (GTE), but was sold to AT&T in 1989 to form part of AT&T Technologies. In 1996, AT&T spun that company off to form Lucent, which was sold to Alcatel (France) in 2006. GTE is now part of Verizon. SxS technology was considered state of the art until the appearance of the crossbar (Xbar) switch in 1938. Large numbers of SxS switches remained in service into the 1970s and even 1980s, and some likely remain in service to this day. See also *cordboard*, *panel switch*, and *Xbar*.

symbol **1.** A sign that has a specific meaning in a specific context, such as mathematics. For example, the Greek letter λ (lambda) is used in physics to mean wavelength, which is the inverse of frequency, represented by the Latin letter f. **2.** Something that represents or suggests something else, usually something abstract. **3.** In digital communications, the smallest amount of data transmitted at one time. In a purely digital system, such as a fiber optic transmission system (FOTS), a symbol is an individual bit. In a digital system involving modulation of an analog carrier waveform, a symbol is an individual baud, or signal change, which may represent multiple bits. In Fiber Distributed Data Interface (FDDI), a broadband LAN standard, a five-bit symbol represents a four-bit nibble. See also *baud*, *bit*, *FDDI*, *intersymbol interference*, *LAN*, *modulation*, and *nibble*.

symmetric **1.** Balanced or proportional. **2.** In telecommunications, a link that supports equal bandwidth in both directions. Symmetric digital subscriber line (SDSL), for example, supports equal bandwidth downstream and upstream. Bluetooth supports an asynchronous data channel that can operate in symmetric mode at speeds of up to 432.6 kbps. Alternatively, the Bluetooth data channel can operate in asymmetric mode at up to 721 kbps in either direction and 57.6 kbps in the reverse direction. See also *asymmetric*, *asynchronous*, *bandwidth*, *Bluetooth*, *channel*, *downstream*, *SDSL*, and *upstream*. **3.** In compression, a process that is equally time-consuming and processor-intensive in terms of compression and decompression. See also *compression*.

symmetric digital subscriber line (SDSL) See *SDSL*.

symmetric DSL (SDSL) See *SDSL*.

symmetric high-bit-rate digital subscriber line (SHDSL) See *SHDSL*.

sync (synchronize or synchronization) **1.** Devices in synchronization are said to be in sync. If it is necessary to synchronize two devices, it may be said that it is necessary to get them in sync or to sync them up. See also *synchronize*. **2.** A control character in some polled protocols for multidrop lines. See also *Bisync* and *synchronize*.

synchronization Referring to the coordination in time between a transmitter and receiver. In video communications, synchronization includes vertical and horizontal sync. Vertical sync keeps the picture from scrolling, or flipping. Horizontal synch keeps the picture from twisting. See also *quadbit*.

synchronize To cause objects or events to move together or occur at the same time.

synchronizing bit A binary digit (bit) used to synchronize devices connected by a circuit. See also *synchronous* and *synchronous transmission*.

synchronous From Latin and Greek origins, synchronous translates as *together with time*. Referring to events that occur at the same instant of a coordinated time scale. If the events are repetitive, the instant of one event bears a fixed time relationship with the instant of a corresponding event, e.g., event a is followed 10 milliseconds later by event b. Synchronous processes in separate, networked devices depend on a common clocking source, on clocking pulses emitted by the transmitting device, or on synchronizing bits or bit patterns embedded in a set of data.

synchronous code division multiple access (S-CDMA) See *S-CDMA*.

synchronous connection-oriented (SCO) See *SCO*.

Synchronous Data Link Control (SDLC) See *SDLC*.

Synchronous Digital Hierarchy (SDH) See *SDH*.

Synchronous Optical Network (SONET) See *SONET*.

synchronous transmission Data transmission in which a relatively large set of data is organized into a frame or block, with one or more synchronization bits or bit patterns used to identify the beginning and end of a logical block of data. T1 transmission, for example, is synchronized through framing bits that occur at the beginning of each frame. E-1 transmission is synchronized through the use of a separate time slot zero (0). Synchronous modems coordinate the receiving terminal on the rate of transmission of the data from the sending terminal. Synchronous data protocols such as Synchronous Data Link Control (SLDC) and High-Level Data Link Control (HDLC) use a specific bit pattern to form synchronizing characters that are integral to each frame. Through the receipt of the synchronizing bits or characters, a receiving device can match its speed of data receipt to the rate of data transmission across the circuit. Thereby, each bit of data and control information can be distinguished at the physical layer. Higher layer protocols sort out when to expect what information, in which data fields, and in what sequence, based on an agreed upon protocol such as frame relay or Internet Protocol (IP). Synchronous transmission is much more efficient than asynchronous transmission, as only a few framing bits and synchronizing bits surround a large block of data. See also *asynchronous transmission*, *block*, *E-1*, *frame*, *frame relay*, *HDLC*, *IP*, *protocol*, *SDLC*, *SYNTRAN*, and *T1*.

Synchronous Payload Envelope (SPE) See *SPE*.

Synchronous Transport Signal (STS) See *STS*.

SYNTRAN (SYNchronous TRANsmission) A standard developed at Bellcore for synchronous add/drop multiplexing (ADM) at rates up to 45 Mbps (T3) on the basis of a single master clocking source, which allowed the elimination of stuff bits and, thereby, reduced overhead. SYNTRAN also allowed DS-0s and DS-1s to be added to, i.e., multiplexed directly into, and dropped from, i.e., demultiplexed directly from, a DS-3 frame, thereby eliminating the intermediate DS-2 level. SYNTRAN formed the basis for the Synchronous Optical Network (SONET) standard. See also *ADM*, *Bellcore*, *DS-0*, *DS-1*, *DS-2*, *multiplexer*, *overhead*, *SONET*, *stuff bit*, *synchronous*, and *T3*.

system **1.** A combination or assembly of components that forms a complex whole entity that functions as single unit, such as a computer system, PBX system, or transmission system. **2.** An established, orderly method or procedure for doing something.

Systems Network Architecture (SNA) See *SNA*.

T **1.** Tera. From the Greek *teras*, meaning *monster*, translates to *trillion*. **2.** In terms of the electromagnetic spectrum, THz (terahertz) is a trillion (10^{12}) hertz, which is in the range of infrared and visible light. Infrared light has application in fiber optic and free space optics (FSO) transmission systems. A Tbps is a trillion (10^{12}) bits per second. In transmission systems, therefore, a trillion is exactly 1,000,000,000,000 since the measurement is based on a base 10, or decimal, number system. **3.** In computing and storage systems, a TB (terabyte) is actually 1,099,511,627,776 (2^{40}) bytes, as the measurement of internal computer memory is based on a base 2, or binary, number system. The term TB comes from the fact that 1,099,511,627,776 is nominally, or approximately, 1,000,000,000,000. See also *byte*, *electromagnetic spectrum*, and *hertz*. **4.** T interface or Reference Point T in ISDN. See *Reference Point T*.

T1 (Terrestrial 1) Corresponds to *DS-1* (Digital Signal level One) in the North American digital signal hierarchy. The fundamental level of the T-carrier digital carrier system. A T1 system comprises circuit-terminating equipment in the form of a combination of a channel service unit (CSU) and data service unit (DSU) that jointly serve to interface a device to a full-duplex (FDX) four-wire digital circuit and to perform various signal-formatting, signal-timing, monitoring, and diagnostic functions. T1 operates at a signaling rate of 1.544 Mbps, which supports a frame rate of 8,000 frames per second (fps), with each frame comprising a framing bit followed by 192 bits of user payload, at least potentially. The framing bits are used for synchronization and, in some cases, for monitoring, diagnostic, and other network management purposes. The 192 bits of user payload are organized into 24 time-division multiplexed (TDM) time slots, each of which is eight bits wide. See Figure T-1. At a rate of 8,000 fps, each time slot is repeated 8,000 times per second, which translates into a DS-O channel at 64 kbps (8 bits × 8,000 per second = 64,000 bps). Taken together, the 24 8-bit TDM channels at 8,000 fps yield an aggregate payload transmission rate of 1.536 Mbps. Adding the 8,000 framing bits (one per frame) per second, yields the aggregate signaling rate of 1.544 Mbps. Actually, different generations of CSUs, DSUs, and channel banks operate on different framing conventions (D1, D2, D3, D4, and ESF). In some cases, a process of bit robbing reduces the amount of user payload to seven bits per eight-bit time slot of each frame, thereby restricting user payload to 56 kbps. Another convention bit robs only certain frames, and yet another bit robs not at all.

T1 was designed to operate over an unshielded twisted-pair (UTP) circuit comprising two two-wire pairs, each of which operates in simplex mode. One pair supports transmission in one direction, and the other pair in the opposite direction. In the aggregate, the physical four-wire circuit supports full-duplex (FDX) transmission. The line coding technique employed in traditional T1 is alternate mark inversion (AMI), which yields 1.544 Mbps at a nominal carrier frequency of 784 kHz, which is exactly half the T1 bit rate, plus some overhead for error control. At such a high frequency, issues of attenuation are significant, and mutual interference between cable pairs must be considered, so repeaters must be placed every 6,000 feet. Contemporary T1 circuits typically are provisioned using high-bit-rate digital subscriber line (HDSL) technology, which mitigates these issues. T1 was initially designed to operate over a physical four-wire twisted-pair copper circuit. The interface, more correctly known as DSX-1, is medium-independent and will run over coaxial cable, optical fiber, microwave, satellite, and free space optics (FSO) just as well. T1 generally is used in local loops and other short haul applications. In long-haul applications, T3 and other, higher speed, standards generally are employed. The *T* for *Terrestrial* was to distinguish the system from satellite transmission as Bell Laboratories both activated the first T1 system and launched Telstar I, the first communications satellite, in 1962. See also *AMI*, *attenuation*, *bit robbing*, *carrier*, *channel bank*, *CSU*, *D1*, *D2*, *D3*, *D4*, *DS-0*, *DS-1*, *DSU*, *error control*, *ESF*, *FDX*, *four-wire circuit*, *fractional T1*, *frame*, *framing bit*, *frequency*, *HDSL*, *line coding*, *overhead*, *payload*, *signaling rate*, *simplex*, *synchronization*, *T3*, *T-carrier*, *TDM*, *time slot*, *transmission rate*, *two-wire circuit*, and *UTP*.

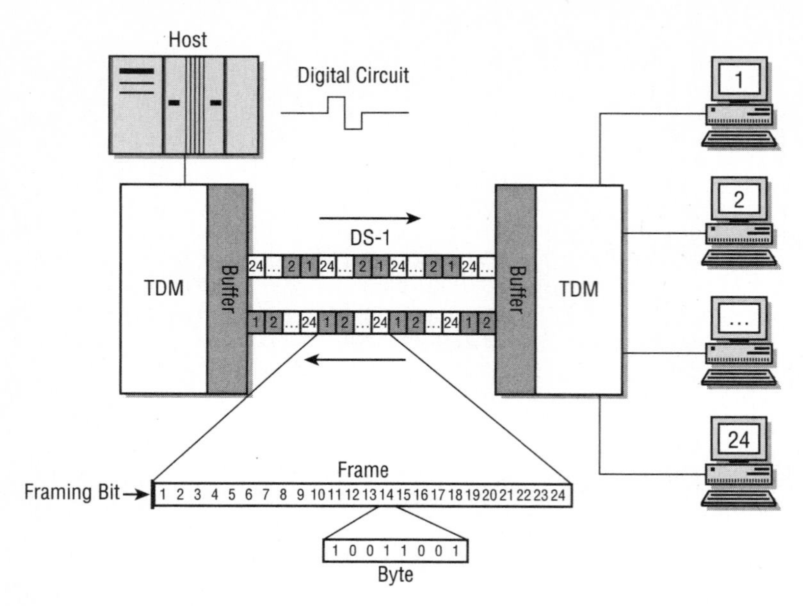

Figure T-1

T1C (T1 Concatenated) Corresponds to *DS-1C* (Digital Signal level One Concatenated) in the North American digital hierarchy. T1C links together two T1s to run at a signaling rate of 3.152 Mbps in support of 48 DS-0 channels. T1C runs at twice the signaling rate and supports twice the number of channels as T1, but with the same repeater spacings as T1. T1C was not deployed widely and is considered obsolete. See also *concatenation*, *DS-0*, *DS-1C*, *repeater*, *T1*, and *T-carrier*.

T.120 The ITU-T Recommendation for the multipoint transport of multimedia data, which can include whiteboarding or binary files. The T.120 series of recommendations supports a broad range of underlying network technologies, and can work either independently or under the H.320 umbrella. T.120 is entirely platform-independent, and can run in a variety of network environments, involving either reliable or unreliable data transport. T.120 supports both unicast and multicast modes of operation. See also *binary*, *data*, *H.320*, *multicast*, *multipoint*, *platform independent*, *transport*, *unicast*, and *whiteboarding*.

T2 Corresponds to *DS-2* (Digital Signal Two), the second level in the North American T-carrier digital hierarchy. T2 was designed for intercity transmission at distances up to 500 miles. T2 runs at 6.312 Mbps and comprises four T1s at 1.544 Mbps each, plus 132 kbps of overhead and justification, or bit stuffing, to adjust for variations in the clocking rates of the incoming T1s. Multiplexing is performed by M12 (Multiplex T1-to-T2) terminals that multiplex four T1 signals, which yields 96 DS-0 channels at 64 kbps per channel. T2 is unusual, although some does remain in place in the local loop, where Bell System companies used it in digital loop carrier (DLC) applications. The Subscriber Line Carrier-96 (SLC-96) system introduced by Western Electric (now Lucent) in 1979, for example, essentially is a remote line shelf and time division multiplexer (TDM mux) that allows a single four-wire twisted-pair or optical fiber circuit to serve as many as 96 channels. See also *bit stuffing*, *digital hierarchy*, *DLC*, *DS-0*, *DS-2*, *local loop*, and *T-carrier*.

T.2 A set of standards for facsimile (fax) machines published by the ITU-T in 1966, and based on the EIA RS-328 specification published by the Electronic Industries Alliance (EIA). Machines conforming to T.2 later became known as Group I. See also *EIA*, *facsimile*, *Group I*, *ITU-T*, *T.3*, *T.4*, and *T.6*.

T3 Corresponds to *DS-3* (Digital Signal Three), the third level in the North American T-carrier digital hierarchy. Designed for long-haul transmission in support of interoffice trunking in the public switched telephone network (PSTN), T3 runs at a signaling rate of 44.736 Mbps. A T3 actually begins as four T1s multiplexed into a T2, by an M13 (Multiplex T1 to T3) mux, which then multiplexes seven T2s to yield a signaling rate of 42.924 Mbps. Stuff bits are added to adjust for variations in the clocking rates of the incoming T2s, bringing the signaling rate up to 44.736 Mbps, comprising 672 DS-0 channels at 64 Kbps. T3 will run over a four-wire twisted-pair circuit, but for no more than 50 feet, due to issues of signal attenuation. Other media, such as coaxial cable, microwave, and optical fiber, are more appropriate. See also *bit stuffing, channel, digital hierarchy, DS-0, four-wire circuit, long haul, multiplexer, PSTN, signaling rate, T-carrier,* and *trunk.*

T.3 A set of standards for facsimile (fax) machines published by the ITU-T in 1978. Machines conforming to T.2 later became known as Group II. See also *facsimile, Group II, ITU-T, T.2, T.4,* and *T.6.*

T.30 An ITU-T specification, published in 1996, describing the handshaking protocol used between Group III/IV facsimile machines to establish and maintain communications. T.30 also provides for routing faxes to users via subaddresses or fax mailboxes. Message security is included, so that only those responsible for certain manual routing processes can view even the cover page. See also *facsimile, Group III, Group IV, handshaking,* and *protocol.*

T.37 An ITU-T Recommendation issued in June 1998 for Fax over Internet Protocol (FoIP) in store-and-forward mode via e-mail, through the use of SMTP (Simple Mail Transfer Protocol) and MIME (Multipurpose Internet Mail Extensions) protocols. The ITU-T and Internet Engineering Task Force (IETF) jointly developed the standard, which the IETF mirrors in RFCs 2301-2306. T.37 specifies the attachment of fax image documents to e-mail headers and their encoding in the Tagged Image File Format-Fax (TIFF-F) data format using Modified Huffman (MH) compression. In simple-mode, T.37 uses the TIFF-F S-profile and restricts fax transmission to the most popular fax machine formats (e.g., standard or fine resolution, and standard page size). Simple-mode provides no confirmation of delivery. Full-mode extensions include mechanisms for ensuring call completion through negotiation of capabilities between transmit and receive devices. Full-mode also provides for delivery confirmation. Extensions have been developed for color fax, as well as black-and-white and grayscale.

T.38 Also known as *Internet Fax Protocol* (IFP). An ITU-T Recommendation originally issued in June 1998 for store-and-forward Fax over Internet Protocol (FoIP) via e-mail. T.38 addresses IP fax transmissions for IP-enabled fax devices and fax gateways, and defines the translation of T.30 fax signals and Internet Fax Protocol (IFP) packets. The specific methods for various T.38 implementations include fax relay and fax spoofing. Fax relay, also known as demod/remod, addresses the demodulation of standard analog fax transmissions from originating machines equipped with modems, and their remodulation for presentation to a matching destination device. Fax relay depends on a low latency IP network in order to avoid session time outs. Fax spoofing is used for fax transmissions over IP networks characterized by longer and less predictable levels of packet latency that could cause the session with the conventional fax machines to time out. T.38 provides for two transport protocols, User Datagram Protocol (UDP) and Transmission Control Protocol (TCP), and several optional means for error control. See also *fax relay, fax spoofing, FoIP, T.30, T.37, TCP,* and *UDP.*

T4 Corresponds to *DS-4* (Digital Signal Four), the fourth level in the North American T-carrier digital hierarchy. T4 was designed primarily as a metropolitan area transmission system, but could operate over distances of up to 500 miles in the backbone of the public switched telephone network (PSTN). Coaxial cable was originally specified, although optical fiber was later preferred. T4 operates at a signaling rate of 274.176 Mbps, which supports 4,032 DS-0 channels at 64 kbps. Little T4 was installed, and it is now considered obsolete, having been superseded by the SONET fiber optic transmission system (FOTS). See also *backbone, channel, digital hierarchy, DS-0, DS-4, FOTS, PSTN, signaling rate, SONET,* and *T-carrier.*

T.4 A set of standards for facsimile (fax) machines published by the ITU-T in 1980. Machines conforming to T.4 later became known as Group III. See also *facsimile, Group III, ITU-T, T.2, T.3,* and *T.6.*

T.434 An ITU-T specification, published in 1999, for binary file transfer (BFT) that permits compliant facsimile devices to send any image file type (e.g., .eps, .pcx, and .bmp) in the form of an editable file that retains its specific file attributes. T.434 offers facsimile the additional advantages of increased throughput and reduced document storage requirements through data compression. T.434 works with computer-based facsimile systems and Group IV fax machines, and supports the linking of fax systems to photocopiers, scanners, e-mail gateways, and PCs. See also *facsimile*, *Group IV*, and *throughput*.

T5 Corresponds to DS-5 (Digital Signal Five), the fifth level in the North American T-carrier digital hierarchy. T5 was designed to operate at a signaling rate of 400.352 Mbps, which supports 5,760 DS-0 channels at 64 kbps. Little, if any, T5 was installed, and it is now considered obsolete, having been superseded by the SONET fiber optic transmission system (FOTS). See also *channel*, *digital hierarchy*, *DS-0*, *DS-4*, *FOTS*, *signaling rate*, *SONET*, and *T-carrier*.

T.6 A set of standards for facsimile (fax) machines published by the ITU-T in 1984. Machines conforming to T.6 later became known as Group IV. See also *facsimile*, *Group IV*, *ITU-T*, *T.2*, *T.3*, and *T.4*.

T.81 The ITU-T Recommendation for Joint Photographic Experts Group (JPEG), a graphics file format for storing highly compressed images. JPEG is a joint standard of the International Telecommunications Union (ITU-T T.81) and the International Organization for Standardization (ISO). See also *compression*, *ISO*, *ITU-T*, and *JPEG*.

TA (Terminal Adapter) Synonymous with *ISDN modem*. An interface adapter for connecting one or more non-ISDN devices (e.g., telephone sets or PCs) to an ISDN network. A TA is ISDN data communications equipment (DCE) that performs protocol conversion for equipment that is not ISDN-compatible. See also *DCE*, *ISDN*, *modem*, and *protocol*.

TACS (Total Access Communications System) A 1G analog cellular radio derivate of the Advanced Mobile Phone Service (AMPS) technology, TACS was developed for use in the United Kingdom. TACS operates in the 900 MHz band and supports either 600 or 1,000 channels of 25 kHz, compared with the 666/832 channels of 30 kHz supported by AMPS. A number of variations were developed, including Narrowband TACS (NTACS), Extended TACS (ETACS), and Japanese Total Access Communications System (JTACS). TACS found acceptance in very few nations, largely has been replaced by GSM, and is considered obsolete in the United Kingdom. See also *1G*, *AMPS*, *analog*, *cellular radio*, *channel*, *ETACS*, *GSM*, and *NTACS*.

Tagged Image File Format (TIFF) See *TIFF*.

Tagged Image File Format-Fax (TIFF-F) See *TIFF-F*.

tail circuit A circuit at the tail end, rather than the headend of a fantail circuit or multi-drop circuit, more formally known as a point-to-multipoint circuit. The tail circuits connect to the main circuit through a simple bridge. See also *bridge*, *drop*, *fantail circuit*, *headend*, *point-to-multipoint circuit*, and *WAN*.

talk battery Referring to the 48V DC current that provides loop current, i.e., supports voice communications over an electrified copper local loop. In a typical single-line residence or business application, the talk battery is provided from a common battery located in the central office exchange (CO or COE), across the local loop, to the telephone set. See also battery, CO, common *battery*, *current*, *DC*, *local loop*, and *V*.

talk path Transmission path, i.e., circuit. See *circuit*.

tandem switch Also known as a *backbone switch* and a *core switch*, a tandem switch is a high-capacity switch positioned in the physical core, or backbone, of a Public Switched Telephone Network (PSTN), where it serves to interconnect edge switches, or Central Office (CO) switches. In the traditional PSTN hierarchy, a tandem might be a Class 1 regional toll center, Class 2 sectional toll center, Class 3 primary toll center, or Class 4 tandem toll center. An access tandem switch serves to connect local exchange carriers (LECs), i.e., local telephone companies, to the interexchange carriers (IXCs), i.e., long distance carriers, over dedicated interoffice trunks, known as access trunks. In a contemporary PSTN, a tandem switch commonly is a hybrid Class 4/5, functioning as both a tandem and a CO (Class 5). See also *IXC*, *LEC*, *PSTN*, and *switch*.

tap **1.** A temporary physical connection to a metallic circuit. See also *bridged tap* and *circuit*. **2.** A wiretap, or secret temporary connection to a circuit for purposes of monitoring the information being transmitted across it. See also *wiretap*. **3.** In Power over Ethernet (PoE), a picker that acts as a splitter, picking off the 48V DC voltage, making it available to the PoE-compliant device at 5V, 6V, or 12V DC, for example. See also *DC*, *PoE*, *splitter*, and *voltage*.

TAPI (Telephony Application Programming Interface) A specification for a computer telephony API developed jointly by Microsoft and Intel in response to the problems associated with Telephony Services Application Programming Interface (TSAPI). As an integral part of Microsoft's Windows Open Services Architecture (WOSA), TAPI runs in Microsoft Windows Server 2003, Windows XP, and Windows NT/2000 environments. See also *API*, *computer telephony*, *TSAPI*, and *WOSA*.

taps The number of horizontal lines or pixels considered in the filtering process, which is one step in video compression. See also *compression*, *filtering*, and *pixel*.

tariff A document that a carrier files with a regulatory authority, describing the services the carrier intends to offer in that domain, the proposed rates and charges, and the proposed obligations, rights, and responsibilities of both the carrier and the customer. The proposal is subject to regulatory review, which generally includes public hearings. See also *carrier*.

TASI (Time-Assignment Speech Interpolation) A technique used on some high capacity, long haul frequency-division multiplexed (FDM) analog voice circuits to improve the efficiency of bandwidth utilization. Through a technique known as silence suppression, TASI senses periods of inactivity in a voice conversation, and inserts the conversation of another speaker into that period of silence. When the first speaker again becomes active, TASI inserts that conversation into another channel where it has detected a period of silence, and so on. If too many speakers are active, voice signals are clipped and quality drops. TASI is no longer used, because all, or nearly all, long haul voice circuits are digital. TASI did, however, form the basis for digital speech interpolation (DSI), which is widely used in contemporary voice networks. See also *analog*, *bandwidth*, *channel*, *digital*, *DSI*, and *FDM*.

TB (TeraByte) One trillion bytes. In computing and storage systems, a TB (terabyte) is actually 1,099,511,627,776 (2^{40}) bytes, since the measurement is based on a base 2, or binary, number system. The term TB comes from the fact that 1,099,511,627,776 is nominally, or approximately, 1,000,000,000,000. See also *byte* and *T*.

Tbps Terabit per second, or trillion (10^{12}) bits per second. A measure of the bandwidth of a digital transmission system. See also *bandwidth*, *bps*, and *T*.

TC (Transmission Convergence) In the ATM reference model, a Physical Layer sublayer that handles frame generation, frame adaption, cell delineation, header error control (HEC), and cell rate decoupling. The frame generation function receives the frame of data presented by the transmitting device across the Physical Medium (PM) sublayer for presentation to the ATM Layer and subsequent segmentation into cells. On the receive side, the TC sublayer receives data in cells and decouples it to reconstitute the frame of data, checking for header errors before presenting the data to the PM sublayer, which passes the data to the end-user device. See also *ATM*, *ATM Layer*, *ATM reference model*, *cell*, *frame*, *HEC*, *Physical Layer*, and *PM*.

TC (Transmit Clock) **1.** A pin on a serial interface that pulses to indicate each bit time when the receiver should sample the data circuit to read a bit value. **2.** A clock that resides in a transmitter and provides a clocking source on which both the transmitting and the receiving data communications equipment (DCE) can synchronize. A TC embedded in a transmitting modem, for example, provides a clocking pulse on which both the transmitting modem and receiving modem can synchronize in order to distinguish between blocks of data. See also *DCE*, *modem*, and *synchronous*.

T-carrier The United States Bell System activated the first commercial digital carrier system in 1962 in Chicago, Illinois, where electrical noise from high-tension lines and automotive ignitions interfered with analog systems. The system was designated T1, with the T standing for Terrestrial to distinguish the land transmission from satellite transmission. (Bell Laboratories also launched Telstar I, the first communications satellite, in 1962.) T-carrier was designed for a four-wire twisted-pair circuit, although the DSX-1 interface is medium-independent, i.e., can be provisioned over any of the transmission media, at least at the T1 rate of 1.544 Mbps. At the T3 rate of 44.736 Mbps, twisted pair is unsuitable over distances greater than 50 feet due to issues of signal attenuation. As the first digital carrier system, T-carrier set the standards for digital transmission and switching, including the use of pulse code modulation (PCM) for digitizing analog voice signals. (Note: T-carrier uses the μ-law (mu-law) companding technique for PCM.) T-carrier not only set the basis for the North American digital hierarchy, but also led to the development of E-carrier in Europe and J-carrier in Japan. The fundamental building block of T-carrier is a 64-kbps channel, referred to as DS-0 (Digital Signal level Zero). Through time-division multiplexing (TDM), T-carrier interleaves DS-0 channels at various signaling rates to create the services that comprise the North American digital hierarchy, as detailed in Table T-1.

Table T-1: North American Digital Hierarchy: T-carrier

Digital Signal (DS) Level	*Data Rate*	*64-Kbps Channels (DS-0s)*	*Equivalent T1s*
DS-0	64 Kbps	1	1/24
DS-1 (T1)	1.544 Mbps	24	1
DS-1C (T1C)	3.152 Mbps	48	2
DS-2 (T2)	6.312 Mbps	96	4
DS-3 (T3)	44.736 Mbps	672	28
DS-4 (T4)	274.176 Mbps	4,032	168
DS-5 (T5)	400.352 Mbps	5,760	250

See also *analog, carrier, digital, DS-0, E-carrier, fractional T1, J-carrier, μ-law, signaling rate, T1, T1C, T2, T3, T4, T5, TDM,* and *transmission rate.* See also *digital signal hierarchy* for a side-by-side comparison of the North American, European, and Japanese digital hierarchies.

TCM (Trellis-Coded Modulation) A modulation scheme based on quadrature amplitude modulation (QAM), but adds a forward error correction (FEC) mechanism to overcome the increased susceptibility to signal impairments. TCM is so named because the plotting of the signal points resembles the latticework of a trellis, such as that used in a rose garden, only four-dimensional. TCM employs a convolutional (i.e., error-correcting) that involves adding an extra bit to every symbol for error control purposes. For example, the 128-QAM technique yields 128 possible signal combinations, with each symbol representing seven bits ($2^7 = 128$). As TCM uses one bit for error control, only six payload bits remain ($2^6 = 64$). Therefore, a modem employing TCM accepts six bits at a time. The two least significant bits (LSBs) are separated from the six-bit payload, are analyzed, and a parity bit is added that describes the mathematical value (odd or even) of the sum of the LSBs. The resulting three bits and the other original four bits are recombined into a seven-bit symbol prior to transmission. The receiving modem reverses the process, analyzes the parity bit describing the LSBs, and either accepts the data as correct, adjusts the data to correct for an error if possible, or requests a retransmission. The LSBs, which are the rightmost bits in a byte, change rapidly if the total byte value changes even slightly. Therefore, they are highly sensitive to errors and very telling in the event that errors occur. When the symbols are plotted onto the logical trellis by the receiver, there are only 64 ($2^6 = 64$) legitimate states, or positions, plus the two for the error control bit, for a total of 66 states. If the indicated plot point is one of the other 62 ($2^7 = 128 - 66 = 62$), the received symbol is assumed to have been errored in transit. TCM is specified by the ITU-T Recommendations for modems at speeds of 19.2 kbps and higher. ITU-T Recommendations for dial-up modems (and maximum speeds) specifying TCM currently include V.32 (9600 bps), V.32bis (14.4 kbps), V.32ter (19.2 kbps), V.34 (28.8 kbps), and V.34bis (33.6 kbps), aka V.34+. See also *amplitude, bit, byte, error control, FEC, ITU-T, LSB, modem, modulation, payload, QAM, symbol, V.32, V.34,* and *V.34+.*

TCP (Transmission Control Protocol) A Transport Layer protocol in the TCP/IP protocol suite, TCP is a connection-oriented protocol designed to provide reliable transmission across inherently unreliable Internet Protocol (IP) networks such as the Internet. Defined in IETF RFC 793, TCP evolved from the ARPANET Network Control Protocol (NCP), which was developed to provide reliable transmission across the analog links of unshielded twisted pair (UTP) and packet radio (e.g., AlohaNet). TCP is a connection-oriented protocol that employs virtual circuits in support of byte-stream-oriented communications. TCP provides for file segmentation into packets prior to transmission, and for reassembly upon receipt. TCP also provides for packet sequencing, end-to-end flow control, and error control, thereby guaranteeing delivery. Each packet in a stream of packets received by the destination device is either acknowledged as having been received correctly, or is retransmitted.

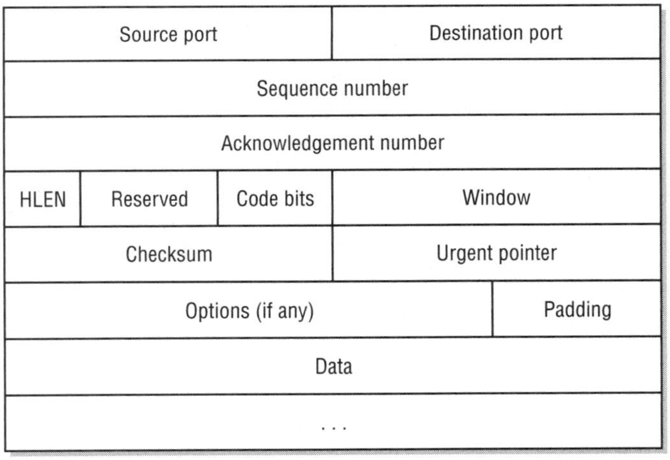

<table>
<tr><td colspan="3">Source port</td><td colspan="2">Destination port</td></tr>
<tr><td colspan="5">Sequence number</td></tr>
<tr><td colspan="5">Acknowledgement number</td></tr>
<tr><td>HLEN</td><td>Reserved</td><td>Code bits</td><td colspan="2">Window</td></tr>
<tr><td colspan="3">Checksum</td><td colspan="2">Urgent pointer</td></tr>
<tr><td colspan="4">Options (if any)</td><td>Padding</td></tr>
<tr><td colspan="5">Data</td></tr>
<tr><td colspan="5">. . .</td></tr>
</table>

Figure T-2

The standard size of the TCP header is 20 octets, as illustrated in Figure T-2, although 4 additional octets may be used to accommodate options. The header fields are as follows:

- **Source Port:** 16 bits that define the TCP port number used by the source application program. TCP ports are logical points of connection.
- **Destination Port:** 16 bits that define the TCP port number used by the destination application program.
- **Sequence Number:** 32 bits that identify the position of the data in the TCP segment relative to the entire originating byte stream.
- **Acknowledgment Number:** 32 bits that identify the acknowledgment number of the octet that the source expects to receive next. The acknowledgment number explicitly acknowledges that all previous data octets associated with all previous segments were received correctly.
- **HLEN:** 4 bits that specify the segment header length.

- **Reserved:** 6 bits reserved for future use.
- **Code Bits:** 6 bits that define the purpose and contents of the segment, e.g., acknowledgment, connection reset, and end of byte stream.
- **Window:** 16 bits that advertise the size of the sender's sliding receive window, that is, how much data the host computer is willing to accept, based on its buffer size.
- **Checksum:** A 16-bit cyclic redundancy check (CRC) used for error control in the data field, as well as the header.
- **Urgent Pointer:** 16 bits that identify urgent out-of-band data (i.e., data not part of the information stream). Such data is treated on a high-priority basis, in advance of data-stream octets that might be awaiting consumption by the destination hosts. Urgent data, for example, might include a keyboard sequence to interrupt or abort a program.
- **Options, If Any:** 24 bits that address a variety of options, such as *maximum segment size* (*MSS*).
- **Padding:** 8 bits in an optional field used only when necessary to ensure that the TCP header extends to an exact multiple of 32 bits. This field is used only when the Options, If Any field is used.
- **Data:** A variable-length field that contains the actual data content, or payload. When TCP is used in conjunction with IPv4, the minimum size of the data field is 536 octets, which is the minimum size of the IPv4 datagram, less 20 octets each for the standard IP and TCP headers.

See also *AlohaNet*, *application*, *byte*, *connection-oriented*, *CRC*, *datagram*, *error control*, *flow control*, *header*, *host*, *IETF*, *Internet*, *IP*, *NCP*, *octet*, *packet*, *payload*, *port*, *protocol*, *protocol suite*, *segment*, *segmentation*, *stream-oriented*, *TCP/IP*, *Transport Layer*, *UTP*, *virtual circuit*, *well-known port*, and *window*.

TCPA (Telephone Consumer Protection Act) See *Telephone Consumer Protection Act*.

TC-PAM (Trellis-Coded Pulse Amplitude Modulation) Also known as *trellis-coded modulation* (TCM). See *TCM*.

TCP/IP (Transmission Control Protocol/Internet Protocol) Referring to the ARPA protocol suite that provides the basis for what has evolved into the Internet and, therefore, often is referred to as the Internet protocol suite. This is a four-layer protocol suite that maps into the OSI Reference Model that followed a few years later. Table T-2 includes the major protocols that comprise the TCP/IP protocol suite as they relate to the ARPA and OSI models.

Table T-2: TCP/IP Protocol Suite

OSI Layer	*ARPA Layer*	*Protocol*			
7 Application	Process/ Application	FTP File Transfer Protocol	SMTP Simple Mail Transfer Protocol	SNMP Simple Network Management Protocol	TELNET Telecommunications Network
6 Presentation					
5 Session					
4 Transport	Host-to-Host	TCP Transmission Control Protocol		UDP User Datagram Protocol	
3 Network	Internet	ARP Address Resolution Protocol	ICMP Internet Control Message Protocol	IP Internet Protocol	
2 Data Link	Network Interface	LAN, MAN, WAN Network Interface Card (NIC)			
1 Physical		Transmission Media			

See also *ARP, ARPA, FTP, ICMP, Internet, IP, LAN, MAN, NIC, OSI Reference Model, protocol, SMTP, SNMP, TCP, TELNET, transmission media, UDP,* and *WAN.*

TCP/IP protocol suite See *TCP/IP.*

TDEA (Triple Data Encryption Algorithm) See *Triple DES.*

TDES (Triple Data Encryption Standard) See *Triple DES.*

TDD **1.** Time-Division Duplex. A means of providing bidirectional communications in digital wireless networks, such as cellular networks. TDD typically is used in conjunction with frequency-division duplex (FDD), with the forward and backward time division multiplexed (TDM) channels riding over separate frequency channels. Further, the time slots are staggered so the frequency-specific transceivers are not asked to transmit and receive at the same exact moments in time. See also *cellular, digital, FDD, TDM, time slot,* and *transceiver.* **2.** Telecommunications Device for the Deaf. In the United States, a printing telegraph service widely used by those with hearing or speech impairments. Also generically known as *TTY* (TeleTYpewriter), *textphone* in Europe, and *minicom* in the United Kingdom. TDDs use the Baudot code, also known as International Telegraph Alphabet #2 (ITA #2). See also *Baudot code, telegraph,* and *teletype.*

TDM (Time-Division Multiplexing) A multiplexing method by which multiple low-speed incoming transmissions can share a single high speed outgoing digital circuit. An analog voice conversation requires bandwidth of 4 kHz. Although there are a considerable number of methods for converting an analog voice signal into a digital signal, the fundamental standard is pulse code modulation (PCM), which requires 64 kbps. A voice grade digital channel, therefore, is 64 kbps wide, which forms the fundamental building block for digital switching and transmission of not only voice, but all forms of data. A typical digital voice application is multi-channel in nature and involves a four-wire circuit with a TDM multiplexer, or mux, placed on each end of the circuit, as illustrated in Figure T-3.

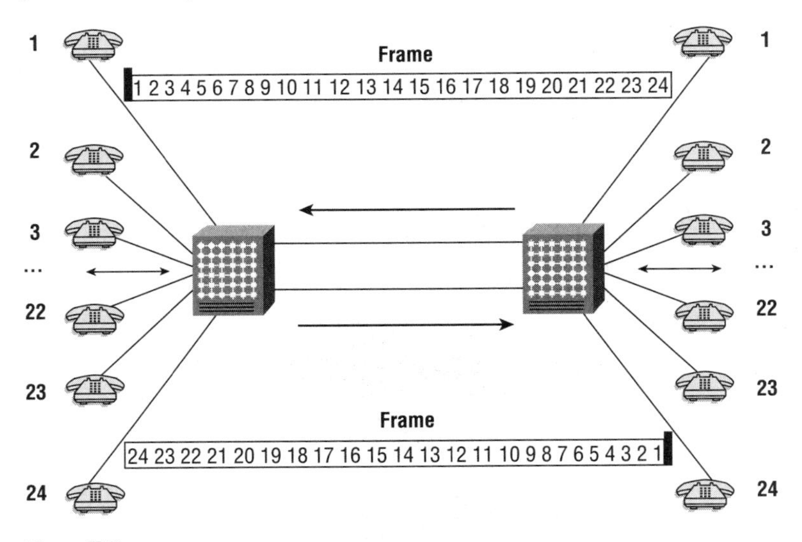

Figure T-3

At the transmitting end of the circuit, the mux scans the buffers associated with the ports to which individual devices are attached. Each device port is allocated a channel in the form of a time slot on the aggregate line for transmission of data. Using T1 as an example, the transmitting TDM mux relieves each buffer of an 8-bit sample of data, beginning with buffer/port #1 and proceeding in sequence through buffer/port #24, and transmits the bytes, in sequence, across the circuit, interleaving them into a frame of data. The mux prepends the frame with a framing bit that delineates that frame from another, and is used by the multiplexers and other intermediate devices for purposes of synchronization and, in some cases, for various signaling and control purposes. This process occurs 8,000 times a second at the precise pace of 125μs (microseconds), with the pace driven by the demands of PCM voice. At the receiving end of the circuit, the process is reversed. As the paired muxes are tightly synchronized, each channel in each frame is identified, the individual transmissions are de-multiplexed, and each is forwarded over the port to which the intended receiving terminal device is attached. See also *FDM* and *STDM*.

TDMA (Time Division Multiple Access) **1.** A multiplexing technique used in digital wireless systems that divides each frequency channel into multiple time slots, each of which supports an individual conversation. The total available bandwidth, the bandwidth of the individual channels, and the number of time slots per channel vary according to the particular standard, as well as the specific coding technique employed. TDMA is the wireless variant of time division multiplexing (TDM) in the wireline domain. A TDMA system, such as GSM, employs both frequency division multiplexing (FDM) and TDMA. FDM derives multiple carrier channels from a wider band of assigned spectrum. Within each frequency channel, TDMA derives multiple time slots (i.e., digital channels), for which incoming and outgoing calls contend. Alternative multiplexing techniques employed in various cellular radio networks are code division multiple access (CDMA) and frequency division multiple access (FDMA). See also *bandwidth*, *carrier*, *CDMA*, *channel*, *digital*, *E-TDMA*, *FDM*, *FDMA*, *frequency*, *GSM*, *multiplexing*, *TDM*, *time slot*, *wireless*, and *wireline*. **2.** The vernacular name often applied to the North American IS-54 and IS-136 standards for cellular radio networks, more correctly known as Digital Advanced Mobile Phone System (D-AMPS). TDMA actually refers to Time Division Multiple Access, the access technique first specified in IS-54 and used in most 2G digital cellular networks, including those conforming to the pan-European GSM standard. See also *2G*, *cellular radio*, *D-AMPS*, *digital*, *GSM*, *IS-54*, and *IS-136*.

TDM MUX or TDM mux (Time Division Multiplexer) A device that performs time division multiplexing (TDM). See also *mux* and *TDM*.

TD-SCDMA (Time Division-Synchronous Code Division Multiple Access) A 3G cellular radio standard under development by the Chinese Academy of Telecommunications Technology (CATT), an agency of the People's Republic of China (PRC), and adopted by the 3rd Generation Partnership Project (3GPP) as UTRA TDD 1.28Mcps Option (UMTS Terrestrial Radio Access Time Division Duplex 1.28 Mega chips per second Option). TD-SCDMA is based on code division multiple access (CDMA), but uses time division multiple access (TDMA), as well, with the uplink signal synchronized by the base station (BS). Like European versions of W-CDMA, TD-SCDMA uses time-division duplex (TDD) to support bidirectional communications by derived uplink and downlink time slots within the same frequency band. TD-SCDMA specifications also include a frequency division duplex (FDD) mode. The modulation scheme on the uplink is quaternary phase shift keying (QPSK) and on the downlink either binary phase shift keying (BPSK) or QPSK. The synchronization of the signal at the base station improves the orthogonality of the coded transmissions, which serves to reduce interference between users and, therefore, allow increased capacity. TD-SCDMA runs in the 2000 MHz (2 GHz) band. Each 5 MHz carrier supports 120 voice channels in TDD mode (one carrier for both uplink and downlink transmission) and 250 channels in FDD mode (one uplink carrier and one downlink carrier). Asynchronous data rates range from 1.2 kbps to 2 Mbps, symmetrical. See also *3GPP*, *BPSK*, *carrier*, *CDMA*, *channel*, *chip*, *chip rate*, *downlink*, *FDD*, *frequency band*, *interference*, *orthogonal*, *QPSK*, *symmetric*, *synchronize*, *TDD*, *TDMA*, *time slot*, *uplink*, and *W-CDMA*.

TE1 (Terminal Equipment 1) ISDN-compatible equipment. In other words, terminal equipment that can interface directly to an ISDN circuit via a four-wire twisted-pair S or T interface. Switches and routers must be natively ISDN-compatible in order to connect to ISDN circuits, typically at the U reference point on the local loop. Telephone sets and data terminals need not be ISDN-compatible. Those not compatible are classified as TE2 and connect through a terminal adapter (TA). See also *compatible*, *four-wire circuit*, *ISDN*, *Reference Point U*, *router*, *S Interface*, *switch*, *TA*, *TE2*, *T Interface*, and *twisted pair*.

TE2 (Terminal Equipment 2) Terminal equipment that is not ISDN-compatible. TE2 equipment must connect through a terminal adapter (TA), also known as an ISDN modem, that resolves issues of incompatibility. Telephone sets and data terminals generally are classified as TE2, as the cost of ISDN compatibility generally is considered prohibitive. See also *compatible*, *ISDN*, *TA*, and *TE1*.

techie A person with technical expertise in an area, especially computer technology. A good techie is invaluable, and can even be a guru. A techie gone bad is a nerd, but may still be invaluable. See also *empty suit*, *geek*, *guru*, *nerd*, and *suit*.

technology From the Greek *tekhnologia*, meaning *systematic treatment* or *science of craft*. Applied science. Practical arts. The application of scientific devices, machines, and techniques for manufacturing and other productive processes. See *Hellenologophobia*, *technophilia*, and *technophobia*.

technophilia The love of technology. See also *technophobia*.

technophobia The fear of technology. See also *technophilia*.

TEI (Terminal Endpoint Identifier) In ISDN, a one-octet address field that identifies the address of the destination terminal device. The TEI and the service access point identifier (SAPI) jointly comprise the data link connection identifier (DLCI), which is the two-octet address field in an LAPD frame. See also *DLCI*, *ISDN*, and *LAPD*.

telco (telephone company) A company in the business of providing local telephone service, i.e., a local exchange carrier (LEC).

Telcordia Technologies Originally Bell Communications Research (Bellcore), the research and development arm of the Regional Bell Operating Companies (RBOCs). SAIC acquired the company 1988 and changed the name to Telcordia Technologies in April 1999, with the stated focus of emerging technologies. Telcordia is now a private, standalone organization involved in the development of OSSs and network management software, as well as consulting, testing, and research services. See also *Bellcore*, *network management*, *OSS*, and *RBOC*.

telco return A hybrid communications technique for videoconferencing in lecture mode, telco return combines satellite communications in one direction and the public switched telephone network (PSTN) in the other. This technique involves satellite transmission of the downstream video and audio signal from the lecturer's location to the multiple locations where the audience members are located. Audience participation is via the PSTN, either on a dial-up basis or over dedicated circuits. This hybrid approach provides sufficient bandwidth at reasonable cost for the point-to-multipoint presentation. Although the downstream satellite transmission involves propagation delay of approximately 0.32 seconds, the lecture is all one-way, so the delay is not an issue, as long as it is consistent, i.e., there is no jitter. (Even at roughly the speed of light, it takes approximately .25 seconds for a radio signal to travel from the transmitting Earth station to the satellite at an altitude of 22,300 miles, and back to a receiving Earth station. Processing time at the Earth stations and onboard the satellite takes another 0.07 seconds or so.) If, however, the interactive question-and-answer portion of the presentation were to take place over the satellite, the propagation time would triple (0.32 seconds for the presentation point, 0.32 seconds for the question, and 0.32 seconds for the response), and the flow of the presentation would suffer irreparably. Telco return overcomes this issue, as the audience interaction takes place over the PSTN, which offers negligible delay. Telco return also eliminates the requirement for two-way satellite antennas and bandwidth, which considerably lowers the costs of the satellite links. Some providers of direct broadcast satellite (DBS) services use this same hybrid approach in support of interactive Internet access. See also *jitter*, *propagation delay*, *satellite*, and *speed of light*.

telecommunication management network (TMN) See *TMN*.

telecommunications From the Greek *tele*, meaning *far off*, and Latin *communicare*, meaning *to share* or *impart*, literally *to make common*. The science and technology of transmitting voice, audio, facsimile, image, video, computer data, and multimedia information over significant distances by the use of electromagnetic energy in the form of electricity, radio, or optics.

Telecommunications Act of 1996 An act of the United States Congress that effectively superseded the 1982 Modified Final Judgement (MFJ), removing line-of-business restrictions and promising to permit full and open competition in virtually every aspect of communications, from radio broadcasting to CATV to local exchange service and long distance service. The local exchange networks were opened to competition, and the incumbent local exchange carriers (ILECs) were required to lease cable pairs, switch ports, central office (CO) space, and other elements of their networks to competitive local exchange carriers (CLECs). The act specified three ways by which CLECs could provide competing local phone service:

- **Build and Interconnect:** CLECs could build their own wireline or wireless local loops and interconnect with the ILEC and interexchange carrier (IXC) networks.
- **Bundled Wholesale Purchase:** CLECs could purchase bundled local telephone service from the ILECs at government-controlled wholesale prices, which typically were 15–25 percent below retail prices.
- **Unbundled Service:** The CLECs could purchase the very same network elements on an unbundled basis, which essentially were a menu of Unbundled Network Elements (UNEs) from which the CLECs could choose to lease only what they needed, on a case-by-case basis.

In exchange for yielding their monopoly status in the local exchange, the Regional Bell Operating Companies (RBOCs) were to be permitted to compete in the interstate long-distance market once they demonstrated that they were no longer the dominant carriers in their local exchange markets. The RBOCs demonstrated, over time, that this was case, and all have now been granted interexchange carrier (IXC) status. The act also, in large part, lifted ownership restrictions, enabling the telecommunications carriers to invest, relate, merge, and acquire. The act also effectively deregulated CATV, opened spectrum for broadcast TV stations to introduce high-definition television (HDTV), and formally established a Universal Service Fund (USF) to keep basic telecommunications service rates low in rural areas and to subsidize telecommunications and Internet services to schools and libraries. See also *bundled service*, *CATV*, *CLEC*, *CO*, *HDTV*, *ILEC*, *IXC*, *local loop*, *MFJ*, *RBOC*, *spectrum*, *UNE*, *UNE-P*, and *USF*.

Telecommunications Device for the Deaf (TDD) See *TDD*.

Telecommunications Industry Association (TIA) See *TIA*.

Telecommunications Network (TELNET) See *TELNET*.

Telecommunications Relay Service (TRS) See *TRS*.

telecommunications organization (TO) See *TO*.

telecommunications service As defined in the U.S. Telecommunications Act of 1996, "the offering of telecommunications for a fee directly to the public, or to such classes of users as to be effectively available directly to the public, regardless of the facilities used." As interpreted by the Federal Communications Commission (FCC), broadband wireline services such as digital subscriber line (DSL) and cable modem service are information services, rather than telecommunications services. See also *broadband*, *cable modem*, *DSL*, *FCC*, *information services*, *Telecommunications Act of 1996*, and *wireline*.

telecommuter A businessperson who works and communicates largely from home through the use of information and communications technologies, rather than traveling to a company office. The terms *telecommuting* and *telework* were coined in the early 1970s by Jack Nilles, who was working at the University of Southern California (USC) on projects aimed at eliminating rush-hour traffic by allowing employees to work closer to home or at home, linked to the central workplace via telecommunications networks. See also *teleworker*.

telegram A message transmitted by telegraph. See also *telegraph.*

telegraph From the Greek *tele*, meaning *far off*, and *graphos*, meaning *written*. See also *Hellenologophobia*. **1.** An apparatus or process for communicating information over a distance by coded signals. Simple telegraphs employ smoke signals, drums, mirrors, flags, fires, lanterns, and mechanical semaphores. **2.** The electric telegraph was invented by Samuel F.B. Morse (1791–1872) and Alfred Vail (1807–1859), and the first intercity telegraph line in the United States began operation in 1844. That system involved a transmitter, or sender, in the form of a manually operated telegraph key that the operator used to open and close an electric circuit, thereby sending coded alphanumeric data using Morse code. The receiver recorded the code symbols with an armature that scratched a paper tape. The manual key has been improved, and now operates on the basis of a side-to-side motion, which is more natural than an up-and-down motion. The receiver eliminated the armature, and became a simple sounder that emits audible clicks, for which the human operator must listen to decode the transmission. *Note:* In 1838, Samuel Morse offered to give his invention of the telegraph to the newly formed (1836) Republic of Texas. After having received no reply for 22 years, Morse withdrew his offer in a letter to Texas governor Sam Houston in 1860. See also *ham*; *Morse Code*; *Morse, Samuel F.B.*; *Pony Express*; *teletype*; *Telex*; *Vail, Alfred*; and *Western Union*.

tele-immersion A developing application of Internet2 intended to allow multiple, geographically distributed users to collaborate in real time in a shared, simulated hybrid environment through a synthesis of media technologies such as three-dimensional (3D) environment scanning, projection and display, tracking, audio, video, robotics, and haptics (i.e., touch) technologies. In other words, tele-immersion creates a multimedia virtual meeting space in cyberspace. See also *cyberspace*, *Internet2*, *multimedia*, *real time*, and *virtual*.

telemanagement software A suite of application software used to manage a telecommunications system. Telemanagement software systems generally are modular and rely on a single, integrated relational database management system (RDBMS). Modules can include call accounting, cost allocation, traffic analysis, network optimization, asset management (system and component inventory management, and wire and cable management), directory management, trouble ticket management, work order management, and security management (intrusion detection and fraud detection).

telemarketing The use of the telephone to solicit prospective customers to sell products or services. Telemarketing dates at least to the 1970s, when AT&T promoted long distance telemarketing under the name Phone Power to promote WATS service. Under pressure from consumers and consumer groups, the United States Congress passed the Telephone Consumer Protection Act (TCPA) in 1991. The TCPA specifically mentions automatic telephone dialing systems, or auto dialers, and prerecorded messages, and includes substantial penalties for telemarketers found guilty of violating the privacy of residential, fax, and certain other categories of users. In 2003, the FCC and Federal Trade Commission (FTC) established the National Do-Not-Call Registry, which allows residential landline customers to opt out of unsolicited telemarketing calls. The Do-Not-Call Registry was extended to cellular phones in late 2004. Telemarketers must honor the opt-out, although there are exceptions for political campaigns and nonprofit organizations. The penalties for violations are substantial. See also *cellular* and *landline*.

telemedicine A videoconferencing or multimedia conferencing application that supports remote medical consultation and even remote diagnosis and treatment.

telemetry From the Greek *tele*, meaning *far off*, and *meter*, meaning *measure*, and translating literally as *measure far off*. The branch of science or process of remote measurement and collection of variable data, such as pressure, temperature, flow, or radiation, and transmission of the data to a distant location for analysis and interpretation.

telephone From the Greek *tele*, meaning *far off*, and *phone*, meaning *a sound*, and translating literally as *a sound far off*. See also *Hellenologophobia*. **1.** A device comprising a transmitter, receiver, and dialing mechanism (dial or keypad), and used for transmitting speech over distances by converting acoustical signals into electrical signals. A typical telephone can be analog or digital in nature, and be equipped with a dial or keypad. See also *acoustics*, *electricity*, *receiver*, *signal*, *speech*, and *transmitter*. **2.** An invention of the devil which abrogates some of the advantages of making a disagreeable person keep his distance, at least according to Ambrose Bierce (1842–1914).

Telephone Consumer Protection Act (TCPA) A United States federal law (1991) that led to Federal Communications Commission (FCC) rules prohibiting telephone solicitation calls to a residence before 8:00 am or after 9:00 pm. Anyone making a telephone solicitation call to a residence must provide his or her name, the name of the person or entity on whose behalf the call is being made, and a telephone number or address at which that person or entity may be contacted. The TCPA specifically mentions automatic telephone dialing systems, or auto dialers, and prerecorded messages, and includes rather substantial penalties for telemarketers found guilty of violating the privacy of residential, fax, and certain other categories of users. In 2003, the FCC revised its rules implementing the TCPA and, together with the Federal Trade Commission (FTC), established a national Do-Not-Call Registry. The FCC also adopted restrictions on the number of abandoned calls that are permissible. See also *Do-Not-Call Registry*, *FCC*, and *telemarketing*.

telephone number Also known as the *directory number* (DN). The logical address assigned to a subscriber, a telephone number comprises a series of digits conforming to a format defined by the ITU-T at the international level in E.164 and comprises a maximum of 15 digits, including a country code. Within that format, a national or regional authority sets the format for that domain. The North American Numbering Council (NANC) sets the format for the North American Numbering Plan (NANP), for example, including a country code, an area code, a central office prefix, and a line number. In the traditional landline context of the public switched telephone network (PSTN), each telephone number is assigned to a circuit. In cellular networks, each telephone number is assigned to a cellular telephone. See also *E.164*, *ITU-T*, *NANC*, and *NANP*.

Telephony Application Programming Interface (TAPI) See *TAPI*.

Telephony Services Application Programming Interface (TSAPI) See *TSAPI*.

Telepoint An early public wireless telephony service offered in the United Kingdom. Telepoint was based on the CT2 standard for digital cordless telephony. See also *cordless telephone*, *CT2*, and *digital*.

teleport From the Greek *tele*, meaning *far off*, and the Latin *porta*, meaning *door*, and translating literally as *far off door*. A contraction of telecommunications port. A telecommunications hub where satellite service providers cluster satellite antennas. A teleport acts as a centralized physical gateway between terrestrial and nonterrestrial networks. A teleport generally is located in proximity to a major population center, but far enough away, or high enough, to have an unobstructed view of orbiting satellites in order to ensure line-of-sight (LOS) connectivity and, therefore, high-quality communications.

teleportation The transportation of matter through space by converting it into pure energy at the source and then reconverting it to matter at the sink. Teleportation is pure theory and will never work. The same was true of VoIP in 1990 or so, but people are still working on perfecting it. Beam me up, Scotty.

teletype From the Greek *tele*, meaning *far off*, and *typos*, meaning *mark*, and translating literally as *mark far off*. A printing telegraph system that replaced the sending key with a typewriter-like keyboard and the receiving sounder with a teleprinter. Western Union introduced teletypewriter service in 1923 so that companies could link branches and even join other companies in private text messaging over leased private-line networks. Teletype service was heavily used by banks, telephone companies, electric utilities, and others into the early 1970s. Teletypewriter (TTY) service, also known as Telecommunications Device for the Deaf (TDD) in the United States, textphone in Europe, and minicom in the United Kingdom, is heavily used by those with hearing or speech impairments. The teletype was based on the Baudot Distributor, an automatic telegraph system that involved pairs of synchronized electromechanical machines. Both the Baudot Distributor and the teletype used the five-bit Baudot coding scheme. See also *Baudot code*, *telegraph*, and *telex*.

teletypewriter See *teletype*.

television (TV) See *TV.*

teleworker A businessperson who works and communicates largely through the use of information and communications technologies, rather than traveling and conducting business face-to-face at various physical company offices and client sites. The terms *telecommuting* and *telework* were coined in the early 1970s by Jack Nilles, who was working at the University of Southern California (USC) on projects aimed at eliminating rush-hour traffic by allowing employees to work closer to home or at home, linked to the central workplace via telecommunications networks. See also *telecommuter.*

telewriter See *teletypewriter.*

telex (*tel*etypewriter *ex*change) Beginning in about 1910, telegraph companies began to use telex, a rotary dialing system for routing telegraph calls, much like that used in telephone networks. Telex initially ran at the amazing signaling speed of 45.5 bits per second (bps) and, later, at 50 bps, or 66 words per minute (wpm). At that signaling rate, one analog voice grade channel could support 24 or 25 telex transmissions through frequency division multiplexing (FDM). In 1958, Western Union introduced its Telex® service, a direct-dial consumer-to-consumer teleprinter service. In 1930, AT&T released its TWX (TeletypeWriter eXchange) service, a high speed telex service that ran at 75 bps, and later 150 bps. Western Union operated the TWX service in the United States for many years until AT&T acquired Western Union's Telex network in 1990. Telex services eventually reached some 190 countries and 3.5 million machines, by some estimates. Telex was the first true electronic mail service and is still in use, largely in developing countries. Telex and TWX were used in business extensively at least in part because the networks provided the equivalent of caller ID, as each user was attached to a known port on a switch. This gave transmissions the property of nonrepudiation. See also *AT&T, FDM, telegraph, voice grade,* and *Western Union.*

TELNET (Telecommunications Network) A TCP/IP protocol functioning at the Application Layer, TELNET provides terminal emulation service over a Transmission Control Protocol (TCP) connection, enabling the user to assume control over the applications that reside on a remote system. Virtual network terminal services permit the data terminal equipment (DTE) to emulate other terminal devices, transparently. TELNET is defined in IETF RFC 854 (May 1983). See also *Application Layer, DTE, IETF, TCP, TCP/IP, terminal emulation,* and *virtual.*

Telstar **1.** The first active telecommunications satellite. Launched in 1962 by the United States National Aeronautics and Space Administration (NASA), Telstar was developed by Bell Telephone Laboratories (Bell Labs) and owned by AT&T. Telstar was placed in an elliptical low-earth orbit (LEO), and had a dwell time of only 20 minutes or so. Its fragile transistors soon failed, and the first Telstar went out of service in 1963, but remains in orbit to this day. Many other generations of Telstars have since been launched, and many remain in service. See also *dwell time, LEO,* and *satellite.* **2.** An instrumental composition written by Joe Meek and first recorded by the British group The Tornadoes in 1962. Inspired by the Telstar satellite, "Telstar" was the first tune by a British group to reach No. 1 on the U.S. Billboard pop music charts. The Beatles finally made it to No 1 in 1964 with "Can't Buy Me Love."

tenant services See *multi-tenant services.*

tensile strength The maximum lengthwise stretching force that a cable can withstand without breaking. Tensile strength must be considered during the installation process, because a pulling load is placed on an optical fiber, in particular. Riser cables must be manufactured and installed with tensile strength in mind, because they hang vertically in place for long periods of time. Aerial cables run horizontally, attached to and suspended from poles placed at intervals along a cable route. Aerial and riser cables commonly are manufactured with strength members, not only to increase the overall tensile strength of the cable, but also to relieve the wires and fibers, themselves, from bearing the load. See also *aerial cable, break strength, flex strength, riser cable,* and *strength member.*

ter From the Latin *ter,* meaning thrice, or threefold. In standards terminology, ter refers to the third version, e.g., V.27ter.

tera- (T) See *T.*

terabit per second (Tbps) See *Tbps.*

terabyte (TB) See *TB.*

terahertz (THz) See *THz.*

terminal **1.** A device that constitutes a point of termination of a communications circuit or channel, i.e., a transmitter or receiver, also known respectively as a source or sink. **2.** In telecommunications, a voice terminal is a telephone set, which can take many forms, including a simple analog rotary dial set, a digital smartphone, a digital softphone, and a cellular telephone. **3.** In data communications, a device comprising a keyboard, video adapter, and monitor. A data terminal is a dumb terminal, i.e., it does no independent processing, but relies on the computational resources of a computer to which it is connected over a dedicated circuit or through a network. A data terminal essentially is an input/output (I/O) device. See also *DTE* and *terminal emulation.*

terminal adapter (TA) See *TA.*

terminal emulation The process by which a microcomputer imitates a dumb terminal in order to communicate with a mainframe computer or other device. See also *dumb terminal, emulation, mainframe, microcomputer,* and *TELNET.*

terminal endpoint identifier (TEI) See *TEI.*

terminal equipment **1.** A device that constitutes a point of termination of a communications circuit or channel. Terminal equipment includes all customer premises equipment (CPE), including voice terminal equipment and data terminal equipment (DTE). See also *CPE, DTE,* and *terminal.* **2.** TE. In ISDN, TE includes all customer premises equipment (CPE), including voice terminal equipment and data terminal equipment (DTE). ISDN TE also includes premises-based switching equipment, such as PBXs and routers. There are two types of ISDN TE: TE1 is ISDN-compatible, and TE2 is not. See also *CPE, DTE, TE1,* and *TE2.*

terminate-and-stay-resident (TSR) See *TSR.*

terminating multiplexer (TM) See *TM.*

text Synonymous with *plain text.* Data comprising standard characters (e.g., letters, numbers, and punctuation marks), with no formatting codes. Such unformatted text usually is coded according to the ASCII standard, which adequately represents the characters and symbols used to form the words used in most spoken languages and which, therefore, lends itself to text-to-speech (TTS) conversion. See also *ASCII, data, rich text, texting,* and *TTS.*

text field Also known as the *data field* and the *payload.* The user data field contained within a block, cell, frame, or packet. The text may be preceded by start-of-text (STX) and succeeded by end-of-text (ETX) control characters so the receiving device can determine the location of the message data. See also *block, cell, data format* (illustration), *frame, header, packet,* and *trailer.*

text messaging See *texting.*

texting Referring to the process of communicating via plain text messages over a cellular radio network using short message service (SMS). See also *cellular radio, SMS,* and *text.*

textphone In Europe, a printing telegraph service widely used by those with hearing or speech impairments. Also generically known as *TTY* (TeleTYpewriter) and *TDD* (Telecommunications Device for the Deaf) in the United States, and *minicom* in the United Kingdom. See also *telegraph* and *teletype.*

text-to-speech (TTS) See *TTS.*

theory A formulation of relationships or principles based on considerable factual evidence objectively analyzed to explain the operation of certain phenomena. A theory is in sharp contrast to pure speculation, which is an opinion based on incomplete evidence or information. See also *law.*

ThickNet A contraction of *thick Ethernet.* See *10Base5.*

thin access point (thin AP) See *thin AP.*

thin AP (thin access point) In wireless local area networks (WLANs), an AP intended to act under the supervision of a centralized controller that configures, manages, and secures the environment. The centralized controller provides a single point of administration for all APs. See also *fat AP* and *WLAN.*

thin client In a client/server network, client software or a client node that has little in the way of resources such as memory, disk space, and processing power and, therefore, relies on the server to provide those resources. A browser, for example, provides client access to server-hosted applications, data, and processing. A diskless workstation is an example of a thin client node. See also *browser, client, client/server, dumb terminal, fat client, node, server,* and *software.*

ThinNet A contraction of *thin Ethernet.* See *10Base2.*

throughput **1.** The total amount of data that can be processed by, passed through, or otherwise put through a system or system element when operating at maximum capacity. The measurement is in data units per unit of time, such as bits, bytes, blocks, cells, frames, or packets per second. **2.** The amount of useful data, user data, or payload that can be processed by, passed through, or otherwise put through a system or system element when operating at maximum capacity. In this sense, overhead, i.e., signaling and control data, is of no relevance except for the fact that it reduces the payload and, therefore, the throughput. Throughput is always less than bandwidth. In other words, the transmission rate, or data rate, is always less than the signaling rate. See also *bandwidth, goodput,* and *overhead.*

THz (TeraHertz) One trillion (10^{12}) hertz. A measure of the frequency of an analog signal. The infrared light spectrum is 300 GHz–400THz, and the visible light spectrum is 400 THz–1 PHz (petahertz, or one quadrillion ((10^{15})) Hz). At this level, bandwidth is quantified in GHz to describe channel widths, and in bps to describe transmission rates. See also *analog, bandwidth, bps, frequency spectrum,* and *T.*

TIA (Telecommunications Industry Association) An industry association that represents information and communications technology manufacturers and service providers in standards development, domestic and international policy advocacy, and facilitating business opportunities for its membership. TIA represents the communications sector of the Electronics Industry Alliance (EIA). See Appendix A for contact information. See also *EIA.*

tie line A line that directly interconnects, or ties together, two key telephone systems (KTSs) in a private network configuration, a tie line is a dedicated circuit that generally is leased from a public carrier. A tie line allows users to avoid toll charges for long-distance calls between two geographically separated offices. If a key system sits behind (i.e., is used in conjunction with) a PBX, one or more tie lines are used to interconnect them. See also *line, nailed-up circuit,* and *tie trunk.*

tie trunk A trunk that directly interconnects, or ties together, two PBXs in a private network configuration, a tie trunk is a dedicated circuit that generally is leased from a public carrier. Through the use of optional automatic route selection (ARS) software, PBX systems can automatically route calls between offices over an available tie trunk rather than over the public switched telephone network (PSTN), thereby avoiding toll charges. See also *nailed-up circuit, tie line,* and *trunk.*

.tiff The file extension identifying files in the Tagged Image File Format (TIFF) format. See also *TIFF.*

TIFF (Tagged Image File Format) A standard bitmapped file format commonly used for scanning, storage, and transfer of image files, including photographs and line art. The T.37 standard for Fax over Internet Protocol (FoIP) transmission, for example, specifies TIFF-F. TIFF files are identified by the .tif or .tiff file extension. See also *FoIP*.

TIFF-F (Tagged Image File Format-Fax) A compressed data format specified by the T.37 standard for Fax over Internet Protocol (FoIP) transmission. See also *compression*, *FoIP*, and *T.37*.

tight buffered cable A type of fiber optic cable with a protective material extruded directly on the acrylate coating of an optical fiber to further allow individual fibers to be handled easily during installation, while protecting them from physical damage. A tight buffered cable is used for short jumper cables and many other indoor applications where the temperature is controlled and the differences in thermal expansion and contraction are not so great between the buffer and fiber as to cause bending, which ultimately can lead to cracking and breaking of the fiber. See also *acrylate coating*, *cable*, and *loose tube cable*.

time-assignment speech interpolation (TASI) See *TASI*.

time compression **1.** A compression technique that speeds up the playback of audio and audio-video content. Linear time compression does so by uniformly compressing and speeding up the playback of all speech segments, but can affect the pitch of the speech. Non-linear time compression does so without affecting the pitch, by selectively speeding up certain speech segments, eliminating or reducing pauses and periods of silence, and uniformly speeding up the remaining speech. More sophisticated approaches attempt to mimic the human compression strategies that people use when they talk faster. Time compression, both linear and non-linear, has been used for some years in voice mail systems, and more recently in streaming media products used in Web-based applications. See also *compression* and *voice mail*. **2.** A compression technique that supports the transfer of data between devices at a rate faster than the playback rate, depending on bandwidth availability. For example, a 30-frame file of 56.7 Mbits that represents 1 second of broadcast-quality video in raw, uncompressed content form might blow up to 65+ Mbits to accommodate overhead. The transmission of that file in real time, i.e., 30 frames per second (fps) to create the illusion of full motion for a human input device (eyeball) and processor (brain) would require a transmission facility capable of a rate of 65 Mbps to support that one audiovisual signal. If, on the other hand, one can compress that content by 90%, the file is 5.67 Mbits of raw content and 6.5 Mbits, including overhead. If the transmission facility is capable of the same 65 Mbps and the content is stored and ready for transmission, 1 second of real-time content can be transmitted from source to sink in one-tenth of a second. The if transmission facility is an OC-192 backbone SDH/SONET fiber optic transmission system (FOTS) running at 10 Gbps in the carrier backbone, then 1,538 seconds (25.64 hours) of full-motion video theoretically can be transmitted in one second. Ultimately, of course, the slowest link determines the maximum speed of transmission, end-to-end, and, therefore, whether or not even the most sophisticated compression technique yields the benefits of time compression. See also *backbone*, *bandwidth*, *broadcast quality*, *compression*, *FOTS*, *overhead*, *persistence of vision*, *phi phenomenon*, *real time*, *SDH*, and *SONET*.

time diversity Referring to the fact that elements of a radio signal transmitted at the same moment in time can arrive at a receiver at different moments in time if they travel different physical paths of different lengths. Through the use of receiving antenna technology known variously as rake receivers and multiple-input-multiple-output (MIMO), multiple spatially diverse receiving antennas can combine and correlate multiple incoming signals received at different times, to yield a stronger signal. See also *antenna*, *radio*, *rake receiver*, *spatial diversity*, and *MIMO*.

time division duplex (TDD) See *TDD*.

time division multiple access (TDMA) See *TDMA*.

time division multiplexer (TDM MUX) A device that performs time division multiplexing (TDM). See also *MUX* and *TDM*.

time-division multiplexing (TDM) See *TDM.*

Time Division-Synchronous Code Division Multiple Access (TD-SCDMA) See *TD-SCDMA.*

time-sharing A mode of operation that allows multiple independent users to share the resources of a multiuser computer system, including the CPU, bus, and memory. Time-sharing generally is accomplished by interleaving computer usage, with the independent applications or users taking turns using computer resources, although the appearance may be of concurrent usage.

time slice See *time slot.*

time slot A defined interval, or slot, of time recognizable by devices and designated for a specific purpose or operation. For example, a T1 facility has a signaling rate of 1.544 Mbps. Therefore, the transmitting time-division multiplexer (TDM mux) creates a signal with every bit occupying a slot of time of $1/1{,}544{,}000$ second, and the receiving mux monitors the circuit to extract the bits at the same pace. The term is also applied to multibit intervals devoted to a single connection, as in a DS-0 time slot, which is an eight-bit slot in an E/T-carrier voice grade channel. See also *channel, circuit, DS-0, E-carrier, multiplexer, signal, signaling rate, T1, T-carrier, TDM,* and *voice grade.*

Time to Live (TTL) See *TTL.*

T interface See *Reference Point T.*

tip The electrically positive (+) side of a cable pair. The tip side serves as the return path from the telephone to the central office (CO). See *tip and ring.*

tip and ring The term tip and ring originated in the cordboard plugs that establish the connection between the copper wire pairs. The tip of the plug connects to the electrically positive (+) wire, and the ring, or seat of the plug, to the electrically negative (–) wire. The ring supports the return path to the central office, for example. If the pair is shielded or if there is a ground wire, the ground is the sleeve and the correct term is *tip ring sleeve.* The term is still used extensively to identify the positive and negative wires in a pair. See also *ground.*

tip ring sleeve See *tip and ring.*

TIR (Total Internal Reflection) See *total internal reflection.*

TLD (Top Level Domain) In the Internet Domain Name System (DNS), the rightmost portion of the address, the TLD identifies the type of entity owning or sponsoring the address, or the country in which the address is located. The several types of Top Level Domains (TLDs) include generic Top Level Domains (gTLDs) and country codes (ccTLDs). The gTLD is always used. If a country code is also used in the address, it becomes the TLD, and the gTLD becomes the secondary domain. See also *ccTLD, DNS, domain, gTLD, Internet,* and *secondary domain.*

TM (Terminating Multiplexer) In SDH and SONET networks, a multiplexer at the end of a fiber link and providing access to one or more end users. See also *link, SDH,* and *SONET.*

TMN (Telecommunication Management Network) Established by the ITU-T in 1985 in Recommendation M.3000, TMN is an architectural framework for the interconnection of operations support system (OSS) components and network elements (NEs), and defines the standard interfaces and protocols for the exchange of information between them. TMN also describes the functional requirements for network management. See also *NE, network management,* and *OSS.*

TMR (Trunk Mobile Radio) Also known as *specialized mobile radio* (SMR), and marketed as *Improved Mobile Telephone Service* (IMTS). See *SMR.*

TO (Telecommunications Organization) A generic term referring to the government agency formed in many countries to assume responsibility for telegraph and telecommunications services from the Post, Telegraph, and Telephone (PTT) agency. Most developed nations have since fully or partially privatized telecommunications. See also *PTT.*

TOH (Transport OverHead) In a SONET or SDH frame, the overhead dedicated to control of data transport between network elements (NEs). TOH comprises Section Overhead (SOH) and Line Overhead (LOH). SOH of nine octets is dedicated to the transport of status, messages, and alarm indications for the maintenance of links between add/drop multiplexers (ADMs). LOH of 18 octets controls the reliable transport of payload data in the Synchronous Payload Envelope (SPE) between any two NEs. See also *ADM, frame, link, NE, octet, overhead, payload, SDH,* and *SONET.*

token **1.** In authentication systems, a hardware device (e.g., smart card, calculator-line device, or flash drive), or software on a client computer, that stores the user's credentials and generates a one-time password. The authentication process commonly involves a challenge-response dialogue between the token and a dedicated authentication server. In conjunction with a personal identification number (PIN) and supplemental password, a token is an excellent security mechanism. See also *authentication, password,* and *security.* **2.** In token passing local area networks (LANs), a special signal in the form of a certain bit pattern that circulates among the connected stations. A station cannot transmit data unless it has possession of the token. See also *LAN, signal, token bus, token passing,* and *token ring.*

Token Bus A local area network (LAN) standard defined in IEEE 802.4, token bus is a token passing protocol based on a physical bus topology. Token Bus is considered an orphaned standard, as the 802.4 committee disbanded in 2004 due to lack of interest. See also *bus, Ethernet, LAN,* and *token passing.*

token passing A deterministic, i.e., noncontentious, access method used in Token-Passing Ring, Token Ring, and Token Bus local area networks (LANs). The token, which consists of a specific bit pattern, indicates the status of the network — available or unavailable. The token is generated by a centralized master control station and transmitted across the network. The station in possession of the token controls the access to the network. That station may either transmit or require other stations to respond. Transmission is in the form of a data packet of a predetermined maximum size, determined by the number of nodes on the ring and the traffic to be supported; oversized transmissions are segmented, or fragmented. After transmitting, the station passes the token to a successor station, in a predetermined sequence. While the process is complex and overhead-intensive, its high level of control over the network avoids data collisions and increases throughput. See also *deterministic, LAN, node, packet, throughput, token,* and *Token Ring.*

Token Ring (TR) A local area network (LAN) technology invented in 1967 by Swedish inventor Olof Soderblom (1940–), who worked for IBM at the time. Soderblom subsequently filed for a patent that was finally awarded in 1981. IBM developed and heavily promoted the technology, a variation of which was standardized by the IEEE as 802.5. The 802.5 standard is for a token-passing ring access method and Physical Layer specifications supporting data rates of 4 Mbps and 16 Mbps. Through a multistation access unit (MSAU or MAU), as many as 250 stations can connect in a physical star topology that operates as a logical ring, and multiple MAUs can be interconnected in a physical ring topology. Token Ring specifies differential Manchester coding and a deterministic access method known as token passing. As is the case with token-based networks, in general, Token Ring offers a high level of access control, which is centralized. Access delay is measured and ensured, with priority access supported. Throughput is very close to raw bandwidth, as data collisions are avoided and retransmissions, therefore, are avoided. Throughput also improves under load, although absolute overhead is higher than with nondeterministic access techniques.

A Token Ring network elects one active monitor (AM), which is the node with the highest medium access control (MAC) address. The AM is responsible for ring administration, including clocking, token management, and removal of circulating frames. All other nodes are standby monitors (SMs), and can assume control in the event of an AM failure. If a station detects a network problem, it invokes a beaconing process that involves the continuous transmission of small control frames that identify the transmitting station, the nearest active upstream neighbor, and everything in between. This triggers a process of autoreconfiguration, in which nodes within the failure domain automatically initiate diagnostic measures in an attempt to identify, isolate, and bypass the point of failure.

A Token Ring moves a small frame, known as a token, around the network. The token, as illustrated in Figure T-4, comprises three fields, as follows:

- **Start Delimiter (SD):** A field of one octet that denotes the beginning of the token. The bit pattern JK0JK000 intentionally violates the Manchester coding scheme used elsewhere in the frame, thereby firmly establishing its identity. (Note: J indicates the intentional violation of a 1 bit, and K indicates the intentional violation of a 0 bit.)
- **Access Control (AC):** A field of one octet in the bit sequence PPPTMRRR. The P bits are Priority bits. The T bit is a Token bit that, when set at a 1 value, indicates that the frame is a token frame rather than a data or command frame. The M bit is a Monitor bit set by the active monitor (AM) station when it sees the frame and subsequently can be used by the AM to identify a frame that is endlessly circling the ring. The R bits are Reservation bits used by a station to reserve the token for a future transmission.
- **End Delimiter (ED):** A field of one octet that variously denotes the ending of the token. The bit pattern JK1JK1IE intentionally violates the Manchester coding scheme used elsewhere in the frame, thereby firmly establishing its identity. (*Note:* J indicates the intentional violation of a 1 bit, and K indicates the intentional violation of a 0 bit. The I bit is an Intermediate frame bit, and the E bit is an Error detection bit.)

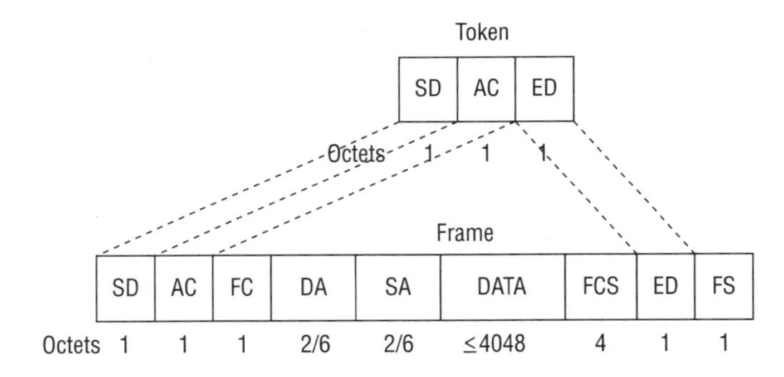

Figure T-4

If a station has no data to transmit, it passes the token to the next station on the ring, perhaps after holding it for a brief period of time established and limited by the master control station. If a station has data to transmit, and its priority level meets or exceeds the priority level established in the Access Control field, it seizes the token and changes one bit of the Start Delimiter field to a start-of-frame sequence, fills out the balance of the frame control fields, inserts the payload into the data field, and transmits the frame to the next station on the ring. A Token Ring command/data frame is formatted as follows:

- **Start Delimiter (SD):** A field of one octet that denotes the beginning of the token. The bit pattern JK0JK000 intentionally violates the Manchester coding scheme used elsewhere in the frame, thereby firmly establishing its identity. (*Note:* J indicates the intentional violation of a 1 bit, and K indicates the intentional violation of a 0 bit.)
- **Access Control (AC):** A field of one octet in the bit sequence PPPTMRRR. The P bits are Priority bits. The T bit is a Token bit that, when set at a 1 value, indicates that the frame is a token frame rather than a data or command frame. The M bit is a Monitor bit set by the active monitor (AM)

station when it sees the frame and subsequently can be used by the AM to identify a frame that is endlessly circling the ring. The R bits are Reservation bits used by a station to reserve the token for a future transmission.

- **Frame Control (FC):** A field of one octet establishing whether the frame is a control frame or a data frame. If the frame is a control frame, the FC field identifies the specific type.
- **Destination Address (DA):** A field of two or six octets identifying the MAC address of the target node.
- **Source Address (SA):** A field of two or six octets identifying the Medium Access Control (MAC) address of the originating node.
- **Data:** A field of variable length containing up to 4,048 octets of payload.
- **Frame Check Sequence (FCS):** A field of four octets containing a cyclic redundancy check (CRC) for use by the receiving node for error control.
- **End Delimiter (ED):** A field of one octet that variously denotes the ending of the token. The bit pattern JK1JK1IE intentionally violates the Manchester coding scheme used elsewhere in the frame, thereby firmly establishing its identity. (*Note:* J indicates the intentional violation of a 1 bit, and K indicates the intentional violation of a 0 bit. The I bit is an Intermediate frame bit, and the E bit is an Error detection bit.)
- **Frame Status (FS):** A field of one octet terminating the frame.

Each downstream station examines the destination address to see if the frame is addressed to it. If so, it copies the data. If not, it passes the frame to the next downstream station. If a station wishes to reserve the token for the next opportunity and has a priority level that meets or exceeds that of the data frame, as set by the transmission station, it sets the priority bits. When the next token is generated, it includes the priority level of the reserving station. Any station that raises the priority level of a frame must reinstate the original priority level once it completes its transmission.

Token Ring is a sophisticated technology that offers excellent *congestion* control and, therefore, excellent throughput. As it also is complicated and expensive to administer, it has been overwhelmed by the simpler, less expensive, and much faster Ethernet (802.3). See also *802.3*, *bandwidth*, *beaconing*, *bit*, *CRC*, *deterministic*, *domain*, *downstream*, *error control*, *Ethernet*, *frame*, *IEEE*, *LAN*, *MAC*, *Manchester coding*, *node*, *non-deterministic*, *octet*, *overhead*, *Physical Layer*, *ring topology*, *star topology*, *throughput*, and *token passing*.

toll call A telephone call for which there is a specific monetary charge. The term generally describes a long distance call, with the charge being sensitive to the time of day (prime time or non-prime time), total elapsed time, day of year (weekday, weekend day, or holiday), and distance between point of origin and point of termination.

toll free service A PSTN service that is free of any toll (i.e., long distance) charges to the caller. The call actually is not toll-free, as the charges are automatically billed to the called party. In the United States, the service originally was called *INWATS* (INward Wide Area Telecommunications Service), although that term has long since fallen out of favor. In the United States, toll free services generally are known as *800 services*, referring to the 800 area code prefix associated with such numbers. United States toll free prefixes now include 800, 866, 877, and 888. Toll free services elsewhere in the world commonly are known as *Freephone* or *Greenphone*, and the associated dial prefixes include 020, 0500, 00800, 0800, and 900.

toll quality A voice call having quality comparable to that of an ordinary long distance call, originally placed over the analog circuit-switched public switched telephone network (PSTN). More recently, the quality measurement technique applied to evaluate various implementations of voice encoding techniques, such as pulse code modulation (PCM). In quantitative terms, a voice connection rated as having a mean opinion score (MOS) of 4.0 or better, on a scale of 1.0 (bad) to 5.0 (excellent). The MOS is a mean average of the subjective evaluations of a group of listeners. The ITU-T standardized the MOS evaluation process in the Conversation Opinion Test, which is documented in Annex A of Recommendation P.800. A companion, and more objective, standard is P.862, the Perceptual Evaluation of Speech Quality (PESQ), which addresses the effects of filters, jitter, and coding distortions. Both P.800 and P.862 can be implemented in hardware. See also *distortion*, *filter*, *ITU-T*, *jitter*, *MOS*, *P.800*, *P.862*, *PCM*, and *PESQ*.

toll restriction A restriction placed on a telephone station or telephone account that prevents outgoing toll calls from being placed. See also *toll call*.

Top Level Domain (TLD) See *TLD*.

topology The physical and logical structure of a network. Physical topology refers to the physical layout of a network, specifically the physical positioning of the nodes and the circuits that interconnect them. Logical topology refers to the manner in which devices logically interconnect in a network, and may differ considerably from the physical topology. For example, an Ethernet LAN segment may comprise a number of workstations and peripheral devices that interconnect through a hub, with each device connecting directly to a hub port. The physical topology is that of a star, but the logical topology is that of a bus. That is to say that, although the devices connect to the hub over circuits that emanate from the hub like the rays of a star, they interconnect through a collapsed bus, or common electrical path, housed within the hub. LAN and WAN topologies variously include bus, mesh, partial mesh, ring, star, and tree. See also *bus*, *Ethernet*, *hub*, *LAN*, *logical*, *mesh topology*, *physical*, *ring*, *segment*, *star topology*, *tree*, and *WAN*.

toroidal Ring-shaped, or donut-shaped. See *loading coil*.

ToS (Type of Service) Referring to a field in the Internet Protocol version 4 (IPv4) header that indicates the quality of service (QoS) requested for the datagram. Although TCP/IP networks currently do not provide guaranteed QoS, the networks will attempt to honor QoS requests in terms of parameters that include packet precedence (i.e., priority), low delay, high throughput, and high reliability. See also *datagram*, *DiffServ*, *header*, *IPv4*, *QoS*, and *TCP/IP*.

Total Access Communications System (TACS) See *TACS*.

total internal reflection (TIR) The complete reflection of a light ray as it strikes the interface between the medium in which it is traveling and a medium with a lower refractive index, or index of refraction (IOR) at an angle greater than the critical angle, which is measured from the perpendicular at the point of reflection. Depending on the specific nature of the glass optical fiber (GOF) and its manner of construction, oblique light rays striking the interface between the core and cladding variously are reflected or refracted back into the core, which is the primary light-conducting medium. If the optical fiber is a step-index fiber, and the angle of incidence is greater than the critical angle, the light rays are reflected back into the core. If the angle is less than the critical angle, the light rays penetrate the core/cladding interface, where they are lost. If the fiber is a graded-index fiber, the light rays also reflect off the core/cladding interface if the angle of incidence is greater than the critical angle. As the angle decreases below the critical angle, the light rays enter the cladding, where they gradually gain velocity and bend, or refract, back into the core. The lesser the angle, the greater the penetration and the greater the associated increases in velocity and refraction. If the angle of the incident light ray is too severe, the light ray will penetrate the core/cladding interface and be lost in the cladding of either type of cable. Total internal reflection essentially confines the optical signal to the core conducting material, thereby maintaining signal strength over a distance. Total internal reflection is the fundamental principle that makes fiber optic transmission possible. See also *angle of incidence*, *cladding*, *core*, *critical angle*, *graded-index fiber*, *IOR*, *reflection*, *refraction*, and *step-index fiber*.

TPC (Transmission Power Control) DFS (Dynamic Frequency Selection) A protocol that controls power levels in an IEEE 802.11a wireless LAN (WLAN). Because 802.11a competes for spectrum with HiperLAN, the standard developed and promoted by the European Telecommunications Standards Institute (ETSI), 802.11a implementations in Europe must use TPC and dynamic frequency shifting (DFS), which dynamically shifts frequency channels. In combination, these protocols serve to eliminate interference issues with incumbent signals. See also *802.11a*, *channel*, *ETSI*, *frequency*, *HiperLAN/1*, *IEEE*, *interference*, *protocol*, and *WLAN*.

TPDDI (Twisted Pair Distributed Data Interface) Also known as *CDDI* (Copper Distributed Data Interface). A standard for extending Fiber Distributed Data Interface (FDDI) connections to workstations via unshielded twisted pair cable (UTP) over distances of 100 meters or less. See also *FDDI* and *UTP*.

TR (Token Ring) See *Token Ring*.

trademark (™ or ®) Intellectual property comprising a word, phrase, logo or other graphic symbol, sounds, or colors used by a manufacturer or seller to distinguish its products. In order to receive federal legal protection, a trademark must be distinctive, affixed to a product in the marketplace, and registered with the United States Patent and Trademark Office. Trademarks are perpetual, as long as they are in continued use. The terms *trademark* and *mark* are often used to refer to both trademarks and service marks. See also *intellectual property* and *service mark*.

trade secret Intellectual property such as a formula, process, recipe, or other information that is kept confidential in order to establish and maintain a competitive advantage in the marketplace.

traffic The total volume of cells, blocks, frames, packets, calls, messages, or other units of data carried over a circuit or network, or processed through a switch, router, or other system.

traffic analysis The process of analyzing traffic volumes, patterns, and rates for the purpose of improving system and network performance, while lowering costs. See also *configuration management* and *network management*.

traffic capacity **1.** The maximum volume of cells, blocks, frames, packets, calls, messages, or other units of data that a circuit or network can support, or that a switch, router, or other system can process. **2.** The measure of the number of simultaneous conversations that a key telephone system (KTS) or PBX can support. Traffic capacity depends on processor capacity and bus capacity. Processor capacity is measured in terms of the maximum number of Busy Hour Call Attempts (BHCAs) and Busy Hour Call Completions (BHCCs) the system can support. The busy hour is the busiest hour of the day, ideally as determined by empirical traffic studies. The bus capacity, i.e., the capacity of the switch matrix, usually is evaluated in terms of Centum Call Seconds (CCS) capacity, with a centum call second being 100 call seconds. One hour contains 3,600 call seconds (60 seconds times 60 minutes), or 36 CCS.

traffic contract In asynchronous transfer mode (ATM), the specification of all characteristics of a connection negotiated between a source endpoint and an ATM network. Traffic parameters are descriptions of the traffic characteristics of a source endpoint, including peak cell rate (PCR), sustainable cell rate (SCR), maximum burst size (MBS), maximum frame size (MFS), and minimum cell rate (MCR). A traffic descriptor is the entire set of traffic parameters associated with a source endpoint. See also *ATM*, *endpoint*, *MBS*, *MCR*, *MFS*, *PCR*, *SCR*, and *source*.

traffic descriptor In asynchronous transfer mode (ATM), the entire set of traffic parameters associated with a source endpoint. See also *ATM*, *endpoint*, *source*, and *traffic parameter*.

traffic engineering **1.** The inexact science of predicting the volume of future traffic and equipping a network with enough switching and transmission resources to handle that volume, thereby providing end users with a satisfactory grade of service (GoS). See also *Erlang*, *Erlang B*, *Erlang C*, *Equivalent Queue Extended Erlang B*, *Extended Erlang B*, *GoS*, *Poisson distribution*, and *traffic*.

traffic manager **1.** Synonymous with *traffic shaper.* See *traffic shaper.* **2.** The equally inexact practice of assigning virtual circuits to paths, and priorities to traffic on those circuits, in order to provide various levels of Quality of Service (QoS). See also *path, QoS, traffic,* and *virtual circuit.*

traffic parameter In asynchronous transfer mode (ATM), the description of a traffic characteristic of a source endpoint. Traffic parameters include peak cell rate (PCR), sustainable cell rate (SCR), maximum burst size (MBS), maximum frame size (MFS), and minimum cell rate (MCR). A traffic descriptor is the entire set of traffic parameters associated with a source endpoint. See also *ATM, endpoint, MBS, MCR, MFS, PCR, SCR,* and *source.*

traffic shaper Synonymous with *traffic manager.* A bandwidth optimizer that classifies packet traffic by analyzing application information contained deep within the packet. Traffic shapers essentially are Application Layer (OSI Layer 7) switches with buffering capability for traffic queuing and partitioning. Traffic shaping is also known as *policing.* See also *application, Application Layer, bandwidth, network optimization, OSI Reference Model,* and *queue.*

trailer **1.** Something pulled by or following something else, e.g., a boat trailer or house trailer is pulled by a car or truck. **2.** Also known as the *tail* or *trace* portion of the data set. The portion of a data block, cell, frame, or packet that succeeds the text field and contains information relative to message analysis, including message tracking and diagnostics. The trailer often includes an error detection and correction mechanism to manage the integrity of the transmitted data. See also *block, cell, data format* (illustration), *frame, header, packet,* and *text field.* **3.** An advertisement for a feature-length motion picture, generally consisting of short extracts of the film, and shown either in conjunction with another feature film or on entertainment television. A trailer originally was attached to the end of a film, hence the term *trailer.* When studio executives discovered that most audience members left after the conclusion of the feature films, rather than staying to watch the trailers, they moved the trailers to the beginning of the films, but the name stuck. Mystery solved. You are most welcome.

transducer A device capable of transforming a signal from one form of energy to another, usually in such a way that the fidelity of the original information is maintained, or substantially so. A microphone, for example, is a transducer that converts acoustic energy into electric current, and a speaker is a transducer that converts electric current into acoustic energy. The transmission of sound through these paired transducers certainly results in some loss of fidelity, the level of which depends on their quality, their proper matching, and the quality of the circuit that interconnects them. A diode laser light source and an Avalanche PhotoDiode (APD) light detector are paired transducers that, in combination, convert electrical energy to optical energy and back to electrical energy in an optical network such as fiber optics or Free Space Optics (FSO).

transceiver (transmitter/receiv*er*) A device that serves as both a transmitter and receiver, with both sets of circuitry contained within a common housing. In a local area network (LAN) context, a media access unit (MAU) is a form of transceiver that serves to physically connect a device such as a workstation or printer to the network. A modem is a transceiver, as well. See also *portmanteau, receiver,* and *transmitter.*

transcode To convert from one format to another. In video processing, it might be necessary to convert a file from VHS to Moving Pictures Experts Group (MPEG) format. In voice, it might be necessary to convert a speech signal from adaptive differential pulse code modulation (ADPCM) format to pulse code modulation (PCM) format, or vice versa. Transcoding typically is performed in a digital signal processor (DSP). See also *ADPCM, DSP, MPEG,* and *PCM.*

transform Referring to a step in the video compression process that converts the native three-dimensional video signal into a two-dimensional form comprising for subsequent compression. Transform compression is based on the premise that the low frequency components of a signal are more important than the high frequency components. Therefore, a substantial reduction in the number of bits used to represent a high frequency component will degrade the quality of the image only slightly. The various approaches include *discrete cosine transform (DCT)*, *fractal transform*, *vector quantization*, and *wavelet compression*. See also *compression*, *DCT*, *fractal transform*, *signal*, *vector quantization*, *video*, and *wavelet compression*.

transformer A static device that couples two circuits for the purpose of transferring alternating current (AC) from one circuit to another by electromagnetic induction. The current remains at the same frequency, but may change in voltage, phase, or impedance values. A transformer comprises two or more wire coils, or windings, but no moving parts. The voltage induced in a coil on a transformer is proportional to the number of turns around the core. See also *AC*, *frequency*, *impedance*, *inductance*, *phase*, and *voltage*.

transistor A contraction of trans-resistor, a transistor is a solid-state active device that controls current flow. A transistor comprises a semiconducting material, such as silicon or germanium, in three electrode regions with two junctions. The regions are alternately doped positive-negative-positive or negative-positive-negative in a semiconducting sandwich, so to speak. One outer region serves as the collector, the inner region as the base, and the other outer region as the emitter. The collector circuit collects power from the external power source, the base acts like a control electrode, and the emitter emits the outbound signal. Small signals applied between the base and the emitter control the larger currents and power from the collector, with a small change in the signal applied to the base producing a large and rapid change in the current flowing through the entire component. A transistor can operate linearly, like an audio amplifier, or like a switch, rapidly opening and closing an electronic gate. A transistor can act on a signal to perform a variety of functions such as amplification, rectification, modulation or demodulation, and buffering. The transistor was invented by William Shockley, John Bardeen, and Walter Britain of AT&T Bell Telephone Laboratories, in 1947, and quickly replaced the electron tube, or vacuum tube. As a result of the 1956 Consent Decree, AT&T was forced to license the patented transistor technology to any company for $25,000. General Electric, IBM, Sony, and Texas Instruments are but a few of the companies that wrote a check. Large numbers of transistors are frequently interconnected with microcircuits and baked into a single integrated circuit, many of which can exist on a single circuit board in an electronic device such as a computer, switch, or router. See also *amplifier*, *buffer*, *Consent Decree*, *current*, *modulation*, *patent*, *power*, *rectifier*, and *signal*.

transition band In CATV, synonymous with *guard band*. See also *CATV* and *guard band*.

translator **1.** A component of a crossbar (Xbar) switch, a translator translates dialed digits into routing instructions, which it forwards to a sender in order that a call can be connected over a trunk. See also *sender* and *Xbar*. **2.** A device that changes the frequency or voltage of a signal without altering the baseband data or modulation.

translucent optical cross-connect (translucent OXC) See *translucent OXC*.

translucent OXC (translucent Optical CROSS-Connect) A hybrid that can function as both an opaque OXC and a transparent OXC. A transparent OXC is a digital cross-connect (DXC) that can be characterized as optical-optical-optical (OOO), as it performs the switching function without converting the optical signal to electronic format. An opaque OXC requires that the optical signal be converted into electrical format before being switched by the electronic DXC, then converted back into optical format before being placed back on the fiber optic transmission system (FOTS). See also *DXC*, *OEO*, *OOO*, *opaque OXC*, *OXC*, and *transparent OXC*.

Transmission Control Protocol (TCP) See *TCP*.

Transmission Convergence (TC) See *TC*.

transmission grating A type of diffraction grating comprising groves etched into a transparent material such as glass. As the elements of light in the incident spectrum strike the grooves at a certain angle, they are diffracted and, therefore, separated to various degrees, with blue light diffracted the least and red light the most. This approach is used effectively in fiber optic transmission systems (FOTS) employing wavelength division multiplexing (WDM). See also *diffraction*, *diffraction grating*, *FOTS*, and *WDM*.

transmission medium Something that passively supports or allows the conveyance or transmission of a signal, a transmission medium is not necessarily a tangible thing that can be touched. Transmission media can be categorized a number of ways:

- **Wired** media involve tangible conductors that conduct electromagnetic energy. Metallic conductors support the transmission of electrical energy, with copper, aluminum, and copper-clad aluminum as examples. Metallic conductors are used in twisted-pair (copper) and coaxial cable (aluminum, copper, and copper-clad aluminum) media, for example. Glass and plastic conductors support the transmission of infrared photonic energy in fiber optic networks. See also *conductor*.
- **Wireless** media do not involve conductors at all. Actually, "media" is a misnomer, for not only are there no wires, but also there is no physical medium of any sort. Rather, the electromagnetic signal propagates through space between a transmitter and a receiver, with any physical matter (e.g., atomic, molecular, and particulate matter) serving only to attenuate, impede, diffuse, distort, and otherwise interfere with the radio or optical signal. Wireless transmission is synonymous with airwave transmission and includes radio and free space optics (FSO). See also *attenuation*, *diffusion*, *distortion*, *FSO*, *impedance*, and *radio*.
- **Guided** media incorporate a waveguide, i.e., an enclosed physical structure that contains and guides the signal. Conducted media fit this definition, as insulation and shields variously serve more or less effectively to confine the signal to the conductors and guide it along a physical path. Some radio systems make use of waveguides comprising hollow metal pipes that generally are rectangular, although sometimes circular, in form. Such radio waveguides are restricted to use in very high power or very high frequency applications over short distances, due to their size, weight, and cost. See also *waveguide*.
- **Bounded** media incorporate some form of twisting, shielding, cladding, and/or insulating material that serves to more or less effectively bind the signal within the core medium, thereby improving signal strength over a distance and enhancing the performance of the transmission system in the process. Twisted pair, coaxial cable, and fiber optic cable fall into this category. See also *coaxial cable*, *fiber optic cable*, and *twisted pair*.

transmission power control (TPC) See *TPC*.

transmission rate The theoretical rate at which user data can be transmitted across a circuit. The signaling rate is the total data rate, including overhead, i.e., signaling and control information. Therefore, the transmission rate is always less than the signaling rate. For example, a typical T1 circuit has a signaling rate of 1.544 Mbps, but a transmission rate of only 1.536 Mbps, as 0.008 Mbps is required for overhead. If the T1 equipment engages in bit robbing, that technique further reduces the transmission rate. If a circuit is used to its maximum capacity, throughput equals the transmission rate. See also *bit robbing* and *throughput*.

transmission speed See *transmission rate*.

transmission system A combination or assembly of components that forms a complex whole entity that functions as a single unit to accomplish the transmission of signals. A transmission system always includes circuit terminating equipment on both the transmit and receive ends of a transmission medium. The system may also include amplifiers, repeaters, and other intermediate devices that variously serve to amplify, reshape, and retime signals. See also *amplifier*, *circuit terminating equipment*, *repeater*, *system*, and *transmission medium*.

transmit clock (TC) See *TC*.

transmitter Also known as a *sender* or *source*, a transmitter is a device that originates, or generates, an information transfer to one or more receivers. Transmitters include telephones, facsimile machines, data terminals, host computers, and video cameras. See also *receiver* and *transceiver*.

transparent Something easily seen through and even invisible, in some cases. A DSL modem and the functions it performs, for example, are transparent to an end user.

transparent optical cross-connect (transparent OXC) See *transparent OXC*.

transparent OXC (transparent Optical CROSS-connect) Also known as a *photonic cross-connect* (PXC). A digital cross-connect (DXC) that can be characterized as optical-optical-optical (OOO), because it performs the switching function without converting the optical signal to electronic format. An opaque OXC requires that the optical signal be converted into electrical format before being switched by the electronic DXC, then converted back into optical format before being placed back on the fiber optic transmission system (FOTS). A translucent OXC is a hybrid that includes both electronic and optical switching modules. See also *DXC*, *OEO*, *OOO*, *opaque OXC*, *OXC*, and *translucent OXC*.

transponder (transmitter/responder) A radio or radar transmitter-receiver. **1.** In a telecommunications context, a transponder is a satellite repeater, which accepts the weak incoming signals from the Earth stations, boosts them, shifts them from the uplink to the downlink frequency, and retransmits them to the Earth stations in what is known as a bent pipe network configuration. Contemporary satellites commonly support as many as 28–46 transponders. Current generations of broadband satellites are replacing the relatively dumb bent pipe transponder approach with onboard processors capable of switching and routing traffic over intersatellite links in support of IP, Frame Relay, and ATM traffic. **2.** Transponders also are used in aviation applications. Commercial and general aviation aircraft are fitted with transponders that automatically respond to queries from ground stations and military aircraft with preprogrammed IFF (Identification Friend or Foe) information, including altitude. The transponder can be tracked to determine direction, speed, rate of ascent or descent, and so on. See also *bent pipe* and *satellite*.

transport To carry from one place to another, especially over long distances. See also *transport circuit*.

transport circuit A circuit employed for long-haul carriage within the core or backbone of a wide area network (WAN). See also *access circuit*, *long haul circuit*, and *transport*.

Transport Layer Layer 4 of the seven-layer Open Systems Interconnection (OSI) Reference Model. Software at the Transport Layer is responsible for maintaining the end-to-end integrity and control of the session. Data is accepted from the Session Layer (Layer 5) and passed through to the Network Layer (Layer 3). Software at this layer fragments long message blocks into shorter packets for transmission and reassembles them at the receiver, manages flow control, adds sequence numbers, calculates and appends checksums, generates retransmissions in the event of errored message blocks or timeouts, and adds security measures. Example protocols that can be used at this layer include Transmission Control Protocol (TCP), User Datagram Protocol (UDP), and the five classes of the OSI Transport Protocol (TP). The X.25 packet-switching protocol operates at the Transport, Network, Data Link, and Physical Layers. See also *block*, *code*, *compression*, *flow control*, *layer*, *network architecture*, *OSI Reference Model*, *packet*, *protocol*, and *software*.

Transport Overhead (TOH) See *TOH*.

transverse At a right angle, i.e., 90 degree angle. Crosswise, or perpendicular. See also *normal*.

.travel Pronounced *dot travel*. The generic Top Level Domain (gTLD) reserved exclusively for those whose primary area of activity is in the travel industry. This domain was created in 2005 under the sponsorship of Tralliance Corporation. See also *gTLD*, *Internet*, and *sponsored domain*.

tree topology A network structure in the form of a multipoint electrical circuit, with multiple branches off the trunk of the central, or root, bus, as illustrated in Figure T-5. The tree topology is a variation of the bus topology and provides only one path between any two nodes. See also *bus topology* and *topology*.

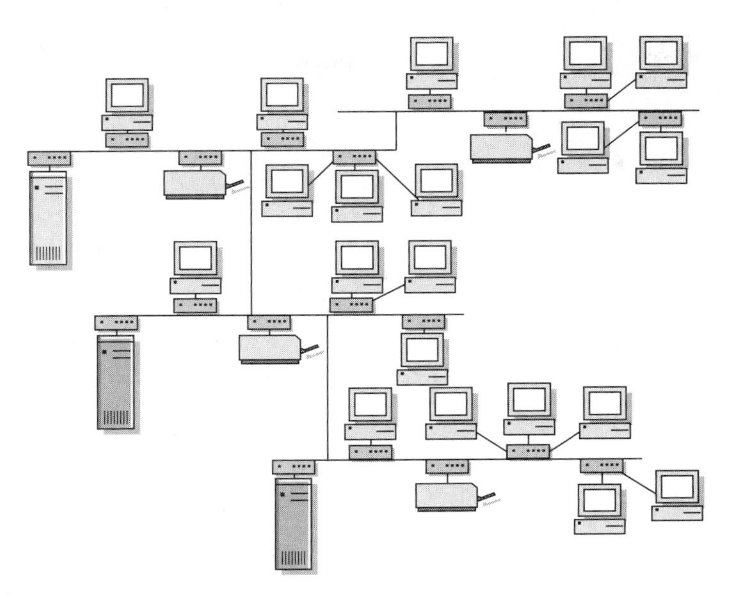

Figure T-5

trellis-coded modulation (TCM) See *TCM.*

trellis-coded pulse amplitude modulation (TC-PAM) Also known as *trellis-coded modulation* (TCM). See *TCM.*

tribit Referring to a modulation technique that impresses three bits on a baud, so that the bit rate is triple the baud rate. Such a technique employs eight signal states. Quadrature amplitude modulation (QAM) is a tribit technique achieved by defining two amplitude values for each of four phase shifts. See also *amplitude, amplitude modulation, baud, baud rate, bit, bit rate, dibit, modulation, phase, QAM, quadbit, signal,* and *unibit.*

Tributary Unit (TU) See *TU.*

Triple Data Encryption Algorithm (TDEA) See *Triple DES.*

Triple Data Encryption Standard (Triple DES) See *Triple DES.*

Triple DES (Triple Data Encryption Standard) More correctly known as *Triple Data Encryption Algorithm* (TDEA). A block cipher encryption method that uses the DES encryption algorithm to encrypt message blocks three times, using three different 56-bit keys. Triple DES is highly secure, but slow, and largely has been superseded by Advanced Encryption Standard (AES). See also *AES, algorithm, block, DES, encryption,* and *key.*

triple play A marketing term used by broadband service providers to describe the combination of voice, high-speed data, and television services over a single local loop. The term is a baseball analogy, referring to the very rare act in which the defense makes three outs on the same play. Some service providers use the term *quadruple play* to describe a triple play plus wireless services, although the baseball analogy breaks down at that point. Actually, the analogy broke down with the marketer who invented the term, as a triple play is a defensive play. A more appropriate term would be *triple,* which is a rare offensive play. Also, that would make the nonexistent defensive quadruple play a *home run.* See also *broadband, local loop,* and *quadruple play.*

Trojan horse A program that allows a remote user secretly to gain control of a computer. A Trojan horse masquerades as a legitimate computer game, utility, or application. Once activated, the program positions itself to load automatically, report its presence to a server, and listen for commands. A Trojan horse often contains hidden malicious code that does something to harm the victim's computer, perhaps severely. Trojan horses often are downloaded over the Internet by unsuspecting end users. Unlike a virus, a Trojan horse cannot replicate itself. The term comes from a Greek story of the Trojan War, in which the Greeks pretended to retreat from the battlefield and left a giant wooden horse at the gates of Troy, ostensibly as a peace offering. But after the Trojans dragged the horse inside the city walls, Greek soldiers sneaked out of the hollow belly of the horse and opened the city gates to the waiting Greek army, which captured Troy. Screensavers and emoticons are common Trojan horses. See also *application*, *emoticon*, *Internet*, *malware*, *utility*, *virus*, and *worm*.

TRS (Telecommunications Relay Service) A service that allows persons with hearing or speech disabilities to place and receive telephone calls through human operators who serve as facilitators. A speech- or hearing-disabled person initiating a TRS call uses a teletypewriter (TTY) or other text input device to call the TRS relay center and gives the communications assistant (CA) the destination telephone number, which the CA uses to place a traditional telephone call. The CA then serves as a translator, relaying the text of the calling party in voice to the called party, and converting the voice of the called party into text for the benefit of the caller. There are a number of variations on the basic text-to-voice and voice-to-text theme of TRS:

- **Voice Carry Over (VCO):** A service that allows a person with a hearing disability to use his or her own voice to speak directly to the called party, but receive responses in text from the CA.
- **Hearing Carry Over (HCO):** A service that allows a person with a speech disability to use his or her own hearing to listen to the called party, but to respond in text through the CA.
- **Speech-to-Speech (STS):** A service that allows a person with a speech disability to use his or her own voice to speak to the called party through a CA specially trained to understand speech affected by a variety of disorders. The CA repeats the spoken words in a clear and understandable manner.
- **Shared Non-English Relay Service:** The FCC requires interstate TRS providers to offer Spanish-to-Spanish traditional TRS service, i.e., TTY-based service. Providers of other non-English, e.g., French-to-French, TRS service can be compensated from the federal TRS fund.
- **Captioned Telephone Service:** Involves a special telephone with a text display. Rather than using TTY technology, the called party's speech is re-voiced by the CA, converted into text by a voice recognition system, and transmitted directly to the caller's display telephone.
- **Video Relay Service (VRS):** A service that allows persons whose primary language is American Sign Language (ASL) to communicate with the CA in sign language using videoconferencing equipment and an Internet Protocol (IP) based connection over the Internet.
- **IP Relay:** A text-based service that uses the Internet rather than a traditional public switched telephone network (PSTN) connection to communicate with the CA.
- **IP Captioned Telephone Service:** A hybrid service that uses the Internet for the text-based portion of the call, but allows the CA to simultaneously listen to one party and read the text of the other party.

TRS is funded through the Universal Service Fund (USF) and is administered by the National Exchange Carrier Association (NECA). See also *Internet*, *IP*, *NECA*, *PSTN*, *TTY*, and *USF*.

truck roll Referring to the need to dispatch a technician in a truck to install, move, or somehow reconfigure an item of equipment or a wire and cable system, or perhaps to respond to a service call or network outage. Because a truck roll is expensive, service providers tend to prefer networks and network elements that are self-healing or that can be configured and reconfigured remotely.

trunk A communications circuit that interconnects switches. As such, multiple users and multiple transmissions can share a trunk on a pooled basis, with contention for trunk access managed by an intelligent switching device. There are many types of trunks. Tie trunks connect Private Branch Exchange (PBX) switches in a private network, Central Office Exchange (COE) trunks connect PBXs to telephone company central office exchange switches, and interoffice trunks interconnect central office exchange switches. Trunk groups are groups of trunks serving the same special purpose. Trunks traditionally are directional in nature, with the options being one-way outgoing (originating), one-way incoming (terminating), or two-way (combination). The term *trunk* comes from the Latin *truncus*, literally meaning something *cut off*. *Trunk* distinguishes the main body of a circuit from the lesser subsidiary lines that come off it, much as the main channel of a river is apart from its lesser tributaries. See also *line* and *trunk group*.

trunk card A printed circuit board (PCB), or card, that fits into a card slot, of which there may be many on a shelf that fits into a common equipment cabinet, of which there may be many in a switch or router. See also *trunk* and *trunk interface*.

trunk group A group of trunks serving the same special purpose. The term commonly is applied to voice Private Branch Exchange (PBX) trunks. Multiple Direct Inward Dial (DID) trunks commonly are grouped together in a trunk group, for example. See also *trunk*.

trunk interface The total of hardware and firmware that serves to interconnect a trunk and a switch, router, or other device, and to facilitate their interoperation. Such interfaces are specific to the Physical Layer protocols, which address such factors as transmission medium, physical dimensions of the medium, signal frequency and wavelength, signal format, and signaling speed. See also *firmware*, *hardware*, *Physical Layer*, *protocol*, and *trunk*.

trunk mobile radio (TMR) Also known as *specialized mobile radio* (SMR), and marketed as *Improved Mobile Telephone Service* (IMTS). See *SMR*.

trunk side In telco terminology, referring to the internal telco network, or telco-facing portion of the Public Switched Telephone Network (PSTN). This is in contrast to the line side of the network, which is the local loop, or customer-facing portion of the PSTN. The demarcation point is at the local loop interface to the central office (CO). See also *CO*, *line side*, *local loop*, and *PSTN*.

TSAPI (Telephony Services Application Programming Interface) An API developed jointly by AT&T and Novell in the early phases (1994) of computer telephony (CT). Strongly oriented toward PC platforms, the TSAPI specification is a lesser protocol stack that does not require that the switch manufacturer fully open the switch interface. Although a number of PBX manufacturers supported TSAPI, the cost of TSAPI PBX drivers was quite high and, as a result, its widespread use has been discouraged. See also *computer telephony*, *driver*, *PBX*, and *TAPI*.

T-span Vernacular for a T1 twisted-pair circuit in the outside plant (OSP). See *circuit*, *OSP*, *T1*, and *twisted pair*.

TSR (Terminate-and-Stay-Resident) A utility program that loads and remains in random access memory (RAM) even when not running, in order to be available instantly when required. TSR programs were popular as extensions to DOS, but largely have been replaced by more sophisticated modules in more recent operating systems (OSs.) See also *DOS*, *OS*, *program*, *RAM*, and *utility*.

TTL (Time To Live) A field in the Internet Protocol version 4 (IPv4) header that specifies the length of time in seconds that the datagram can live in the Internet system. The maximum length of time is 255 seconds (2^8–1, with 0 not considered, as it is the official time of death), or 4.25 minutes. From the instant the IP datagram enters the Internet, each gateway and host that acts on the datagram decrements the TTL by at least one second, although the time it has possession of the datagram generally is much less. When the TTL reaches 0, the datagram is declared dead and is discarded. The TTL mechanism prevents packets from wandering the Internet for eternity, at which point they would have no value, and would only contribute to overall network congestion. Over time, the TTL field has been redefined to indicate, as an option, the number of hops (i.e., routers) through which the packet travels. In effect, the TTL is a hop count, anyway. The default TTL is 64. See also *datagram*, *gateway*, *header*, *host*, *Internet*, *IPv4*, *MPLS*, and *packet*.

TTS (Text-To-Speech) The process of converting digital text into speech output in order to support interactive voice response (IVR) applications. See also *IVR*, *speech*, and *text*.

TTY (TeleTYpewriter) A printing telegraph service widely used by those with hearing or speech impairments. Also known as *Telecommunications Device for the Deaf* (TDD) in the United States, *textphone* in Europe, and *minicom* in the United Kingdom. TTY uses the Baudot code, also known as International Telegraph Alphabet #2 (ITA #2). See also *Baudot code*, *telegraph*, and *teletype*.

TU (Tributary Unit) In SONET specifications, a Virtual Tributary (VT) along with a pointer that identifies the location of the VT for purposes of switching and cross-connecting. See also *cross-connect*, *pointer*, *SONET*, *switch*, and *VT*.

tunneling A transmission method used in internetworks whereby a packet of one type is encapsulated in a packet of another type, in essence becoming the payload of the encapsulating packet. For example, an IPv6 packet can be encapsulated in an IPv4 packet in order to transit an intermediate IPv4 network, which otherwise would not accept it due to numerous incompatibilities. Tunneling is used extensively in IP-based virtual private networks (VPNs). In this application, the payload packet is encrypted and encapsulated in another packet. The encapsulating packet insulates the payload packet, shielding it from view and protecting it as it transits the network. See also *encapsulate*, *encryption*, *IPv4*, *IPv6*, *packet*, *payload*, and *VPN*.

Tuvalu A small, poor island country located in the western Pacific Ocean. Tuvalu comprises nine low-lying coral atolls with a total land mass of about 10 square miles, and the highest point of land is approximately 16 feet above sea level. The main wild animals are rats, lizards, and turtles. The only source of water is rainwater, which is collected in catchment basins. Exports include copra, i.e., dried coconut meat, and postage stamps. In terms of population (estimated at 11,810 as of July 2006), Tuvalu is the smallest member of the United Nations. Its closest neighbors are the Fiji Islands and Samoa, both of which are about 650 miles away. In April 2000, the country of Tuvalu licensed its .tv Internet top level domain (TLD) to Dot TV Corp., a California-based company since acquired by VeriSign, in a deal that guarantees Tuvalu a payment of at least $4 million per year for at least ten years. Tuvalu also received a significant minority position in Dot TV. Dot TV markets .tv as a master portal for TV-related content providers, commanding fees in the range of $25 to $100,000 for Web site registrations. The Tuvalu philatelic bureau issued a series of four stamps to commemorate the .tv deal. The population enjoys little in the way of modern conveniences or infrastructure, as the country is listed by the United Nations as one of the least developed in the world. So, a TLD is of relatively little value to the citizens of Tuvalu. In fact, there is great concern among the citizens of Tuvalu that, if global warming continues at the current pace, the entire nation will be under water within 50 years or so. See also *Internet*, *portal*, *TLD*, and *WWW*.

TV (TeleVision) From the Greek *tele*, meaning *far off*, and *visio*, meaning *seen*, and translating literally as *seen far off*. A system in which series of still images are communicated across a distance. The TV transmitter scans a view by means of a camera, converts light rays into electrical signals used to modulate a carrier, adds audio converted from acoustical to electrical by a microphone, and adds control signals. The receiver uses the electrical signal to modulate and control a display device and speakers that reconstruct the video and audio signals. Broadcast TV is transmitted over the airwaves. Community antenna television (CATV), more commonly known as cable TV, is transmitted over networks of coaxial cable or hybrid fiber/coax (HFC) networks of coax and optical fiber. Analog TV standards include National Television Standards Committee (NTSC), Phase Alternate Line (PAL), and **S**équential **C**ouleur **A**vec **M**émoire (SECAM). Digital TV (DTV) standards include high-definition television (HDTV) and standard definition television (SDTV). See also *acoustics*, *audio*, *broadcast TV*, *carrier*, *CATV*, *coaxial cable*, *DTV*, *electricity*, *HDTV*, *HFC*, *light*, *microphone*, *NTSC*, *optical fiber*, *PAL*, *receiver*, *SDTV*, *SECAM*, *transmitter*, and *video*.

twinax (twinaxial cable) See *twinaxial cable*.

twinaxial cable (twinax) A cable system containing two thin coaxial cables, perhaps contained within a single cable sheath. Twinaxial cables once were used extensively with IBM System 36, System 38, and AS 400 computer systems. More recently (February 2006), the IEEE 802.3ak task force finalized the 10GBase-CX4 standard in support of 10 Gigabit Ethernet (10 GbE). The specification provides for twinax patch cord assemblies operating over distances up to 50 feet for interconnecting switches and aggregating servers in data centers. See also *10GBase-CX4*.

twist pitch The level or degree of twisting in a twisted pair. See also *lay length* and *twisted pair*.

twist ratio The number of twists per inch, foot, or meter in a twisted pair. See also *lay length*, *twisted pair*, and *twist pitch*.

twisted pair A twisted pair comprises two copper conductors, separately insulated by a dielectric material, and smoothly twisted in a helix with a constant pitch or distance to make a 360° twist. Shielded twisted pair (STP) and screened twisted pair (ScTP) incorporate special shields for protection from electromagnetic interference (EMI). Unshielded twisted pair (UTP), the most common configuration, has no such shielding. See also *EMI*, *ScTP*, *STP*, and *UTP*.

Twisted-Pair Distributed Data Interface (TPDDI) See *TPDDI*.

two-way paging (TWP) See *TWP*.

two-wire circuit A circuit that carries information signals in both directions over the same physical link or path. A two-wire circuit commonly is provisioned through the use of a single twisted pair copper wire, within which two wires are required to complete the electrical circuit, with the current in one wire opposite to the current in the other, and with both wires carrying the information signal. The most common example of a two-wire circuit is an electrically-based unshielded twisted pair (UTP) local loop access circuit between a telephone company central office (CO) switching center and a residential or small business premises, where the loop ultimately terminates in an individual single-line or multiline telephone set, data terminal, or key service unit (KSU) associated with a key telephone system (KTS). Two-wire circuits generally span a relatively short distance, and are analog, narrowband, and single channel in nature. Two-wire UTP local loops, for example, typically are designed to be less than 18,000 feet (6 kilometers) in length, analog voice grade, and single channel. However, the same loop will support ISDN BRI (Basic Rate Interface) service, a digital transmission technology. See Figure T-6. See also *BRI*, *channel*, *four-wire circuit*, *ISDN*, *local loop*, *narrowband*, and *voice grade*.

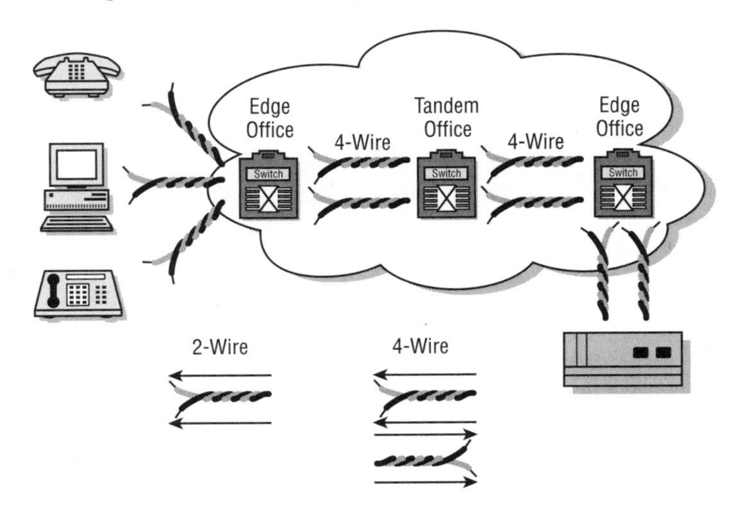

Figure T-6

TWP (Two-Way Paging) A paging system that supports duplex communications. The simplest and most common version is known as 1.5-way paging, which supports guaranteed message delivery, as the network does not attempt to download messages until such time as the pager is within range, turned on, and has enough memory to support the download. The general location of the pager is communicated upstream, so the messages can be downloaded to the antennas supporting that particular geographic area, rather than being broadcast across the entire paging network. Once downloaded successfully, the pager acknowledges to the network the receipt of the page. Full two-way paging enables the recipient of the page to select and transmit a return message, which most commonly is selected from a small set of predetermined messages, although some pagers support user-definable return messages. See also *download*, *duplex*, *pager*, *paging system*, and *upstream*.

TWX (TeletypeWriter eXchange) Pronounced *twix*. A high-speed telex service that AT&T introduced in 1930. TWX ran at the blazing speed of 75 bps, and later 150 bps. See also *AT&T*, *bps*, and *telex*.

type of service (ToS) See *ToS*.

U **1.** In electricity, the symbol for potential energy difference, which is the work done to move a unit charge between two points in an electric field. In other words, the difference in voltage level between two points in an electric field. **2.** U interface or Reference Point U in ISDN. See *Reference Point U.*

UADSL (Universal Asymmetric Digital Subscriber Line Lite) See *G.lite.*

U-Band (Ultralong Wavelength) Band) The ITU-T standard optical transmission window in the wavelength range of 1,625–1,675 nm. See also *wavelength* and *window.*

UBR (Unspecified Bit Rate) In asynchronous transfer mode (ATM), a best-effort class of traffic. Traffic parameters include peak cell rate (PCR) and cell delay variation tolerance (CDVT). No quality of service (QoS) commitment is made. ATM also defines available bit rate (ABR), constant bit rate (CBR), non real-time Variable Bit Rate (nrt-VBR), real-time Variable Bit Rate (rt-VBR), and variable bit rate (VBR) traffic classes. UBR traffic includes traditional computer applications such as file transfer and e-mail. See also *ABR, ATM, CBR, CDVT, compression, nrt-VBR, PCR, QoS, real-time, rt-VBR,* and *VBR.*

UCAID (University Corporation for Advanced Internet Development) A not-for-profit consortium formed specifically to advance the development of Internet2. All nomenclature since has transitioned to Internet2. See also *Internet2.*

UCD (Uniform Call Distributor) A standard feature of many PBXs that automatically routes incoming calls to the next available agent, scanning agent ports in a predetermined, rigid hunt pattern. A UCD serves many of the same functions as an automatic call distributor (ACD), but lacks the same level of intelligence and, therefore, is much less capable. See also *ACD.*

UCS (Universal Character Set) A joint initiative of the International Organization for Standardization (ISO) and the Unicode Consortium to develop a 32-bit character set capable of encoding all written languages. ISO 10646 defines the Universal Character Set (UCS), into which all Unicode Transformation Format (UTF) code sets map. UCS-4 is a four-octet code set into which UTF-32 maps and UCS-2 is a two-byte code set into which UTF-16 maps. UCS-1 encodes all characters in byte sequences varying from one to five bytes. See also *bit, byte, code set, octet,* and *Unicode.*

UDP **1.** Uniform Dialing Plan. See *uniform dialing plan.* **2.** User Datagram Protocol. A Transport Layer host-to-host protocol in the TCP/IP protocol suite, UDP is a connectionless protocol for datagram-oriented applications. Like the Transmission Control Protocol (TCP), UDP uses the Internet Protocol (IP) for addressing and routing purposes. Unlike TCP, UDP provides no sequencing, error control, or flow control mechanisms. An application program that uses UDP assumes full responsibility for all issues of reliability, including data loss, data integrity, packet latency, data sequencing, and loss of connectivity. UDP is used extensively in voice over Internet Protocol (VoIP) and stream-oriented multimedia applications, where compression techniques are designed to mitigate such issues over a highly shared packet network. UDP also works well where transactions are of such short duration that connection setup overhead comprises a large proportion of overall transaction traffic, as exemplified by Simple Network Management Protocol (SNMP) exchanges.

The standard size of the UDP header is eight octets, as illustrated in Figure U-1. The UDP header comprises the following fields:

- **Source Port:** 16 bits that define the port number used by the source application program.
- **Destination Port:** 16 bits that define the port number used by the destination application program.

- **UDP Message Length:** 16 bits that identify the length of the message in the Data field.
- **Header Checksum:** 16 bits used for error control in the header, only. The checksum process is that of cyclic redundancy check (CRC).
- **Data:** A variable-length field that contains the actual data content, or payload,

UDP source port	UDP destination port
UDP message length	UDP checksum
Data	
. . .	

Figure U-1

See also *application*, *compression*, *connectionless*, *CRC*, *datagram*, *error control*, *flow control*, *header*, *host*, *IP*, *latency*, *multimedia*, *octet*, *overhead*, *packet*, *payload*, *protocol suite*, *SNMP*, *stream-oriented*, *TCP/IP*, *Transport Layer*, and *VoIP*.

UGS (Unsolicited Grant Service) A WiMAX quality of service (QoS) level involving a polling schedule designed for services that periodically generate fixed units of data. T1, E-1, and other services based on time division multiplexing (TDM) require UGS polling. See also *E-1*, *QoS*, *T1*, *TDM*, and *WiMAX*.

UHF (Ultra High Frequency) UHF radio is in the frequency range of 300 MHz – 3 GHz and has a wavelength of 1 m – 10 cm. UHF radio has applications in microwave and satellite radio, UHF TV, paging systems, cordless telephony, cellular and PCS telephony, and wireless LANs. See also *electromagnetic spectrum*, *frequency*, *Hz*, and *wavelength*.

UIFN (Universal International Freephone Number) A toll-free telephone number for international calling. A UIFN charges the cost of the call to the called party. Standardized by the ITU-T, the numbering convention for UIFNs is 800 + 8 digits. See also *ITU-T* and *toll free service*.

U interface See *Reference Point U*.

UL (Underwriters Laboratories) An independent, for-profit laboratory that has developed a large number of safety standards against which it offers testing and certification services for potentially hazardous products, including electrical equipment, marine products and life-saving devices, fire suppression and fire containment equipment, and heating, air conditioning, and refrigeration equipment. See Appendix A for contact information.

ultra high frequency (UHF) See *UHF*.

ultralong wavelength band (U-Band) See *U-Band*.

ultraviolet (UV) light See *UV*.

UMA (Unlicensed Mobile Access) The original term for the Generic Access Network (GAN) standards. See *GAN*.

UMTS (Universal Mobile Telecommunications System) Also known as Wideband CDMA (W-CDMA). A 3G digital cellular radio technology that is an upgrade to the pan-European Global System for Mobile Communications (GSM). UMTS is based on code division multiple access (CDMA) and operates over a carrier 5 MHz wide, compared to the 200 kHz carrier used for narrowband CDMA. UMTS specifications provide for both time division duplex (TDD) and frequency division duplex (FDD) modes of operation, with TDD largely used in Europe and FDD in the United States. The FDD specifications call for the downlink to run in the 2110–2200 MHz range and the uplink in the 1885–2025 MHz range. As is the case with all true 3G systems, UMTS specifications include 128 kbps for high-mobility applications, 384 kbps for pedestrian speed applications, and 2 Mbps (1.920 Mbps) for fixed in-building applications. In reality, UMTS currently caps the transmission rate at a theoretical 384 kbps. UMTS was first deployed in Japan (2000), where it is known as Freedom of Mobile Multimedia Access (FOMA). High Speed Downlink Packet Access (HSDPA), sometimes characterized as a 3.5G technology, is an upgrade to UMTS that increases theoretical downlink data rates to 14.4 Mbps. High Speed Uplink Packet Access (HSUPA), which remains in development, is expected to increase uplink speeds to a maximum of 5.76 Mbps. UMTS standards are published by the 3rd Generation Partnership Project (3GPP). See also *3G, 3GPP, carrier, CDMA, cellular radio, digital, downlink, FDD, GSM, HSDPA, HSUPA, narrowband, TDD, uplink*, and *wideband.*

UMTS Terrestrial Radio Access Time Division Duplex 1.28 Mega chips per second Option (UTRA TDD 1.28Mcps Option) See *UTRA TDD 1.28Mcps Option.*

unbalanced Referring to a lack of electrical symmetry. A coaxial cable (coax) comprises two conductors. The center conductor(s) carries the carrier signal. The outer conductor generally is used only for electrical grounding, and is maintained at zero volts (0V). Therefore, coax is described as an electrically unbalanced medium. A balun (balanced/unbalanced) connector is used to connect (balanced) twisted pair and (unbalanced) coax cables. See also *balanced* and *twisted pair.*

unbundled service The Telecommunications Act of 1996 specifies three ways that competitive local exchange carriers (CLECs) in the United States can provide competing local telephone service: build and interconnect, bundled wholesale purchase, and unbundled service. Securing unbundled service from the incumbent LECs (ILECs) essentially was a matter of selecting, on a case-by-case basis, from a menu of unbundled network elements (UNEs), including local loops, local exchange and tandem switches (including software features), interoffice transmission facilities, signaling and call-related database facilities, operations support systems (OSSs) and information, and operator and directory assistance facilities. When all elements from the menu are chosen, this approach is known as Unbundled Network Elements-Platform (UNE-P) and the effect is the same as the bundled service, but the total price typically is at a discount of 40–60 percent from retail prices. See also *bundled service, CLEC, ILEC*, and *Telecommunications Act of 1996.*

underground cable Cable designed to be place underground, either directly buried or in conduit. Underground cable must be specially designed to protect against moisture, rodents, cable-seeking backhoes, and other earth-moving equipment. See also *conduit* and *direct bury cable.*

UN (United Nations) An association of nations with stated goals, including uniting the strength of nations to maintain international peace and security, and promoting the economic and social advancement of all peoples. With respect to telecommunications, the UN system of organizations includes the International Telecommunications Union (ITU), which establishes standards in the wireline and radio sectors, manages the global frequency spectrum, and works to improve infrastructure in developing nations. See also *ITU.*

Unbundled Network Element (UNE) See *UNE.*

Underwriters Laboratories (UL) See *UL.*

undocumented feature A euphemism for bug. See *bug.*

UNE (Unbundled Network Element) The Telecommunications Act of 1996 specified three ways that competitive local exchange carriers (CLECs) in the United States can provide competing local telephone service: build and interconnect, bundled wholesale purchase, and unbundled service. Securing unbundled service from the incumbent LECs (ILECs) essentially was a matter of selecting from a menu of network elements, including local loops, local exchange and tandem switches (including software features), interoffice transmission facilities, signaling and call-related database facilities, operations support systems (OSSs) and information, and operator and directory assistance facilities. See also *CLEC, CO, database, ILEC, local loop, OSS, signaling and control, tandem switch,* and *Telecommunications Act of 1996.*

Unbundled Network Elements Platform (UNE-P) See *UNE-P.*

UN/EDIFACT (United Nations/Electronic Data Interchange For Administration, Commerce, and Transport) The international standard for Electronic Data Interchange (EDI), UN/EDIFACT is a United Nations recommendation that has been adopted by the International Organization for Standardization (ISO) and is predominant outside of North America. The competing X12 standard is dominant in North America. EDI standards specify data formats, character sets, and data elements. See also *EDI, standard, UN,* and *X12.*

UNE-P (Unbundled Network Elements-Platform) The Telecommunications Act of 1996 specified three ways that competitive local exchange carriers (CLECs) in the United States can provide competing local telephone service: build and interconnect, bundled wholesale purchase, and unbundled service. Securing unbundled service from the incumbent LECs (ILECs) essentially was a matter of selecting from a menu of network elements, including local loops, local exchange and tandem switches (including software features), interoffice transmission facilities, signaling and call-related database facilities, operations support systems (OSSs) and information, and operator and directory assistance facilities. When all elements from the menu are chosen, this approach is known as Unbundled Network Elements-Platform (UNE-P) and the effect is the same as the bundled service, but the total price typically is at a discount of 40–60 percent from retail prices. See also *CLEC, CO, database, ILEC, local loop, OSS, signaling and control, tandem switch,* and *Telecommunications Act of 1996.*

UNI (User Network Interface) The boundary or point of interaction between a network and the end user, the UNI is both a physical and logical point of demarcation. The UNI serves the technical boundary where protocol issues are resolved and as the point of division between the responsibilities of the service provider and those of the end user. In ISDN, for example, the UNI is defined as basic rate interface (BRI) for low-speed access and primary rate interface (PRI) for high-speed access, and is implemented in customer premises equipment (CPE). In frame relay, the UNI is implemented in the frame relay access device (FRAD), which is CPE, and the frame relay network device (FRND) located at the edge of the carrier network. In asynchronous transfer mode (ATM), a Private UNI is employed in a private ATM network, such as an ATM-based LAN. A Public UNI the interface between an ATM endpoint or private ATM switch and an ATM switch in a public network. See also *BRI, CPE, FRAD, frame relay, FRND, ISDN, logical, physical, PRI,* and *protocol.*

unibit Referring to a modulation technique that impresses a single bit on a baud, so that the bit rate is the same as the baud rate. Such a technique employs only two signal states. Binary phase-shift keying (BPSK) is an example of a unibit modulation scheme. Basic amplitude modulation (AM) and frequency modulation (FM) also are unibit in nature. See also *AM, BPSK, dibit, FM, quadbit,* and *tribit.*

unicast Referring to the transmission of a signal or packet from a single device to another single device over a circuit or network. See also *anycast, anycast address, broadcast, global unicast address,* and *multicast.*

unicast address In Internet Protocol version 6 (IPv6), an address associated with a single interface that is associated with a single node, and can, in effect, identify the node. A unicast address can be of several types and here are also special purpose unicast subtypes, including IPv6 addresses with embedded IPv4 addresses. A global unicast address is a conventional, publicly routable address that can be used in the Internet or any public domain. A link-local address is similar to an IPv4 private IP address, as it is not meant to be routed, but confined to a single segment. A site–local address is used by an organization that has not yet connected to the Internet. A loopback address is used when a host needs to send a packet back to itself. See also *domain, global unicast address, interface, Internet, IPv4, IPv6, link-local address, loopback address, node, packet, router,* and *site–local address.*

Unicode (Universal code) A set of standard coding schemes intended to replace the multiple coding schemes currently used, worldwide. The Unicode Consortium developed the original standard, Unicode Transformation Format-16 (UTF-16), in 1991 as a standard coding scheme to support multiple complex alphabets such as Chinese, Devanagri (Hindi), Japanese, and Korean. In the Japanese language, for example, even the abbreviated Kanji writing system contains well over 2,000 written ideographic characters; the Hirigana and Katakana alphabets add considerably to the complexity. As 7- and 8-bit coding schemes cannot accommodate such complex alphabets, computer manufacturers traditionally have taken proprietary approaches to this problem through the use of two linked 8-bit values. UTF-16 supports 65,536 (2^{16}) characters, which accommodates the most complex alphabets. Unicode accommodates pre-existing standard coding schemes, using the same byte values for consistency. For example, Unicode mirrors ASCII in UTF-7 and EBCDIC in UTF-EBCDIC, specifically for IBM mainframes. UTF-8 supports any universal character in the Unicode range, using one-to-four octets (eight-bit bytes) to do so, depending on the symbol. UTF-32 uses four octets for each symbol, but is rarely used due to its inherent inefficiency. The Unicode Standard has been adopted by most, if not all, major computer manufacturers and software developers, and is required by modern standards such as CORBA 3.0, ECMAScript (JavaScript), Java, LDAP, WML, and XML. Unicode is developed in conjunction with the International Organization for Standardization (ISO) and Internet Engineering Consortium (IEC), which also define the Universal Character Set (UCS), into which the UTF code sets map. UCS-4 is a four-octet code set into which UTF-32 maps and UCS-2 is a two-byte code set into which UTF-16 maps. UCS-1 encodes all characters in byte sequences varying from one to five bytes. See also *ASCII, code set, EBCDIC, IEC,* and *ISO.*

Unicode Consortium A not-for-profit consortium founded to develop and promote the use of the Unicode Standard. The consortium operates with the International Organization for Standardization (ISO), Internet Engineering Consortium (IEC) and World Wide Web Consortium (W3C) to refine the Unicode specifications and expand the character set. See Appendix A for contact information. See also *IEC, ISO, Unicode,* and *W3C.*

Unicode Transformation Format (UTF) See *Unicode.*

unified communications The ability to communicate in real time in the preferred mode (i.e., landline, cellular telephone, e-mail, or fax), unified communications incorporates presence technology, thereby allowing the user to indicate availability (e.g., available, unavailable, or out to lunch) and communications mode preference (e.g., business phone, cellular phone, text message, or instant message) to prospective callers. See also *unified messaging.*

unified messaging Synonymous with integrated messaging and multimedia messaging. The integration of voice, audio, text, facsimile, image, and video messaging, ideally on a single messaging platform or on a suite of platforms interrelated in such a way as to be transparent to the end user. Unified messaging is intended to provide an end user with a single interface to a messaging system or suite of messaging systems that support all types of messages in all formats, and that can adapt the message format to match the limitations of the terminal device and network that the end user employs to access the messaging system. For example, a unified messaging system could convert a fax message into a voice message if the user calls from a conventional telephone over a landline, or into a text message if the user has access to a cellular telephone. Ideally, unified messaging works in both directions, enabling the recipient of the message to respond in an inverse manner. If, for example, a unified messaging system converted a fax message into a text message for a user who has access to a cellular telephone, unified communications allows the user to respond with a text message that the system will convert into a fax that it will send to the originator of the message. See also *unified communications*.

uniform call distributor (UCD) See *UCD*.

uniform dialing plan (UDP) Also known as *coordinated dialing plan*. An optional PBX feature for large, complex multisite PBX networks, UDP supports simplified station-to-station dialing, allowing multiple PBXs to share a single numbering plan wherein each station has a unique number comprising four or five digits. The end user need only dial those digits to be connected to another user, regardless of where the two are located. If the two are served by the same PBX, the call connection process is simple. If the two are served by separate PBXs, the call will be sent over a private facility such as a tie trunk if one is available. If the call is sent over a central office (CO) trunk for connection through the PSTN, the originating PBX will automatically insert the necessary country code, area code and central office prefix, or otherwise modify the dialed digits as necessary to, for example, direct the call to the DID number of the called party.

Uniform Resource Identifier (URI) See *URL*.

Uniform Resource Locator (URL) See *URL*.

uninterruptible power supply (UPS) See *UPS*.

unipolar A digital signaling technique that makes use of a positive (+) voltage and a null, or zero (0), voltage to represent data in binary form, i.e., ones (1s) and zeroes (0s). See Figure U-2. See also *binary* and *bipolar*.

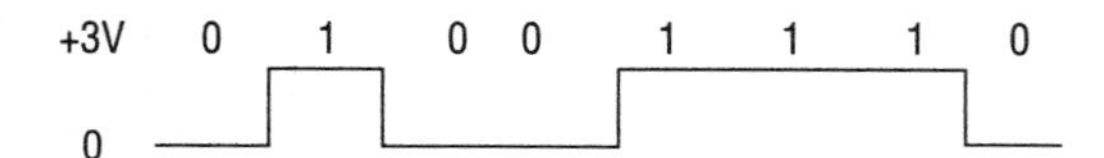

Figure U-2

United Nations (UN) See *UN*.

United Nations/Electronic Data Interchange For Administration, Commerce, and Transport (UN/EDIFACT) See *UN/EDIFACT*.

United States Computer Emergency Readiness Team (US-CERT) See *US-CERT*.

United States Digital Cellular (USDC) A digital cellular radio standard better known as Digital Advanced Mobile Phone Service (D-AMPS). See *D-AMPS*.

United States Independent Telephone Association (USITA) See *USTA*.

United States Telecom Association (USTA) See *USTA*.

United States Time Division Multiple Access (US TDMA) A digital cellular radio standard better known as Digital Advanced Mobile Phone Service (D-AMPS). See *D-AMPS*.

U-NII (Unlicensed National Information Infrastructure) Radio spectrum in the 5 GHz range made available for use on an unlicensed basis in the United States. Indoor applications include wireless LANs (WLANs) such as 802.11a (Wi-Fi5) and HiperLAN. Outdoor applications include wireless local loop (WLL). See also *802.11a*, *HiperLAN*, *Wi-Fi-5*, *WLAN*, and *WLL*.

universal ADSL (universal Asymmetric Digital Subscriber Line Lite) See *G.lite*.

Universal Character Set (UCS) See *UCS.*

Universal Code (Unicode) See *Unicode.*

Universal Coordinated Time (UTC) See *UTC.*

Universal Mobile Telecommunications System (UMTS) See *UMTS.*

Universal Plug and Play (UPnP) See *UPnP.*

Universal International Freephone Number (UIFN) See *UIFN.*

University Corporation for Advanced Internet Development (UCAID) See *UCAID.*

Unlicensed National Information Infrastructure (U-NII) See *U-NII.*

universal serial bus (USB) See *USB.*

Universal Service Administrative Company (USAC) See *USAC.*

universal service The concept of making affordable basic telephone service available to everyone everywhere within a nation, state, or other governmental jurisdiction. The United States Communications Act of 1934 first established the concept and led to the formation of a fund known as the Universal Service Fund (USF), which was finally codified in the Telecommunications Act of 1996. In some cases, the concept has been widened to include other telecommunications services classified as information services, most especially Internet access. That is the case in the United States, where the Universal Service Administrative Company (USAC) administers programs that provide support for incumbent local exchange carriers (ILECs) in high-cost areas, assistance for low-income subscribers, discounts for telecommunications service to schools and libraries, and discounts to rural health care providers. See also *Communications Act of 1934*, *Internet*, *Telecommunications Act of 1996*, *USAC*, and *USF.*

Universal Service Fund (USF) See *USF.*

UNIX A powerful multi-tasking, multi-user computer operating system (OS). UNIX was developed at AT&T Bell Telephone Laboratories during the years 1969 to 1973 by Ken Thompson and Dennis Ritchie for minicomputer application. As UNIX is written in the highly portable C programming language, it is used on a wide variety of computers, from mainframes to PDAs. A number of UNIX variations have been developed, some of which are available as freeware, and are known as Linux. See also *Bell Labs*, *freeware*, *Linux*, *mainframe computer*, *minicomputer*, *OS*, *PDA*, and *portability*.

Unlicensed Mobile Access (UMA) The original term for the Generic Access Network (GAN) standards. See *GAN.*

unshielded twisted pair (UTP) See *UTP.*

unsolicited grant service (UGS) See *UGS.*

unspecified bit rate (UBR) See *UBR.*

unsponsored domain In the Internet Domain Naming System (DNS), the original Top Level Domains (TLDs) that operate under policies established by the global Internet community, directly through the administration process of the Internet Corporation for Assigned Names and Numbers (ICANN). See also *DNS*, *ICANN*, *Internet*, *sponsored domain*, and *TLD.*

update **1.** A manipulation involving adding, modifying, or deleting data to bring a file or database up-to-date. **2.** A relatively minor release or version upgrade to an existing software product that adds minor features or corrects bugs. An update generally is denoted by a decimal fraction notation in the version number, such as 6.1 as an update to 6.0. See also *bug*, *bug fix*, *patch*, and *upgrade*.

upgrade A new, enhanced, or more powerful version or release of a product. A software upgrade generally is denoted by a new version number, such as 7.0 as an upgrade to 6.0. See also *bug*, *bug fix*, *patch*, and *update*.

uplink **1.** The microwave radio link from the terrestrial transmit antenna to the satellite. See also *downlink*. **2.** In a cellular network, the radio link from the mobile station (MS) to the base station (BS). See also *downlink*.

upload To transfer a file copy from a local computer to a remote computer over a network. See also *download*.

UPnP (Universal Plug and Play) A Microsoft initiative that extends the PnP specification to networks. UPnP enables a PC to automatically recognize the presence of a new peripheral hardware device as it is connected to a network, locate the necessary support software (e.g., driver), and configure the device interface. UPnP is aimed primarily at home and small office users who need to network devices such as personal digital assistants (PDAs) and digital cameras. See also *PnP*.

UPS (Uninterruptible Power Supply) A device or system that provides electrical power without interruption in the event that commercial power drops to an unacceptable voltage level. A UPS comprises circuitry and batteries that may provide power just long enough to shut down a computer or other system gracefully, without loss of data, or perhaps for many hours of normal operation in the event of a catastrophic commercial power failure. A typical UPS system operates in a hot standby, or offline, mode, continuously charging its batteries from a commercial power source and constantly prepared to assume responsibility within a few milliseconds for powering the client system. A more expensive online UPS actively filters commercial power, running it through the battery packs and an inverter, smoothing out the electrical waveforms and correcting for any power spikes and dips. See also *inverter* and *waveform*.

upstream The signal direction from the customer premises to the network edge. See also *downstream*.

uptime The time during which a machine or system is functioning properly. See also *downtime*.

URI (Uniform Resource Identifier) See *URL*.

URL (Uniform Resource Locator) A type of uniform resource identifier (URI) that consists of a uniform address that both identifies an abstract or physical resource on the World Wide Web (WWW) and indicates how to locate it. As specified in IETF RFC 3986, the syntax follows a standard convention:

scheme://authority/path?query#fragment

Consider the example:

http://info.cern.ch/hypertext/WWW/MarkUp/MarkUp.html

where http = hypertext transport protocol and html = hypertext markup language.

The method, or scheme name, indicates the network protocol used to assign identifiers. Examples of schemes include the following:

- **http:** Hypertext Transfer Protocol (HTTP)
- **https:** Hypertext Transfer Protocol (HTTP) over Secure Sockets Layer (SSL)
- **ftp:** File Transfer Protocol (FTP)
- **news:** Usenet newsgroups
- **Telnet:** Telecommunications Network protocol (TELNET)

The authority is preceded by a double slash (//) and is terminated by the next slash (/), question mark (?), or number sign (#), or the end of the URI. The authority can contain user information followed by a commercial at sign (e.g., ray@) and host information in the form of an Internet Protocol (IP) address (e.g., IPv4 dotted decimal notation) or registered domain name (e.g., contextcorporation.com). The authority also may contain an optional port number, which is unnecessary if the number is the same as the scheme's default (e.g., 80 would be redundant with http).

The path component contains data, usually organized in a hierarchical form that serves to identify a resource within the scope of the URI's scheme and naming authority. Within the path, a slash (/) is used as a delimiter between components. The path is terminated by the first question mark (?), number sign (#), or the end of the URI.

The query component contains non-hierarchical data that, along with data in the hierarchical path component, serves to identify a resource within the scope of the URI's scheme and naming authority.

The fragment component allows the indirect identification of a secondary resource that may be some portion or subset of the primary resource, some views or representations of the primary resource, or some other resource defined or described by those representations. See also *domain name*, *dotted decimal notation*, *HTML*, *HTTP*, *IETF*, *IP*, *IP address*, *IPv4*, *port*, *protocol*, *SSL*, *TELNET*, and *WWW*.

USAC (Universal Service Administrative Company) A subsidiary of the National Exchange Carrier Association (NECA), USAC administers programs that provide support for incumbent local exchange carriers (ILECs) companies in high-cost areas, assistance for low-income subscribers, discounts for telecommunications service to schools and libraries, and discounts to rural health care providers. USAC accepts funds collected by the ILECs and the interexchange carriers (IXCs) and distributes designated funds in support of the following programs and organizations set up under the Telecommunications Act of 1996:

- **Schools and Libraries Program:** Also known as the E-rate Program. Administered by the Schools and Libraries Corporation (SLC) to make telecommunications service, Internet access, and internal connections affordable for eligible schools and libraries.
- **Rural and High Cost Program:** Subsidizes basic telephone service in high-cost areas of the United States and its territories.
- **Rural Health Care Program:** Administered by the Rural Health Care Corporation (RHC) to subsidize telecommunications and Internet service for eligible rural health care providers in high-cost areas.
- **Low Income Program:** Lifeline subsidies to reduce the installation and monthly costs of basic telephone service for low-income consumers.

See also *ILEC*, *IXC*, *LEC*, *Telecommunications Act of 1996*, *universal service*, and *USF*.

USB (Universal Serial Bus) A serial bus with a data transfer rate of 12 Mbps for connecting a microcomputer to peripherals such as keyboards, mice, printers, and digital cameras through a single, general purpose port. USB is a considerable improvement over the RS-232 interface, but does not compare favorably with IEEE 1394, also known as FireWire in the Apple computer domain. See also *1394*, *bus*, *FireWire*, *IEEE*, *microcomputer*, *peripheral*, *port*, *RS-232*, and *serial*.

US-CERT (United States Computer Emergency Readiness Team) A team within the Department of Homeland Security charged with protecting the nation's Internet infrastructure by coordinating defense against and response from cyberattacks. See also *CERT*, *cyberspace*, and *Internet*.

USDC (United States Digital Cellular) A digital cellular radio standard better known as Digital Advanced Mobile Phone Service (D-AMPS). See *D-AMPS*.

U.S. Digital Cellular (USDC) A digital cellular radio standard better known as Digital Advanced Mobile Phone Service (D-AMPS). See *D-AMPS.*

user **1.** End user. The living, breathing human being who actually uses a computer or application to perform processes that yield end results. **2.** The person, organization, process, device, program, system, or other entity that exploits another person, organization, process, device, program, system, or other entity.

User Datagram Protocol (UDP) See *UDP.*

user ID (user IDentification) The user name, or username, by which a person is identified to a computer system or network. A user commonly must enter both a user ID and a password as an authentication mechanism during the logon process. If the system or network is connected to the Internet, the username typically is the leftmost portion of the e-mail address, which is the portion preceding the @ sign. In the e-mail address ray@contextcorporation.com, for example, ray is the username. User ID is synonymous with username. See also *password.*

user interface (UI) The portion of a computer program with which the user interacts, i.e., the interface between a user and a computer program. There are command-line interfaces, menu-driven interfaces, and graphical user interfaces (GUIs). See also *GUI.*

username Synonymous with user ID. See *user ID.*

user network interface (UNI) See *UNI.*

user plane In the ATM reference model, the functions that deal with issues of user-to-user information transfer and associated controls such as flow control and error control mechanisms. See also *ATM reference model, control plane, error control, flow control,* and *management plane.*

USF (Universal Service Fund) A fund established in the United States to support the concept of universal service, as first described in the Communications Act of 1934. The USF traditionally was intended to subsidize the cost of providing service to high-cost areas, defined as areas where the cost of providing service is at least 115 percent of the national average. Thereby, the USF ensured that even the most remote, sparsely populated, and impoverished areas of the United States had access to good quality basic voice telephone service at reasonable cost. In effect, it was a national cost-averaging scheme designed for the benefit of society. The USF extended over time to support the provisioning of lifeline service to those end users who cannot afford the cost of basic telephone service. The Telecommunications Act of 1996 codified the USF, extending its benefits to subsidize Internet access to schools and libraries, and telecommunications networks to link rural health care providers to urban medical centers in order to provide access to advanced diagnostic and other medical services. Under the direction of the Federal Communications Commission (FCC), the National Exchange Carrier Association (NECA) governs the USF, which actually is administered by the Universal Service Administrative Company (USAC), a NECA subsidiary. USAC accepts the collected funds from the local exchange carriers (LECs) and interexchange carriers (IXCs). The LECs collect USF fees from both IXCs and subscribers. Those fees are embedded in the access charges to the carriers, and generally are billed to subscribers as a separate line item. The LECs net out their approved USF requirements, retain a percentage as reimbursement for billing and administrative costs, and pass any remaining monies to USAC. Cellular providers contribute to the USF, usually on the basis of an average assumed percentage of interstate and international traffic. Voice over Internet Protocol (VoIP) providers also contribute to the fund for VoIP to PSTN calls; peer-to-peer VoIP-to-VoIP calls remain untaxed. In the United States, each of the individual states also has a USF, or something similar. Many other countries do, as well. For example, Australia formally established a Universal Service Obligation (USO) in 2001 and South Africa established the Universal Service Agency (USA) in the Telecommunications Act of 1996. See also *access charges, cellular radio, Communications Act of 1934, FCC, IXC, LEC, lifeline service, NECA, PSTN, Telecommunications Act of 1996, universal service, USAC,* and *VoIP.*

USITA (United States Independent Telephone Association) See *USTA*.

USTA (United States Telecom Association) A trade association of United States telecommunications carriers and suppliers, the USTA lobbies Congress, regulatory agencies, and other policy makers on behalf of its membership. USTA has its roots in the National Telephone Association (1897), which later changed its name to the United States Independent Telephone Association (USITA) and represented the interests of independent (i.e., non-Bell) telephone companies. When the break-up of the Bell System was decided in 1982, USITA admitted the newly divested Baby Bells as full members and changed its name to USTA. See Appendix A for contact information.

UTC (Universal Coordinated Time) Also known as Zulu (Z) Time. The International Telecommunications Union (ITU) struggled to develop an initialization for a universal time coordinated among all nations of the Earth. English speakers preferred CUT (Coordinated Universal Time). French speakers preferred TUC (Temps Universel Coordonné). So, the ITU managed to reach a consensus that totally satisfied neither group, but resolved the stalemate. UTC is based on the average of multiple atomic clocks, many of which are cesium clocks, which are the standard on which the SI second is based. UTC sometimes is referred to as GMT (Greenwich Mean Time), which is archaic. See also *ITU*, *second*, *SI*, and *Zulu Time*.

US TDMA (United States Time Division Multiple Access) A digital cellular radio standard better known as Digital Advanced Mobile Phone Service (D-AMPS). See *D-AMPS*.

UTF (Unicode Transformation Format) See *Unicode*.

utility **1.** A computer program dedicated to the management of a computer system, a utility is narrowly focused on a task such as maintaining a database, improving the efficiency of computer storage, searching for viruses, encrypting and decrypting data, and compressing and decompressing data. **2.** A company that performs or provides an essential public service, such as electric power, natural gas, sewer, telephone, or water.

utilize A frilly word that means to make use of something. People who want to sound smarter than they really are utilize utilize. I just use use.

Utopia From the Greek *ou*, meaning *not*, and *topos*, meaning *place*, and translating literally as *no place*. The word was first used by Sir Thomas More (1516) in his book *Utopia* as the name of an imaginary island that was the home of a perfect political and social system. In contemporary usage, utopia refers to an ideal place, state of being, or situation. (*Note:* Utopia sounds like no place I've ever been. If there were such a place, someone surely would foul it up. If not it would get so crowded that nobody would go there any more.)

UTP (Unshielded Twisted Pair) A pair of copper conductors, separately insulated by a dielectric material and smoothly twisted in a helix with a constant pitch or distance to make a 360° twist, as illustrated in Figure U-3. The conductors generally are solid core, although stranded wires are employed where additional flex strength is required. The insulating material is polyethylene, polyvinyl chloride, flouropolymer resin, Teflon®, or some other low-smoke, fire-retardant substance. The insulation separates the conductors so that the electrical circuit is not shorted, and protects the conductors from physical damage. Twisted pair is known as a balanced medium as both conductors serve for signal transmission and reception, and as each conductor carries a similar electrical signal with identical direct and return current paths. At any given point in the cable, the signals are equal in voltage to ground but opposite in polarity, which has the effect of reducing radiated energy and, therefore, reducing attenuation. The result is increased signal strength over a distance. Reducing the radiated energy also serves to minimize the impact on adjacent pairs in a multipair cable configuration. Generally speaking, the tighter the twist, i.e., the more twists per foot, the better the performance of the circuit. Unlike shielded twisted pair (STP) and screened twisted pair (ScTP), UTP has no shield to protect transmissions from electromagnetic interference (EMI). See also *dielectric*, *EMI*, *flex strength*, *ScTP*, *shield*, *STP*, and *UTP*.

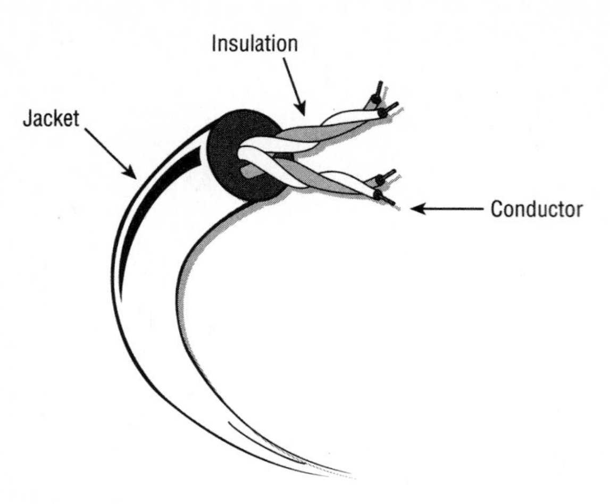

Figure U-3

UTRA TDD 1.28Mcps Option (UMTS Terrestrial Radio Access Time Division Duplex 1.28 Mega chips per second Option) The 3rd Generation Partnership Project (3GPP) designation for Time Division-Synchronous Code Division Multiple Access (TD-SCDMA), a 3G cellular radio standard under development by the Chinese Academy of Telecommunications Technology (CATT), an agency of the People's Republic of China (PRC). See also *3G*, *3GPP*, *cellular radio*, and *TD-SCDMA*.

UV (UltraViolet) The portion of the electromagnetic spectrum in the frequency range of 1–30 PHz and with a wavelength of 380–10 nm. UV light has no applications in telecommunications other than to make it possible for you to enjoy the full impact of psychedelic posters from the 1960s while you are contemplating what to do next. UV is the most dangerous portion of the spectrum in causing sunburn because the energy of a photon is proportional to its frequency ($e = h \times f$), where h is Planck's constant (6.6×10^{-34} joule-second). See also *electromagnetic spectrum*, *frequency*, *Hz*, and *wavelength*.

V **1.** Volt. See *volt.* **2.** Vertical. From the Latin *vertex*, translating literally as *turning point*, and meaning *top, zenith*, or *summit*. At a right angle to the horizon. The y-axis on a two- or three-dimensional grid, chart, or graph in a Cartesian coordinate system. See also *Cartesian coordinates*, *V & H*, *y-axis*, and *z-axis*. **3.** The V series of ITU-T Recommendations. The V series specifies protocols relating to modems and modem communications protocols over the public switched telephone network (PSTN). See *V series.* **4.** V interface or Reference Point V in ISDN. See *Reference Point V.*

V.32 The ITU-T Recommendation for full-duplex (FDX) asynchronous dial-up modems operating over the public switched telephone network (PSTN) at transmission rates up to 9600 bps, with the capability to fall back to 4800 bps, over a 4-kHz analog voice grade channel. See also *analog*, *asynchronous*, *channel*, *FDX*, *modem*, *PSTN*, *transmission rate*, and *voice grade.*

V.32bis The ITU-T Recommendation for full-duplex (FDX) asynchronous dial-up modems operating over the public switched telephone network (PSTN) at transmission rates up to 14.4 kbps, with the capability to fall back to rates as low as 4800 bps, over a 4-kHz analog voice grade channel. See also *analog*, *asynchronous*, *channel*, *FDX*, *modem*, *PSTN*, *transmission rate*, and *voice grade.*

V.34 The ITU-T Recommendation for full-duplex (FDX) asynchronous dial-up modems operating over the public switched telephone network (PSTN) at transmission rates up to 28.8 kbps over a 4-kHz analog voice grade channel. See also *analog*, *asynchronous*, *channel*, *FDX*, *PSTN*, *transmission rate*, and *voice grade.*

V.34+ The vernacular for V.34bis. See *V.34bis.*

V.34bis The ITU-T Recommendation for full-duplex (FDX) asynchronous dial-up modems operating over the public switched telephone network (PSTN) at transmission rates up to 33.6 kbps over a 4-kHz analog voice grade channel. See also *analog*, *asynchronous*, *channel*, *FDX*, *PSTN*, *transmission rate*, and *voice grade.*

V.35 Long deprecated standard for a modem that survived only in the form of a connector, the shape and pinout of which were defined for the modem. The V.35 connector is big, bulky, and slow by today's standards.

V.42 The ITU-T Recommendation for an error correction protocol that addresses error detection and correction through packet retransmission. V.42 is implemented in modems with asynchronous-to-synchronous conversion capability. As a default error correction method, V.42 includes the link access procedure for modems (LAPM), which is based on the High-Level Data Link Control (HDLC) protocol. In the event that LAPM fails, modems implementing V.42 fall back to the Microcom Networking Protocol version 4 (MNP4). See also *error control*, *HDLC*, *LAPM*, *MNP 4*, *modem*, and *protocol.*

V.42bis The ITU-T Recommendation specifying procedures for data compression in modems and other devices using error correction. V.42bis yields transmission rates up to 28.8 kbps. See also *error control*, *HDLC*, *modem*, and *protocol.*

V.90 The ITU-T Recommendation (1999) for full-duplex (FDX) asynchronous dial-up modems operating over the public switched telephone network (PSTN) at transmission rates up to 56 kbps downstream and 33.6 kbps upstream over a 4-kHz analog voice grade channel. In order to achieve this level of performance, a 56 kbps modem configuration requires that only one transmission link be analog. The end user dial-up connection is through a V.90 modem over an analog local loop to the PSTN. The connection through the PSTN must be entirely digital, including the originating and terminating central offices (COs), all tandem offices, and all transmission facilities. The local loop connection at the corporate intranet site, Internet service provider (ISP), or other site also must be digital, such as T-carrier, E-carrier or ISDN.

This configuration limits the transmissions to only a single D-to-A-to-D (digital-to-analog-to-digital) conversion process, which limits the amount of quantizing noise associated with the PCM encoding process. The V.90 standard was known as *V.last*, as it was expected to be the last conventional modem standard developed by the ITU-T. V.92 was developed shortly thereafter, in 2000. *Note:* Although V.90 modem technology is fully capable of achieving a downstream transmission rate of 64 kbps, the T-carrier systems in the North American PSTN use a bit-robbing convention that reduces the reliable maximum transmission rate to 56 kbps. Further, FCC limitations on amplitude levels restrict the downstream rate to 53.5 kbps in the United States. See also *analog, asynchronous, bit robbing, channel, CO, dial-up circuit, digital, downstream, E-carrier, encode, FDX, Internet, intranet, ISDN, ISP, ITU-T, modem, PCM, PSTN, quantizing noise, tandem switch, T-carrier, transmission rate, upstream, V.92, voice grade,* and *V series.*

V.92 The ITU-T Recommendation that improved on the V.90 standard in several ways. The QuickConnect feature reduces the handshaking time by approximately 50 percent, to about 10–15 seconds. QuickConnect trains the modem on the first call and remembers the characteristics of the circuit. Assuming that the circuit is the same on the next call the circuit characteristics do not have to be relearned, which results in faster connect times. V.92 increased the upstream transmission speed from 33.6 kbps to 48 kbps under optimum conditions, using a variation of PCM that allows the upstream datastream to use the same clocking source as the downstream datastream. V.90 also replaces the V.42bis compression algorithm with V.44, a string-coding algorithm that offers compression in the range of 6:1, improving throughput by 20–60 percent, and as much as 200 percent for certain kinds of highly compressible data. That yields theoretical downstream throughput rates as high as 300 kbps, compared with the maximum rates of 150–200 kbps possible with V.90 modems. A V.92 modem can put a data session on hold when it detects a voice call, either incoming or outgoing, through a call waiting indication, and gracefully resume that session when the voice call is terminated, thereby allowing a single analog line to be used for both voice and data applications. See also *circuit, compression, downstream, handshaking, modem, PCM, session, throughput, upstream, V.42,* and *V.90.*

V.110 The ITU-T Recommendation that specifies support for data terminal equipment (DTE) with asynchronous or synchronous serial interfaces over an ISDN network through rate adaption. See also *asynchronous, ISDN, rate adaption,* and *synchronous.*

V.120 The ITU-T Recommendation that specifies support for data terminal equipment (DTE) with asynchronous or synchronous serial interfaces over an ISDN network through data encapsulation. V.120 includes specifications for allowing multiple terminals to share a 64-kbps bearer channel (B channel) through statistical time division multiplexing (STDM). See also *asynchronous, B channel, ISDN, STDM,* and *synchronous.*

V.last The term once applied to V.90, as it was expected to be the last conventional modem standard developed by the ITU-T. V.92 was developed shortly thereafter. See also *V.90* and *V.92.*

V & H (Vertical and Horizontal) **1.** Geographical V & H coordinates traditionally are used to determine the straight-line distance between toll centers for purposes of rating long distance calls. **2.** Information oftentimes is graphically represented along the two dimensions of the vertical (V) and the horizontal (H). The graphic representation of an electromagnetic waveform, for example, plots amplitude (A), or signal strength, on the vertical axis, and frequency (f), or the periodic variation in value over time (t), on the horizontal axis.

vacuum A space completely void of matter. Although a complete vacuum is unachievable on earth, outer space is theoretically a vacuum to within a few molecules per cubic inch.

Vail, Alfred (1807–1859) A machinist and inventor who collaborated with Samuel F.B. Morse (1791–1872) on the invention and subsequent development of the electric telegraph. Vail is credited by some with the invention of Morse code, an improvement on Morse's original coding scheme, although Morse filed the patent on the code and, therefore, owned the rights to it. Vail left the telegraph industry in 1848, as he found he could not make a living in it. In fact, his last job was as superintendent of a telegraph line from Washington, D.C., to Columbia, South Carolina, at an annual salary of $900. In a letter to

Morse, he said, "I have made up my mind to leave the Telegraph to take care of itself, since it cannot take care of me." See also *Morse, Samuel F.B.*; *patent*; and *telegraph*.

value-added network (VAN) A network that offers a value-added service, i.e., services that alter the form, content, or nature of the information, thereby adding value to it. Packet-switched networks, and specifically X.25, were the first VANs. Specifically, X.25 added value through error correction. See also *packet switch*, *value-added service*, and *X.25*.

value-added reseller (VAR) See *VAR*.

value-added service Also known as *enhanced service*. Services that alter the form, content, or nature of the information, thereby adding value to it. Examples include store-and-forward services such as voice mail, e-mail, and fax mail. Voice mail systems, for example, cannot only store messages for subsequent retrieval, but also often allow a user to annotate a message before forwarding or archiving it. Some voice messaging systems even perform language translation. Voice-to-text capability allows the user to request that the system convert a voice message to text format and then send it via e-mail. Similarly, unified messaging systems can convert e-mail to voice format, fax mail to e-mail or voice format, and so on. See also *basic service*, *e-mail*, *facsimile*, *store-and-forward*, and *voice mail*.

VAR (Value-Added Reseller) A company that builds products or systems incorporating components or products that are manufactured by another company commonly referred to as an original equipment manufacturer (OEM). A manufacturer of laptop or tablet computers, for example, might incorporate an Ethernet network interface card (NIC), built by an OEM to its specifications.

variable bit rate (VBR) See *VBR*.

variable quantizing level (VQL) See *VQL*.

vBNS (very-high-speed Backbone Network Service) A broadband optical network provided under a cooperative agreement between the National Science Foundation (NSF) and Worldcom (now MCI, which is part of Verizon) in support of NSF-approved institutions of higher learning. vBNS initially (1995) ran over an ATM/SONET backbone at 155 Mbps (OC-3), and later was upgraded to 2.5 Gbps (OC-48). When the NFS contract expired, the vBNS largely transitioned to a federal government network from virtual private networks (VPNs) and the universities and research institutions transitioned to Internet2. See also *Abilene Project*, *ATM*, *backbone*, *broadband*, *Internet2*, *OC*, *SONET*, and *VPN*.

VBR (Variable Bit Rate) In asynchronous transfer mode (ATM), a class of traffic that requires access to time slots at a rate that can vary significantly from time to time. Real-time compressed voice and video and time-sensitive bursty data traffic are examples of VBR traffic. ATM also defines constant bit rate (CBR), non real-time Variable Bit Rate (nrt-VBR), non real-time Variable Bit Rate (nrt-VBR), real-time Variable Bit Rate (rt-VBR), and unspecified bit rate (UBR) traffic classes See also *ABR*, *ATM*, *CBR*, *compression*, *nrt-VBR*, *real-time*, *rt-VBR*, *time slot*, and *UBR*.

VC **1.** Virtual Container Synchronous Digital Hierarchy (SDH) terminology for Virtual Tributary (VT). See also *SDH* and *VT*. **2.** Virtual Channel. In asynchronous transfer mode (ATM), a unidirectional channel for transporting cells between two consecutive ATM entities across a link. See also *ATM*, *cell*, *channel*, and *link*. **3.** Virtual Connection. A logical connection to a virtual circuit.

VCI (Virtual Channel Identifier) In asynchronous transfer mode (ATM), 16 bits in the cell header that identify the virtual channel (VC), which is established each time a call is set up. See also *ATM*, *ATM reference model*, *call*, *cell*, *header*, and *VC*.

VCO (Voice Carry Over) An offering of Telecommunications Relay Service (TRS) that allows a person with a hearing disability to use his or her own voice to speak directly to the called party, but receive responses in text from the communications assistant (CA), who acts as a facilitator. See also *TRS*.

VCSEL (Vertical Cavity Surface-Emitting Laser) A laser with the lasing cavity running vertically through the layers of the semiconductor chip, which is mirrored at the bottom in order to maximize signal emission power at the top surface.VCSELs have capabilities somewhere between LEDs and other lasers. VCSELS have a spectral width somewhere between the two.VCSELs can couple effectively to a multimode fiber (MMF) with a narrower core than LEDs (50 microns versus 62.5 microns), but not as narrow (5–10 microns) as the single-mode fiber (SMF) to which a distributed feedback (DFB) laser connects.VCSELs also have a faster cycle time than LEDs, if somewhat slower than DFB lasers, and, therefore, have bandwidth capabilities somewhere between the two.The first generation ofVCSELs operates in the 850 nm window. The second generation (2005) ofVCSELs can run in the 1300 nm and 1310 nm regions.VCSELs are used primarily in high-speed LAN applications. See also *cycle time*, *DFB laser*, *laser*, *LED*, *MMF*, *SMF*, and *window*.

VDSL (Very-high-data-rate Digital Subscriber Line) Specified by the ITU-T in Recommendation G.993.1 (June 2004),VDSL is a high speed DSL variant that provides for downstream data rates up to 55 Mbps and upstream rates up to 15 Mbps over distances up to 1,000 feet (300 meters), sensitive to factors such as local loop characteristics.VDSL operates in a frequency range up to approximately 8.8 MHz divided amongst 2,048 subcarriers.As attenuation is a considerable issue such high frequencies, performance drops precipitously beyond 1,000 feet.VDSL2 was published in February 2006. See also *attenuation*, *downstream*, *frequency*, *ITU-T*, *subcarrier*, *upstream*, and *VDSL2*.

VDSL2 (Very-high-data-rate Digital Subscriber Line version 2) Specified by the ITU-T in Recommendation G.993.2 (February 2006), is the specification for two versions ofVDSL2. The long reach version runs at 12 MHz, divided among 2,872 subcarriers. Bandwidth is asymmetric, with transmission rates up to 55 Mbps downstream and 30 Mbps upstream over local loops up to 1,000 feet in length. Considerably reduced rates are achievable at distances up to 4,000–5,000 feet. The short reach version runs variously at 17.6 MHz and up to 30 MHz, divided among as many as 4,096 and 3,478 subcarriers. Bandwidth is symmetric, with downstream and upstream transmission rates as high as 100 Mbps over loops up to 500 feet. Considerably reduced rates are achievable at distances up to 4,000–5,000 feet.The short reach version will run in asymmetric mode, as well.VDSL2 employs the same discrete multitone (DMT) modulation scheme as ADSL.Trellis-coded modulation (TCM) yields higher throughput on long loops where the signal-to-noise ratio (SNR) is low, although data rates drop considerably beyond 500–1,000 feet. VDSL2 defines eight profiles for services, including asynchronous transfer mode (ATM) and Ethernet, and quality of service (QoS) features are integrated into the specification. See also *ADSL*, *asymmetric*, *ATM*, *bandwidth*, *DMT*, *downstream*, *Ethernet*, *ITU-T*, *local loop*, *modulation*, *QoS*, *SNR*, *subcarrier*, *symmetric*, *TCM*, *throughput*, *transmission rate*, *upstream*, and *VDSL*.

vector **1.** A mathematical expression of a quantity, such as velocity, that possesses both magnitude (i.e., amplitude) and direction, and that may or may not be a function of time. See also *amplitude*. **2.** A directed line segment of such an expression. See also *HCV* and *VQC*. **3.** A set of numbers in an order that has meaning when each position is mapped to a corresponding dimension. **4.** In video, a frequency or series of frequencies associated with a video signal. See also *vector quantization*.

vector quantization A lossy video compression technique that analyzes blocks of video pixels to determine their vectors, or frequencies. Prior to transmission, the video codec consults a codebook that contains a number of standard abbreviated vector descriptions in the form of codewords.The codec selects the codeword that produces the lowest level of distortion and outputs that to the channel. A matching codec associated with the receiver reverses the process. See also *channel*, *codec*, *frequency*, *lossy compression*, *pixel*, *vector*, and *video*.

vector quantizing code (VQC) See *VQC*.

vector sum excited linear predictive coding (VSELP) See *VSELP*.

velocity of light See *Vp*.

velocity of propagation (Vp) See *Vp.*

Veronica (Very Easy Rodent-Oriented Net-wide Index to Computerized Archives) A variation on the Archie Internet browser that searches Gopherspace titles and creates a menu with the results of the search. See also *Archie, browser, Gopher, Gopherspace, Internet,* and *JUGHEAD.*

vertical cavity surface-emitting laser (VCSEL) See *VCSEL.*

vertical redundancy check (VRC) See *VRC.*

very-high-data-rate digital subscriber line (VDSL) See *VDSL.*

Very High Frequency (VHF) See *VHF.*

very-high-speed Backbone Network Service (vBNS) See *vBNS.*

Very Low Frequency (VLF) See *VLF.*

very small aperture terminal (VSAT) See *VSAT.*

vestigial sideband (VSB) See *VSB.*

VHF (Very High Frequency) VHF radio is in the frequency range of 30–300 MHz and has a wavelength of 10–1 m.VHF radio has applications in amateur (Ham) radio,VHF TV, FM radio, mobile satellite systems (MSS), and mobile radio and fixed wireless. See also *electromagnetic spectrum, frequency, Ham radio, Hz, MSS,* and *wavelength.*

video **1.** The visual component of a television signal, which actually comprises a set of still images presented in rapid succession. See also *image* and *television.* **2.** Relating to the display of image data on a television set, computer monitor, cellular telephone, or other display device.

videoconference A video telephone call involving more than two parties.Videophones originated with the AT&T Picturephone, which was demonstrated at the New York World's Fair in 1964. Never intended for practical application, the Picturephone required bandwidth of about 90 MHz, and weighing about 26 pounds. During the 1980s, AT&T, British Telecom, and others developed videophones that sold for less than $1,000. As the cost was high, as each party was required to have a videophone of the same manufacture, and as the picture quality was poor at 2 frames per second (fps), videophones were stunning failures. A contemporary videoconferencing system can be in the form of an expensive and complex room system, a portable and less expensive rollabout system, or a desktop PC–based system. Regardless of its specific nature, a videoconferencing system consists of cameras, monitors, video boards, microphones, speakers, and software. Additional, specialized equipment includes the following:

- **Codecs** accomplish the process of digitizing, or coding, the analog signal on the transmit side and decoding it on the receive end. The codecs also accomplish the process of data compression and decompression, according to the specifics of the compression algorithm used.
- **Inverse Multiplexers (Inverse MUXs)** are used in commercial videoconferencing systems where sufficient dedicated bandwidth is not available over a single circuit.
- **Multipoint Control Units (MCUs)** are digital switching and bridging devices that support multipoint videoconferencing.

See also *analog, bandwidth, bridge, codec, compression, digital, fps, inverse multiplexing, MCU,* and *switch.*

video dial tone Also known as *visual dial tone.* Referring to the notion of a broadband network that is available to process a videoconference on demand. See also *broadband* and *dial tone.*

video-on-demand (VOD) See *VOD.*

Video Relay Service (VRS) See *VRS.*

V interface See *Reference Point V.*

virtual Being in essence or effect, although not in reality. A virtual circuit, for example, is not a physical circuit, but behaves as though it were, at least in some respects. Virtual is virtually the opposite of transparent, as something transparent exists but appears almost as though it does not. See also *transparent, virtual Centrex, virtual circuit, virtual path, virtual WATS,* and *VPN.*

virtual Centrex **1.** A Centrex technique that networks multiple geographically distributed CO Centrex systems, thereby creating the effect that all users are collocated and served by a single CO switch. See also *Centrex.* **2.** IP Centrex, which is Centrex service based on a softswitch platform and delivered over the Internet, rather than based on a central office (CO) circuit switch and delivered over the public switched telephone network (PSTN). See also *Centrex, CO, Internet, PSTN,* and *softswitch.*

virtual channel (VC) See *VC.*

virtual channel identifier (VCI) See *VCI.*

virtual circuit A circuit that exists in essence or effect, although not in the reality of dedicated components. In other words, a virtual circuit is a logical, rather than a physical, circuit. A virtual circuit commonly exists in the form of channel capacity provided over high-capacity, multichannel physical circuits, such as fiber optic transmission facilities, in a packet network. Virtual circuits are established end-to-end through a packet network, such as X.25 and Frame Relay, based on options and instructions defined in software routing tables. Permanent Virtual Circuits (PVCs) are permanently defined in routing tables, until such time as the carrier permanently redefines them. Switched Virtual Circuits (SVCs) are determined at the moment in time the connection is requested, with the specific path selection made in consideration of factors such as the level of congestion, level of error performance, geographic distance, and number of hops. A virtual circuit provides connectivity much as though it were a physical circuit, with all data traveling the same path. See also *circuit, PVC,* and *SVC.*

Virtual Container (VC) Synchronous Digital Hierarchy (SDH) terminology for Virtual Tributary (VT). See also *SDH* and *VT.*

Virtual LAN (VLAN) See *VLAN.*

virtual memory In a computer, disk space pretending to be random access memory (RAM). See also *computer, memory,* and *RAM.*

virtual path (VP) See *VP.*

virtual path identifier (VPI) See *VPI.*

Virtual Private Network (VPN) See *VPN.*

virtual routing and forwarding (VRF) See *VRF.*

virtual tributary (VT) See *VT.*

virtual tributary group (VTG) See *VTG.*

virtual WATS See *WATS.*

virus A type of intrusive malware that replicates itself and inserts copies of itself in legitimate programs, where it carries out unwanted and often damaging operations. Viruses initially were spread through infected floppy disks, which users frequently exchanged to share data and software. The most common contemporary methods of propagation are through attachments to Internet e-mail and programs downloaded from Websites. Viruses can be prevented if users open attachments only from trusted correspondents, visit only trusted websites, and purchase anti-virus software that they keep current. The term *virus*, in the contemporary context, was first used by Fred Cohen in his paper "Experiments with Computer Viruses" (1984). According to Cohen, the term was coined by Len Adleman; however, the term was in common usage long before. The science fiction novel *When HARLIE was One* (1972), by David Gerrold, describes a computer program named VIRUS, which could be countered by a program named VACCINE. See also *malware*, *spyware*, *Trojan horse*, and *worm*.

visible light The portion of the electromagnetic spectrum in the frequency range of 400 THz – 1 PHz and has a wavelength of 750–380 nm. Visible light has no applications in telecommunications other than to make it possible for you to see what is going on and what you are doing about it. See also *electromagnetic spectrum*, *frequency*, *Hz*, and *wavelength*.

visible light-emitting diode (VLED) See *VLED*.

visual dial tone Also known as *video dial tone*. Referring to the notion of a broadband network that is available to process a videoconference on demand. See also *broadband* and *dial tone*.

VLAN (Virtual Local Area Network) A software-defined LAN that groups users by logical addresses into a virtual, rather than physical, LAN through a switch or router. Users within a VLAN traditionally are grouped by physical ports, Transmission Control Protocol (TCP) port address, medium access control (MAC) address, or Internet Protocol (IP) address. A LAN switch or router can support many VLANs, which operate as subnets. See also *IP*, *LIS*, *MAC*, *physical*, *port*, *subnet*, *TCP*, and *virtual*.

VLED (Visible Light-Emitting Diode) An LED that emits light in the visible spectrum. See also *LED* and *visible light*.

VLF (Very Low Frequency) VLF radio is in the frequency range of 3–30 kHz and has a wavelength of 100–10 km. VLF radio has applications in navigation and weather science, and submarine communications. See also *electromagnetic spectrum*, *frequency*, *Hz*, and *wavelength*.

VoATM (Voice over Asynchronous Transfer Mode) Referring to voice communications over an ATM network. Uncompressed VoATM traffic is constant bit rate (CBR) traffic based on pulse code modulation (PCM) and time division multiplexing (TDM). CBR traffic requires the presentation of time slots on a regular and unswerving basis. Compressed voice is variable bit rate (VBR) traffic and requires access to time slots at a rate that can vary dramatically from time to time. See also *ATM*, *CBR*, *PCM*, *TDM*, and *VBR*.

V band The portion of the radio spectrum in the range of 46–56 GHz, as specified by the ITU-R. Current applications are limited to inter-satellite links. See also *electromagnetic spectrum* and *ITU-R*.

vocoder (voice coder) The equivalent of a codec in cellular wireless networks, a device that interfaces an analog device to a digital circuit or channel. Codecs operate in operate in balanced and symmetrical pairs, with one at each end of the communications circuit and with both having the same capabilities, at least at a minimum level. On the transmit side of the connection, a codec accepts an incoming analog signal, encodes it, i.e., converts it into digital form, and places it on a digital circuit. On the receive side of the connection, a codec with matching capabilities accepts the digital signal and decodes it to, i.e. recreates an approximation of the original analog signal. Many codecs are capable of operating in full duplex (FDX) , simultaneously encoding signals as they transmit them and decoding signals as they receive them. See also *analog*, *channel*, *circuit*, *codec*, *digital*, *encode*, *FDX*, and *signal*.

VOD (Video-On-Demand) A system that allows a user to select and access stored video content as desired. CATV providers, for example, commonly offer VOD, storing large numbers of movies and previously aired television programs on video servers. When the subscriber selects content, the movie or program is accessed on the server, decompressed and streamed over the network, and begins to play almost immediately on the user's TV set. See also *CATV*, *compression*, and *server*.

VoDSL (Voice over Digital Subscriber Line) Referring to nonstandard techniques for supporting packet voice over symmetric digital subscriber line (SDSL) and symmetric high-bit-rate digital subscriber line (SHDSL). See also *SDSL* and *SHDSL.*

VoFR (Voice over Frame Relay) Referring to techniques for transmitting real-time voice over a frame relay network. The standards for VoFR were set forth in the Frame Relay Forum's FRF11.1, Voice over Frame Relay Implementation Agreement (December 1998). As frame relay is a packet data network intended for LAN-to-LAN internetworking rather than isochronous traffic, levels of latency, loss, and error are variable and unpredictable in nature, which creates issues for real-time voice communications. However, a number of major domestic and international carriers offer, and even promote, VoFR as part of an integrated network solution and a managed service offering. The design may include separate VoFR permanent virtual circuits (PVCs) and oversized committed information rates (CIRs).

Still, VoFR must contend with issues of latency, jitter, loss and error, and does so through the use of various low bit-rate compression algorithms, the most popular of which are in the CELP (Code-Excited Linear Prediction) family. CELP and other compression algorithms support very reasonable business quality voice, at bit rates as low as 8 kbps, under conditions of low network congestion or where the voice traffic remains within its CIR. In order to mitigate the inherent difficulties of VoFR, some manufacturers and carriers offer various priority management techniques. Some service providers also offer PVCs of varying levels of delay/priority, usually by mapping the frame relay connection to an ATM connection with these properties. Priority levels generally are defined as follows:

- **Real-Time Variable Frame Rate:** Top priority; suited to delay-sensitive, mission-critical applications such as voice and SNA.
- **Non Real-Time Variable Frame Rate:** No-priority designation; suited to LAN and business class Internet and intranet IP traffic.
- **Available/Unspecified Frame Rate:** Low-priority designation; suited to Internet access, e-mail, file transfer, monitoring, and other low-priority applications.

See also *CELP*, *CIR*, *compression*, *congestion*, *frame relay*, *Frame Relay Forum*, *isochronous*, *LAN*, *latency*, *managed service provider*, *packet*, *PVC*, *real time*, *voice*, and *VoIP*.

voice Sounds made through the mouth by humans while talking, singing, or otherwise audibly communicating through the use of vocal organs. Although human voice frequencies mostly fall in the range of 100–8,000 Hz, the energy in the speech spectrum peaks at approximately 500 Hz, with most articulation at higher frequencies. Human hearing can distinguish signals as low as 20 Hz and as high as 20 kHz, and is most sensitive in the range of 1,0003,000 Hz. Human-to-human voice communications seldom requires technical support over short distances. Voice communications over distances of more than a few meters, however, requires that the acoustical energy be converted into some form of electromagnetic energy and sent over a transmission system of some description. See also *transmission system* and *voice grade.*

voice activity detection (VAD) Also known as *digital speech interpolation* (DSI). See *DSI.*

voice band The frequency band, or range, specified for voice communications in the public switched telephone network (PSTN). The total bandwidth of a voice grade channel is nominally 4 kHz. So, a single-channel voice grade circuit, supports a frequency band of 0–4,000 Hz. The bandwidth in the range 0–300 Hz generally is ignored, as the equipment is unable to deal those low frequencies. The voice band is approximately 3.0 kHz wide, running at 300–3,300 Hz. Signaling and control functions take place in the band 3,300–3,700 Hz. The lower band of 0–300 Hz and the upper band of 3,700–4,000 Hz are used as guard bands for maintaining separation between information channels, each of which is supported over a separate carrier frequency range, when analog voice channels are multiplexed using Frequency Division Multiplexing (FDM). See also *analog*, *bandwidth*, *FDM*, *frequency*, *guard band*, *multiplexing*, *PSTN*, and *signaling and control.*

Voice Carry Over (VCO) See *VCO.*

Voice Extensible Markup Language (VoiceXML) See *VoiceXML.*

voice grade Bandwidth sufficient to support voice communications. In analog transmission systems, a standard voice grade narrowband channel has nominal bandwidth of 4,000 Hz (4 kHz), which is the standard for analog voice. Within that channel and as illustrated in Figure V-1, the 0–300 Hz range generally is ignored, suppressed by the equipment's lack of ability to deal with voice at those low frequencies, which also avoids picking up hum from AC power lines. The active voice band is approximately 3.0 kHz wide, running at 300–3,300 Hz. As most speech activity takes place within this range, the level of fidelity is considered quite acceptable. Signaling and control functions take place in the 3,300–3,700 Hz band. In analog multiplexers, the lower band of 0–300 Hz and the upper band of 3,700–4,000 Hz are used as guard bands, i.e., for maintaining separation between information channels; on local loops these frequencies are filtered out. Multiple voice grade channels can coexist on an analog transmission facility, each running in a distinct carrier frequency range, and multiplexed using Frequency Division Multiplexing (FDM). Bandlimiting filters employed in carrier networks constrain the amount of bandwidth provided for a voice application, which conserves bandwidth without overly compromising fidelity. Capping the bandwidth at 3,300 Hz also prevents aliasing, a phenomenon that occurs when different continuous signals overlap and become indistinguishable when encoded into digital format. In digital systems, a narrowband channel is 64 kbps, based on G.711, which is the fundamental standard for digitized voice. There are numerous other standards for voice encoding, most of which involve considerable compression to improve bandwidth efficiency. Multiple voice grade channels can coexist on a digital transmission facility, each running in a distinct time slots, and multiplexed using Time Division Multiplexing (TDM). See also *aliasing, bandwidth, carrier, G.711, guard band,* and *narrowband.*

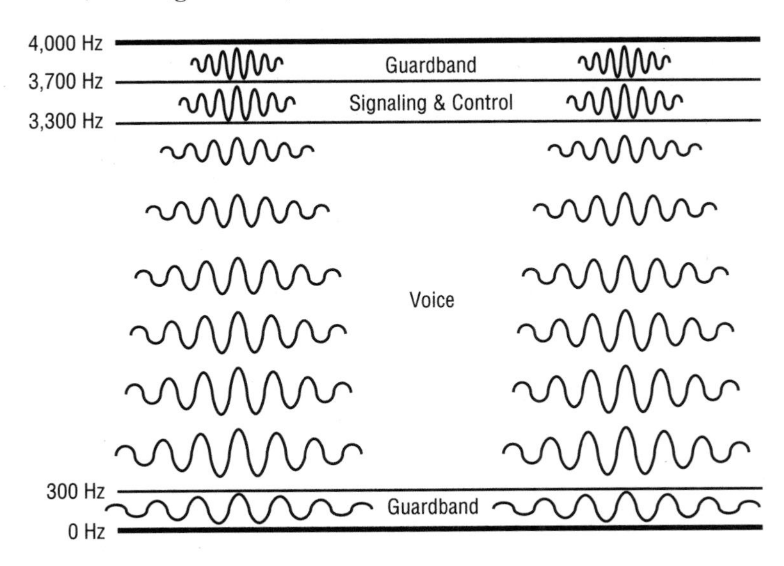

Figure V-1

voice mail The intervention of a voice processor in order that a caller can leave a voice message in the event that the incoming call encounters a busy signal or a no answer condition. In such an event, the telephone system (KTS, PBX, Centrex, or CO) directs the call to the voice processor, which answers the call with a default or customized greeting and directs the incoming call to a voice mailbox associated with a particular user or application. The voice processor digitizes, compresses, and stores the voice incoming message in the mailbox, and then advises the user of the fact that a message is waiting, most commonly through message-waiting lamp indication or stuttered dial tone. When the user accesses the system and enters the proper command and password, the message is restored to analog form and played back. Features may include annotation, archiving, attendant access, broadcasting, certification, class of service (CoS), find-me, forwarding, off-site notification, personalized greeting, prioritization, privacy, and purge. See the features for more detail. See also *audiotex*, *automated attendant*, *dial tone*, *time compression*, and *voice processor*.

voice mail jail The predicament a caller finds himself in when reaching an automated attendant that does not provide a menu option for reaching a live, human attendant. As the automated attendant feature is programmable, the network administrator failing to enable the feature can be characterized as being in violation of the Geneva Conventions dealing with the treatment of prisoners and civilians, and should be prosecuted as such. Fortunately, there are means by which a prisoner can organize a voice mail jailbreak. Unfortunately, the means vary by voice mail system and company. Usually, a prisoner can zero out, i.e., depress the zero (0) button on the keypad, sometimes repeatedly, to reach a human being. Sometimes, the hash (#) button works. Sometimes, it is the asterisk (*) or the nine (9) button. Sometimes, the process is more complicated. A simple search on the Web for "voice mail jail" will lead to a Web site with jailbreak instructions for many companies' voice mail jails. See also #, *automated attendant*, *Geneva Convention*, *human*, and *voice mail*.

voice over asynchronous transfer mode (VoATM) See *VoATM*.

voice over DSL (voice over Digital Subscriber Line) See *VoDSL*.

voice over frame relay (VoFR) See *VoFR*.

Voice over Internet Protocol (VoIP) See *VoIP*.

Voice over IP (Voice over Internet Protocol) See *VoIP*.

voice over packet Referring to voice communications over packet network technologies, as opposed to voice over the circuit-switched public switched telephone network (PSTN). In the generic sense, packet refers to the manner in which data are organized into discrete units for transmission and switching through a data network. The data unit can be in the form of a block, frame, cell, or packet, depending on the protocol specifics. In a technology-specific sense, voice over packet includes Voice over Asynchronous Transfer Mode (VoATM), Voice over Frame Relay (VoFR), Voice over Internet Protocol (VoIP), and Voice over Wi-Fi (VoWiFi). See also *block*, *cell*, *frame*, *packet*, *protocol*, *PSTN*, *VoATM*, *VoFR*, *VoIP*, and *VoWiFi*.

voice over Wi-Fi (VoWiFi) See *VoWiFi*.

voice processor A system, generally a specialized computer system, running application software that may perform a number of functions, including audiotex, automated attendant, and voice mail. In a large application, a voice processor is in the form of a standalone computer system linked to a central office (CO) or PBX. In an application serving a small-to-medium size enterprise (SME), a voice processor commonly is in the form of a set of application-specific integrated circuits (ASICs) on a printed circuit board (PCB) that fits in the chassis of a key telephone system (KTS) or PBX. The voice processor was invented by Gordon Mathews. See also *ASIC*; *audiotex*; *automated attendant*; *Mathews, Gordon*; and *voice mail*.

Voice Profile for Internet Mail (VPIM) See also *VPIM*.

voice recognition See *speech recognition*.

VoiceXML (Voice eXtensible Markup Language) A set of specifications for a high-level programming interface to speech and telephony resources for the development of speech recognition technology. Voice XML is based on the industry standard XML developed by the World Wide Web Consortium (W3C) for use on the World Wide Web (WWW). See also *speech recognition* and *XML*.

VoIP (Voice over Internet Protocol) Referring to voice communications over the public Internet or any packet network employing the TCP/IP protocol suite. Specifically, VoIP operates in datagram mode, employing the Internet Protocol (IP) for addressing and routing, the User Datagram Protocol (UDP) for host-to-host data transfer between application programs, and the Real Time Transport Protocol (RTP) for end-to-end delivery services. VoIP also typically employs sophisticated predictive compression algorithms, such as low delay code excited linear prediction (LD-CELP), to mitigate issues of latency and jitter over a packet-switched network. See also *IP, jitter, latency, LD-CELP, RTP,* and *UDP.*

volt (V) The unit of electric potential difference, or electromotive force (emf). A volt is equal to the difference in electric potential between two points in a circuit carrying one ampere (A) of constant current (I) across a resistance of 1 ohm, and thus dissipating one watt (W) of power. Voltage essentially is electrical pressure — the higher the voltage, the greater the pressure forcing electrons to flow through a metallic circuit. The volt is named for Count Alessandro Volta (1745–1827), the Italian physicist who invented the first electric battery. See also *ampere, current, emf, sine wave,* and *watt.*

voltage (E) Electric potential, expressed in volts (V). Voltage is the push, or pressure, behind current flow. See also *current* and *volt.*

VoWiFi (voice over Wi-Fi) The transmission of packet voice over an IEEE 802.11 wireless LAN (WLAN). VoWiFi takes advantage of the bandwidth offered by 802.11g (54 Mbps) and 802.11n (108 Mbps) to support voice over Internet Protocol (VoIP) over networks of fast Layer 2 Ethernet switches controlling handoffs among large number of thin access points (APs). Quality of service (QoS) issues were resolved in 802.11e (2005) through a coordination function that provides a station with high priority traffic such as voice with more frequent network access than a station with low priority traffic such as e-mail. Furthermore, the station with the high priority traffic is granted a longer transmit opportunity. See also *802.11, 802.11e, 802.11g, 802.11n, AP, bandwidth, handoff, QoS, thin AP, VoIP,* and *WLAN.*

Vp (Velocity of propagation) The speed with which a signal travels through a medium or through free space, expressed as a percentage of the speed of light in a vacuum. All electromagnetic signals propagate, or travel, in a vacuum at the speed of light, which is 299,792.458 kilometers per second (km/s), or 186,282.397 miles per second. (*Note:* This actually is a definition rather than a measurement, as the meter is formally defined as the distance traveled by light in a vacuum in $1/299,792,458$ of a second.) The speed of light typically is expressed in nominal, or approximate, terms as 300,000 km/s, or 186,000 miles per second, for ease of expression and calculation. As signals propagate in a vacuum unimpeded by any physical matter, this speed is the base number for indexing the velocity of propagation in all media. Table V-1 (Electromagnetic Signal Propagation Velocity) compares nominal signal velocities in various media, some of which are relevant and some of which simply serve as points of reference. The signal velocity in a vacuum applies to the space segment of a microwave satellite transmission, while the velocity in air applies to the propagation of the signal in the few kilometers of atmosphere it encounters. Terrestrial microwave systems, cellular systems, and all other varieties of terrestrial radio systems all operate in the atmosphere, of course. The signal velocity in copper cable refers to the propagation of electricity in a copper conductor, such as those in a twisted pair cable or a coaxial cable. The signal velocity in polyethylene, polypropylene, and polyvinyl chloride insulating materials is significant as much of the electrical signal energy is in the form of an electromagnetic field that propagates through the insulation surrounding the metallic conductor, rather than traveling through it. The signal velocity in air also is significant as air is introduced into the insulation of some high speed, multi-pair data cables in order to compensate for delay skew, i.e., differences in propagation delay across pairs. Propagation delay is the fundamental factor impacting latency. The inverse of Vp is the index of refraction (IOR), which is used to describe the Vp in a given fiber optic medium relative to the speed of light in a vacuum. See also *delay skew, IOR, latency, meter,* and *vacuum.*

Table V-1: Electromagnetic Signal Propagation Velocity (Approximate)

Medium	*Signal Velocity (km/s)*	*Velocity of Propagation (Vp)*
Vacuum	300,000	100.00
Air	299,890	99.97
Copper Cable	180,000–240,000	60.00–80.00
Water	226,000	75.33
Teflon	210,000	70.00
Optical Fiber	205,000	68.33
Polyethylene; Polypropylene	200,000	66.67
Polyvinyl Chloride	135,000–180,000	45.00–60.00

VP (Virtual Path) **1.** In SDH and SONET specifications, an end-to-end path that is temporary in nature, which is to say that the path between two endpoints can change from one connection to another. The VP also is shared among multiple calls. See also *path*, *SDH*, and *SONET*. **2.** In packet networks, a group of virtual circuits (VCs) sharing the same endpoints. See also *packet-switching* and *VC*.

VPI (Virtual Path Identifier) In asynchronous transfer mode (ATM), 8 bits in the cell header that identify the virtual path (VP), which is determined at the input port and is fixed for each call, and is shared among multiple calls. The path is from switch input port, through the switching matrix, to the output port, and then across a link between any two consecutive ATM entities and, therefore, has only local significance. See also *ATM*, *ATM reference model*, *cell*, *header*, *link*, *port*, *switch matrix*, and *VP*.

VPIM (Voice Profile for Internet Mail) A specification from the Internet Engineering Task Force (IETF) for transporting voice mail over TCP/IP networks, including the public Internet, VPIM superseded the Audio Messaging Interchange Specification (AMIS). VPIM wraps ADPCM-encoded voice messages in Multipurpose Internet Mail Extensions (MIME) message parts and uses Simple Mail Transfer Protocol (SMTP) to send them over TCP/IP networks. The first specification for VPIM was RFC 1911 (1996), which was followed by RFCs 2421 and 2421v2 (1998). See also *ADPCM*, *AMIS*, *IETF*, *Internet*, *IP*, *MIME*, *RFC*, *SMTP*, *TCP*, and *voice mail*.

VPN (Virtual Private Network) **1.** Also known as *SDN* (Software-Defined Network). A voice network service that creates the effect of a private, leased line network, but without the associated issues of design complexity, long deployment time, high recurring cost, and vulnerability to failure. Rather than interconnecting the various sites with dedicated circuits, the carrier routes the traffic through the public switched telephone network (PSTN) facilities on a priority basis over pre-determined paths, thereby ensuring that the level of service provided is roughly equivalent to that of a true private network. VPN services generally are interexchange, and often international, in nature, as they find their greatest application in very large multisite, national, or multinational enterprises. **2.** Also known as *SDN* (Software-Defined Network) and *SDDN* (Software-Defined Data Network). A digital data network service that operates much like a voice VPN or Switched 56, although the level of bandwidth provided can be much greater in support of intensive data communications, videoconferencing, or multimedia conferencing. Such a VPN creates the effect of a private, leased-line network, but without the associated issues such as design complexity and lengthy deployment. Rather than interconnecting the various sites with dedicated circuits, the carrier routes the traffic through the public switched telephone network (PSTN) facilities on a priority basis over pre-determined paths that are digital from end to end, thereby ensuring that the level of service provided is roughly equivalent to that of a true private data network. VPN services generally are interexchange, and often international, in nature, as they find their greatest application in very large multisite, national or multinational enterprises. Depending on the carrier, VPNs support bandwidth of 56/64 kbps, N × 64 kbps, 384 kbps, 768 kbps, 1.544 Mbps (T1) or 2.048 Mbps (E-1), and 44.736 Mbps (T3) or 34.368 Mbps (E-3). See also *dedicated circuit*, *PSTN*, and *Switched 56*.

VQC (Vector Quantizing Code) A voice compression technique that encodes analog voice signals based on a series of samples represented as a bit string, which is termed a vector. The vector is compared to a set of standard vectors stored in a codebook, the standard vector closest to the actual is selected, and an abbreviated identifier (i.e., code) is transmitted to the target station. VQC supports high-quality voice at rates of 32 kbps and 16 kbps, compression ratios of 2:1 and 4:1 as compared to pulse code modulation (PCM). VQC also does a good job of supporting modem transmission at speeds up to 9600 bps. VQC is an early form of code-excited linear prediction (CELP), which is now a dominant compression mechanism in packet voice applications such as voice over Internet protocol (VoIP). See also *analog, CELP, compression, encode, HCV, modem, PCM, quantize, vector,* and *VoIP.*

VQL (Variable Quantizing Level) A voice compression technique that encodes analog voice signals in blocks, adjusting the size of the quantizing steps based on the highest amplitude value in each block. VQL supports high-quality voice at a rate of 32 kbps, a 2:1 compression ratio, as compared to pulse code modulation (PCM). See also *amplitude, analog, compression, encode, PCM,* and *quantize.*

VRC (Vertical Redundancy Check) A simple parity checking error control method used in asynchronous transmission and primary storage. See Table V-2. VRC entails the appending of a parity bit at the end of each character or value to create an odd or even total mathematical bit value. The letter V, for example, in ASCII, is coded as a bit sequence of 0110101, which is an even number of marks, or 1 bits. If the network is set for the default odd parity, the parity bit would be a 1, as that would create an eight-bit byte with the sequence 01101011, thereby creating an odd parity value. Alternatively, the parity bit would be a 0 if the network is set for even parity, as that would create an eight-bit byte with the sequence 01101010, thereby retaining an even parity value. The receiving device executes the same mathematical process to verify that the correct total bit value was received, hence the use of the terms redundancy and checking. Speaking in terms of the logical manner in which humans add numbers physically positioned in columns, the two devices sum the bit values vertically, as represented in the following table, hence the use of the term vertical. VRC is easily and inexpensively implemented in computers employing asynchronous transmission, but is highly unreliable, as two errored bits in a character yield an undetectable error. Further, VRC provides no inherent means of error correction. VRC often is characterized as send and pray.

Table V-2: Vertical Redundancy Check (VRC)

Bit/Value	*C*	*O*	*N*	*T*	*R*	*O*	*L*
1	1	1	0	0	0	1	0
2	1	1	1	0	1	1	0
3	0	1	1	1	0	1	1
4	0	1	1	0	0	1	1
5	0	0	0	1	1	0	0
6	0	0	0	0	0	0	0
7	1	1	1	1	1	1	1
8 (Odd Parity)	0	0	1	0	0	0	0

See also *asynchronous, error control, LRC, parity bit,* and *parity check.*

VRF (Virtual Routing and Forwarding) In multiprotocol label switching (MPLS) or a router-based network, a routing table instance associated with a virtual private network (VPN) or frame relay access port. There may be multiple VRFs (one per VPN) in a single router serving multiple VPNs, a common situation within carriers and Internet service providers (ISPs). In a complex multi-site enterprise, there may be several routers, perhaps one serving the headquarters site and one or more serving the remote sites, and multiple VRFs, with each VRF serving one VPN. See also *carrier, frame relay, ISP, MPLS, router,* and *VPN.*

VRS (Video Relay Service) A Telecommunications Relay Service (TRS) that allows persons whose primary language is American Sign Language (ASL) to communicate through a communications assistant (CA) in sign language using videoconferencing equipment and an Internet Protocol (IP) based connection over the Internet. The signing person communicates in ASL to the CA, who voices the message to the hearing party, who responds to the CA, who then signs to the hearing-impaired caller. See also *Internet*, *IP*, *TRS*, and *videoconference*.

VSAT (Very Small Aperture Terminal) A satellite Earth station, or terrestrial terminal, characterized as having a dish antenna with an aperture, or opening, of very small diameter, at least in relative terms. A typical VSAT dish is 0.9, 1.2, 1.8 or 2.4 meters (approximately 3 to 8 feet) in diameter, with the specific size sensitive to the position of the antenna within the satellite footprint. The smallest dishes work well in the center of the footprint, where the signal is strongest. As antenna placements move farther from the center and closer to the fringes of the footprint contour, larger dishes are required to collect more signal and, thereby to improve the quality of reception. VSATs are associated with digital satellite systems operating in the C-band and Ku-band, and are designed primarily to support data communications applications such as retail inventory management, credit authorization, and general transaction processing. Bandwidth commonly is in channel increments of 56/64 kbps, generally up to an aggregate bandwidth of 512 kbps, which is the equivalent of 8 channels. Some newer systems support much higher levels of bandwidth that can be asymmetric in nature. See also *antenna*, *C-band*, *footprint*, *Ku-band*, and *satellite*.

VSB (Vestigial SideBand) A technique, used with amplitude modulation (AM), involving the transmission of the carrier, one complete sideband and only a portion of the other (vestigial) sideband. (Note: The process of amplitude modulation results in the creation of two sidebands. An upper sideband is above the carrier frequency and a lower sideband is below the carrier frequency.) The vestigial sideband assists in demodulation of the signal. See also *8-VSB*, *AM*, *carrier*, *DSB*, *frequency*, *Group II*, *sideband*, and *SSB*.

VSELP (Vector Sum Excited Linear Predictive Coding) A proprietary voice encoding standard developed by Motorola for digital cellular radio communications in the United States. VSELP is a type of code excited linear predictive (CELP) coding algorithm that typically encodes voice at an average rate of 7.95 kbps and adds overhead of 5.05 kbps for error control and synchronization for a total rate of 13 kbps, that actually can burst up to 48 kbps. Other versions of VSELP support voice at rates as low as 4.8 kbps. CELP uses a stochastically overlapped codebook, with each entry sharing all but two samples with its neighboring entries. VSELP offers the advantage of a codebook structure that allows a more efficient search procedure by using the sum of two basis vectors, although the computational complexity is greater. VSELP is specified in IS-54 and IS-136, better known as *D-AMPS*, in the pan-European GSM standard, and the Japanese PDC standard. See also *cellular radio*, *CELP*, *D-AMPS*, *digital*, *encode*, *GSM*, and *PDC*.

V Series The series of ITU-T Recommendations specifying protocols relating to data communication over the over the public switched telephone network (PSTN). V-series specifications largely relate to modems, data interfaces on modems (e.g., V.35, discontinued as a modem recommendation but still alive as a connector), and modem communications protocols, over dial-up and leased lines in both full duplex (FDX) and half-duplex (HDX). See Table V-3 for selected V-series Recommendations. For a full listing of ITU-T Recommendations, see the contact information in Appendix A.

Table V-3: Selected ITU-T V-Series Recommendations

Recommendation Number	*Description*
V.17	14.4 kbps two-wire modems for facsimile transmission over dial-up PSTN circuits
V.21	300 bps FDX modems for use over dial-up PSTN circuits
V.22	1200 bps FDX modems for use over dial-up PSTN circuits and point-to-point two-wire leased lines

Table V-3: Selected ITU-T V-Series Recommendations *(continued)*

Recommendation Number	*Description*
V.22bis	2400 bps FDX modems using the frequency division multiplexing (FDM) technique for use over dial-up PSTN circuits and point-to-point two-wire leased lines
V.23	600/1200-baud synchronous and asynchronous HDX modems for use over dial-up PSTN circuits and point-to-point two-wire leased lines
V.24	Definitions for interchange circuits between data terminal equipment (DTE) and data circuit-terminating equipment (DCE)
V.26	2400 bps FDX modems for use on four-wire leased lines
V.26bis	1200/2400 bps FDX modems for use over dial-up PSTN circuits
V.26ter	2400 bps FDX modems with echo cancellation for use over dial-up PSTN circuits and point-to-point two-wire leased lines
V.27	4800 bps FDX modems with manual equalization for use on leased lines
V.27bis	2400/4800 bps FDX modems with automatic equalization for use on leased lines
V.27ter	2400/4800 bps FDX modems for use for use over dial-up PSTN circuits
V.28	Electrical characteristics for unbalanced double-current interchange circuits
V.29	9600 bps HDX and FDX modems for use on point-to-point four-wire leased lines
V.32	9600 bps FDX modems with echo cancellation for use over dial-up PSTN circuits and point-to-point two-wire leased lines
V.32bis	4800/7200/9600/12,000/14,4000 bps FDX modems with echo cancellation for use over dial-up PSTN circuits and point-to-point two-wire leased lines
V.33	12,000/14,400 bps synchronous FDX modems for use on point-to-point four-wire leased lines
V.34	28.8 kbps modems for use over dial-up PSTN circuits and point-to-point two-wire leased lines
V.34bis	33.6 kbps modems for use over dial-up PSTN circuits and point-to-point two-wire leased lines
V.35	All that remains of this modem specification is a clunky connector for serial data.
V.41	Code-independent error-control system
V.42	Error-correcting procedures for DCE using asynchronous-to-synchronous conversion
V.42bis	Data compression procedures for DCE using error correction procedures
V.43	Data flow control
V.44	Data compression procedures
V.54	Loop test devices for modems
V.56	Comparative tests of modems for use over two-wire voice grade PSTN connections
V.56bis	Network transmission model for evaluating modem performance over two-wire voice grade PSTN connections
V.56ter	Test procedure for evaluation of two-wire 4 kHz voiceband FDX modems

Table V-3: Selected ITU-T V-Series Recommendations *(continued)*

Recommendation Number	*Description*
V.61	4800 bps simultaneous voice plus data modems, with optional automatic switching to 14.4 kbps data-only signaling rates of up to 14 400 bit/s for use over dial-up PSTN circuits and point-to-point two-wire leased lines
V.90	56 kbps downstream and 33.6 kbps upstream digital and analog modem pairs for use over dial-up PSTN circuits
V.91	64 kbps digital modems for use over four-wire circuit-switched connections and point-to-point leased four-wire digital circuits
V.92	Enhancements to Recommendation V.90
V.110	Support by an ISDN of DTE with V-Series type interfaces
V.120	Support by an ISDN of DTE with V-Series type interfaces with provision for statistical multiplexing
V.130	ISDN terminal adaptor framework
V.140	Procedures for establishing communication between two multiprotocol audiovisual terminals using digital channels at a multiple of 64 kbps or 56 kbps
V.150.0	Modem-over-IP networks: Foundation
V.150.1	Modem-over-IP networks: Procedures for the end-to-end connection of V-series DCE
V.151	Procedures for end-to-end connection of analog PSTN text telephones over an IP network utilizing text relay
V.152	Procedures for supporting voice-band data over IP networks
V.300	A 128 (144) kbps DCE for use on digital point-to-point leased circuits

VT (Virtual Tributary) In SONET standards, a bit-transparent time division multiplexed (TDM) connection that carries one form of signal (e.g., DS-1, DS-2, or DS-3) within a byte-interleaved frame. VTs are sized to accommodate the originating signal. A VT1.5, for example, operates at 1.544 Mbps (T1), VT2 at 2.048 Mbps (E-1), VT3 at 3.152 Mbps (T1C), and VT6 at 6.312 Mbps (T2). Individual VTs are distinguished by the use of a pointer, which identifies the position of the VT within a Virtual Tributary Group (VTG). A Tributary Unit (TU) is a VT, along with a pointer. SDH specifications refer to a VT as a Virtual Container (VC). See also *bit transparent*, *DS-1*, *DS-2*, *DS-3*, *SDH*, *SONET*, *TDM*, *TU*, and *VTG*.

VT1.5 (Virtual Tributary 1.5 Mbps) In SDH and SONET standards, a Virtual Tributary (VT) that operates at 1.544 Mbps, which is the signaling rate of a T1. See also *SDH*, *signaling rate*, *SONET*, *T1*, and *VT*.

VT2 (Virtual Tributary 2 Mbps) In SDH and SONET standards, a Virtual Tributary (VT) that operates at 2.048 Mbps, which is the signaling rate of an E-1. See also *E-1*, *SDH*, *signaling rate*, *SONET*, and *VT*.

VT3 (Virtual Tributary 3 Mbps) In SDH and SONET standards, a Virtual Tributary (VT) that operates at 3.152 Mbps, which is the signaling rate of a T1C. See also *SDH*, *signaling rate*, *SONET*, *T1C*, and *VT*.

VT6 (Virtual Tributary 6 Mbps) In SDH and SONET standards, a Virtual Tributary (VT) that operates at 6.312 Mbps, which is the signaling rate of a T2. See also *SDH*, *signaling rate*, *SONET*, *T2*, and *VT*.

VTG (Virtual Tributary Group) In SDH and SONET standards, group of Virtual Tributaries (VT), each of which carries the same form of signal (e.g., DS-1, DS-2, or DS-3) within a byte-interleaved frame. SONET can map as many as four VTs into a VTG, and as many as seven VTGs into a single Virtual Path (VP). See also *DS-1*, *DS-2*, *DS-3*, *SDH*, *signaling rate*, *SONET*, and *VT*.

W Symbol for watt. See *watt.*

WACS (Wireless Access Communication System) A U.S. specification for cordless telephony that was modified to an industry standard known as Personal Access Communications Services (PACS) and ultimately evolved into Personal Communications Services (PCS). See also *cordless telephone* and *PCS.*

WAIS (Wide Area Information Service) A UNIX-based system of servers that enables users to specify the databases requested for search, and to conduct a subject-matter search on the basis of keywords. See also *browser, server,* and *UNIX.*

Walkabout An early wireless telephony trial conducted in Canberra, Australia, in support of pedestrian traffic. Walkabout was based on CT2+ digital cordless telephony standards. See also *cordless telephone, CT2+,* and *digital.*

walkie talkie A portable handheld radio transmitter-receiver invented in 1938 by Al Gross while a high school student in Cleveland, Ohio (United States). The device caught the attention of the Office of Strategic Services (OSS), predecessor to the Central Intelligence Agency (CIA). The OSS recruited Goss, who then lead the effort to develop the walkie-talkie for clandestine and military uses. Code-named Joan/Eleanor, the first walkie talkie system comprised a ground unit, Joan, and an airborne unit, Eleanor. The system allowed OSS agents behind enemy lines to communicate with aircraft in a manner that virtually defied detection at the time. See also *Gross, Al.*

WAN (Wide Area Network) A network that covers a wide geographic area such as a state, province, region, or country. The Public Switched Telephone Network (PSTN) is a voice-oriented WAN. The Internet is a data-oriented public WAN. WANs can serve to interconnect LANs and MANs. See also *LAN* and *MAN.*

WAN Interface Sublayer (WIS) See *WIS.*

WAP (Wireless Access Protocol) A carrier-independent, device-independent, transaction-oriented protocol employed in Web-enabled cellular networks in support of text, graphics, and audio. The best performance is achieved when accessing Web sites written in WML (Wireless Markup Language), which is similar to HTML. The alternative is transcoding from HTML to WML, which is accomplished through gateways. A much simpler, but much less attractive technique is Web clipping, which strips the graphic content out of Web pages. Security over the wireless link is provided through Wireless Transport Layer Security (WTLS). WAP is employed outside of Japan, where the i-Mode microbrowser technology is employed. See also *browser, carrier, cellular, i-Mode, protocol, Web,* and *WML.*

WARC (World Administrative Radio Conferences) Now known as the World Radio Conferences (WRC). See *WRC.*

warrior's code A code of conduct that defines what warriors can and cannot do if they wish to continue to be regarded as warriors, rather than murderers or cowards. For the warrior who adheres to such an informal code or to more formal rules of engagement, certain actions are unthinkable, even in the most dire or extreme circumstances. Within a decade of the introduction of firearms to Japan in 1543, the Japanese were arguably the best gun makers in the world, and there were more guns per capita in Japan than in any other country in the world. The warrior code of the Samurai, however, viewed the use of guns in warfare as dishonorable. Centuries of tradition demanded that Samurai warriors engage in elaborate rituals prior to combat, which was conducted man-to-man and hand-to-hand between gentleman warriors. Under pressure from the Samurai, the Emperor of Japan gradually reduced the number of authorized gun factories to zero and, over time, subsequently reduced the number of gun repair shops to zero. By the time that Commodore Perry visited in the 1840s, there was not a single gun left in Japan, which left the Japanese at a decided disadvantage against the superior weaponry of the Europeans. See also *Geneva Convention* and *rules of engagement.*

water-blocking gel A soft, gooey, gelatinous, hydrophobic substance used to flood outside plant (OSP) cables to protect them from moisture and even standing water. Moisture can cause electrical noise and short circuits in copper cables, ice crush in fiber optic cables, and can cause cable sheaths to crack in subfreezing temperature conditions. An alternative to the unpleasantly gooey and sticky gel is a dry powdered compound that becomes a water-blocking gel on contact with moisture. See also *icky-pic* and *OSP.*

water peak A peak in attenuation in optical fibers caused by contamination from hydroxyl (OH) ions that are residuals of the manufacturing process. Water peak causes wavelength attenuation and pulse dispersion in the general regions of 950 nm, 1380 nm and 2730 nm. Low-water-peak fiber (LWPF) and zero-water-peak fiber (ZWPF) resolves water peak issues in the 1380 region (1383 nm), thereby opening the entire spectrum from 1260 to 1625 nm for high-performance optical transmission technologies employing coarse wavelength division multiplexing (CWDM). See also *attenuation, CWDM, hydroxyl, LWPF, pulse dispersion, wavelength,* and *ZWPF.*

watermark **1.** A translucent mark or image in paper produced by pressing the paper in a mold or on a processing roll during the manufacturing process. The watermark is visible when the paper is held to a light. A watermark is used as a sign of authenticity in order to make the counterfeiting of currency and postage stamps more difficult, for example. **2.** Digital watermarking is the process by which visible or invisible copyright notices or other messages are embedded in audio, image, or video signals or files, program files, or Web pages, thereby providing a tracking mechanism and discouraging copyright violations. The term is derived from the practice of marking paper, especially currency and postage stamps, to discourage counterfeiting. A hidden digital watermark is a form of steganography. See also *steganography.*

watt (W) Watt is the fundamental unit of electrical power, and is a rate unit, rather than a quantity. The wattage is determined by multiplying the voltage (E), as measured in volts (V), by the current (I), as measured in amperes (A). ($W = V \times A$). A watt is the amount of power required to do work at the rate of one joule per second. The watt is named for James Watt (1736–1819), who invented the steam engine and, in collaboration with Matthew Boulton, also invented a pumping engine and a rotative engine. See also *current, sine wave,* and *voltage.*

WATS (Wide Area Telecommunications Service) A PSTN offering for discounted bulk long distance service provided over special trunks in the United States and Canada. In the 1970s and 1980s, WATS was highly attractive to medium and large business enterprises that placed large volumes of long distance calls. WATS services were either full-time or measured. Full-time WATS service could be used 24 × 7 at a flat rate. Measured WATS was metered and billed in increments of 0.1 hour above a monthly usage threshold of 10 hours. WATS service was banded, meaning that it was organized, and rated, in crude concentric bands that radiated from the subscriber's home state. Band 0 WATS was intrastate, Band 1 included the band of contiguous states, Band 2 included the next concentric band of contiguous states, and so on to Band 5, which included the entire United States. As the service was directional in nature, WATS service was for outgoing calls, only, and INWATS was for incoming calls, only. Virtual WATS was organized in the same manner, but only logically so, with no need for special purpose circuits. WATS service is highly unusual in a contemporary context, having been made obsolete by discounted long distance billing plans that are independent of trunk facilities.

wave Something that moves up and down, back and forth, in and out, left and right, or otherwise in a gradual, curving, or undulating motion. See also *waveform.*

waveform The geometric shape of a wave. A waveform is used to graphically represent some recurring characteristic of a wave over time. Electromagnetic energy is commonly plotted in two dimensions as a sinusoidal waveform that varies in amplitude (A), or signal strength, on the vertical (V) axis at a periodic rate, or frequency (f), over time (t) on the horizontal (H) axis. See also *sine wave.*

waveguide An enclosed physical structure that contains and guides a signal. Conducted transmission media fit this definition, as insulation and shields variously serve more or less effectively to confine a signal to the electrical or optical conductor and guide it along a physical path. Unshielded twisted pair (UTP), shielded twisted pair (STP), coaxial cable, and optical fiber are all waveguides. Some radio systems make use of waveguides comprising hollow metal pipes made of a good conductor such as copper, aluminum, brass, or silver, surrounding a dielectric region, usually of air. The pipes generally are rectangular in form, although they sometimes are circular and can take a variety of other shapes, as well. Such radio waveguides are restricted to use in very high power or very high frequency applications over short distances due to their size, weight, inflexibility, and cost. See also *transmission medium*.

waveguide dispersion **1.** A type of dispersion caused by the different refractive indexes of the core and cladding of an optical fiber. Regardless of the nature of the light source and optical fiber, some light travels in the cladding, as well as the core. Assuming a step-index fiber, the core is of one highly consistent index of refraction (IOR), and the cladding is of another, although sometimes the cladding is of several layers of glass, each with a sharp step in IOR. As the IOR of glass varies as the wavelength varies, with longer wavelengths propagating at higher velocities, as no light pulse has a perfectly narrow spectral width, and as multiple layers of glass of different properties make up the core and cladding, different wavelengths of light propagate at different velocities in the different layers. So, the optical pulse can disperse, or spread, over a distance, which clearly can confuse the light detector at the far end of the fiber. Waveguide dispersion is one factor contributing to chromatic dispersion, both of which are issues in long haul fiber optic transmission systems (FOTS) employing single-mode fiber (SMF) of step-index construction. Multimode graded-index fibers suffer so much from modal dispersion over short distances that material dispersion and chromatic dispersion never become factors. See also *chromatic dispersion, dispersion, IOR, SMF, spectral width, step-index fiber,* and *waveguide dispersion*. **2.** A type of dispersion attributable to the relationship of the physical dimensions of the waveguide and the optical signal, specifically, the diameter of the fiber in cross-section and the length of the optical wave, i.e., the wavelength of the signal. The closer the relationship is to 1:1, the less the waveguide dispersion. As the waveguide increases in size from 50 microns to 62.5 microns, for example, waveguide dispersion increases at a given wavelength, such as 1300 nm. The diameter of the waveguide determines the number of modes, or physical paths, along which the signals are allowed to propagate. As the wavelength decreases from 1300 nm to 850 nm, for example, waveguide dispersion increases in fiber of a given core diameter, such as 62.5 microns. This is due to the increased frequency of the signal and, therefore, the increased opportunity for the signal to interact with the waveguide. This type of waveguide dispersion affects only multimode fiber (MMF) as single-mode fiber (SMF) supports only a single mode. See also *MMF, mode, propagate, SMF,* and *wavelength*.

wavelength The length of an electromagnetic waveform, wavelength (λ) is inversely proportional to frequency (f). As the frequency of the signal (number of cycles per second) increases, the wavelength (length of the electromagnetic waveform) of the signal decreases. In other words, the more waveforms transmitted per second, the shorter the length, or cycle, of each individual wave. Figure W-1 illustrates the relationship between frequency and wavelength — as the frequency doubles, the wavelength halves.

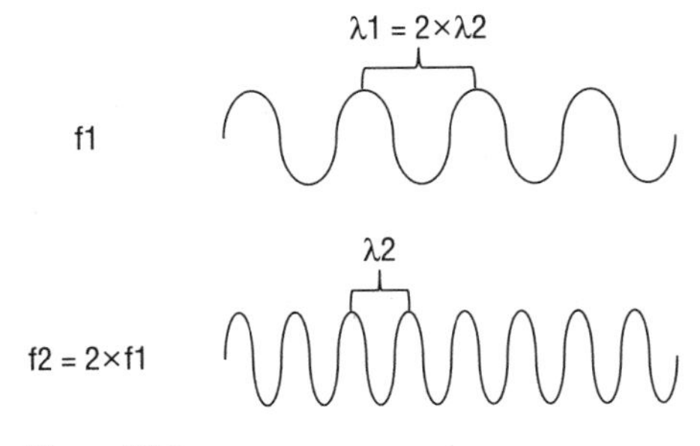

Figure W-1

Signals in electrical and radio networks are defined in terms of frequency and described in terms of cycles per second (cps) traditionally, and Hertz (Hz) in a contemporary context. Once the frequency of the electromagnetic signals exceeds the extremely high frequency (EHF) of level of 300 GHz and crosses into the infrared light (IR) range of the optical spectrum, however, Hz no longer has relevance as either a bandwidth measurement or a signal descriptor. The numbers are just too large and difficult to express. Rather, wavelength is used in the optical domain to describe the nature of the signal. By way of illustration, consider that the upper range of an analog voice channel is 4 kHz. The velocity of propagation of all electromagnetic energy in a vacuum is roughly that of the speed of light, or 300,000 kilometers per second. The velocity of electromagnetic energy through a copper wire is approximately 60–80 percent of that, or 180,000–240,000 kilometers per second. Therefore, at a frequency of 4 kHz (4,000 cycles per second) and assuming that the speed of the electrical current through the twisted pair copper circuit is 200,000 kilometers per second, each cycle is 50 kilometers in length over a typical single-channel, voice grade analog local loop. See the mathematical equation below:

$$\frac{200{,}000{,}000km/s}{4{,}000cycles/s} = 50{,}000m$$

Now consider that an infrared optical signal in a fiber optic cable at a commonly used wavelength of 1550 nm (.000001550 meters) has a nominal frequency of 193 THz (193,548,387,096,774 Hz). So, the same voice conversation over a typical long haul fiber optic system is carried by an optical signal that has a wavelength of .000001550 meters, not considering the Vp. In consideration of the fact that adjacent signals can be spaced at intervals of 200 GHz, or 1.6 nm (at 1550 nm), it is fairly obvious that it is a lot easier to discuss optical signals in terms of wavelength than frequency (Hz). See the math below:

$$\frac{300{,}000km/s}{193{,}548{,}387{,}096{,}774cycles/s} = .000001550m$$

wavelength band A continuous group, or range, or wavelengths with an upper limit and a lower limit. In analog terms, bandwidth and channel width are defined as a range of frequencies. The ITU-T defines standard optical transmission windows in bands of wavelengths. See also *wavelength* and *window*.

wavelength division multiplexer (WDM mux) A device that performs wavelength division multiplexing (WDM). See *WDM*.

wavelength division multiplexing (WDM) See *WDM*.

wavelet compression A technique for video compression that uses wavelet transforms to compress data. Wavelet compression treats the entire image as a series of small waveforms, or signals, known as wavelets, with one per color channel, e.g., red, green, and blue. A wavelet transform is applied that quantizes the wavelets by measuring the distances between the zero line and points along each wavelet and records theses distances as coefficients, with one coefficient for each pixel in the image. The coefficients of adjacent images are averaged to produce a simplified version of the wave, which process effectively halves the size of the image description. The process is repeated again and again and so on, producing progressively smaller waves, therefore, and smaller data files in a process known as decomposition. At each step of the averaging process, the difference between the coefficients is noted. Wavelet compression is used effectively to compress transient images, such as twinkling stars in a night sky. The U.S. Federal Bureau of Investigation (FBI) uses wavelet compression to store fingerprints. Smooth, periodic images are best compressed using other methods. Wavelet compression can be either lossless or lossy in nature. See also *compression*, *lossy compression*, *transform*, and *video*.

W band The portion of the radio spectrum in the range of 75–110 GHz, as specified by the ITU-R. The W band is employed in radar and scientific applications. See also *electromagnetic spectrum* and *ITU-R*.

W-CDMA (Wideband Code Division Multiple Access) Also known as Universal Mobile Telecommunications System (UMTS). See *UMTS*.

WD-40 (Water Displacement, 40th attempt) Arguably one of the two greatest inventions of the twentieth century (duct tape is the other), WD-40 was invented by Norm Larsen on his 40th attempt to develop a formula to prevent corrosion by displacing water. In addition to displacing water, WD-40 is a multipurpose problem solver that variously cleans, removes, dissolves, degreases, penetrates, and lightly lubricates various other items. The formula for WD-40 remains a closely guarded secret. As the saying goes, "If it is stuck and shouldn't be, WD-40 it. It if is unstuck and shouldn't be, duct tape it." The WD-40 Company did not pay me for this definition, but they should. See also *duct tape*.

WDM (Wavelength Division Multiplexing) A multiplexing technique by which multiple wavelengths of light, or lambdas, share a single optical fiber. Wavelength division multiplexing is essentially frequency division multiplexing (FDM) at the optical level. Much as multiple electrical frequencies can coexist on an electrified copper circuit in support of multiple, simultaneous conversations in a FDM transmission system, multiple wavelengths can coexist on a single fiber of the appropriate type in a WDM system. A number of carriers now routinely deploy dense wavelength division multiplexing (DWDM) on fiber optic systems, introducing eight or more lambdas into an optical fiber through the use of tunable, cooled lasers firing through windows, or wavelength ranges. The ITU-T has defined 160 wavelengths at spacings of 100 GHz (at 1550 nm) and manufacturers currently offer DWDM systems that multiplex as many as 80 lambdas. Coarse wavelength division multiplexing (CWDM) is defined by the ITU-T as 18 wavelengths in the 1270–1610 nm range, with spacing of 2500 GHz (at 1550 nm). See also *CWDM*, *DWDM*, *FDM*, *lambda*, *laser*, *optical fiber*, *wavelength*, and *window*.

Web (World Wide Web) See *WWW*.

Web 2.0 A term coined by O'Reilly Media (2003) and referring to a second generation of the World Wide Web (WWW) as an enabling platform for Web-based communities of interest, collaboration, and hosted services. Web 2.0 supports *mashing*, which is a process of building new services from reusable components of other services. Applications include social bookmarking, calendaring, and VoIP. Really Simple Syndication (RSS) is one of the most popular tools associated with Web 2.0. See also *RSS*, *VoIP*, and *WWW*.

Web browser See *browser*.

Web log (blog) See *blog*.

Web clipping Referring to the technique of stripping the graphic content out of Web pages for use in microbrowsers built into terminals used in cellular networks running the Wireless Access Protocol (WAP). Web clipping is simpler, but much less aesthetically pleasing than accessing Web sites written in WML (Wireless Markup Language), which is similar to HTML. The alternative is transcoding from HTML to WML, which is accomplished through gateways. The latter approach is commonly employed in contemporary networks, as the cost of such gateway technology has dropped considerably in the recent past, while the performance of the gateways has increased. See also *cellular radio*, *gateway*, *HTML*, *microbrowser*, *WAP*, and *WML*.

Web page See *home page*.

WECA (Wireless Ethernet Compatibility Alliance) See *Wi-Fi Alliance*.

well-known port A port reserved for the use of system (root) processes or of programs executed by privileged users for specific applications or services. Well-known ports are numbered 0 through 1023, and include 25 for SMTP (Simple Mail Transfer Protocol), 80 for HTTP (HyperText Transport Protocol), and 107 for Remote TELNET Service. See also *dynamic port*, *port*, *private port*, and *registered port*.

WEP (Wired Equivalent Privacy) In IEEE 802.11b wireless LAN (WLAN) specifications, an optional security mechanism that uses a stream cipher with a 40-, 64- or 128-bit (WEP2) encryption key to protect data in transit. Since compromised by hackers in 2001, WEP largely has been replaced by Wi-Fi Protected Access (WPA). See also *802.11b*, *cipher*, *encryption*, *hacker*, *key*, *WLAN*, and *WPA*.

Western Union The Mississippi Valley Printing Telegraph Company, formed in 1851, became the Western Union Telegraph Company in 1856 through a series of acquisitions. Western Union completed the first transcontinental telegraph line in 1861, which put the Pony Express out of business. Western Union introduced the stock ticker in 1866, a standardized time service in 1870, and a money transfer business in 1871. In the fall of 1876, the American Bell Telephone Company offered to sell the Bell telephone patents to Western Union for $100,000. Western Union President Carl Orton reportedly replied, "What use would this company make of an electric toy?" In 1877, however, Western Union became convinced of the telephone's importance, and began to compete fiercely against AT&T until it lost a patent lawsuit and exited the business in 1879. Western Union later was a leader in intercity fax service, microwave and satellite communications, and Telex service. On January 27, 2006, the company discontinued telegram service, and is now a financial services company. The last 10 telegrams included birthday wishes, condolences on the death of a loved one, notification of an emergency, and several people trying to be the last to send a telegram. See also *fax*, *microwave*, *Pony Express*, *satellite*, *telegram*, *telegraph*, and *Telex*.

Wide Area Telecommunications Service (WATS) See *WATS*.

whiteboarding A graphical conferencing technology that allows multiple users to collaborate over a network on a graphic basis as though they were drawing on a physical whiteboard. Participants can create and modify a document by clicking and dragging with a mouse, with each participant assigned a different color marker. Participants also have the use of a pointer and highlighter. See also *collaborative computing*.

white light Visible light that is a combination of all frequencies or wavelengths in the visible light spectrum. See also *white noise*.

white noise The background noise that is continuously present on electrical circuits or radio circuits due to the thermal agitation of electrons. White noise has a flat power spectral density, which is to say that it has equal power at any frequency in any given frequency band. The term white noise comes from the fact that it is analogous to white light, which is a combination of all frequencies or wavelengths in the visible light spectrum. At acceptable levels of signal-to-noise ratio (SNR), white noise takes the form of a background hiss or mild level of static noise. White noise is even desirable and often added to digital circuits, which can be so quiet during periods of voice inactivity as to fool a listener into thinking that the connection has been dropped. People find the addition of this comfort noise to be reassuring at mild levels. At unacceptable levels of SNR, white noise can overwhelm an audio signal or cause bit errors in a data transmission. White noise is often referred to as Gaussian noise, although the two are not necessarily the same, as Gaussian noise refers to the distribution of the signal values. See also *Gaussian noise*, *noise*, and *SNR*.

Whois **1.** A utility on UNIX systems that provides information about other users who are logged on to the same system. See also *UNIX* and *utility*. **2.** An Internet database service that provides identity and contact information about owners of domain names. See also *database*, *Domain Name System*, and *Internet*.

WiBro (Wireless Broadband) A wireless local loop (WLL) specification developed by the government of South Korea. WiBro allocates 100 MHz of spectrum set in the 2.3 GHz band, offers aggregate throughput of 30–50 Mbps, and has a reach of 1–5 kilometers. See also *BRAN*, *WiMAX*, and *WLL*.

Wicked Witch of the West (WWW) A wicked witch played by Margaret Hamilton in the 1939 film "The Wizard of Oz." Ms. Hamilton often used *WWW* in autographing photos of herself in costume. The most oft-quoted line of the Wicked Witch of the West, whose real name was never revealed, is "I'll get you, my pretty — and your little dog, too!" Dorothy, however, was protected by Glinda, the Good Witch of the North. Alas, Good does not always triumph over Wickedness in the real world. See also *WWW*.

Wide Area Network (WAN) See *WAN.*

wideband **1.** A circuit or channel with capacity greater than narrowband. See also *narrowband.* **2.** A circuit or channel with bandwidth wider than normal for operation. See also *bandwidth.* **3.** A radio channel covering a relatively wide range of frequencies. Ultra-Wideband (UWB) for example, is a radio system with occupied bandwidth (i.e., the difference between the highest and lowest frequencies in the radio channel) greater than 25 percent of the center frequency. See also *bandwidth.* **4.** Wideband sometimes is used interchangeably with broadband. See also *broadband.* (I promise that I don't make these things up. I just explain them as they are.)

Wideband CDMA (W-CDMA) Also known as Universal Mobile Telecommunications System (UMTS). See *UMTS.*

Wi-Fi (Wireless Fidelity) The name given IEEE 802.11 by the Wireless Ethernet Compatibility Alliance (WECA, now the Wi-Fi Alliance), Wi refers to the wireless nature of the LAN and Fi to the fidelity (i.e., faithfulness, or integrity) of the signal. The term also has been attributed to the IEEE 802.11 Working Group, with Wi referring to the fact that a wire traditionally served as the physical medium for LANs, and the homonym Fi referring to PHY, the PHYsical Layer of the OSI Reference Model. So, Wireless PHY became Wi-Fi. One way or another, or perhaps both ways, Wi-Fi became the vernacular for 802.11 and especially 802.11b. See also *802.11*, *802.11a*, *802.11b*, *802.11g*, *802.11n*, *IEEE*, *OSI Reference Model*, *Physical Layer*, and *VoWiFi.*

Wi-Fi5 (Wireless Fidelity 5 GHz) The vernacular for IEEE 802.11a. See *802.11a* and *Wi-Fi.*

Wi-Fi Alliance Previously the Wireless Ethernet Compatibility Alliance (WECA). A special interest group comprising manufacturers and vendors that work to develop specifications for and promote the worldwide adoption of products based on the IEEE 802.11 standards for wireless local area networks (WLANs). See Appendix A for contact information. See also *802.11*, *IEEE*, and *WLAN.*

Wi-Fi Multimedia Extension (WMM or WME) The term the Wi-Fi Alliance uses for the priority classes specified in IEEE 802.11e for quality of service (QoS) over 802.11 wireless LANs (WLANs). See also *802.11*, *802.11e*, *IEEE*, *QoS*, *Wi-Fi Alliance*, and *WLAN.*

Wi-Fi Protected Access (WPA) See *WPA.*

Wi-Fi TV A term sometimes used for the wireless regional area network (WRAN) project of the IEEE 802.22 Working Group. Wi-Fi TV is intended to operate in the UHF and VHF broadcast TV bands. See also *802.22* and *WRAN.*

wiki **1.** Quick or fast, in the Hawaiian language. **2.** A type of authoring software that enables users to easily and quickly create and edit Web server content using any browser.

WiMAX (Worldwide Interoperability for Microwave Access) A broadband wireless access (BWA) solution based on the standards recommendations from the IEEE 802.16 Working Group and the European Telecommunications Standards Institute (ETSI) HiperMAN group. WiMAX is promoted by the WiMAX Forum, a special interest group with members from the manufacturing, carrier, service provider, and consulting communities. Where line of sight (LOS) can be achieved, the WiMAX cell radius is as much as 50 kilometers (31 miles). Under non-line of sight (NLOS) conditions, the maximum cell radius is approximately 9 kilometers (5.6 miles). WiMAX standards provide for aggregate raw bandwidth up to about 70 Mbps per base station (BS), although the throughput is much less due to overhead, as well as issues of LOS, link distance, air quality, electromagnetic interference (EMI), radio frequency interference (RFI), and other signal impairments. Mobile network deployments described in 802.16e are expected to provide up to 15 Mbps of aggregate raw bandwidth within a cell radius of up to 3 kilometers. WiMAX supports a maximum signaling rate of 70 Mbps and the maximum throughput of approximately 40 Mbps over the shortest distance between the BS and the user antenna under LOS conditions. Over the maximum distance of 50 kilometers under LOS conditions, or the maximum distance of 9 kilometers under NLOS conditions, throughput drops considerably. The transmission rate is symmetrical, i.e., the same for the uplink (upstream), i.e., the link from the remote terminal back to the BS, as for the downlink (downstream). The sole exception to this symmetry is in the case of full-featured CPE at the cell edge, where uplink transmission rates are constrained by power limitations.

WiMAX employs orthogonal frequency division multiplexing (OFDM), which subdivides the spectrum into a number of independent, narrowband subcarriers, across which it sends the signal in parallel fashion. Through sub-channelization on the uplink, WiMAX concentrates signal power into fewer OFDM subcarriers, thereby extending the reach of the system, mitigating the effects of physical obstructions in an NLOS environment and reducing CPE power consumption. Multiple-input multiple-output (MIMO) antennas employ space/time coding to compensate for multipath fading over long loops. At the customer premises is an adaptive, passive array antenna known as a pizza box, as it is about the size and shape of a pizza box. Rate-adaptive modulation dynamically adjusts the signal modulation technique of each carrier to compensate for variations in signal quality at that carrier frequency. Reed-Solomon forward error correction (FEC) is employed to deal with issues of signal quality and automatic repeat request (ARQ) is employed to request retransmission of any remaining errored frames. 802.16 specifications include several multiplexing options. Frequency division duplex (FDD) supports both half-duplex (HDX) and full duplex (FDX) communications, and time division duplex (TDD) supports half-duplex (HDX), only. The 802.16 security protocol is built on enhancements to the privacy key management (PKM) developed for cable modem communications. The protocol uses X.509 digital certificates with Rivest-Shamir-Adleman (RSA) encryption for authentication and key exchange. Traffic encryption options are data encryption standard (DES) and advanced encryption standard (AES). 802.16 specifications include convergence sublayers designed for mapping services to and from 802.16 connections. The ATM convergence sublayer is for ATM services and the packet convergence sublayer is for packet services such as IPv4, IPv6, Ethernet, and Virtual LAN (VLAN).

WiMAX offers differential quality of service (QoS) based on four polling schedules:

- Unsolicited grant service (UGS) is designed for services such as T1 and E-1.
- Real-time polling service is designed for services such as voice over Internet Protocol (VoIP) and IP-based streaming audio and video.
- Non-real-time polling service is designed for services such as Internet access.
- Best effort service provides neither throughput nor latency guarantees.

See also *802.16, AES, ARQ, ATM, authentication, base station, broadband, BWA, cable modem, carrier, DES, digital certificate, E-1, EMI, encryption, Ethernet, ETSI, FDD, FDX, FEC, HDX, HiperMAN, IPv4, IPv6, latency, LOS, MIMO, modulation, narrowband, NLOS, OFDM, overhead, pizza box, PKM, protocol, rate adaption, Reed-Solomon, RFI, RSA, signal, subcarrier, T1, TDD, throughput, VLAN, VoIP, WiMAX Forum,* and *X.509.*

WiMAX Forum A not-for-profit organization formed to promote and certify the compatibility and interoperability of broadband wireless access (BWA) products based on the IEEE 802.16 and ETSI HiperMAN specifications. For contact information, see Appendix A. See also *802.16, BWA, ETSI, HiperMAN, IEEE,* and *WiMAX.*

window **1.** An opening or opportunity for passage of data frames or packets without the requirement for an acknowledgement from the receiving device. See modulo and TCP. **2.** An opening or opportunity for passage of a range of wavelengths in a fiber optic transmission system (FOTS). For example, a laser diode might fire at 1550 nm, referring to a range of wavelengths with a nominal center point of 1550 nm. A light-emitting diode (LED) might fire at 850 nm, and a vertical cavity surface-emitting laser (VCSEL) at 1300 nm or 1310 nm. The ITU-T has established a number of standard windows, as detailed in Table

W-1. Generally speaking, the higher the transmission window (i.e., the longer the wavelength and lower the frequency), the less the signal attenuation, but the more expensive the associated electronics. See also *attenuation*, *FOTS*, *frequency*, *laser diode*, *LED*, *VCSEL*, and *wavelength*.

Table W-1: ITU-T Transmission Windows

Band Designation	*Wavelength Window*
850 Band	810–890 nm
O-Band (Original Band)	1,260 nm–1,360 nm
E-Band (Extended Band)	1,360 nm–1,460 nm
S-Band (Short Wavelength Band)	1,460 nm–1,530 nm
C-Band (Conventional Band)	1,530 nm–1,565 nm
L-Band (Long Wavelength Band)	1,565 nm–1,625 nm
U-Band (Ultralong Wavelength Band)	1,625 nm–1,675 nm

wiper Also known as a *selector*, a wiper is a component of a step-by-step (SxS) electromechanical circuit switch. See *selector*.

wire A current-carrying metal conductor, generally encased in a dielectric insulating material. A solid core conductor comprises a single wire. A stranded conductor comprises a number (usually 7 or 17, because they pack neatly) of small wires. Telecommunications wires generally are made of copper to conduct electrical current, although tinned copper, copper-clad aluminum, and other metals and metal combinations also can be used. Stranded, rather than solid core, conductors are used in applications requiring high flex strength. The wires generally are separately insulated with polyethylene, polyvinyl chloride (PVC), flouropolymer resin, Teflon, or some other low-smoke, fire-retardant, dielectric material. Two wires then typically are twisted in a helix with a constant pitch or distance to make a 360-degree twist to form a twisted pair. One or more pairs then are formed into a cable, which is covered in a protective sheath of dielectric material. See also *cable*, *conductor*, *current*, *dielectric*, *flex strength*, *insulation*, and *twisted pair*.

wire center A central point where physical circuits are interconnected, a wire center generally is housed in a Central Office (CO) owned by an Incumbent Local Exchange Carrier (ILEC). See also *CO* and *ILEC*.

wired equivalent privacy (WEP) See *WEP*.

wireless Referring to a link, circuit, or network that employs either radio frequency (RF) or infrared (IR) transmission medium, rather than a wired technology such as coaxial cable, twisted pair, or optical fiber. See also *coaxial cable*, *IR*, *RF*, *optical fiber*, *transmission medium*, and *twisted pair*.

Wireless Access Communication System (WACS) See *WACS*.

Wireless Access Protocol (WAP) See *WAP*.

Wireless Broadband (WiBro) See *WiBro*.

Wireless Ethernet Compatibility Alliance (WECA) See *Wi-Fi Alliance*.

wireless fiber A term sometimes applied to free space optics (FSO) systems, which are optical airwave systems operating in the infrared (IR) spectrum and offering bandwidth up to the Gbps range. See also *bandwidth*, *infrared*, *FSO*, and *spectrum*.

wireless local area network (WLAN) See *WLAN*.

wireless local loop (WLL) See *WLL*.

wireless local number portability (WLNP) Synonymous with wireless number portability (WNP). See *WNP.*

Wireless Markup Language (WML) See *WML.*

wireless media See *transmission medium.*

Wireless Medical Telemetry Service (WMTS) See *WMTS.*

wireless number portability (WNP) See *WNP.*

wireless office telecommunications system (WOTS) See *WOTS.*

wireless personal area network (WPAN) See *WPAN.*

wireless regional area network (WRAN) See *WRAN.*

Wireless Transport Layer Security (WTLS) See *WTLS.*

wireline Referring to a service that connects to the public switched telephone network (PSTN) through a local loop of copper wire or glass fiber that terminates in a fixed location at a customer premises. A wireline service is in contrast to a wireless local loop (WLL) and a wireless service such as cellular. See also *cellular radio, fiber, PSTN, wire, wireless,* and *WLL.*

wireline carrier Also known as a B Carrier. A provider of traditional landline telecommunications services. Such services involve connections to the public switched telephone network (PSTN) by wire (or fiber) local loops that terminate in fixed locations at customer premises. The distinction between wireline and non-wireline carriers was made primarily for purposes of segregating bidders for radio spectrum assignment during the FCC cellular radio spectrum auctions. The initial approach toward spectrum assignment was designed to ensure that there was one wireline (i.e., telephone company) and one non-wireline carrier per market. See also *carrier, cellular radio, FCC, landline, local loop, premises, PSTN, radio,* and *spectrum.*

wiretap In historical terms, a temporary physical connection secretly placed on a metallic circuit in order to monitor the information being transmitted across it. In contemporary terms, a wiretap need not be a physical connection on a circuit, but can take many forms, including the interception of a radio signal. In the United States and many other countries, wiretaps are illegal unless authorized by court order, or perhaps the order of a federal agency or of the executive branch of government in times of war or in the interests of national security. See also *eavesdrop* and *Echelon.*

WIS (WAN Interface Sublayer) A protocol sublayer that enables compatibility between 10 Gigabit Ethernet (10GbE) equipment and SONET long-haul equipment in a LAN-to-WAN interface scenario. See also *10GbE, LAN, long haul circuit, SONET,* and *WAN.*

WLAN (Wireless Local Area Network) A LAN that employs radio frequency (RF) or perhaps infrared (IR) transmission rather than a wired technology such as coaxial cable, twisted pair, or optical fiber. A typical WLAN comprises fixed-location transceivers known as *access points* (APs) to which client workstations and peripherals connect via RF technology. The access points typically are hard wired to switches and routers that interconnect them and provide access to servers. The APs are fitted with radio transceivers and omnidirectional antennas. The client transceivers, or network adapters, may be in the form of PCMCIA cards, although major computer manufacturers now offer laptops with built-in transceivers. The APs are located at central points where there is good line of sight (LOS) to the workstations and link quality, therefore, is best. Most WLANs are standards-based versions from the IEEE 802.11 Working Group. At the Physical Layer, the RF specifications include both direct sequence spread spectrum (DSSS) and frequency-hopping spread spectrum (FHSS). At the Data Link Layer, the medium access control (MAC) protocol is carrier sense multiple access with collision avoidance (CSMA/CA). Most WLANs operate in the 2.4 GHz and 5 GHz unlicensed ISM (Industrial, Scientific, and Medical) bands, which approach avoids the expensive and lengthy licensing process, but carries with it the potential for interference from other systems in proximity. As power levels are low, distances generally are limited to 500–800 feet or so. See also *802.11*, *AP*, *coaxial cable*, *CSMA/CA*, *Data Link Layer*, *DSSS*, *FHSS*, *hardwire*, *IR*, *ISM*, *LOS*, *MAC*, *RF*, *router*, *switch*, *optical fiber*, *Physical Layer*, *switch*, *transceiver*, *transmission medium*, and *twisted pair*.

WLL (Wireless Local Loop) Also known as *fixed wireless*. A group of airwave transmission technologies designed to support communications from the edge of a public network to the customer premises. These fixed wireless technologies include both radio frequency (RF) and infrared (IR) options. RF solutions include Local Multipoint Distribution Service (LMDS), Multichannel Multipoint Distribution Services (MMDS), and Worldwide Interoperability for Microwave Access (WiMAX). Free space optics (FSO) systems are wireless optical transmissions using wavelengths in the infrared range. The technologies supporting broadband performance to an individual user are sometimes known as *broadband wireless access* (BWA) and include all of the above. See also *airwave transmission*, *BWA*, *FSO*, *IR*, *LMDS*, *local loop*, *MMDS*, *RF*, and *WiMAX*.

WLNP (Wireless Local Number Portability) Synonymous with wireless number portability (WNP). See *WNP*.

WMBTOTCITBWTNTALI (We May Be The Only Telephone Company In Town, But We Try Not To Act Like It) Southwestern Bell Telephone Company used this initialism in an advertising campaign during the 1970s. The campaign backfired, of course, because a substantial number of people felt very strongly that the company did try to act like it. Sometimes they were right. (I worked for the company in management at the time, so I know.) The ad campaign was short-lived, but the company was not. Southwestern Bell later changed its name to SBC, and merged with and acquired a number of companies outside the southwest region of the United States. One of those acquisitions was the remnants of the once great AT&T, which had been Southwestern Bell's parent company until the breakup of the Bell System in 1984. SBC assumed the name in 2006, and is now known as AT&T.

WME (Wi-Fi Multimedia Extension) See *Wi-Fi Multimedia Extension*.

WML (Wireless Markup Language) A tag-based notation language used to format documents for microbrowsers used in Web-enabled cellular networks employing the Wireless Access Protocol (WAP). WML is similar to but incompatible with Hypertext Markup Language (HTML). See also *browser*, *cellular radio*, *HTML*, *WAP*, and *Web*.

WMM (Wi-Fi MultiMedia extension) See *Wi-Fi Multimedia Extension*.

WNP (Wireless Number Portability) Also known as *full mobile number portability* (FMNP) and *wireless local number portability* (WLNP). Referring to the ability to port, i.e., move, a cellular telephone number from one carrier to another. Local number portability (LNP) is the equivalent term in the fixed-line, or wireline, domain. See also *carrier* and *LNP*.

WMTS (Wireless Medical Telemetry Service) In the United States, a service that supports the remote monitoring of a patient's health for medical purposes. All types of communications other than voice and video are permitted as long as they relate to the provision of medical care. WMTS operates in the 608–614 MHz, 1395–1400 MHz, and 1427–1432 MHz bands. The Federal Communications Commission (FCC) regulates WMTS, which is in the family of personal radio services. See also *FCC* and *personal radio services*.

workgroup switch In a local area network (LAN), a workgroup switch is a relatively low capacity switch that serves the needs of a workgroup, or small group of workers who generally are geographically clustered. A workgroup switch is the LAN equivalent of an edge switch in a public wide area network (WAN). See also *backbone switch*, *LAN*, *switch*, and *WAN*.

World Administrative Radio Conferences (WARC) Now known as the *World Radio Conferences* (WRC). See *WRC*.

World Radio Conferences (WRC) See *WRC*.

World Trade Organization (WTO) See *WTO*.

Worldwide Interoperability for Microwave Access (WiMAX) See *WiMAX*.

World Wide Web (WWW) Also known as the *Web*. See *WWW*.

World Wide Web Consortium (W3C) See *W3C*.

worm A type of malware that replicates itself across a computer network by making copies of itself, which it sends to other computers. A worm embeds itself in memory and may replicate itself so many times that it causes the host to crash. Note that a worm is neither a Trojan horse nor a virus. See also *malware*, *spyware*, *Trojan horse*, and *virus*.

WOTS (Wireless Office Telecommunications System) A wireless telephone system generally in the form of one or more adjuncts that provide cordless telephony communications capabilities behind PBXs, electronic key telephone systems (EKTS), or Centrex systems. WOTS generally are limited to voice applications, although some also support low-speed data. WOTS systems involve a wireless master controller, which is hardwired to special ports on the voice communications system. See also *adjunct*, *Centrex*, *cordless telephone*, *EKTS*, *hardwire*, *PBX*, and *port*.

WPA (Wi-Fi Protected Access) A security mechanism based on IEEE 802.11i, WPA was designed by the Wi-Fi Alliance to replace the flawed Wired Equivalent Privacy (WEP) for 802.11b, aka Wi-Fi, wireless LANs (WLANs). WPA2 includes an encryption algorithm based on Advanced Encryption Standard (AES), which employs a 128-bit block cipher that is considered to be completely secure. See also *802.11b*, *802.11i*, *AES*, *block cipher*, *encryption*, *WEP*, and *WLAN*.

WPAN (Wireless Personal Area Network) A wireless network defined by personal operating space (POS), which simply is the area in the near vicinity of a device or individual. 802.15 is the IEEE specification for WPANs and 802.15.1 is the specification for Bluetooth. See also *802.15* and *Bluetooth*.

WRAN (Wireless Regional Area Network) A technology under development by the IEEE 802.22 Working Group directed toward the development of a cognitive radio air interface for use by license-exempt radios on a non-interfering basis in spectrum currently allocated to television broadcast service. See also *air interface*, *cognitive radio*, *IEEE*, and *spectrum*.

wrap A restoral mechanism employed in Resilient Packet Ring (RPR). In the event of a node or link failure, wrap calls for data to travel around the fiber optic ring until it reaches the node nearest the break. That node turns the traffic around and sends it in the reverse direction over the counter-rotating ring. See also *dual counter-rotating ring*, *fiber optics*, *link*, *node*, *RPR*, and *steer*.

WRC (World Radio Conferences) Previously known as the *World Administrative Radio Conferences* (WARC). A group of conferences at which the various national regulatory authorities meet to sort out national and international spectrum allocation issues. WRC are sponsored by the ITU-R every two years. See also *ITU-R*.

w^3 (WWW, or World Wide Web) See *WWW*.

W3C (World Wide Web Consortium) A cooperative venture of the l'Conseil Européen pour la Recherche Nucléaire (CERN), The Massachusetts Institute of Technology (MIT), and l'Institut National de Recherche en Informatique et en Automatique (INRIA). Formed in 1994, W3C's primary focus is that of leading the technical evolution of the Web by promoting interoperability and providing an open forum for discussion. See also *CERN*, *INRIA*, and *WWW*.

WTLS (Wireless Transport Layer Security) Pronounced witless. The security layer that provides authentication services for the Wireless Access Protocol (WAP) used in many Web-enabled cellular networks. See also *authentication*, *cellular radio*, *WAP*, and *WWW*.

WTO (World Trade Organization) An international organization that deals with the global rules of trade between nations. The WTO was formed in 1995 as the successor to the Global Agreement on Tariffs and Trade (GATT), which was established in 1955 in the wake of World War II (WWII). In the context of telecommunications, the WTO's involvement largely has been in the realm of international long distance rates.

WWW **1.** World Wide Web. Also known as w^3 and the Web. A global interlinked hypertext system that uses the Internet infrastructure to network client workstations and servers all around the world based on the Hypertext Transport Protocol (HTTP). Documents on the WWW — known as *pages*, *home pages*, or *Web pages* — are written in Hypertext Markup Language (HTML), are identified by Uniform Resource Locators (URLs), and are transmitted over the Internet through the use of HTTP. The WWW also incorporates hypermedia, which is hyperlinked multimedia, including not only text, but also audio, graphics, animations, and video. Tim Berners-Lee developed the WWW at l'Conseil Européen pour la Recherche Nucléaire (CERN), which translates from French as The European Council for Nuclear Research and is generally known as the *European Laboratory for Particle Physics*, in Geneva, Switzerland. The home of the WWW is now the World Wide Web Consortium (W3C), a cooperative venture of CERN, The Massachusetts Institute of Technology (MIT), and l'Institut National de Recherche en Informatique et en Automatique (INRIA), which translates from French as the National Institute for Research in Computer Science and Control. See also *CERN*, *client*, *HTML*, *HTTP*, *hypermedia*, *hypertext*, *INRIA*, *multimedia*, *server*, *URL*, and *W3C*. **2.** Wicked Witch of the West. Margaret Hamilton, who played the Wicked Witch of the West in the 1939 film "The Wizard of Oz," often used WWW in autographing photos of herself in costume. That was long before the World Wide Web was invented. The WWW's most oft-quoted line is, "I'll get you, my pretty — and your little dog, too!" Dorothy, however, was protected by Glinda, the Good Witch of the North. Alas, Good does not always triumph over Wickedness in the real world.

WYSIWYG (What You See Is What You Get) Pronounced *wizzywig*. Referring to a program that allows the user to see the document on screen just as it will appear in final form. Working with a word processing program in print view is WYSIWYG. Programming in a markup language such as HTML definitely is not.

x **1.** In mathematics, the symbol for an unknown quantity, person, factor, or thing. X is the first in a set of two or three unknowns comprising x, y, and z. In telecommunications, for example, xDSL refers to any of a group of Digital Subscriber Line services. See also *x-axis*, *xDSL*, *y*, and *z*. **2.** In mathematics, *times*. For example, 3 × 4 = 12 means 3 times 4 equals 12. **3.** In measurements, by. For example, a 2 × 4 (2 by 4) is a piece of lumber measuring 2 inches wide by 4 inches deep, or at least it used to be. A contemporary 2 × 4 actually measures 1½″ by 3½″ (38 by 89 mm), with the difference being due to planing (i.e., shaping) and shrinking as the lumber is dried. Once upon a time, lumber was cut so that a 2 × 4 was actually 2″ by 4″ after planing and shrinking. Once upon a time a lot of things were different. In olden times, a can of coffee (yes, people actually used to buy ground coffee in cans) actually weighed one pound, comprising 16 ounces. Contemporary cans of coffee variously weigh 12 oz., 12.5 oz, or perhaps 14 oz., but rarely 16 oz. The reason, of course, is marketing, which translates into perceptual selling, which at the very least borders on deceptive selling. Many marketers would rather reduce the size of the product while holding the price firm rather than holding the size of the product firm while increasing the price. Ultimately, it is a matter of ethics. See also *ethics*.

X **1.** The symbol for reactance. See *reactance*. **2.** X. The X series of ITU-T Recommendations. The X series addresses data networks, open systems communication, and security. See *X Series*.

X12 A standard for Electronic Data Interchange (EDI) from the American National Standards Institute (ANSI) Accredited Standards Committee (ASC). X12 is popular in North America. The competing UN/EDIFACT international standard is predominant outside of North America. EDI standards specify data formats, character sets, and data elements. See also *ANSI*, *EDI*, *standard*, and *UN/EDIFACT*.

X.21bis The Physical Layer specification within the X.25 protocol suite. X.21bis defines call control procedures, synchronization mechanisms, and the mechanical and electrical parameters for cables and connectors, which are similar to RS-232. X.21bis supports full-duplex (FDX) transmission at speeds from 9600 bps to 64 kbps. See also *FDX*, *Physical Layer*, *synchronization*, and *X.25*.

X.25 The ITU-T Recommendation for the interface into a packet-switched network, X.25 was published in 1976, and subsequently revised several times, most recently in 1993. X.25 actually is a protocol suite comprising three layers that map into the bottom three layers of the OSI Reference Model:

- **packet layer protocol** (PLP) maps into Layer 3, the Network Layer
- **link access procedure, balanced** (LAPB) maps into Layer 2, the Data Link Layer
- **X.21bis** maps into Layer 1, the Physical Layer

Figure X-1 illustrates the relationship between the PLP, LAPB, and X.21bis.

X.25 originally was intended for interactive time-sharing, which involves long connect times and low data volumes over error-prone circuits. Although X.25 still supports such applications effectively, contemporary applications include online interactive processing (e.g., reservations systems and credit card processing), electronic messaging (e.g., e-mail), batch file transfer (e.g., data backup), and Internet access. X.25 largely has been replaced by frame relay and IP networks, although X.25 remains heavily used in developing countries and in areas where error prone analog circuits must be used in data communications applications. See also *frame relay*, *IP*, *ITU-T*, *LAPB*, *packet switch*, *PLP*, *protocol suite*, *X.21bis*, and *X Series*.

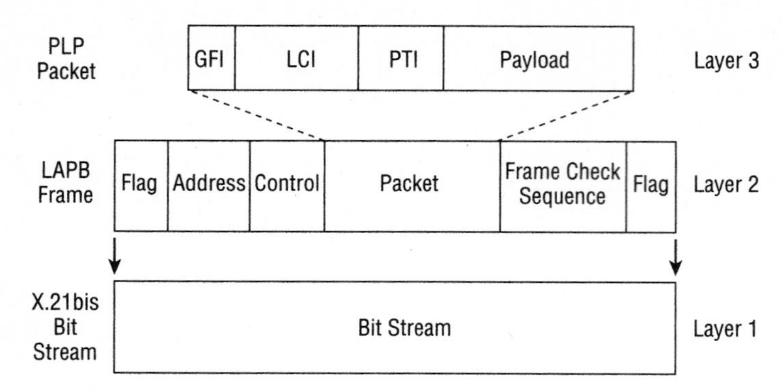

Figure X-1

X.28 The ITU-T Recommendation for the interface between data terminal equipment (DTE) and a packet assembler/disassembler (PAD) in an X.25 public data network (PDN). See also *DTE, PAD, PDN,* and *X.25.*

X.29 The ITU-T Recommendation for communications formats between data terminal equipment (DTE) and an X.3 packet assembler/disassembler (PAD) in an X.25 public data network (PDN). See also *DTE, PAD, PDN, X.3,* and *X.25.*

X.3 The ITU-T Recommendation for a packet assembler/disassembler (PAD) in a public data network (PDN). The PAD organizes the user data into packets according to the X.25 packet layer protocol (PLP) and encapsulates each in an LAPB frame before presenting it to the network. See also *packet, PAD, PDN, LAPB,* and *X.25*

X.31 The ITU-T Recommendation for X.25 packet-mode services over ISDN. See also *ISDN* and *X.25.*

X.32 The ITU-T Recommendation for access to an X.25 packet network in synchronous dial-up mode through a public switched telephone network (PSTN), an integrated services digital network (ISDN), or a circuit-switched public data network (PDN). See also *dial-up access, PDN, PSTN, synchronous,* and *X.25.*

X3T9-3 The ANSI (American National Standards Institute) specification for High Performance Parallel Interface (HIPPI). See also *ANSI* and *HIPPI.*

X3T9-5 The ANSI (American National Standards Institute) specification for Fiber Distributed Data Interface (FDDI). See also *ANSI* and *FDDI.*

X.75 The ITU-T Recommendation for the interconnection of X.25 packet-switched networks. Through an X.75 gateway that serves as a network-to-network interface, issues of packet size are resolved. The X.75 gateway examines packets for errors, and either segments the payloads of large packets into multiple smaller packets or combines the payloads of multiple smaller packets into larger ones. The gateway then encapsulates each with the necessary control data in the form of a header and trailer, modifies the addressing scheme as required, and presents each to the target network.

X.200 The ITU-T Recommendation that mirrors the Open Systems Interconnection (OSI) Reference Model. See *ITU-T, OSI Reference Model,* and *X Series.*

X.400 The ITU-T standard Message Handling Service Protocol (MHS or MHSP) for e-mail and document exchange, X.400 is a complex gateway protocol that permits disparate e-mail systems to interoperate at a minimal level, over either packet networks or asynchronous dial-up circuits. Created in 1984 and updated in 1988 and again in 1992, X.400 functions at Layer 7, the Application Layer, of the OSI

Reference Model. X.400 gained considerable popularity in Europe, largely for in-house implementations. In the United States, however, the preference developed for the Simple Mail Transfer Protocol (SMTP) of the TCP/IP protocol suite, upon which the Internet is based. Although most carriers do not use X.400 in native implementations, they commonly use it as a gateway protocol for X.400-to-SMTP gateways, particularly for international networks. See also *Application Layer, gateway, ITU-T, Layer 7, OSI Reference Model, SMTP, TCP/IP protocol suite*, and *X Series*.

X.500 An international standard (1988) developed jointly by the International Organization for Standardization (ISO) and the ITU-T to support the e-mail name and address lookup requirements of X.400. X.500 provides for global directory services that theoretically enable network managers to store information about all users, machines, and applications in a distributed fashion. X.500 is a very robust global directory standard that requires significant computational resources to implement and, therefore, is criticized for being over-engineered. See also *e-mail, ISO, ITU-T, LDAP, X.400*, and *X Series*.

X.509 An ITU-T Recommendation (1988) for a public key infrastructure (PKI). X.509 establishes a hierarchical structure of certificate authorities (CAs) that issue digital certificates, which are electronic credentials that authenticate the identity of users on the Internet and intranets. See also *authentication, CA, digital certificate, Internet, intranet*, and *PKI*.

XAUI (10 Gigabit Attachment Unit Interface) The physical layer chip interface specified for 10 Gbps Ethernet (10 GbE) in 802.3ae. The AUI defines the manner in which an Ethernet cable physically attaches to a network interface card (NIC).

x-axis The horizontal (H), or nearest horizontal, plane on a two- or three-dimensional grid, chart, or graph in a Cartesian coordinate system. See also *Cartesian coordinates, y-axis*, and *z-axis*.

X band Frequencies in the 10 GHz range of the microwave spectrum. X band is used in some radar and satellite systems. X Band Radar (XBR) is the primary fire control sensor, providing surveillance, acquisition, tracking, discrimination, fire control support and kill assessment for the United States National Missile Defense (NMD) system affectionately known as Star Wars. See also *electromagnetic spectrum, frequency, microwave, radar, satellite*, and *spectrum*.

Xbar (CROSSbar) Electromagnetic, common control circuit switches that replaced the electromechanical step-by-step (SxS) switches. AT&T Bell Telephone Laboratories accomplished most of the conceptual work on the Xbar, but the Ericsson company (Sweden) was responsible for much of the early practical development. The first Xbar switch installed in the United States was a central office (CO) exchange in Brooklyn, New York in 1938. Xbar technology quickly became preferred over that of the electromechanical SxS switch, which dates to 1891. In an Xbar switch, a marker recognizes a request for dial tone, directs a sender to store the dialed digits, and directs a translator to route the call, reserving a path through a switching matrix. Once the call connects, these various components become available to serve other calls. Compared to the SxS switch, the Xbar has relatively few moving parts. Xbar switches offer the advantages of increased intelligence, common control, faster connection speed, smaller physical footprint, lower maintenance costs, and greater traffic capacity. Xbar switches were considered state of the art for nearly 30 years, until the appearance of the electronic common control (ECC) switch in 1965. See also *ECC, switch matrix*, and *SxS*.

xDSL (generic Digital Subscriber Line) Referring to any of a group of broadband digital access technologies operating over embedded unshielded twisted pair (UTP) telco local loops. xDSL technologies employ sophisticated compression algorithms and multiplexing techniques to derive performance that often exceeds 1 Mbps over a voice grade local loop. Most DSL technologies support simultaneous voice and high speed Internet access, and a number support video, as well. Most of the technologies involve centralized splitters, also called *modems* or *filters*, on the customer premises side of the loop. All DSL technologies support always-on data access, as the circuit is always available from the PC through the on-premises splitter to the centralized splitter and DSL access multiplexer (DSLAM) in the central office, or other centralized location, and to the Internet. Therefore, there are no dial-up delays such as those experienced when using a modem to establish a circuit-switched connection to the Internet over the public switched telephone network (PSTN). Table X-1 provides a comparative listing of a number of DSL technologies.

Table X-1: DSL Technology Comparisons

DSL Type	*ITU-T Standard*	*Max Rate Downstream*	*Max Rate Upstream*	*Max Reach*	*Applications*
ADSL	G.992.1 (1999)	7 Mbps	800 kbps	18,000 ft	Consumer-class Internet
G.lite	G.992.2 (1999)	1.544 Mbps	512 Mbps	18,000 ft	Consumer-class Internet
ADSL2	G.992.3, G.992.4 (2002)	12 Mbps	1 Mbps	18,600 ft	Consumer-class Internet
ADSL2+	G.992.5 (2003)	24.5 Mbps	1 Mbps	18,000 ft	Consumer-class, SME Internet
ADSL2-RE	G.992.3 (2003)	8 Mbps	1 Mbps	20,700 ft	Consumer-class Internet
IDSL	N/A	144 kbps	144 kbps	18,000 ft	Consumer-class Internet
SHDSL	G.991.2 (2003)	4.6 Mbps	4.6 Mbps	6,000 ft.	Business-class Internet
VDSL	G.993.1 (2004)	55 Mbps	15 Mbps	1,000 ft	Voice, Data, Video
VDSL2: 12 MHz	G.993.2 (2005)	55 Mbps	30 Mbps	1,000 ft	Voice, Data, Video
VDSL2: 30 MHz	G.993.2 (2005)	100 Mbps	100 Mbps	500 ft	Voice, Data, Video

See also *ADSL, ADSL2, ADSL2+, always on, broadband, compression, downstream, DSLAM, G.lite, HDSL, high speed, IDSL, ITU-T, local loop, multiplexer, SDSL, SHDSL, splitter, telco, upstream, UTP, VDSL,* and *x.*

xerography From the Greek *xeros*, meaning *dry*, and *graphos*, meaning *written*. Also known as *electrophotography*. A process for copying material through the production of photographic images. The latent image is transferred by the action of light on a photoconductive insulated drum to which the image attracts oppositely charged dry ink particles that are then fused in place on paper. The process was invented and patented in 1938 by Chester Carlson (1906–1968) and subsequently was commercialized by the Haloid Corporation, which later became Xerox Corporation. *Note:* It is a violation of federal law to xerograph copyrighted material such as this book. Just buy a copy. You'll be a better person for it and you'll feel better about yourself. I'll feel better about it, as well.

Xerox PARC (Xerox Palo Alto Research Center) The Xerox research center where a number of significant inventions were made in the 1970s. The first graphical user interface (GUI) was invented there, as was Ethernet, laser printing, and object-oriented programming (OOP). Xerox PARC is now PARC (Palo Alto Research Center, Inc.), an independent research and development company.

XML (eXtensible Markup Language) A language used by Web developers and designers for creating declarative markup languages like Hypertext Markup Language (HTML), only more flexible in that documents written in XML can be shared across different information systems, particularly the Internet, and can adapt to different presentation style sheets and applications. A condensed form of Standard Generalized Markup Language (SGML), XML is published and maintained by the World Wide Web Consortium (W3C). See also *W3C* and *Web*.

XMODEM A public domain file transfer protocol used in asynchronous data communications, XMODEM organizes data into 128-byte blocks and employs a cyclic redundancy check (CRC) for error control. XMODEM was developed by Ward Christensen in 1977 and quickly became popular in the bulletin board system (BBS) community for file downloads. XMODEM has largely been replaced by ZMODEM. See also *asynchronous*, *BBS*, *block*, *CRC*, *error control*, *Kermit*, *modem*, *YMODEM*, and *ZMODEM*.

XMPP (eXtensible Messaging and Presence Protocol) The Internet Engineering Task Force (IETF) specification (RFCs 3920-3923) for an open XML technology for real-time communications, including instant messaging (IM), presence, and whiteboarding. XMPP is based on Extensible Markup Language (XML). See also *IETF*, *IM*, *presence*, *whiteboarding*, and *XML*.

XO The abbreviation for crystal oscillator. See *crystal oscillator*.

X-rays The portion of the electromagnetic spectrum in the frequency range of 30 PHz – 30 EHz and with a wavelength of 10–.01 nm. X-rays have no applications in telecommunications (unless you count their use in the printing of semiconductor chips used in telecommunications), but they are useful as diagnostic tools in medicine and science. X-rays can penetrate solids and ionize gas. See also *electromagnetic spectrum*, *frequency*, *Hz*, and *wavelength*.

X Series The series of ITU-T Recommendations specifying protocols relating to data networks, open systems communications, and security. See Table X-2 for selected X-series Recommendations by 100s group, and Table X-3 for specific recommendations of interest. For a full listing of ITU-T Recommendations, see the contact information in Appendix A.

Table X-2: ITU-T X Series Recommendation Groups

Recommendation Group	*Description*
100	Data communications and networks
200	Open systems interconnection
300	Data transmission services between public networks, and between public networks and other networks
400	Message Handling Services and Systems (MHS)
500	Open Systems Interconnection (OSI): Directory services
600	Open Systems Interconnection (OSI): Multi-peer communications
700	Open Systems Interconnection (OSI): Systems Management
800	Open Systems Interconnection (OSI): Security
900	Open Distributed Processing

Table X-3: Select X Series Recommendations

Recommendation	*Description*
X.3	Packet assembly/disassembly facility (PAD) in a public data network (PDN)
X.20bis	Use on public data networks (PDNs) of Data Terminal Equipment (DTE) designed for interfacing to asynchronous duplex V-Series modems
X.21	Interface between Data Terminal Equipment (DTE) and Data Circuit-terminating Equipment (DCE) for synchronous operation on public data networks

continued

Table X-3: Select X Series Recommendations *(continued)*

Recommendation	*Description*
X.21bis	Use on public data networks of Data Terminal Equipment (DTE) which is designed for interfacing to synchronous V-Series modems
X.25	Interface between Data Terminal Equipment (DTE) and Data Circuit-terminating Equipment (DCE) for terminals operating in the packet mode and connected to public data networks (PDNs) by dedicated circuit
X.28	DTE/DCE interface for a start-stop mode Data Terminal Equipment accessing the Packet Assembly/Disassembly (PAD) in a public data network (PDN) situated in the same country
X.29	Procedures for the exchange of control information and user data between a Packet Assembly/Disassembly (PAD) and a packet mode DTE or another PAD
X.31	Support of packet mode terminal equipment by an ISDN
X.32	Interface between Data Terminal Equipment (DTE) and Data Circuit-terminating Equipment (DCE) for terminals operating in the packet mode and accessing a packet-switched public data network through a public switched telephone network (PSTN) or an integrated services digital network (ISDN) or a circuit-switched public data network (PDN)
X.75	Packet-switched signaling system between public networks providing data transmission services
X.121	International numbering plan for public data networks (PDNs)
X.122	Numbering plan interworking for the E.164 and X.121 numbering plans
X.140	General quality of service parameters for communication via public data networks (PDNs)
X.141	General principles for the detection and correction of errors in public data networks (PDNs)
X.142	Quality of service metrics for characterizing frame relay/ATM service interworking performance
X.144	User information transfer performance parameters for public frame relay data networks
X.145	Connection establishment and dis-engagement performance parameters for public frame relay data networks providing SVC services
X.146	Performance objectives and quality of service classes applicable to frame relay
X.147	Frame relay network availability
X.148	Procedures for the measurement of the performance of public data networks (PDNs) providing international frame relay service
X.149	Performance of IP networks when supported by public frame relay data networks
X.151	Frame relay operations and maintenance — Principles and functions
X.200	Information technology — Open Systems Interconnection (OSI) - Basic Reference Model: The basic model
X.300	General principles for interworking between public networks and between public networks and other networks for the provision of data transmission services
X.400	Message handling services: Message handling system and service overview

Table X-3: Select X Series Recommendations *(continued)*

Recommendation	*Description*
X.500	Information technology — Open Systems Interconnection (OSI) - The Directory: Overview of concepts, models and services
X.601	Multi-peer communications framework
X.700	Management framework for Open Systems Interconnection (OSI) for CCITT applications
X.800	Security architecture for Open Systems Interconnection for CCITT applications
X.901	Information technology - Open Distributed Processing - Reference Model: Overview

XT (CROSSTalk) The abbreviation for crosstalk. See *crosstalk*.

.xxx In May 2006, the Board of Directors of the Internet Corporation for Assigned Names and Numbers (ICANN) reversed itself and voted against a proposed agreement for an .xxx generic Top Level Domain (gTLD) reserved for pornography. Some argued that .xxx would serve as a positive, if voluntary, means of segmenting the Internet. ICANN tentatively approved the new TLD before receiving an unprecedented level of correspondence in opposition. The .xxx domain has since been rejected multiple times. See also *gTLD*, *ICANN*, and *Internet*.

y (year) **1.** In mathematics, the symbol for an unknown quantity, person, factor, or thing. Y is the second in a set of two or three unknowns comprising x, y, and z. See also *x*, *y-axis*, and *z*. **2.** The abbreviation for year. See *yy* and *yyyy*.

Y **1.** Yotta. A septillion (10^{24}), or 1,000,000,000,000,000,000,000,000. See *YB*. **2.** The symbol for yttrium. See *yttrium*.

Y2K (Year 2000) See *yy*.

yada yada See *yada yada yada*.

yada yada yada Urban slang for and so on, and often used to suggest that the specifics are boring and unworthy of repetition.

YAG/LED (Yttrium Aluminum Garnet/Light-Emitting Diode) A light source used at one time in fiber optic transmission systems (FOTS). See also *FOTS*, *LED*, and *yttrium*.

Yagi antenna Also known as *Yagi-Uda antenna*. A VHF or UHF vertical antenna array comprising a basic radiator (dipole) antenna supplemented by a slightly longer reflector element mounted directly behind the dipole and a slightly shorter director element mounted directly in front of the dipole. A Yagi antenna is used to improve the directional gain of a television antenna when reception is otherwise weak. Ham radio enthusiasts commonly use homemade Yagi-Uda antennas. The Yagi-Uda antenna was invented at Tohuku University in 1926 by Hidetsugu Yagi (1886–1976), a Japanese electrical engineer, and Shitaro Uda (1896–1976), an assistant professor. See also *amateur radio service*, *antenna*, and *gain*.

Yagi-Uda antenna See *Yagi antenna*.

Yahoo! (Yet another hierarchically officious oracle!) A Web directory developed in the early 1990s by Stanford graduate students David Filo and Jerry Yang. Yahoo! has expanded into a full-featured Web portal, including a search engine, chat groups, instant messaging (IM), and e-mail. The word *Yahoo* was coined by Jonathan Swift in his book *Gulliver's Travels* (1726), referring to a race of foul, uncultivated, loutish, brutish creatures in the form of men. The term has since evolved to refer to a coarse, unrefined, unruly, crudely materialistic person. Filo and Yang reportedly selected the name because they considered themselves yahoos. See also *endianess*.

y-axis The vertical (V), or nearest vertical, plane on a two- or three-dimensional grid, chart, or graph in a Cartesian coordinate system. See also *Cartesian coordinates*, *x-axis*, and *z-axis*.

Yb The symbol for ytterbium. See also *ytterbium*.

YB (YottaByte) A septillion (10^{24}) bytes. In computing and storage systems, a YB (YottaByte) is exactly 1,208,925,819,614,629,174,706,176 (2^{80}) bytes, since the measurement is based on a base 2, or binary, number system. The term YB comes from the fact that 1,208,925,819,614,629,174,706,176 is nominally, or approximately, 1,000,000,000,000,000,000,000,000, which suggests that mathematicians consider a difference of 208,925,819,614,629,174,706,176 to be a rounding error. See also *byte*.

Year 2000 (Y2K) See *yy*.

Year To Date (YTD) See *YTD*.

YMODEM An improvement on the XMODEM file transfer protocol that transfers data in blocks of 1,024 bytes, performs a cyclic redundancy check (CRC) on each block, supports the transfer of multiple files in a batch transmission, and supports file transfer abort. XMODEM and YMODEM have largely been replaced by ZMODEM. See also *CRC*, *protocol*, *XMODEM*, and *ZMODEM*.

yotta- See *Y*.

yottabyte (YB) See *YB*.

YTD (Year To Date) A term used in accounting to identify the fact that the revenue or expense category in question reflects financial from January 1 to the present date.

ytterbium (Yb) A rare divalent or trivalent, silvery, soft, malleable, and ductile metallic rare-earth element. Ytterbium has little practical application. Number 70 in the Periodic Table of Elements, ytterbium is named for the village of Ytterby, Sweden, where it was discovered. So were erbium, yttrium, and terbium. See also *erbium* and *television*.

Ytterby Pronounced *Iterbe*. The village in Sweden where the element erbium (Er) was discovered. So were the elements ytterbium, yttrium, and terbium. Also discovered in Ytterby was the element holmium, but it was named for Stockholm, which is some 10 miles distant. Further discovered in Ytterby was the element gadolinium, which was named for Johan Gadolin (1760–1852), the Finnish scientist who discovered them all in 1794. There probably have been other things discovered in Ytterby, but you'll have to discover them in another book. Erbium, by the way is used in erbium-doped fiber amplifiers (EDFAs), which are amplifiers used in fiber optic transmission systems (FOTS). See also *EDFA*, *erbium*, and *FOTS*.

yttrium (Y) A rare trivalent, silvery, metallic chemical element used in various alloys and to make the red color phosphors in cathode ray tubes (CRTs). Yttrium also is used in YAG/LED (Yttrium Aluminum Garnet/Light-Emitting Diode) light sources. Number 39 in the Periodic Table of Elements, yttrium is named for the village of Ytterby, Sweden, where it was discovered. So were erbium, ytterbium, and terbium. See also *erbium*, *LED*, *television*, *YAG/LED*, and *Ytterby*.

yy Abbreviation for the two-digit storage and display of a year, as in 07. Older computer systems stored dates in the format xx/xx/yy, which represents variously represents month/day/year or day/month/year, depending on whether one is using the American system or the little-endian notation used in most of the rest of the world. In either case, before the year 2000 (Y2K), the possibility existed that a two-digit date would be interpreted as 1900 and disrupt the operation of older computers that the manufacturers had fully expected would be replaced long before 00 became an issue. See *little-endian*.

yyyy Abbreviation for the four-digit display of a year, as in 2007.

Z **1.** Zeta. The sixth letter of the Greek alphabet, written in the English alphabet as Z. **2.** The symbol for impedance. See *impedance*. **3.** The symbol for Zulu time. See *Zulu time*. **4.** Zetta sextillion (10^{21}), or 1,000,000,000,000,000,000,000. See *ZB*. **5.** In mathematics, the third of a set of unknown variables, with x and y being the first two unknowns. See also *x*, *y*, and *z-axis*. **6.** One of the two letters, along with Q, that traditionally did not appear on a telephone dial or keypad. The thought was that Z could be confused with 2 and that Q could be confused with O. Z now appears with W, X, and Y on number 9. Alphanumeric dialing was, and remains, a North American practice. Telephones in most other countries do not sport letters.

z-axis The third axis, usually representing depth, of a three-dimensional grid, chart, or graph in a Cartesian coordinate system. The z-axis is perpendicular to both the x-axis and y-axis and is used to plot the value of z, the third unknown in mathematics. See also *Cartesian coordinates*, *x-axis*, *y-axis*, and *z*.

ZB (ZettaByte) A sextillion (10^{21}) bytes. In computing and storage systems, a ZB is exactly 1,180,591,620,717,411,303,424 (2^{70}) bytes, since the measurement is based on a base 2, or binary, number system. The term ZB comes from the fact that 1,180,591,620,717,411,303,424 is nominally, or approximately, 1,000,000,000,000,000,000,000, which suggests that mathematicians consider a difference of 180,591,620,717,411,303,424 to be a rounding error. See also *byte* and *Z*.

ZC (ZigBee Coordinator) A ZigBee device that initializes the network, coordinates its operation, and is responsible for security. There is one ZC per network, but there can be a great many ZigBee End Devices (ZEDs) and ZigBee Routers (ZRs). See also *ZED*, *ZigBee*, and *ZR*.

ZED (ZigBee End Device) A ZigBee terminal device, such as a sensor, that can perform only a single monitoring or control function. A ZED comprises a low-cost microprocessor, RAM and ROM, a long-life battery, and a low-power radio and controller. A ZED is small enough to be embedded in a light switch, smoke or carbon dioxide detector, thermostat, security sensor, utility meter, or medical sensor. A ZED can communication only with a ZigBee Router (ZR), and not other ZEDs. See also *ZigBee* and *ZR*.

ZDSF (Zero Dispersion-Shifted Fiber) A type of dispersion-shifted, single-mode fiber (DSF SMF) that shifts the point of zero dispersion by increasing material dispersion to the point that it cancels out chromatic dispersion at 1550 nm, rather than 1310 nm. Dense Wavelength Division Multiplexing (DWDM) and Erbium-Doped Fiber Amplifiers (EDFAs) both work in this higher window, which can create yet another noise problem in the form of four-wave mixing (FWM), a phenomenon by which wavelengths interact to create additional wavelengths. The EDFAs amplify those signals, and superimpose them on the DWDM channels. Non Zero Dispersion-Shifted Fiber (NZDF) addresses this issue by shifting the optimal dispersion point slightly above the range in which EDFAs operate. See also *chromatic dispersion*, *cladding*, *core*, *dispersion*, *DSF*, *DWDM*, *EDFA*, *material dispersion*, *noise*, *NZDF*, *refractive index*, *SMF*, *wavelength*, and *window*.

zero dispersion-shifted fiber (ZDSF) See *ZDSF*.

zero-water-peak fiber (ZWPF) See *ZWPF*.

ZigBee A specification from the ZigBee Alliance for a set of high-level communications protocols based on the IEEE 802.15.4 standard for a low-data-rate wireless personal area network (WPAN) comprising devices of low complexity and long battery life. ZigBee is designed for connecting devices in ad hoc networks over very short distances with very low power consumption. ZigBee specifies star, peer-to-peer and mesh network topologies, with mesh being the preferred approach for reasons of redundancy and resiliency. ZigBee runs in the ISM band using direct sequence spread spectrum (DSSS) transmission. A ZigBee Coordinator (ZC) initializes the network, coordinates its operation, and is responsible for security. ZigBee End Devices (ZEDs) are terminal devices that are very limited in function, and can communicate only with ZigBee Routers (ZRs), which function to pass messages to other ZRs or to the ZC. In beacon enabled networks, ZigBee Routers (ZRs) periodically beacon their presence and, therefore, need power up only during the beaconing cycle, which conserves battery life. In non–beacon enabled networks, all devices remain active all the time and employ the carrier sense multiple access with collision avoidance (CSMA/CA) medium access control (MAC) mechanism.

Although most devices run in the 2.4 GHz range, some run at 915 MHz (Americas) and 868 MHz (Europe) ranges, as those bands offer better signal propagation through physical obstructions such as walls, floors, ceilings, and windows. Depending on the frequency band selected, raw data rates are 20 kbps (1 channel at 868 MHz), 40 kbps (10 channels at 915 MHz), and 250 kbps (16 channels at 2.4 GHz). Specified modulation techniques are binary phase-shift keying (BPSK) in the 868 and 915 MHz bands, and quaternary phase-shift keying (QPSK) in the 2.4 GHz band. Distances range from 10 meters to 100+ meters, depending on frequency, power output, and environmental characteristics. Devices generally transmit at a maximum power level of approximately 1 mW. Security features include access control and encryption based on the advanced encryption standard (AES). ZigBee is theoretically scalable up to 65,536 devices. The term *ZigBee* refers to the technique, known as the ZigBee principle, that a domestic honeybee uses to communicate the location of a new food source to other members of the colony. See also *ad hoc*, *AES*, *BPSK*, *CSMA/CA*, *DSSS*, *frequency band*, *ISM*, *MAC*, *mesh topology*, *peer-to-peer*, *propagation*, *QPSK*, *star topology*, *WPAN*, *ZC*, *ZED*, *ZigBee principle*, and *ZR*.

ZigBee Alliance An association of companies working to develop and promote a set of open, global standards for a low-power, cost effective wireless personal area network (WPAN) technology based on IEEE 802.15.4. For contact information, see Appendix A. See also *WPAN* and *ZigBee*.

ZigBee Coordinator (ZC) See *ZC*.

ZigBee End Device (ZED) See *ZED*.

ZigBee Router (ZR) See *ZR*.

ZigBee principle The technique that a domestic honeybee uses to communicate the location of a new food source to other members of the colony. The bee dances in a zigzag pattern that communicates information such as distance and direction, at least according to the ZigBee Alliance, which developed and promotes the ZigBee specification for a wireless personal area network (WPAN). My personal research, as well as the research of many others, suggests there is no such thing, at least not until the ZigBee Alliance coined the term. Oh, well. See also *WPAN*, *ZigBee*, and *ZigBee Alliance*.

ZMODEM A public domain file transfer protocol used in asynchronous data communications. ZMODEM offers a number of improvements to the earlier XMODEM protocol, including an expanded 32-bit cyclic redundancy check (CRC) for purposes of improved error control. ZMODEM also supports larger block sizes (512 bytes), allows a transmission to resume where it left off in the event of a communications failure, and reduces latency through a sliding window protocol that allows a modem to transfer a series of blocks without waiting for individual acknowledgements from the receiving modem. ZMODEM was developed by Chuck Forsberg in 1986 for use in the Telenet X.25 network, and quickly became popular in the bulletin board system (BBS) community for file downloads. See also *asynchronous*, *BBS*, *block*, *CRC*, *error control*, *Kermit*, *modem*, *X.25*, *XMODEM*, and *YMODEM*.

ZigBee Router (ZR) A ZigBee device that functions as a router to pass messages from ZigBee End Devices (ZEDs) to other ZRs or to the ZigBee Coordinator (ZC). A ZR also can function as a ZED. See also *ZC*, *ZED*, and *ZigBee*.

Zulu Time (Z) Also known as *Universal Coordinated Time* (UTC) and *Greenwich Mean Time* (GMT), the latter of which is used for civilian purposes. The world is divided into 25 integer time zones, each of which is identified by a letter in the English alphabet. (*Note:* There is no time zone J.) The clock at Greenwich, England, is the standard clock for international time reference, and is referred to in military and aviation parlance as the Zulu (Z) Time, as it is in the time zone designated by the letter Z. On a personal note, I live in the Mount Vernon, Washington area of the United States. Mount Vernon is on the West Coast, which is in the Pacific Standard Time (PST) zone, which is time zone U (Uniform), which is Z -8 hours, or, more correctly, UTC –8 hours, so we are eight hours behind Zulu Time. Denver, Colorado is in the Mountain Standard Time (MST) zone, which is time zone T (Tango), which is Z –7. Hawaii is in the Hawaii Standard Time (HST) zone, which is time zone W (Whiskey), which is Z –10. See also *UTC*.

ZWPF (Zero-Water-Peak Fiber) Single-mode fiber (SMF) manufactured without hydroxyl (OH) ion contamination in order eliminate the attenuation peak in the 1400 nm window, which is in the E-band (1360–1460 nm). The traditional SMF manufacturing process introduces hydroxyl (OH) ions into the fiber core. Wavelengths in the region around 1400 nm attenuate about 2 dB/km as a result of their interaction with those ions. As traditional single-wavelength fiber optic transmission systems (FOTS) employing SMF operate in the 1310 nm or 1550 nm window, water peak attenuation does not affect them. However, 4 of the 18 channels in coarse wavelength division multiplexing (CWDM) systems fall within the E-band and, therefore, are rendered unusable by water peak attenuation. Low–water-peak fiber (LWPF) has low levels of hydroxyl ion contamination and, therefore, suffers low water peak attenuation. ZWPF is the contemporary industry standard for all SMF. See also *attenuation*, *CWDM*, *dB*, *dB/km*, *E-band*, *FOTS*, *hydroxyl*, *LWPF*, *SMF*, *water peak*, *wavelength*, and *window*.

Appendix A: Standards Organizations and Special Interest Groups (SIGs)

Formal Standards Organizations

There is no useful rule without an exception. Thomas Fuller, Gnomlogia (1732)

3GPP

3rd Generation Partnership Project
ETSI
Mobile Competence Centre
650, route des Lucioles
06921 Sophia-Antipolis Cedex
Tel: +33 (0) 4-92-94-42-00
Fax: +33 (0) 4-93-65-47-16
www.3gpp.org

3GPP2

3rd Generation Partnership Project 2
2500 Wilson Boulevard, Suite 300
Arlington, VA 22201
Telephone: 703.907.7700
Fax: 703.907.7728
www.3gpp2.org

ANSI

American National Standards Institute
1819 L Street, NW, Suite 600
Washington, DC 20036
Tel: 202.293.8020
Fax: 202.293.9287
www.ansi.org

ATSC

Advanced Television Systems Committee
1750 K Street NW, Suite 1200
Washington, DC 20006
Tel: 202.872.9160
Fax: 202.872.9161
www.atsc.org

ATIS

Alliance for Telecommunications Industry Solutions
(nee ECMA, Exchange Carriers Standards Association [ECSA])
1200 G St. NW, Suite 500
Washington, DC 20005
Tel: 202.628.6380
Fax: 202.393.5453
www.atis.org

CableLabs

Cable Television Laboratories, Inc.
858 Coal Creek Circle
Louisville, CO 80027-9750
Tel: 303.661.9100
Fax: 303.661.9199
www.cablelabs.com

CEN

Comité Européen de Normalisation
European Committee for Standardization
36 rue de Stassart, B
1050 Brussels, Belgium
Tel: 32-2-550-08-11
Fax: 32-2-550-08-19
www.cenorm.be

CENELEC

Comité Européen de Normalisation Electrotechnique
European Committee for Electrotechnical Standards
35, rue de Stassartstraat
B-1050 Brussels, Belgium
Tel: 32-2-519-68-71
Fax: 32-2-519-69-19
www.cenelec.org

CERN

l'Conseil Européen pour la Recherche Nucléaire
The European Organization for Nuclear Research
CERN CH-1211 Genève 23
Switzerland
Tel: +41 22 76 761 11
Fax: +41 22 76 765 55
www.cern.ch

CSA International

Canadian Standards Association International
5060 Spectrum Way
Mississauga, ONT L4W 5N6
Canada
Tel: 416.747.4000, or 800-463-6727
Fax: 416.747.2473
www.csa.ca

DISA

Data Interchange Standards Association
7600 Leesburg Pike, Suite 430
Falls Church, VA 22043
Tel: 703.970.4480
Fax: 703. 970.4488
www.disa.org

Ecma International

(nee ECMA, European Computer Manufacturers Association)
114 Rue du Rhone CH-1204
Geneva, Switzerland
Tel: 41-22-849-6000
Fax: 41-22-849-6001
www.ecma-international.org

EIA

Electronic Industries Alliance
2500 Wilson Blvd.
Arlington, VA 22201
Tel: 703.907.7500
Fax: 703.907.7501
www.eia.org

ETSI

European Telecommunications Standards Institute
650, route des Lucioles
06921 Sophia Antipolis Cedex
France
Tel: +33 (0) 4-92-94-42-00
Fax: +33 (0) 4-93-65-47-16
www.etsi.org

FCC

Federal Communications Commission
445 12th Street, S.W.
Washington, DC 20554
Tel: 888.225.5322
Fax: 202.418.0232 or 866-418-0232
www.fcc.gov

ICEA

Insulated Cable Engineers Association
P.O. Box 1568
Carrollton, GA 30112
www.icea.net

IEC

International Electrotechnical Commission
3, Rue de Varembe
P.O. Box 131
CH-1211 Geneva 20
Switzerland
Tel: 41-22-919-02-11
Fax: 41-22-919-03-00
www.iec.ch

IEEE

Institute of Electrical and Electronics Engineers
445 Hoes Lane
Piscataway, NJ 088541331
Tel: 732.981.0060
Fax: 732.981.1721
www.ieee.org

ISO

International Organization for Standardization
1, Rue de Varembe
Case postale 56
CH-1211 Geneva 20
Switzerland
Tel: 41-22-749-01-11
Fax: 41-22-733-34-30
www.iso.ch

ISOC

Internet Society
1775 Wiehle Avenue, Suite 102
Reston, VA 20190-5108
Tel: 703.326.9880
Fax: 703.326.9881
www.isoc.org

ITU

International Telecommunication Union
Place des Nations
CH-1211 Geneva 20
Switzerland
Tel: 41-22-730-5111
Fax: 41-22-733-7256
www.itu.ch

NIST

National Institute of Standards and Technology
(nee National Bureau of Standards, NBS)
100 Bureau Drive
Gaithersburg, MD 20899
Tel: 301.975.2000
www.nist.gov

NTIA

National Telecommunications and Information Administration
U.S. Department of Commerce
1401 Constitution Avenue, N.W.
Washington, DC 20230
Tel: 202.482.7002
www.ntia.doc.gov

Telcordia Technologies

(nee Bellcore)
1 Telcordia Drive
Piscataway, NJ 08854-4157
Tel: 732.699.2000
Fax: 732.336.2320
www.telcordia.com

TIA

Telecommunications Industry Association
2500 Wilson Boulevard, Suite 300
Arlington, VA 22201-3837
Tel: 703.907.7700
Fax: 703.907.7727
www.tiaonline.org

UL

Underwriters Laboratories Inc.
333 Pfingsten Road
Northbrook, IL 60062-2096
Tel: 847.272.8800
Fax: 847.272.8129
www.ul.com

W3C

World Wide Web Consortium
Massachusetts Institute of Technology
32 Vassar Street, Room 32-G515
Cambridge, MA 02139
Tel: 617.253.2613
Fax: 617.258.5999
www.w3.org

Consortia, Fora, and Special Interest Groups (SIGs)

A complex society is not necessarily more advanced than a simple one; it has just adapted to conditions in a more complicated way. Peter Farb, Man's Rise to Civilization (1968)

ACM

Association for Computing Machinery
1515 Broadway
New York, NY 10036
Tel: 212.626.0500 or 800.342.6626
www.acm.org

ATM Forum

See MFA Forum

BICSI

Building Industry Consulting Service International
8610 Hidden River Parkway
Tampa, FL 33637-1000
Tel: 813.979.1991 or 800.242.7405
Fax: 813.971.4311
www.bicsi.org

Bluetooth Special Interest Group

500 108th Avenue N.E., Suite 250
Bellevue, WA 98004
Tel: 425.691.3535
www.bluetooth.com

CDG

CDMA Development Group
575 Anton Boulevard, Suite 560
Costa Mesa, CA 92626
Tel: 714.545.5211 or 888.800.2362
Fax: 714.545.4601
www.cdg.org

CEA

Consumer Electronics Association
2500 Wilson Blvd.
Arlington, VA 22201-3834
Tel: 703.907.7600 or 866.858.1555
Fax: 703.907.7675
www.ce.org

CEPT

European Conference of Postal and Telecommunications Administrations
Bezuidenhoutseweg 30
P.O.Box 20101
2500 EC The Hague
The Netherlands
Tel: 31 70 379 8164
Fax: 31 70 379 8267
www.cept.org

CERT

Computer Emergency Response Team
Software Engineering Institute
Carnegie Mellon University
Pittsburgh, PA 15213-3890
www.cert.org

CompTIA

Computer Technology Industry Alliance
1815 S. Meyers Road, Suite 300
Oakbrook, IL 60181-5228
Tel: 630.678.8300
Fax: 630.628.1384
www.comptia.org

CTIA

The Wireless Association
(Previously Cellular Telecommunications & Internet Association)
1400 16th Street, NW, Suite 600
Washington, DC 20036
Tel: 202.785.0081
Fax: 202.785.0721
www.ctia.org

DSL Forum

39355 California Street, Suite 307
Fremont, CA 94538
Tel: 510.608.5905
Fax: 510.608.5917
www.dslforum.org

EFF

Electronic Frontier Foundation
454 Shotwell Street
San Francisco, CA 94110-1914
Tel: 415.436.9333
Fax: 415.436.9993
www.eff.org

EMA

Electronic Messaging Association
See The Open Group

Ethernet Alliance

P.O. Box 200757
Austin, TX 78720
Tel: 512.363.9932
Fax: 512.532.6894
www.ethernetalliance.org

ETNO

European Telecommunications Network Operators' Association
Avenue Louise 54
1050 Brussels, Belgium
Tel: (32) 2 219-3242
Fax: (32) 2 219-6412
www.etno.be

Fibre Channel Industry Association

www.fibrechannel.org

Frame Relay Forum

See MFA Forum

IEC

International Engineering Consortium
300 W. Adams Street, Suite 1210
Chicago, IL 60606-5114
Tel: 312.559.4100
Fax: 312.559.4111
www.iec.org

IMC

Internet Mail Consortium
127 Segré Place
Santa Cruz, CA 95060
Tel: 831.426.9827
Fax: 831.426.7301
www.imc.org

IMTC

International Multimedia Telecommunications Consortium, Inc.
Bishop Ranch 6
2400 Camino Ramon, Suite 375
San Ramon, CA 94583
Tel: 925.275.6600
Fax: 925.275.6691
www.imtc.org

IPv6 Forum

www.ipv6forum.com

IrDA

Infrared Data Association
P.O. Box 3883
Walnut Creek, CA 94598
Tel: 925.943.6546
Fax: 925.943.5600
www.irda.org

MFA Forum

(merged MPLS, Frame Relay and ATM Fora)
39355 California Street, #307
Fremont, CA 94538
Tel: 510.608.5910
Fax: 510.608.5917
www.mfaforum.org

MPLS Forum

See MFA Forum

MSF

Multiservice Switching Forum
39355 California Street, #307
Fremont, CA 94538
Phone: 510.608.5922
Fax: 510.608.5917
www.msforum.org

NAB

National Association of Broadcasters
1771 N Street, NW
Washington, DC 20036-2891
Tel: 202.429.5300
Fax: 202.429.4199
www.nab.org

NARTE

National Association of Radio and
Telecommunications Engineers
167 Village Street
Medway, MA 02053
Tel: 508.533.8333 or 800.896.2783
Fax: 508.533.3815
www.narte.org

NARUC

National Association of Regulatory and Utility Commissioners
1101 Vermont Avenue, N.W.
Washington, DC 20005
Tel: 202.898.2200
Fax: 202.898.2213
www.naruc.org

NCTA

National Cable & Telecommunications Association
1724 Massachusetts Avenue, N.W.
Washington, DC 20036
Tel: 202.775.3550
www.ncta.com

NECA

National Exchange Carrier Association
80 South Jefferson Road
Whippany, NJ 07981-1009
Tel: 973.884.8000, or 800.228.8597
Fax: 973.884.8469
www.neca.org

NTIS

National Technical Information Service
Technology Administration
U.S. Department of Commerce
5285 Port Royal Road
Springfield, VA 22161
Tel: 703.605.6000
Fax: 703.321.8547
www.ntis.gov

OMA

Open Mobile Alliance
4275 Executive Square, Suite 240
La Jolla, CA 92037
Tel: 858.623.0740
Fax: 858.623.0743
www.openmobilealliance.org

OMG

Object Management Group, Inc.
1410 Kendrick Street
Building A, Suite 300
Needham, MA 02494
Tel: 781.444.0404
Fax: 781.444.0320
www.omg.org

The Open Group

Previously the Electronic Messaging
Association (EMA)
44 Montgomery Street, Suite 960
San Francisco, CA 94104-4704
Tel: 415.374.8280
Fax: 415.374.8293
www.opengroup.org

PCIA

Personal Communications Industry Association
500 Montgomery Street, Suite 700
Alexandria, VA 22314-1561
Tel: 703.739.0300, or 800.759.0300
Fax: 703.836.1608
www.pcia.com

PCMCIA

Personal Computer Memory Card
International Association
2635 North First Street, Suite 218
San Jose, CA 95134
Tel: 408.433.2273
Fax: 408.433.9558
www.pcmcia.org

SAI

Satellite Industry Association
1730 M Street, N.W., Suite 600
Washington, DC 20036
Tel: 202.349.3650
Fax: 202.349.3622
www.sia.org

SBCA

Satellite Broadcasting and Communications Association
1730 M Street, N.W., Suite 600
Washington, DC 20036
Tel: 202.349.3620 or 800.541.5981
Fax: 202.349.3621
www.sbca.com

SCTE

Society of Cable Telecommunications Engineers
140 Philips Road
Exton, PA 19341-1318
Tel: 610.363.6888, or 800.542.5040
Fax: 610.363.5898
www.scte.org

SIIA

Software & Information Industry Association
1090 Vermont Ave NW, Sixth Floor
Washington, DC 20005-4095
Tel: 202.289.7442
Fax: 202.289.7097
www.siia.net

SIP Forum

Session Initiation Protocol Forum
Tel: 978.824.0111
www.sipforum.org

SMPTE

Society of Motion Picture & Television Engineers
3 Barker Avenue
White Plains, NY 10601
Tel: 914.761.1100
Fax: 914.761.3115
www.smpte.org

SNIA

Storage Networking Industry Association
500 Sansome Street, Suite 504
San Francisco, CA 94111
Tel: 415.402.0006
Fax: 415.402.0009
www.snia.org

TM Forum

TeleManagement Forum
240 Headquarters Plaza
East Tower, 10th Floor
Morristown, NJ 07960-6628
Tel: 973.944.5100
Fax: 973.944.5110
www.nmf.org

Unicode Consortium

Attn: Magda Danish
1065 L'Avenida Street
Microsoft Building 5
Mountain View, CA 94043
Tel: 650.693.3921
Fax: 650.693.3010
www.unicode.org

USTA

United States Telecom Association
607 14th Street NW, Suite 400
Washington, DC20005
Tel: 202.326.7300
Fax: 202.326.7333
www.ustelecom.org

WCA International

Wireless Communications Association International
1333 H Street, N.W.
Suite 700 West
Washington, DC 20005-4754
Tel: 202-452-7823
Fax: 202-452-0041

Wi-Fi Alliance

Previously Wireless Ethernet Compatibility Alliance (WECA)
3925 West Braker Lane
Austin, TX 78759
Tel: 512.305.0790
Fax: 512.305.0791
www.wi-fi.com

WiMAX Forum

2495 Leghorn Street
Mountain View, CA 94043
Tel: 503.712.2206
www.wimaxforum.org

WTO

World Trade Organization
Centre William Rappard
Rue de Lausanne 154
CH-1211 Geneva 21
Switzerland
Tel: 41-22-739-51-11
Fax: 41-22-731-42-06
www.wto.org

ZigBee Alliance

2400 Camino Ramon, Suite 375
San Ramon, CA 94583
Tel: 925-275-6607
Fax: 925.886.3850
www.zigbee.org